New Perspectives in the Definition/Evaluation of Seismic Hazard through Analysis of the Environmental Effects Induced by Earthquakes

New Perspectives in the Definition/Evaluation of Seismic Hazard through Analysis of the Environmental Effects Induced by Earthquakes

Editors

Sabina Porfido
Giuliana Alessio
Germana Gaudiosi
Rosa Nappi

MDPI • Basel • Beijing • Wuhan • Barcelona • Belgrade • Manchester • Tokyo • Cluj • Tianjin

Editors

Sabina Porfido
Institute of Food Sciences
(ISA) National Research
Council (CNR), Italy and
Istituto Nazionale diGeofisica
e Vulcanologia,sezione di
Napoli Osservatorio Vesuviano
Italy

Giuliana Alessio
Istituto Nazionale di Geofisica e
Vulcanologia,sezione di Napoli
Osservatorio Vesuviano
Italy

Germana Gaudiosi
Istituto Nazionale di Geofisica e
Vulcanologia, sezione di Napoli
Osservatorio Vesuviano
Italy

Rosa Nappi
Istituto Nazionale di Geofisica e
Vulcanologia, sezione di Napoli
Osservatorio Vesuviano
Italy

Editorial Office
MDPI
St. Alban-Anlage 66
4052 Basel, Switzerland

This is a reprint of articles from the Special Issue published online in the open access journal *Geosciences* (ISSN 2076-3263) (available at: https://www.mdpi.com/journal/geosciences/special_issues/seismic_hazard).

For citation purposes, cite each article independently as indicated on the article page online and as indicated below:

LastName, A.A.; LastName, B.B.; LastName, C.C. Article Title. *Journal Name* **Year**, *Volume Number*, Page Range.

ISBN 978-3-0365-1877-0 (Hbk)
ISBN 978-3-0365-1878-7 (PDF)

Cover image courtesy of Chiara Frigerio (UninSubria).
Mt. Vettore coseismic rupture of the 2016-2017 Amatrice-Visso-Norcia seismic sequence: area of the maximum vertical displacement measured

Contents

About the Editors

Sabina Porfido

Sabina Porfido is a researcher of ISA-CNR. She is engaged in research for the evaluation and reduction of the natural hazards. The research activities is mainly focused on the study of recent and historical seismicity, macroseismic studies, seismic microzonation; Seismotectonic processes and definition of seismogenic areas; Study of induced seismic primary and secondary ground effects: new macroseismic scale (INQUA scale, 2003-2004; ESI scale, 2007). Natural hazards connected to hydrogeological events: historical floods in the Campania Region; coastal hazards; land use and resilience. She has participated in numerous national and international research projects (INQUA-TERPRO, EEE Catalogue ISPRA; Cooperazione Bilaterale Italia-Ungheria (CNR-MTA); IODP-Campi Flegrei; Proyecto de red interuniversitaria regional Italo-Centroamericana en análisis de los fenómenos naturales para la evaluación de la peligrosidad (Ministero Affari Esteri); Accordo bilaterale Italia-Moldavia.

Giuliana Alessio

Giuliana Alessio is a researcher of INGV-OV. She deals mainly with seismicity and seismotectonics of active faults in the Central-Southern Apennines (Irpinia; Sannio-Beneventano; Matese chain; Salento peninsula) and in the volcanic areas of Vesuvius, Campi Flegrei and Ischia island, studying their geometry and activity by space-time correlation of local seismicity, field data recognition, recent deformation data, historical seismicity. She participated to field surveys for the seismic emergencies of L'Aquila, 2009; Emilia-Romagna, 2012; Central Italy (Amatrice and Norcia) 2016; the island of Ischia, 2017; also for the Stromboli volcanic eruptions of 2002/2003 and 2019. She also has applied multiparameter approach for hydrogeologic hazard assessment in the Somma-Vesuvius and Sorrento peninsula areas. She operates for mitigation of seismic and volcanic risk with scientific dissemination activities for various national and international Institutions and geoarcheological campaigns.

Germana Gaudiosi

Germana Gaudiosi is a researcher of INGV-OV. She studies the geodynamical models of Western Mediterranean, shallow structure of volcanic areas of southern Italy from joined analysis of active and passive seismic data, deep crustal structure of Italian peninsula from active seismological techniques (Campanian volcanoes, Campanian Plain, Eolie, Etna, southern-central Appennines ,Eastern Sicily, Ionian sea). Through the integrated analysis of tectonic, seismic and gravimetric data, she performs the geometry of active/silent outcropping/buried faults of central-southern Italy (campanian Plain, Mt. Matese). She re-evaluating the macroseismic effects in MCS scale and ground effects in natural environment according to the ESI 2007 scale. She contributed in numerous national and international research projects and collaborate with Universities of Naples Federico II, University of Cosenza, University of Palermo, University of Milano, University of Trieste, University of Paris, CNR-ISA.

Rosa Nappi

Rosa Nappi is a researcher of INGV-OV. She studies the fault process in tectonic and volcanic area in the seismotectonic framework, earthquake geology and induced primary and secondary seismic ground effects. Through integrated geological and geophysical data, in GIS environment, she performs the quantitative geomorphic analysis and ground field deformation for identifying the seismogenetic source, with the aim to contribute of the evaluation and reduction of the seismic hazards. Also she re-evaluating the macroseismic effects in MCS scale and ground effects in natural environment according to the ESI 2007 scale. The recent study areas are: Salento Peninsula (Apulia Region, Southern Italy); Central Apennines (l'Aquila earthquake, Amatrice-Visso-Norcia 2016/17 seismic sequences, 2017, Casamicciola, Ischia island). She has participated in numerous national and international research projects and collaborate with Universities of Naples Federico II and University of Insubria, CNR-ISA.

geosciences

MDPI

Editorial

New Perspectives in the Definition/Evaluation of Seismic Hazard through Analysis of the Environmental Effects Induced by Earthquakes

Sabina Porfido [1,2], Giuliana Alessio [2], Germana Gaudiosi [2] and Rosa Nappi [2,*]

1 Consiglio Nazionale delle Ricerche - ISA - Via Roma 64, 80100 Avellino, Italy; sabina.porfido@cnr.it
2 Istituto Nazionale di Geofisica e Vulcanologia - Sezione di Napoli Osservatorio Vesuviano. Via Diocleziano, 328, 80124 Napoli, Italy; giuliana.alessio@ingv.it (G.A.); germana.gaudiosi@ingv.it (G.G.)
* Correspondence: rosa.nappi@ingv.it

Received: 31 January 2020; Accepted: 31 January 2020; Published: 4 February 2020

Abstract: The application of the Environmental Seismic Intensity (ESI) scale 2007 to moderate and strong earthquakes, in different geological context all over the word, highlights the importance of Earthquake Environmental Effects (EEEs) for the assessment of seismic hazards. This Special Issue "New Perspectives in the Definition/Evaluation of Seismic Hazard through Analysis of the Environmental Effects Induced by Earthquakes" presents a collection of scientific contributions that provide a sample of the state-of-the-art in this field. Moreover the collected papers also analyze new data produced with multi-disciplinary and innovative methods essential for development of new seismic hazard models.

Keywords: ESI scale; earthquake environmental effects; seismic hazard

---

The devastating effects caused by the recent catastrophic earthquakes that took place all over the world from Japan, New Zealand, to Chile, as well as those occurring in the Mediterranean basin, have once again shown that the ground motion, although a serious source of direct damage, is not the only parameter to be considered, with most damage being the result of coseismic geological effects that are directly connected to the earthquake source or caused by ground shaking.

The primary environmental effects induced by earthquakes such as surface faulting, regional uplift, and subsidence, as well as the secondary effects such as tsunami, liquefaction, ground failure, and landslides (sensu Environmental Seismic Intensity (ESI) 2007 scale) [1], must be considered for a more correct and complete evaluation of seismic hazard, at both regional and local scales.

This Special Issue aims to collect all contributions that, using different methodologies, integrate new data produced with multi-disciplinary and innovative methods. These methodologies are essential for the identification and characterization of seismically active areas, and for the development of new hazard models, obtained using different survey techniques.

The topic attracted a lot of interest; moreover, different areas of the world have been analyzed through these methodologies (Italy, USA, Spain, Australia, Ecuador, Guatemala, South Korea, Kyrgyzstan, Mongolia, Russia, China, Japan, Nepal).

This Special Issue is composed of 18 peer-reviewed articles, out of 27 papers that had been received; different approaches to the evaluation of seismic hazard are presented in the volume, from laboratory tests on liquefaction phenomena and landslides triggered by earthquakes, to the analysis of the environmental effects (primary and secondary coseismic effects) induced by the earthquakes.

In detail, this Special Issue contains 5 review papers and 13 articles. Below is a brief description of the contents of all articles.

- The review paper "History of the Environmental Seismic Intensity Scale ESI-07" by L. Serva [1,2] presents a brief note aiming to describe the history, from its early original idea, to the present of the new Environmental Seismic Intensity Scale 2007 (ESI 2007, Michetti et al., 2007). This innovative macroseismic scale can be used together with other existing scales or alone, if necessary, for measuring the intensity of an earthquake on the basis of the primary and secondary effects caused by the seismic event on the natural environment. The ESI 2007 scale was promoted by several geologists, seismologists, and engineers from different countries, forming a multidisciplinary international research group. This scale has been tested worldwide on several modern earthquakes, historical earthquakes, and paleoearthquakes. The review by Serva also presents a wide and comprehensive bibliography with the most important examples of ESI-07 scale applications carried out worldwide.
- The review paper "Paleoliquefaction Studies and the Evaluation of Seismic Hazard" by M.P. Tuttle et al. [3] presents recent and historical studies of earthquake-induced liquefaction, as well as paleoliquefaction studies, in order to demonstrate the potential usefulness of liquefaction data in the assessment of the earthquake hazard due to seismic sources. This paper presents background information on earthquake-induced liquefaction and the resulting soft-sediment deformation features that may be preserved in the geologic record, best practices used in paleoliquefaction studies, and the application of paleoliquefaction data in earthquake source characterization. The paper shows two examples of regional paleoliquefaction studies—in the Charleston seismic zone and the New Madrid seismic zone in the United States, respectively—which contributed to seismic source models used in earthquake hazard assessment.
- The review paper "Catalogue of the Geological Effects of Earthquakes in Spain Based on the ESI-07 Macroseismic Scale: A New Database for Seismic Hazard Analysis" by P.G. Silva et al. [4] summarizes the content and scope of the "Catalogue of Earthquake Geological Effects in Spain", published by the Geological Survey of Spain, and constitutes the first official publication on seismic hazard, containing geological information for this nation. The catalogue offers a variety of parametric information, quality indexes, and seismic intensity (ESI-07), with a description of environmental damage structured in individual "event files". Sixteen events present full information files with individualized analyses of the geological and geo-archaeological data as well as graphic information with hybrid Environmental Seismic Intensity-European Macroseismic Scale (ESI–EMS) intensity maps, shake maps, and complementary kmz files (Google Earth) among which is the well-known AD 1755 Lisbon earthquake-tsunami.
- The review paper "Surface-Rupturing Historical Earthquakes in Australia and Their Environmental Effects: New Insights from Re-Analyses of Observational Data" by T.R. King et al. [5] presents the digitization of surface rupture maps and compiles observational data from 67 publications on ten of eleven historical, surface-rupturing earthquakes in Australia, in order to analyze the prevailing characteristics of surface ruptures and other environmental effects in this crystalline basement-dominated intraplate environment. The authors have analyzed the earthquakes that occurred between 1968 and 2018, with a magnitude (Mw) from 4.7 to 6.6, and collected environmental effects including primary surface ruptures, secondary fracture/cracks, fissures, rockfalls, ground-water anomalies, vegetation damage, sand-blows/liquefaction, displaced rock fragments, and holes from collapsible soil failure, assessing the relative ESI-07 seismic intensity. This paper represents an important contribution that highlights Mw/ESI-07 relations in geologically different environments.
- The review paper "Post Seismic Catalog Incompleteness and Aftershock Forecasting" by E. Lippiello et al. [6] shows how standard models for earthquake forecasting can be modified to take into account this incompleteness, in particular, the authors focus on forecasting methods based on the data available in real time, in which many events are missing and the uncertainty in hypocenter location is considerable. Furthermore, the authors present retrospective tests that demonstrate the

usefulness of these novel methods compared with traditional ones, which implement average values of parameters obtained from previous sequences.

- The paper "Landslides Triggered by the 2016 Mw 7.8 Pedernales, Ecuador Earthquake: Correlations with ESI-07 Intensity, Lithology, Slope and PGA-h" by K. Chunga et al. [7] presents a dataset of the landslides induced by the 2016 Pedernales megathrust earthquake in Ecuador (Mw 7.8, focal depth of 20 km). The authors compare landslides spatial distribution with mapped bedrock lithology and horizontal peak ground acceleration (PGA-h) in order to evaluate the macroseismic intensity taking into account these earthquake-induced environmental effects (ESI-07). Moreover, the authors underline that coseismically-triggered landslides are among the most common environmental effects occurring during large subduction events, and they can be effectively used to properly evaluate the earthquake macroseismic field.
- The paper "The 1976 Guatemala Earthquake: ESI Scale and Probabilistic/Deterministic Seismic Hazard Analysis Approaches" by M. Caccavale et al. [8] shows how the hazard assessment of the 1976 Guatemala earthquake (M = 7.5) based on the environmental effects had effectively contributed to estimating the high destructive impact of that event. The results evidence that the probabilistic/deterministic hazard analysis procedures may result in very different indications on the PGA distributions, and PGA values often display significant discrepancy from the macroseismic intensity values calculated with the ESI scale. Therefore, the incorporation of the environmental geological effects into the probabilistic/deterministic hazard analysis appears to be mandatory in order to achieve a more accurate seismic hazard estimation.
- The paper "Geological and Structural Control on Localized Ground Effects within the Heunghae Basin during the Pohang Earthquake (MW 5.4, 15 November 2017), South Korea" by S.P. Naik et al. [9] presents the results of a systematic survey of the secondary ground effects, that is, soil liquefaction and ground cracks developed during the Pohang earthquake in South Korea (M_W = 5.4) on 15 November 2017. The liquefaction phenomenon associated with the 2017 Pohang earthquake emphasizes that there is an urgent need of liquefaction potential mapping for Pohang city and other areas with a similar geological setting. The authors highlight that this phenomenon represents a hazard that may cause significant societal and economic threats in the future.
- The paper "Earthquake Environmental Effects of the 1992 M_S = 7.3 Suusamyr Earthquake, Kyrgyzstan, and Their Implications for Paleo-Earthquake Studies" by C. Grützner et al. [10] presents the application of the ESI-07 scale to the 1992 M_S = 7.3 Suusamyr Earthquake in the Kyrgyz Tien Shan. The author shows that the ESI-2007 intensity values distribution differs somewhat from traditional intensity assessments Medvedev-Sponheuer-Karnik (MSK) and Modified Mercalli Intensity scale (MMI), because of the sparse population in the epicentral area and the spatial distribution of primary and secondary Environmental Earthquake Effects (EEEs). However, the ESI-2007 scale captures a similar overall pattern of the intensity distribution. The final results highlight the applicability of this scale, even for earthquakes with complex and unusual primary surface rupture patterns.
- The paper "Earthquake-Induced Landslide Risk Assessment: An Example from Sakhalin Island, Russia" by Konovalov et al. [11] presents a new probabilistic technique for earthquake-induced landslide risk assessment. A fully probabilistic technique suggests a multi-stage hazard assessment; that is, an example from Sakhalin Island, Russia. The given approach follows the rational risk management idea that handles well all possible ground motion scenarios, slope models, and parameters. The authors suggest that the given approach can improve geotechnical studies of slope stability.
- The paper "Site Effect Assessment in Ulaanbaatar, Mongolia through Inversion Analysis of Microtremor H/V Spectral Ratios" by Z. Tumurbaatar et al. [12] presents the evaluation of a site effect, which is one of the essential parts of the earthquake hazard estimation, in Ulaanbaatar city, Mongolia, through inversion analysis of microtremor horizontal-to-vertical (H/V) spectral ratios. In this study, the V_S models are estimated using microtremor data at 50 sites and inversion

analysis is applied to the observed data in order to evaluate site amplifications in Ulaanbaatar city. In particular, the joint inversion technique based on a diffuse field approach is applied to estimate the V_S structures at three sites using the observed horizontal-to-vertical (H/V) spectral ratios and surface wave phase velocities obtained by Odonbaatar (2011). The result shows that the largest site amplification zone is computed along the Tuul river in the southeastern part of Ulaanbaatar.

- The paper "Bayesian Variable Selection for Pareto Regression Models with Latent Multivariate Log Gamma Process with Applications to Earthquake Magnitudes" by H-C. Yang et al. [13] presents a Bayesian spatial variable selection for Pareto regression based on Bradley et al. and Hu et al. to tackle the variable selection issue in generalized linear regression models with spatial random effects. The authors demonstrate the applicability of the proposed method in an analysis of the earthquake data obtained from the United States Geological Survey (USGS).
- The paper "Machine Learning Methods for Seismic Hazards Forecast" by V.G. Gitis and A.B. Derendyaev [14] presents two machine learning methods for seismic hazard forecast. The first method is used for the spatial forecasting of maximum possible earthquake magnitudes (*Mmax*), whereas the second one is used for space-time forecasting of strong earthquakes. The authors show the results of testing the approach on earthquake prediction in the Mediterranean and Californian regions.
- The paper "Suitability Analysis for the Emergency Shelters Allocation after an Earthquake in Japan" by T. Akamatsu and K. Yamamoto [15] presents an applied statistical method and public open data related to population and emergency shelters, with the aim to quantitatively conduct a suitability analysis for the allocation of emergency shelters after an earthquake in Japan, in a GIS environment. The results show the districts that lack emergency shelters, and visually show the places where such facilities should be newly established on the digital map of GIS. Additionally, the assessment method is reproducible in the spatial and temporal dimension. However, it is necessary to create more original data related to emergency shelters to raise the reliability of the results, as the present research has the limitation of data availability.
- The paper "Use of Macroseismic Intensity Data to Validate a Regionally Adjustable Ground Motion Prediction Model" by Y. Tang et al. [16] presents a regionally adjustable ground motion prediction equation (GMPE), known as the component attenuation model (CAM), by which a diversity of crustal conditions can be covered in one model. In addressing the challenge of validating a GMPE for use in an area where instrumental data are scarce, Modified Mercalli intensity (MMI) data inferred from peak ground velocity values predicted by CAM are compared with the records of MMI of past earthquake events, as reported in historical archives. South-Eastern Australia and South-Eastern China are the two study regions used in this article for demonstrating the viability of CAM as a ground motion prediction tool in an intraplate environment.
- The paper "Comparison of Earthquake-Triggered Landslide Inventories: A Case Study of the 2015 Gorkha Earthquake, Nepal" by S.R. Meena and S. Tavakkoli Piralilou [17] presents an overview of the impact of methodology selection and outlines the limitations and advantages of different remote sensing and mapping techniques for landslide inventorying: a case study of the 2015 Gorkha earthquake, Nepal. After the main event on 25 April 2015, researchers around the world mapped the landslides induced by this earthquake. In this research, the authors compared four of these published inventories qualitatively and quantitatively using different techniques. Two principal methodologies, namely the cartographical degree of matching and frequency area distribution (FAD), were optimized and applied to evaluate inventory maps.
- The paper "Dynamics of the Zones of Strong Earthquake Epicenters in the Arctic-Asian Seismic Belt" by L.P. Imaeva [18] presents a comprehensive study of the Russian Arctic region to clarify the features and types of seismotectonic deformation of the crust in the Arctic-Asian Seismic Belt, specifically in the zones of strong earthquakes in the Laptev Sea Segment, the Kharaulakh Segment, and the Chersky Seismotectonic Zone. The authors have analyzed modern tectonic structures and active fault systems, as well as tectonic stress fields reconstructed by tectonophysical analysis of

the Late Cenozoic faults and folds. A set of models has been constructed for the studied segments of plate boundaries with account of the dynamics of the regional geological structures. The models can give a framework for the assessment of potential seismic risks of seismicity-generating structures in the Russian Arctic region.

- The paper "Undrained Cyclic Laboratory Behavior of Sandy Soils" by Castelli et al. [19] presents the complex cyclic shear stress path experienced by the soil during an earthquake, which could also induce the liquefaction phenomena. The authors performed a detailed geological and geotechnical characterization of the area through in situ and laboratory tests, including seismic dilatometer Marchetti tests (SDMTs), the combined resonant column (RCT) and cyclic loading torsional shear tests (CLTSTs), and undrained cyclic loading triaxial tests (CLTxTs). The paper presents the results of cyclic triaxial tests carried out on isotropically consolidated specimens of a sandy soil.

Acknowledgments: The Guest Editors thank all the Authors, the *Geosciences'* Editors, and the reviewers for their appreciable contributions and commitment to this Special Issue. Special thanks go to Richard Li, *Geosciences'* Assistant Editor, for his dedication to this project and his valuable collaboration in the setup, promotion, and management of the Special Issue.

Conflicts of Interest: The authors declare no conflict of interest.

References

1. Michetti, A.M.; Esposito, E.; Guerrieri, L.; Porfido, S.; Serva, L.; Tatevossian, R.; Vittori, E.; Audemard, F.; Azuma, T.; Clague, J.; et al. Environmental Seismic Intensity scale—ESI 2007. *Mem. Descr. Carta Geol. D'Ital.* **2007**, *74*, 7–23.
2. Serva, L. History of the Environmental Seismic Intensity Scale ESI-07. *Geosciences* **2019**, *9*, 210. [CrossRef]
3. Tuttle, M.P.; Hartleb, R.; Wolf, L.; Mayne, P.W. Paleoliquefaction Studies and the Evaluation of Seismic Hazard. *Geosciences* **2019**, *9*, 311. [CrossRef]
4. Silva, P.G.; Rodríguez-Pascua, M.A.; Giner Robles, J.L.; Élez, J.; Pérez-López, R.; Davila, M.B.B. Catalogue of the Geological Effects of Earthquakes in Spain Based on the ESI-07 Macroseismic Scale: A New Database for Seismic Hazard Analysis. *Geosciences* **2019**, *9*, 334. [CrossRef]
5. King, T.R.; Quigley, M.; Clark, D. Surface-Rupturing Historical Earthquakes in Australia and Their Environmental Effects: New Insights from Re-Analyses of Observational Data. *Geosciences* **2019**, *9*, 408. [CrossRef]
6. Lippiello, E.; Cirillo, A.; Godano, C.; Papadimitriou, E.; Karakostas, V. Post Seismic Catalog Incompleteness and Aftershock Forecasting. *Geosciences* **2019**, *9*, 355. [CrossRef]
7. Chunga, K.; Livio, F.A.; Martillo, C.; Lara-Saavedra, H.; Ferrario, M.F.; Zevallos, I.; Michetti, A.M. Landslides Triggered by the 2016 Mw 7.8 Pedernales, Ecuador Earthquake: Correlations with ESI-07 Intensity, Lithology, Slope and PGA-h. *Geosciences* **2019**, *9*, 371. [CrossRef]
8. Caccavale, M.; Sacchi, M.; Spiga, E.; Porfido, S. The 1976 Guatemala Earthquake: ESI Scale and Probabilistic/Deterministic Seismic Hazard Analysis Approaches. *Geosciences* **2019**, *9*, 403. [CrossRef]
9. Naik, S.P.; Kim, Y.-S.; Kim, T.; Su-Ho, J. Geological and Structural Control on Localized Ground Effects within the Heunghae Basin during the Pohang Earthquake (MW 5.4, 15th November 2017), South Korea. *Geosciences* **2019**, *9*, 173. [CrossRef]
10. Grützner, C.; Walker, R.; Ainscoe, E.; Elliott, A.; Abdrakhmatov, K. Earthquake Environmental Effects of the 1992 MS7.3 Suusamyr Earthquake, Kyrgyzstan, and Their Implications for Paleo-Earthquake Studies. *Geosciences* **2019**, *9*, 271. [CrossRef]
11. Konovalov, A.; Gensiorovskiy, Y.; Lobkina, V.; Muzychenko, A.; Stepnova, Y.; Muzychenko, L.; Stepnov, A.; Mikhalyov, M. Earthquake-Induced Landslide Risk Assessment: An Example from Sakhalin Island, Russia. *Geosciences* **2019**, *9*, 305. [CrossRef]
12. Tumurbaatar, Z.; Miura, H.; Tsamba, T. Site Effect Assessment in Ulaanbaatar, Mongolia through Inversion Analysis of Microtremor H/V Spectral Ratios. *Geosciences* **2019**, *9*, 228. [CrossRef]
13. Yang, H.-C.; Hu, G.; Chen, M.-H. Bayesian Variable Selection for Pareto Regression Models with Latent Multivariate Log Gamma Process with Applications to Earthquake Magnitudes. *Geosciences* **2019**, *9*, 169. [CrossRef] [PubMed]

14. Gitis, V.G.; Derendyaev, A.B. Machine Learning Methods for Seismic Hazards Forecast. *Geosciences* **2019**, *9*, 308. [CrossRef]
15. Akamatsu, T.; Yamamoto, K. Suitability Analysis for the Emergency Shelters Allocation after an Earthquake in Japan. *Geosciences* **2019**, *9*, 336. [CrossRef]
16. Tang, Y.; Lam, N.; Tsang, H.-H.; Lumantarna, E. Use of Macroseismic Intensity Data to Validate a Regionally Adjustable Ground Motion Prediction Model. *Geosciences* **2019**, *9*, 422. [CrossRef]
17. Meena, S.R.; Tavakkoli Piralilou, S. Comparison of Earthquake-Triggered Landslide Inventories: A Case Study of the 2015 Gorkha Earthquake, Nepal. *Geosciences* **2019**, *9*, 437. [CrossRef]
18. Imaeva, L.P.; Imaev, V.S.; Koz'min, B.M. Dynamics of the Zones of Strong Earthquake Epicenters in the Arctic–Asian Seismic Belt. *Geosciences* **2019**, *9*, 168. [CrossRef]
19. Castelli, F.; Cavallaro, A.; Grasso, S.; Lentini, V. Undrained Cyclic Laboratory Behavior of Sandy Soils. *Geosciences* **2019**, *9*, 512. [CrossRef]

geosciences

MDPI

Review

History of the Environmental Seismic Intensity Scale ESI-07

Leonello Serva †

Independent Consultant, Via della Resistenza 171, 00185 Cantalice (Rieti), Italy; lserva@alice.it

† Former General Director of the Geological Survey of Italy-Institute for Environmental Protection and Research Agency (ISPRA-Italy); Member of the European Academy of Sciences and Arts-Class V-Environmental and Technological Sciences.

Received: 11 March 2019; Accepted: 6 May 2019; Published: 10 May 2019

Abstract: This brief note aims to describe the history, from its early original idea, of the new macroseismic scale: The Environmental Seismic Intensity Scale 2007 (ESI 2007). It can be used together with other existing scales or alone when needed for measuring the intensity of an earthquake on the basis of the primary and secondary effects of a seismic event on the natural environment. These effects could be the major sources of earthquake hazards, as recently proved. This note also aims to contribute to the understanding of processes that induced the researcher to develop an idea, to pursue it, and bring it to its end, first through the help of valuable Italian researchers and then through the constructive exchange of ideas with researchers of different cultural backgrounds operating almost everywhere in the world. This note is sponsored and approved by the International Union for Quaternary Research (INQUA), and the Environmental Seismic Intensity scale (ESI-07) was published in 2007 after a revision process of about eight years.

Keywords: earthquake hazards; ground effects; ESI scale 2007; EEE database

---

1. Introduction

Following the kind invitation of the Editors, herein I introduce some personal considerations on the state of the Environmental Seismic Intensity scale (ESI 2007 scale) [1] starting from its creation to the present and its perspective for the future. This brief note aims to contribute to an understanding of the processes that induced the researcher to develop an idea, pursue it, and bring it to an end, first through the help of valuable Italian researchers and then through the constructive exchange of ideas with researchers of different cultural backgrounds operating almost everywhere in the world.

The Environmental Seismic Intensity scale (ESI 2007 scale was published in 2007 after a revision process of about eight years as a new intensity scale based only on the Earthquake Environmental Effects (EEEs). This scale integrates the traditional macroseismic scales, of which it represents an evolution, allowing the intensity parameter to also be assessed where buildings are absent and when diagnostic damage-based elements have saturated, exclusively on the basis of environmental effects. Actually, the ground effects have recently proved to be major sources of hazard in addition to vibratory ground motion. In fact, although a serious source of direct damage, the ground motion is not the only parameter to be considered since most damage is due to coseismic geological effects that are directly connected to the earthquake's source or caused by ground shaking. Primary effects such as surface faulting, regional uplift, and subsidence, and secondary effects such as tsunamis, liquefaction, ground failure, and landslides (sensu ESI 2007 scale) must now be taken into account for a more correct and complete evaluation of seismic hazards at both regional and local scales.

The ESI 2007 scale is a 12-degree scale: each degree reflects the corresponding strength of an earthquake and provides a measure of its intensity on the basis of its characteristics. The main advantage of the ESI 2007 scale is the classification, quantification, and measurement of several known geological, hydrological, and geomorphological features that are associated with each intensity degree, and therefore it allows for the definition of seismic intensity based on the entire scenario of geological ground effects.

The ESI 2007 scale was promoted by several geologists, seismologists, and engineers coordinated by the Servizio Geologico d'Italia of the Institute for Environmental Protection and Research (ISPRA). This scale has been tested worldwide on several modern earthquakes, historical earthquakes, and paleoearthquakes [2–7].

2. History and Considerations on the Scale and Associated EEEs Database

In the early 1990s, I was working at the Italian Nuclear Regulatory Commission (ENEA-DISP), responsible for the safety of nuclear power plants (NPPs) regarding the effects of natural phenomena, with special attention devoted to those caused by earthquakes. During those years, I realized that the earthquake intensity parameter—a significant parameter for evaluating the whole set of potential earthquake effects—lacked its initial and fundamental characteristics, mainly because the people in charge of evaluating intensity were not making proper use of the traditional intensity scales (Mercalli–Cancani–Sieberg, MCS; Medvedev–Sponhouer–Karnik, MSK; and Modified Mercalli, MM). In fact, they were steadily neglecting the effects of earthquakes on the ground (and more generally on the natural environment) which, as is now clearly accepted by all (differently from the 1990s), are responsible for quite a large percentage of earthquake damage.

A clear example of this case is the Japanese Fukushima NPP during the 2011 earthquake, where the effects of ground shaking on the built environment were negligible as opposed to the catastrophic tsunami generated by the surface faulting of the ocean bottom [8].

As a consequence of this bad approach, the publication of the European Macroseismic Scale (EMS-98) [9]) appeared, where the ground effects were relegated to an appendix of no real value. As stated above, the non-utilization of ground effects resulted in a complete alteration of the meaning and value of the intensity parameter. For example, assessing intensity without considering ground effects would have made it impossible to compare the intensity associated with past earthquakes to the intensity assigned to recent ones, and this could have led to erroneous conclusions about the seismic hazard in a given territory and therefore for the facilities located therein, with a special emphasis on NPPs. At that time, these intensity evaluators claimed that the use of the ground effects was not possible due to their extreme variability in both time and space, and the absence of a proper ground-effects database from which to derive a reliable description of the effects for each intensity degree. This was partly true considering the incompleteness of the database, but in my opinion was completely wrong with regard to the great variability of the effects, because this was fully comparable to that of man-made structures. Furthermore, the great variability of the ground effects enables the proper estimation of the strongest earthquakes (mainly intensities between X and XII degrees where in most cases the effects on built structures saturate).

It is important to remark that between 1980 and 1990, paleoseismology was developing considerably all over the world, and I was lucky to meet Professor D. Burt Slemmons, one of the few great fathers of earthquake geology, in 1982 when I was engaged in a training course at the United States Nuclear Regulatory Commission in Washington, DC (USNRC). He was a teacher at the Mackay School of Mines in Reno (Nevada, USA). We became good friends and accomplished many missions together in several countries for the International Atomic Energy Agency (IAEA). Through contact with him and his assistants, in particular with Robert E. Whitney, I started, as a pioneer in Europe, to study and publish papers on paleoseismology [10–22]. These papers, some of which are in Italian [10–12,16,17], show how throughout the years the study of paleoseismicity and its application to Italian earthquakes have grown with the aim of carrying out seismic hazard assessments from paleoseismic evidence. The pathway goes from the first studies concerning the Rieti basin [18,19] or the Fucino Plain [20] to the evidence of strong seismic paleoevents in the "aseismic" zone of Pollino [21], proving the state-of-the-art of paleoseismology in those years.

I also need to say that my interest in paleoseismology was facilitated by the lessons of Giorgio Magri, my Italian supervisor at the Comitato Nazionale Energia Nucleare/Direzione Sicurezza e Protezione (CNEN-DISP). He always repeated to me that it is better to say that mountains grow because of earthquakes as opposed to tectonics.

With the support of paleoseismology, I decided to adopt the approach of producing something useful for the proper reuse of ground effects in intensity evaluation and restoring the intensity parameter to its proper meaning.

The first step was the production of two articles [22,23]. The first was published in Terra Nova. In that article, I made an extraction of the ground effects according to the various degrees as reported in the MCS, MSK, MM, and Japanese scales. The main goal was to remind the intensity evaluators that ground effects are massively included in the historically most-applied intensity scales (MCS, MSK, and MM) from the beginning of their conception. The second one was written to present the idea to a very broad international audience during the special session of the International Union of Geological Sciences (IUGS) Congress held in Beijing in 1996. The simple database, presented in [22], was the "seed" of the huge plant that later gave rise to the ESI scale. I must remind current users that it should be used in conjunction with one of the other traditional scales in order to avoid the same mistake made in the 1990s by the intensity evaluators concerning built structures, that is, the lack of consideration for the entire set of earthquake effects when assessing intensity. It can be used alone only for higher degrees and in the case of sparsely populated areas. It can also be used to verify the intensity of past earthquakes in cases where the intensity was evaluated not using the ground effects properly. As said above, the seeds started to sprout, and the two roadmaps were the initial drafts of the ESI scale and the construction of the associated database.

Regarding the ESI scale, the steps have been as follows:

1. The collection and careful reading of published and unpublished papers and documents dealing with the same objective, first among them: [24]. I think we have to be very grateful to these authors because they opened the road for using ground effects in earthquake intensity evaluation.
2. The presentation of the drafted ESI content at conferences at national and international levels (i.e., [25–28]) and the continuous discussion of the content at national and international levels with the involvements of numerous well-known international experts, to whom I am truly grateful. Among them, special mentions (also due to our friendship) are to be given to:

    - Bagher Mohammadioun and his wife Jody. Bagher is a well-known seismologist [27] who for many years headed the Bureau d'Évaluation du Risque Sismique pour la Sûreté des Installations Nucléaires at the Institut Radioprotetion Surete Nucléaire (IPSN) of France;
    - Ruben Tatevossian (seismologist) and Eugene Roghozin (geologist)—two well-known Russian scientists with great experience concerning macroseismic data and paleoseismology [24];
    - Aybars Gurpinar, a civil engineer who for many years has been the Director of the Nuclear Installation Safety Division at the IAEA, with great worldwide experience on seismic hazards in relation to NPP sites [29];
    - Frank Audemard, a Venezuelan paleoseismologist with great experience, mainly in seismically-active South American countries [30];
    - Shmulik Marco, a well-known Israeli paleoseismologist and author of significant papers on the subject [31];
    - James McCalpin, author of the international benchmark book titled *Paleoseismology* [32];
    - Nils-Axel Mörner, a Swedish geologist with a broad field of experience, including in paleoseismology [33];
    - John Clague, a Canadian authority in quaternary and environmental earth sciences [34];
    - Yoko Ota, also a geologist with great knowledge of the earthquake geology of Japan [35];

- Takashi Azuma, a geologist with significant knowledge of paleoseismology in Japan [35].

The first version of the new macroseismic scale was published in 2004 [36], and the second in 2007 [1]. The scale was very well accepted worldwide, and a huge number of papers dealing with the use of this scale have been produced since then across the five continents [1–7,37–48].

Thanks to the kind insistence of my collaborators, the version integrated with three historical cases and with my name as the first author was published in 2015 [49]. Therefore, based on several years of worldwide application in the field, in this paper: (a) we introduced the ESI scale to the community of earth scientists (geologists, geophysicists, and seismologists) and civil engineers as a survey instrument to better characterize a seismic event, also in terms of local effects and attenuation with distance, and (b) we provided insurers, civil protection agencies, and administrators with an integrated tool to assess the potential damage deriving from geological effects during a future earthquake in an area, to be added to the damage directly associated to seismic shaking. Presently, this scale is also going to be used officially by some institutions (e.g., in Georgia) dealing with seismic hazard evaluation, and I hope this practice will also be adopted by other institutions. In fact, a seismic hazard is often underestimated because it is based on lower intensities than the ground environmental effects (e.g., earthquakes of Mw > 7 have been associated to I = X), which are instead considered in the ESI scale.

I also think it is important to remark that seismic hazard assessment (SHA) would benefit from a comprehensive consideration of all earthquake-related effects, including environmental ones. The key role of the ESI scale use to SHA is the improved intensity assessment. It is also necessary in order to preserve the consistency between the source parameters assessed for historical earthquakes and for recent ones. The basic message is that despite the advent of magnitude, earthquake intensity persists as a fundamental seismic parameter for reliable SHA, especially when EEEs are properly taken into account.

The EEE Catalogue is a database containing information on earthquake effects on the environment [50]. Such information includes the effects of recent earthquakes but also data derived from historical and paleo-earthquakes. The objective of this database is to procure data to be used for the present and future updating of the ESI scale. This worldwide database was initially created in the framework of the International Union for Quaternary Research (INQUA) activities. The web infrastructures were developed by ISPRA—the Geological Survey of Italy. In 2007, the first access-type database including around 20 events was created. This first online structure was presented in the Bern (Switzerland) INQUA Congress in 2011 [50]. Thanks to its online structure, recent, historical, and paleo-earthquakes' environmental effects data were subsequently uploaded to this database, provided by worldwide authors. It currently includes data on around 200 events (http://193.206.192.211/wfd/eee_catalog/viewer.php). Since 2012, the EEE catalogue has been accessible online on the websites of the IAEA and the International Seismic Safety Center (ISSC). It is important to note that the IAEA, in the wake of the disastrous Tohoku 2011 tsunami, has recommended that member states should carefully consider earthquake ground effects—and, more generally, paleoseismological data—in order to achieve a better definition of the seismic hazards in areas where nuclear installations are already installed.

Finally, it is important to recall the numerous papers published worldwide recognizing the importance of this catalogue. Among them is the ISPRA volume published in 2015 [51], a significant monograph that reports the description of the ESI scale in ten languages (English, Italian, Spanish, French, German, Japanese, Russian, Greek, Dutch, and Korean).

Currently, the Italian National Institute of Geophysics and Volcanology (INGV) has also finally recognized the importance of the environmental effects induced by earthquakes by including them in their new seismic catalogs [52].

Together with the EEE database, the ITaly HAzards from CApable faults (ITHACA) database born from an idea of Eutizio Vittori. It includes the current knowledge on capable faulting in Italy, to which paleoseismological studies widely contribute. At present, this continuously updated catalogue is used for the revision or planning of new infrastructure and microzonation purposes in the whole territory of Italy [53,54].

It is important to underline that, globally, there are very well-organized similar catalogues of active faults, such as the one proposed for the USA by the United States Geological Survey (USGS) (https://usgs.maps.arcgis.com/apps/webappviewer/index.html?id=5a6038b3a1684561a9b0aadf88412fcf) and another for Central Asia organized by the University of Tubingen (https://esdynamics.geo.uni-tuebingen.de/faults/).

In conclusion, I strongly hope that the institution where I worked (currently named ISPRA) will continue our efforts to maintain the databases quoted above and soon release an update of the ESI scale based on the wealth of information gathered in recent seismic events worldwide.

Funding: This research received no external funding.

Acknowledgments: I take the opportunity here to express my sincere gratitude to my collaborators, since without their help the scale would not have been published. First, to Alessandro Maria Michetti, who led the group responsible for the first publication in 2007. Then, Eutizio Vittori, who was of considerable help concerning the draft of the scale and its final publication in 2015. Next, Luca Guerrieri, especially for the work accomplished for the EEE database. Fourth, Sabina Porfido and Eliana Esposito, for the enthusiasm they always expressed for this initiative and for the considerable effort they have made for its use at a global scale, and finally for Valerio Comerci, who arrived later but in that short period was still able to make a significant contribution.

Conflicts of Interest: The authors declare no conflict of interest.

References

1. Michetti, A.M.; Esposito, E.; Guerrieri, L.; Porfido, S.; Serva, L.; Tatevossian, R.; Vittori, E.; Audemard, F.; Azuma, T.; Clague, J.; et al. Intensity Scale ESI 2007. *Mem. Descr. Carta Geol. D'Italia* **2007**, *74*, 11–20.
2. Silva, P.G.; Rodríguez-Pascua, M.A.; Pérez-López, R.; Bardaji, T.; Lario, J.; Alfaro, P.; Martínez-Díaz, J.J.; Reicherter, K.; Giménez García, J.; Giner, J.; et al. Catalogacion de los efectos geologicos y ambientales de los terremotos en Espana en la Escala ESI 2007 y su aplicacion a los estudios paleosismologicos. *Geotemas* **2008**, *6*, 1063–1066. (In Spanish)
3. Lekkas, E. The 12 May 2008 Mw 7.9 Wenchuan, China, Earthquake: Macroseismic Intensity Assessment using the EMS-98 and ESI 2007 Scales and their correlation with the geological structure. *Bull. Seism. Soc. Am.* **2010**, *100*, 2791–2804. [CrossRef]
4. Reicherter, K.; Michetti, A.M.; Silva, P.G. (Eds.) *Paleoseismology: Historical and Prehistorical Records of Earthquake Ground Effects for Seismic Hazard Assessment*; The Geological Society of London: London, UK, 2009; pp. 544–552. [CrossRef]
5. Papanikolau, I.D. Uncertainty in intensity assignment and attenuation relationships: How seismic hazard maps can benefit from the implementation of the Environmental Seismic Intensity scale (ESI 2007). *Quat. Int.* **2011**, *242*, 42–51. [CrossRef]
6. Giles, D. Intensity scales. In *Encyclopedia of Natural Hazards*; Bobrowsky, P.T., Ed.; Springer: Dordrecht, The Netherlands, 2013.
7. Porfido, S.; Nappi, R.; De Lucia, M.; Gaudiosi, G.; Alessio, G.; Guerrieri, L. The ESI scale, an ethical approach to the evaluation of seismic hazards. In Proceedings of the EGU General Assembly 2015, Vienna, Austria, 12–17 April 2015. Geophysical Research Abstracts Vol. 17, EGU2015-11732-2.
8. IAEA. *Report on the Fukushima Daiichi Accident*; IAEA: Vienna, Austria, 2015.
9. Grunthal, G. (Ed.) *European Macroseismic Scale 1998 (EMS-98)*; Centre Européen de Géodynamique et de Séismologie: Luxembourg, 1998.
10. Magri, G.; Serva, L. Approccio metodologico per la valutazione del terremoto di riferimento per la progettazione di impianti nucleari. *Mem. Soc. Geol. Ital.* **1986**, *35*, 507–513.
11. Serva, L.; Blumetti, A.M.; Michetti, A.M. Gli effetti sul terreno del terremoto del Fucino (13.11.1915). Tentativo di interpretazione della evoluzione tettonica recente di alcune strutture. *Mem. Soc. Geol. Ital.* **1986**, *35*, 893–907.
12. Serva, L. Un metodo per una migliore comprensione della sismicítà di un'area (Conca del Fucino). In *Notiziario ENEA-DISP: Sicurezza e Protezione*; ENEA: Roma, Italy, 1987; pp. 55–63.

13. Calderoni, G.; Lorenzoni, P.; Ortolani, F.; Pagliuca, S.; Serva, L. Paleoseismological evidences at Rivisondoli. Central Apennines, Italy. *Rend. Soc. Geol. Ital.* **1990**, *13*, 27–32.
14. Gurpinar, A.; Serva, L. Paleoseismological evidences in a railroad cut near Baiji (Iraq). *Rend. Soc. Geol. Ital.* **1990**, *13*, 33–36.
15. Vittori, E.; SylosLabini, S.; Serva, L. Paleoseismicity: Critical review of the state of the art. *Tectonophysics* **1991**, *193*, 9–32. [CrossRef]
16. Blumetti, A.M.; Brunamonte, F.; Michetti, A.M.; Serva, L.; Vittori, E. Dislocazioni cosismiche e magnitudo: Una proposta di classificazione basata sull'analisi paleosismologica di strutture tettoniche dell'Appennino Centrale. *Il Quaternario* **1993**, *6*, 338–339.
17. Brunamonte, B.; Michetti, A.M.; Serva, L.; Vittori, E. Evidenze paleosismologic nell'Appennino Centrale ed implicazioni neotettoniche. *Studi Geol. Camerti* **1992**, *2*, 265–270.
18. Brunamonte, F.; Michetti, A.M.; Serva, L.; Whitney, R.E. Seismic hazard assessment from paleoseismological evidence in the Rieti Region (Central Italy). In *Perspectives in Paleoseismology*; Serva, L., Slemmons, D.B., Eds.; AEG Special Publications: Seattle, WA, USA, 1995; pp. 63–82.
19. Michetti, A.M.; Serva, L. New data on seismotectonic potential of the Leonessa fault area (Rieti, Central Italy). *Rend. Soc. Geol. Ital.* **1990**, *13*, 37–46.
20. Michetti, A.M.; Brunamonte, F.; Serva, L.; Vittori, E. Trench investigations alongthe 1915 Fucino earthquake fault scarps (Abruzzo, Central Italy): Geological evidence of large historical events. *J. Geophys. Res.* **1996**, *101*, 5921–5936. [CrossRef]
21. Michetti, A.M.; Ferreli, L.; Serva, L.; Vittori, E. Geological evidence for strong historical earthquakes in an "aseismic" region: The Pollino case (Southern Italy). *J. Geodyn.* **1997**, *24*, 67–86. [CrossRef]
22. Serva, L. Ground effects in intensity scales. *Terra Nova* **1994**, *6*, 414–416. [CrossRef]
23. Esposito, E.; Porfido, S.; Mastrolorenzo, G.; Nikonov, A.A.; Serva, L. Brief review and preliminary proposal for the use of ground effects in the macroseismic intensity assessment. In Proceedings of the 30th IUGS Congress, Beijing, China, 4–14 August 1996; Contemporary Lithospheric Motion Seismic Geology International Science Publishers: Zeist, The Netherlands, 1997; pp. 233–243.
24. Tatevossian, R.; Roghozin, E.; Guerrieri, L.; Michetti, A.M.; Serva, L.; Vittori, E. *Effektyzemletryaseniya v Prirodnoysrede (EEE) Iotsenkaintensivnosti: Proyektshkaly INQUA*; Earthquake Effects in Natural Environment (EEE) for Earthquake Intensity Calibration: The INQUA Scale Project. Issledovaniya po Seismotektonike i Sovremennoy Geodinamike (Studies on Seismotectonics and Modern Geodynamics); IFZ RAN Publ, 2006. Special Issue (In Russian)
25. Dengler, L.A.; McPherson, R. The 17 August 1991 Honeydew earthquake: A case for revising the Modified Mercalli scale in sparsely populated areas. *Bull. Seism. Soc. Am.* **1993**, *83*, 1081–1094.
26. Abdel Aziz, M.A.H.; Serva, L.; Michetti, A.M.; Giardina, F.; Swedan, A.H.; Gad Elhaq, G.; Soliman, H. Assessing intensity values for earthquakes affecting low populated areas: The case of 22 November 1995 Nuweiba earthquake. *Ann. Geol. Surv. Egypt* **2007**, *29*, 321–334.
27. Mohammadioun, B.; Serva, L. Stress Drop, Slip Type, Earthquake Magnitude, and Seismic Hazard. *Bull. Seism. Soc. Am.* **2001**, *91*, 694–707. [CrossRef]
28. Cisternas, A.; Philip, H.; Bousquet, J.C.; Cara, M.; Deschamps, A.; Dorbath, L.; Dorbath, C.; Haessler, H.; Jimenez, E.; Nercessian, A.; Rivera, L.; Romanowicz, B.; et al. The Spitak (Armenia) earthquake of 7 December 1988: Field observations, seismology and tectonics. *Nature* **1989**, *339*, 675–679. [CrossRef]
29. Gürpinar, A. The importance of paleoseismology in seismic hazard studies for critical facilities. *Tectonophysics* **2005**, *408*, 23–28. [CrossRef]
30. Audemard, F.A.M. Contribucions of paleoseimology to historical seismicity: The 1610 and 1894 earthquakes in the southern Venezuelan Andes. *Rev. Geogr. Venez.* **1998**, *39*, 87–105.
31. Marco, S.; Stein, M.; Agnon, A.; Ron, H. Long-term earthquake clustering: A 50,000-year paleoseismic record in the Dead Sea Graben. *J. Geophys. Res. Solid Earth* **1996**, *101*, 6179–6191. [CrossRef]
32. McCalpin, J.P. *Paleoseismology*, 2nd ed.; Academic Press: San Diego, CA, USA, 2009.
33. Mörner, N.A. *Paleoseismicity of Sweden: A Novel Paradigm*. 2003. ISBN9163140721, 9789163140723- 320. Available online: https://www.researchgate.net/publication/286457401_Paleoseismicity_of_Sweden_-_a_novel_paradigm (accessed on 9 May 2019).
34. Clague, J.J. Evidence for large earthquakes at the Cascadia subduction zone. *Rev. Geophys.* **1997**, *35*, 439–460. [CrossRef]

35. Ota, Y.; Azuma, T.; Nina Lin, Y. Application of INQUA Environmental Seismic Intensity Scale to recent earthquakes in Japan and Taiwan. *Geol. Soc. Lond.* **2009**, *316*, 55–71. [CrossRef]

36. Michetti, A.M.; Esposito, E.; Gurpinar, A.; Mohammadioun, B.; Mohammadioun, G.; Porfido, S.; Roghozin, E.; Serva, L.; Tatevossian, R.; Vittori, E.; et al. The INQUA Scale. An innovative approach for assessing earthquake intensities based on seismically-induced ground effects in natural environment. *Mem. Descr. Carta Geol. D'Italia* **2004**, *67*, 1–118.

37. Mosquera-Machado, S.; Lalinde-Pulido, C.; Salcedo-Hurtado, E.; Michetti, A.M. Ground effects of the 18 October 1992, Murindo earthquake (NW Colombia), using the Environmental Seismic Intensity Scale (ESI 2007) for the assessment of intensity. *Geol. Soc. Lond.* **2009**, *316*, 123–144. [CrossRef]

38. Berzhinskii, Y.A.; Ordynskaya, A.P.; Gladkov, A.S.; Lunina, O.V.; Berzhinskaya, L.P.; Radziminovich, N.A.; Radziminovich, Y.B.; Imayev, V.S.; Chipizubov, A.V.; Smekalin, O.P. Application of the ESI-2007 Scale for Estimating the Intensity of the Kultuk Earthquake, August 27, 2008 (South Baikal). *Seism. Instr.* **2010**, *46*, 307–324. [CrossRef]

39. Gosar, A. Application of Environmental Seismic Intensity scale (ESI 2007) to Krn Mountains 1998 Mw = 5.6 earthquake (NW Slovenia) with emphasis on rockfalls. *Nat. Hazards Earth Syst. Sci.* **2012**, *12*, 1659–1670. [CrossRef]

40. Fountoulis, I.G.; Mavroulis, S.D. Application of the Environmental Seismic Intensity scale (ESI 2007) and the European Macroseismic Scale (EMS-98) to the Kalamata (SW Peloponnese, Greece) earthquake (Ms = 6.2, 13 September 1986) and correlation with neotectonic structures and active faults. *Ann. Geophys.* **2013**, *56*, S0675. [CrossRef]

41. Ahmad, B.; Sana, H.; Alam, A. Macroseismic intensity assessment of 1885 Baramulla Earthquake of northwestern Kashmir Himalaya, using the Environmental Seismic Intensity scale ESI 2007. *Quat. Intern.* **2014**, *321*, 59–64. [CrossRef]

42. Garduño-Monroy, V.H. Una propuesta de escala de intensidad sísmica obtenida del códice náhuatl Telleriano Remensis. *Arqueol. Iberoam.* **2016**, *31*, 9–19.

43. Heddar, A.; Beldjoudi, H.; Aurhemayou, C.; SiBachir, R.; Yelles-Chaouche, A.A.; Boudiaf, A. Use of the ESI-2007 scale to evaluate the 2003 Boumerdès earthquake (North Algeria). *Ann. Geophys.* **2016**, *59*, 0211. [CrossRef]

44. Sanchez, J.J.; Maldonado, R.F. Application of the ESI 2007 Scale to Two Large Earthquakes: South Island, New Zealand (2010 Mw 7.1), and Tohoku, Japan (2011 Mw 9.0). *Bull. Seism. Soc. Am.* **2016**, *106*, 1151–1161. [CrossRef]

45. Papanikolaou, I.; Melaki, M. The Environmental Seismic Intensity Scale (ESI 2007) in Greece, addition of new events and its relationship with magnitude in Greece and the Mediterranean; preliminary attenuation relationships. *Quat. Int.* **2017**, *451*, 37–55. [CrossRef]

46. Nappi, R.; Gaudiosi, G.; Alessio, G.; De Lucia, M.; Porfido, S. The environmental effects of the 1743 Salento earthquake (Apulia, southern Italy): A contribution to seismic hazard assessment of the Salento Peninsula. *Nat. Hazards* **2017**, *86*, 295–324. [CrossRef]

47. King, T.R.; Mark, C.Q.; Clark, D. Earthquake environmental effects produced by the Mw 6.1, 20th May 2016 Petermann earthquake, Australia. *Tectonophysis* **2018**, *747*, 357–372. [CrossRef]

48. Chunga, K.; Livio, F.; Mulas, M.; Ochoa-Cornejo, F.; Besenzon, D.; Ferrario, M.F.; Michetti, A.M. Earthquake Ground Effects and Intensity of the 16 April 2016 Mw 7.8 Pedernales, Ecuador, Earthquake: Implications for the Source Characterization of Large Subduction Earthquakes. *Bull. Seism. Soc. Am.* **2018**, *108*, 3384–3397. [CrossRef]

49. Serva, L.; Vittori, E.; Comerci, V.; Esposito, E.; Guerrieri, L.; Michetti, A.M.; Mohammadioun, B.; Mohammadioun, G.C.; Porfido, S.; Tatevossian, R.E. Earthquake Hazard and the Environmental Seismic Intensity (ESI) Scale. *Pure Appl. Geophys.* **2015**, *173*, 1479–1515. [CrossRef]

50. Guerrieri, L. The EEE Catalogue: A global catalogue of Earthquake Environmental Effects. *Quat. Int.* **2012**, *279*, 179–180. [CrossRef]

51. Audemard, F.; Azuma, T.; Baiocco, F.; Baize, S.; Blumetti, A.M.; Brustia, E.; Clague, J.; Comerci, V.; Esposito, E.; Guerrieri, L.; et al. Earthquake Environmental Effect for seismic hazard assessment: The ESI intensity scale and the EEE Catalogue. *Mem. Desc. Carta Geol. D'Italia* **2015**, *97*. [CrossRef]

52. Guidoboni, E.; Ferrari, G.; Mariotti, D.; Comastri, A.; Tarabusi, G.; Sgattoni, G.; Valensise, G. *CFTI5Med, Catalogodei Forti Terremoti in Italia (461 a.C.-1997) e Nell'area Mediterranea (760 a.C.-1500)*; Istituto Nazionale di Geofisica e Vulcanologia (INGV): Roma, Italy, 2018.
53. Comerci, V.; Blumetti, A.M.; Di Manna, P.; Fiorenza, D.; Guerrieri, L.; Lucarini, M.; Serva, L.; Vittori, E. ITHACA Project and Capable Faults in the Po Plain (Northern Italy). *Ingegneria Sismica* Special Issue "Seismic Risk in the Po Plain". **2013**, 36–50. Available online: http://ingegneriasismica.org/2013/01/ (accessed on 9 May 2019).
54. Guerrieri, L.; Blumetti, A.M.; Comerci, V.; Di Manna, P.; Michetti, A.M.; Vittori, E.; Serva, L. Surface Faulting Hazard in Italy: Towards a First Assessment Based on the ITHACA Database. In *Engineering Geology for Society and Territory*; Springer: Cham, Switzerland, 2015.

geosciences

MDPI

Review

Paleoliquefaction Studies and the Evaluation of Seismic Hazard

Martitia P. Tuttle [1,*], Ross Hartleb [2], Lorraine Wolf [3] and Paul W. Mayne [4]

1 M. Tuttle & Associates, P.O. Box 345, Georgetown, ME 04548, USA
2 Lettis Consultants International, Inc., Santa Clarita, CA 91355, USA
3 Department of Geosciences, Auburn University, Auburn, AL 36849, USA
4 Civil & Environmental Engineering, Georgia Institute of Technology, 790 Atlantic Drive, Atlanta, GA 30332, USA
* Correspondence: mptuttle@earthlink.net; Tel.: +01-207-371-2007

Received: 7 June 2019; Accepted: 7 July 2019; Published: 13 July 2019

Abstract: Recent and historical studies of earthquake-induced liquefaction, as well as paleoliquefaction studies, demonstrate the potential usefulness of liquefaction data in the assessment of the earthquake potential of seismic sources. Paleoliquefaction studies, along with other paleoseismology studies, supplement historical and instrumental seismicity and provide information about the long-term behavior of earthquake sources. Paleoliquefaction studies focus on soft-sediment deformation features, including sand blows and sand dikes, which result from strong ground shaking. Most paleoliquefaction studies have been conducted in intraplate geologic settings, but a few such studies have been carried out in interplate settings. Paleoliquefaction studies provide information about timing, location, magnitude, and recurrence of large paleoearthquakes, particularly those with moment magnitude, **M**, greater than 6 during the past 50,000 years. This review paper presents background information on earthquake-induced liquefaction and resulting soft-sediment deformation features that may be preserved in the geologic record, best practices used in paleoliquefaction studies, and application of paleoliquefaction data in earthquake source characterization. The paper concludes with two examples of regional paleoliquefaction studies—in the Charleston seismic zone and the New Madrid seismic zone in the southeastern and central United States, respectively—which contributed to seismic source models used in earthquake hazard assessment.

Keywords: paleoliquefaction; paleoearthquake; earthquake hazard

1. Introduction

Paleoseismology is the study of prehistoric earthquakes as preserved in the geologic record, and it improves our understanding of the long-term behavior of fault zones and seismic sources (e.g., Reference [1]). Paleoseismology is especially useful in regions where strain rates are relatively low and recurrence times of large earthquakes are longer than the historical record. In such regions, the seismicity catalog is often insufficient to characterize the expected rates of large events, a critical issue for estimating uncertainty in seismic hazard assessments.

The paleoliquefaction approach to paleoseismology focuses on soft-sediment deformation structures and related ground failures resulting from liquefaction induced by earthquakes. This field of study developed over the past 40 years and provides important information about timing, source areas, magnitudes, and recurrence times of large paleoearthquakes during the Late Quaternary. Paleoliquefaction studies are especially useful in intraplate and interplate regions where seismogenic faults may not rupture the surface or are otherwise difficult to identify [2,3]. Paleoliquefaction studies have been conducted in seismically active regions of central and eastern North America, including the New Madrid seismic zone in the central United States (US) (e.g., References [4–8]), the Charleston seismic zone in the southeastern US (e.g., References [9–12]), and the Charlevoix seismic zone in southeastern Canada [13], where large historical earthquakes are known to have induced liquefaction (Figure 1). They have been carried out in the Wabash Valley (e.g., References [14,15]) and the Eastern Tennessee [16] seismic zones, where only small to moderate earthquakes occurred during the historical period. In addition, paleoliquefaction studies were conducted in interplate settings like the Dominican Republic and Puerto Rico in the northeastern Caribbean and the Pacific Northwest in the US, where subduction zones and crustal faults pose a significant seismic hazard (e.g., References [3,17–20]). Paleoliquefaction studies have been conducted in a lacustrine setting in eastern Turkey [21] and a volcanic setting in southern Italy [22]. Studies focusing on soft-sediment deformation structures in lacustrine deposits were reported for southern Italy [23], Mexico [24], and Argentina [25]. Recently, several paleoliquefaction studies were carried out in the Canterbury region of New Zealand, where a system of crustal faults, some of which did not rupture the surface, produced the 2010–2011 sequence of earthquakes and caused extensive and recurrent liquefaction (e.g., References [26–32]).

Paleoliquefaction data were used to develop seismic source models for the US national seismic hazard maps [33,34] and for the central and eastern US (CEUS) seismic source characterization for nuclear facilities [35]. Paleoliquefaction studies, along with other paleoseismic investigations such as fault studies, supplement modern seismicity studies and provide critical information on the long-term behavior and earthquake potential of seismic sources, which is essential for probabilistic seismic hazard assessments. This review paper presents background information on earthquake-induced liquefaction and resulting soft-sediment deformation features that may be preserved in the geologic record, best practices used in paleoliquefaction studies, and application of paleoliquefaction data in earthquake source characterization.

Figure 1. Photograph of large sand blow that formed as the result of liquefaction during the 1811–1812 New Madrid earthquake sequence [36].

2. Earthquake-Induced Liquefaction

The 1964 Alaska (US), 1964 Niigata (Japan), and 1967 Caracas (Venezuela), earthquakes caused catastrophic liquefaction-related failures (e.g., Reference [37]). More recent earthquakes, including the 1995 Hyogo-ken Nanbu (Japan) and 2010–2019 Canterbury (New Zealand) earthquakes also caused large economic losses due to liquefaction-related damages to the built environment (e.g., References [38,39]). These and other events spurred research in earthquake-induced liquefaction and earthquake soil dynamics in the geological and geotechnical engineering communities. Progress was made in better understanding the process of liquefaction, the conditions contributing to liquefaction, and ground motions that initiate liquefaction; however, many uncertainties remain regarding liquefaction triggering assessment, including the influence of fines and gravel content of soils, age effects on liquefaction susceptibility, the degree of saturation below the water table, and methods for predicting liquefaction triggering [40]. Readers are referred to the National Academies of Sciences [40] for a thorough review of the art and practice in the assessment of earthquake-induced liquefaction and its consequences. Aspects of earthquake-induced liquefaction especially pertinent to paleoliquefaction studies are summarized below.

2.1. Process of Liquefaction

There are two primary types of liquefaction: (a) flow liquefaction, associated with static monotonic loading; and (b) cyclic liquefaction, caused by repeated (earthquake) loading (Figure 2 [41,42]). In cases of both cyclic and flow liquefaction, a significant loss of soil strength occurs. Cyclic liquefaction results from the accumulation of excess porewater pressure in granular soils (sands and gravels) as repeated load cycles are applied by earthquake ground motions. Flow liquefaction occurs during first-time (monotonic static) loading of unstable soils, including sands, silts, and clays.

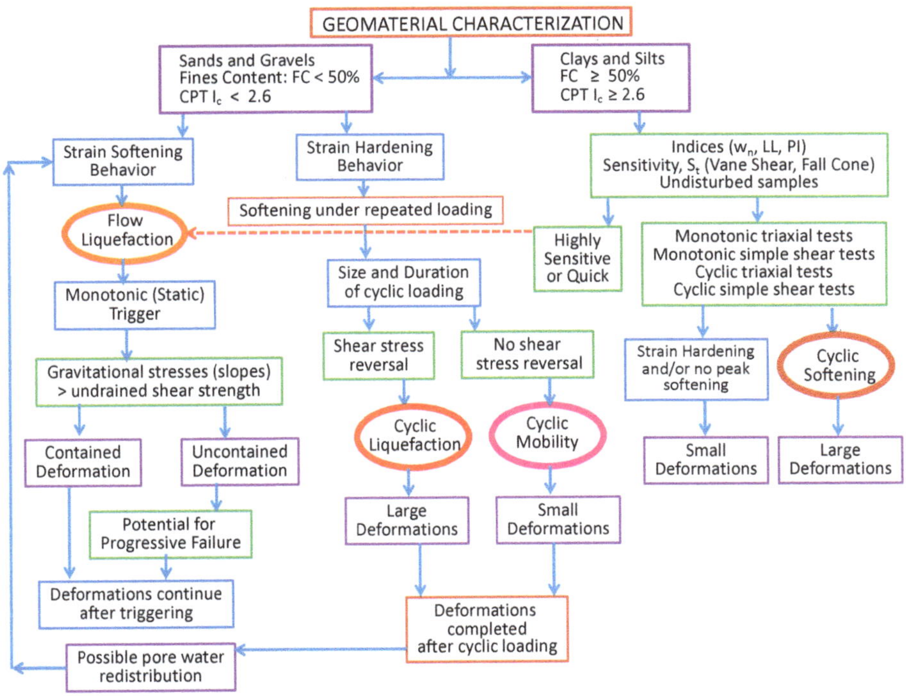

Figure 2. Flow chart showing different types of soil liquefaction: flow liquefaction, cyclic liquefaction, cyclic mobility, and cyclic softening (modified from References [41,42]).

At most documented liquefaction sites, the more common mechanism is cyclic liquefaction. Cyclic liquefaction occurs when the structure of water-saturated, loose, granular sediment breaks down due to rapidly applied and repetitive loading caused by earthquakes [37,41–43]. During cyclic loading, the loosely packed particles attempt to move into a denser configuration (termed contraction). During an earthquake, however, there is not enough time for the water in the pores of the sediment to be squeezed out. Instead, the water is trapped and prevents the particles from moving closer together. This is accompanied by an increase in porewater pressure that reduces the contact forces between the individual particles, thereby softening and weakening the sediment. In loose contractive sand, the development of positive porewater pressure reduces the effective stress state and, when the effective strength envelope is reached, the onset of cyclic liquefaction occurs (e.g., Reference [41]). During large earthquakes, the increase in porewater pressure can lead to upward flow of water and entrainment of sediment through the process of fluidization [43–45]. The upward flow of water and entrained sand can lead to the formation of liquefaction features, including sand dikes, diapirs, sills, and blows (Figure 3). These and other soft-sediment deformation structures are discussed further in Section 2.4 below. For additional information on the process of liquefaction, see References [40,41,46–50].

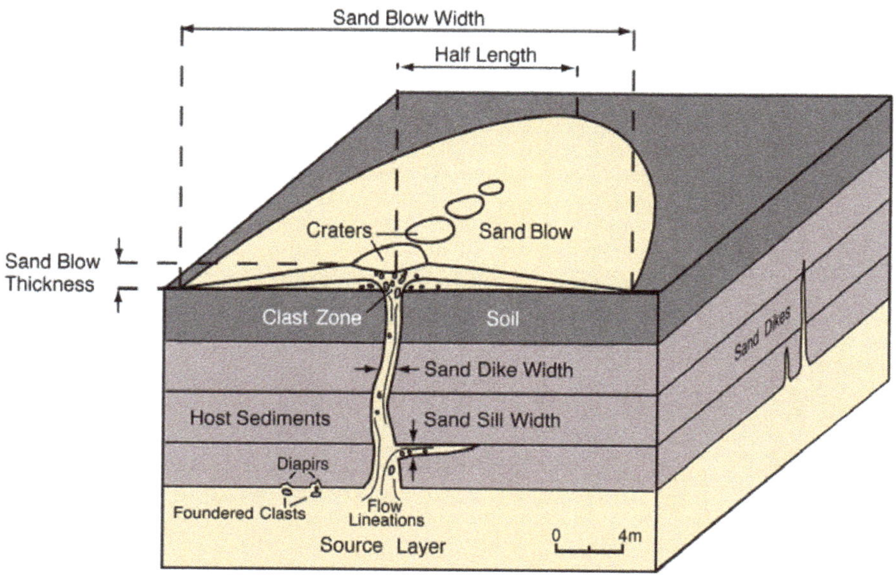

Figure 3. Block diagram showing relationship between the liquefied layer and overlying liquefaction features, including intrusive sand dikes and sills and an extrusive sand blow [51].

2.2. Conditions That Influence the Formation of Liquefaction Features

Earthquake-induced liquefaction features commonly form in alluvial, coastal, deltaic, and lacustrine deposits of the Holocene age (0.01 Ma to present), where sand is interbedded with silt and clay and shallow groundwater conditions prevail (e.g., References [28–32,52–61]). Sedimentological and hydrological conditions in these environments are conducive to the formation of liquefaction features. Liquefaction features also formed in similar deposits of the Late Pleistocene age (0.126 to 0.01 Ma) (e.g., References [13,14,52]) and as old as 200,000–240,000 years before present (BP) [61].

The physical properties of sediment that influence liquefaction susceptibility, or their tendency to liquefy during earthquakes include the size, shape, and packing arrangement of the sediment grains [37,40,49,62]. Well-rounded, well-sorted, loose to moderately loose, fine sand with a high void ratio or low relative density would be especially susceptibility to liquefaction, so long as it is saturated. Liquefaction susceptibility decreases with increasing fines content, especially fines with a high plasticity index [40,63–65]. Layering in sediment plays an important role in liquefaction and the formation of liquefaction features. Layers of clay or silt form impermeable barriers that prevent or retard the upward flow of water, thereby promoting the buildup of porewater pressure in underlying sandy layers (e.g., References [66–68]). In addition, layers of clay or silt above sandy layers can lead to the formation of a water layer, foundering of the overlying layer, and formation of sand diapirs and dikes (e.g., References [67,69]).

Aging can have a significant effect on the strength of soil and, thus, may influence its liquefaction susceptibility (e.g., References [70,71]). Due largely to aging effects, Pleistocene (2.58 to 0.01 Ma) [72] deposits tend to be less susceptible to liquefaction than Holocene (0.01 Ma to present) [72] deposits, and pre-Pleistocene (before 2.58 Ma) deposits tend to have a very low liquefaction susceptibility [54,73]. Over time, sediment consolidates, especially if buried by subsequent deposits, becoming more densely packed and less susceptible to liquefaction. In addition, cementation of sand grains by clay, calcium carbonate, and iron and manganese oxides will strengthen the sediment and reduce its liquefaction susceptibility [74]. A fluctuating water table can promote the precipitation and concentration of clays, carbonates, and oxides. On the other hand, considerable groundwater flow can flush potential cementing agents out of sandy deposits [75]. Research in the Charleston seismic zone of the southeastern US suggests that repeated liquefaction by large earthquakes every few hundred years can counter the effects of aging or "reset the clock" (e.g., References [40,76]).

2.3. Ground Motions That Cause Liquefaction

The key index parameters of ground motion include ground acceleration, ground velocity, ground displacement, and duration of ground shaking [37,77,78]. These factors, combined with the frequency characteristics of ground motion, account for the intensity of ground shaking. Characteristics of earthquake ground motions at a particular site, which may or may not suffer liquefaction, will depend on several seismological and geological factors. These factors include magnitude of the earthquake, the source mechanism of the earthquake, distance to the site from the earthquake source, directivity of seismic energy related to the direction and speed of fault rupture, characteristics of the rocks along the wave path from source to site, and local soil conditions at the site [37].

Earthquake-induced liquefaction is caused by the buildup of excess porewater pressure due to cyclic shear stresses imparted by ground motions. In particular, the amplitude of the cyclic shear stresses and the number of cycles of shearing contribute to liquefaction [37]. Maximum shear stress is related to maximum ground acceleration. The number of cycles depends on the duration of the earthquake and also, implicitly, its frequency content. Both maximum ground acceleration and duration of ground shaking generally increase with earthquake magnitude and may also increase as a result of seismotectonic setting, site conditions, and bedrock topography (i.e., basin effects) (e.g., Reference [37]).

During the 1976 **M** 7.5 Guatemala, 1976 **M** 7.6 Tangshan (China), 1978 **M** 6.5 Miyagiken-Oki (Japan), 1983 **M** 6.8 and 7.7, Nihonkai-Chuba (Japan), 1989 **M** 6.9 Loma Prieta (California), 2016 **M** 5.8 Pawnee (Oklahoma), and **M** 7.1 Darfield and **M** 6.2 Christchurch (New Zealand) earthquakes, liquefaction was triggered by peak ground accelerations (PGAs) of ≤0.1 g [26,79–81]. The US National Research Council judged that a typical peak ground acceleration (PGA) threshold value of liquefaction is 0.1 g and that smaller accelerations associated with long-duration earthquakes also can induce liquefaction in very susceptible soils [82]. More recently, de Magistris et al. [83] estimated a threshold value of 0.09 g based on a statistical analysis of case histories of earthquake-induced liquefaction, primarily from Japan and the US, including earthquakes ranging from **M** 5.9 to **M** 8+. The peak acceleration for triggering liquefaction ranged from 0.08 to 0.84 g, with the most cases of liquefaction between 0.16 and 0.32 g.

Liquefaction triggering is assessed by using simplified empirical models based on in situ testing (e.g., standard penetration test or SPT, cone penetration test or CPT, shear-wave velocity) to correlate penetration resistance to relative soil density [37,41,42,49,84–86]. From in situ measurements at sites of liquefaction and non-liquefaction, relations are developed for predicting resistance to liquefaction triggering, or the cyclic resistance ratio (CRR). The simplified empirical models evaluate the seismic loading that initiates the liquefaction process. The seismic loading is characterized as the magnitude- and stress-corrected cyclic stress ratio (CSR_{M,σ'_v}), which is expressed as follows [87]:

$$CSR_{M,\sigma'_v} = 0.65 \cdot \left(\frac{a_{max}}{g}\right) \cdot \left(\frac{\sigma_v}{\sigma'_v}\right) \cdot r_d \cdot \frac{1}{MSF}, \tag{1}$$

where a_{max} is the peak (horizontal) ground acceleration (or PGA), g is the gravitational acceleration constant (g = 9.8 m/s^2 = 32 ft/s^2), σ_v and σ'_v are the total and effective vertical stresses, respectively, r_d is a stress reduction coefficient that accounts for the flexibility of the model soil column ($r_d \leq 1.0$), and MSF is a magnitude scaling factor, which is a function of **M** and is a proxy for the duration of loading [88]). Liquefaction triggering potential is often expressed as a factor of safety against liquefaction (FS_L) as follows:

$$FS_L = \frac{resistance}{loading} = \frac{CRR}{CSR_{M,\sigma'_v}}. \quad (2)$$

The characterization of seismic loading in terms of a_{max} and **M** is a critical part of empirical liquefaction triggering assessment [89].

2.4. Earthquake-Induced Liquefaction Features

There is a large body of literature on earthquake-induced liquefaction features, including articles about laboratory experiments (e.g., References [90,91]), post-earthquake surveys and studies (e.g., References [6,26–32,52,53,58,92–96]), and paleoliquefaction studies (e.g., References [6,8,9,14,15,21–25,97–99]). In addition, there are reviews on earthquake-induced liquefaction features and criteria for distinguishing them from non-seismic features (e.g., References [2,23,44,51,91,97–100]). Earthquake-induced liquefaction features are also important in the assessment of intensity fields and epicentral intensity of modern and historical earthquakes [101–104]. In the environmental seismic intensity scale (ESI) based on effects of ground shaking in the natural environment, liquefaction features along with other environmental effects are used to define earthquake intensity levels IV–XII [105–107]. This overview is not intended as a comprehensive discussion of all the relevant literature, but rather as an introduction for new investigators to the topic, drawing on several classic studies, our own experience, and several recent studies that advanced the field of study.

As described above in Section 2.1, strong ground shaking can induce liquefaction and fluidization of water-saturated, loose, sandy sediment, and it can lead to the formation of liquefaction features. From a geological perspective, deformation of unconsolidated sand is related to a trigger (groundwater movements, wave action, and seismic shaking), a deformation mechanism (liquefaction and fluidization), and a driving force (gravitational body force, unevenly distributed loads, unstable density gradients, and shear forces) [44]. Seismic shaking is the likely trigger for regionally extensive liquefaction and can result in the formation of sand dikes and sand blows if fluidization is involved. In addition, dish structures, load casts, pseudonodules, and ball-and-pillow structures can form in muddy and sandy sediments in response to seismic shaking; however, these features may also form as the result of other triggers. Many of the liquefaction features that are observed following earthquakes were reproduced during laboratory experiments, including sand blows with a central crater and a thin surface coating of finer sand, and loss of stratification in the feeder area or source zone [91].

Earthquake-induced liquefaction features can be divided into two categories: (1) features related to deformation extending beyond the layer that liquefies; and (2) features related to deformation within the sedimentary layer that liquefies. Features that extend beyond the liquefied layer include intrusive dikes, sills, and diapirs, and extrusive sand blows or volcanoes (e.g., References [6,21–24,26,28–31,92,100,108,109]). Features that form within the liquefied layer include disturbed bedding, dish structures, ball-and-pillow structures, load casts and related folds, pseudonodules, convolute bedding and lamination, and folds related to slumping (e.g., References [21–25,32,44,54,91,97–100,110,111]).

2.4.1. Blows, Dikes, Sills, and Diapirs

Sand blows result from venting of water and entrained sediment onto the ground surface at the time of the earthquake, or soon thereafter in cases of delayed effects (Figures 3–5; e.g., References [51,92,99,100]). Sand blows are also called sand volcanoes because they resemble small volcanic cones, some with small craters at the surface, aligned along ground fissures. The opening at the ground surface through which the slurry of water and sediment flows is referred to as the vent. As vented water flows across the ground surface, the entrained sediment is deposited to form constructional cones or to fill topographically low or subsided areas (see photographs of sand blows that formed during the 2010–2011 Canterbury (New Zealand) earthquake sequence in the Supplementary Materials of Reference [30] and during the 2012 Emilia (Italy) earthquake sequence in Reference [112]. In plan view, the shape of a sand blow is related to the shape of the conduit through which the slurry of water and sand vented. Linear to elliptical blows result from venting through fissures [26,28,30,69,92,95]. Circular blows result from venting through tubular-shaped conduits such as root casts (e.g., Reference [113]).

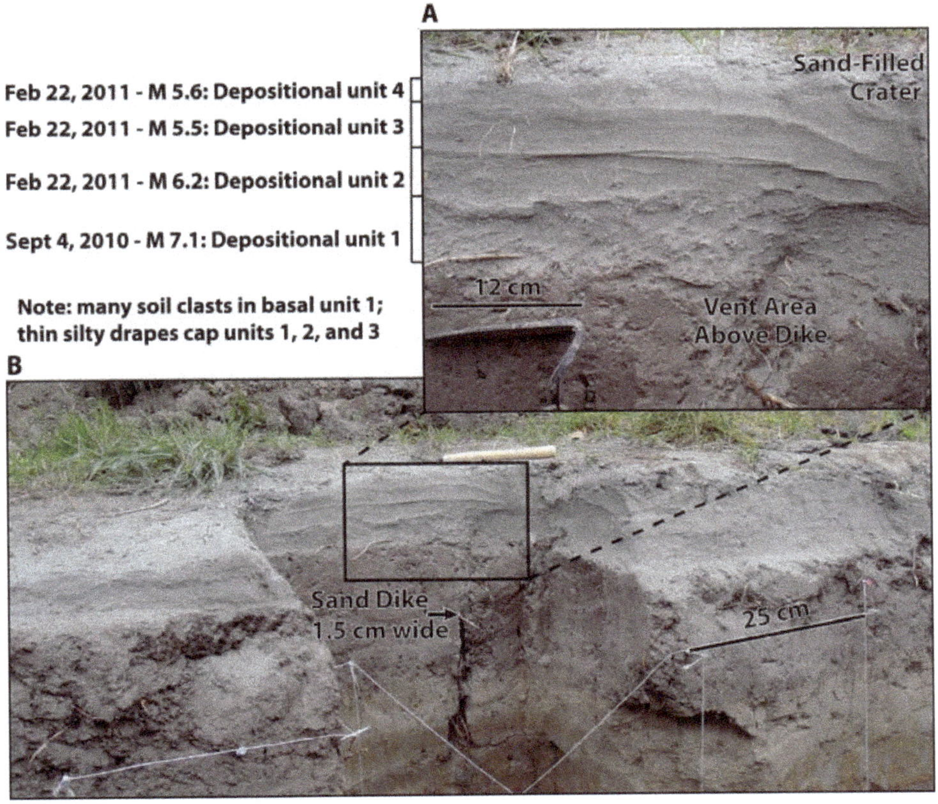

Figure 4. A compound sand blow that formed during September 2010 Darfield and February 2011 Christchurch mainshocks and two aftershocks as exposed in a trench at the Hardwick site south of Christchurch, New Zealand [30]. Photographs by M. Tuttle. (**A**) Closeup of sand blow shows vent area above sand dike, four depositional units of the compound sand blow, and portion of a sand-filled crater. Depositional unit 1 contains soil clasts eroded from the dike wall by venting water and deposited with entrained sand on the ground surface. Depositional units 1, 2, and 3 are composed of sand overlain by silt drapes that stand out in relief. Upper contacts of units 2, 3, and 4 dip toward the sand-blow crater. Upper contacts of units 1 and 4 are irregular, consistent with bioturbation during periods of subaerial exposure. (**B**) Sand blow and related feeder dike; area of closeup in (**A**) indicated by black rectangle.

Generally, sand blows are thicker and coarser-grained above the vent area and thin and fine away from the vent [6,8]. Several investigators noted that some sand blows also fine upward and are capped by silt or clay (e.g., Reference [22]), including depositional units of compound sand blows resulting from multiple earthquakes in a sequence (e.g., References [6–8,26,30,92]. For example, compound sand blows that formed during the 2010–2011 Canterbury (New Zealand) earthquake sequence were composed of multiple fining-upward units capped by silt drapes (Figure 4 [26,30]). Each of these fining-upward units formed during a different earthquake in the sequence. In the New Madrid seismic zone of the central US, compound sand blows that formed during the 1811–1812 earthquake sequence, as well as paleo-sand blows in the region, are composed of several depositional units that generally fine upward and are capped by silt or clay layers (Figure 5 [6–8]). The silt and clay layers capping the sandy units are thought to be related to low flow velocity during the waning stage of venting and, in some locations, to settling of fines out of standing water following an event.

Sand blows that formed during a single event, involving ground oscillation and pulsing of vented water and sediment, should not be confused with compound sand blows (e.g., References [58,114]). These single-event sand blows are composed of multiple sand layers that are not capped by silt or clay.

Craters form in the surface of the sand blow above the most recently active vents. Following an event, the craters are often coated with silt or clay (see examples of craters in sand blows that formed during the 2010–2011 Canterbury earthquake sequence in the Supplementary Materials of Reference [30]) and are eventually filled with wind- and water-borne sediment [92]. It was also noted that sand blows may contain clasts of the underlying deposits through which the slurry of water and sand flowed [6,8,92]. The clasts tend to be larger and more abundant in close proximity to the vent. In cases of compound sand blows that formed as a result of recurrent liquefaction, clasts are often more abundant in the lowest and earliest unit to be deposited (Figures 4 and 5). The sedimentological characteristics of sand blows vary from site to site and region to region and likely depend on a number of factors such as grain-size distribution of the source bed, the flow rate and duration of venting water and entrained sediment, and variations in the flow rate related to ground oscillation and pulsing (e.g., References [5,77,92]). In cases where a large volume of subsurface sediment vented to the surface or where lateral spreading or faulting occurred, the soil horizon buried beneath the blow may dip toward the vent or be displaced downward across the vent (Figure 5 [22,69]).

Sand-blow craters that formed in the ground surface were studied in the Charleston seismic zone of the southeastern US [10,109]. Vented sediment, including clasts of soil, was deposited around the crater rims and in the base of the craters. Over time, the craters filled with reworked sand-blow deposits, slack-water deposits, and organic material [10,109]. Sand-blow craters were studied during the post-earthquake survey of the **M** 7.7 Bhuj (India) earthquake [58]. In at least one location, a sand-blow crater was formed by explosive deformation. Crater formation with forceful ejection of soil clasts post-date the vented sand and silt surrounding the crater, suggesting that explosive ground deformation resulted from delayed effects of soil liquefaction [115].

Sand dikes are sheet-like or tabular-shaped intrusive bodies that crosscut bedding in the host deposits (Figures 5a and 6 [51,92,99,100]). Dikes usually have well-defined margins and can be differentiated from the host deposit by differences in grain size and weathering characteristics. Dikes typically originate in a layer of sandy sediment (i.e., the source layer) and are composed of sediment derived from the source layer (Figure 6). The dikes may contain clasts of the intruded host deposits and exhibit flow structure or lineations. They often become narrower and more fine-grained up section, and sometimes branch upward. Dikes may pinch out or terminate within the stratigraphic section or extend through the entire section to the ground surface. Tree root casts, animal burrows, desiccation cracks, and other voids or weak spots near the ground surface can be utilized as pathways to the surface (e.g., References [5,20,113]). Sand diapirs are similar to dikes but are relatively small intrusions of sediment extending from the layer that liquefied into the base of the overlying layer (e.g., Reference [13]). In contrast to dikes and diapirs, sills are intruded parallel to bedding of host deposits and usually take the form of lenses intruded below low-permeability layers (Figures 3 and 6). The source layer of dikes, diapirs, and sills may contain foundered clasts of the overlying layer and exhibit deformation structures related to liquefaction and fluidization, such as disturbed bedding, dish structures, ball-and-pillow structures, load casts and related folds, and flow structure or lineations (e.g., References [5,13,22,97]). Alternatively, the source layer may be massive if bedding is completely destroyed.

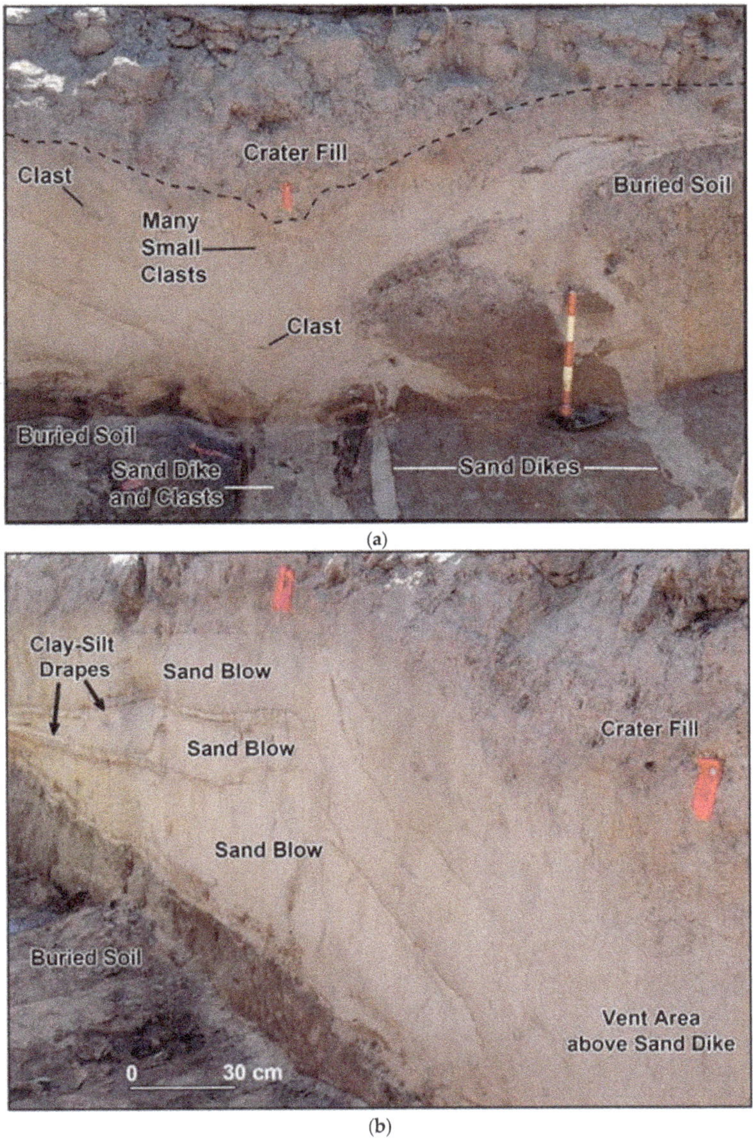

Figure 5. A compound sand blow that formed during the 1811–1812 New Madrid earthquake sequence as exposed in a paleoseismic trench near Blytheville, Arkansas [116]. Photographs by M. Tuttle. (**a**) Vent area of sand blow and related dikes. Brown soil horizon is crosscut by two sand dikes, displaced downward ~1 m, and buried by the sand blow. Clasts of the soil horizon occur within dikes and the overlying sand blow. Each colored interval on shovel handle represents 10 cm. (**b**) Sand blow is composed of three sandy depositional units with the lower two units capped by clay–silt drapes. The top of the upper unit was subjected to soil-forming processes, whereas the lower two units were protected and preserved by the upper unit.

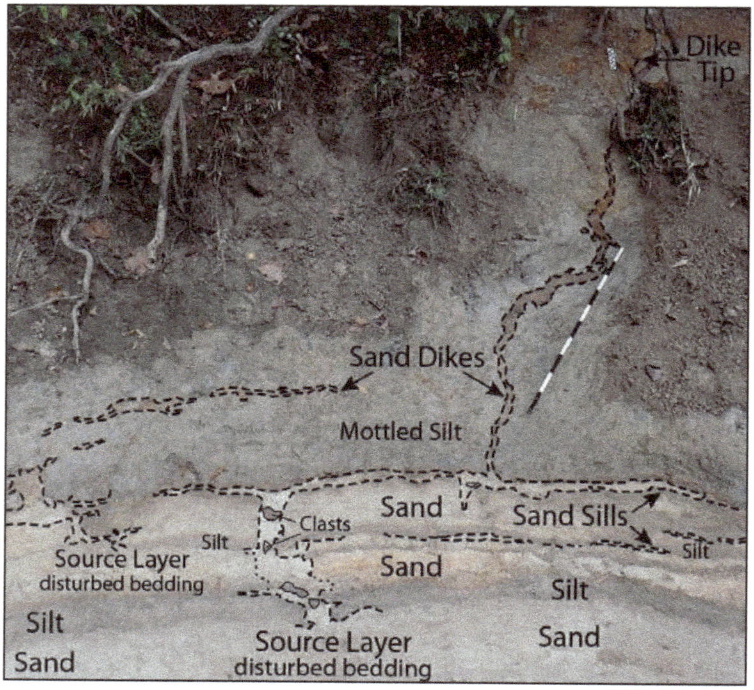

Figure 6. Annotated photograph of liquefaction features exposed in a river cutbank in northwestern Mississippi southeast of the New Madrid seismic zone in the central United States (US) [117]. Source layers that liquefied occur at the base of the cutbank, sand dikes crosscut overlying interbedded silt and sand layers and mottled silt, and sills intrude along the basal contacts of silt layers. The black and white intervals on the meter stick represent 10 cm. Photograph by M. Tuttle.

2.4.2. Soft-Sediment Deformation Structures within the Layer That Liquefied

Deformation structures within the layers that liquefied include disturbed bedding, dish structures, ball-and-pillow structures, load casts and related folds, pseudonodules, convolute bedding and lamination, and folds related to slumping (e.g., References [5,13,22,24,25,32,44,54,91,97–99,111,117]). In a review of the physical conditions under which soft-sediment deformation occurs, Owen [44] concluded that dish structures, load casts, pseudonodules, and ball-and-pillow structures can form in muddy and sandy sediments in response to seismic shaking, but that these features may also form as the result of other triggers. Sims [97], who pioneered the use of these types of features in paleoliquefaction studies, developed recognition criteria by studying earthquake-induced structures that formed in different environments during modern and historical earthquakes. It was noted that these liquefaction features typically form in interbedded fine- and coarse-grained deposits close to the sediment-water interface as a result of bearing strength failure due to liquefaction of the coarse-grained layers. In a more recent review, Sims [98] provided a detailed discussion of the characteristics of liquefaction features and the conditions under which these features form. In the case of load casts, fine-grained layers sag into the liquefied coarse-grained layer but without completely detaching (Figure 7). If the sagging layers detach, convolute laminations and ball-and-pillow structures form as they sink into the liquefied coarse-grained layer. Pseudonodules form when the coarse-grained layer separates into domains or irregular masses. Disturbed bedding, dish structures, and foundered clasts were also described in source layers of sand dikes and sand blows that formed during earthquakes (e.g., References [5,13,117]). In his early research, Sims [53] found that load casts form at lower modified Mercalli intensities of about VI, whereas convolute laminations and pseudonodules form at higher intensities of VIII–IX. Therefore, the occurrence of the different types of strata-bound liquefaction features may help constrain the locations and magnitudes of paleoearthquakes.

Figure 7. Load cast that formed in interbedded and laminated sediment of Van Norman Lake during the 1952 **M** 7.5 Kern County (California) earthquake [97].

Given that strata-bound soft-sediment deformation can form as a result of a variety of triggers, it may be best to use these structures in combination with sand dikes and sand blows [13,110]. A recent study by Uner [21] was a good example of this approach. In this study, fluid-escape structures, including sand volcanoes, flame structures, and dish and pillar structures, were found in combination with disturbed layers, ball-and-pillow structures, convolutions, and slump structures in Quaternary lacustrine deposits of Lake Van in eastern Turkey. The lacustrine deposits are composed of sand, silt, and clay and contain the soft-sediment deformation structures at various stratigraphic levels. The most recent deformation was likely caused by the 2011 **M** 7.2 Van-Tabanlı earthquake, and the earlier episodes of deformation are attributed to earlier earthquakes of magnitude 5 or greater [21]. Moretti and Ronchi [25] made a compelling case that a 1.5-m-thick and 500-m-long deformed layer in Pleistocene fluvio-lacustrine deposits of the Neuquén Basin in northern Patagonia (Argentina) is the product of earthquake-induced liquefaction. They performed facies analysis of the Pleistocene deposits and morphological analysis of the soft-sediment deformation structures, primarily large load structures, and excluded non-seismic triggers, such as overloading and wave action, for the deformation of the layer.

2.4.3. Diagnostic Criteria for Earthquake-Induced Liquefaction Features

Criteria used to identify and utilize paleoliquefaction features in paleoseismology studies were developed over the past forty years. Focusing primarily on intrusive and extrusive features extending beyond the layer that liquefies, the following general criteria were recommended for identifying earthquake-induced liquefaction features [2,99]:

(1) Sedimentary characteristics consistent with case histories of earthquake-induced liquefaction;
(2) Sedimentary characteristics indicative of a sudden, strong, upwardly directed hydraulic force of short duration;
(3) Occurrence of more than one type of liquefaction feature and of similar features at multiple nearby locations;

(4) Occurrence in geomorphic settings where hydraulic conditions described in (2) would not develop under non-seismic conditions;
(5) Age data to support both contemporaneous and episodic formation of features over a large area.

Criteria (1)–(3) are required and (4) and (5) are preferred to attribute a soft-sediment deformation structure to earthquake-induced liquefaction. The more criteria that are satisfied, the greater the confidence is in the interpretation. Lunina and Gladkov [108] proposed criteria for recognizing earthquake-induced soft-sediment deformation structures, specifically dikes, in regions of past and present permafrost where cryogenic processes commonly deform sediment. Sims [97] suggested criteria for distinguishing seismically from non-seismically induced strata-bound soft-sediment deformation structures as follows:

(1) Liquefiable sediment is present or potentially present;
(2) Deformational structures observed are similar to those formed experimentally or are shown to have formed during seismic events;
(3) Structures are restricted to or originate from a single stratigraphic interval;
(4) Zones of structures are correlated over large areas;
(5) Absence of detectable influence by slopes, slope failures, or other sedimentological, biological, or deformational processes.

Wheeler [110] later expanded the criteria and further evaluated limitations related to determining the origin of these deformation structures (Table 1).

Table 1. Summary of tests for determining seismic origin of soft-sediment structures (from References [100,110]).

Test Name	Observation	Limitation
Sudden formation	Structure formed more suddenly, and perhaps more violently, than any non-seismic alternative	May be unable to rule out some nonseismic origins without additional evidence
Synchronous formation	Nearby structures of same type formed at times indistinguishable from each other	May be unable to rule out some nonseismic origins; dating and correlation lack resolution to distinguish synchronous from near-synchronous formation
Zoned distribution	Size of structure decreases away from a central area	Cannot rule out earthquake origin
Size	Structure not larger than similar structures formed by historical earthquakes	Maximum size may be unknown; cannot rule out an earthquake origin for small structures
Tectonic setting structure	Seismic shaking strong enough to form the structure occurs more frequently than nonseismic alternatives in modern analog settings	Threshold magnitudes and accelerations for formation are only generally known
Depositional setting	Seismic shaking by itself forms the structure in similar modern deposits	Difficulty in recognizing some newly formed structures in the field

Having performed detailed analysis of deformation structures, including deformed laminations, load structures, large water-escape structures, slumps, and Neptunian dykes, in a Pleistocene lacustrine succession of the Sant'Arcangelo Basin of southern Italy, Moretti and Sabato [23] proposed criteria to distinguish seismic from non-seismic soft-sediment deformation structures. They differentiated various deformation structures by driving force system and trigger mechanism and found that all the studied deformation structures may have been triggered by earthquakes, except for small-scale load structures and Neptunian dikes, which they attributed to overloading and extensional tectonics, respectively.

2.4.4. Preservation of Liquefaction Features

Whether or not liquefaction features are preserved in the geologic record depends on their position on the landscape or in the stratigraphic section, the size of the features, and the geologic processes that affect the features over time. Within weeks to months of their formation, sand blows can show signs of erosion and bioturbation. For example, following the 2010–2011 Canterbury earthquake sequence [26,28,30], sand blows were incised by water that continued to flow from the vents following the earthquakes. In addition, sand blows were somewhat eroded by rainfall and bioturbated by plant roots within several months. Over time, erosion can completely remove liquefaction features, and bioturbation can totally rework sand blows and near-surface portions of dikes. If the features are small, they are likely to be destroyed by erosion and bioturbation within decades or centuries, leaving truncated and weathered sand dikes below [51,60]. Liquefaction features can also be destroyed by other geologic processes such as river channel migration and mass movements. In addition, human modification of the landscape can destroy liquefaction features, especially if the features are small and the landscape modification is severe. If sand blows are large (1–2+ m thick), they may persist on the ground surface for thousands of years as in the New Madrid seismic zone and nearby Marianna area in the central US despite decades of plowing [6,8,118]. Sand blows exposed to subaerial weathering develop soil characteristics indicative of their age (Figure 8; e.g., References [5,118,119]). Alternatively, sand blows may be buried and preserved by subsequent deposits (e.g., References [22,120]. For example, a paleo-sand blow and related dikes that formed between anno Domini (AD) 890 and 1400 were found in a trench of sand blows that formed during the 2010–2011 Canterbury earthquake sequence [29,30]. The paleoliquefaction features, weathered and partially eroded, were buried and preserved by a crevasse splay deposit. Even burial by sand blows that form during aftershocks, later earthquakes in a sequence, or earthquakes hundreds of years later can help preserve earlier sand blows (Figure 4, Figure 5, and Figure 8; e.g., References [26,116,118]). Lacustrine basins may be a near-ideal setting for the formation and preservation of earthquake-related soft-sediment deformation structures [120], as demonstrated by several recent studies [21,23,25].

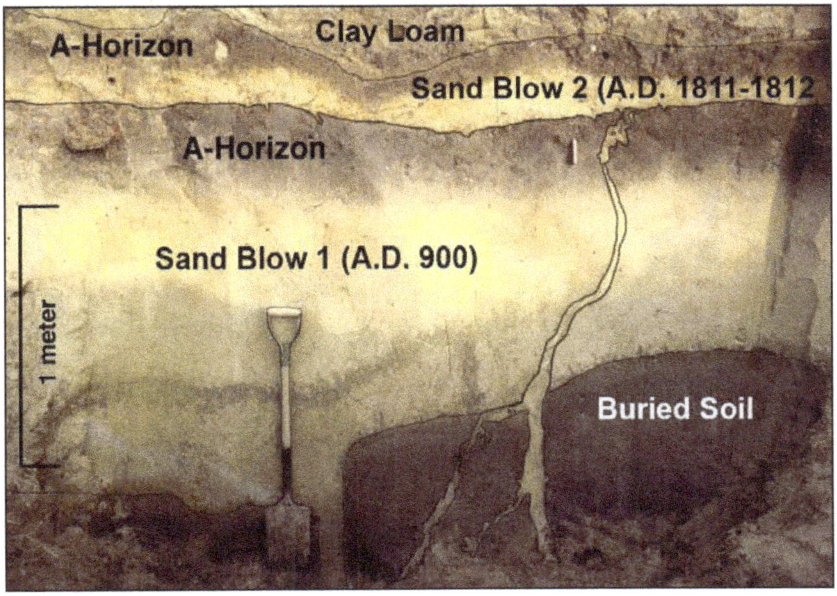

Figure 8. Two generations of sand blows and related feeder dikes exposed in a drainage ditch in the northern portion of the New Madrid seismic zone [119]. The large lower sand blow was subjected to soil-forming processes for about 900 years resulting in the formation of a ~37-cm-thick soil A-horizon.

The sand blow was later buried by a sand blow that formed during anno Domini (AD) 1811–1812 New Madrid earthquakes. Photograph by M. Tuttle.

3. Paleoliquefaction Studies

Paleoliquefaction studies are interdisciplinary and employ methods used in paleoseismology, Quaternary geology, and the affiliated fields of sedimentology, structural geology, geochronology, geomorphology, geophysics, geotechnical engineering, soil science, and sometimes archeology (e.g., References [2–32,97–100,108–111,116–158]). This section discusses various aspects of paleoliquefaction studies, including searching for and documenting liquefaction features and interpreting liquefaction data in terms of past earthquakes.

3.1. Selection of Study Area

Study areas should be selected where sedimentological and hydrological conditions are conducive to the formation and preservation of liquefaction features (e.g., References [2,97,100]). For liquefaction features to form, loose to moderately dense sandy sediment must be present within about 15 m of the ground surface and the sandy sediment must be saturated. A relatively impermeable capping layer will promote the increase of porewater pressure in and liquefaction of saturated sandy sediment during ground shaking (e.g., References [67–69]). Liquefaction features are most likely to be preserved in geologic settings that are relatively free from erosion or are experiencing sediment accumulation. Sand blows may endure for millennia on surfaces of abandoned floodplains or high terraces, or they may be buried and preserved by subsequent deposits in coastal, fluvial, and lacustrine environments.

In selecting a study area, it is helpful to review a variety of information. This information might include eyewitness accounts of ground failure indicative of liquefaction during modern and historical earthquakes, aerial photographs and satellite imagery on which sand blows and related ground failures can be identified, Quaternary geology and geomorphology, geotechnical properties of soil or sediment, depth of the water table, and natural and human-made exposures of Quaternary deposits. As was noted during studies of modern liquefaction features in Ferland (Quebec) [56] and in the Canterbury region of New Zealand [24], sites of liquefaction during modern or historical earthquakes provide good targets for paleoliquefaction studies since liquefaction can occur time and time again at the same site (e.g., References [54,151]). It was recognized for decades that alluvial, deltaic, and coastal sediment, especially fluvial channel and point bar, dune, estuarine, and lagoonal deposits, are susceptible to earthquake-induced liquefaction (e.g., References [5,11,54,57,59,73,95]). Thus, surficial geology maps showing the occurrence of these types of deposits are used for identifying areas where liquefaction features may have formed during past earthquakes. Many geologic maps are now available in digital and georeferenced formats and can be downloaded over the internet. Georeferenced maps can be imported into a geographical information system to be displayed and analyzed in combination with other datasets such as satellite imagery, soils maps, and topography for the purpose of selecting study areas.

With the advent of light detection and ranging (LiDAR) and the growing availability of high-resolution elevation data, LiDAR-derived digital elevation models (DEMs) coupled with geomorphic mapping are being used to study the environments in which liquefaction features formed during recent earthquakes, thus identifying areas where earthquake-induced liquefaction features may have formed during past earthquakes. Using LiDAR-derived DEMs, geomorphic mapping, and other techniques (i.e., geophysical surveys, trenching, and coring) in the Canterbury region, strong spatial correlations were found between liquefaction induced by the 2010–2011 Canterbury earthquakes and river channel and point-bar deposits in the alluvial setting (Figure 9) [29,31] and interdune deposits in the coastal setting [32]. Following the 2012 Emilia earthquake sequence that induced liquefaction in the Po River alluvial plain of northern Italy, LiDAR-derived DEMs and geomorphic mapping found that many of the liquefaction features correlated with and were aligned with abandoned riverbeds, alluvial ridges, levees, crevasse splays, and out-flow channel and fans [95,159]. In the Canterbury and Po River regions, LiDAR-derived DEMs improved the resolution of mapping and helped identify geologic deposits and geomorphic elements prone to liquefaction that can be targeted during paleoliquefaction studies [29,31,95,159].

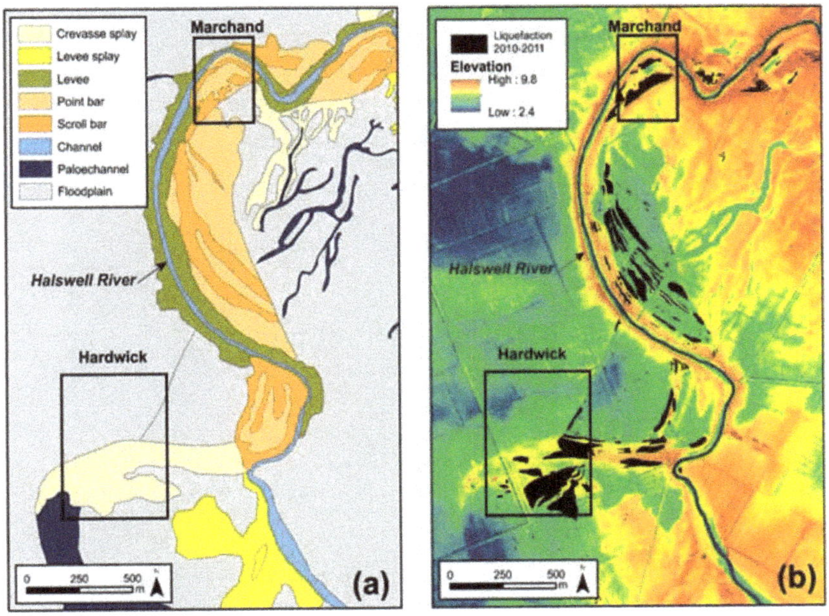

Figure 9. (**a**) Geomorphic map of the Halswell River study area and (**b**) digital elevation model (DEM) from light detection and ranging (LiDAR) with the locations of the 2010–2011 liquefaction sand blows (black polygons) [29]. The aerial photograph shown in Figure 10 is of the Hardwick site outlined by a black rectangle.

Aerial photographs and satellite imagery can provide a bird's eye view of surficial sand-blow deposits. As long as the deposits are not obscured by forest canopy and other vegetation, sand blows appear as circular, elliptical, and linear, light-colored patches on the ground surface (Figure 10) [5,9,29–31,95,112]. Historical aerial photographs that predate modern development may be especially useful in identifying and mapping paleo-sand blows. Sand blows that form on river floodplains or old lake beds may be more apparent since the coarse-grained sand blows have a lower moisture-holding capacity than the surrounding soils developed in fine-grained overbank and lacustrine deposits [5]. The lower moisture-holding capacity of sand blows also may be reflected in vegetation type and in crop growth.

3.2. Field Studies

Paleoliquefaction studies are commonly designed to characterize the ages, sizes, and spatial distribution of paleoliquefaction features across a region. Paleoliquefaction studies at several sites can provide valuable information about paleoearthquakes; however, confidence grows in the interpretation of the occurrence of such an event and in the estimates of its timing, location, and magnitude by studying numerous liquefaction sites across a broad region. As mentioned above, it is important to conduct field work in areas where liquefaction features are likely to form, to be preserved, and to be exposed. Fieldwork is conducted at times of the year, and even at times of the day in coastal areas when exposure is optimal, in order to minimize the chance that liquefaction features are missed due to high water or vegetative cover. In areas where liquefaction features are small and sparse, including areas far from the epicenters of causative earthquakes, more extensive reconnaissance is often required. For the purpose of estimating the locations and magnitudes of paleoearthquakes, it is important to document where paleoliquefaction features did not form, as well as where they did form. Paleoliquefaction field studies often include three phases: initial reconnaissance, site investigations, and river surveys.

Figure 10. Aerial photograph showing gray-colored sand blows that formed at the Hardwick site on the floodplain of the Halswell River south of Christchurch during the 4 September 2010, **M** 7.1, Darfield earthquake (modified from Reference [160]).

3.2.1. Initial Reconnaissance

A reconnaissance plan is developed based on background information described in Section 3.1 above. Locations are selected for reconnaissance to assess whether or not conditions are favorable for formation and preservation of liquefaction features, and whether the exposure is adequate to warrant searches for liquefaction features. Reconnaissance often involves inspection of sites where possible sand blows were identified on aerial photographs and/or satellite imagery. After permissions are arranged with property owners, sites are walked, and possible sand blows are identified on the ground surface. If the possible sand blow appears sandier on the surface than the surrounding soils, test pits are dug by hand to observe the sedimentary characteristics of the deposit. Typically, sand blows are sandier than, deposited on top of, and contain clasts of the surrounding soil, and are characterized by irregular bedding or flow structure. Possible sand blows found in this manner are evaluated, and those deemed the most likely to provide crucial information about paleoearthquakes are selected for further investigation (see Section 3.2.2 below). Reconnaissance also includes inspection of borrow pits and river cutbanks. Those portions of borrow pits and rivers with the most suitable conditions and best exposure are selected for systematic searches for liquefaction features (see Section 3.2.3 below).

3.2.2. Site Investigations

Detailed investigations are conducted at sites where sand blows are identified and that hold promise for providing information about the timing, locations, and magnitudes of paleoearthquakes.

Sand blows that occur in association with cultural horizons and features (e.g., pits, post molds, wall trenches) are often selected for site investigations because cultural artifacts and abundant organic material found in these horizons and features can be used to estimate the ages of the liquefaction features. For example, many of the better-constrained age estimates of sand blows in the New Madrid seismic zone in the central US came from investigations at archeological sites (e.g., References [5,7,161]). If investigations are to be conducted at archeological sites, it is imperative to involve professional archeologists in the endeavor, to take steps to minimize impacts to the site, and to comply with any national, provincial, or local rules and regulations pertaining to the documentation and preservation of cultural sites.

Site investigations often include geophysical surveys followed by trenching. As explained below in more detail, geophysical techniques are used to map the extent of the sand blow, to locate the main feeder dikes of the sand blow, and to identify cultural features that may occur in association with the sand blow. The results of geophysical surveys help to position trenches to reveal critical relationships between the liquefaction features and cultural horizons and features, while minimizing the impact to sites. Excavation of trenches is necessary to document and study characteristics of liquefaction features and their relationships with host sediment and cultural horizons and features, as well as to collect samples for dating the liquefaction features and estimating the timing of their formation.

Geophysical Techniques

The ability of geophysical techniques to detect vertical and lateral changes in sediment properties makes them an effective mapping tool in paleoliquefaction studies. These techniques offer a non-invasive tool for mapping the three-dimensional (3D) morphology of sand blows and locating sand dikes and sand blows in the subsurface. Although sand-blow deposits typically are identified by careful examination of surface deposits and natural exposures, geophysical techniques can provide a better understanding of subsurface relationships and can help guide trench excavations necessary to collect samples for dating sand blows and their causative earthquakes. The effectiveness of geophysical techniques is largely dependent on the physical properties of surficial deposits and their 3D geometries.

Electrical resistivity, ground-penetrating radar, seismic methods, magnetic methods, and ground conductivity surveys are all used in paleoliquefaction investigations. All methods are useful for locating subsurface features or structures, although each has its advantages and its limitations. These relative assets and disadvantages must be carefully considered when choosing the technique or combination of techniques to be employed at a given site. All methods derive their success from their sensitivity to variations in the physical properties of the sediment, and all are best used following surface reconnaissance to determine and narrow the general survey area. The most commonly used of these geophysical methods for paleoliquefaction surveys are electrical resistivity and ground-penetrating radar.

Although electrical methods were in use for resource exploration since Schlumberger's early work in the last century [162], the method was first applied to studies of earthquake-induced liquefaction in the mid-1990s [124,125]. The time required to perform surveys and their imaging capability has since improved with the use of multichannel equipment with automated switching capability and 3D inversion algorithms for estimating the true subsurface distribution of resistivities [121]. For alluvial sediments, resistivity (or conductivity) varies largely as a function of porosity, the degree of saturation, the type of fluid in the pore space, and clay content. Saturated sediment or sediment containing salt or clay, because of its ionic content, increases the ability for current to flow and thus decreases the resistivity of the material.

The factors affecting the measured values are the input current, the type of electrode array used, and the sediment properties. If 3D data acquisition is not employed, profile lines should be oriented perpendicular to the strike of elongated sand blows or sand fissures, if they are observed at the surface. In map view, sand dikes and related sand blows often form en echelon patterns and parallel profile lines help to determine their orientation. Important to all geophysical surveys are the depth and scale of the imaging desired. For electrical resistivity (ER) surveys, the depth of imaging is greatly affected by the conductive properties of the near-surface layers and the arrangement or spacing of the electrodes. Tighter spacing will resolve smaller features, but there will be a trade-off in the maximum depth imaged.

When done correctly in suitable environs, the imaging capability and resolution of the ER method is excellent. Figure 11 shows an example of an electrical resistivity profile that was correlated with a log of a wall of an excavated trench. The location for the excavation was chosen based on the results of the site reconnaissance (e.g., archeological surveys, soil test pits, etc.) and the geophysical survey. Warm colors in the geophysical cross-section denote more resistive material (e.g., fine to coarse sand and higher porosity associated with sand-blow deposits and dikes). Cooler colors reflect sediments with higher percentages of fines, such as silts and clays. Important in the interpretation of the geophysical data is the recognition of patterns and geometries commonly associated with liquefaction deposits, because interbedded fine- and coarse-grained sediment is common in fluvial environments.

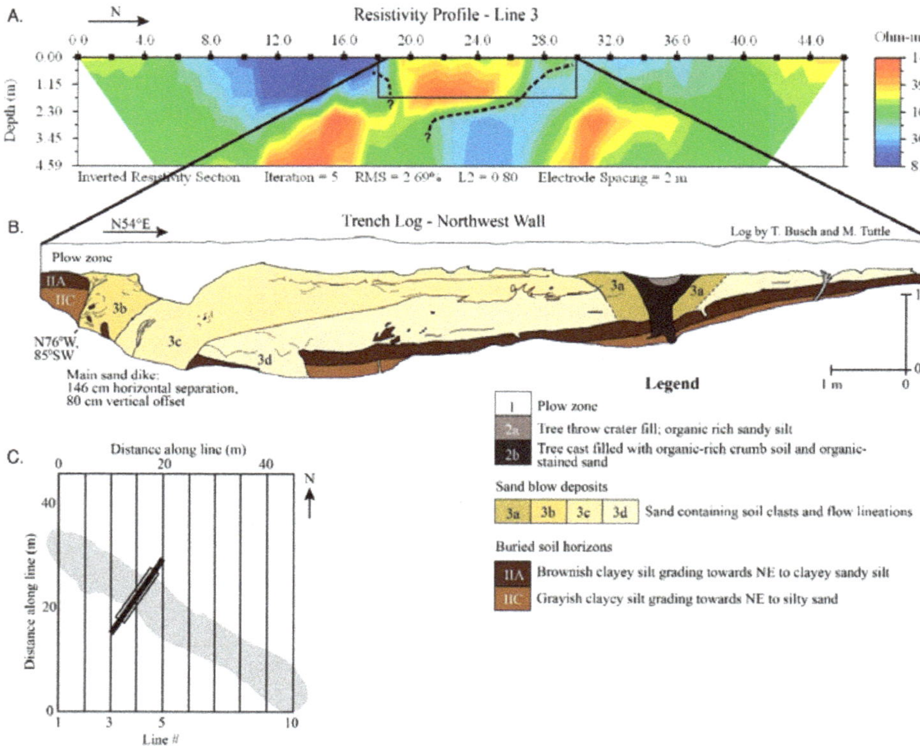

Figure 11. (**A**) Example of a resistivity profile at a liquefaction site in the New Madrid seismic zone in northeastern Arkansas [116]. Dotted line outlines approximate contact between sand-blow deposit and host sediments. Warm colors reflect sediment associated with sand blow and sand-filled dikes, which typically have higher resistivities than surrounding fine-grained sediment. Blue colors reflect material with higher percentages of clay and moisture, which are more conductive. The resistivity profile is correlated with a trench log (**B**) that is oriented at an angle to the resistivity profile. The trench log corresponds with the small portion of the resistivity section outlined by the black rectangle in (**A**). Dark-brown colors on the log correlate with green to blue colors in the resistivity profile, while yellow to red colors correlate to sand-blow deposits and the top of the feeder dike. (**C**) Inset map shows location of the trench (heavy black line) and trench log (open rectangle) in relation to the north–south-oriented geophysical profiles and northwest-oriented sand blow (shaded gray).

In addition to the factors mentioned above, success in imaging is determined by the contrast in physical properties between the liquefaction deposit or feature, and the surrounding host material. The contrast in grain sizes between the more coarse-grained, sandy liquefaction features and more fine-grained host deposits provides a good indication of the potential success of the survey. Wolf et al. [121] compared the grain-size distribution of sediment sampled from liquefaction deposits with that of the surrounding host sediment at seven sites in the New Madrid seismic zone. Results showed that, when the contrast in grain size is large, imaging by resistivity surveys is excellent. However, at sites where the contrast is low (<0.4 mm), results are relatively poor and success is achieved only through decreasing the electrode spacing, which can increase the time needed to perform the survey.

Ground-penetrating radar (GPR) methods involve the transmission and reflection of electromagnetic energy. Like seismic methods, transmitted waves (in this case electromagnetic fields) are scattered, reflected, and partitioned at interfaces in the subsurface due to contrasts in impedance such that some energy is returned to the surface. This partitioning is determined by the contrast in physical properties at stratigraphic and structural boundaries. GPR systems operate over a range of frequencies, and the ability for waves to propagate is a function of the dielectric permittivity of the medium, the operating frequency of the system (antenna choice), and the propagation dispersion (radar response to small-scale heterogeneities in the subsurface). GPR is best used in areas of dry, sandy soils, as the signal penetration through clayey soils and saturated soils can be poor. An important decision to be made before beginning GPR surveys is the target depth and size of the features being imaged and the appropriate antenna to achieve this depth and resolution. An excellent source for a thorough discussion of GPR theory and the relations affecting GPR practice can be found in Annan [126]. With the proper choice of antenna frequency and suitable soils, GPR methods offer a fast, efficient tool for mapping sand blows and locating subsurface feeder dikes at depths less than 5 m.

The use of GPR methods for earthquake-related liquefaction studies was pioneered by Liu and Li [127] and later further developed by Al-Shukri et al. [122]. Liu and Li [127] used both a 400- and 100-MHz antenna to study three sites in the New Madrid seismic zone. As with ER methods, there is a trade-off between depth of signal penetration and feature resolution; lower frequencies penetrate deeper, but lose their resolving capability. Similarly, the success of the method is dependent on the physical properties of the near-surface sediment. Like electrical methods, GPR works best when there is a contrast in grain size between the sandy liquefaction features and the surrounding host sediment. However, a key limitation of GPR surveys is the loss of signal penetration associated with conductive overburden. Clays and clayey soils are very good conductors and effectively prevent electromagnetic energy from penetrating to strata below. Penetration depths in such conditions can be less than one meter. Because true 3D data acquisition is not common in most GPR surveys, parallel profile lines are oriented perpendicular to elongated sand blows. The profiles can later be combined to form a pseudo-3D dataset.

Al-Shukri et al. [122] had good success using a 400-MHz antenna in their work on paleoliquefaction deposits in east-central Arkansas south of the New Madrid seismic zone. They noted that, because the sand blow thickness was less than 4 m, this antenna frequency provided the necessary penetration depth, as well as the spatial resolution needed to image sand dikes crosscutting silty host deposits, contacts between buried soils and overlying sand blows, and displacements of those contacts due to liquefaction-related ground failure. Resolution was also improved by post-processing the data (e.g., filtering for noise reduction, removal of direct wave arrival) following acquisition. Results showed clear reflectors similar to what one might achieve with seismic reflection surveys (Figure 12 [122]). Nobes et al. [128], in a post-earthquake liquefaction study, combined GPR imaging (200-MHz antenna) with electrical imaging and demonstrated good agreement between the two methods. A key advantage to GPR surveys is the speed with which surveys can be accomplished relative to seismic surveys. In addition, GPR surveys require only minimal personnel (e.g., References [122,127,129]) and preliminary data processing can be accomplished in the field.

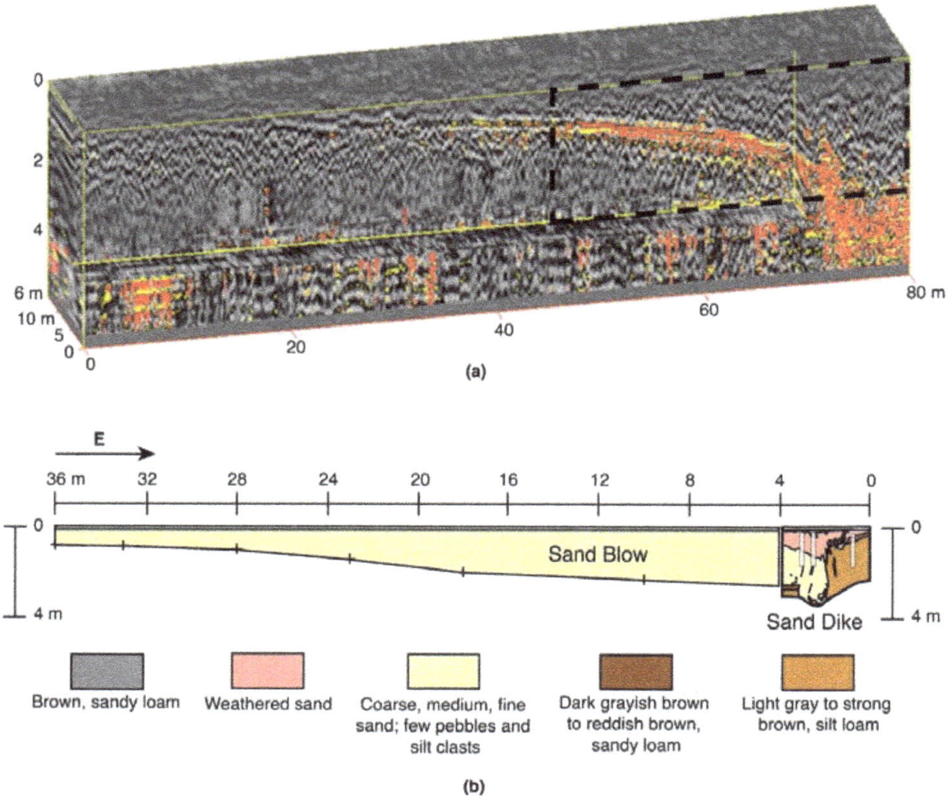

Figure 12. (**a**) Pseudo-three-dimensional (3D) ground-penetrating radar (GPR) image correlated with (**b**) log of paleoseimic trench showing large sand blow and related sand dike near Marianna, Arkansas [122]. Area of trench log is outlined on (**a**) with a dashed rectangle.

Paleoseismic Trenches

Trenches are excavated in surface sand blows to gather information about the ages and sizes of earthquake-induced liquefaction features used to estimate the timing, locations, and magnitudes of past earthquakes. Working in trenches, especially in loose sand prone to collapse or cave-ins, can be very dangerous. Many countries developed excavation standards in order to prevent or greatly reduce risks of collapse and other excavation-related incidents (e.g., References [163–165]. It is recommended that investigators follow safe excavation standards to prevent injury and deaths. For example, the Occupational Safety and Health Administration in the US (OSHA) [163] recommends that trenches in sandy soils have sloped walls at a horizontal to vertical ratio of 1.5:1, or that they be shored if walls are vertical. Stable trenches deeper than 2 m are not feasible at sites where the water table depth is close (<2 m) to the ground surface.

Trenches are typically 1.5 to 2 m deep, depending on the thickness of the sand blow, and usually excavated roughly perpendicular to the trend of the liquefaction features, which is determined from aerial photography and/or satellite imagery, geophysical surveys, and field observations. It is recommended that the trenches be dug with a backhoe or excavator with a smooth-blade bucket, sometimes referred to as a sand bucket or mop-out bucket. A smooth-blade bucket creates a clean cut, making it easier to examine the vertical and horizontal surfaces of the excavation and recognize features of interest, such as sand dikes and cultural features. If there is a plow zone at the site, it is removed with the backhoe in thin, 2–5-cm-thick cuts until the base of the plow zone is reached. The contact with the underlying sand blow is carefully cleaned and examined for intruding root casts or cultural features such as post molds, pits, and wall trenches (Figure 13). Any intruding features are documented, photographed, and sampled. The excavation continues through the sand blow, with frequent cleaning and examination of the walls and floor, and documentation of biological, cultural, and geological features encountered. This procedure is followed until the soil or sediment layer buried beneath the sand blow is reached. The contact of the buried soil or sediment layer is examined for, and samples collected of, buried leaves, tree debris, tree trunks, and other organic materials. If cultural artifacts are found in the buried soil, backhoe excavation ceases and an archeological excavation, referred to as a test unit, is planned and carried out. If no cultural artifacts are found in the buried soil, backhoe excavation continues as needed to reveal the feeder dikes of the sand blow. In projects funded by the US government, there is an obligation to follow the Section 106 process of the National Historic Preservation Act, to consult with state and tribal historic preservation officers, and to involve an archaeologist in the excavation, collection, and analysis of cultural artifacts [116,166].

Figure 13. Paleoseismic investigation at an archeological site in the New Madrid seismic zone [130]. The contact between the plow zone and the underlying sand blow is examined for features, such as cultural pits and tree-root casts, that may help to constrain the minimum age of the sand blow. Photograph by M. Tuttle.

After trench excavation is completed, trench walls are cleaned and logged at an appropriate scale (e.g., 1 inch = 25 cm or 50 cm) depending on the complexity of the features and their relationships (Figure 14). Logging, a fundamental technique used in paleoseismic studies, requires careful observations and recording of crucial relationships. Logs are created of the trench walls, and sometimes the trench floor, by gridding the surfaces and measuring points of interest relative to the grid, by surveying points of interest using a total station, or by photographing the exposures and digitally assembling the images into mosaics with photographic software, using the grid to orthorectify the images. During logging, liquefaction features and their sedimentological, stratigraphical, and structural relationships are studied. Liquefaction features are photographed and described in terms of size (i.e., width of dikes, and lateral extent and thickness of sand blows and sills), orientation, sedimentary structures, crosscutting relationships, soil development, and stratigraphic context. Liquefaction-related ground failure and the amount of vertical and horizontal displacements are measured and described. The characteristics of the host sediment are also documented, including sediment type, bedding, thickness, lateral continuity, and soil development.

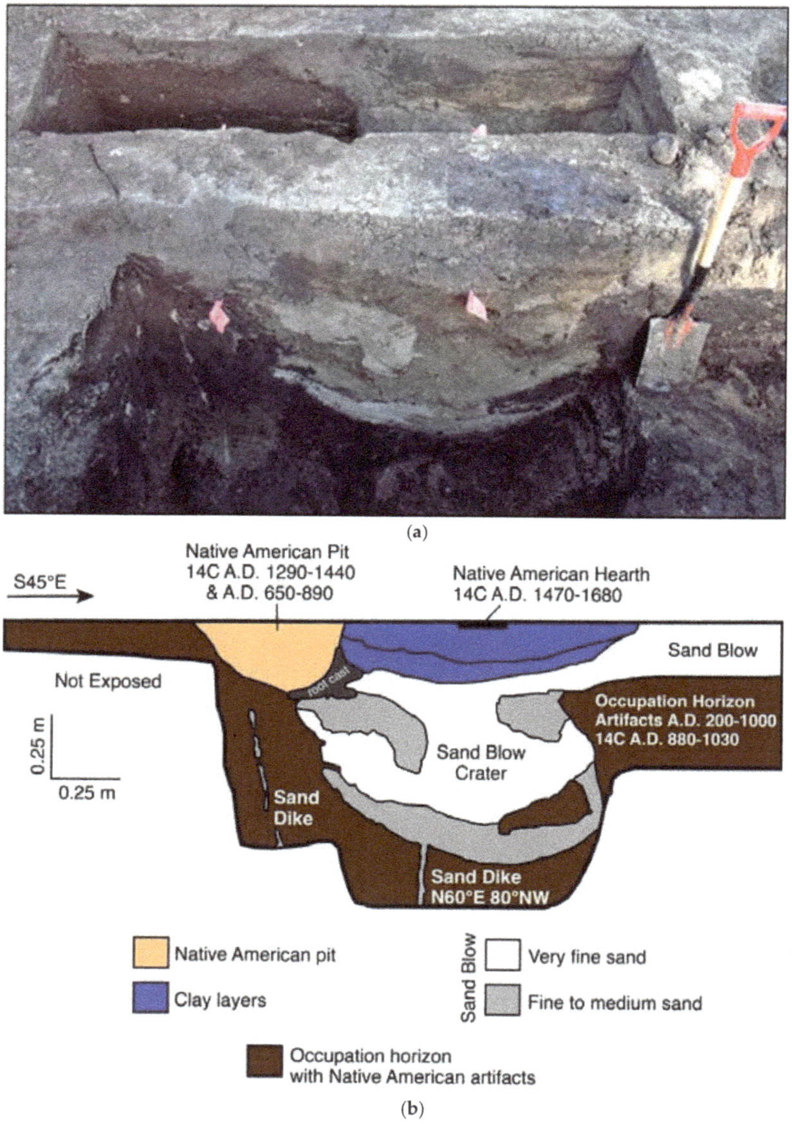

Figure 14. (**a**) Photograph of excavations and (**b**) log of trench wall at archeological site 3MS306 in northeastern Arkansas, showing relationships between the sand-blow crater, sand blow, sand dikes, and Native American hearth and pit above the sand-blow crater, and occupation horizon below the sand blow [130]. Also shown are two-sigma calibrated radiocarbon dates of organic samples collected from cultural features and horizon. On the basis of the artifact assemblage and radiocarbon dates, earthquake-induced liquefaction features are estimated to have formed in AD 900 ± 100 years.

Organic samples are collected for radiocarbon dating (e.g., References [5–18,27–32,61,109,116–119,123,125,130–138]) and sediment samples are collected for optically stimulated luminescence (OSL) dating (e.g., References [28,118,139,140]). Organic samples are sought in close association with liquefaction features, such as in soils buried by or developed in sand blows. OSL samples are collected of the contacts between the sand blows and the underlying soils or sediment in order to date the time of burial. The locations of the samples are noted on the trench log. Samples collected in the field are later reviewed in the office, and selected samples are submitted to reputable radiocarbon and luminescence laboratories for dating. Artifacts collected by an archeologist are bagged and tagged according to provenance. Cultural features that intrude or are intruded by liquefaction features are excavated and feature fill collected for analysis. Later in the archeological laboratory, feature fill is processed by flotation to recover organic materials for radiocarbon dating and small artifacts. Artifacts are described in terms of sherd sizes and types, lithic types and sources, and tool types.

Results of radiocarbon and OSL dating and information about the artifact assemblage of a site are used to estimate the ages of liquefaction features. It is advisable to date multiple samples at each site in order to have confidence in the results and to narrowly constrain the age estimates of the liquefaction features. This will facilitate correlation of similar-age liquefaction features across the region and interpretation of the source areas and magnitudes of the causative earthquakes.

3.2.3. Surveys of Rivers and Other Exposures

Eroding river cutbanks, active borrow pits, and recently excavated drainage ditches can provide exposures of Holocene and Late Pleistocene deposits in which liquefaction features may be preserved. Surveys of selected rivers, borrow pits, and ditches should be conducted when water levels are low and exposure is optimum. During the surveys, exposures are examined for the presence or absence of liquefaction features and other deformation related to earthquakes. This often involves scraping subvertical features, such as cracks and soil discontinuities that might be sand dikes and sandy lenses that might be sand blows. If sand dikes are found, they are traced upsection to determine if they terminate within the host deposit or if they connect with sand blows above (Figures 6 and 15). Sandy lenses are examined to see if they exhibit characteristics of sand blows, such as flow structure and clasts, and if they are connected to a sand dike.

As with trench studies, liquefaction features are photographed and described and organic and sediment samples collected for radiocarbon and OSL dating, respectively. The locations of liquefaction sites are measured with a global positioning system and marked on topographic maps. In addition, it is important to document the amount of exposure and sedimentary conditions where liquefaction features are found and not found. This information helps assess and define the limit of liquefaction for particular events, which in turn helps estimate the location and magnitude of past earthquakes.

3.3. *Dating Liquefaction Features*

Estimating the ages of liquefaction features is a critical part of any paleoliquefaction investigation because this information is used to estimate the timing of the earthquakes that were responsible for their formation. It is important to constrain the ages of liquefaction features as narrowly as possible to correlate similar-age features across a region and to differentiate closely timed earthquakes. Paleoliquefaction studies use a variety of dating methods, and the decision regarding which techniques to use is often based on the types of dateable material available at a site, the time period of interest, and limitations that may be imposed by project budget and schedule. The subsections below provide strategies for dating liquefaction features and overviews of methods commonly used to date liquefaction features, which include but are not limited to radiocarbon dating, OSL dating, soil development, stratigraphic context, archeological context, and dendrochronology. These overviews are not intended as comprehensive discussions of the dating methods, but rather as high-level summaries that point out some of the advantages and disadvantages of the methods. Where possible, it is recommended that paleoliquefaction studies employ more than one of these dating methods to corroborate the results.

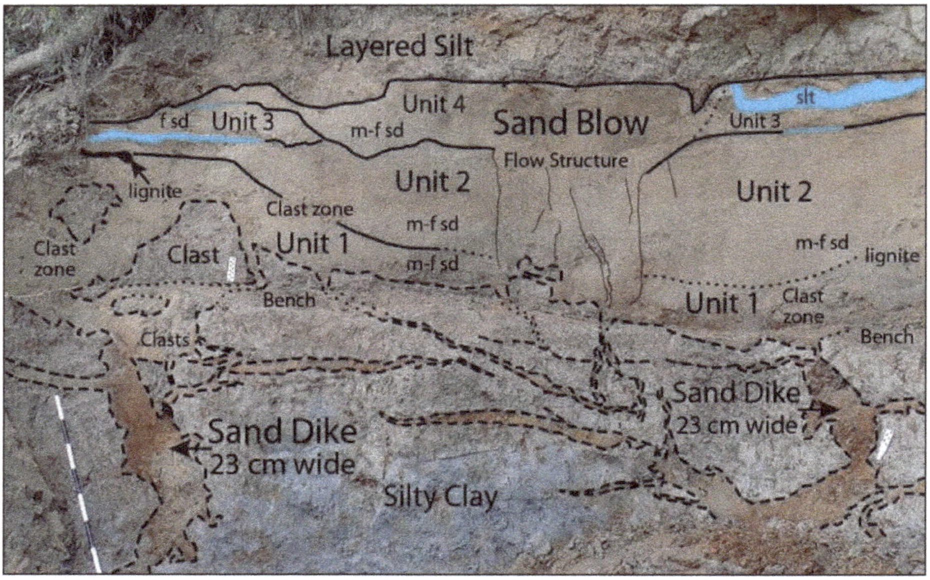

Figure 15. Annotated photograph of liquefaction features exposed in a river cutbank in the vicinity of the New Madrid seismic zone in the central US [117]. Two 23-cm-wide sand dikes with multiple splays crosscut silty clay and connect with the base of a large compound sand blow, composed of four depositional units, that formed during four earthquakes in a sequence. Dashed lines represent clear contacts; dotted lines represent inferred contacts; solid lines represent clear contacts of sand blow units; blue represents silt layers at top of sand blow units. The black and white intervals on the meter sticks represent 10 cm. Photograph by M. Tuttle.

3.3.1. Dating Strategies

Sand blows typically provide the best opportunity for estimating the ages of paleoearthquakes with relatively small uncertainties, because it is often possible to determine both maximum and minimum age constraints for sand blows and, thus, bracket their age of formation (Figure 16) [2]. Close maximum age constraints can be determined by radiocarbon dating of plant material, such as twigs and leaves, and OSL dating of sediment that was at or near the ground surface and buried by the sand blows at the time of the earthquake. Similarly, plant material derived from surface soils and incorporated in the vented deposits of sand blows and sand-blow craters also provides close maximum age constraints. Close minimum age constraints can be achieved by dating plant material and sediment that accumulated in craterlets in the upper surface of sand blows soon after they formed. More commonly, minimum age constraints are derived by dating plant material in soils that developed in the sand blows over time and from tree roots and cultural pits that extend down into sand blows from above. Similarly, well-dated tephra beds in volcanically active areas can provide maximum and minimum age constraints for sand blows [22].

Estimating the ages of sand dikes and sand sills usually involves greater uncertainty than estimating the ages of sand blows and sand-blow craters, because dikes and sills may terminate several meters below the ground surface at the time of the paleoearthquake (Figure 16) [2]. Maximum age constraints can be determined by dating the uppermost stratigraphic units that the dikes crosscut or the sills overlie, but these ages may be hundreds to thousands of years older than the liquefaction feature [125]. Minimum age constraints of dikes and sills can be determined by dating roots, animal burrows, and cultural pits that clearly intrude and postdate the liquefaction features or by dating deposits that overlie unconformities truncating the liquefaction features. However, it is fairly uncommon to find circumstances such as these that help constrain the minimum ages of dikes and sills [2]. Therefore, age estimates of sand dikes and sills often have large uncertainties. Some investigators make educated guesses as to the ages of these types of liquefaction features based on weathering characteristics of the features themselves or the approximate age of the deposits in which they occur. These estimates, however, can have large uncertainties on the order of thousands of years.

Pseudonodules, load casts, and related folds typically form close to the sediment–water interface at the time of sediment deposition [97]. Age estimates and related uncertainties for causative earthquakes can be derived by dating the deformed sediment itself or by dating plant material above and below the deformed sediment. There often are much larger uncertainties in estimating the ages of sand diapirs and foundered clasts because these soft-sediment deformation structures may have formed anytime following deposition of the stratigraphic units involved. Maximum age constraints can be established by dating the deformed deposits, but the deformation may be hundreds or thousands of years younger than the deposits.

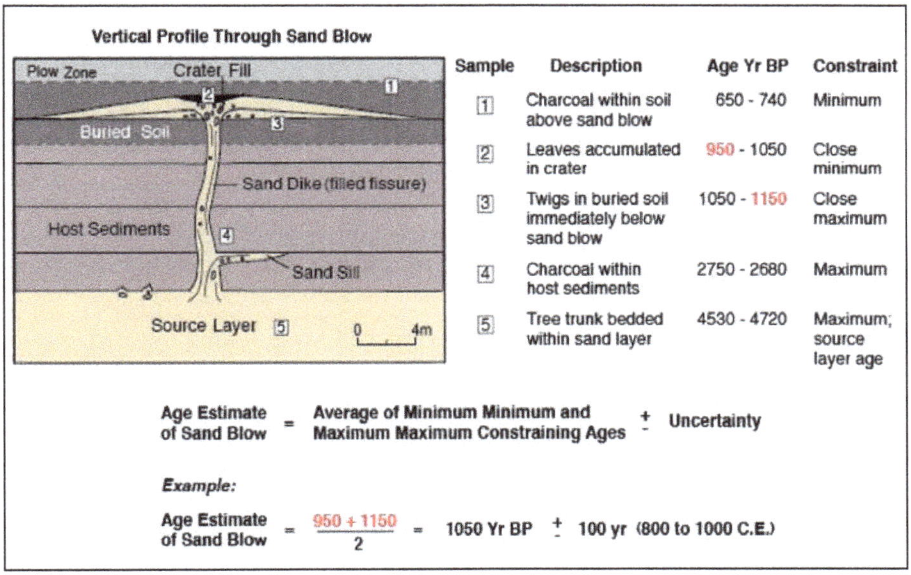

Figure 16. Diagram illustrating sampling strategies and age estimation of liquefaction features (modified after References [35,116]).

3.3.2. Radiocarbon Dating

Radiocarbon (^{14}C) dating is commonly used in paleoliquefaction studies. Although reliable for only the past 50,000 years, radiocarbon dating is useful for the time period of interest for most paleoseismic studies. In addition, dateable organic material is common in many environments, and the method is relatively inexpensive. The type and location of samples collected for radiocarbon dating affect the uncertainty of the age estimate of the liquefaction features. Plant remains that occur in close stratigraphic position to a sand blow will fairly closely reflect its age. For example, leaves or seeds that occur at the contact of a buried soil horizon and an overlying sand blow would provide a close maximum age constraint for the sand blow. Similar materials incorporated into the top of a sand blow would provide a close minimum age constraint (Figure 16). Samples collected for radiocarbon analysis should be inspected in the field to ensure that their origin is understood. For example, a piece of detrital charcoal in the sediment underlying a sand blow would provide a maximum age constraint, but could be hundreds or even thousands of years older than the sand blow, depending on its transport time. Likewise, plant roots can return radiocarbon ages that are much younger than the sediment from which they are collected. Bulk samples of soils buried by or developed in sand blows also can be dated.

However, radiocarbon dates of soils reflect the mean residence time of carbon in those samples [167,168]. Also, contamination by young (e.g., modern humic acids) and old (e.g., lignite and calcium carbonate) carbon can be a significant problem in dating soils. Therefore, radiocarbon dating of bulk soil samples is usually a last resort and requires a sampling strategy to help minimize the uncertainties [5].

3.3.3. Luminescence Dating

Luminescence dating methods are also commonly used in paleoliquefaction studies. These numerical methods are used to determine the amount of time that passed since sediment was last exposed to light. Luminescence dating can be used to estimate the ages of sand blows and, thus, their causative earthquakes, by dating sediment that pre- and post-dates sand blows [120,139,140]. Luminescence dating can be employed in a variety of terrestrial stratigraphic settings, can date sediment on the order of several years to several hundreds of thousands of years old [169], and is particularly useful for dating sediments that receive brief exposure to sunlight prior to deposition [169–171]. A particular advantage of luminescence dating methods is that they rely on abundant minerals such as quartz and feldspars. Quartz is commonly used in optically stimulated luminescence (OSL) dating, but saturation effects of the quartz luminescence signal with increasing dose generally restrict the method to dating sediments younger than about 1500,000 to 200,000 years old [172,173]. However, where environmental dose rates are low, OSL dating of quartz may be extended to approximately 400,000 years [173]. Feldspars can be dated by OSL and infrared stimulated luminescence (IRSL) methods. Because the IRSL signal from feldspar typically saturates at higher dose levels relative to quartz OSL, the dateable range for IRSL on feldspar is on the order of 500,000 years [174], and this range may be extended further using relatively recent and ongoing developments in post-IR IRSL (pIR-IRSL) dating of potassium feldspars [172,173].

There are multiple sources of uncertainty that can limit the accuracy of luminescence dating. The most common problem is that the sediment did not receive enough sunlight exposure prior to burial in order to rid the sample of previously acquired luminescence [175]. In addition, silt and sand grains that are coated with clay during their erosional travel path may be shielded from the bleaching effects of sunlight. Because cosmic radiation often fluctuates and attenuates quickly with depth, uncertainties in the depositional elevation of the sediment sample also may contribute to the uncertainty in the resulting age estimates. In addition, the accuracy of luminescence dating may be limited by estimates of past water content in the deposit because of the radiation-absorbing characteristic of water. To minimize the impact of the sources of uncertainties described above, selection and sampling of sediment are crucial, as well as early guidance on sampling techniques and sample handling from the luminescence dating facility.

3.3.4. Soil Development and Weathering Characteristics

In many paleoliquefaction studies, soil development is used as a relative dating technique to distinguish young, unweathered features from significantly older, weathered features. In the New Madrid seismic zone, for example, the thickness of soil A-horizons developed in sand blows was used to distinguish paleoliquefaction features from those that formed during the 1811–1812 earthquake sequence and to estimate the age of prehistoric sand blows (Figure 8) [141]. The age estimates are based on a rate of A-horizon development derived from measurements of A-horizon thickness of sand blows whose ages were determined by radiocarbon dating. Age estimates derived in this manner typically have uncertainties on the order of 100–200 years, similar to the uncertainties of radiocarbon dating on which the rate of A-horizon development is based. In a few regions where sand dikes terminated within the stratigraphic section, and where organic material and cultural artifacts were not available for constraining the ages of the features, soil characteristics such as iron staining and accumulation of fine-grained sediment are used to correlate features over large distances. This practice is not recommended unless the soil characteristics are regionally calibrated and the uncertainties associated with their rates of development quantified. Otherwise, the spatial correlation of features and the interpretations related to the spatial distribution of those features may be erroneous [2].

3.3.5. Stratigraphic Context

Stratigraphic context and relationships can be used as a means to estimate the relative ages of buried sand blows and sand dikes, and to correlate paleoliquefaction features between exposures. The age of the hosting sediments provides a maximum age for the liquefaction features that intrude therein. For example, stratigraphic context and relationships can be used to place maximum ages on sand dikes that terminate upward at a stratigraphic level that may be lower than the paleo-ground surface at the time of the causative earthquake (Figure 16). Sand dikes that terminate below the event horizon, or below a horizon that represents the ground surface at the time of the earthquake, are commonly encountered outside of active seismic zones and at greater distances from the seismic source than sand blows. By numerical or relative dating of the host deposits, it is possible to place at least a maximum age constraint on the timing of dike formation (Figure 16).

In addition, the law of superposition, crosscutting relationships, and identification of paleosurface indicators preserved in the stratigraphic record can be used to help determine the relative ages of paleoliquefaction features. In an area with laterally continuous stratigraphy or prominent marker beds, age equivalence can be established between different exposures or sites. If the ages of some or all of these continuous strata are determined by numerical or other means at one exposure, these ages can be extrapolated to other nearby exposures. Vitale at al. [22] reported evidence of faults, as well as seismically induced sand blows, clastic dikes, and other soft sediment deformation structures preserved in the stratigraphic record, in the area of the active Campi Flegrei caldera in southern Italy. The volcanic sedimentary sequence in which these features are preserved is well dated and, thus, provides the means to constrain the ages of liquefaction-inducing, moderate-magnitude earthquakes between 4.55 and 4.28 ka BP [22].

3.3.6. Archeological Context

Cultural artifacts found at liquefaction sites can help estimate the ages of the liquefaction features [4,14,142–144,161]. Due to the common abundance of organic-rich material, archeological sites often provide good opportunities for finding samples suitable for radiocarbon dating as well. Where possible, it is desirable to conduct both archeological analyses and radiocarbon dating to provide a means of independently verifying, and adding confidence in, age estimates of liquefaction features. Some artifact types are narrowly tied to specific cultural periods while others are not. Because cultural periods are based on radiocarbon dating of archaeological contexts of artifacts, age estimates of liquefaction features based on their archeological context will have uncertainties at least as great as those based solely on radiocarbon dating [2].

The archaeological context of a sand blow or sand dike is defined by the presence of artifacts and/or cultural features that may occur stratigraphically above, below, or within the geologic feature. The stratigraphic relationships between liquefaction features and cultural features and horizons, as well as the assemblage of artifacts (especially if diagnostic artifact types are present) can help constrain the ages of the liquefaction features (e.g., References [7,144,161]). For example, the assemblage of artifacts within a soil A-horizon buried by a sand blow can provide an estimate of the maximum age of the liquefaction feature [8,125]. The assemblage of artifacts within an occupation horizon developed in a sand blow or cultural features, such as a storage pit or wall trench dug into a sand blow, can provide an estimate of the minimum age of the sand blow. It is important to study assemblages of artifacts at a site since there are still many uncertainties regarding the temporal and geographical ranges of artifact types.

3.3.7. Dendrochronology

Dendrochronology is the dating of past events through the study of the tree ring records. As trees grow, they add annual growth bands, or rings, to their circumference. Tree ring width varies from species to species and from year to year depending on the amount of annual rainfall and other environmental factors. Therefore, tree-ring chronologies are both species- and region-specific. Master chronologies are constructed by starting with tree-ring sequences for modern trees and adding antecedent and overlapping sequences back through time [176–178]. Dating a sample of an ancient tree is accomplished by matching the sequence of rings with the master chronology. It is desirable to have a sequence of 100 rings for dating purposes. Accurate dating of the death of a tree requires having the exterior ring of the specimen.

Dendrochronology has the potential to date paleoearthquakes to the year and even the season [178,179]. For example, trees killed by co-seismic subsidence along the coast of Washington State helped provide exact dates of megathrust earthquakes along the Cascadia subduction zone [145]. In the New Madrid seismic zone, bald cypress in Reelfoot Lake in westernmost Tennessee showed a dramatic increase in ring width following the 1811–1812 earthquakes due to increased water depth resulting from co-seismic subsidence, whereas bald cypress in the St. Francis sunkland in northeastern Arkansas showed decreased ring widths for about 45 years following the earthquakes due to tissue damage inflicted by earthquake shaking [180].

To date, dendrochronology was used very little in paleoliquefaction studies but it has the potential to better constrain age estimates of paleoearthquakes, especially in regions where liquefaction-related ground failures were severe. Abrupt changes in soil-moisture conditions due to liquefaction-related subsidence of the ground surface and/or burial by thick sand blows, as well as disruption of tree root systems by lateral spreading, may affect tree ring growth and even lead to tree death [5]. Trees buried and preserved below sand blows may provide accurate dates of paleoearthquakes. Before dendrochronology can be used to date paleoliquefaction features, however, regional chronologies that extend beyond the historic period must be developed for long-lived tree species.

3.4. *Interpreting Liquefaction Features*

During field studies and dating of field samples, information is gained about the locations, sizes, sedimentary and structural characteristics, archeological and stratigraphic context, and ages of liquefaction features. This information is interpreted in terms of the number, timing, source areas, and magnitudes of paleoearthquakes, as well as the recurrence times of large earthquakes and long-term behavior of earthquake sources. There are numerous factors that contribute to uncertainties in the earthquake parameters interpreted from liquefaction features (Table 2 [51]). These uncertainties can be reduced by conducting a broad and balanced field program, by making detailed and thorough field observations, by collecting samples that will narrowly constrain the age estimates of liquefaction features, and by wisely employing analysis tools in the interpretation of liquefaction data.

3.4.1. Correlation of Features

Correlation of liquefaction features is an important and necessary step in the interpretation of the timing, location, and magnitude of paleoearthquakes. Correlation of liquefaction features is based on available information, including one or more of the following (from Reference [51]):

- Chronological control: Paleoearthquakes are identified based on grouping of paleoliquefaction features that have overlapping age estimates (e.g., References [5–8,12,29,109,117,118]. As described above in Section 3.3.1, sand blows usually provide the best chronological control because the event horizons (e.g., soil horizons buried by sand blows) are more easily identified and their age estimates are usually better constrained, whereas the event horizon and age estimates associated with sand dikes are often poorly constrained.
- Size distribution: In general, the size of liquefaction features diminishes as ground shaking decreases (e.g., References [2,6,8,10,30,58,59,100,181,182]). Therefore, the size distribution of liquefaction features relates to magnitude and distance from the causative earthquake. The size distribution of features is also important for interpreting whether similar-age features formed during a single large earthquake or multiple smaller earthquakes.

- Stratigraphic control: Paleoearthquakes are distinguished based on grouping of paleoliquefaction features found in deposits of similar age (see caveats described in Sections 3.3.2 and 3.3.3).
- Pedologic or weathering characteristics: Paleoearthquakes are distinguished based on grouping of paleoliquefaction features with similar soil or weathering characteristics (see caveats described in Section 3.3.4).

If the different types of information described above provide conflicting correlations for a specific field study, the investigator must assess the relative quality of the information and provide a preferred interpretation. Additional field reconnaissance may be necessary to identify areas for targeted studies that might be necessary to resolve these conflicts.

Table 2. Uncertainties related to interpretation of liquefaction features [51].

Earthquake Parameter	Range in Uncertainty	Factors that Contribute to Uncertainty	Observations and Analyses that Reduce Uncertainty
Timing	10s–1000s of years	(1) Dating of liquefaction features (2) Use of sand dikes in absence of sand blows	(1) Well-constrained age estimates of liquefaction features (2) Space–time diagrams (3) Statistical analysis of uncertainty range of age estimates of multiple liquefaction features
Location	Few–100s of km	(1) Same as above (2) Correlation of features across region (3) Size and spatial distribution of contemporaneous features a. Style of faulting b. Earthquake source characteristics c. Directivity of seismic energy d. Attenuation & amplification of ground motion e. Relative density of sediment f. Distribution of liquefiable sediment g. Water table depth (4) Field sampling and exposure	(1) through (3) above (4) Size distribution of features (5) Information regarding uncertainty factors (3a) through (3g). (6) Field studies conducted where sedimentary and hydrologic conditions suitable for formation and preservation of liquefaction features, and when adequate exposure available to find features, if present (7) Comparative study with calibration event in same region (8) Relationship to active fault
Magnitude	0.25–1+ unit	(1) through (4) above (5) Epicentral distance to farthest sand blow unlikely to be known (6) Changes in source sediment due to liquefaction or to postliquefaction effects such as cementation and compaction	(1) through (8) above (9) Empirical relations based on global database of earthquakes that induced liquefaction (10) Evaluation of scenario earthquakes using liquefaction potential analysis
Recurrence time	10s–1000s of years	(1) Uncertainty in timing of paleoearthquakes (2) Completeness of paleoearthquake record in space and time	(1) Well-constrained age estimates of paleoearthquakes (2) Space–time diagrams (3) Consideration of history of sedimentation and erosion, as well as of changes in water table

3.4.2. Timing of Paleoearthquakes

Given the uncertainties related to dating as discussed in Section 3.3 above, it is prudent to interpret the timing of paleoearthquakes from the age estimates of at least several individual liquefaction features and possibly other paleoseismic deformation structures (Table 2). The more well-constrained age estimates of liquefaction features are used in the interpretation of event timing, the higher the confidence level in the result is. If ages of liquefaction features cannot be constrained within a few hundred years, it may not be possible to resolve the timing of paleoearthquakes with confidence or to correlate features chronologically across a region.

Clustering of age estimates of liquefaction features that can be reasonably correlated across a region is thought to reflect the timing of paleoearthquakes (Figures 17 and 18). For a particular cluster, the union of well-constrained age estimates of liquefaction features represents the time period during which the paleoearthquake is likely to have occurred. It is not uncommon for this time period to have a range of 100s to 1000s of years (e.g., References [6,12,51]. The intersection of overlapping age estimates of sand blows may provide an event time with a range of 10s to 100s of years, as long as there is a high degree of confidence in the accuracy of the age estimates. Statistical analysis of data clusters that includes several well-constrained age estimates can lead to a more narrowly defined range of several to tens of years and, thus, smaller uncertainties in the estimated timing of the paleoearthquakes (Figure 18).

Due to the effects of soil development, weathering, and bioturbation, it is more difficult to recognize older liquefaction features and to narrowly constrain their ages. Liquefaction features that are more than 50,000 years old are beyond the reach of radiocarbon dating. For these paleoliquefaction features, age estimates may be provided by stratigraphic context and OSL dating of the host deposit, but the uncertainties may be on the order of 1000s of years. Ancient liquefaction features were identified in rocks 100s of Ma in age (e.g., Reference [146]) but are rarely relevant to the current tectonic setting and seismic hazard.

3.4.3. Location and Magnitudes of Paleoearthquakes

As demonstrated by numerous case studies of modern earthquakes that induced liquefaction, the distributions of liquefaction features or liquefaction fields are roughly centered on their earthquake epicenters or reflect the earthquake source area. Sand blows generally decrease in size and frequency with increasing distance from the epicenter (e.g., References [59,181,183]). In addition, the size of sand blows near the epicenter, as well as the maximum distance at which sand blows form, generally increase with earthquake magnitude. Therefore, the size and spatial distribution of sand blows can help to estimate the locations and magnitudes of paleoearthquakes (e.g., References [2,100]). However, there are examples of modern earthquakes that produced liquefaction fields not fitting these general patterns. For example, the 2010 **M** 7.1 Darfield (New Zealand) earthquake, produced a liquefaction field skewed toward the coast due to regional differences in liquefaction susceptibility of sediment and water table depth [30,184–190]. The 2002 **M** 7.9 Denali fault (Alaska) earthquake produced an extensive liquefaction field (at least 100 km from the epicenter) [94] that increased in severity from west to east, or from the main shock to the third subevent [93,94]. Harp et al. [94] attributed this unusual distribution of liquefaction to differences in fault length, duration of shaking, and lower frequency content of the third subevent. Therefore, factors that can influence the size and distribution of liquefaction features, such as regional tectonics and earthquake characteristics (e.g., style of faulting, directivity of seismic energy, and attenuation and amplification of ground motion), as well as local site conditions (e.g., grain-size distribution and relative density of sediment, distribution of liquefiable sediment, and water table depth), should be considered when making interpretations about paleoearthquakes from paleoliquefaction features (e.g., References [30,51,56,75,95,100,147]). In the case of seismically triggered soft-sediment deformation structures in lacustrine deposits, the spatial distribution of the features and the intensity of deformation are used to construct isoclines of deformation and to identify the probable paleoearthquake source area [24].

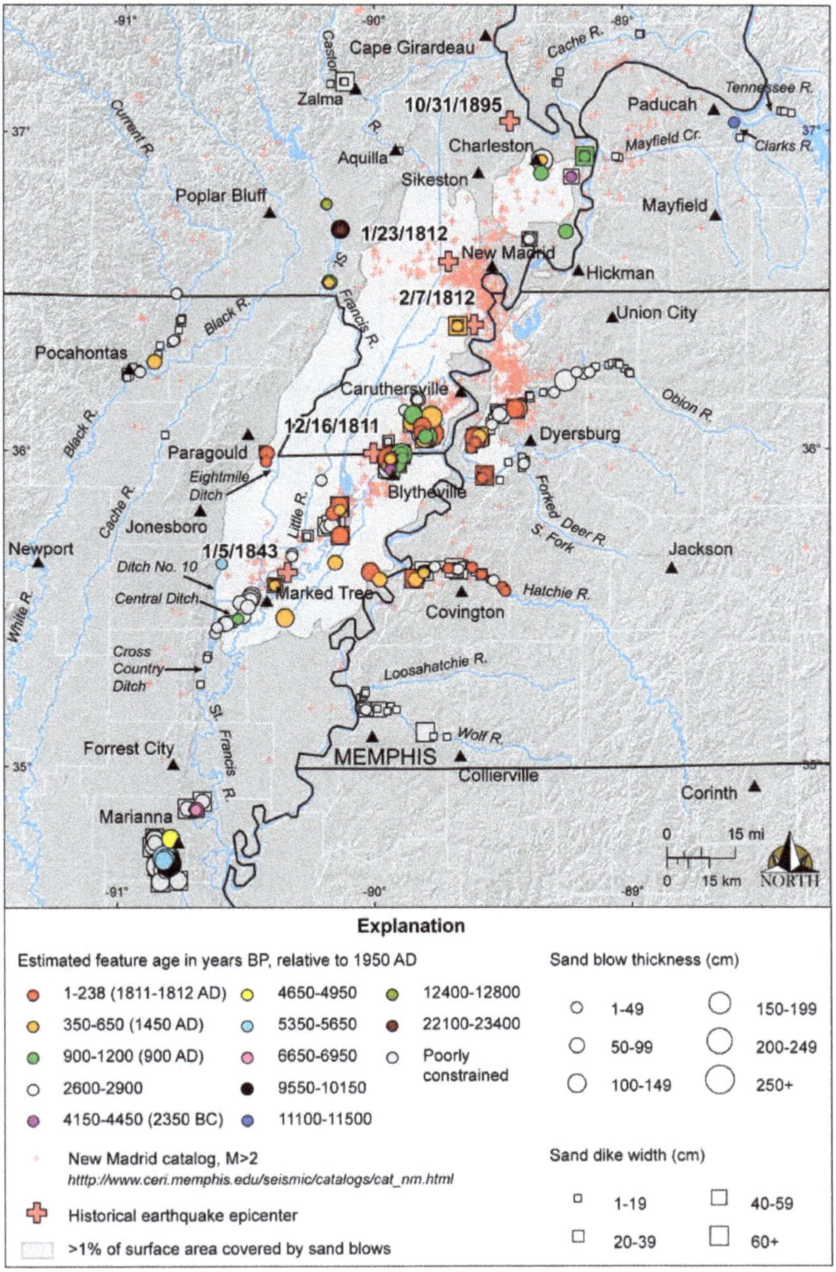

Figure 17. Shaded relief map of New Madrid seismic zone (NMSZ) and surrounding region showing ages and measured sizes of earthquake-induced liquefaction features, previously recognized liquefaction field, inferred locations of historical earthquakes, and instrumental located earthquakes [148].

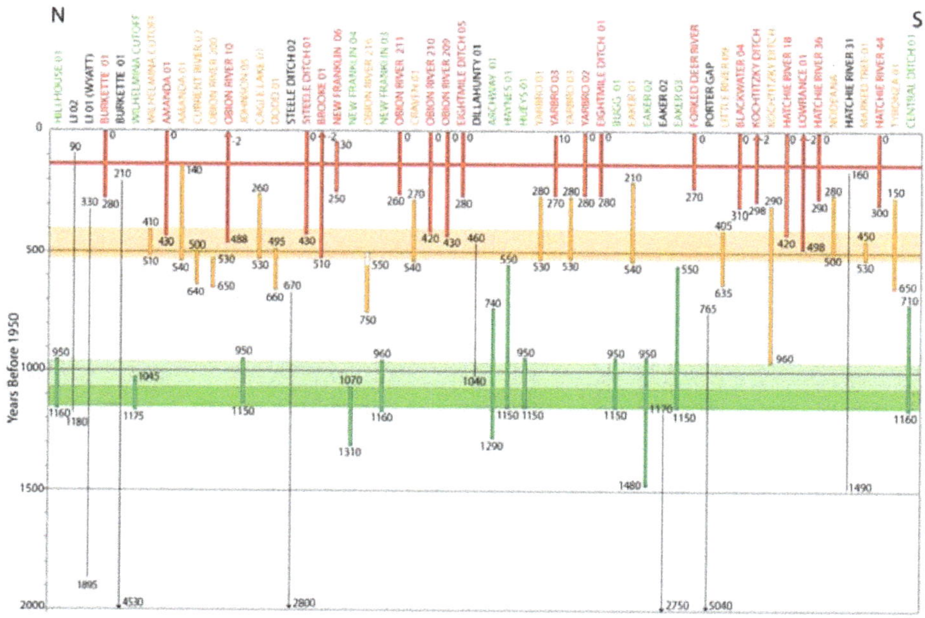

Figure 18. Earthquake chronology of the New Madrid seismic zone for the past 2000 years based on paleoliquefaction studies (e.g., References [6,7,51]). Vertical bars represent age estimates of individual sand blows, and horizontal bars represent inferred event times based on intersection of overlapping age estimates: 138 years before present (BP) (AD 1811–1812); 500 years BP (AD 1450) ± 150 years; 1050 years BP (AD 900) ± 100 years. Statistical analysis resolved event times with narrower uncertainty ranges of 503 years BP ± 8 years and 1110 years BP ± 40 years as indicated by darker portions of the horizontal bars [35].

When sand dikes are apparently the only earthquake-induced liquefaction features that formed in a region, special care must be taken in interpreting them. Sand dikes indicate that ground motions were strong enough to induce liquefaction in subsurface sediment. Therefore, distributions of similar-age sand dikes can be used to define liquefaction fields and to make interpretations about the locations and magnitudes of paleoearthquakes. However, dike width appears to be strongly influenced by proximity to river cutbanks or other abrupt changes in topography (e.g., References [28,52,55,100,111] and, therefore, the sizes of dikes may be misleading. Other earthquake-induced liquefaction features such as sand diapirs and foundered clasts may be indicative of the threshold of liquefaction, or lowest levels of ground shaking at which liquefaction occurs. These features can help delineate the outer limits of liquefaction fields, which is useful information for interpreting locations and magnitudes of paleoearthquakes, as long as they can be correlated with sand blows and sand dikes in the region. Several approaches to interpreting paleoliquefaction in terms of locations and magnitudes of paleoearthquakes are summarized below.

Comparison with Modern or Historical Analogues

Modern and historical earthquakes that induced liquefaction can be used as calibration events for interpreting paleoliquefaction features. If the size and spatial distribution of liquefaction features generated by a paleoearthquake are similar to those for a modern or historical earthquake in the same region, the paleoearthquake can be inferred to have a similar source area and magnitude to that of the modern or historical earthquake. For example, the source area of the 1886 Charleston (South Carolina) earthquake is thought to have produced several large paleoearthquakes during the past

5500 years on the basis of similar spatial distributions of historical and prehistoric sand-blow craters (e.g., Reference [12]). Similarly, the New Madrid seismic zone is thought to be the source of **M** 7 to 8 earthquakes about AD 1450 (500 years BP) and AD 900 (1050 years BP) on the basis of similar size and spatial distributions of historical and prehistoric sand blows (Figure 19 [6,7]). If a paleoearthquake can be directly related to movement on a particular fault, such as with the Reelfoot fault in the New Madrid seismic zone, the location of the event can be further constrained (Table 2; e.g., References [149,191]).

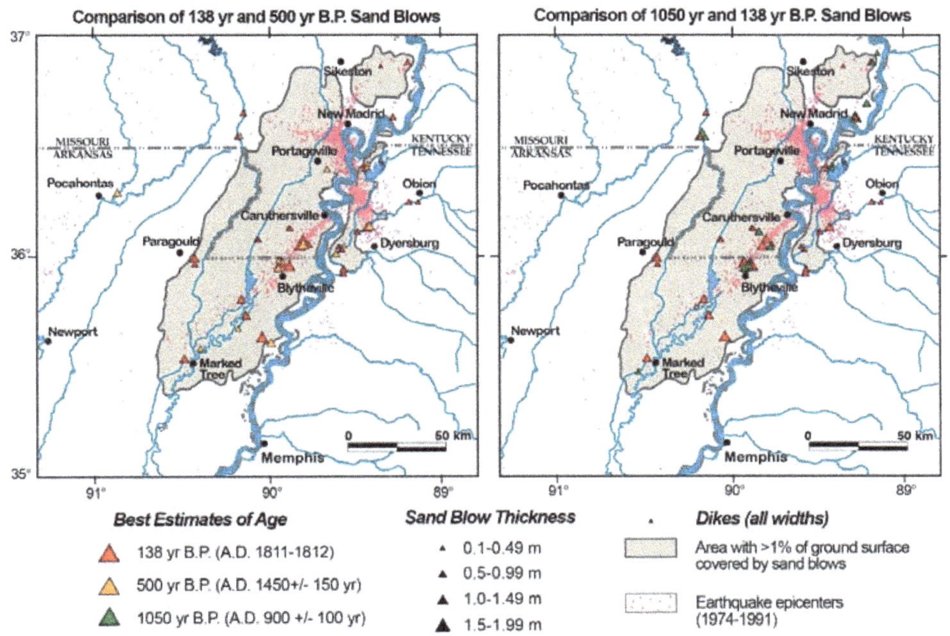

Figure 19. Maps of the New Madrid seismic zone showing comparison of sizes and spatial distributions of sand blows and dikes that formed during the 138 years BP (AD 1811–1812) event with those attributed to the events 500 years BP (AD 1450) and 1050 years BP (AD 900) (modified from Reference [6]).

Empirical Relations

Empirical relations used in paleoliquefaction studies include (1) the trend between earthquake magnitude and maximum distance of surface manifestations of liquefaction, or sand blows; and (2) the relations between the liquefaction severity index (LSI) and distance of liquefaction from the seismic energy source. Relations were also developed between earthquake intensity and epicentral distance of liquefaction features (e.g., References [97,192]. The magnitude–distance and intensity–distance relations are based on modern and historical cases of earthquake-induced liquefaction for which the locations and magnitudes of the earthquakes are fairly well known (Figure 20; e.g., References [181,183,192–195]). Recent studies found that style of faulting, directivity of seismic energy, and other factors influence the size and distance of liquefaction features, but that the maximum distance of surface manifestations of liquefaction is still a useful indicator of earthquake magnitude [95,101,181]. The ESI, with its specifics regarding the abundance and size of sand blows, as well as the severity of related ground failures and other environmental effects, has the potential to be useful in estimating the intensity field and epicentral intensity of paleoearthquakes (e.g., References [102,103,105–107,192,196]). Efforts to develop catalogs and databases of liquefaction features and other environmental effects, such as through the International Union for Quaternary Research and the US Nuclear Regulatory Commission (e.g., References [51,107,112]), are ongoing and will enhance comparisons between modern and historical earthquakes and improve empirical relations between various earthquake parameters and liquefaction (e.g., References [101,104]).

To use magnitude–distance relations in paleoliquefaction studies, also known as the magnitude-bound method, the source area of a paleoearthquake is first interpreted from the size and spatial distribution of contemporaneous liquefaction features; then, the magnitude of the causative earthquake is estimated from the maximum distance of sand blows from the inferred epicenter. There are several obvious factors that contribute to uncertainties in magnitude estimates for paleoearthquakes based on the magnitude-bound method. The epicentral location may be poorly defined and the farthest sand blow is unlikely to be known. In addition, the magnitude–distance relations themselves have some inherent uncertainties since the epicentral distance to the farthest sand blow may not be known even for recent instrumentally recorded earthquakes. Due to the sparse data for infrequent very large earthquakes, the relations are poorly constrained for earthquakes greater than magnitude 7.5 (Figure 20). Attempts were made to reduce uncertainties in the magnitude estimates of paleoearthquakes by performing regional calibration of the magnitude–distance relations [123,150,151,192]. Calibration of the relations may be most fruitful for regions, such as the Canterbury region of New Zealand and the Po Plain in northern Italy, which recently experienced earthquakes that were instrumentally recorded and for which a post-earthquake survey of liquefaction features was conducted (e.g., References [26,95,151]). Given the uncertainties in the locations and magnitudes of historical earthquakes, calibrations that rely heavily on historical earthquakes may not significantly improve the magnitude–distance relations developed from the worldwide database of instrumentally recorded earthquakes [35].

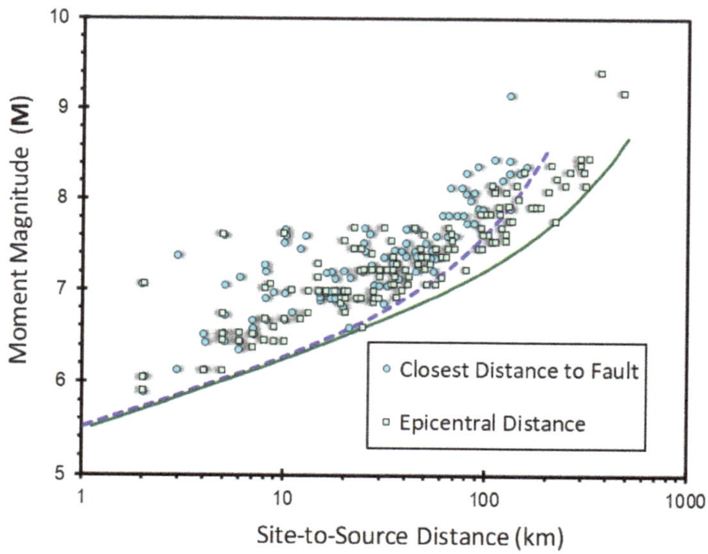

(a)

Figure 20. *Cont.*

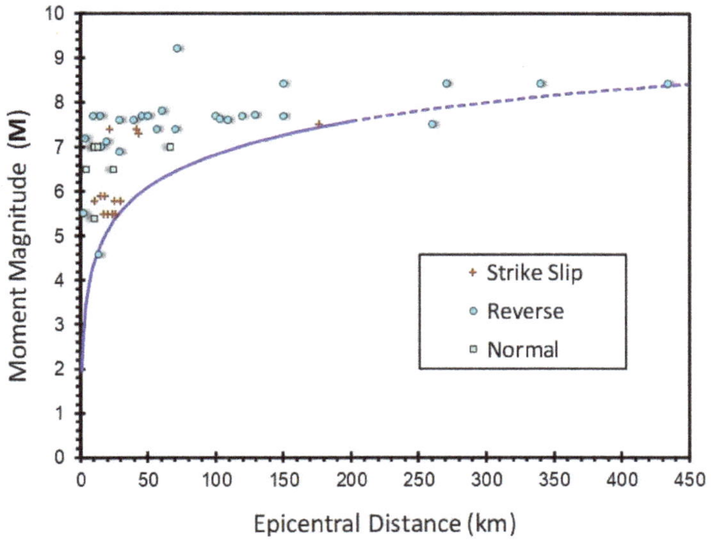

(b)

Figure 20. Empirical relations between moment magnitude, **M**, and distance to farthest surface expression of liquefaction. (**a**) Relation between magnitude and fault distance, as well as epicentral distance of liquefaction (after Reference [195]). (**b**) Relation between magnitude and epicentral distance to farthest known sand blows that formed during earthquakes produced by strike slip, reverse, and normal faulting (after Reference [181]).

LSI is a measure of ground failure displacement related to lateral spreading on gently sloping late Holocene fluvial deposits [73]. LSI represents the maximum observed severity of ground failure at a given locality, with displacements greater than 2.5 m receiving the limiting value of 100. Values of LSI were determined for several earthquakes, ranging in magnitude from **M** 5.2 to 9.2 in the western US, and plotted against horizontal distance from fault rupture [73]. Relations were developed for one modern and two historical earthquakes in the eastern US and Canada that show liquefaction at greater distances in this region compared to similar-size earthquakes in the west [197]. For example, the 1988 **M** 5.9 Saguenay (Quebec) earthquake induced liquefaction 25–30 km from its epicenter, a much greater distance than reported for liquefaction triggered by similar-magnitude earthquakes in the western US [52,73]. Liquefaction at greater distances is probably due in part to lower attenuation of ground motion in the crystalline rocks of eastern North America. Although they are employed only rarely in paleoliquefaction studies, LSI–distance relations allow for the use of liquefaction features, including sand blows, in a meizoseismal area and do not rely on distal liquefaction features that may be difficult to find and to date.

Geotechnical Approach

The geotechnical approach for evaluating paleoearthquakes involves performing liquefaction potential analysis for scenario earthquakes of various locations and magnitudes (e.g., References [75,152]). Liquefaction potential analysis is usually performed both for sites that did and did not liquefy across a region. Predicted liquefaction is then compared with observed liquefaction in order to identify the locations and magnitudes of earthquakes most likely to have produced the observed paleoliquefaction field [116]. The cyclic-stress method, also known as the simplified procedure (e.g., References [37,49,84–86,198]), is usually applied in paleoearthquake studies, although the seismic energy method (e.g., Reference [153]) was also used. Peak ground accelerations used in the analysis are estimated for the scenario earthquakes, employing regionally appropriate ground motion prediction equations (e.g., Reference [199]).

A variation of this approach involves determining combinations of peak ground acceleration and earthquake magnitude required to induce liquefaction for an individual site (Figure 21 [75]). Once the boundary line (i.e., $FS_L = 1$ line) is determined which separates combinations of $a_{max} - M$ that can and cannot induce liquefaction (dashed orange line in Figure 21a), regional attenuation relations are used to define credible $a_{max} - M$ combinations for the site (dotted blue line in Figure 21b). The intersection of the boundary line and the attenuation relation is the lower bound $a_{max} - M$ combination for liquefaction at the site. A regional assessment of $a_{max} - M$ combinations for all liquefaction sites is then performed in order to identify the location of the energy center and to estimate the probable magnitude of the paleoearthquake [75].

The geotechnical approach may suggest more than one source area and a range of magnitudes for a given paleoearthquake. For example, paleoliquefaction features that formed about 6500 years BP in southeastern Missouri and southwestern Illinois, north of the New Madrid seismic zone, in the central US, can be explained by one very large (**M** > 7) regional earthquake or two smaller earthquakes located in close proximity to the clusters of features (Figure 22 [131]). This is largely due to the prevalence of sand dikes and soft-sediment deformation structures in this region whose ages were not well constrained.

There are numerous sources of uncertainty associated with evaluating the locations and magnitudes of paleoearthquakes from geotechnical data including the following: (1) identification of the sediment layer that liquefied during a particular event; (2) measurements of the geotechnical properties of the layer that liquefied; (3) changes in geotechnical properties of the layer due to liquefaction and to post-liquefaction effects related to aging and groundwater conditions; (4) seismic parameters (e.g., amplitude, duration, frequency, and directivity), regional ground motion attenuation, and local site effects; and (5) interpretation of site-specific results in a regional context [2,51,75,116,150]. Improving the understanding of these sources of uncertainty and quantifying their influence on location and magnitude estimates of paleoearthquakes is an area of ongoing research.

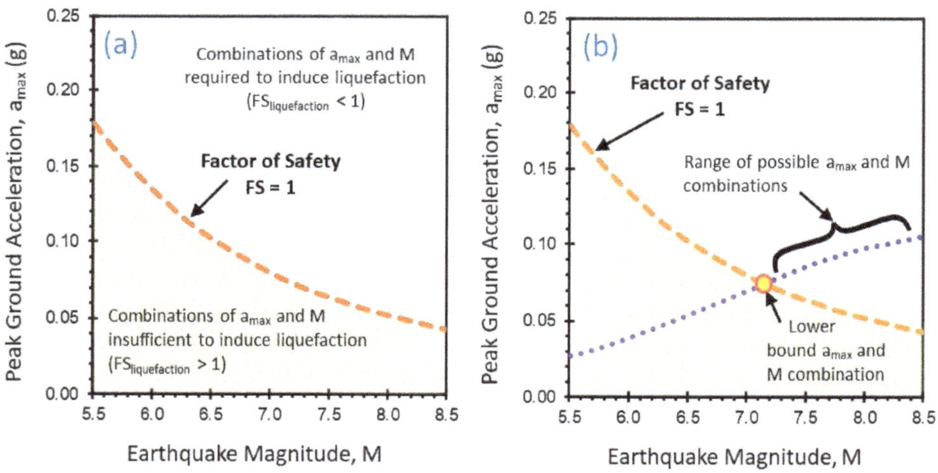

Figure 21. (**a**) a_{max}—**M** combination required to induce liquefaction at a hypothetical site; (**b**) lower bound a_{max}—**M** combination determined for the same hypothetical site (modified from Reference [75]).

Geotechnical studies significantly advanced through the utilization of electronic CPT, compared with more conventional rotary drilling, sampling, and SPT. Whereas the older SPT collects a small disturbed sample and penetration resistance every 1.5 m, the newer CPT collects at least three continuous readings with depth: (a) cone tip resistance; (b) sleeve friction; and (c) porewater pressure. In addition, the inclusion of a biaxial or triaxial geophone sensor allows the direct procurement of downhole shear wave velocity measurements with depth. Readings taken every 1 cm with depth permit much more detailed profiling of soil strata, liquefiable layers, and layer boundaries. Considerable research showed that much more reliable results are obtained with CPT soundings than the traditional SPT when evaluating liquefaction triggering for a specific site [42,190,200]. A new approach to liquefaction triggering that uses yield stress evaluations of soil from CPT and critical state soil mechanics may help better constrict the PGA and **M** of paleoearthquakes when combined with the more traditional approaches that utilize cyclic stress ratio (*CSR*), cyclic resistance ratio (*CRR*), and normalized cone resistance, as discussed by Mayne and Styler [200].

3.4.4. Recurrence of Paleoearthquakes

A primary contribution of paleoliquefaction studies is a longer view of the behavior of seismic source areas and fault zones than is afforded by the historical period. However, the paleoearthquake record may be incomplete for a particular region or for a certain time period for a variety of reasons, including spatial variability in sedimentological and hydrological conditions, changes in liquefaction susceptibility of sediment over time, and availability of exposures or inadequate sampling [51]; thus, these factors must be evaluated cautiously. The degree of completeness of the paleoearthquake record contributes to uncertainty in the locations, magnitudes, and recurrence times of paleoearthquakes. For cases in which the paleoearthquake record may be incomplete, due to a lack of exposure of certain ages of sediment or to periods of low water table, for example, paleoliquefaction features can still be used to determine the minimum number of earthquakes of a particular magnitude range for a given period of time. In this way, a minimum recurrence rate can be estimated. For cases in which the paleoearthquake record appears to be reasonably complete based on the sedimentological and hydrological conditions in a study region, thoroughness and quality of paleoliquefaction studies, and observed versus predicted seismicity rates, recurrence times for liquefaction-inducing earthquakes can be more reliably estimated.

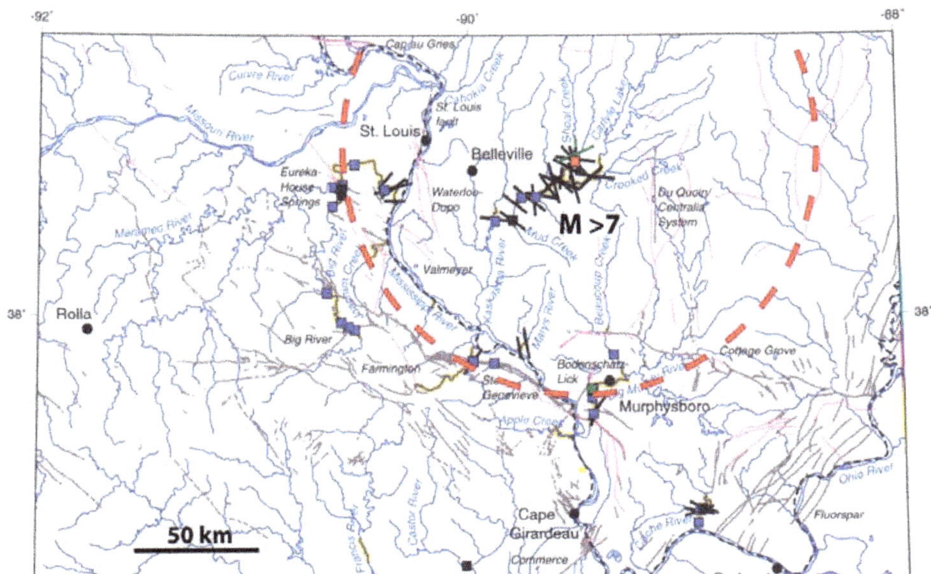

(a)

Figure 22. *Cont.*

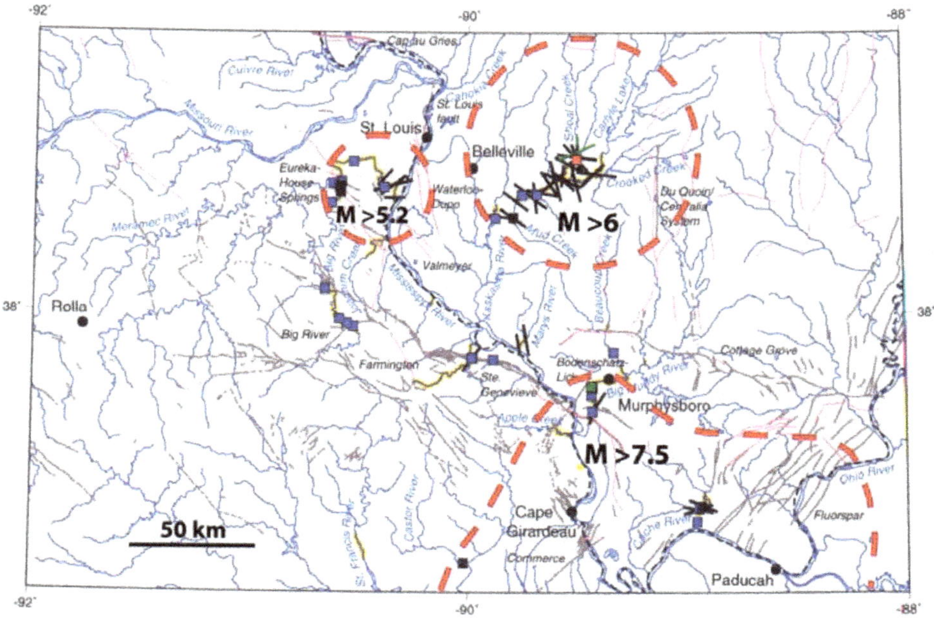

(b)

Figure 22. Alternative earthquake scenarios that can explain the distribution of paleoliquefaction features in the greater St. Louis region (modified from [131]); dashed red lines represent possible liquefaction fields: (**a**) **M** > 7 earthquake centered near Germantown (Illinois) in vicinity of largest liquefaction features on Shoal Creek; (**b**) **M** > 6 earthquake centered near Germantown, **M** > 5.2 near St. Louis, and **M** > 7.5 centered near New Madrid (Missouri). Solid black lines indicate location and strike of dikes, where length of lines represents relative size of dike width. Blue filled squares indicate soft-sediment deformation structures. Red- and green-filled squares represent sand blows and sand sills, respectively.

In paleoliquefaction studies, space–time diagrams and time series of paleoliquefaction data aid in the correlation of similar-age features across a region, reducing the uncertainty associated with the timing of a paleoearthquake, the identification of paleoearthquake source areas, and the calculation of recurrence times for each source (Figures 18 and 23) [1,8,35]. The precision with which recurrence times can be calculated depends on the precision of the estimated timing of the paleoearthquakes. Well-constrained age estimates of liquefaction features and paleoearthquakes contribute to well-constrained estimates of recurrence times (Table 2), provided that multiple earthquake cycles are recognized.

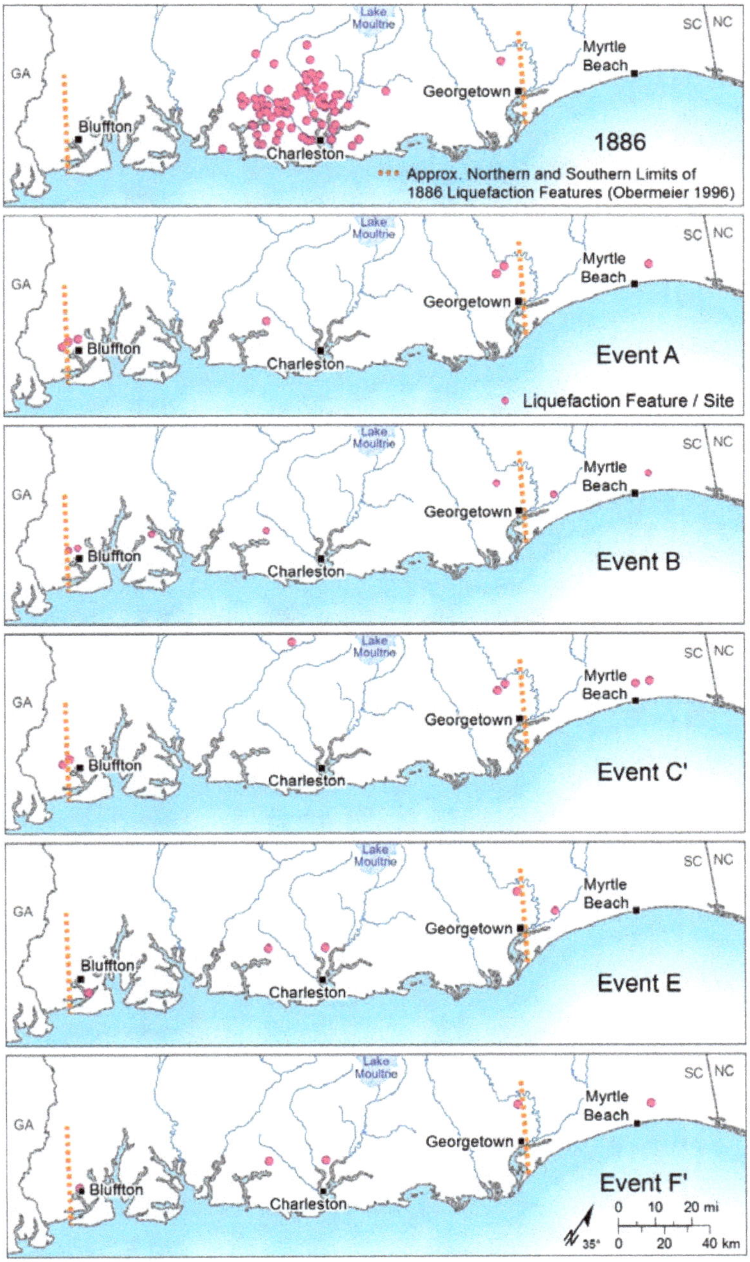

Figure 23. Time series of maps showing the distribution of earthquake-induced liquefaction features in the vicinity of Charleston (South Carolina) and the approximate location of the AD 1886 earthquake. Each panel represents a different earthquake: (top row) AD 1886 and Event A (~546 years BP); (middle row) Events B (~1021 years BP) and C′ (~1683 years BP); (bottom row) Events E (~3548 years BP) and F (~5038 years BP) (modified from References [11,12]). A recurrence time of 500–600 years was estimated from the timing of the past three events (AD 1886 and Events A and B), in which there was more confidence than the earlier three events [12]. Red dashed lines show approximate northern and southern limits of 1886 liquefaction features [100].

4. Earthquake Source Characterization

Information about paleoearthquakes resulting from paleoliquefaction studies is useful in the development of seismic source models, seismic hazard maps, and in site-specific seismic hazard assessments (e.g., References [13,33–35]. More specifically, paleoliquefaction studies provide information about source areas, magnitudes, and recurrence times of paleoearthquakes. In regions with fairly long and complete paleoearthquake records, paleoliquefaction data also may provide insights regarding clustered versus non-clustered behavior of earthquake sources and migration of seismicity within and between fault systems.

4.1. Development of Seismic Source Models

In the US, two recent national-level seismic hazard assessments used paleoliquefaction data in the development of seismic source models. These assessments include the US National Probabilistic Seismic Hazard Maps by the US Geological Survey and the Central and Eastern US Seismic Source Characterization for Nuclear Facilities (CEUS SSC) Project sponsored by the US Nuclear Regulatory Commission, the US Department of Energy, and the Electric Power Research Institute. The US National Probabilistic Seismic Hazard Maps display the ground-motion hazard component of the seismic provisions of national building codes and are used as input to design maps. Paleoliquefaction and other paleoseismic data were used in the regional fault models of the seismic hazard model [33,34]. Paleoliquefaction data figured largely in characterizing seismic source parameters for the Charleston, Charlevoix, New Madrid, and Wabash Valley source models.

The CEUS SSC Project made unprecedented use of paleoliquefaction data in regional seismic source models developed for use in probabilistic hazard analyses for nuclear facilities [35]. Using a master logic-tree framework, the seismic source models provide a hierarchical approach to the identification and characterization of seismic sources. Sources of repeated large-magnitude earthquakes (RLMEs) with moment magnitudes greater than 6.5, in this case, are included as a branch of the master logic tree and help define the maximum magnitude (Mmax) zones for assessing the spatial and temporal characteristics of future earthquakes. Paleoliquefaction data, collected over the past 30 years, were used to evaluate whether the RLME sources were in or out of an earthquake cluster and to define the range and weights of magnitudes and recurrence times of RLMEs.

4.2. Charleston Seismic Zone, Southeastern United States

On 31 August 1886 (local time), the largest historical earthquake ever recorded in the eastern US struck near Charleston (South Carolina) (Figure 24). Magnitude estimates vary, but typically are in the high-6 to mid-7 range, with more recent estimates generally near the lower end of this range (e.g., References [201–205]). This earthquake produced modified Mercalli intensity (MMI) X shaking in the epicentral area near Charleston and was felt as far away as Chicago [202]. Strong ground shaking during the 1886 Charleston earthquake resulted in extensive liquefaction, which was expressed primarily as sand-blow craters at the ground surface [206]. Liquefaction features attributed to the 1886 earthquake extend along coastal South Carolina from Bluffton on the southwest to approximately Georgetown on the northeast (Figure 24), a distance of approximately 200 km. Concern about the possible repeat of an 1886-type earthquake motivated paleoliquefaction studies in the Charleston region.

4.2.1. Summary of Paleoliquefaction Studies

Systematic searches for paleoliquefaction features in the Charleston region began in the early 1980s [132,133]. Since that time, several studies documented liquefaction features from the 1886 earthquake that are preserved in geologic deposits in the South Carolina coastal region (e.g., References [9–12,61,109,134,135]. Additionally, documentation of sand-blow craters and other paleoliquefaction features throughout coastal South Carolina provides evidence for prior strong ground motions during prehistoric large earthquakes (e.g., References [10–12,61,134,136,154].

Talwani and Schaeffer [12] combined data from previous studies with data from their own work to estimate the ages of sand blows in the Charleston region, primarily on the basis of radiocarbon dating, supported by stratigraphic context, soil development, and weathering characteristics. Their database included 121 radiocarbon dates from 54 sand blows at 14 sites. By correlating the ages of sand blows across the region, and comparing the areal extents of paleoliquefaction with the extent of 1886 liquefaction, they interpreted between three and four large-magnitude earthquakes in the past approximately 2000 years, and between five and seven large-magnitude earthquakes in the past approximately 5800 years [12].

Dike widths and other dimensional data for paleoliquefaction features are generally lacking for the Charleston region; thus, magnitude estimates for prehistoric Charleston earthquakes are based on in situ soil properties and the spatial distribution of paleoliquefaction features. Geotechnical estimates for the magnitudes of prehistoric earthquakes in the Charleston seismic zone vary widely from approximately the high-5 to mid-7 range [70,155–157].

Reconnaissance-level searches for paleoliquefaction features did not find paleoliquefaction features much beyond the Charleston region. This suggests a stationary source of repeated, large-magnitude earthquakes located near Charleston.

4.2.2. Use of Paleoliquefaction Data in Seismic Source Model of the Charleston Seismic Zone

Paleoliquefaction studies from the Charleston region were used to help constrain source parameters describing the geometry, magnitude, and recurrence for the Charleston seismic zone. In the 2012 CEUS SSC Project [35], Talwani and Schaeffer's [12] paleoliquefaction compilation was updated with data from other studies (e.g., References [154,158]) (Figure 24). Also, the conventional radiocarbon ages were recalibrated and reported with two-sigma error bands for use in the CEUS SSC Project. These data indicate a total of four liquefaction-inducing, large-magnitude earthquakes in the past approximately 2000 years BP (events 1886, A, B, and C in Figure 25). In the older portion of the record that extends to approximately 5500 years BP, there is evidence for two additional earthquakes (events D and E in Figure 25). Neither the 1886 nor the prehistoric earthquakes in the Charleston area can be definitively attributed to any specific fault or fault zone at the present time. As such, the alternative geometries for the Charleston seismic source in the CEUS SSC Project are influenced by the spatial distribution of observed liquefaction features.

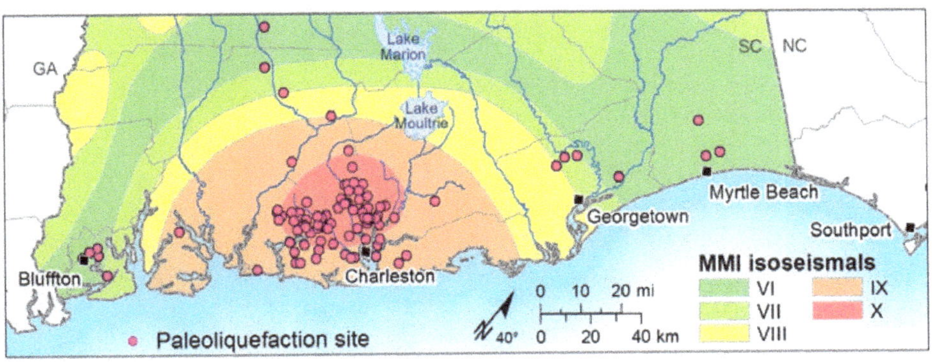

Figure 24. Map of the Charleston seismic zone showing distribution of paleoliquefaction features in the Central and Eastern US Seismic Source Characterization for Nuclear Facilities (CEUS SSC) Project database (from Reference [35]) and modified Mercalli intensity (MMI) isoseismals from the AD 1886 earthquake [207].

In the CEUS SSC Project, the modeled maximum magnitude distribution for future large earthquakes in the Charleston seismic zone is based on estimates of the magnitude 1886 earthquake, geotechnical evaluations of historical and paleoliquefaction features, and interpretations of the extents of paleoliquefaction fields relative to that produced by the 1886 earthquake. Given the large uncertainties in the magnitude estimates for the 1886 and prehistoric earthquakes, the modeled Mmax distribution spans a wide range from **M** 6.7 to **M** 7.5, with a weighted mean of **M** 7.1 (Figure 26).

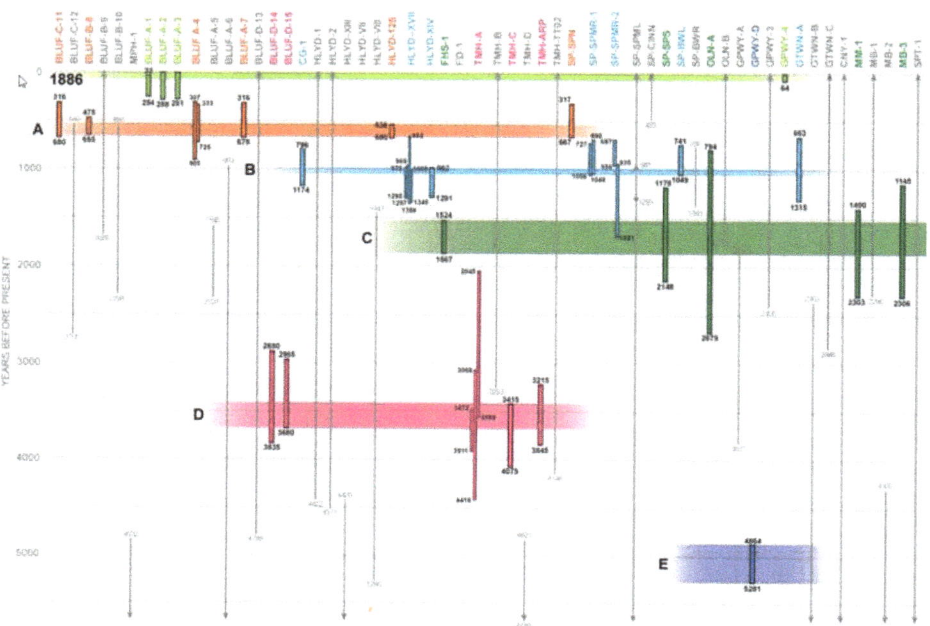

Figure 25. Earthquake chronology of the Charleston seismic zone for the past ~5500 years based on paleoliquefaction studies [35]. Text across top of figure indicates paleoliquefaction site name from Bluffton, South Carolina (BLUF) on the southwest to South Port, North Carolina (SPT) on the northeast (not to scale) [12]. Vertical bars represent radiocarbon age estimates of individual sand blows, and horizontal bars represent inferred event times based on intersection of overlapping age estimates: 64 years BP (AD 1886); 536–655 years BP (Event A); 982–1046 years BP (Event B); 1524–1867 years BP (Event C); 3415–3680 years BP (Event D); and 4864–5281 years BP (Event E).

For estimating the recurrence of large Charleston seismic zone earthquakes in the CEUS SSC Project, greater weight was placed on recurrence estimated from the younger portion of the record because, although it is based on fewer events, it was judged to be more complete. Moreover, there is some uncertainty regarding whether older events D and E were large earthquakes similar in magnitude to the 1886 earthquake, or whether they could have been more moderate-magnitude events [35]. This is reflected in the logic tree (Figure 26), in which greater weight (0.8) is placed on recurrence estimates from the paleoliquefaction record for the past ~2000 years and lesser weight (0.2) is placed on recurrence estimates from the past ~5500 years.

The logic tree from the CEUS SSC Project for the Charleston seismic source also includes a node describing whether the source is "in" or "out" of a cluster of activity (Figure 26). This node addresses the possibility that recurrence behavior of large earthquakes in the Charleston seismic zone may be highly variable through time, such that periods of activity alternate with periods of quiescence. Geomorphic observations suggest that the relatively high rate of recurrence of large earthquakes in the Charleston seismic zone inferred from mid-Holocene to recent liquefaction and from modern seismicity may not be indicative of the longer-term behavior of the zone. Evidence for such clustered behavior is based largely on the observed lack of obvious geomorphic expression of faulting. A protracted period of large earthquakes recurring on the order of hundreds to a few thousand years apart would produce tectonic landforms with clear geomorphic expression, such as those present in regions of the world with comparably high rates of moderate to large earthquakes. Based largely on the lack of such geomorphic expression, the high rate of mid-Holocene to modern seismicity in the Charleston seismic zone is not interpreted to be a long-lived phenomenon. However, because no data definitively demonstrate whether the Charleston seismic zone is currently in or out of a temporal cluster, the Charleston seismic source is modeled as "in" a temporal cluster with a weight of 0.9 and "out" of a temporal cluster with a weight of 0.1 in the 2012 CEUS SSC Project (Figure 26 [35]).

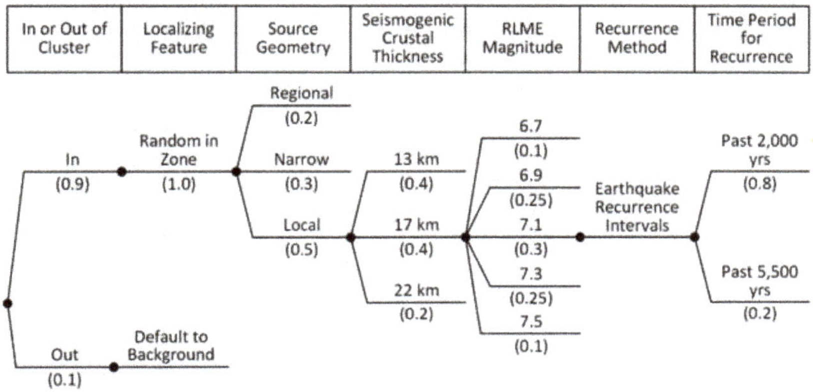

Figure 26. Charleston seismic zone logic tree of seismic source model in CEUS SSC Project (modified from Reference [35]). Paleoliquefaction data are reflected in clustered behavior and in recurrence data used in seismic source model.

The characterization of the Charleston seismic source from the 2012 CEUS SSC Project [35] was adapted with minor changes for use in the 2014 US National Probabilistic Seismic Hazard Maps [34]. One difference is that the CEUS SSC Project allows time-dependent earthquake behavior with a low weight, whereas the National Seismic Hazard Maps use only time-independent behavior for the Charleston seismic zone [34]. Both the 2012 CEUS SSC Project and the 2014 US National Probabilistic Seismic Hazard Maps demonstrate that the Charleston seismic zone contributes significantly to hazard in the southeastern US, although there are several outstanding questions. As such, continued research on Charleston liquefaction is warranted to refine our understanding of the geometry, magnitude, and recurrence for this seismic zone.

4.3. *New Madrid Seismic Zone, Central United States*

In 1811–1812, a major earthquake sequence including three mainshocks with moment magnitudes **M** 7 to 8 were centered in the New Madrid seismic zone (NMSZ) in the central US (Figure 27; e.g., References [202,203,208]). The earthquakes induced severe liquefaction and related ground failures over a ~10,000 km^2 area in the New Madrid region [36,100,209]. They also induced liquefaction at distant locations more than 240 km from their inferred epicenters [210,211], supporting the interpretation that the 1811–1812 mainshocks were very-large-magnitude earthquakes [6,181,211]. Concern about a possible recurrence of an earthquake sequence like the 1811–1812 event motivated paleoliquefaction studies in the NMSZ.

4.3.1. Summary of Paleoliquefaction Studies

Over the past 30 years, various investigators conducted paleoliquefaction studies in the NMSZ and surrounding region (e.g., References [4–8,125,137,144]). Earthquake-induced liquefaction features in Holocene and Pleistocene fluvial deposits of the Mississippi River and its tributaries were identified on floodplains and in cutbanks of rivers and drainage ditches. As of 2019, more than 850 liquefaction features, of which 166 are sand blows, were studied at 218 sites (Figures 28 and 29) [51,117]. At the sites, liquefaction features were documented, described, and measured, and samples, where available, were collected for dating the liquefaction features. Primarily on the basis of radiocarbon dating and supported by archaeological context, stratigraphic context, soil development, and weathering characteristics, the age estimates of sand blows and related sand dikes are well constrained at more than 100 sites (Figures 27 and 28; e.g., References [6–8]).

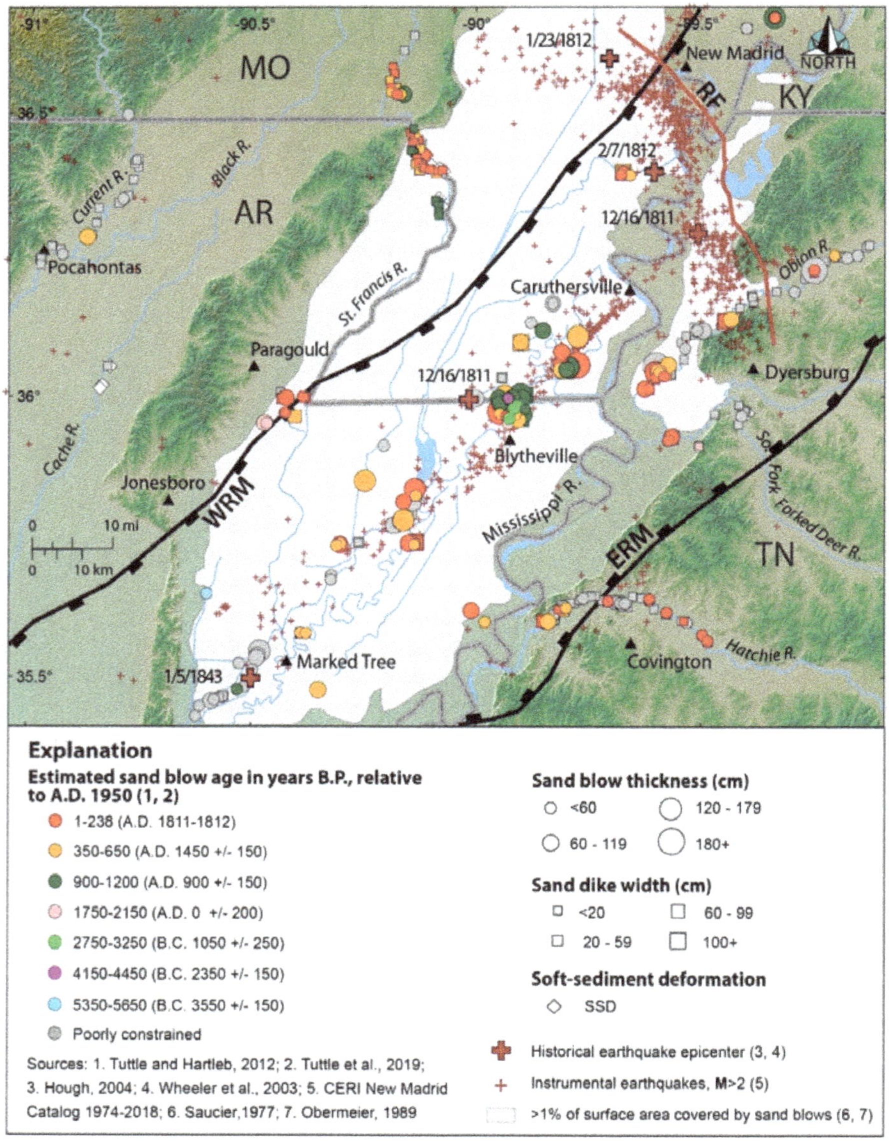

Figure 27. Map of NMSZ (defined by small red crosses), historical earthquakes epicenters [110,212], and surrounding region showing ages and measured sizes of studied liquefaction features (from Reference [8], after Reference [51]), mapped surface area of sand blows [209,213], and geologic structures including Reelfoot fault (RF) and Eastern and Western Reelfoot Rift margins (ERM and WRM).

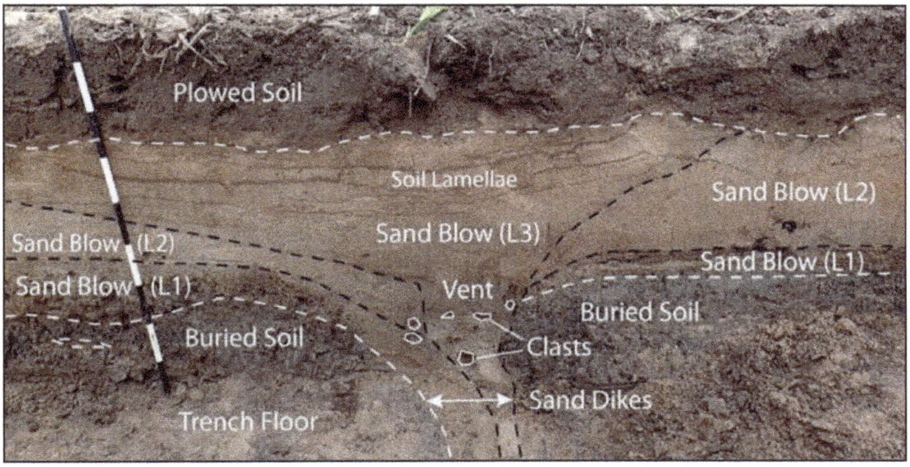

Figure 28. Example of a compound sand blow composed of three depositional units (L1, L2, and L3) and related feeder dikes that formed during three earthquakes in a sequence [8]. Radiocarbon dating of the buried soil immediately below the sand blow provides close maximum calibrated date of before Christ (BC) 1010. Note the soil lamellae in the sand blow which takes hundreds to thousands of years to form. Dashed lines represent contacts. Black and white intervals on meter stick are 10 cm long.

On the basis of the age estimates of the paleoliquefaction features, large paleoearthquakes are estimated to have occurred in the NMSZ in AD 1450 ± 150 years, AD 900 ± 100 years, AD 0 ± 200 years, before Christ (BC) 1050 ± 250 years, and 2350 BC ± 200 years (Figure 29) [6–8]. The AD 1450 and AD 900 events are supported by studies of the Reelfoot fault [149,214], and the BC 1050 and BC 2350 events are corroborated by a channel morphology study of the Mississippi River upstream from the Reelfoot fault [191]. Recently, the AD 0 event was interpreted from liquefaction features at two archeological sites in northeastern Arkansas [8].

The size, spatial distribution, and compound nature of sand blows that formed during the AD 1450 and AD 900 events are strikingly similar to those that formed during the 1811–1812 New Madrid sequence (Figure 27, Figure 28, and Figure 30) [5,6,8]. In addition, there is a close spatial correlation of both historical and prehistoric sand blows with the NMSZ, interpreted as the likely source of the earthquakes responsible for the liquefaction features. The similarity of the historical and prehistoric sand blows also suggests that the AD 1450 and AD 900 events had similar magnitudes to the mainshocks of the 1811–1812 sequence. Using magnitude–distance relations developed from case studies of liquefaction (e.g., References [181,183]), a lower-bound moment magnitude of 7.6 was estimated for the 1811–1812 event given a maximum distance of 240 km for surface manifestation of liquefaction. Lower-bound moment magnitudes of 6.7 and 6.9 were estimated for the AD 1450 and AD 900 events, respectively, based on the distance of observed sand blows. These magnitude estimates represent minimum values since the limits of the liquefaction fields, or the greatest epicentral distances of liquefaction, are yet to be defined for these events. Several studies used the geotechnical approach to estimate the magnitude of New Madrid events with similar results. Using electronic cone penetration soundings collected at liquefaction sites in the NMSZ, these studies estimated moment magnitudes for the 11 December 1811 earthquake that ranged from **M** 7.4 to **M** 8.0 [215–217]. In a more recent study that used geotechnical data collected near distal sites of liquefaction, a combination of scenario earthquakes with locations of the 16 December 1811, 23 January 1812, and 23 January 1812 mainshocks and with moment magnitudes of **M** 7.6, **M** 7.5, and **M** 7.8, respectively, best predicted the observed liquefaction [117]. Although their associated uncertainties are large (0.25–0.5), these magnitudes for the three 1811–1812 mainshocks agree with estimates based on modeling of felt reports [203].

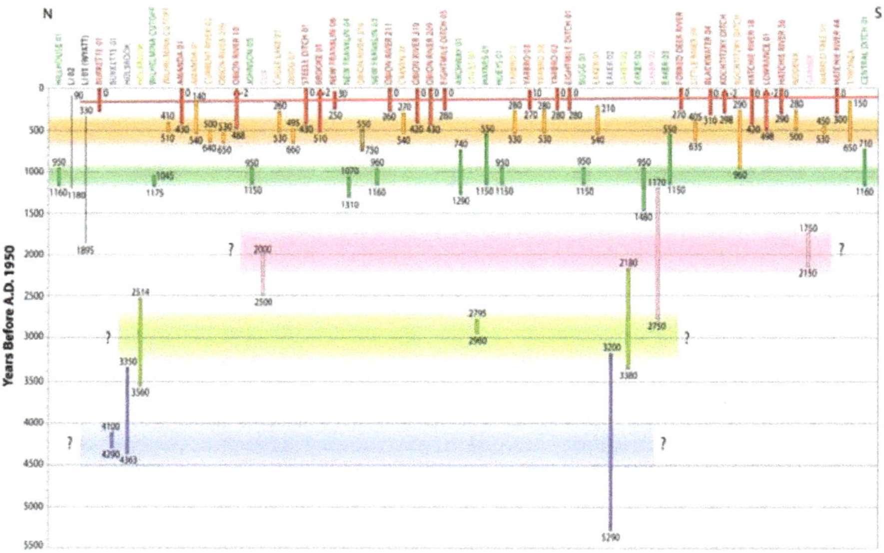

Figure 29. Diagram illustrating earthquake chronology of New Madrid seismic zone for past 5500 years based on dating and correlation of liquefaction features at sites (listed at top) across region from northeast to southwest. Vertical bars represent age estimates of individual sand blows, and horizontal bars represent event times of AD 1811–1812 (138 years BP); AD 1450 ± 150 years (500 years BP); AD 900 ± 100 years (1050 years BP); AD 0 ± 200 years (1950 years BP), BC 1050 ± 250 years (3000 years BP), and BC 2350 ± 200 years (4300 years BP) (modified from References [8,51]).

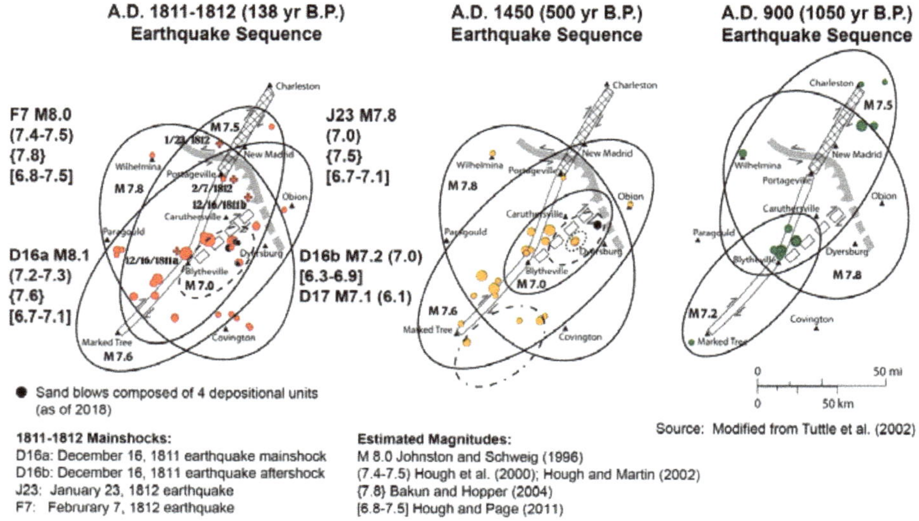

Figure 30. Liquefaction fields for the past three New Madrid earthquake sequences as interpreted from spatial distribution and stratigraphy of sand blows (modified from Reference [6]). The colored circles represent sand blows that formed during the AD 1811–1812 (red), AD 1450 (orange), and AD 900 (green) sequences. Black circles indicate sand blows composed of four depositional units. Sand blows are shown relative to the preferred fault rupture scenario for the AD 1811–1812 earthquake sequence (from Reference [211]). Earthquake magnitudes shown within 1811–1812 ellipses are from Reference [203]. Other magnitude estimates are from References [208,218,219].

To date, the paleoliquefaction findings suggest that the NMSZ generated earthquakes sequences, including **M** 7 to 8 earthquakes, every ~500 years on average during the past 1200 years and every ~1100 years on average during the previous 3300 years (Figure 31 [6,8]). This apparent variability in recurrence times may be due to temporal clustering of seismicity with active and inactive periods [191] or to an incomplete paleoearthquake record prior to AD 900 [8].

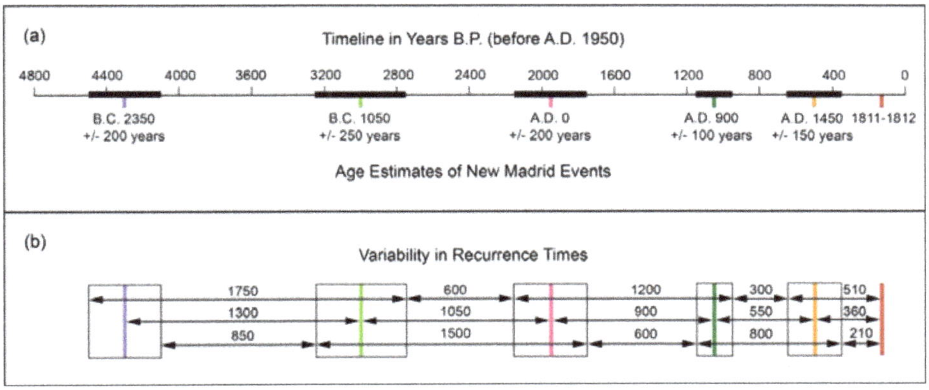

Figure 31. Estimated ages of New Madrid paleoearthquakes and resulting recurrence times [8] (modified from Reference [6]): (**a**) timeline showing timing of New Madrid events during past 4500 years; (**b**) uncertainties in timing of New Madrid events lead to variability in estimates of recurrence times. Average recurrence time for two most recent earthquake cycles is ~500 years and ~1100 years for three earlier earthquake cycles.

4.3.2. Use of Paleoliquefaction Data in Seismic Source Model of the NMSZ

Paleoliquefaction studies made great strides in deciphering the prehistorical earthquake record of the NMSZ and improving understanding of the hazard it poses. As of 2012, it was clear from paleoliquefaction studies that the NMSZ (1) is a source of RLMEs; (2) produced three earthquake sequences during the past 1200 years and a fourth earthquake sequence about 3100 years earlier; and (3) is likely to be within an active period, or cluster, with events occurring on average every 500 years. There was and still is uncertainty in magnitude estimates of the New Madrid paleoearthquakes, as well as the 1811–1812 mainshocks; however, most magnitude estimates fall within the 6.9–7.9 range.

Results of the New Madrid paleoliquefaction studies, including insights on timing, source area, temporal clustering, and recurrence times of large-magnitude earthquakes, were incorporated into seismic source models developed for seismic hazard assessments. In the logic tree of the 2012 CEUS SSC project, a heavier weight (0.9) is placed on clustered earthquakes than on non-clustered earthquakes (0.1), and the dates of the last three New Madrid events (1811–1812, AD 1450 and AD 900) are included in the recurrence data (Figure 32 [35]). In the logic tree of the 2014 US National Probabilistic Seismic Hazard Maps, weightings on earthquake sequences and on a recurrence time of 500 years are increased over the values used for the 2008 maps (Figure 33 [34]). In addition, the 2014 logic tree includes the 2012 CEUS SSC project fault model as an alternative branch.

The NMSZ is recognized as having the highest hazard in the CEUS, but large uncertainties in the seismic hazard model spurred considerable debate and controversy. Sensitivity analysis of sources associated with the NMSZ suggests that the earthquake rate has the greatest effect on hazard estimates [35]. Therefore, additional paleoliquefaction information regarding timing and recurrence of New Madrid paleoearthquakes is needed to further reduce uncertainties in the hazard model.

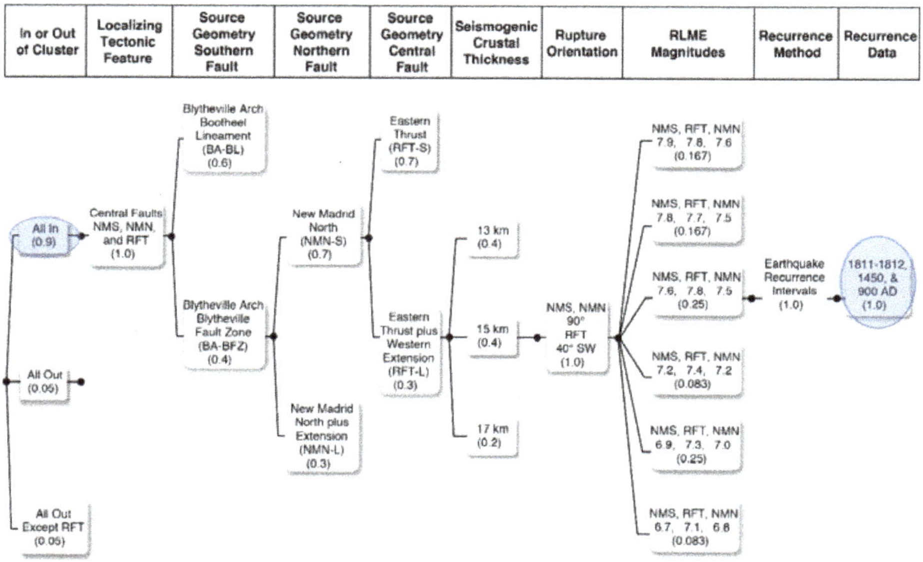

Figure 32. NMSZ logic tree of seismic source model in CEUS SSC Project (modified from Reference [35]). Paleoliquefaction data are reflected in weighting of clustered behavior and in recurrence data used in seismic source model.

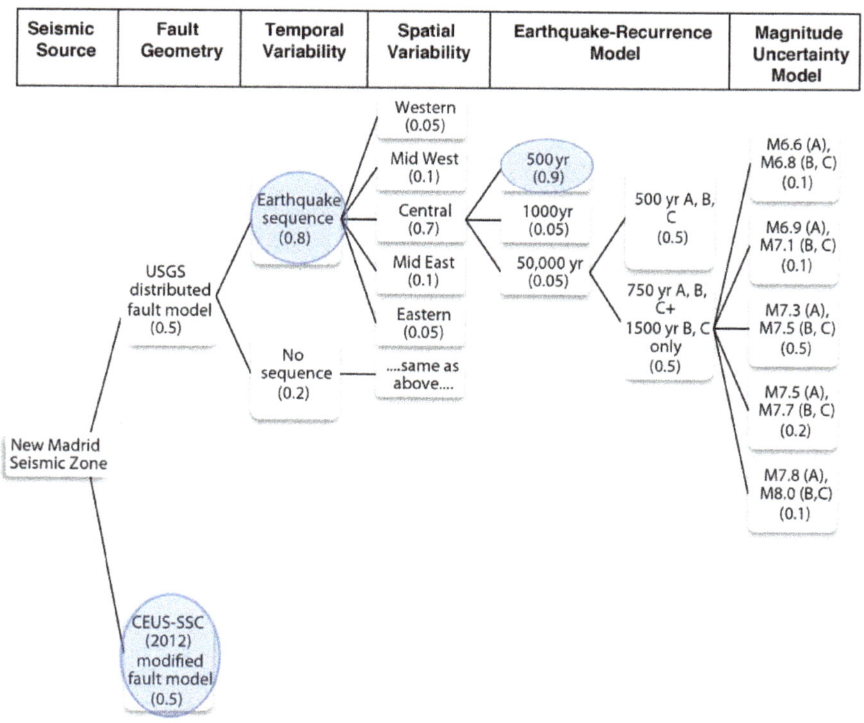

Figure 33. Logic tree of NMSZ seismic source model in US National Probabilistic Seismic Hazard Maps (modified from Reference [34]). Paleoliquefaction data are reflected in increased weighting of earthquake sequence and 500-year recurrence time compared with the earlier model.

5. Conclusions and Recommendations for Future Research

Paleoliquefaction studies, a relatively new and interdisciplinary field of inquiry, provides a unique perspective on past earthquakes that advances knowledge of the earthquake potential of both intraplate and interplate regions. Case studies of modern and historical earthquakes that induced liquefaction inform our understanding of this phenomenon and contribute to the use of liquefaction features in paleoseismology. In paleoliquefaction studies, earthquake-induced liquefaction features, especially sand blows and sand dikes, that result from strong ground shaking are interpreted in terms of the timing, location, magnitude, and recurrence of large earthquakes. Results from these studies supplement information about historical and instrumental seismicity and extend the earthquake record back in time. This valuable insight about past earthquakes is being used in the development of seismic source models for earthquake hazard assessments and is reducing uncertainties regarding clustered seismicity and recurrence intervals. In addition, paleoliquefaction has the potential to address questions regarding the long-term behavior of fault zones.

Despite the important contributions that paleoliquefaction studies are making to earthquake hazard assessments, there remain large uncertainties associated with the derived seismic geoparameters, and there are many regions around the world where paleoliquefaction studies have yet to be conducted. With additional research on earthquake-induced liquefaction, it may be possible to reduce uncertainties associated with estimates of earthquake hazard and to advance the usefulness and application of this field of study. Below are recommendations for future research.

Case Studies: As was done for the 2010–2011 Canterbury earthquake sequence in New Zealand and the 2012 Emilia earthquake sequence in Italy, documentation of liquefaction induced by modern earthquakes, with well-constrained locations, magnitudes, and other earthquake parameters (e.g., source mechanism, directivity, acceleration, frequency content, and intensity), is recommended and would help to advance paleoliquefaction studies in several ways. Detailed information about the sedimentary characteristics of sand blows, dikes, and sills, as well as the size and spatial distributions of liquefaction features, would be useful for direct comparisons between paleo and modern liquefaction features for interpretations of paleoearthquakes. The information would further enhance empirical relations of earthquake and liquefaction parameters, especially for great (**M** > 8) earthquakes for which there are fewer data. Also, ESI could be applied to a paleoliquefaction dataset to define the intensity field and epicentral intensity of a paleoearthquake and the results compared with those of other approaches for estimating the locations and sizes of past events. It is strongly encouraged that liquefaction information gathered during cases studies be made available through online catalogs and databases for research and seismic hazard assessment.

To gain a better understanding of both the processes of liquefaction and the effects on the source layers that liquefied, instrumentation of liquefaction-prone sites in seismically active regions is encouraged, as is pre- and post-event geological sampling and measurement of geotechnical properties. This information may help determine whether strata-bound soft-sediment deformation structures can provide reliable magnitude estimates and help reduce uncertainties related to back-calculating magnitudes of paleoearthquakes using post-event geotechnical measurements. Also, it would be beneficial to better understand differences between characteristics, such as frequency content and attenuation of ground motion, of both intraplate and interplate earthquakes as they relate to liquefaction, so that results of field experiments could be applied to both settings.

Mapping Using Remotely Sensed Data: Modern technology provides new data and tools that make it possible to identify and map geologic and geomorphic features over large areas. Satellite imagery, LiDAR data, unmanned aerial vehicles or drones, image-processing software, and geographical information systems provide the opportunity to combine datasets within a geographical framework and classify and map sand blows over areas large and small. It behooves us to develop methodologies that utilize these data and tools for mapping earthquake-induced liquefaction features and related ground failures, especially given the increasing human disturbance of the landscape and its impact on the geologic record of past earthquakes.

Dating Techniques: Dating paleoliquefaction features is a critical aspect of any paleoliquefaction study. If the ages of paleoliquefaction features are not well constrained, it is difficult to estimate the timing of paleoearthquakes, to correlate similar-age features across a region, and to interpret the locations and magnitudes of paleoearthquakes with confidence. Therefore, research to improve dating of paleoliquefaction features is a high priority. Radiocarbon and OSL are commonly used dating techniques in these studies. Because they are often collected stratigraphically above and below sand blows, samples for radiocarbon and OSL dating provide minimum and maximum constraining dates for liquefaction features and, thus, for the earthquakes that caused them. Although the individual dates may have precisions of ±20–80 years, the age estimates of liquefaction features based on the combination of the minimum and maximum constraining dates will have uncertainties of about 100 years in the best of circumstances. Dating techniques and sampling strategies that provide more precise and accurate results would help improve the usefulness of liquefaction features in earthquake source characterization.

Geophysical Techniques: Geophysical surveys proved a useful tool for locating sand dikes, mapping sand blows, and selecting locations for paleoseismic trenches. However, distinguishing between earthquake-induced liquefaction features and non-seismic sedimentary features and deposits in remotely sensed data can be difficult, particularly where sand blows are buried and have no surface expression. A better understanding of the morphological and spatial differences between features and sedimentary deposits could increase the usefulness of geophysical methods as a reconnaissance tool and as a means for tracing features beyond the reach of trench excavations. Although geophysical methods yield results sensitive to the physical properties of sediment, better knowledge of empirical relations between these properties and the geophysical observations would increase confidence in data interpretations. This information could be used to understand better the soil and sediment characteristics that contributed to the occurrence of liquefaction in a particular environment. Specifically, the influence of stratigraphic and sedimentological properties in the development of porewater pressure at liquefaction sites could lead to a better understanding of the process of liquefaction and the conditions that contribute to liquefaction susceptibility. Further work on calibrating geophysical surveys with sedimentological and geotechnical analyses in a variety of environments is needed to extend the utility of site-specific geotechnical measurements and improve the ability to relate laboratory studies of liquefaction to actual field studies.

Geotechnical Techniques: Additional research is needed to further evaluate and reduce the sources of uncertainties associated with the geotechnical approach for estimating the magnitudes of paleoearthquakes. The development and application of new techniques, devices, and procedures for in situ testing of soil properties in paleoliquefaction studies could help reduce some of these uncertainties. The traditional sampling interval of soil borings may be too gross (~1.5 m) to accurately reflect the properties of the layers that liquefied and adjacent layers that govern the buildup of porewater pressure. The conventional seismic cone test and seismic dilatometer test improve significantly on this as penetrometer data are collected on 1- to 2-cm intervals and dilatometer data are procured every 20 cm. Downhole shear wave velocity data by traditional approaches are collected at 1-m intervals. Newer techniques using either a seismic piezocone or seismic dilatometer allow for more detailed profiling of shear-wave velocities and small-strain stiffness at 0.2-m intervals that may help reduce uncertainties. The continuous-interval seismic piezocone test is now available and permits profiling at 0.1-m intervals, although the data are noisier and less robust. Also, it may be possible to use a mini-cone to construct higher-resolution stratigraphic profiles of soil layering, including lenses. This could facilitate and complement trench investigations and permit rapid characterization of sand blows at many sites across a region. Development of this technique, which will require field testing, has the potential to deliver a more accurate and detailed characterization of sand blows that have limited layer thickness, or that are composed of multiple depositional units related to recurrent liquefaction during an earthquake sequence. In addition, the technique may make it possible to identify sand blows buried below the reach of trenches.

Paleoearthquake Chronologies: In regions where paleoearthquake records exist but may not be fully developed, paleoseismic studies could be designed that would improve the completeness and extend the length of the paleoearthquake chronologies in order to improve recurrence estimates of large earthquakes and understanding of earthquake sources. It is also recommended that paleoliquefaction studies be conducted in regions of historically low seismicity that share geologic and tectonic characteristics with known seismogenic zones to better understand the earthquake potential of those regions and to test the hypothesis that inherited geologic structures (e.g., faults that were active during the Mesozoic) control seismicity.

Author Contributions: M.P.T., R.H., and L.W. have conducted research in paleoseismology, and P.W.M. has studied earthquake-induced liquefaction, each for more than 20 years. M.P.T. focuses primarily on paleoliquefaction studies and R.H. on fault studies. L.W. developed geophysical techniques for imaging earthquake-induced liquefaction features. Mayne developed geotechnical techniques for in situ testing of soils, primarily concerning the cone penetration test. During the Central and Eastern US Seismic Source Characterization Project, M.P.T. and R.H. developed guidance for the project to use paleoliquefaction data in the development of seismic source models. Later, M.P.T., L.W., and P.W.M. further developed guidance for conducting paleoliquefaction investigations for the US Nuclear Regulatory Commission. Those two efforts led to the conceptualization of this paper. All four co-authors participated in writing the paper and descriptions of various methodologies. M.P.T. took the lead summarizing paleoliquefaction studies in the New Madrid seismic zone, and R.H. did the same for the Charleston seismic zone. L.W. wrote the section on geophysical techniques and P.W.M. contributed primarily to the sections on earthquake-induced liquefaction and the use of geotechnical data to estimate the locations and magnitudes of paleoearthquakes. All four co-authors edited the manuscript and R.H. compiled the reference list.

Funding: Although no particular grant or contract funded the writing of this review paper, we credit the Electric Power Research Institute and the US Nuclear Regulatory Commission for past support through agreement EP-P38320/C17094 and contract NRC-HQ-11-C-04-0041, respectively, which contributed to the development of this paper. The CESU SSC Project [27] described in this review paper was sponsored by the Electric Power Research Institute (EPRI), US Department of Energy (US DOE) under Award Number DE-FG07-08ID14908, and the US Nuclear Regulatory Commission (US NRC) under Award Number NCR-04-09-144. The views and conclusions presented in this paper are those of the authors and should not be interpreted as necessarily representing the official policies, either expressed or implied, of the US Government.

Acknowledgments: We thank Mark Quigley, Sabina Porfido, and two anonymous reviewers for their constructive comments and suggestions that helped improve this manuscript.

Conflicts of Interest: The authors declare no conflicts of interest.

References

1. McCalpin, J.P.; Nelson, A.R. Introduction to paleoseismology. In *Paleoseismology*, 2nd ed.; McCalpin, J., Ed.; Academic Press: Burlington, MA, USA, 2009; pp. 1–27.
2. Tuttle, M.P. The use of liquefaction features in paleoseismology: Lessons learned in the New Madrid seismic zone, central United States. *J. Seismol.* **2001**, *5*, 361–380. [CrossRef]
3. Tuttle, M.P.; Prentice, C.S.; Dyer-Williams, K.; Pena, L.R.; Burr, G. Late Holocene liquefaction features in the Dominican Republic: A powerful tool for earthquake hazard assessment in the northeastern Caribbean. *Bull. Seismol. Soc. Am.* **2003**, *93*, 27–46. [CrossRef]
4. Saucier, R.T. Geoarchaeological evidence of strong prehistoric earthquakes in the New Madrid (Missouri) seismic zone. *Geology* **1991**, *19*, 296–298. [CrossRef]
5. Tuttle, M.P. Late Holocene Earthquakes and Their Implications for Earthquake Potential of the New Madrid Seismic Zone, Central United States. Ph.D. Thesis, University of Maryland, College Park, MD, USA, 1999.
6. Tuttle, M.P.; Schweig, E.S.; Sims, J.D.; Lafferty, R.H.; Wolf, L.W.; Haynes, M.L. The earthquake potential of the New Madrid seismic zone. *Bull. Seismol. Soc. Am.* **2002**, *92*, 2080–2089. [CrossRef]
7. Tuttle, M.P.; Schweig, E., III; Campbell, J.; Thomas, P.M.; Sims, J.D.; Lafferty, R.H., III. Evidence for New Madrid earthquakes in AD 300 and 2350 BC. *Seismol. Res. Lett.* **2005**, *76*, 489–501. [CrossRef]
8. Tuttle, M.P.; Wolf, L.W.; Starr, M.E.; Villamor, P.; Mayne, P.W.; Lafferty, R.H., III; Morrow, J.E.; Scott, R.J., Jr.; Forman, S.L.; Hess, K.; et al. Evidence for large New Madrid earthquakes about AD 0 and BC 1050, Central United States. *Seismol. Res. Lett.* **2019**, *90*, 1393–1406. [CrossRef]
9. Talwani, P.; Cox, J. Paleoseismic evidence for recurrence of earthquakes near Charleston, South Carolina. *Science* **1985**, *228*, 379–381. [CrossRef]

10. Obermeier, S.F.; Weems, R.E.; Jacobson, R.B.; Gohn, G.S. Liquefaction evidence for repeated Holocene earthquakes in the coastal region of South Carolina. *Ann. NY Acad. Sci.* **1989**, *558*, 183–195. [CrossRef]
11. Amick, D.; Maurath, G.; Gelinas, R. Characteristics of seismically induced liquefaction sites and features located in the vicinity of the 1886 Charleston, South Carolina earthquake. *Seismol. Res. Lett.* **1990**, *61*, 117–130. [CrossRef]
12. Talwani, P.; Schaeffer, W.T. Recurrence rates of large earthquakes in the South Carolina Coastal Plain based on paleoliquefaction data. *J. Geophys. Res.* **2001**, *106*, 6621–6642. [CrossRef]
13. Tuttle, M.P.; Atkinson, G.M. Localization of large earthquakes in the Charlevoix seismic zone, Quebec, Canada, during the past 10,000 years. *Seismol. Res. Lett.* **2010**, *81*, 140–147. [CrossRef]
14. Munson, P.J.; Obermeier, S.F.; Munson, C.A.; Hajic, E.R. Liquefaction evidence for Holocene and latest Pleistocene seismicity in the southern halves of Indiana and Illinois: A preliminary overview. *Seismol. Res. Lett.* **1997**, *68*, 521–536. [CrossRef]
15. Obermeier, S.F. Liquefaction evidence for strong earthquakes of Holocene and Latest Pleistocene ages in the States of Indiana and Illinois, USA. *Eng. Geol.* **1998**, *50*, 227–254. [CrossRef]
16. Hatcher, R.D., Jr.; Vaughn, J.D.; Obermeier, S.F. Large earthquake paleoseismology in the eastern Tennessee seismic zone–results of an 18-month pilot study. *Geol. Soc. Am. Spec. Pap.* **2012**, *491*, 111–142. [CrossRef]
17. Tuttle, M.; Dyer-Williams, K.; Schweig, E.S.; Prentice, C.; Moya, J.; Tucker, K. Liquefaction induced by historic and prehistoric earthquakes in western Puerto Rico. In *Active Tectonics and Seismic Hazards of Puerto Rico, the Virgin Islands, and Offshore Areas*; Mann, P., Ed.; Geological Society of America: Boulder, CO, USA, 2005; pp. 263–276.
18. Obermeier, S.F.; Dickenson, S.E. Liquefaction evidence for the strength of ground motions resulting from late Holocene Cascadia subduction earthquakes, with emphasis on the event of 1700 A.D. *Bull. Seismol. Soc. Am.* **2000**, *90*, 876–896. [CrossRef]
19. Bourgeois, J.; Johnson, S.Y. Geologic evidence of earthquakes at the Snohomish delta, Washington, in the past 1200 yr. *Geol. Soc. Am. Bull.* **2001**, *113*, 482–494. [CrossRef]
20. Martin, M.E.; Bourgeois, J. Vented sediments and tsunami deposits in the Puget Lowland, Washington–differentiating sedimentary processes. *Sedimentology* **2012**, *59*, 419–444. [CrossRef]
21. Uner, S. Seismogenic structures in Quaternary lacustrine deposits of Lake Van (eastern Turkey). *Geologos* **2014**, *20*, 79–87. [CrossRef]
22. Vitale, S.; Isaia, R.; Ciarcia, S.; Di Giuseppe, M.G.; Iannuzzi, E.; Prinzi, E.P.; Tramparulo, F.D.A.; Troiano, A. Seismically induced soft-sediment deformation phenomena during the volcano-tectonic activity of Campi Flegrei caldera (southern Italy) in the last 15 kyr. *Tectonics* **2019**, *38*. [CrossRef]
23. Moretti, M.; Sabato, L. Recognition of trigger mechanisms for soft-sediment deformation in the Pleistocene lacustrine deposits of the Sant'Arcangelo Basin (Southern Italy): Seismic shock vs. overloading. *Sediment. Geol.* **2007**, *196*, 31–45. [CrossRef]
24. Rodríguez-Pascua, M.A.; Garduño-Monroy, V.H.; Israde-Alcantara, I.; Perez-Lopez, R. Estimation of the paleoepicentral area from the spatial gradient of deformation in lacustrine seismites (Tierras Blancas Basin, Mexico). *Quat. Int.* **2010**, *219*, 66–78. [CrossRef]
25. Moretti, M.; Ronchi, A. Liquefaction features interpreted as seismites in the Pleistocene fluvio-lacustrine deposits of the Neuquén Basin (Northern Patagonia). *Sediment. Geol.* **2011**, *235*, 200–209. [CrossRef]
26. Quigley, M.C.; Bastin, S.; Bradley, B.A. Recurrent liquefaction in Christchurch, New Zealand, during the Canterbury earthquake sequence. *Geology* **2013**, *41*, 419–422. [CrossRef]
27. Villamor, P.; Giona-Bucci, M.; Almond, P.; Tuttle, M.; Langridge, R.; Clark, K.; Ries, W.; Vandergoes, M.; Barker, P.; Martin, F.; et al. *Exploring Methods to Assess for Paleoliquefaction in the Canterbury Area*; GNS Science Consultancy Report 2014/183; University of Canterbury: Christchurch, New Zealand, 2014.
28. Bastin, S.H.; Quigley, M.C.; Bassett, K. Paleoliquefaction in Christchurch, New Zealand. *Geol. Soc. Am. Bull.* **2015**, *127*, 1348–1365. [CrossRef]
29. Villamor, P.; Almond, P.; Tuttle, M.; Giona Bucci, M.; Langridge, R.; Clark, K.; Ries, W.; Bastin, S.; Eger, A.; Vandergoes, M.C.; et al. Liquefaction features produced by the 2010–2011 Canterbury earthquake sequence in southwest Christchurch, New Zealand, and preliminary assessment of paleoliquefaction features. *Bull. Seismol. Soc. Am.* **2016**, *106*, 1747–1771. [CrossRef]

30. Tuttle, M.P.; Villamor, P.; Almond, P.; Bastin, S.; Giona Bucci, M.; Langridge, R.; Clark, K.; Hardwick, C. Liquefaction induced during the 2010–2011 Canterbury, New Zealand, earthquake sequence and lessons learned for the study of paleoliquefaction features. *Seismol. Res. Lett.* **2017**, *88*, 1403–1414. [CrossRef]
31. Giona Bucci, M.; Villamor, P.; Almond, P.; Tuttle, M.; Stringer, M.; Smith, C.; Ries, W.; Watson, M. Associations between sediment architecture and liquefaction susceptibility in fluvial settings: The 2010–2011 Canterbury earthquake sequence, New Zealand. *Eng. Geol.* **2018**, *237*, 181–197. [CrossRef]
32. Giona Bucci, M.; Almond, P.; Villamor, P.; Tuttle, M.; Stringer, M.; Smith, C.; Ries, W.; Bourgeois, J.; Loame, R.; Howarth, J.; et al. Controls on patterns of liquefaction in a coastal dune environment. *Sediment. Geol.* **2018**, *377*, 17–33. [CrossRef]
33. Petersen, M.D.; Frankel, A.D.; Harmsen, S.C.; Mueller, C.S.; Haller, K.M.; Wheeler, R.L.; Wesson, R.L.; Zeng, Y.; Boyd, O.S.; Perkins, D.M.; et al. *Documentation for the 2008 Update of the United States National Seismic Hazard Maps*; NO. 2008–1128; US Geological Survey: Reston, VA, USA, 2008.
34. Petersen, M.D.; Moschetti, M.P.; Powers, P.M.; Mueller, C.S.; Haller, K.M.; Frankel, A.D.; Zeng, Y.; Rezaeian, S.; Harmsen, S.C.; Boyd, O.S.; et al. *Documentation for the 2008 Update of the United States National Seismic Hazard Maps*; US Geological Survey Open-File Report: Reston, VA, USA, 2014.
35. Electric Power Research Institute (EPRI); US Department of Energy (DOE); US Nuclear Regulatory Commission (NRC). *Central and Eastern United States Seismic Source Characterization for Nuclear Facilities*; NUREG-2115; EPRI: Palo Alto, CA, USA, 2012.
36. Fuller, M.L. *The New Madrid Earthquake*; US Geological Survey, US Government Printing Office: Washington, DC, USA, 1912.
37. Seed, H.B.; Idriss, I.M. *Ground Motions and Soil Liquefaction during Earthquakes*; Earthquake Engineering Research Institute: Berkeley, CA, USA, 1982.
38. Chang, S.E. Disasters and transport systems: Loss, recovery and competition at the Port of Kobe after the 1995 earthquake. *J. Trans. Geogr.* **2000**, *8*, 53–65. [CrossRef]
39. Parker, M.; Steenkamp, D. The economic impact of the Canterbury earthquakes. *Reserve Bank N. Z. Bull.* **2012**, *75*, 13–25.
40. National Academies of Sciences; Engineering, and Medicine. *State of the Art and Practice in the Assessment of Earthquake-Induced Soil Liquefaction and its Consequences*; The National Academies Press: Washington, DC, USA, 2016.
41. Robertson, P.K.; Wride, C.E. Evaluating cyclic liquefaction potential using the cone penetration test. *Can. Geotech. J.* **1998**, *35*, 442–459. [CrossRef]
42. Idriss, I.M.; Boulanger, R.W. *Soil Liquefaction During Earthquakes*; Earthquake Engineering Research Institute: Oakland, CA, USA, 2008.
43. Allen, J.R.L. *Sedimentary Structures: Their Character and Physical Basis, Vol. II*; Elsevier: New York, NY, USA, 1982.
44. Owen, G. Deformation processes in unconsolidated sands. In *Deformation of Sediments and Sedimentary Rocks*; Jones, M.E., Preston, R.M.F., Eds.; Geological Society of London Special Publication: London, UK, 1987; Volume 29, pp. 11–24.
45. Kramer, S.L.; Hartvigsen, A.J.; Sideras, S.S.; Özener, P.T. Site response modeling in liquefiable soil deposits. In Proceedings of the 4th IASPEI/IAEE International Symposium on Effects of Surface Geology on Seismic Motion, Santa Barbara, CA, USA, 23–26 August 2011.
46. Seed, H.B.; Idriss, I.M. Evaluation of liquefaction potential of sand deposits based on observations of performance in previous earthquakes. In Proceedings of the Session on In-Situ Testing to Evaluate Liquefaction Susceptibility, ASCE National Convention, St. Louis, MO, USA, 26–30 October 1981.
47. Ishihara, K. Stability of natural soils during earthquakes. In Proceedings of the 11th International Conference on Soil Mechanics and Foundation Engineering, San Francisco, CA, USA, 12–16 August 1985.
48. Castro, G. On the behavior of soils during earthquakes-liquefaction. In *Soil Dynamics and Liquefaction*; Cakmak, A.S., Ed.; Elsevier: Amsterdam, The Netherlands, 1987.
49. Youd, T.L.; Idriss, I.M.; Andrus, R.D.; Arango, I.; Castro, G.; Christian, J.T.; Dobry, R.; Finn, W.D.L.; Harder, L.F., Jr.; Hynes, M.E.; et al. Liquefaction resistance of soils: Summary report from the 1996 NCEER and 1998 NCEER/NSF workshops on evaluation of liquefaction resistance of soils. *J. Geotech. Geoenviron. Eng.* **2001**, *127*, 817–833. [CrossRef]

50. Jefferies, M.; Been, K. *Soil Liquefaction: A Critical State Approach*, 2nd ed.; CRC Press: Boca Raton, FL, USA, 2015.
51. Tuttle, M.P.; Hartleb, R.D. *Appendix E: Central and Eastern US Paleoliquefaction Database, Uncertainties Associated with Paleoliquefaction Data, and Guidance for Seismic Source Characterization*; Technical Report; Electric Power Research Institute: Palo Alto, CA, USA, 2012.
52. Tuttle, M.; Law, K.T.; Seeber, L.; Jacob, K. Liquefaction and ground failure in Ferland, Quebec, triggered by the 1988 Saguenay earthquake. *Can. Geotech. J.* **1990**, *27*, 580–589. [CrossRef]
53. Sims, J.D. Earthquake-induced structures in sediments of Van Norman Lake, San Fernando California. *Science* **1973**, *182*, 161–163. [CrossRef]
54. Youd, T.L.; Hoose, S.N. Liquefaction susceptibility and geologic setting. In Proceedings of the 6th World Conference on Earthquake Engineering, New Delhi, India, 10–14 January 1977; pp. 37–42.
55. Youd, T.L. *Geologic Effects—Liquefaction and Associated Ground Failure*; Open File Report for US Geological Survey: Reston, VA, USA, 1984; pp. 84–760.
56. Tuttle, M.P.; Cowie, P.; Wolf, L. Liquefaction induced by modern earthquakes as a key to paleoseismicity: A case study of the 1988 Saguenay earthquake. In Proceedings of the Nineteenth Water Reactor Information Meeting, US Nuclear Regulatory Commission, Bethesda, MA, USA, 28–30 October 1991.
57. Tinsley, J.C.; Egan, J.A.; Kayen, R.E.; Bennet, M.J.; Kropp, A.; Holzer, T.L. Appendix: Maps and descriptions of liquefaction and associated effects. In *The Loma Prieta, California, Earthquake of October 17 1989–Liquefaction*; Holzer, T.L., Ed.; US Geological Survey Professional Paper; US Government Printing Office: Washington, DC, USA, 1998; pp. B287–B314.
58. Tuttle, M.P.; Hengesh, J.; Tucker, K.B.; Lettis, W.; Deaton, S.L.; Frost, J.D. Observations and comparisons of liquefaction features and related effects induced by the Bhuj earthquake. *Earthq. Spectra* **2002**, *18*, 79–100. [CrossRef]
59. Holzer, T.L.; Noce, T.E.; Bennett, M.J. Liquefaction probability curves for surficial geologic deposits. *Environ. Eng. Geosci.* **2010**, *17*, 1–21. [CrossRef]
60. Reid, C.M.; Thompson, N.K.; Irvine, J.R.M.; Laird, T.E. Sand volcanoes in the Avon-Heathcote estuary produced by the 2010–2011 Christchurch earthquakes: Implications for geological preservation and expression. *N. Z. J. Geol. Geophys.* **2012**, *55*, 249–254. [CrossRef]
61. Obermeier, S.F.; Jacobson, R.B.; Smoot, J.P.; Weems, R.E.; Gohn, G.S.; Monroe, J.E.; Powars, D.S. *Earthquake-Induced Liquefaction Features in the Coastal Setting of South Carolina and in the Fluvial Setting of the New Madrid Seismic Zone*; US Geological Survey Professional Paper; US Government Printing Office: Washington, DC, USA, 1990.
62. Iai, S.; Tsuchida, H.; Koizumi, K. *A New Criterion for Assessing Liquefaction Potential Using Grain Size Accumulation Curve and N-Value*; Report of the Port and Harbour Research Institute: Nagase, Yokosuka, Japan, 1986; pp. 125–234.
63. Seed, H.B.; Idriss, I.M.; Arango, I. Evaluation of liquefaction potential using field performance data. *J. Geotech. Geoenviron. Eng.* **1983**, *109*, 458–482. [CrossRef]
64. Boulanger, R.W.; Idriss, I.M. Closure to "Liquefaction Susceptibility Criteria for Silts and Clays" by Ross W. Boulanger and IM Idriss. *J. Geotech. Geoenviron. Eng.* **2008**, *134*, 1027–1028. [CrossRef]
65. Bray, J.D.; Sancio, R.B. Assessment of the liquefaction susceptibility of fine grained soils. *J. Geotech. Geoenviron. Eng.* **2006**, *132*, 1165–1177. [CrossRef]
66. Elgamal, A.W.; Dobry, R.; Adalier, K. Small scale shaking table tests of saturated layered sand-silt deposits. In Proceedings of the 2nd US-Japan Workshop on Soil Liquefaction, Buffalo, NY, USA, 1989.
67. Fiegel, G.L.; Kutter, B.L. Liquefaction mechanism for layered soils. *J. Geotech. Eng.* **1994**, *120*, 737–755. [CrossRef]
68. Özener, P.; Özaydın, K.; Berilgen, M. Investigation of liquefaction and pore water pressure development in layered sands. *Bull. Earthq. Eng.* **2009**, *7*, 199–219. [CrossRef]
69. Tuttle, M.; Barstow, N. Liquefaction-related ground failure: A case study in the New Madrid seismic zone, central United States. *Bull. Seismol. Soc. Am.* **1996**, *86*, 636–645.
70. Leon, E.; Gassman, S.L.; Talwani, P. Effect of soil aging on assessing magnitudes and accelerations of prehistoric earthquakes. *Erthq. Spectra* **2005**, *21*, 737–759. [CrossRef]

71. Andrus, R.D.; Mohanan, M.P.; Piratheepan, P.; Ellis, B.S.; Holzer, T.L. Predicting shear-wave velocity from cone penetration resistance. In Proceedings of the 4th International Conference on Earthquake Geotechnical Engineering, Thessaloniki, Greece, 25–28 June 2007.
72. Gibbard, P.L.; Head, M.J.; Walker, M.J.C.; Subcommission on Quaternary Stratigraphy. Formal ratification of the Quaternary System/Period and the Pleistocene Series/Epoch with a base at 2.58 Ma. *J. Quat. Sci.* **2010**, *25*, 96–102. [CrossRef]
73. Youd, T.L.; Perkins, D.M. Mapping of liquefaction severity index. *J. Geotech. Engn.* **1987**, *113*, 1374–1392. [CrossRef]
74. Sowers, G.F. *Introductory Soil Mechanics and Foundations: Geotechnical Engineering*, 4th ed.; Macmillan Publishing: New York, NY, USA, 1979.
75. Green, R.A.; Obermeier, S.F.; Olson, S.M. Engineering geologic and geotechnical analysis of paleoseismic shaking using liquefaction effects: Field examples. *Eng. Geol.* **2005**, *76*, 263–293. [CrossRef]
76. Andrus, R.D.; Hayati, H.; Mohanan, N. Correcting liquefaction resistance of aged sands using measured to estimated velocity ratio. *J. Geotech. Geoenviron. Eng.* **2009**, *135*, 735–744. [CrossRef]
77. Holzer, T.L.; Youd, T.L.; Hanks, T.C. Dynamics of liquefaction during the Superstition Hills, California, earthquake. *Science* **1989**, *244*, 56–59. [CrossRef]
78. Kayen, R.; Moss, R.E.S.; Thompson, E.M.; Seed, R.B.; Cetin, K.O.; Der Kiureghian, A.; Tanaka, Y.; Tokimatsu, K. Shear wave velocity-based probabilistic and deterministic assessment of seismic soil liquefaction potential. *J. Geotech. Geoenviron. Eng.* **2013**, *139*, 407–419. [CrossRef]
79. Cetin, K.O.; Seed, R.B.; Der Kiureghian, A.K.; Tokimatsu, K.; Harder, L.F., Jr.; Kayen, R.E.; Moss, R.E.S. Standard penetration test-based probabilistic and deterministic assessment of seismic soil liquefaction potential. *J. Geotech. Geoenviron. Eng.* **2004**, *130*, 1314–1340. [CrossRef]
80. Boulanger, R.W.; Wilson, D.W.; Idriss, I.M. Examination and reevaluation of SPT-based liquefaction triggering case histories. *J. Geotech. Geoenviron. Eng.* **2012**, *138*, 898–909. [CrossRef]
81. Clayton, P.; Zalachoris, G.; Rathje, E.; Bheemasetti, T.; Caballero, S.; Yu, X. *The Geotechnical Aspects of the September 3, 2016 M5.8 Pawnee, Oklahoma Earthquake*; Geotechnical Extreme Events Reconnaissance Association: Berkeley, CA, USA, 2016.
82. National Research Council. *Liquefaction of Soils during Earthquakes*; National Academy Press: Washington, DC, USA, 1985.
83. De Magistris, F.S.; Lanzano, G.; Forte, G.; Fabbrocino, G. A database for PGA threshold in liquefaction occurrence. *Soil Dyn. Erthq. Eng.* **2013**, *54*, 17–19. [CrossRef]
84. Boulanger, R.W.; Idriss, I.M. *CPT and SPT Based Liquefaction Triggering Procedures*; Report No. UCD/CGM-14/01; Center for Geotechnical Modeling, University of California: Davis, CA, USA, 2014.
85. Robertson, P.K. Evaluating soil liquefaction and post-earthquake deformations using the CPT. In *ISC-2 Proceedings on Geotechnical and Geophysical Site Characterization*; Millpress: Rotterdam, The Netherlands, 2004; pp. 233–252.
86. Robertson, P.K. Interpretation of cone penetration tests: A unified approach. *Can. Geotech. J.* **2009**, *46*, 1335–1355. [CrossRef]
87. Seed, H.B.; Idriss, I.M. Simplified procedure for evaluating soil liquefaction potential. *J. Geotech. Eng. Div.* **1971**, *97*, 1249–1273.
88. Boulanger, R.W.; Idriss, I.M. Probabilistic standard penetration test-based liquefaction: Triggering procedure. *J. Geotech. Geoenviron. Eng.* **2012**, *138*, 1185–1195. [CrossRef]
89. Franke, K.W.; Lingwall, B.N.; Youd, T.L.; Blonquist, J.; Liang, J.H. Overestimation of liquefaction hazard in areas of low to moderate seismicity due to improper characterization of probabilistic seismic loading. *Soil Dyn. Erthq. Eng.* **2019**, *116*, 681–691. [CrossRef]
90. Kuenen, P.H. Experiments in geology. *Trans. Geol. Soc. Glasg.* **1958**, *23*, 1–28. [CrossRef]
91. Owen, G. Experimental soft-sediment deformation: Structures formed by the liquefaction of unconsolidated sands and some ancient examples. *Sedimentology* **1996**, *43*, 279–293. [CrossRef]
92. Sims, J.D.; Garvin, C.D. Recurrent liquefaction at Soda Lake, California, induced by the 1989 Loma Prieta earthquake and 1990 and 1991 aftershocks: Implications for paleoseismicity studies. *Bull. Seismol. Soc. Am.* **1995**, *85*, 51–65.

93. Eberhart-Phillips, D.; Haeussler, P.J.; Freymueller, J.T.; Frankel, A.D.; Rubin, C.M.; Craw, P.; Ratchkovski, N.A.; Anderson, G.; Carver, G.A.; Crone, A.J.; et al. The 2002 Denali fault earthquake, Alaska: A large magnitude, slip-partitioned event. *Science* **2003**, *300*, 1113–1118. [CrossRef] [PubMed]
94. Harp, E.L.; Jibson, R.W.; Kayen, R.E.; Keefer, D.K.; Sherrod, B.S.; Carver, G.A.; Collins, B.D.; Moss, R.E.S.; Sitar, N. Landslides and liquefaction triggered by the M 7.9 Denali Fault earthquake of 3 November 2002. *GSA Today* **2003**, *13*, 4–10. [CrossRef]
95. Emergeo Working Group. Liquefaction phenomena associated with the Emilia earthquake sequence of May–June 2012 (Northern Italy). *Nat. Hazards Earth Syst. Sci.* **2013**, *13*, 935–947. [CrossRef]
96. Naik, S.P.; Kim, Y.-S.; Kim, T.; Su-Ho, J. Geological and structural control on localized ground effects within the Heunghae Basin during the Pohang earthquake (M_W 5.4, 15th November 2017), South Korea. *Geosciences* **2019**, *9*, 173. [CrossRef]
97. Sims, J.D. *Determining Earthquake Recurrence Intervals from Deformational Structures in Young Lacustrine Sediments*; Elsevier: Amsterdam, The Netherlands, 1975.
98. Sims, J.D. Earthquake-induced load casts, pseudonodules, ball-and-pillow, and convolute lamination: Additional deformation structures for paleoseismic studies. In *Recent Advances in North American Paleoseismology and Neotectonics East of the Rockies*; Cox, R.T., Tuttle, M.P., Boyd, O.S., Locat, J., Eds.; Geological Society of America Special Paper: Boulder, CO, USA, 2012; pp. 191–202.
99. Obermeier, S.F. Use of liquefaction-induced features for paleoseismic analysis—an overview of how seismic liquefaction features can be distinguished from other features and how their origin can be used to infer the location and strength of Holocene paleo-earthquakes. *Eng. Geol.* **1996**, *44*, 1–76. [CrossRef]
100. Obermeier, S.F. Using liquefaction-induced features for paleoseismic analysis. In *Paleoseismology*, 2nd ed.; McCalpin, J., Ed.; Academic Press: Burlington, MA, USA, 2009; pp. 497–564.
101. Porfido, S.; Esposito, E.; Guerrieri, L.; Vittori, E.; Tranfaglia, G.; Pece, R. Seismically induced ground effects of 1805, 1930 and 1980 earthquakes in the Southern Apennines (Italy). *Boll. Soc. Geol. It.* **2007**, *126*, 333–346.
102. Serva, L.; Esposito, E.; Guerrieri, L.; Porfido, S.; Vittori, E.; Comerci, V. Environmental effects from some historical earthquakes in Southern Apennines (Italy) and macroseismic intensity assessment: Contribution to INQUA EEE scale project. *Quat. Int.* **2007**, *173*, 30–44. [CrossRef]
103. Serva, L.; Vittori, E.; Comerci, V.; Esposito, E.; Guerrieri, L.; Michetti, A.M.; Mohammadioun, B.; Porfido, S.; Tatevossian, R.E. Earthquake hazard and the environmental seismic intensity (ESI) scale. *Pure Appl. Geophys.* **2015**, *173*, 1479–1515. [CrossRef]
104. Blumetti, A.M.; Guerrieri, L.; Porfido, S. Cataloguing the EEEs induced by the 1783 5th February Calabrian earthquake: Implications for an improved seismic hazard. *Mem. Descr. Carta Geol. D'Ital.* **2015**, *97*, 153–164. [CrossRef]
105. Michetti, A.M.; Esposito, E.; Guerrieri, L.; Porfido, S.; Serva, L.; Tatevossian, R.; Vittori, E.; Audemard, F.; Azuma, T.; Clague, J.; et al. Environmental seismic intensity scale—ESI 2007. *Mem. Descr. Carta Geol. D'Ital.* **2007**, *74*, 7–23.
106. Guerrieri, L.; Tatevossian, R.; Vittori, E.; Comerci, V.; Esposito, E.; Michetti, A.M.; Porfido, S.; Serva, L. Earthquake environmental effects (EEE) and intensity assessment: The INQUA scale project. *Boll. Soc. Geol. Italy* **2007**, *126*, 375–386.
107. Audemard, F.; Azuma, T.; Baiocco, F.; Baize, S.; Blumetti, A.M.; Brustia, E.; Clague, J.; Comerci, V.; Esposito, E.; Guerrieri, L.; et al. Earthquake Environmental Effect for Seismic Hazard Assessment: The ESI Intensity Scale and the EEE Catalogue. *Mem. Descr. Carta Geol. D'Ital.* **2015**, *97*, 184. [CrossRef]
108. Lunina, O.V.; Gladkov, A.S. Soft-sediment deformation structures induced by strong earthquakes in southern Siberia and their paleoseismic significance. *Sediment. Geol.* **2016**, *344*, 5–19. [CrossRef]
109. Amick, D.C. Paleoliquefaction Investigations along the Atlantic Seaboard with Emphasis on the Prehistoric Earthquake Chronology of Coastal South Carolina. Ph.D. Thesis, University of South Carolina, Columbia, SC, USA, 1990.
110. Wheeler, R.L.; Omdahl, E.M.; Dart, R.L.; Wilkerson, G.D.; Bradford, R.H. *Earthquakes in the Central United States, 1699–2002*; Version 1; US Geological Survey: Denver, CO, USA, 2003.
111. Counts, R.; Obermeier, S. Seismic signatures: Small-scale features and ground fractures. In *Recent Advances in North American Paleoseismology and Neotectonics East of the Rockies*; Cox, R.T., Tuttle, M.P., Boyd, O.S., Locat, J., Eds.; Geological Society of America Special Paper: Boulder, CO, USA, 2012; pp. 203–220.

112. EMERGEO Working Group. Technologies and new approaches used by the INGV EMERGEO Working Group for real-time data sourcing and processing during the Emilia Romagna (northern Italy) 2012 earthquake sequence. *Ann. Geophys.* **2012**, *55*, 689–695. [CrossRef]
113. Audemard, F.; de Santis, F. Survey of liquefaction structures induced by recent moderate earthquakes. *Bull. Int. Assoc. Eng. Geol.* **1991**, *44*, 5–16. [CrossRef]
114. Rodriguez-Pascua, M.A.; Silva, P.G.; Perez-Lopez, R.; Giner-Robles, J.L.; Martin-Gonzalez, F.; Del Moral, B. Polygenetic sand volcanoes: On the features of liquefaction processes generated by a single event 2012 Emilia Romagna 5.9 Mw earthquake. *Quat. Int.* **2015**, *357*, 329–335. [CrossRef]
115. Rydelek, P.A.; Tuttle, M.P. Explosive craters and soil liquefaction. *Nature* **2004**, *427*, 115–116. [CrossRef] [PubMed]
116. Tuttle, M.P.; Wolf, L.W.; Mayne, P.W.; Dyer-Williams, K.; Lafferty, R.H. *Guidance Document: Conducting Paleoliquefaction Studies for Earthquake Source Characterization;* US Nuclear Regulatory Commission: Washington, DC, USA, 2018.
117. Tuttle, M.P.; Wolf, L.W.; Dyer-Williams, K.; Mayne, P.W.; Lafferty, R.H.; Hess, K.; Starr, M.E.; Haynes, M.H.; Morrow, J.; Scott, R.; et al. *Paleoliquefaction Studies in Moderate Seismicity Regions with a History of Large Earthquakes;* US Nuclear Regulatory Commission: Washington, DC, USA, 2019.
118. Tuttle, M.P.; Al-Shukri, H.; Mahdi, H. Very large earthquakes centered southwest of the New Madrid seismic zone 5,000–7,000 years ago. *Seismol. Res. Lett.* **2006**, *77*, 755–770. [CrossRef]
119. Li, Y.; Schweig, E.S.; Tuttle, M.P.; Ellis, M.A. Evidence for large prehistoric earthquakes in the northern New Madrid seismic zone, central United States. *Seismol. Res. Lett.* **1998**, *69*, 270–276. [CrossRef]
120. Moretti, M.; Alfaro, P.; Owen, G. The environmental significance of soft-sediment deformation structures: Key signatures for sedimentary and tectonic processes. *Sediment. Geol.* **2016**, *344*, 1–4. [CrossRef]
121. Wolf, L.W.; Tuttle, M.P.; Browning, S.; Park, S. Geophysical surveys of earthquake-induced liquefaction deposits in the New Madrid seismic zone. *Geophysics* **2006**, *71*, 223–230. [CrossRef]
122. Al-Shukri, H.; Mahdi, H.; Tuttle, M. Three-dimensional imaging of earthquake-induced liquefaction features with ground penetrating radar near Marianna, Arkansas. *Seismol. Res. Lett.* **2006**, *77*, 505–513. [CrossRef]
123. Obermeier, S.F.; Pond, E.C.; Olson, S.M. *Paleoliquefaction Studies in Continental Settings–Geologic and Geotechnical Factors in Interpretations and Back-Analysis;* US Geological Survey Open-File Report: Washington, DC, USA, 2001.
124. Wolf, L.W.; Collier, J.; Tuttle, M.; Bodin, P. Geophysical reconnaissance of earthquake-induced liquefaction features in the New Madrid seismic zone. *J. Appl. Geophys.* **1998**, *39*, 121–129. [CrossRef]
125. Tuttle, M.P.; Collier, J.; Wolf, L.W.; Lafferty, R.H. New evidence for a large earthquake in the New Madrid seismic zone between AD 1400 and 1670. *Geology* **1999**, *27*, 771–774. [CrossRef]
126. Annan, A.P. The principals of ground penetrating radar. In *Near Surface Geophysics;* Butler, D.K., Ed.; Society of Exploration Geophysicists: Tulsa, OK, USA, 2005; pp. 357–438.
127. Liu, L.; Li, Y. Identification of liquefaction and deformation features using ground penetrating radar in the New Madrid seismic zone, USA. *J. Appl. Geophys.* **2001**, *47*, 199–215. [CrossRef]
128. Nobes, D.C.; Bastin, S.; Cook, G.C.R.; Gallagher, M.; Graham, H.; Grose, D.; Hedley, J.; Sharp-Heward, S.; Templeton, S. Geophysical imaging of subsurface earthquake-induced liquefaction features at Christchurch Boys High School, Christchurch, New Zealand. *J. Environ. Eng. Geophys.* **2013**, *18*, 255–267. [CrossRef]
129. Salvi, S.; Cinti, F.R.; Colini, L.; D'addezio, G.; Doumaz, F.; Pettinelli, E. Investigation of the active Celano-L'Aquila fault system, Abruzzi (central Apennines, Italy) with combined ground-penetrating radar and palaeoseismic trenching. *Geophys. J. Int.* **2003**, *155*, 805–818. [CrossRef]
130. Tuttle, M.; Lafferty, R.H., III; Schweig, E.S., III. *Dating of Liquefaction Features in the New Madrid Seismic Zone and Implications for Earthquake Hazard;* US Nuclear Regulatory Commission: Washington, DC, USA, 1998.
131. Tuttle, M.; Chester, J.; Lafferty, R.; Dyer-Williams, K.; Cande, B. *Paleoseismology Study Northwest of the New Madrid Seismic Zone;* US Nuclear Regulatory Commission: Washington, DC, USA, 1999; p. 98.
132. Cox, J.; Talwani, P. Paleoseismic studies in the 1886 Charleston earthquake meizoseismal area. *Geol. Soc. Am. Abstr. Prog.* **1983**, *16*, 130.
133. Cox, J.H.M. Paleoseismology Studies in South Carolina. M.S. Thesis, University of South Carolina, Columbia, SC, USA, 1984.
134. Amick, D.; Gelinas, R.; Maurath, G.; Cannon, R.; Moore, D.; Billington, E.; Kemppinen, H. *Paleoliquefaction Features along the Atlantic Seaboard;* US Nuclear Regulatory Commission: Washington, DC, USA, 1990.

135. Amick, D.; Gelinas, R. The search for evidence of large prehistoric earthquakes along the Atlantic seaboard. *Science* **1991**, *251*, 655–658. [CrossRef]
136. Weems, R.E.; Obermeier, S.F. The 1886 Charleston earthquake—An overview of geological studies. In Proceedings of the US Nuclear Regulatory Commission Seventeenth Water Reactor Safety Information Meeting, Rockville, MD, USA, 23–25 October 1990.
137. Vaughn, J.D. *Paleoseismological Studies in the Western Lowlands of Southeast Missouri*; Technical Report to U.S. Geological Survey: NEHRP Award 14-08-0001-G1931; U.S. Geological Survey: Reston, VA, USA, 1994; p. 27.
138. Tuttle, M.P. *Earthquake Potential of the Central Virginia Seismic Zone*; Final Technical Report to US Geological Survey: NEHRP Award G13AP00045; Reston, VA, USA, 2016; p. 32. Available online: https://earthquake.usgs.gov/cfusion/external_grants/reports/G13AP00045.pdf (accessed on 11 July 2019).
139. Mahan, S.A.; Crone, A.J. Luminescence dating of paleoliquefaction features in the Wabash River Valley of Indiana. In *4th New World Luminescence Dating and Dosimetry Workshop, Denver, Colorado*; Wide, R.A., Ed.; US Geological Survey Open-File Report: Reston, VA, USA, 2006; p. 12.
140. Mahan, S.; Counts, R.; Tuttle, M.; Obermeier, S. Can OSL be used to date paleoliquefaction events? In *Abstracts Volume from Meeting of Central and Eastern U.S. (CEUS) Earthquake Hazards Program, October 28-29, 2009*; Tuttle, M.P., Boyd, O., McCallister, N., Eds.; US Geological Survey Open-File Report: Reston, VA, USA, 2013; p. 50.
141. Tuttle, M.P.; Sims, J.D.; Dyer-Williams, K.; Lafferty, R.H., III; Schweig, E.S., III. *Dating of Liquefaction Features in the New Madrid Seismic Zone*; NUREG/GR-0018; US Nuclear Regulatory Commission: Washington, DC, USA, 2000.
142. Tuttle, M.P.; Schweig, E.S. Archeological and pedological evidence for large earthquakes in the New Madrid seismic zone, central United States. *Geology* **1995**, *23*, 253–256. [CrossRef]
143. Lafferty, R.H., III. Archeological techniques of dating ancient quakes. *Geotimes* **1996**, *41*, 24–27.
144. Tuttle, M.P.; Lafferty, R.H., III; Guccione, M.J.; Schweig, E.S., III; Lopinot, N.; Cande, R.F.; Dyer-Williams, K.; Haynes, M. Use of archaeology to date liquefaction features and seismic events in the New Madrid seismic zone, central United States. *Geoarchaeology* **1996**, *11*, 451–480. [CrossRef]
145. Atwater, B.F.; Tuttle, M.P.; Schweig, E.S.; Rubin, C.M.; Yamaguchi, D.K.; Hemphill-Haley, E. Earthquake recurrence inferred from paleoseismology. In *The Quaternary Period in the United States, Developments in Quaternary Science 1*; Gillespie, A.R., Porter, S.C., Atwater, B.F., Eds.; Elsevier: New York, NY, USA, 2004; pp. 331–350.
146. Loope, D.B.; Elder, J.F.; Zlotnik, V.A.; Kettler, R.M.; Pederson, D.T. Jurassic earthquake sequence recorded by multiple generations of sand blows, Zion National Park, Utah. *Geology* **2013**, *41*, 1131–1134. [CrossRef]
147. Quigley, M.; Hughes, M.; Bradley, B.; van Ballegooy, S.; Reid, C.; Morgenroth, J.; Horton, T.; Duffy, B.; Pettinga, J. The 2010–2011 Canterbury earthquake sequence: Environmental effects, seismic triggering thresholds, and geologic legacy. *Tectonophysics* **2016**, *672*, 228–274. [CrossRef]
148. Tuttle, M.P. *Search for and Study of Sand Blows at Distant Sites Resulting from Prehistoric and Historic New Madrid Earthquakes: Collaborative Research between M. Tuttle & Associates and Central Region Hazards Team*; Final Technical Report to US Geological Survey: NEHRP Award 02HQGR0097; Reston, VA, USA, 2010; p. 48. Available online: https://earthquake.usgs.gov/cfusion/external_grants/reports/02HQGR0097.pdf (accessed on 11 July 2019).
149. Kelson, K.I.; Simpson, G.D.; Van Arsdale, R.B.; Harris, J.B.; Haradan, C.C.; Lettis, W.R. Multiple Holocene earthquakes along the Reelfoot fault, central New Madrid seismic zone. *J. Geophys. Res.* **1996**, *101*, 6151–6170. [CrossRef]
150. Olson, S.M.; Green, R.A.; Obermeier, S.F. Revised magnitude bound relation for the Wabash Valley seismic zone of the central United States. *Seismol. Res. Lett.* **2005**, *76*, 756–771. [CrossRef]
151. Maurer, B.; Green, R.; Quigley, M.; Bastin, S. Development of magnitude-bound relations for paleoliquefaction analyses: New Zealand case study. *Eng. Geol.* **2015**, *197*, 253–266. [CrossRef]
152. Olson, S.M.; Obermeier, S.F.; Stark, T.D. Interpretation of penetration resistance for back-analysis at sites of previous liquefaction. *Seismol. Res. Lett.* **2001**, *72*, 46–59. [CrossRef]
153. Pond, E.C. Seismic Parameters for the Central United States based on Paleoliquefaction Evidence in the Wabash Valley. Ph.D. Dissertation, Virginia Polytechnic Institute, Blacksburg, VA, USA, 1996.

154. Talwani, P.; Dura-Gomez, I.; Gassman, S.; Hasek, M.; Chapman, A. Studies related to the discovery of a prehistoric sandblow in the epicentral area of the 1886 Charleston SC earthquake: Trenching and geotechnical investigations. *Program Abstr. East. Sect. Seismol. Soc. Am.* **2008**, 50.
155. Hu, K.; Gassman, S.L.; Talwani, P. In-situ properties of soils at paleoliquefaction sites in the South Carolina coastal plain. *Seismol. Res. Lett.* **2002**, *73*, 964–978. [CrossRef]
156. Hu, K.; Gassman, S.L.; Talwani, P. Magnitudes of prehistoric earthquakes in the South Carolina coastal plain from geotechnical data. *Seismol. Res. Lett.* **2002**, *73*, 979–991. [CrossRef]
157. Gassman, S.; Talwani, P.; Hasek, M. Maximum magnitudes of Charleston, South Carolina earthquakes from in-situ geotechnical data. In *Abstracts Volume from Meeting of Central and Eastern U.S. Earthquake Hazards Program*; University of Memphis: Memphis, TN, USA, 2009; p. 19.
158. Noller, J.S.; Forman, S.L. Luminescence geochronology of liquefaction features near Georgetown, South Carolina. In *Dating and Earthquakes: Review of Quaternary Geochronology and Its Application to Paleoseismology*; Sowers, J.M., Noller, J.S., Lettis, W.R., Eds.; US Nuclear Regulatory Commission: Washington, DC, USA, 1998; pp. 49–57.
159. Civico, R.; Brunori, C.A.; De Martini, P.M.; Pucci, S.; Cinti, F.R.; Pantosti, D. Liquefaction susceptibility assessment in fluvial plains using airborne lidar: The case of the 2012 Emilia earthquake sequence area (Italy). *Nat. Hazard Earth Syst. Sci.* **2015**, *15*, 2473–2483. [CrossRef]
160. Almond, P.; Wilson, T.; Shanhun, F.L.; Whitman, Z.; Eger, A.; Moot, D.; Cockcroft, M.; Nobes, D. Agricultural land rehabilitation following the 2010 Darfield (Canterbury) earthquake: A preliminary report. *Bull. N. Z. Soc. Earthq. Eng.* **2010**, *43*, 432–438.
161. Tuttle, M.P.; Lafferty, R.H., III; Cande, R.F.; Sierzchula, M.C. Impact of earthquake-induced liquefaction and related ground failure on a Mississippian archeological site in the New Madrid seismic zone, central USA. *Quat. Int.* **2011**, *242*, 126–137. [CrossRef]
162. Oristaglio, M.; Dorozynski, A. *A Sixth Sense: The Life and Science of Henri-Georges Doll, Oilfield Pioneer and Inventor*, 1st ed.; Gerald Duckworth & Co. Ltd.: London, UK, 2009.
163. Occupational Safety and Health Administration (OSHA). *Trenching and Excavation Safety*; US Department of Labor: Washington, DC, USA, 2015; p. 28.
164. Work Safe. *Excavation Safety: Good Practice Guidelines*; New Zealand Government: Wellington, New Zealand, 2016.
165. Canadian Centre for Occupational Health and Safety (CCOHS). Trenching and Excavation Fact Sheet. Available online: https://www.ccohs.ca/oshanswers/hsprograms/trenching_excavation.html (accessed on 25 June 2019).
166. Advisory Council on Historic Preservation. Section 106 regulations: National historic preservation act of 1966. In *Code of Federal Regulations, 36 CFR Part 800*; Advisory Council on Historic Preservation: Washington, DC, USA, 1966.
167. Trumbore, S.E. AMS ^{14}C measurements of fractionated soil organic matter: An approach to deciphering the soil carbon cycle. *Radiocarbon* **1989**, *31*, 644–654. [CrossRef]
168. Walker, M. *Quaternary Dating Methods*; John Wiley & Sons Ltd.: London, UK, 2005.
169. Rhodes, E.J. Optically stimulated luminescence dating of sediments over the past 200,000 years. *Ann. Rev. Earth Planet. Sci.* **2011**, *39*, 461–488. [CrossRef]
170. Lian, O.B.; Roberts, R.G. Dating the Quaternary: Progress in luminescence dating of sediments. *Quat. Sci. Rev.* **2006**, *25*, 2449–2468. [CrossRef]
171. Duller, G.A.T. *Luminescence Dating: Guidelines on Using Luminescence Dating in Archaeology*; English Heritage: Swindon, UK, 2008.
172. Li, B.; Jacobs, Z.; Roberts, R.G.; Li, S.-H. Review and assessment of the potential of post-IR IRSL dating methods to circumvent the problem of anomalous fading in feldspar luminescence. *Geochronometria* **2014**, *41*, 178–201. [CrossRef]
173. Wallinga, J.; Cunningham, A. Luminescence dating, uncertainties and age range. In *Encyclopedia of Scientific Dating Methods*; Rink, W.J., Thompson, J.W., Eds.; Springer: New York, NY, USA, 2015; pp. 440–444.
174. Kars, R.H.; Busschers, F.S.; Wallinga, J. Validating post IR-IRSL dating on K-feldspars through comparison with quartz OSL ages. *Quat. Geol.* **2012**, *12*, 74–86. [CrossRef]
175. Lian, O.B. Luminescence dating—Optically stimulated luminescence. In *Encyclopedia of Quaternary Science*; Elias, S.A., Ed.; Elsevier: New York, NY, USA, 2007; pp. 1491–1505.

176. Douglass, A.E. Crossdating in dendrochronology. *J. For.* **1941**, *39*, 825–831.
177. Stahle, D.W.; Cook, E.R.; White, J.W.C. Tree-ring dating of baldcypress and the potential for millennia-long chronologies in the southeast. *Am. Antiquity* **1985**, *50*, 796–802. [CrossRef]
178. Pierce, K.L. Dating methods. In *Active Tectonics: Impact on Society*; The National Academies Press: Washington, DC, USA, 1986; pp. 195–214.
179. Stahle, D.W.; Fye, F.K.; Therrell, M.D. Interannual to decadal climate and streamflow variability estimated from tree rings. In *The Quaternary Period in the United States, Developments in Quaternary Science 1*; Gillespie, A.R., Porter, S.C., Atwater, B.F., Eds.; Elsevier: New York, NY, USA, 2004; pp. 491–504.
180. Van Arsdale, R.B.; Stahle, D.W.; Cleaveland, M.K.; Guccione, M.J. Earthquake signals in tree-ring data from the New Madrid seismic zone and implications for paleoseismicity. *Geology* **1998**, *26*, 515–518. [CrossRef]
181. Castilla, R.A.; Audemard, F.A. Sand blows as a potential tool for magnitude estimation of pre-instrumental earthquakes. *J. Seismol.* **2007**, *11*, 473–487. [CrossRef]
182. Allen, J.R.L. Earthquake magnitude-frequency, epicentral distance, and soft-sediment deformation in sedimentary basins. *Sediment. Geol.* **1986**, *46*, 67–75. [CrossRef]
183. Ambraseys, N.N. Engineering seismology. *Erthq. Eng. Struct. Dyn.* **1988**, *17*, 1–105. [CrossRef]
184. Mandel, S. *The Groundwater Resources of the Canterbury Plains*; New Zealand Agricultural Engineering Institute: Canterbury, New Zealand, 1974; p. 59.
185. Elder, D.M.G.; McCahon, I.F.; Yetton, M.D. *The Earthquake Hazard in Christchurch: A Detailed Evaluation*; Report funded by the Earthquake Commission; Soils and Foundations Ltd.: Christchurch, New Zealand, 1991.
186. Tonkin, T. *Darfield Earthquake Recovery Geotechnical Factual Report–Kaiapoi North*; Report Prepared for the Earthquake Commission; Tonkin & Taylor: Christchurch, New Zealand, 2011; p. 9.
187. Brackley, H.L. *Review of Liquefaction Hazard Information in Eastern Canterbury, Including Christchurch City and Parts of Selwyn, Waimakariri and Hurunui Districts*; GNS Science Consultancy Report 2012/218; University of Canterbury: Christchurch, New Zealand, 2012.
188. Cubrinovski, M.; Green, R.A.; Allen, J.; Ashford, S.; Bowman, E.; Bradley, B.; Cox, B.; Hutchinson, T.; Kavazanjian, E.; Orense, R.; et al. Geotechnical reconnaissance of the 2010 Darfield (Canterbury) earthquake. *Bull. N. Z. Soc. Earthq. Eng.* **2010**, *43*, 243–320.
189. Cubrinovski, M.; Bradley, B.; Wotherspoon, L.; Green, R.A.; Bray, J.; Wood, C.; Pender, M.; Allen, J.; Bradshaw, A.; Rix, G.; et al. Geotechnical aspects of the 22 February 2011 Christchurch earthquake. *Bull. N. Z. Soc. Earthq. Eng.* **2011**, *44*, 205–226.
190. Taylor, M.L. The Geotechnical Characterisation of Christchurch Sands for Advanced Soil Modeling. Ph.D. Dissertation, University of Canterbury, Christchurch, New Zealand, 2015.
191. Holbrook, J.; Autin, W.J.; Rittenour, T.M.; Marshak, S.; Goble, R.J. Stratigraphic evidence for millennial-scale temporal clustering of earthquakes on a continental-interior fault: Holocene Mississippi River floodplain deposits, New Madrid seismic zone, USA. *Tectonophysics* **2006**, *420*, 431–454. [CrossRef]
192. Galli, P. New empirical relationships between magnitude and distance for liquefaction. *Tectonophysics* **2000**, *324*, 169–187. [CrossRef]
193. Kuribayashi, E.; Tatsuoka, F. Brief review of liquefaction during earthquakes in Japan. *Soils Found.* **1975**, *15*, 81–92. [CrossRef]
194. Youd, T.L. Brief review of liquefaction during earthquakes in Japan. *Soils Found.* **1977**, *17*, 81–92.
195. Papadopoulos, G.A.; Lefkopoulos, G. Magnitude-distance relations for liquefaction in soil from earthquakes. *Bull. Seismol. Soc. Am.* **1993**, *83*, 925–938.
196. Boulton, S.J. Paleoseismology. In *Encyclopedia of Earthquake Engineering*; Springer: Berlin, Germany, 2014.
197. Youd, T.L.; Perkins, D.M.; Turner, W.G. *Liquefaction Severity Index Attenuation for the Eastern United States, Proceedings from the Second U.S.-Japan Workshop on Liquefaction, Large Ground Deformation and their Effects on Lifelines*; Technical Report NCEER-89-0032; National Center for Earthquake Engineering Research: Buffalo, NY, USA, 1989.
198. Cetin, K.O.; Turkoglu, M.; Unsal Oral, S.; Nacar, U. *Van Tabanli Earthquake (Mw = 7.1) October 23, 2011 Preliminary Reconnaissance Report*; Report Number GEER-028; Geotechnical Extreme Events Reconnaissance Association, 2011; Available online: http://www.geerassociation.org/administrator/components/com_geer_reports/geerfiles/Van_EQ_Preliminary_Report_KOC.pdf (accessed on 11 July 2019).
199. Atkinson, G.; Adams, J. Ground motion prediction equations for application to the 2015 Canadian national seismic hazard maps. *Can. J. Civ. Eng.* **2013**, *40*, 988–998. [CrossRef]

200. Mayne, P.W.; Styler, M. Soil liquefaction screening using CPT yield stress profiles. In *Geotechnical Earthquake Engineering and Soil Dynamics V: Liquefaction Triggering, Consequences, and Mitigation*; Brandenberg, S.J., Manzari, M.T., Eds.; Curran Associates Inc.: Red Hook, NY, USA, 2018; pp. 605–616.
201. Martin, J.R.; Clough, G.M. Seismic parameters from liquefaction evidence. *J. Geotech. Eng.* **1994**, *120*, 1345–1361. [CrossRef]
202. Johnston, A.C. Seismic moment assessment of stable continental earthquakes, Part III: 1811–1812 New Madrid, 1886 Charleston, and 1755 Lisbon earthquakes. *Geophys. J. Int.* **1996**, *126*, 314–344. [CrossRef]
203. Bakun, W.H.; Hopper, M. Magnitudes and locations of the 1811–1812 New Madrid, Missouri and the 1886 Charleston, South Carolina earthquakes. *Bull. Seismol. Soc. Am.* **2004**, *94*, 64–75. [CrossRef]
204. Boyd, O.S.; Cramer, C.H. Estimating earthquake magnitudes from reported intensities in the central and eastern United States. *Bull. Seismol. Soc. Am.* **2014**, *104*, 1709–1722. [CrossRef]
205. Cramer, C.H.; Boyd, O.S. Why the New Madrid earthquakes are M 7–8 and the Charleston earthquake is ~M 7. *Bull. Seismol. Soc. Am.* **2014**, *104*, 2884–2903. [CrossRef]
206. Dutton, C.E. The Charleston Earthquake of August 31, 1886. In *US Geological Survey 9th Annual Report 1887–1888*; US Government Printing Office: Washington, DC, USA, 1889; pp. 203–528.
207. Bollinger, G.A. Reinterpretation of the intensity data for the 1886 Charleston, South Carolina, earthquake. In *Studies Related to the Charleston, South Carolina, Earthquake of 1886—A Preliminary Report*; Rankin, D.W., Ed.; US Geological Survey Professional Paper; US Government Printing Office: Washington, DC, USA, 1977; pp. 17–32.
208. Hough, S.E.; Armbruster, J.G.; Seeber, L.; Hough, J.F. On the modified Mercalli intensities and magnitudes of the 1811–1812 New Madrid earthquakes. *J. Geophys. Res.* **2000**, *105*, 23839–23864. [CrossRef]
209. Saucier, R.T. *Effects of the New Madrid Earthquake Series in the Mississippi Alluvial Valley*; US Army Corps of Engineers Waterways Experiment Station: Vicksburg, MS, USA, 1977.
210. Street, R.; Nuttli, O. The central Mississippi Valley earthquakes of 1811–1812. In *Proceedings of the Symposium on the New Madrid Seismic Zone*; Gori, P.L., Hays, W.W., Eds.; US Geological Survey Open-File Report: Reston, VA, USA, 1984; pp. 33–63. [CrossRef]
211. Johnston, A.C.; Schweig, E.S. The enigma of the New Madrid earthquakes of 1811–1812. *Ann. Rev. Earth Planet. Sci.* **1996**, *24*, 339–384. [CrossRef]
212. Hough, S.E. Scientific overview and historical context of the 1811–1812 New Madrid earthquake sequence. *Ann. Geophys.* **2004**, *47*, 523–537. [CrossRef]
213. Obermeier, S. *The New Madrid Earthquakes: An Engineering-Geologic Interpretation of Relict Liquefaction Features*; US Geological Survey Professional Paper; US Government Printing Office: Washington, DC, USA, 1989.
214. Russ, D.P. Style and significance of surface deformation in the vicinity of New Madrid, Missouri. In *Investigations of the New Madrid, Missouri, Earthquake Region*; McKeown, F.A., Pakiser, L.C., Eds.; US Geological Survey Professional Paper; US Government Printing Office: Washington, DC, USA, 1982; pp. 94–114.
215. Schneider, J.A.; Mayne, P.W. Liquefaction response of soils in Mid-America by seismic cone tests. In *Innovations and Applications in Geotechnical Site Characterization (GSP 97)*; American Society of Civil Engineers: Reston, VA, USA, 2000; pp. 1–16.
216. Schneider, J.A.; Mayne, P.W.; Rix, G.J. Geotechnical site characterization in the greater Memphis area using CPT. *Eng. Geol.* **2001**, *62*, 169–184. [CrossRef]
217. Liao, T.; Mayne, P.W.; Tuttle, M.P.; Schweig, E.S.; Van Arsdale, R.B. CPT site characterization for seismic hazards in the New Madrid seismic zone. *Soil Dyn. Erthq. Eng.* **2002**, *22*, 943–950. [CrossRef]
218. Hough, S.E.; Martin, S. Magnitude estimates of two large aftershocks of the 16 December 1811 New Madrid earthquakes. *Bull. Seismol. Soc. Am.* **2002**, *92*, 3259–3268. [CrossRef]
219. Hough, S.E.; Page, M. Toward a consistent model for strain accrual and release for the New Madrid, central United States. *J. Geophys. Res.* **2011**, *116*, B03311. [CrossRef]

geosciences

MDPI

Review

Catalogue of the Geological Effects of Earthquakes in Spain Based on the ESI-07 Macroseismic Scale: A New Database for Seismic Hazard Analysis

Pablo G. Silva [1,*], Miguel A. Rodríguez-Pascua [2], Jorge L. Giner Robles [3], Javier Élez [1], Raúl Pérez-López [2] and M. Begoña Bautista Davila [4]

1 Depto. Geología, Escuela Politécnica Superior de Ávila, Universidad Salamanca, Ávila 05003, Spain
2 Instituto Geológico y Minero de España (IGME), Madrid 28004, Spain
3 Depto. Geoquímica y Geología, Universidad Autónoma de Madrid, Madrid 28049, Spain
4 Escuela Politécnica Superior de Ávila, Universidad de Salamanca, Ávila 05003, Spain
* Correspondence: pgsilva@usal.es

Received: 21 June 2019; Accepted: 24 July 2019; Published: 29 July 2019

Abstract: This paper summarizes the content and scope of the "Catalogue of Earthquake Geological Effects in Spain". The catalogue has been published by the Geological Survey of Spain (IGME) and constitutes the first official publication (in Spain) on seismic hazard containing geological information. The catalogue gathers the 51 stronger earthquakes that have occurred in Spain since the Neolithic period to the present and classifies earthquakes with geological or archaeological seismic records in paleoseismic, ancient, historical and instrumental earthquakes. The catalogue offers a variety of parametric information, quality indexes (Qe, Qi, Qg), and Environmental Seismic Intensity Scale (ESI-07) based description of environmental damage structured in individual "event files". Sixteen of the 51 catalogued events present full information files (full event files), with individualized analyses of the geological and geoarchaeological data as well as graphic information with hybrid ESI-EMS intensity maps, ShakeMaps (seismic scenarios) and complementary kmz files (Google Earth) for each of the sixteen selected earthquakes; among which is the well-known AD 1755 Lisbon earthquake-tsunami. These selected earthquakes present individual environmental earthquake effects (EEE) or earthquake archaeoseismological effects (EAE) files for each catalogued effect containing specific site geo-information and graphic data (photos, graphs, maps, etc.). The second edition of the catalogue record 1027 EEEs and 187 EAEs, of which 322 effects have individual files.

Keywords: ESI-07 scale; earthquake environmental effects (EEEs); earthquake archaeological effects (EAEs); intensity maps; seismic scenarios; earthquake catalogues; seismic hazard; Spain

1. Introduction to the Environmental Seismic Intensity Scale ESI-07

Recent evaluations of the geological and environmental effects of earthquakes (environmental earthquake effects (EEEs)) indicate that such effects can be certainly parameterized and used for relative intensity assessments [1,2]. The introduction of the geological analysis of earthquake effects is primarily based on the application of the Environmental Seismic Intensity Scale (ESI-07) developed by the INQUA Paleoseismology International Focus Group in 2007 [3]. The ESI-07 scale itself constitutes the quantification and parameterization of the environmental effects considered in qualitative terms by the traditional macroseismic scales, such as the Mercalli–Cancani–Sieberg (MCS), the Medvéded–Sponheuer–Karnik (MSK) or the Mercalli Modified Scale (MMS). Unfortunately, recent macroseismic practices in the European Union (European Macroseismic Sacale EMS-98) do not consider earthquake environmental or ground effects for intensity assessments [4]. In this byzantine situation the ESI-07 scale provides tools for intensity assessments independent from building damage and help to understand the contribution of EEEs to the global damage recorded in localities within epicentral areas [1,5].

As with the traditional scales, the ESI-07 scale is a 12-degree intensity scale (Figure 1) only based on analysis and dimensions of EEEs, whose documentation has seen a considerable growth in the past decades, considering secondary (i.e., landslides, liquefaction, tsunamis, etc.) and primary (surface faulting, tectonic uplift) EEEs [3]. However, EEEs in the ESI-07 scale start to be observable, but marginal, from intensities up to IV–V; evident and noticeable from intensities VI–VII; but evident and characteristic triggering permanent changes in the geological and geomorphological record for intensities ≥IX [5] (Figure 1). In this last group EEEs become dominant, characteristic and diagnostic for intensities IX–X, when commonly building damage is nearly saturated in the traditional macroseismic scales. From intensities up to X building damage is saturated in most of the cases and EEEs are the unique tool for an adequate intensity assessment [1]. The ESI-07 scale aims to catalogue EEEs with a multiple purpose [5]:

- To complete historical records of earthquakes for a more fine-tuning of intensity assessments.
- To link historical EEE records with paleoseismological data in the zone/fault of interest.
- To include paleoseismological and geological earthquake records in seismic hazard assessments (SHA).
- To assess intensities in modern earthquakes both in urban and no-urban areas where EEEs occurred and participated in the recorded damage. This point is critical for the future refinement of the scale.

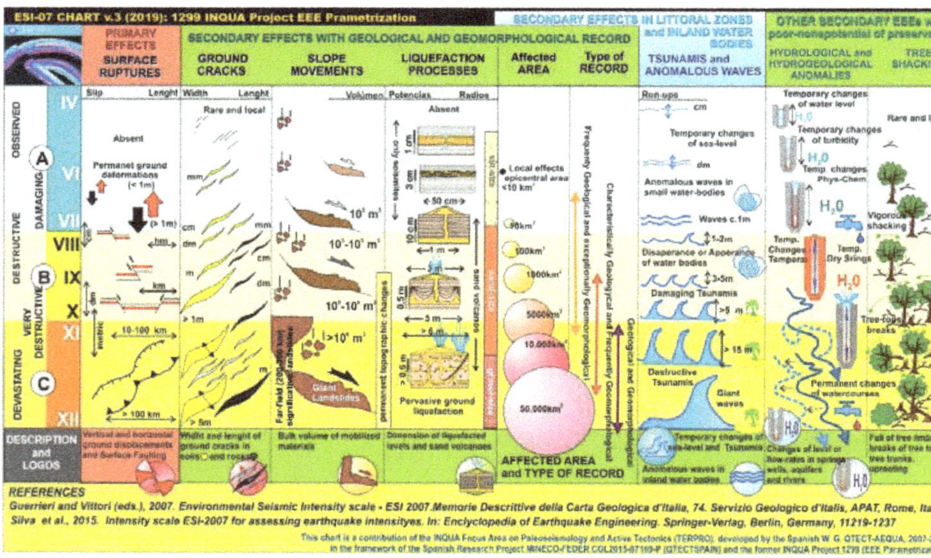

Figure 1. Synoptic chart of the Environmental Seismic Intensity Scale (ESI-07) scale illustrating the earthquake environmental effects modified and updated from Silva et al. [5].

In summary, analyses of EEEs try to enlarge the temporal period of macroseismic observations and data some thousand years ago to the more recent geological past, commonly to the last 10,000–12,000 years BP (i.e., Holocene epoch) [1,5,6]. This fact is essential to perform the scientific analysis of a particular geological process (Seismic Cycle) which normally exceeds the temporal scales commonly considered in the conventional seismic catalogues [5]. In the case of Spain, the official catalogue of historical earthquakes extends to the year 800 BCE [7]. However, as noticed in recent macroseismic reviews, the historical data are scarce and doubtful before the year 1000 CE, and nearly accurate for strong events from the 14th–15th centuries [8–10]. This implies about 500–600 years of useful macroseismic record for Spain, which in turn are the approximate return periods used in SHA for seismic codes [11,12].

2. The Content of the Catalogue

The catalogue includes data for the 51 strongest earthquakes felt in the Iberian Peninsula from the Neolithic period to the present (Figure 2). The reviewed earthquake list also incorporates the well-known AD 1755 Lisbon earthquake-tsunami with 753 catalogued EEEs including the far-field natural anomalies recorded in the British Islands, Central-Northern Europe, Africa and The Caribbean Sea. The catalogue also considers geological information for ancient earthquakes only documented in archaeological sites, differentiating several types of earthquake archeological effects (EAEs) of the classification [13]. Among the most ancient events included in the catalogue are those of the Cueva del Toro in Antequera (Málaga) of Neolithic age (c. 3700 BCE) [14] and the Bronze Age event of La Tira del Lienzo in Totana, Murcia (c. 1550 BCE) [15]. A surface rupture affecting dry masonry walls along the well-known Lorca-Alhama de Murcia left-lateral fault in the Eastern Betic Cordillera documents this Bronze Age earthquake. This fault is also responsible for the last damaging earthquake that occurred in Spain, the 2011 CE Lorca event (5.1 Mw) which produced about 250 EEEs (mainly slope movements) with a maximum intensity of VIII ESI-07 [16] of which the more relevant 49 EEEs have been catalogued [10]. The macroseismic area of this instrumental earthquake was located few kilometers north of Lorca in a nearly depopulated area without EMS assessments, but in which important collapses of the subvertical slopes of dry creeks massively occurred coinciding with the area where the earthquake epicenter was located [10]. This fact illustrates the value of the ESI-07 scale even for instrumental events, since this area of massive slope collapses identified an ESI intensity VIII, whilst the EMS assessments only quoted a maximum intensity of VII EMS in the locality of Lorca [16]. Figure 2 illustrates the distribution of the 51 catalogued events around the Iberian Peninsula identifying the Betic Cordillera (southeast) and the Pyrenees (north) as the two main zones where the convergence between the two large tectonic plates of Eurasia and Africa trigger the ongoing seismotectonic activity. A third important zone is the Gulf of Cádiz and the southern Portugal offshore area, "locus" of nucleation of strong historical earthquake-tsunami events in response to the suspected lithospheric delamination of the Atlantic oceanic floor beneath the Iberian Peninsula (Figure 2) [17,18].

The newly edited catalogue is an official publication of the Geological Survey of Spain (IGME) and does not aim to substitute the running seismological catalogues based on historical data written in EMS language [7]. Furthermore, the published ESI-07 Catalogue [10] incorporates all the available EMS macroseismic data to produce hybrid ESI-EMS Intensity maps as suggested in the guidelines of the ESI-07 scale [3]. Therefore, the variety of geological, geoarchaeological, archaeological and historic information summarized in the catalogue aims to complement the existing EMS seismic catalogues to improve (or refine) seismic scenarios for historical and instrumental earthquakes for their further application to SHA studies [19] and to earthquake drills [20]. The compilation of the information started in the year 2008 [12] leading to the first edition of the catalogue in 2014 [20] and this second edition (revised and expanded) in the present year [9], resulting in the most complete edition of a catalogue on geological effects of the earthquakes at national scale [10]. The next sections explain the content, structure and the variety of graphical information included in the Spanish EEEs catalogue.

Table 1. Parametric list of pre-instrumental (1900–1963) and instrumental earthquakes (after 1963) included in the Spanish catalogue [10] with their corresponding "Quality indexes" discussed in text. Earthquakes in bold corresponds to those with "full event files"; the remainder events have only summarized information in "event files". Two earthquakes are still under study. See Figure 2 for earthquake location.

Year (CE)	Month & Day	Longitude	Latitude	Quality Qe Qi Qg	Magnitude	EMS-98	ESI-07	Location	Code Number
1911	Mar 21	1°15′ O	38°01′ N	AIC	5.7 Ms	VII-VIII		Torres de Cotilla. Murcia	43. PRE 19110321
1911	Apr 03	1°12′ O	38°06′ N	AIC	5.3 Ms	VIII	VIII	Lorquí. Murcia	44. PRE 19110403
1911	May 31	3°42′ O	37°12′ N	AIC	4.9 Ms	VII-VIII	VIII	Santa Fe. Granada	45. PRE 19119531
1917	Jan 28	1°21′ O	38°01′ N	AIC		VI-VII		Torres de Cotilla. Murcia	46. PRE 19170128
1919	Sep 10	0°50′ O	38°05′ N	AIB	5.2 Ms	VIII	VIII	Jacarilla. Alicante	47. PRE 19191010
1930	Jul 05	4°38′ O	37°35′ N	AIB	5.4 Ms	VII-VIII		Montilla. Cordoba	48. Under study
1956	Apr 19	3°41′ O	37°11′ N	IIB	5.0 Ms	VIII	VIII	**Albolote. Granada**	**49. PRE 19560419**
1964	Jun 09	2°34′ O	37°44′ N	IIB	4.8 mb	VIII	VIII	**Galera. Granada**	**50. PRE 19640609**
1993	Dec 23	2°55′ O	36°46′ N	III	5.0 Mw	VII		Adra. Almería	51. Under study
1999	Feb 2	1°50′ O	38°10′ N	III	4.8 Mw	VII	VII	Mula. Murcia	52. INS 19990202
2002	Aug 6	1°51′ O	37°53′ N	III	5.0 Mw	VI	VI	Bullas. Murcia	53. INS 20020806
2003	May 21	3°42′ O	36°541′ N	III	6.8 Mw	I	VI*	Mahón (tsunami*) Balearic Islandas	54. INS 20030521
2005	Jan 29	1°45′ O	37°51′ N	III	4.8 Mw	VII	VII	La Paca. Murcia	55. INS 20050129
2011	May 11	1°42′ O	37°43′ N	III	5.2 Mw	VII	VIII	**Lorca. Murcia**	**56. INS 20110511**

Table 2. Parametric list of historical, ancient and paleoseismic events included in the catalogue with their corresponding "Quality indexes" discussed in text. Earthquakes in bold corresponds to those with "full event files"; the remainder events have only summarized information in "Event files". Two historical earthquakes are still under study. See Figure 2 for earthquake location.

Year (CE)	Month & Day	Longitude	Latitude	Quality Qe Qi Qg	Magnitude	EMS-98	ESI-07	Location	Code Number
4200–3700 BCE		4°32′ O	36°57′ N	CCG	≥ 6.7 Mw		≥ IX	Cueva del Toro. Antequera. Málaga	1. APQ 004AC3700
1900–1550 BCE		1°29′ O	37°47′ N	GGG	6.4 Mw	≥ VIII	≥ VIII	Tira del Lienzo. Totana, Murcia	2. APQ 002AC1550
800 BCE		2°46′ E	42°20′ N	CCC			≥ VIII	Pirineos Orientales Alto Ampurdán. Gerona	3. HST 001AC800
575–500 BCE		2°46′ E	42°20′ N	CCC			≥ VIII	Pirineos Orientales Alto Ampurdán. Gerona	4. HST 001AC 575
570–500 BCE		10°00′ O	36°28′ N	CCC			≥ VIII	Andalucía, Mar Alborán. Mediterranean	5. HST 001AC500
218–209 BCE		9°42′ O	36°38′ N	GGG	≥ 8.0 Mw		≥ IX	*Lacus Ligustinus*: SW Cabo San Vicente, Atlantic Ocean	6. GEO 001AC218
040–060 CE		5°46′ O	36°05′ N	QQG	≥ 5.5 Mw	≥ VIII	≥ VIII	***Baelo Claudia*. Bolonia. Cádiz**	**7. ARQ 004060**
260–280 CE		5°46′ O	36°05′ N	QQG	≥ 5.5 Mw	≥ VIII	≥ VIII	***Baelo Claudia*. Bolonia. Cádiz**	**8. ARQ 0260280**

Table 2. *Cont.*

Year (CE)	Month & Day	Longitude	Latitude	Quality Qe Qi Qg	Magnitude	EMS-98	ESI-07	Location	Code Number
300–400 CE		1°39′ O	38°35′ N	GGG	6.8 Mw		X	**Illunum Tobarra. Albacete**	**9. GEO 0300400**
350–400 CE		3°19′ O	40°30′ N	QQG	≥ 5.5 Mw	≥ VIII	≥ VIII	**Complutum, Alcalá de Henares, Madrid**	**10. ARQ 0350400**
881 CE	May 26	8°00′ O	36°00′ N	DCC	7.2 Mw	XI-X	XI	Golfo de Cádiz. Atlantic Ocean	11. HST 08810000
949 CE				DCC				Zamora - Oporto. Atlantic Ocean	12. HST 09490000
1048 CE		0°55′ O	38°05′ N	DCQ		VIII	VIII	Orihuela. Alicante	13. HST 10480000
1169 CE	Jan 21	4°00′ O	38°00′ N	DCQ	6.0 Mw	VIII-IX	IX	Andújar. Jaén	14. HST 11690121
1356 CE	Agu 24	10°00′ O	36°30′ N	DCC	6.5 Mw	VIII		SW. Cabo San Vicente. Atlantic Ocean	15. HST 13560824
1373 CE	Mar 02	0°45′ O	42°30′ N	CCG	6.2 Mw	VIII-IX	IX	Ribagorça, Lerida	16. HST 13730302
1396 CE	Dec 18	0°13′ O	39°05′ N	BBB	6.5 Mw	VIII-IX	X	**Tavernes de Valldigna. Valencia**	**17. HST 13961218**
1427 CE	May 15	2°30′ E	42°12′ N	BBG		VIII-IX	IX-X	Olot. Gerona	18. Under study
1428 CE	Feb 02	2°10′ E	42°21′ N	BBG		IX-X	IX-X	Queralbs. Gerona	19. Under study
1431 CE	Apr 24	3°38′ O	37°08′ N	CBB	6.7 Mw	VIII-IX	≥ VIII	Sierra Elvira. Granada	20. HST 14310424
1494 CE	Jan 26	4°20′ O	36°35′ N	CBB		VIII	VIII	Málaga. Málaga	21. HST 14940126
1504 CE	Apr 05	5°28′ O	37°23′ N	BAA	6.8 Mw	VIII-IX	IX	Carmona. Sevilla	22. HST 15040405
1518 CE	Nov 09	1°52′ O	37°14′ N	BBA		VIII-IX	IX	Vera. Almaría	23. HST 15181109
1522 CE	Sep 22	2°40′ O	36°58′ N	DBB	6.5 Mw	VIII-IX	IX	W. Alhama de Almería. Almería	24. HST 15220922
1531 CE	Sep 30	2°44′ O	37°32′ N	CBB		VIII-IX	IX	Baza. Granada	25. HST 15310930
1620 CE	Dec 02	0°28′ O	38°42′ N	CBB		VII-VIII	VIII	Alcoy. Alicante	26. HST 16201202
1656 CE	Jun 07	1°14′ O	40°05′ N	BBB		VIII	VIII	Ademuz. Valencia	27. HST 16560607
1644 CE	Jun 19	0°25′ O	38°48′ N	CBB		VIII		Muro de Alcoy. Alicante	28. HST 16440619
1658 CE	Dec 31	2°28′ O	36°50′ N	DBC		VIII	X	Almería. Almería	29. HST 16581231
1674 CE	Agu 28	1°42′ O	37°41′ N	BBB		VIII		Lorca. Murcia	30. HST 16740828
1680 CE	Oct 9	4°36′ O	36°48′ N	BAB	6.8 Mw	VIII-IX	IX	**NW de Málaga. Málaga**	**31. HST 16801009**
1722 CE	Dec 27	7°46′ O	36°24′ N	CBC	6.5 Mw	VIII		Golfo de Cádiz. Atlantic Ocean	32. Under study
1748 CE	Mar 23	0°38 O	39°02′ N	AAA	6.2 Mw	IX	IX	**Estubeny. Valencia**	**33. HST 17480323**
1755 CE	Nov 01	10°00′ O	36°30′ N	DAA	8.5 Mw	XI-XII	XI	**Lisbon Event: SW Cabo San Vicente. Atlantic Oc.**	**34. HST 17551101**
1804 CE	Jan 13	3°35′ O	36°05′ N	CAB	6.7 Mw	VII-VIII	VIII	Almería, Mar Alborán Mediterranean	35. HST 18040113
1804 CE	Agu 25	2°50′ O	36°46′ N	BAB	6.4 Mw	VIII-IX	IX	**Dalías. Almería**	**36. HST 18040825**
1806 CE	Oct 27	3°44′ O	37°14′ N	ABB	5.3 M	VIII		**Pinos-Puente. Granada**	**37. HST 18061027**
1829 CE	Mar 21	0°41′ O	38°05′ N	AAA	6.6–6.9 Mw	IX-X	X	**Torrevieja. Alicante**	**38. HST 18290321**
1845 CE	Oct 07	0°45′ E	41°02′ N	BBA	5.1 Mw	VI-VII	VII	Tivissa. Tarragona	39. HST 18451007
1851 CE	May 15	2°40′ E	39°38′ N	BAA	≥ 4.5 Mw	VII	VII	NE Palma de Mallorca. Balearic Islands	40. HST 18510515
1863 CE	Jun 10	1°56′ O	37°22′ N	AAA	4.2–4.6 Mw	VI-VII	VIII	**Huércal-Overa. Almería**	**41. HST 18630610**
1884 CE	Dec 25	3°59′ O	37°00′ N	AAA	6.5–6.7 Mw	IX-X	X	**Arenas del Rey. Granda**	**42. HST 18841225**

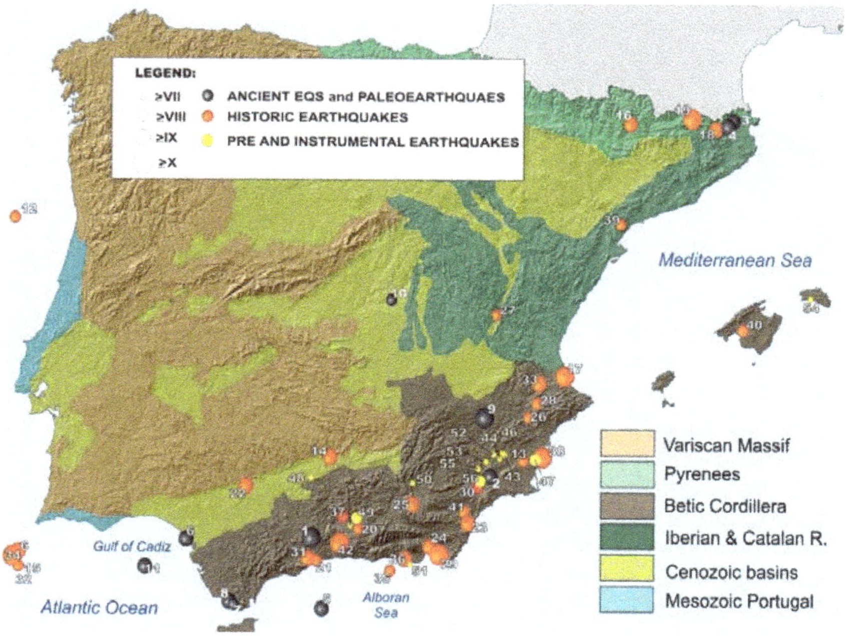

Figure 2. Geological Sketch of the Iberian Peninsula displaying the location of the 56 earthquakes listed in Tables 1 and 2 (51 of them catalogued). For earthquake identification, please see numbered lists of Tables 1 and 2.

3. Type of Catalogued Earthquakes

Conventional seismic catalogues include information for historical and instrumental earthquakes, the last commonly available from the first quarter of the 20th century for most of the European countries. In contrast, the Spanish catalogue differentiates five types (or periods) of seismic events for which the seismic information comes from different sources, with a variety of geological, archaeological or geoarchaeological data. Figure 3 illustrates the provenance and use of the different typology of macroseismic information used for the elaboration of the ESI-07 catalogue [10]. The different types of events are the following:

(a) Instrumental Earthquakes (Code: INS) occurred since AD 1963 to the present and recorded by the National Seismic Network of the "Instituto Geográfico Nacional" (IGN). These types of events have the more precise instrumental records, mainly from 1988 CE [21], complemented with quality macroseismic information. In some cases, there are specific field reports produced by official institutions (i.e., IGN, IGME and IAG) or Civil Protection for those that occurred from 1995 CE. Five events of this type have been catalogued, one of them with complete information (Table 1) which corresponds to the Lorca 2011 CE earthquake—the only one with a specific geological field report produced by the Geological Survey of Spain [22].

(b) Pre-Instrumental Earthquakes (Code: PRE) occurred between the years AD 1900 and 1962. This is a special period in Spain between the installation of the first mechanical seismographs and the implementation of the National Seismic Network ("Red Sísmica Nacional") [21]. These types of events have macroseismic information of different quality from old seismic records and field reports. In general, most of this type of events are the worst documented ones from the point of view of the geological effects. The catalogue includes seven (7) events of this type, two of them with complete macroseismic information (Table 1).

(c) Historical Earthquakes (Code: HST) are documented by written historical sources from the year 800 BCE to 1900 CE. For this type of events the information is of macroseismic character but strongly supported by the analysis of the geological effects (EEEs) reported in historic documents and newly studied archaeoseismological data (EAEs), especially for those occurred after the 1755 CE Lisbon Earthquake-Tsunami for which field earthquake reports are common. Forty-one events of this type have been catalogued, nine of them with complete macroseismic information (Table 2).

(d) Ancient Earthquakes (Code: ARQ). Mainly documented by archaeological and geoarchaeological data (EAEs) and supported by a variety of geological data (EEEs), but still not included in the conventional EMS catalogues [7] or in the updated on-line databases of the National Seismic Network [23]. As aforementioned, within this category the more ancient ones are the Neolithic event documented in "La Cueva del Toro" in Antequera (Málaga) and the Bronze Age event of "La Tira del Lienzo" (Murcia) located just above the trace of the Lorca-Alhama de Murcia fault. However, the more significant events included in this category are those that occurred during the Roman Period in Spain documented in the archaeological sites of "*Baelo Claudia*" [24] in Tarifa (Cádiz) and "*Complutum*" [25] in Alcalá de Henares, 25 km south of the city of Madrid. There are five (5) events catalogued for this category, three of them with complete macroseismic information (Table 2).

(e) Paleoearthquakes or paleoseismic events (Code: GEO) with only geological record, albeit in some cases supported by limited geoarchaeological data. This type of event covers from the most ancient historical periods in Spain (c. 10th century CE) to the onset of the Holocene epoch (c. 11.2 ka BP) and they are not included in the conventional seismic catalogues as commonly occurs with the ancient earthquakes. These events are documented in scientific papers published during the 21st century, and they correspond to fault-trenching analyses (primary surface ruptures) or EEEs analysis (primary and secondary effects). Those related to fault-trenching studies correspond to surface-faulting events (≥7.0 Mw) that occurred during the late Holocene. However, there are fault-trenching data for some ancient earthquakes such as the Tobarra (Albacete) that occurred around the 500 CE [26] or Historical events such as the Arenas del Rey (Granada) that occurred in 1884 CE [27]. Another case of paleoearthquake largely documented by secondary EEEs is the 218 BCE Lacus Ligustinus event recorded in the present "Doñana Marshlands" (an ancient lake by that time) at the Guadalquivir river outlet in the Gulf of Cádiz (Atlantic Ocean) [28]. These authors identify an earthquake-tsunami event similar to the 1755 CE Lisbon event flooding the ancient roman embayment and breaking the existing littoral spit-bars. Other authors document this event in turbidite layers obtained in deep sea cores in the Atlantic ocean near the Gorringe Bank [29] or in other marshlands of the Gulf of Cádiz, such as the Odiel and Tinto river mouths at Huelva or the Guadalete river mouth near Cádiz [30,31]. It is to note that the Lacus Ligustinus event is the unique geological event with historical written records describing the destruction of Cádiz just before the march of Hannibal to Rome across the Alps in the beginning of the Second Punic War in 218 BCE [32]. All these paleoseismic events are supported by a significant number of geochronologic data (^{14}C, OSL and TL data), as is also the case for most of the ancient earthquakes further documented by archaeological data.

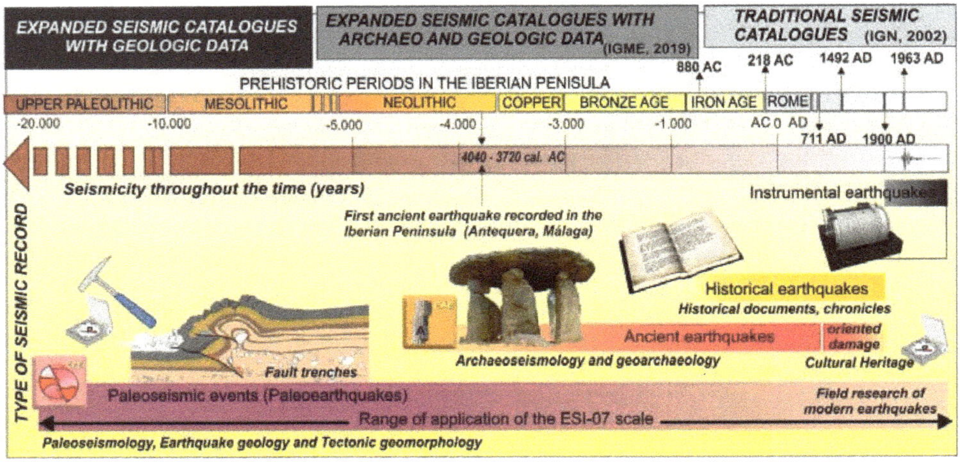

Figure 3. Type of information and temporal extent of the different types of earthquakes (paleo, ancient, historical, pre-instrumental and instrumental) in relation to the time-record for the different prehistoric and historic periods in the Iberian Peninsula. The temporal scale highlights representative historic cut-off dates for the type of information of the catalogued events.

4. Type of Information for the Catalogued Earthquakes.

The written documentary information is unequal for the different historical periods of Spain [8]. The conventional seismic catalogues only include eighteen (18) earthquakes previous to the sixth century CE, most of them suspected strong events that occurred in the Gulf of Cádiz, the Atlantic Portuguese coast or the Pyrenees [7]. The oldest ones (years 800 to 500 BCE) correspond to suspected events that occurred near the first settlements of the Phoenicians in the Pyrenees (*Emporiom*, Gerona) or in the Gibraltar Strait (*Gadir*, Cádiz). However, most of these old historic earthquakes correspond to those that occurred during the Roman period (third century BCE to fifth century CE) in the southern zone of the Iberian Peninsula. More consistent historical data appear from 800 CE by Muslim historians [5], but is not until the Reconquest of the Iberian Peninsula by the Catholic Kings in the 1492 CE, when earthquake data start to be really of quality and representative. Fifty-eight earthquakes are catalogued before this date, half of them (29) for the period between the 11th and 16th centuries CE [10]. The number of historical events catalogued between 1500 and 1800 CE is around 500 which contrast with the about 1200 ones catalogued for the 19th century alone [19]. These numbers help to illustrate the unequal written historical information for seismic events in Spain, even though a majority of the strong earthquakes (≥VIII EMS) occurred after the 10th century can be considered catalogued. This situation is not the same for those historic events that occurred before the 10th century, when the near absence of written seismic information is noticeable including the roman and visigothic periods (218 BCE–711 CE). The low quality of historical data for these period implied that most of these historical earthquakes were removed from the official seismic database of the IGN [23]. After the critical review of Udías [5], only 32 earthquakes remain catalogued before 1500 CE, all them within the present Common Era (CE), and 23 of them occurred in the period 1000–1500 CE (Figure 4). This seismic account contrasts with the about 480 events catalogued for the period 1500–1800 CE and the near 1200 earthquakes catalogued for the 19th century (excluding foreshocks and aftershocks events) still remaining in the official seismic databases (Figure 4) [23].

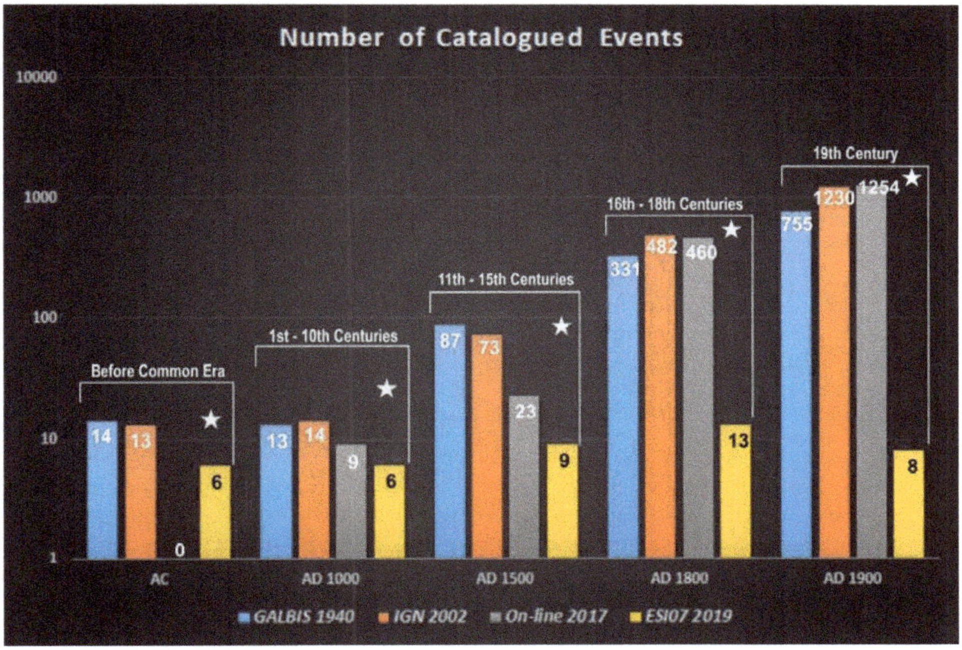

Figure 4. Number of catalogued earthquakes in the different seismic catalogues of Spain for different temporal periods. Stars and yellow bars identify the strong earthquakes included in ESI-07 Catalogue for the Iberian Peninsula [10] reviewed in this paper.

Figure 4 displays the evolution on the number of earthquake events catalogued for the different historic periods of interest discussed here. It is evident than independently of the nature of seismic information (written sources, archaeological or geological), there is a clear incompleteness of the record of seismic events before the year 1500 CE, and especially for those that occurred prior to 1000 CE. This is a typical artifact present in nearly all the seismic catalogs for the Mediterranean region [33].

These modern seismic practices largely contrast with the increasing number of ancient events recovered from paleoseismological and archaeoseismological studies for the visigothic, roman, phoenician periods and earlier ones (i.e., Bronze Age, Neolithic) [10] in Spain. It is for these ancient periods that geological and archaeological information is critical to recuperate old "lost events" no documented by written sources, leading to expand the seismic catalogues to the most recent geological record [34]. As evidence of these trends, since the end of the compilation of information for the published catalogue, three more ancient earthquakes were recovered. This is the case of the roman events of *Idanha* (Portugal) and *Mulva-Munigua* (Sevilla) that occurred during the 4th–5th centuries CE [35,36], as well as the suspected seismic event of Phoenician age (7th century BCE) documented in the Lower Segura Basin at Alicante [37].

Nevertheless, the geological and archaeological research of historical earthquakes has also been largely incorporated in the published catalogue, especially that related to damage to the historic heritage. This ESI-based investigation complements the EMS data allowing upgrading and completing macroseismic information especially in the county-side and depopulated areas, as does the geological investigation of instrumental earthquakes. This last will help to improve and refine data to update the ESI-07 scale in the future, but also to understand the assemblages and patterns of oriented building damage in relation to well-constrained seismic sources by instrumental data, as was the case of the 2011 Lorca earthquake in Spain [22,38].

The different nature of the information incorporated in the catalogue made necessary the use and combination of different types of macroseismic scales or classifications, such as the earthquake environmental effects (EEEs) of the ESI-07 scale, the destructive effects, or the tsunami environmental effects (TEEs) recently classified in the TEE-16 Scale [39]. These last are a complement and refinement to the ESI-07 scale on paleo-tsunami records. In the same way, the archaeological record of earthquake archaeological effects (EAEs) scaled in the works of Rodríguez-Pascua et al. [13,40] have been considered. Regarding EEEs in the ESI-07 scale, the Spanish catalogue [10] differentiates between tsunami effects (TSU) and anomalous waves (AW) in inland water bodies, such lakes, dams, rivers, etc. (Figure 1). In the same way, ESI-07 hydrologic effects have been split in two different categories: (1) hydrological effects (HA) such as changes in level or flow rates in water wells, springs and fountains; and (2) hydrogeological effects (HD) related to changes in physico-chemical properties, temperature, turbidity, etc. The rest of secondary EEEs, such as ground cracks (GK), slope movements (SM), liquefaction (LIQ) and other effects (OT), maintain their status and meaning as originally defined in the ESI-07 scale (Figure 1) [3].

5. Quality of the Seismic and Geological Information for the Catalogued Earthquakes

For all the catalogued earthquakes, source location parameters (date, origin time and geographical coordinates) correspond to those published in the catalogue of historical earthquakes of the IGN [7], except for those events not catalogued by this institution. Likewise, we consider convenient to introduce the indicative quality parameters on the "epicentral source location" (Qe) and "maximum epicentral intensity" (Qi) implemented for the abovementioned IGN catalogue, as well as the evaluation of the related theoretical "Moment Magnitude" (Mw) estimated by means of empirical relationships of magnitude–intensity specifically developed for that catalogue [7]. In spite of the publication of more recent source location and size parametric data on the historical and pre-instrumental seismicity for Spain [41], we preferred to keep the original source parameters listed in the official catalogue and databases of the IGN [7,23] since the new published data are not official data.

5.1. Quality of Source Location Determinations (Qe)

Regarding the source location data, the geographical coordinates of the macroseismic epicenters are given as sexagesimal degrees and full minutes, equivalent to a precision of about ± 1.5 km sufficient for the entire historical catalogued period [7]. However, for the most recent earthquakes that have occurred since the year 1900 more detailed source location data were extracted from the IGN databases [23].

The quality parameter for the source location (Qe) presents five different classes with the following quality indexes:

- Quality A (location error <10 km): macroseismic epicenter surrounded by several macroseismic data-points, whose density allows estimating the source location with an uncertainty ± 5 km. This category, of course includes those earthquakes instrumentally recorded after the year 1963 CE (labelled as Quality I).
- Quality B (location error <20 km): macroseismic epicenter defined by a small number of macroseismic data-points, which density allows estimating the source location with an uncertainty ± 10 km. The epicentral determinations for some of the earthquakes recorded between 1900 and 1963 CE (Spanish pre-instrumental period) are very imprecise and most of the events occurred in that period hold B quality (some even C).
- Quality C (location error 20–50 km): macroseismic epicenter located in a continental area defined by a single macroseismic data-point or by a few ones very distant each other. These can present an uncertainty above ± 25 km, normally within a radius of 50 km.
- Quality D (location error >50 km): macroseismic epicenter located in a coastal locality as the unique macroseismic data-point, or those that occurred offshore which can be poorly defined assuming location uncertainties above ± 25 km. This quality-class also includes very ancient

historical earthquakes recorded by vague and generic written descriptions refereed to very broad regions, such as Gulf of Cádiz, Atlantic Ocean, Mediterranean coast and the Pyrenees. This is the general case for most of the catalogued historical earthquakes before the 10th century BCE around the Iberian Peninsula.

- Quality Q (location error <20 km): Events not recorded by written sources with an only point, or very few points with archaeoseismic and/or paleoseismic information. This quality-class includes those earthquakes with limited archaeoseismic information indicating ground shaking (building fabric EAEs) or occasional paleoseismic information related to secondary earthquake environmental effects (EEEs), avoiding the identification of any reliable epicenter. In these cases, the provided geographical coordinates correspond to the more relevant archaeoseismic or paleoseismic data-points related to the catalogued event. In most cases, these correspond to near-field effects occurring to epicentral distances down to 20 km, and by the acquired experience, the uncertainty would be even down to ±5 km considered for the "Quality A" events.
- Quality G (location error <10 km): The available information is only of geologic nature, providing the geographical coordinates to the investigated points with paleoseismic information regarding primary EEEs of the ESI-07 scale, such as surface ruptures, fault scarps or fault-trenching analysis defining the seismic source. Consequently, this quality-class may hold more accurate locations than the "Quality A" events. In the case of the record of an important number of secondary EEEs the provided geographical coordinates are those corresponding to the zones where EEEs display largest dimensions, frequency or number (consequently intensity). These zones generally define areas no larger than a few squared kilometers wide, where the true macroseismic epicenters are located. This is not valid for strong offshore events or zones with an important proved susceptibility to ground shaking amplification.

These two last quality-classes (Q and G) were introduced in the Catalogue of geological effects of earthquakes by Silva et al. [10,12], whilst the previous classes (A, B, C, and D) are those already developed for the Catalogue of historical earthquakes in Spain [7]. Classes Q and G also include the generic date of the earthquake as provided by the battery of isotopic dates (^{14}C, Th/U, OSL, TL, ESR, etc.) available for the archaeoseismic or paleoseismic data-points, as well as morphometric or archaeological relative dates for particular cases.

5.2. Quality of Earthquake Intensity Assessments (Qi)

The new published Spanish ESI-07 catalogue introduces the maximum intensity value assessed from the ESI-07 scale, but earthquake intensities estimated from other macroseismic scales are also listed and considered. This is the case for the intensity values corresponding to the European Macroseismic Scale (EMS-98) [7] and the Mevdéved–Sponheuer–Karnik Scale (MSK-81) applied in previous Spanish catalogues [42]. As is known, the MSK and EMS scales are nearly equivalent, with the exception that EMS mainly considers building damage, with the exclusion of environmental damage and geological effects of earthquakes for intensity assessments, which is one of the causes for the development of the ESI-07 scale [2].

It is also convenient to indicate that the EMS-98 scale does not consider damage on buildings of historical heritage (Castles, Fortress, Cathedrals, Old Churches, etc.) for intensity assessments [4]. This is a serious inconvenience for evaluating the intensity of historical earthquakes for which the most complete information comes from the description of the damage that occurred in this type of building. This fact has an important impact for earthquakes that occurred before AD 1500, for which written documentation is sparse, and only the analyses of EAEs preserved in these patrimonial buildings or archaeological sites allow estimating the earthquake size [38]. Accordingly, intensity assessments developed for the catalogue display better similarity to those of the ancient catalogues with the MSK scale than the most recent ones based on the EMS scale. In general, an old maximum intensity IX MSK is nowadays converted in a maximum intensity VIII–IX or VIII EMS in the most recent catalogues and databases of the IGN [7,23]. This generally occurs for most of the historical earthquakes between VIII and X intensity in the previous MSK catalogues. This means that the application of the EMS-98 scale in Spain resulted in the reduction of one to half degree with respect to those corresponding to the MSK-81 or ESI-07 scales, which has special negative incidence in the more recent SHA studies in Spain [11]. These evaluate horizontal ground accelerations from the conversion of EMS intensities from historical events (last 450 years) by means of complex combinations of empirical relationships developed for other traditional (i.e, MM or MCS) macroseismic scales, inducing very probable underestimations for the ground movement [19]. As stated in the published catalogue [10], the EMS-98 scale is a nice technical guide for the quantification of building damage produced by recent earthquakes, but of limited application for historical events. Data on old historical events in Spain mainly provide information on environmental damage and significant buildings, now catalogued in the national cultural heritage lists, and the EMS scale does not consider damage on historical buildings and environmental effects for intensity assessments [4]. It is necessary to note that the intensity degrees ≥IX EMS-98 mainly base the intensity assessments on the effects on metallic structure buildings or quake-resistant ones, since the rest of buildings are destroyed at this stage. However, these types of buildings are virtually absent for all the historical periods of Spain, where the first modern seismic code was introduced at the end of the 20th century in the year 1994 (NCS-94). Consequently, the EMS-98 scale has a poor application for historical earthquakes not only in Spain, but also for most of the European territory.

The published catalogue attempts to document and graphically illustrate (photos, diagrams, sketches, drawings, maps, etc.) in the most exhaustive mode possible the large variety of EEEs recorded during the 51 stronger events that occurred in Spain. These have normally intensities ≥VIII for the pre-instrumental period, but some well-documented events of intensity VII that occurred after the year 1900 produced significant EEEs (Table 1). Nevertheless, the catalogue also lists the other important earthquake size parameter: the magnitude. In this case, all the events before the year 1900 have estimated magnitude (Mw) values coming from empirical relationships published in previous Spanish catalogues or recent scientific papers. In many cases the estimated Mw values may appear as bracketed values (i.e., 6.2–6.6) since published data provide different, but similar, magnitude values. This is relevant for those earthquakes only catalogued from paleoseismological data after fault-trenching analyses, where Mw evaluations came specifically from empirical relationships based on the work of Wells and Coppersmith [43]. Instead, events that occurred after the year 1900 are instrumentally recorded and magnitude values are well constrained.

In this way, the catalogue differentiates several quality classes in relation to the maximum known or geologically recorded intensity using the ESI-07 scale. In addition to the intensity points assessed by means of EEEs or EAEs, quality class also considers all macroseismic data-points with EMS information. The quality classes (Qi) are the following:

- Quality A: Density of points with macroseismic information is sufficient to undertake reliable intensity assessment. Normally, this class includes those earthquakes with more than 30 data-points, which generally refers to events that occurred after the 17th century CE.
- Quality C: The available points with macroseismic information are insufficient to undertake reliable intensity assessments. In some cases, there is a single data-point. These earthquakes include old events historically documented, but with little to no geological information, normally occurring before the year 1000 CE.
- Quality B: The accuracy of the intensity assessments is between the two previously mentioned quality classes. This is normally the case of those events that occurred between the 10th and 17th centuries CE. However, for some of them, macroseismic data-points are up to 15–20, including a variety of geological information on secondary EEEs.
- Quality D: EEEs ambiguously or roughly described in historical documents in reference to a region instead to particular localities, which lack modern geological studies. This is normally the case of all the catalogued events for the first millennia of the present era, but also most of the earthquakes that occurred before the present common era, with poor to no geological evidence. Descriptions such as "mountains and plains opened; the earth guzzled and engorged villages and hills; the sea

shrank away, islands and islets disappeared; many villages in the southern and western coasts of Spain were totally ruined", characterize this type of events. This corresponds to the translation of original Arabic text for the 881CE earthquake affecting the old Al-Andalus in South Spain.

- Quality G: EEEs related to historical or prehistoric events, normally not included in the conventional seismic catalogues. They are only documented by geological information coming from recent scientific paleoseismic and archaeoseismic published papers. In general, this quality class includes earthquakes that occurred during the first millennia of the present common era, particularly those that occurred during the Roman and Visigoth period in Spain, but also those only geologically documented.
- Quality I: EEEs linked to recent earthquakes that occurred during the 20th century and the onset of the 21st century. These events are instrumentally recorded and geological effects have been documented "in situ" after the earthquake by scientific commissions or groups of earthquake geologists interested in the event. This quality class normally has the best Qe, Qi and Qg quality indexes. However, as aforementioned some of the pre-instrumental period events (1900–1962 CE) have a poor Qg index with poorly defined descriptions of geological effects, similar to those of Quality B.

5.3. Quality of the Geological Information (Qg)

The published Spanish catalogue introduces a third quality index related to the value of available information on geological and environments effects of the earthquakes (EEEs) listed in the ESI-07 Scale. This quality index was specifically developed for this catalogue and considers the following quality classes:

- Quality A: EEEs widely documented by technical or field reports, scientific papers and journals of the earthquake epoch that have undergone modern geological research published in scientific papers, allowing the quantification of primary and or secondary effects. Generally, this class includes those events that occurred after 1800 CE (19th century), although some 18th century events were also well documented in this time (i.e., 1755 CE Lisbon Earthquake).
- Quality B: EEEs vaguely described in historical documents, but subject of modern published geological research. These events mainly correspond to those that occurred between the 15th and 19th centuries and most of those that occurred during the so-called "pre-instrumental period" in Spain (1900–1962 CE). These normally have good Qe and Qi indexes, but poor geological information, since the progressive implementation of seismographs in Spain during the first half of the 20th century resulted in increasingly worse field-reports.
- Quality C: EEEs ambiguously or roughly described in historical documents in reference to a region instead to particular localities, which lack modern geological studies. This is normally the case of all the catalogued events for the first millennia of the present Common Era, but also most of the earthquakes that occurred before the present Common Era, with poor to no geological evidence, with historical descriptions similar to that transcribed for the Quality C events of the Qi index.
- Quality G: EEEs related to historical or prehistoric events, normally not included in the conventional seismic catalogues. They are only documented by geological information coming from recent scientific paleoseismic and archaeoseismic published papers. In general, this quality class includes earthquakes that occurred during the first millennia of the present common era, particularly those that occurred during the Roman and Visigoth period in Spain, such as those recorded in Baelo Claudia [44], but also those only geologically documented.
- Quality I: EEEs linked to earthquakes that occurred during the 20th century and the onset of the 21st century. These events are instrumentally recorded and geological effects have been documented "in situ" after the earthquake by scientific commissions or groups of earthquake geologists interested in the event. This quality class commonly holds the best Qe, Qi and Qg quality indexes. However, as aforementioned some of the events of the pre-instrumental period

(1900–1962 CE) have a poor Qg index with poorly defined descriptions of geological effects, similar to those of Quality B events.

5.4. Overall Quality for Earthquake Information

Taking into account these three different quality indexes (Qe, Qi, Qg) we have different quality-type events for which macroseismic information allows labelling the earthquakes as credited or poorly credited. In this way, "triple A" (AAA) and "triple I" (III) events can be considered as fully credited events with the most accurate assessments for source location, date and intensity. However, GGG and GQG events are credited with a similar quality than the AAA ones, these are also well-constrained events, but source locations in the near field are still doubtful (Tables 1 and 2). On the contrary, CCC or DCC events are the poorly constrained events. In fact, most of this type of events included in the ESI-07 catalogue [10] have been recently removed from the official macroseismic databases of the IGN [23]. In that exercise, some of the well constrained geologically documented GGG, BGG or QGG earthquakes included in the new published catalogue [10] were also removed from the official macroseismic databases, since these still do not consider geological information [8]. This is the case of the 218 BCE *Lacus Ligustinus* event, which aside from being historically documented, has been identified by tsunamite deposits in all the littoral spit-bars and marshlands of the Gulf of Cádiz, as well as in offshore turbidite deposits near the suspected earthquake source area [28–32]. Whatever the case, all the double AA, double BB, AB or AI–BI events certainly occurred, but source parameters present some uncertainty. Only those labelled as CCC or some DCC are doubtful events (Table 2). These mainly correspond to those that occurred before the present common era (BCE) which are vaguely documented by historical written reports in relation to the first Phoenician and Greek settlements on the Spanish coast, around or fairly before the 500 BCE in the Pyrenees (*Emporium*) and the Gibraltar Strait (*Gadir*).

6. Event Files: Summarized Information for the 51 Catalogued Earthquakes

The information for each catalogued event has been summarized in a two-page file format (Event files) with one-page versions in Spanish and English (Figure 5). These files are available for the 51 catalogued events. They display a numerical summary of the abovementioned source and size parametric data and quality indexes. In the case of instrumental events (INS Code), data on source depth, location errors, recorded magnitude and occurrence of foreshocks and aftershocks is also included in the parametric summary. The maximum intensity in the different scales applied in Spain (MSK-80, EMS-98 and ESI-07) are displayed for all the events, as well as summarized data on the number and size of catalogued EEEs and significant EAEs where pertinent. Event files also include a numerical list of the historical documentation referring to the catalogued earthquake listed in the official catalogue of the IGN [7], as well as five to six key bibliographic references documenting earthquake environmental and building damage.

All the events have a Spanish "resumen" and an English "abstract" summarizing the type and number of the ESI-07 catalogued EEEs, as well as a brief description of the extension/nature of environmental damage (ESI-based) and general information on affected localities and building damage (EMS-based). The info also includes the location of the earthquake within the different "seismotectonic zones" (GM12 Zones) defined for the new "Upgraded Seismic Hazard Maps for Spain" elaborated after the 5.1 Mw 2011 Lorca earthquake to update the present Spanish Seismic Code [11]. Additionally, the "Event files" include a simplified geological map of the Iberian Peninsula displaying the location of the earthquake epicenter within the different structural units of Spain depicted in Figure 2: Alpine Cordilleras (Betics, Pyrenees, Iberian range, Catalan ranges); Palaoezoic-Variscan massif (west Iberia); and Cenozoic sedimentary basins (Tagus, Duero, Guadalquivir, Ebro basins). In the endnote of each "Event file" are displayed the name(s) and affiliation(s) of the author(s) in charge of the compilation and structuring of the macroseismic information.

GEO 001AC 218: *LACUS LIGUSTINUS* (Océano Atlántico). 218 – 209 AC GEOLÓGICO

FECHA AMD	HORA Y MIN.	LONGITUD	LATITUD	MAG (Mw/Ms/M*)	EMS 98	MSK 64	ESI 07	Q_e	Q_i	Q_g	LOCALIZACION TERREMOTO
218 - 209 AC	-	9°42′ O	36°38′ N	8,0 M*	-	-	X	G	G	G	*Lacus Ligustinus.* SW Cabo de San Vicente (O. ATL.)

Método de datacion:	CATALOGADO Ref. Bibliográfica: Galbis (1932), Luque *et al.* (2002), Lario *et al.* (2011), Rodríguez Vidal *et al.* (2011)							Tipo Evento:	GEO (HST)
Prof. km	-	±EZ	-	±EH-	-	M_o	-	Eventos asociados		Agencia:	IGN
EEE Primarios	NO	Longitud Falla	-	Desplazamiento	-			Tipo Desp.:	SIN DATOS
EEE Secundarios	SÍ	Categoria ESI07	☐GK ☒SM ☐LQ ☐HD ☐HA ☐WA ☒TS					Area afectada:	> 2.500 km² (Litoral Español)
Otros EEE Ambientales	NO	☒ OTROS EFECTOS (OT): Las descripciones indican una retirada y posterior inundación marina de envergadura que provocó serios daños y mortalidad en la fauna marina: *"se arrojaron fuera del mar multitud de pescados".*							
EAE Efectos Arqueosismologicos	NO	☒ EAEs: Antiguos asentamientos humanos anteriores al siglo III AC fueron abandonados (Flecha de La Algaida) en el litoral del antiguo *Lacus Ligustinus* romano (Marismas de Doñana), no volviéndose a ocupar hasta el siglo I AD.							

Documentación IGN 2002: 165, 166, 280, 335, 405, 426, 504, 545, 687

Referencias bibliográficas básicas:

Galbis Rodriguez, J. (1932): Catálogo Sísmico de la zona comprendida entre los meridianos 5° E y 20° W de Greenwich y los paralelos 45° y 25° Norte. Instituto Geográfico Catastral y de Estadística, Madrid.
Grácia, E. *et al.* (2011): Holocene earthquake record offshore Portugal (SW Iberia): Testing turbidite palaeoseismology in a slow-convergence margin. Quaternary Science Reviews, 29, 1156-1172.
Lario, J., *et al.* (2011): Holocene paleotsunami catalogue of SW Iberia. Quaternary Intern. 242: 196-200.
Luque, L. *et al.* (2002): Sedimentary record of tsunami during Roman times, Bay of Cadiz, Spain. Journal of Quaternary Science 17 (5–6), 623–631.
Rodriguez-Vidal, J. *et al.* (2011): Geomarkers of the 218-209 BC Atlantic tsunami in the Roman Lacus Ligustinus (SW Spain): A palaeogeographical approach. Quaternary International, 242: 201- 212.
Ruiz, F. *et al.* (2013): Morpho-Sedimentary evidence of Holocene Tsunamis in southwestern Spanish estuaries: A summary. Actas VIII Reunión Cuaternario Ibérico, 145 – 149.
Silva, P.G.*et al.* (2015): Seismic palaeogeography of coastal zones in the Iberian Peninsula: Understanding ancient and historic earthquakes in Spain. Cuaternario y Geomorfologia, 29. 31 – 56

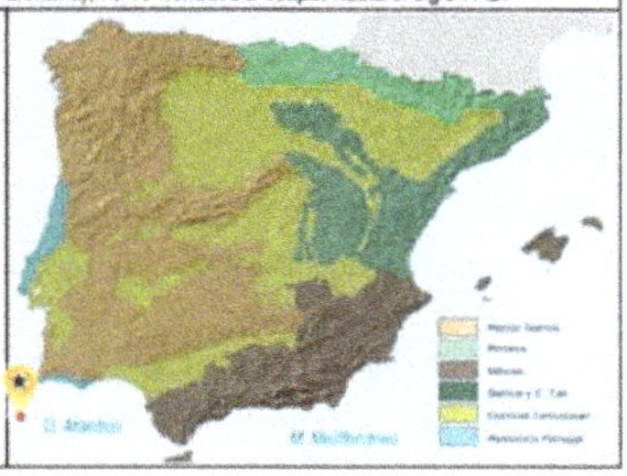

EARTHQUAKE ABSTRACT

EVENT	LONGITUDE	LATITUDE	MAG (Mw/Ms/M*)	EMS 98	MSK 64	ESI 07	Q_g	SEISMOTECTONIC ZONE IGN 2012 (GM12 ZONES)
GEO 001BC 218 *Lacus Ligustinus.* SW Cabo de San Vicente	9°42′ W	36°38′ N	8.0 M*	-	-	X	G	Atlantic Ocean; Gorringe Bank (Zone 52)

EVENT CATALOGED IN THE IGN DATA BASES Ref.: Galbis (1932), Luque *et al.* (2002), Grácia *et al.* (2010), Lario *et al.* (2011), Rodriguez Vidal *et al.* (2011), Ruiz *et al.* (2013)	Event Type:	GEOLOGIC (HST)

GENERAL DAMAGE: Earthquake felt in all the Atlantic littoral of the Gulf of Cadiz (SW Spain) especially affecting to the ancient town of Cádiz (Gades) in that time most important locality of the area located in an island within the estuary of the Guadalete river. The description of the event by Galbis (1932), extract from the historic chronicles of Florián de Ocampo (1553) indicate that: *"The island of Cadiz, but also all the littoral zone of Andalucia suffered strong earthquakes or tremors, inducing the collapse of buildings, the death of the inhabitants and producing terrible wrongs; the sea flooded many places that were first uncovered by the waters, throwing out a multitude of fishes, some of them common but others never seen".* This historical chronicle is in the context of the initiation of the surprising overland journey of Hannibal Barca from Quart Hadast (Carthago Nova) across the Alps to Rome in the beginning of the Second Punic War (218 BC).

ENVIRONMENTAL DAMAGE: The historical descriptions indicate the occurrence of an earthquake generating tsunami, which severely affected the littoral in the Gulf of Cadiz, and especially the old insular emporium of Gades (Cádiz). There are no descriptions associated with earthquake damage, but a relatively good geological record of tsunami deposits associated with both coastal estuarine Tinto-Odiel (Huelva); Guadalete (Cádiz) and the Guadalquivir marshes (Doñana) and in ocean turbiditic levels (submarine landslides) recorded near the Gorringe Bank. Core data from coastal areas document centimeter-thick high energy laminar deposits (tsunamites), with a mixture of estuarine and marine fauna (planktonic and benthic), recorded as far as 12 to 17 km inland from the present coastline. Geological data indicate the occurrence of a tsunami of at least 5 m (run-up) causing significant environmental damage in the aforementioned estuarine zones, such as breakage of spits-bars, changes of drainage patterns within the marshlands and important erosion phenomena on beaches and coastal dune systems (Doñana), comparable to those produced by the AD 1755 Lisbon earthquake-tsunami event. Geological data indicate the record of X ESI-07 intensity, along at least the c. 140 km separating the Tinto-Odiel (Huelva) and the Guadalete (Cádiz) estuaries.

☐ ***Surface ruptures (SR):*** *No Reported.*
☐ ***Tectonic uplift / subsidence (UP):*** *No Reported.*
☐ ***Ground cracks (GK):*** *No Reported.*
☒ ***3 Slope movements (SM):*** *Submarine landslides in the vicinity of the Gorringe Bank (Atlantic Ocean) around the suspect epicentral area. The event is recorded as turbidite levels, up to 20 cm thick, in 3 of the 4 drill-cores conducted in the area.*
☐ ***Liquefaction processes (LQ):*** *No Reported.*
☐ ***Hydrogeological anomalies (HD; HA):*** *No Reported.*
☒ ***7 Anomalous Waves and Tsunamis (7 TS):*** *Record of tsunami deposits, and sedimentary levels produced by erosion-reworking of coastal beaches and spit-bars in the Tinto-Odiel estuaries (2 core-probes) and Guadalquivir marshlands (Doñana, 5 core-probes). Washover fans in the Guadalete estuary and Valdelagrana spit-bar (Cádiz: 2 records). Different effects associated with both the tsunami erosive record and/or backwash under study to be catalogued.*
☒ ***1 Other effects:*** *Significant retreat and subsequent flooding by the sea of coastal areas, causing serious damage and mortality of the marine fauna.*
☒ ***1 Archaeoseismic damage:*** *Some coastal pre-Roman (7th – 3rd centuries BC) human settlements around the ancient* Lacus Ligustinus *(Doñana marshlands), were abandoned (i.e. La Algaida), and later, after the tsunami new Roman saltworks were installed from the 1st century AD.*

☒ ***Geological Data:*** *There is and relevant geological record on the occurrence of a significant tsunami event in the more important estuarine zones of the Gulf of Cádiz. ^{14}C dating indicate the occurrence of a tsunami during roman times with a run-up of 5 m, bracketed in 218-210 BC (Lario et al., 2011; Rodriguez Vidal et al., 2011). Data from offshore drill-cores around the Gorringe Bank (suspect seismic source) record different tsunami-type events triggering submarine landslides of Holocene age (Gràcia et al., 2010). Three of the four cores conducted in the area record the roman event (Event 5: E5) dated in 1,980 – 2,280 yr. BP, with an estimated magnitude ≥ 8.0 Mw.* ***Quality of Geological Information: Class G.***

Research / Data compilation by: *Pablo G. Silva (USAL); Javier Lario (UNED). Updated 2019.*

Figure 5. Example of an Event file (English version) produced for the 51 catalogued events. This corresponds to the 218 AC *Lacus Ligustinus* earthquake [28] with historical, geological and archaeological records.

The 51 catalogued events appear listed and ordered temporally (by date). Each "Event file" displays a code identifying the type of catalogued event (GEO, ARQ, HST, PRE or INS) followed by the corresponding date in numerical format YYYYMMDD (year/month/day); as an example, the most recent event catalogued, the 2011 Lorca earthquake has the code INS 20110511 (Table 1). The code for paleoseismic (GEO), archaeoseismic (ARQ) or historical (HST) events that occurred before the present Common Era (BCE or BC), for which there is only information of the year, the code identifies the millennia and the year (Table 2).

In the case of paleoseismic events with bracketed ages resulting from the uncertainty errors of the applied dating method documenting the earthquake always use the more recent date (year). As an example, the date of the most ancient catalogued earthquake (La Cueva del Toro, Antequera, Málaga) is bracketed by ^{14}C dating in the fourth millennia BC (4200–3700 BCE) and has the code ARQ 04 3700BC (Table 2). Figure 5 displays the structure and format of the English version of the "Event files" included in the published catalogue.

7. Full Event and EEE Files: Full Information for Selected Catalogued Earthquakes

The second edition of the catalogue offers full information for 16 selected events representative of each earthquake type or period (GEO, ARQ, HST, PRE, INS). For these events, a complete analysis of the catalogued environmental (EEEs) or archaeoseismological (EAEs) effects is available. A total amount of 1027 effects has been identified of which 840 are EEEs and 187 EAEs. From this large amount of catalogued effects, about 680 are briefly analyzed in "Event files" and 322 in detailed individual "EEE files" (Figure 6). These last cover all the typologies of EEEs observed in the ESI-07 scale [3], including 84 tsunami effects analyzed by means of the upgraded TEE scale [39]. Some of the earthquakes with a large amount of information over the 75 effects, such as the 2011 Lorca (79 EEEs), 1829 Torrevieja (78 EEEs) and 1755 Lisbon (674 EEEs) have been simplified to a representative number of about 45–50 "EEE files".

Aside from the information summarized in the "event file", earthquakes with complete information have a section in which the original descriptions of the early seismic catalogues for Spain (e.g., Galbis Catalogues) are literally transcribed including the prime references used by Galbis [45]. The "full event files" also include descriptive and interpretative sections in which the source location and maximum intensity of the earthquakes are evaluated in relation to the catalogued EEEs, EAEs, and data from other recent scientific publications. In those cases in which published geological data, ESI-07 parametric evaluations (EEEs) or archaeoseismic oriented damage (EAEs), point to different source locations than those listed in the IGN catalogue [7] the parametric data of the earthquake is disused in the interpretative section. For those recent earthquakes not included in the Galbis catalogues [45], previous to AD 1940, the section of the original descriptions by Galbis have been substituted by a section of "general information" summarized from available seismic reports or field reports developed by different institutions (IGN; IGME; IAG or Civil Protection). For ancient or paleoseismic events not catalogued by Galbis [45], the general information section includes data on the archaeological site or fault-trench site documenting the respective earthquakes. The descriptive and interpretative sections for these full-information events are structured in the style of four–five-page short papers and include a complete reference list of historical sources and recent scientific publications analyzing the corresponding earthquake.

LOCATION	SITE CODE	HST 18841225.14.03 *14.03. Ventas de Zafarraya*	EARTHQUAKE	HST 18841225 ARENAS DEL REY. GR	EEE ESI-07	SURFACE RUPTURE: SR/10

Locality Code	Longitude	Latitude	Altitude	Affected Area	Epicentral Distance	Location of the Epicenter	Macroseismic Data IGN MAXIMUM INTENSITY and MAGNITUDE
14 / Ventas de Zafarraya (Granada)	4° 3'27.97"O	36°56'22.67"N	1145 m	14 km² (nearly linear)	9.6 km	NORTHWEST	IX-X EMS/ 6,5-7,5 M*

DESCRIPTION			
Geomorphologic Context	Southern Slope of a tectonically controlled polje (karstic) depression	Ground Slope	Steep Slope (50-70%)
Geologic description	Steep slope (50-70%) on hard stratified Lower Jurassic limestones in the footwall of the Ventas de Zafarraya Fault, the suspect seismic source of the earthquake. The fault bound a c. 15 km length karstic depression (poljé), related to dextral strike-slip faults with associated normal faulting. The fault is closely related to the limit between the external and internal zones of the Betic Cordillera in the studied zone.		

GROUND				
Bedrock Lithology	**Lithology of affected materials**	**Internal structure**	**Consolidation degree**	**Ground Coefficient NCSE02**
Dolomies and limestones	Thin slope debris deposits	Stratified	Strongly cemented	II (1,3)
Bedrock age (MAGNA)	**Sedimentary cover age (MAGNA)**	**Numeric Age**	**PGA475 IGN 2013 (g)**	**Recorded PGA Local (g)**
Lower Jurassic	Holocene	≤ 2.960 ± 40 yr BP (¹⁴C)	0,23	Historic event (no data)

DAMAGE				
Type of constructions	**Type of structure**	**Affected Amount (quantity)**	**Building Damage**	**INT. LOCAL EMS/ MSK (IGN)**
Masonry, brick	Houses, Church	Multiple (c. 75%)	Ruinous (Ventas de Zafarraya)	IX EMS

GEOLOGICAL OR ENVIRONMENTAL CATALOGUED EFFECT

EEE Category ESI-07	Primary Effect	EEE Type ESI-07	SURFACE RUPTURE: SR/10

Description: Original historical reports (Orueta and Duarte, 1885) indicate the opening of numerous, but discontinuous, ground cracks in the western zone of the southern slope of the Zafarraya Poljé during the earthquake. More recent analyses identify a surface rupture of about 14 km length following the trace of the so-called Ventas de Zafarraya Fault (VZF), which displays a dominant normal fault kinematics (Reicherter et al., 2003; Grützner et al., 2013). At present it is possible to identify discontinuous bedrock fault scarps in the Lower Jurassic limestones which seems to diminish in height from east (c. 60 cm) to west (130 cm) with mean striae pitches around 150° (near vertical). Fault trenches excavated west of the village of Ventas de Zafarraya identified previous surface faulting events on this fault. Paleoseismic and radiocarbon data allow inferring that at least two pre-1884 ruptures occurred on the fault during the last 10 kyr (Grützner et al., 2013).

Interpretation: The southern boundary of the Zafarraya Poljé is delineated by the VZF with dominant normal kinematics during the Holocene. The foot wall is constituted by Jurassic limestones and the hanging wall by cretaceous marls, but the bedrock fault cuts and displaces Holocene colluvial wedges (Grützner et al. 2013. This slope sediments are in addition offset by minor subsidiary synthetic listric normal faults. The trenches excavated west to Ventas de Zafarraya identify coseismic vertical fault displacements of a maximum of 130 cm linked to the AD 1884 Earthquake. This values are reached near the El Cortijo del Barranco increasing east to west.

Quantification: Magnitudes estimated from the empirical relationships of Wells and Coopersmith (1994) result in 6.5 to 7.0 Mw. This will agree with maximum ESI-07intensies along the fault trace of ***intensity X***: metric-scale vertical displacements and kilometric length for surface rupture.

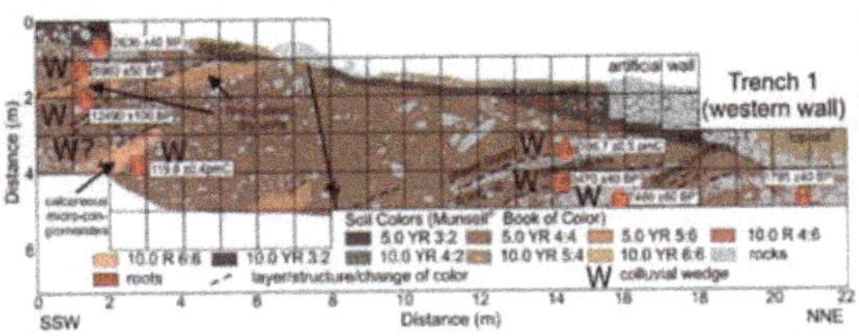

Upper image: natural bedrock fault scarp east of Ventas de Zafarraya (Photo M.A. Rodriguez-Pascua, 2015) Lower image: Exhumed fault plane in the ancient Quarries west of Ventas de Zafarraya. Note the AD 1884 fault scarp (above the red dotted line). The present scarp height is due to excavation works in the ancient quarries (Photo K. Reicherter 2000). To the right trench log of the western wall of T1 excavated in the hanging wall of the fault (Grutzner et al., 2003)

ASSESMENT OF ESI-07 INTENSITY	X	Quality of Geological data (Qg)	G	*Geological/ Environmental effects observed in natural outcrops and fault trenching analyses*

Key References (Bibliography)	*Grützner, C. et al. (2013): Late Holocene rupture history of the Ventas de Zafarraya Fault (Southern Spain). Cuaternario y Geomorfología, 27 (3-4), 51-61.* *Orueta and Duarte, D. (1885): Informe sobre los terremotos ocurridos en el sur de España en diciembre de 1884 y Enero de 1885.: Tip. y Lit. de Fausto Muñoz. Málaga* Reicherter, K.R. *et al.* (2003): Repeated palaeoseismic activity of the Ventas de Zafarraya fault (South Spain) and its relation with the 1884 Andalusian earthquake. Int. Journal Earth Science, 92 (6), 912-922.
Authors	*M. A. Rodriguez-Pascua (IGME) & M.A. Perucha (IGME).* ***Translated and updated by P.G. Silva (USAL)***
File Published in	*Catálogo de los Efectos Geológicos de los Terremotos en España (2nd Edition)* — Vol. Number: 2 — Year: 2019

Figure 6. Example of an "environmental earthquake effects (EEE) file" produced for 16 selected earthquakes included in the Spanish Catalogue of geological effects of earthquakes [10]. The example corresponds to the surface rupture (primary effect) triggered by the AD 1884 Arenas del Rey earthquake (Intensity X). The file has been translated and updated for the present paper by P.G. Silva. Original EEE files are in Spanish.

7.1. EEE Files: Detailed Information for Each Catalogued Event

The detailed information of each catalogued EEEs or EAEs is implemented in individual "EEE files", which offer parametric macroseismic information on geographical coordinates, epicentral distances, epicentral orientation, affected area by the EEE (km² or m²) and the assessed local intensities (EMS/MSK) by building damage coming from previous catalogues. These "EEE files" also include normalized data on the local geomorphology (landscape context and terrain slope), geology (age and nature of affected material and substratum, if the case), but also geotechnical data. This latter information considers the different soil classes observed by the present Spanish Seismic Code [46], as well as normalized PGA values (%g) considered by this seismic code for the affected zone. This geotechnical section also offers normalized descriptions for the construction-types, building-types, amount and damage level that occurred in the locality affected by the particular catalogued EEE in each individual "EEE file". In a last section the "EEE file" displays geological information on the category (primary/secondary), type (liquefaction, slope Movement, etc.) and ESI-07 intensity of the catalogued EEE, as well as individual abridged information on the description, interpretation, quantification (intensity assessment) and the particular Quality index (Qg) of the catalogued effect. This ESI-07 section of the file also includes graphical information of the catalogued EEE, such as photos, drawings, sketches, maps, GoogleEarth imagery, etc., depending on the period of the corresponding earthquake. Finally, the "EEE files" offer four to five key bibliographic references that specifically document, describe or analyze the catalogued EEE. Figure 6 illustrates the above-described structure of the EEE/EAE files.

7.2. Graphic Information Included in the "Full Event Files"

The full event files content a variety of info-graphic data. The primary info is a parametric list of effects (EEEs and EAEs) identified for a specific symbol developed for the scale (see symbol effects in Figure 1) and ordered alphabetically by locality and locality-site when precise (Figure 7). Selected Google Earth images of the kmz files developed for each earthquake (Figure 8) illustrate these parametric lists. The kmz files (and selected views) identify each catalogued EEE or EAE type-effect by its corresponding symbols (Figures 7 and 8). In the future, these kmz event files will be uploaded in a public geo-information database of the Geological Survey of Spain (IGM) for the interactive on-line search of the entire catalogue. Additionally, the "full event files" include two basic graphic databases in form of intensity maps and seismic scenarios (ShakeMaps). When sufficient data hybrid ESI-EMS intensity maps were produced following the guidelines of the ESI-07 scale [3].

Información macrosísmica catalogada (*catalogued macroseismic info*)

Localidad / Sitio *Locality / Site*	Longitud *Longuitude*	Latitud *Latitude*	Efecto *Effect*	Localización *Epicentral location*	EMS-98	MSK-64	ESI-07
01.01 Albojaira, Laguna de	1°55'42,67"O	37°20'29,76"N	GK	N 2,8 km	No data	VI (Zone)	**VIII**
01.02 Albojaira, Laguna de	1°55'42,67"O	37°20'29,76"N	WA	N 2,8 km	No data	VI (Zone)	**VIII**
02.01 Almanzora, Valle	1°53'13,00"O	37°17'50,71"N	SM	NE 1,9 km	No data	VI (Zone)	**VII**
02.02 Almanzora, Valle	1°57'34,48"O	37°23'17,79"N	HA	NE 1,2 km	No data	VI (Zone)	**VIII**
03.01 Cuevas de Almanzora	1°52'46,64"O	37°17'49,30"N	GK	NNO 8,9 km	V-VI	V	**VI**
03.02 Cuevas de Almanzora	1°53'9,08"O	37°17'58,55"N	SM	NNO 8,5 km	V-VI	V (Zone)	**VI**
03.03 Cuevas de Almanzora	1°52'58,57"O	37°17'49,50"N	EAE	NNO9,1 km	V-VI	V	**VI**
04.01 Huércal-Overa	1°56'34,42"O	37°23'8,89"N	HD	SSE 2,6 km	VI-VII	VII	**VII**
04.02 Huércal-Overa	1°56'31,63"O	37°23'9,27"N	GK	SSE 2,3 km	VI-VII	VII	**VI**
04.03 Huércal-Overa	1°56'31,63"O	37°23'9,27"N	OT	SSE 2,4 km	VI-VII	VII	**VII**
04.04 Huércal-Overa	1°56'36,19"O	37°23'20,83"N	EAE	SSE 2,7 km	VI-VII	VII	**VII**
05.01 Huércal, Castillo	1°56'4,41"O	37°23'15,87"N	SM	S 2,3 km	VI-VII	VII (Zone)	**VII**

Figure 7. Example of classified parametric list of the catalogued EEEs and earthquake archaeoseismological effects (EAEs) for the 16 selected earthquakes in the "full event files". The example, translated for this paper, contains macroseismic information for the first five sites of the AD 1863 Huércal-Overa earthquake (Intensity VIII). The full original list has 22 EEEs, 7 EAEs files and 13 complementary EMS data [10].

INFORMACIÓN MACROSÍSMICA HST 18630610: HUÉRCAL – OVERA (Almería). 1863 AD											EFECTOS GEOLÓGICOS
EVENTO	HORA Y MIN.	LONGITUD	LATITUD	MAG (Mw/Ms/M*)	EMS 98	MSK 64	ESI 07	Q_e	Q_i	Q_g	LOCALIZACIÓN TERREMOTO
HST18630610 *AD*	11:10	1°56′ O	37°22′ N	4,2–4,6 M*	VI-VII	VII	VIII	B	B	A	Huércal-Overa. AL.
Tipo de Efectos Catalogados											Número de Fichas 29 — 22 EEEs, 07 EAEs

Imagen Google Earth de la localización de efectos catalogados en la zona epicentral

Figure 8. Example of Google Earth image extracted from the kmz file produced for each of the 16 selected earthquakes with "full event files". The example displays information for the macroseismic area of the AD 1863 Huércal-Overa earthquake [10].

The intensity maps, implemented in shaded-relief models of the affected areas, illustrate the intensity distribution in relation to the Quaternary active faults of the zone and the set of catalogued EEEs for each event (Figure 9). The style of the intensity maps tries to resemble the color-scales used in the instrumental seismicity maps of the USGS Earthquake hazard program (https://earthquake.usgs.gov/) [47]. As mentioned, the hybrid intensity maps incorporate all the EMS information (when available) coming from previous catalogues [7]. The EMS information is also provided at the end of the parametric list illustrated in Figure 7.

In the intensity maps EEEs localities are identified with circles of the same color of the corresponding intensity level (Figure 9). In many cases, the localities have "satellite data-points" represented by smaller colored circles representing other EEEs or EAEs around the stronger (or more important) one that occurred in that locality. The use of "satellite data-points" is based on the recommendations of the International Atomic Energy Agency (IAEA) on the guidelines of paleoseismic analyses to seismic hazard assessments in site evaluation for nuclear installations [48]. The identification and delineation of the geometry of intensity zones was done taking into account the location of the more important Quaternary faults of the region (black lines), but also the geology and geomorphology of the affected areas. In particular, the distribution of poorly consolidated Quaternary deposits and the different topographic contexts on which the documented EEEs occurred was key for the production of intensity maps. As illustrated in Figure 9, colored triangles identify complementary EMS data for different intensity levels. EMS data are especially important for a more accurate definition of the geometry for intensity zones VI to IV in the lower limit of sensitivity of the ESI-07 scale [5]. The intensity maps are not available for ancient or paleoseismic events only documented in one archaeological site of fault-trench excavation. In these cases the intensity maps are substituted by another type of graphic information from geoarchaeological or pure geological sources, such as a maps and sketches of oriented damage (ancient events) or the log of the fault-trench documenting the earthquake (paleoseismic events).

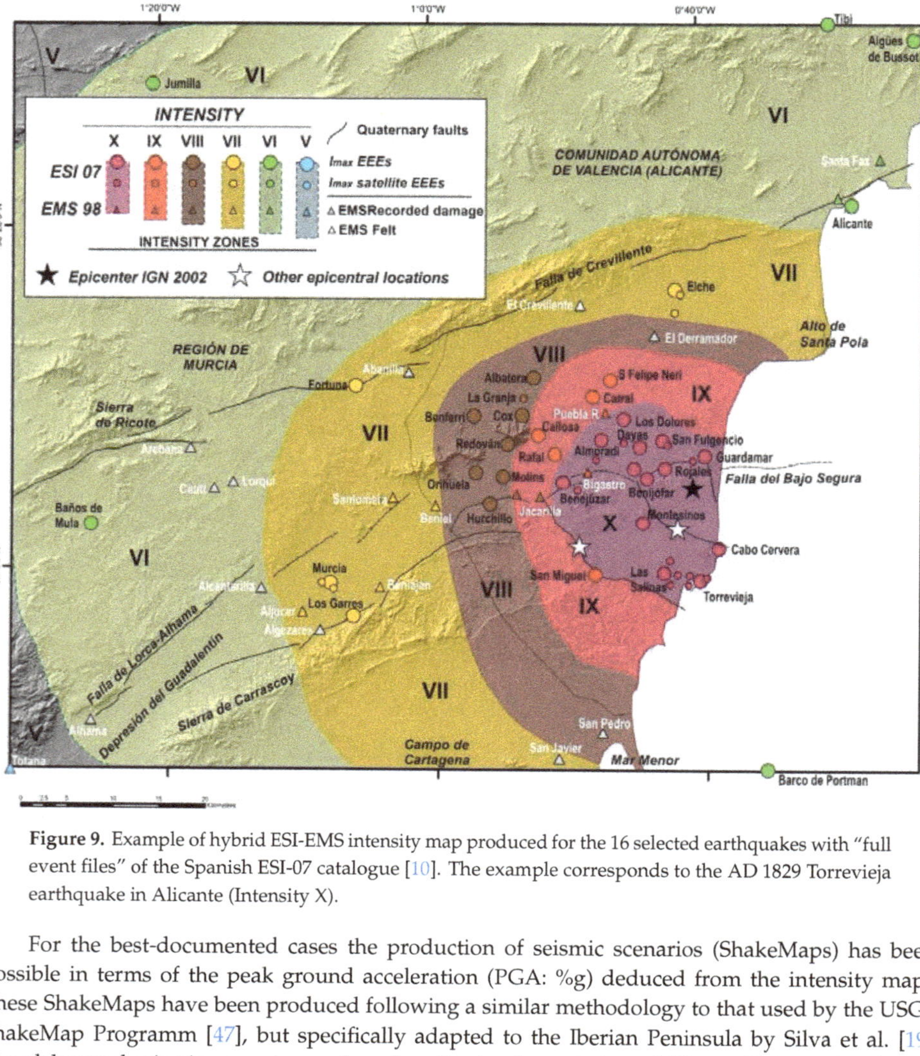

Figure 9. Example of hybrid ESI-EMS intensity map produced for the 16 selected earthquakes with "full event files" of the Spanish ESI-07 catalogue [10]. The example corresponds to the AD 1829 Torrevieja earthquake in Alicante (Intensity X).

For the best-documented cases the production of seismic scenarios (ShakeMaps) has been possible in terms of the peak ground acceleration (PGA: %g) deduced from the intensity maps. These ShakeMaps have been produced following a similar methodology to that used by the USGS ShakeMap Programm [47], but specifically adapted to the Iberian Peninsula by Silva et al. [19]. The elaborated seismic scenarios are based on the implementation of "Ground Motion Prediction Equations" (GMPE) on detailed digital terrain models (5m/pixel DTMs) pixel by pixel. The selected pixel size (5m) is clearly sufficient to identify individual EEEs and to upload detailed slope and geological data into the resulting shake models. This pixel size upgrades the 900 m/pixel DTMs used in the USGS earthquake hazard program [47] and allows one to define the contribution of topography (slope) and geology (loose soils) to the triggered environmental damage [19]. The selected GMPE for the production of seismic scenarios in the Spanish catalogue corresponds to those used in the first versions of the USGS Earthquake Hazard Program [49]. These equations are first generation GMPE, specifically designed for geodynamic contexts dominated by strike-slip faults in the SW EEUU [49], which in turn are similar to the geodynamic framework of southern Spain (Betic Cordillera; Figure 1), where the main strong earthquakes of the catalogue are placed [19].

These GMPE offer broad seismic scenarios in the range of the expected uncertainties for the hypothetical seismic source characterization of the historic, ancient and paleoseismic events analyzed in the catalogue. On the other hand, these basic first generation GMPE allow to implement simple source parameters (i.e. source location, fault type, fault dimensions/geometry and earthquake magnitude) in relation to epicentral distance (Joyner–Boore radius) [49], deduced Vs30 values from proxy data [50] and topographic and geologic correction factors specifically produced for Spain [19]. The modelling of seismic scenarios (shakemaps) for Spain is based on the large set of instrumental, macroseismic and geological data resulting from the last important earthquake that occurred in Spain (2011 Lorca Earthquake; 5.1 Mw; VIII ESI-07), which was used as a check-model in a ArcGis environment [19]. The Figure 10 illustrates the updated shakemap developed for the AD 1755 Lisbon earthquake-tsunami included in the catalogue.

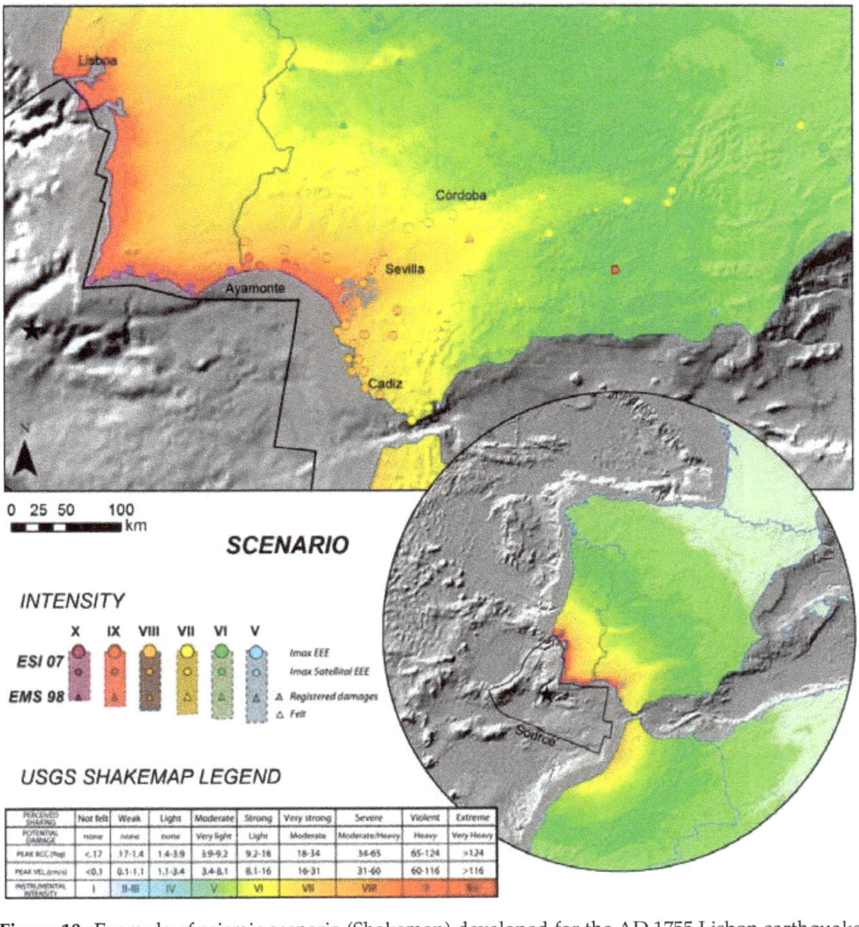

Figure 10. Example of seismic scenario (Shakemap) developed for the AD 1755 Lisbon earthquake and tsunami (XI ESI-07)—one of the 16 selected earthquakes with "full event files" of the Spanish ESI-07 catalogue [10]. The example corresponds to the AD 1755 Lisbon earthquake-tsunami (XI ESI-07). Note the dimensions of the seismogenic source as proposed by Silva et al. [17].

7.3. *Summary on the Structure and Content of the "Full Event Files"*

The information contained in the "full event files" is summarized in the following items:

- Event file (51 earthquakes): These files contain parametric data, earthquake type, quality indexes, summarized information on building (EMS-based) and environmental damage (ESI-based), a list of identified EEEs for the different categories of ESI-07 effects, and archaeoseismic damage (EAEs). Additionally, these files display summarized information on the geological data supporting the earthquake, a geological sketch of the Iberian Peninsula for location (Figure 2), five to six key references documenting the earthquake and a list of the authors compiling the information. Fifty one-page files in Spanish and English (Figure 5).
- Earthquake description in the Galbis Catalogue (16 selected Earthquakes): Transcription of original descriptions and references in these initial seismic catalogues. For those events not listed in the Galbis catalogue this section is substituted by a "general information" section including data from field, technical, scientific reports and scientific publications available for those earthquakes. Sixteen one-page sections.
- Earthquake environmental effects (16 selected earthquakes): Analysis and discussion of the catalogued EEEs in the format of a two-page short paper.
- Earthquake archaeological effects (16 selected earthquakes): Analysis and discussion of the catalogued EAEs in the format of a two-page short paper. In modern earthquakes, the analysis of EAEs is focused on its impact on the cultural heritage of the affected area. As in the previous case, this section has also a two-page format.
- Parametric list of the particular EEEs and EAEs (16 selected earthquakes), identified for each earthquake. For those cases with a great number of effects (above 50) the number of published EEEs/EAEs files was normalized to about 50 representative files (Figure 7).
- Parametric list of EMS data available for each earthquake (16 selected earthquakes) coming from the existing official databases of the IGN [7,23].
- EEE and EAE files (totaling 322) for the 16 selected earthquakes, with detailed information in the geological and geomorphological context, ground conditions, as well as descriptions, interpretation and quantification of the particular EEE or EAE catalogued effect.
- Google Earth image (16 selected earthquakes) representative of the distribution of EEEs and EAEs coming from the kmz files produced for the catalogue which will be implemented in the near future in an online public database of the Geological Survey of Spain (Figure 8).
- ESI-07/EMS-98 hybrid intensity maps (16 selected earthquakes) on shaded relief models, specifically designed for the published catalogue. In the case of earthquakes with few intensity data, like ancient or paleoseismic events, the intensity map is substituted by damage maps/sketches or by the log of the fault-trench documenting the event (Figure 9).
- Seismic Scenarios or ShakeMaps (16 selected earthquakes) in terms of PGA values, based on those developed by the USGS ShakeMap Program but specifically adapted for the Iberian Peninsula (Figure 10).
- Reference list (16 selected earthquakes): Complete bibliographic list with the old historic documents and modern research papers documenting or analyzing the catalogued earthquake.

7.4. *Special "Full Event File" for the AD 1755 Lisbon Earthquake-Tsunami Event*

Due to the special features and size and large area affected for this earthquake its full event file contains additional information. This file has different independent sections on the analysis of tsunami effects, geological effects extracted from the examination of the about 1300 individual locality-reports for this earthquake kept in the Spanish Historical National Archives (Archivo Histórico Nacional; AHN) [51], from which 673 localities reported natural effects [17]. This file also has a section on the large set of far-field hydrological and hydrogeological anomalous effects reported for the northern area of the Iberian Peninsula, British islands, Europe, Africa, the Caribbean Sea and the oceanic islands of Canary,

Madeira and Azores. Though most of the far-field effects are related to the tsunami propagation, some of them occurred in inland water bodies at Reading, Sussex, Plymouth, London (England); Lake district (Cumbria); Lakes Lomod and Ness (Scotland); Lake Binnenalster of Hamburg (Germany); Lakes Léman, Neuchatel and Zürichsee (Switzerland); Telpice Baths (Prague) and the Dal River in Sweden (Baltic Sea). Some of these last effects occurred more than 2000–3000 km away from the epicenter. This event has a total of 753 catalogued EEEs, 673 of them occurring in Spain [10].

All the above mentioned far-field natural effects were compiled in the volume 69 of the "Philosophical transactions of the Royal Society of London" published in December 1755 and reprinted in 1809 [52]. This volume contains an special section entitled: "An Extraordinary and Surprising Agitation of the Waters though without any perceptible Motion of the Earth having been observed in various parts of this Island, both Maritime and Inland, on the same Day and chiefly about the Time that the more Violent Commotion of both Earth and Waters so extensively affected many very distant Parts of the Globe" (pages 646 to 656). This special section collects 17 letters describing tsunami effects in littoral zones of the Atlantic Ocean, but also other 12 letters on the anomalous effects observed in inland water bodies at different parts of Europe. These very far-field effects represent an anomaly in relation to recent strong earthquake-tsunami events of similar size (i.e., 2004 Sumatra, 2010 Chile, 2011 Japan), which did not cause that variety of far-field anomalous effects [17,30].

The Lisbon event also has a particular section regarding other effects not truly considered by the ESI-07 scale [52], but of interest for this particular earthquake [10,51]. Long period waves on standing waters were recorded at 29 localities several hundreds of kilometers away in the whole of Spain, mostly in Central Castilla, but also in Cataluña (1200 km away) where the earthquake was not felt. Perception of an underground noise occurred almost simultaneously to ground shacking in 285 locations covering the whole of mainland Spain, except the north and northeast zones. Perception of ground waving and anomalous tree shaking was observed in many localities throughout the Guadalquivir valley. Finally, luminous effects observed in the sky (51 records) 5–4 hours before the earthquake along the western sector of the Iberian Peninsula throughout a SW–NE band 600 km long (Huelva to Palencia) which have been also catalogued. These effects are interpreted as the occurrence of a comet or meteor crossing the sky [51], eventually exploding over the north of the Duero river basin (north Spain) with a great luminosity and a strong noise [10]. All the far-field EEEs, tree shaking and luminous effects were catalogued totaling 753 effects, but only 49 of them have individual "EEE files" in the recently published catalogue [10]. Taking into account the abovementioned other effects not considered in the ESI-07 (long period waves, underground noise, ground waving, etc.) the total number of effects would amount to over 1000 EEEs for the AD 1755 Lisbon event, which would deserve a separate analysis and catalogue.

8. Brief Numerical Analysis of the Catalogued EEEs

This section introduces a short analysis of the catalogued events for the 51 analyzed earthquakes for the maximum intensity of the events (Figure 11) and for the different intensity levels (Figure 12). As observed in the bar-diagram of Figure 11, the AD 1755 Lisbon event produced a large variety of effects in the entire range of intensity levels, even where the earthquake was not felt. The most conspicuous reported effects were those related to hydrological (HA), hydrogeological (HD) or anomalous waves in inland water bodies (WA), which were more frequent and widely observed for intensities IV to VII (Figure 11). Noticeable ground effects were also liquefaction (LQ), ground cracks (GK) and slope movements (SM) for intensities VI to IX and strong tsunami damage from intensity VIII (Figure 11). Due to the offshore nature of the event, primary effects were few and mostly related to cases of subsidence of the coastal area around the Gulf of Cádiz.

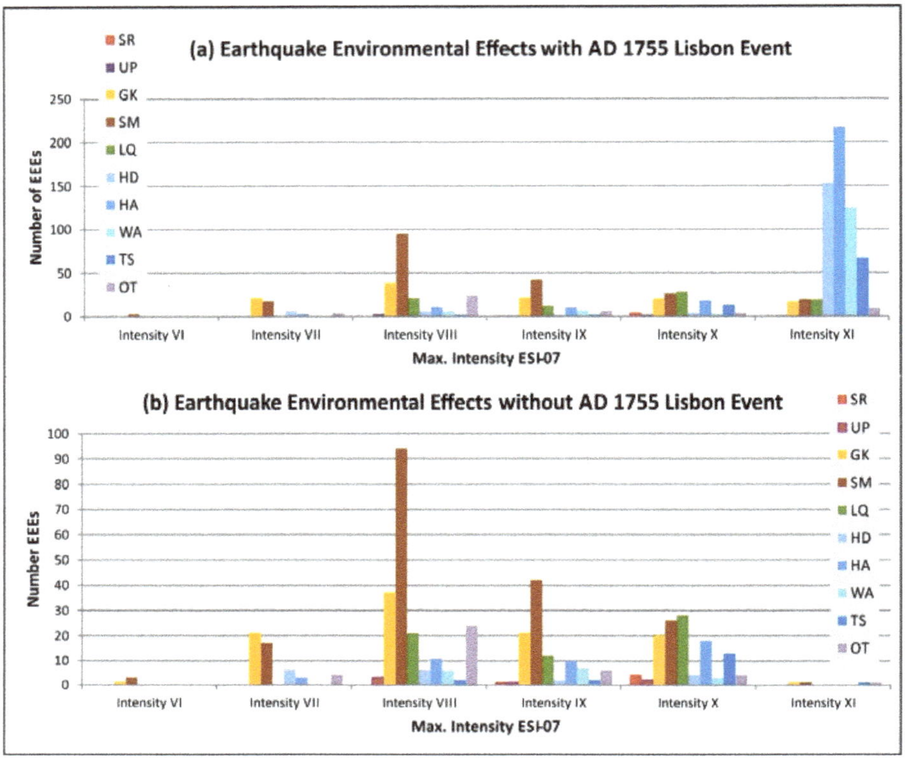

Figure 11. Statistical analysis of EEEs that occurred in the 51 catalogued events considering the maximum intensity (I_{max}) of the triggering event: (**a**) Distribution of EEEs for different intensity earthquakes considering the complete analyzed earthquake list; (**b**) distribution of EEEs for different intensity earthquakes removing the data for the AD 1755 Lisbon event.

The Figure 11 illustrates the distribution of triggered EEEs listed in the "event files" for the whole number of catalogued earthquakes by their maximum intensity. The analyzed data total 976 EEEs including those of the AD 1755 (Figure 11a) but are only 489 without the mentioned event (Figure 11b). The different distribution of EEEs is also noticeable when considering or not the Lisbon Event, which displaces the statistical mode to intensity XI due to the large number of hydrological effects (HA, HD and WA) reported for this earthquake (Figure 11a). On the contrary, removing the Lisbon EEEs, the statistical mode is clearly located in intensity VIII, where the most frequent effects are slope movements (SM), ground cracks (GK) and ground liquefaction (LQ) with an amount of 204 EEEs (Figure 11b), which practically doubles the number of EEEs for intensities IX and X (around 100 each). This is a statistical bias because intensity VIII events are the most numerous among the catalogued Spanish events (Tables 1 and 2). In any case, it seems clear that the minimum I_{max} value to produce secondary earthquake effects with a relative long-lasting geological record is intensity VII.

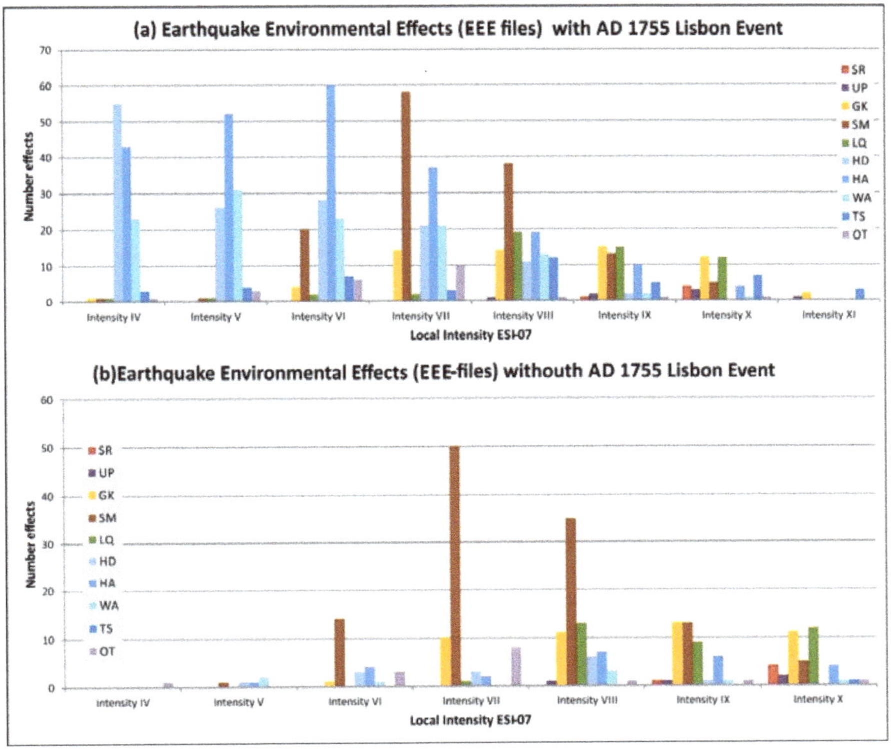

Figure 12. Statistical analysis of EEEs occurred in the 16 selected earthquakes with full event files considering the local intensity for the sixteen analyzed events. (**a**) Distribution of EEEs for different intensity zones considering the sixteen analyzed earthquakes; (**b**) distribution of EEEs for different intensity zones removing the data for the AD 1755 Lisbon event.

Figure 12 displays the catalogued EEEs by their intensity level, but only for the 16 selected earthquakes with "full event files", which sum up to a total of 811 records. Again, the comparison of the diagrams with and without the AD 1755 Lisbon event offers a very different distribution of the number and type of catalogued EEEs. Once the large amount of hydrological and tsunami effects produced by the Lisbon event are removed, again intensities VII and VIII collect the largest number of ground effects (SM, GK and LIQ) (Figure 12b). In many cases, the catalogued ground cracks are related to lateral spreading linked to ground liquefaction or to slope movements. This last category of ESI-07 effects is the more abundant one in the analyzed earthquakes, but the SM peak for intensity VII in Figure 12 is again a statistical bias due to the large amount of slope movements catalogued for the 2011 Lorca earthquake (Figure 12b). It is also noticeable that primary effects start to be observable for intensity VIII, but the few cases of surface ruptures occur for events of intensity ≥IX (Figure 12b). This is the general case for inland Spain, where surface faulting events are very rare, but the occurrence of extensive secondary earthquake effects is common. These mainly include hazardous slope movements, ground liquefaction and ground cracking processes in inland events, but also damaging tsunamis in the Gulf of Cádiz (Atlantic Ocean) and Alborán Sea (Mediterranean Sea) around the Gibraltar strait (Figure 2). In fact, the strongest earthquakes occurred within the Iberian Peninsula have moderate magnitudes around 6.0–7.0 Mw [7,19,23] but they triggered strong intensity levels up to IX–X as a consequence of the participation of geological effects in environmental and building damage. This is of special interest for the dangerous earthquake-tsunami events around the Gibraltar strait zone. In this sense, Figures 11 and 12 display the number of catalogued EEEs with and without the data corresponding to the 1755

Lisbon earthquake-tsunami. In both cases, the noise introduced by this special "Atlantic Event" is clear in relation to the EEEs commonly triggered by the strongest earthquakes that occurred within the Iberian Peninsula.

9. Achievements and Future Perspectives

In spite of the existence of databases with information on earthquake environmental effects—like that developed by the Italian Geological Survey (ISPRA), which includes data on about 200 globally distributed earthquakes (http://193.206.192.211/wfd/eee_catalog/viewer.php) [53], or databases with paleoseismic and active faults information [2]—none of them proceed to a proper ESI-07-based classification and interpretation. In this way, as noted by Serva [2] after the devastating 2011 Japan earthquake-tsunami (Tohoku) the IAEA recommended to the members states to carefully consider secondary earthquake ground effects in order to envisage a more accurate definition of the seismic hazard (in terms of intensity) in zones with nuclear installations [47] and the ESI-07 scale was largely considered in the published recommendations [54]. Some very recent catalogues and on-line databases, like that developed by the "Italian National Geophysics and Volcanologic Survey (INGV)" [55], consider ground effects, but they are treated in a descriptive-informative way and are not used for true intensity assessments following the ESI-07 guidelines. As a consequence, currently there is no similar catalogue like that recently published in Spain [10] considering geological and archaeoseismological data extending the earthquake records to the recent geological past. However, the growing data on earthquake environmental effects, and complementary archaeoseismological analyses, will improve our knowledge on the seismic history of a region in the future, by the recovering of lost earthquakes from the geological record. The future of SHA studies needs in a mandatory way the incorporation of geological and archaeoseismological data to offer to the modern societies more comprehensive data within the temporal range (recurrence periods) of the seismic cycle. In this sense, the Spanish catalogue incorporates seismic scenarios for historical events (i.e., Figure 10) which will help to develop a better definition of the seismic hazard for forthcoming SHA studies but also for the preparation of future earthquake drills [20].

Author Contributions: Project design, management, conceptualization, coordination and funding acquisition, investigation on historical earthquakes, supervision of all Event and EEE files and writing—original draft preparation, P.G.S; project management, administration, and funding acquisition, investigation on paleoseismic and ancient events, supervision and final edition, M.A.R.P.; design of graphic information-software for intensity maps and visualization, investigation on ancient events and cultural heritage, writing—review and editing, J.L.G.R.; design of graphic information-software for seismic scenarios, management of information in ArcGis environment and writing—review and editing, J.E.V.; investigation of historical and ancient events, geochronologic databases, writing—review and editing, R.P.L.; database management, statistical analyses, supervision review and editing of all the Event and EEE files, M.B.B.D.

Funding: This research was funded by the Spanish Research Project MINECO-FEDER CGL2015-67169-P (QTECSPAIN-USAL). This is contribution of the QTECT-AEQUA Working Group.

Acknowledgments: The authors are grateful to the rest of the geologists and scientists contributing to the publication of the 2nd Edition of the Catalogue of the geological effects of earthquakes in Spain: Teresa Bardají (UAH), Pedro Huerta (USAL), M. Ángeles Perucha (IGME), Francisco García-Tortosa (UJAEN), Javier Lario (UNED), Pedro Vicente Gómez (USAL) and Elvira Roquero (UPM).

Conflicts of Interest: The authors declare no conflict of interest.

References

1. Serva, L.; Vittori, E.; Comerci, V.; Esposito, E.; Guerrieri, L.; Michetti, A.M.; Mohammadioun, B.; Mohammadioun, G.C.; Porfido, S.; Tatevossian, R. Earthquake Hazard and the Environmental Seismic Intensity (ESI) Scale. *Pure App. Geophys.* **2016**, *173*, 1479–1555. [CrossRef]
2. Serva, L. History of the Environmental Seismic Intensity Scale ESI-07. *Geosciences* **2019**, *9*, 210. [CrossRef]
3. Michetti, A.M.; Esposito, E.; Guerrieri, L.; Porfido, S.; Serva, L.; Tatevossian, R.; Vittori, E.; Audemard, F.; Azuma, T.; Clague, J.; et al. *Intensity Scale ESI 2007*; ISPRA: Roma, Italy, 2007; Volume 74, ISBN 9788824029032.

4. Grünthal, G. *European Macroseismic Scale 1998: EMS98*; Musée Natioal d´Historie Narurelle: Luxembourg, 1988; p. 99.
5. Silva, P.G.; Guerrieri, L.; Michetti, A.M. Intensity scale ESI 2007 for assessing earthquake intensities. In *Encyclopedia of Earthquake Engineering*; Beer, M., Kougioumtzoglou, I.A., Patelli, E., Au, S.K., Eds.; Springer: Berlin/Heidelberg, Germany, 2015; ISBN 978-3-642-35344-4.
6. IAEA. *Seismic Hazards in Site Evaluation for Nuclear Installations*; IAEA Safety Standards Series No. SSG-9; International Atomic Energy Agency: Vienna, Austria, 2010; p. 193.
7. Martínez Solares, J.M.; Mezcua, J. *Catálogo Sísmico de la Península Ibérica (880 a.C.–1900)*; Instituto Geográfico Nacional: Madrid, Spain, 2002; p. 253. (In Spanish)
8. Udías, A. Historical Earthquakes (before 1755) of the Iberian Peninsula in Early Catalogs. *Seism. Res. Let.* **2015**, *86*, 999–1003. [CrossRef]
9. Silva, P.G.; Rodríguez-Pascua, M.A.; Giner-Robles, J.L.; Pérez-López, R.; Lario, J.; Perucha, M.A.; Bardají, T.; Huerta, P.; Roquero, E.; Bautista Davila, M.B. *Catálogo de los Efectos Geológicos de los Terremotos de España, 1ª Ed*; Riesgos Geológicos y Geotecnia 4. IGME: Madrid, Spain, 2014; p. 352. (In Spanish)
10. Silva, P.G.; Rodríguez-Pascua, M.A.; Giner-Robles, J.L.; Pérez-López, R.E.; García-Tortosa, F.J.; Gómez Vicente, P.; Bardají, T.; Perucha, M.A.; Huerta, P.; Lario, J.J.; et al. *Catálogo de los Efectos Geológicos de los Terremotos de España, 2ª Ed. Revisada y Ampliada*; Riesgos Geológicos y Geotecnia 6. IGME: Madrid, Spain, 2019; p. 804. (In Spanish)
11. IGN. *Actualización de Mapas de Peligrosidad Sísmica en España 2012*; Instituto Geográfico Nacional (IGN): Madrid, Spain, 2013; p. 228. (In Spanish)
12. Silva, P.G.; Rodríguez-Pascua, M.A.; Pérez-López, R.; Bardají, T.; Lario, J.; Alfaro, P.; Martínez-Díaz, J.J.; Reicherter, K.; Giménez García, J.; Giner, J. Catalogación de los efectos geológicos y ambientales de los terremotos en España en la Escala ESI 2007 y su aplicación a los estudios paleosismológicos. *Geotemas* **2008**, *6*, 1063–1066. (In Spanish)
13. Rodríguez-Pascua, M.A.; Pérez-López, R.; Silva, P.G.; Giner-Robles, J.L.; Garduño-Monroy, V.H.; Reicherter, K. A Comprehensive Classification of Earthquake Archaeological Effects (EAE) for Archaeoseismology. *Quat. Int.* **2011**, *242*, 20–30.
14. Bradley, R.; García Sanjuán, L. Sudden time? Natural disasters as a stimulus to monument building, from Silbury Hill (Great Britain) to Antequera (Spain). In *The Neolithic of Europe*; Bickle, P., Cummings, V., Hoffman, D., Pollard, J., Eds.; Oxbow Books: Oxford, UK, 2017; pp. 188–200.
15. Ferrater, M.; Silva, P.G.; Ortuño, M.; Rodríguez-Pascua, M.A.; Masana, E. Archaeoseismologic analysis of a Late Bronze Age site on the Alhama de Murcia Fault: Tira del Lienzo (Murcia, SE Spain). *Geoarchaeology* **2015**, *30*, 151–165. [CrossRef]
16. Silva, P.G.; Pérez-López, R.; Rodríguez-Pascua, M.A.; Roquero, E.; Giner Robles, J.L.; Huerta, P.; Martínez-Graña, A.; Bardají, T. Macroseismic analysis of slope movements triggered by the 2011 lorca earthquake (Mw 5.1): Application of the ESI-07 scale. *Geogaceta* **2015**, *57*, 35–38.
17. Silva, P.G.; Elez, J.; Giner-Robles, J.L.; Gómez-Diego, P.V.; Rodríguez-Pascua, M.A.; Roquero, E.; Martínez-Graña, A.; Bardají, T. The AD 1755 Lisbon Earthquake-Tsunami: Modeling the seismic source from the analysis of environmental and building macroseismic data. *GNS Sci. Misc. Ser.* **2017**, *110*, 358–361.
18. Duarte, J.C.; Rosas, F.M.; Terrinha, P.; Schellart, W.P.; Boutelier, D.; Gutscher, M.-A.; Ribeiro, A. Are subduction zones invading the Atlantic? Evidence from the southwest Iberia margin. *Geology* **2013**, *41*, 839–842. [CrossRef]
19. Silva, P.G.; Elez, J.; Giner-Robles, J.L.; Rodríguez-Pascua, M.A.; Pérez-López, R.; Roquero, E.; Bardají, T.; Martínez-Graña, A.M. ESI-07 ShakeMaps for instrumental and historical events in the Betic Cordillera (SE Spain): An approach based on geological data and applied to seismic hazard. *Quat. Int.* **2017**, *451*, 185–208. [CrossRef]
20. Pérez-López, R.; Elez, J.; Silva, P.G.; Giner-Robles, J.L.; Rodríguez-Pascua, M.A.; Roquero, E.; Bardají, T. Utilización de shakemaps y efectos geológicos como escenarios para simulacros en gestión de desastres. In Proceedings of the 3ª Reunión Ibérica sobre Fallas Activas y Paleosismología IBERFAULT III, Alicante, Spain, 11–15 June 2018. (In Spanish).
21. Martínez Solares, J.M. Sismicidad pre-instrumental: Los grandes terremotos históricos en España. *Enseñ. Cienc. Tierra* **2011**, *19*, 296–304. (In Spanish)
22. IGME. *Informe Geológico Preliminar del Terremoto de Lorca del 11 de Mayo de 2011 (Mw 5,1)*; Instituto Geológico y Minero de España (IGME): Madrid, España, 2011; p. 47. (In Spanish)

23. IGN. *Official web page of the Spanish Seismic Network*; Red Sísmica Nacional. Instituto Geográfico Nacional (IGN): Madrid, España. Available online: http://www.ign.es/web/ign/portal/sis-catalogo-terremotos (accessed on 26 July 2019). (In Spanish)
24. Silva, P.G.; Borja, F.; Zazo, C.; Goy, J.L.; Bardají, T.; De Luque, L.; Lario, J.; Dabrio, C.J. Archaeoseismic record at the ancient Roman City of Baelo Claudia (Cádiz, South Spain). *Tectonophysics* **2005**, *408*, 129–146. [CrossRef]
25. Rodríguez-Pascua, M.A.; Silva, P.G.; Perucha, M.A.; Giner-Robles, J.L.; Heras, C.; Bastida, A.B.; Carrasco García, P.; Roquero, E.; Bardají, T.; Lario, J.; et al. Seismically induced liquefaction structures in La Magdalena archaeological site, the 4th century AD Roman Complutum (Madrid, Spain). *Sediment. Geol.* **2016**, *344*, 34–46. [CrossRef]
26. Rodríguez-Pascua, M.A.; Silva, P.G.; Garduño-Monroy, V.H.; Pérez-López, R.; Israde-Alcántara, I.; Giner-Robles, J.L.; Bischoff, J.; Calvo, J.P. Ancient earthquakes from archaeoseismic evidence during the Visigothic and Islamic periods in the archaeological site of "Tolmo de Minateda" (SE of Spain). In *Special Paper of the Geological Society of America*; Sintubin, M., Stewart, I., Niemi, T., Altunel, E., Eds.; Geological Society of America: Boulder, CO, USA, 2010; Volume 471, pp. 171–184.
27. Grützner, C.; Ruano, P.; Jabaloy, A.; Galindo-Zaldívar, J.; Becker-Heidmann, P.; Sanz de Galdeano, C.; Rudersdorf, A.; Reicherter, K. Late Holocene rupture history of the Ventas de Zafarraya Fault (Southern Spain). *Cuatern. Geomorfol.* **2013**, *27*, 51–61.
28. Rodríguez-Vidal, J.; Ruiz, F.; Cáceres, L.M.; Abad, M.; González-Regalado, M.L.; Pozo, M.; Carretero, M.I.; Monge-Soares, A.M.; Gómez-Toscano, F. Geomarkers of the 218-209 BC Atlantic tsunami in the Roman Lacus Ligustinus (SW Spain): A palaeogeographical approach. *Quat. Int.* **2011**, *242*, 201–212. [CrossRef]
29. Gràcia, E.; Vizcaino, A.; Escutia, C.; Asioli, A.; Rodés, A.; Pallás, R.; García-Orellana, J.; Lebreiro, S.; Goldfinger, C. Holocene earthquake record offshore Portugal (SW Iberia): Testing turbidite palaeoseismology in a slow-convergence margin. *Quat. Sci. Rev.* **2010**, *29*, 1156–1172. [CrossRef]
30. Lario, J.; Zazo, C.; Goy, J.L.; Silva, P.G.; Bardají, T.; Cabero, A.; Dabrio, C.J. Holocene palaeotsunami catalogue of SW Iberia. *Quat. Int.* **2011**, *242*, 196–200. [CrossRef]
31. Silva, P.G.; Bardají, T.; Roquero, E.; Martínez-Graña, A.; Perucha, M.A.; Lario, J.; Giner Robles, J.L.; Rodríguez-Pascua, M.A.; Pérez-López, R.; Cabero, A.; et al. Seismic palaeogeography of coastal zones in the Iberian Peninsula: Understanding ancient and historic earthquakes in Spain. *Cuatern. Geomorfol.* **2015**, *29*, 31–56.
32. Gómez, F.; Arruda, A.M.; Rodríguez-Vidal, J.; Cáceres, L.M.; Ruiz, F. Eventos marinos de alta energía y cambios traumáticos en los asentamientos costeros del Suroeste de la Península Ibérica. *Cuatern. Geomorfol.* **2015**, *29*, 57–74. (In Spanish) [CrossRef]
33. Ambraseys, N. *Earthquakes in the Eastern Mediterranean and the Middle East: A Multidisciplinary Study of Seismicity up to 1900*; Cambridge University Press: New York, NY, USA, 2009; p. 901.
34. Rodríguez-Pascua, M.A.; Silva, P.G.; Giner-Robles, J.L.; Pérez López, R.; Perucha, M.A.; Martín-González, F. Arqueosismología: una nueva herramienta para la sismología y la protección del patrimonio. *Rev. Otarq* **2016**, *1*, 151–169. (In Spanish) [CrossRef]
35. Rodríguez-Pascua, M.A.; Giner-Robles, J.L.; Silva, P.G.; Pérez-López, R.; Perucha, M.A.; Elez, J.; Bardají, T.; Roquero, E.; Sánchez Ramos, I.; Morín de Pablos, J. The record of ancient and historic earthquakes in the archaeological site of Idanha a Vela (Central Portugal; Iberian Peninsula). *GNS Sci. Misc. Ser.* **2017**, *110*, 338–341.
36. Giner-Robles, J.L.; Bardají, T.; Rodríguez-Pascua, M.A.; Silva, P.G.; Roquero, E.; Elez, J.; Perucha, M.A.; Baena, R.; Guerrero, I.; Fernández-Caro, J.J.; et al. Análisis arqueosismológico del conjunto arqueológico romano de Mulva- Munigua (Sevilla, España). *Geotemas* **2016**, *16*, 605–608. (In Spanish)
37. Arteaga, C.; García Menárguez, A.; Prados Martínez, F.; Baudot, E. El Cabezo del Estaño de Guardamar (Alicante, España): Avance Preliminar de Evidencias Arqueosísmicas en un asentamiento Fenicio del Siglo VII a.C. *Rev. Mundo Investig.* **2016**, *2*, 146–155. (In Spanish)
38. Giner-Robles, J.L.; Rodríguez-Pascua, M.A.; Silva, P.G.; Pérez-López, R. Efectos sísmicos en yacimientos arqueológicos: catalogación y cuantificación arqueosismológica. *Bol. Geol. Min.* **2018**, *129*, 445–463. (In Spanish) [CrossRef]
39. Lario, J.; Bardají, T.; Silva, P.G.; Zazo, C.; Goy, J.L. Improving the coastal record of tsunamis in the ESI-07 Scale: Tsunami Environmental Effects Scale (TEE-16 scale). *Geol. Acta.* **2016**, *14*, 179–193.

40. Rodríguez-Pascua, M.A.; Silva, P.G.; Giner-Robles, J.L.; Perucha, M.A.; Roquero, E.; Bardají, T.; Elez, J.; Pérez-López, R. Using earthquake archaeological effects for the identification of seismic damage and intensity assessments in the cultural heritage. In *Scientific Management of Cultural Heritage*; Pechioli, L., D'Amico, S., Venuti, V., Eds.; Springer Handbooks: Berlin, Germany, 2019; in press.
41. Mezcua, J.; Rueda, J.; García Blanco, R.M. Iberian Peninsula Historical Seismicity Revisited: An Intensity Data Bank. *Seism. Res. Let.* **2013**, *84*, 9–18. [CrossRef]
42. Mezcua, J.; Martínez Solares, J.M. *Sismicidad del área Ibero-Mogrebí*; Instituto Geográfico Nacional (IGN). Publicación 203: Madrid, Spain, 1985; p. 301. (In Spanish)
43. Wells, D.L.; Coppersmith, K.J. New empirical relationships among magnitude, rupture length, rupture width, rupture area, and surface displacement. *Bull. Seism. Soc. Am.* **1994**, *84*, 974–1002.
44. Silva, P.G.; Giner-Robles, J.L.; Reicherter, K.; Rodríguez-Pascua, M.A.; Grüztner, C.; García-Jiménez, I.; Carrasco, P.; Bardají, T.; Santos, G.; Roquero, E.; et al. Los terremotos antiguos del conjunto arqueológico romano de Baelo Claudia (Cádiz, Sur de España): Quince años de investigación arqueosismológica. *Estud. Geol.* **2016**, *72*, e050. [CrossRef]
45. Galbis, J. *Catálogo Sísmico de la zona Comprendida Entre los Meridianos 5° E y 20° W de Greenwich y los Paralelos 45° y 25° Norte*; Instituto Geográfico Nacional: Madrid, Spain, 1932 & 1940; Volumes I and II. (In Spanish)
46. NCSE-02. Real Decreto 997/2002, de 27 de septiembre. Boletín Oficial del Estado (BOE). *Norma de la Construcción Sismorresistente Española* **2002**, *244*, 35898–35967. (In Spanish)
47. Wald, D.J.; Worden, B.C.; Quitoriano, V.; Pankow, K.L. *ShakeMap Manual: Technical Manual, User's Guide, and Software Guide*; U.S. Geological Survey: Boulder, CO, USA, 2005; p. 132.
48. IAEA. *The Contribution of Paleoseismology to Seismic Hazard Assessment in Site Evaluation for Nuclear Installations, TEC-DOC 1767*; International Atomic Energy Agency: Vienna, Austria, 2015; p. 206.
49. Boore, D.M.; Joyner, W.; Fumal, T. Equations for estimating horizontal response spectra and peak acceleration from Western North American earthquakes: a Summary of recent Work. *Seismol. Res. Lett.* **1997**, *68*, 128–153. [CrossRef]
50. Wald, D.J.; Allen, T.I. Topographic slope as a proxy for seismic site conditions and amplification. *Bull. Seismol. Soc. Am.* **2007**, *97*, 1379–1395. [CrossRef]
51. Martínez Solares, J.M. *Los Efectos en España del Terremoto de Lisboa*; Monografías IGN, 19. IGN: Madrid, Spain, 2001; p. 756. (In Spanish)
52. Hutton, C.; Shaw, G.; Pearson, R. *Philosophical Transactions of the Royal Society of London 1750–1755*; London: London, UK, 1809; Volume LXIX (69), pp. 646–656.
53. Guerrieri, L. The EEE Catalogue: A global catalogue of Earthquake Environmental Effects. *Quat. Int.* **2012**, *279*, 179–180. [CrossRef]
54. Audemard, F.; Azuma, T.; Baiocco, F.; Baize, S.; Blumetti, A.M.; Brustia, E.; Clague, J.; Comerci, V.; Esposito, E.; Guerrieri, L.; et al. *Earthquake Environmental Effects for Seismic Hazard Assessment: The ESI Intensity Scale and the EEE Catalogue*; ISPRA: Rome, Italy, 2015; Volume 97, ISBN 978-88-9311-007-5.
55. Guidoboni, E.; Ferrari, G.; Tarabusi, G.; Sgattoni, G.; Comastri, A.; Mariotti, D.; Ciuccarelli, C.; Bianchi, M.G.; Valensise, G. CFTI5Med, the new release of the catalogue of strong earthquakes in Italy and in the Mediterranean area. *Sci. Data* **2019**, *6*, 1–15. [CrossRef]

geosciences

MDPI

Review

Surface-Rupturing Historical Earthquakes in Australia and Their Environmental Effects: New Insights from Re-Analyses of Observational Data

Tamarah R. King [1,*], Mark Quigley [1] and Dan Clark [2]

[1] School of Earth Sciences, University of Melbourne, Melbourne 3010, Australia; Mark.quigley@unimelb.edu.au

[2] Geoscience Australia, Canberra 2609, Australia; Dan.clark@ga.gov.au

* Correspondence: tamarah.king@unimelb.edu.au

Received: 2 August 2019; Accepted: 16 September 2019; Published: 20 September 2019

Abstract: We digitize surface rupture maps and compile observational data from 67 publications on ten of eleven historical, surface-rupturing earthquakes in Australia in order to analyze the prevailing characteristics of surface ruptures and other environmental effects in this crystalline basement-dominated intraplate environment. The studied earthquakes occurred between 1968 and 2018, and range in moment magnitude (Mw) from 4.7 to 6.6. All earthquakes involved co-seismic reverse faulting (with varying amounts of strike-slip) on single or multiple (1–6) discrete faults of ≥ 1 km length that are distinguished by orientation and kinematic criteria. Nine of ten earthquakes have surface-rupturing fault orientations that align with prevailing linear anomalies in geophysical (gravity and magnetic) data and bedrock structure (foliations and/or quartz veins and/or intrusive boundaries and/or pre-existing faults), indicating strong control of inherited crustal structure on contemporary faulting. Rupture kinematics are consistent with horizontal shortening driven by regional trajectories of horizontal compressive stress. The lack of precision in seismological data prohibits the assessment of whether surface ruptures project to hypocentral locations via contiguous, planar principal slip zones or whether rupture segmentation occurs between seismogenic depths and the surface. Rupture centroids of 1–4 km in depth indicate predominantly shallow seismic moment release. No studied earthquakes have unambiguous geological evidence for preceding surface-rupturing earthquakes on the same faults and five earthquakes contain evidence of absence of preceding ruptures since the late Pleistocene, collectively highlighting the challenge of using mapped active faults to predict future seismic hazards. Estimated maximum fault slip rates are 0.2–9.1 m Myr^{-1} with at least one order of uncertainty. New estimates for rupture length, fault dip, and coseismic net slip can be used to improve future iterations of earthquake magnitude—source size—displacement scaling equations. Observed environmental effects include primary surface rupture, secondary fracture/cracks, fissures, rock falls, ground-water anomalies, vegetation damage, sand-blows/liquefaction, displaced rock fragments, and holes from collapsible soil failure, at maximum estimated epicentral distances ranging from 0 to ~250 km. ESI-07 intensity-scale estimates range by ± 3 classes in each earthquake, depending on the effect considered. Comparing Mw-ESI relationships across geologically diverse environments is a fruitful avenue for future research.

Keywords: Intraplate earthquake; surface rupture; Australian earthquakes; earthquake environmental effects; reverse earthquake; ESI 2007 scale; historical and recent earthquakes

1. Introduction

In the 50 years between 1968 and 2018 Australia experienced eleven known surface rupturing earthquakes (Table 1, Figure 1). Studies of Australian surface rupturing earthquakes have contributed to improvements in our collective understanding of intraplate earthquake behavior, including rupture recurrence, in stable continental regions (SCR) [1–5] and empirically-derived scaling relationships for reverse earthquakes [6–9]. This paper reviews available published literature on historic surface ruptures (Tables 1 and 2) and collates geological data (Tables 3 and 4, Figures 1 and 2), seismological data and analyses (Table 5), surface rupture measurements (Table 6), environmental damage (Table 7), and paleoseismic data (Table 8) (Figures 3–10). We re-evaluate and reconsider rupture and fault characteristics in light of new data (e.g., geophysical and geological) using modern analysis techniques (e.g., environmental seismic intensity scale (ESI-07) [10]) and new or updated concepts in earthquake science since the time of publication (e.g., paleoseismology, SCR earthquake recurrence).

Table 1. Summary of known historic Australian surface rupturing earthquakes and relevant references.

Name	Fig. 1	Magnitude (Mw) [11]	Date	This Paper: Length (km)	Dip	Avg. Net-Slip (m)	Published Length (km)	Max. Vert. Disp. (m)	Relevant References
Meckering, WA	1	6.59	14/10/1968	40 ± 5	35° ± 10	1.78	37	2.5	[12–37]
Calingiri, WA	8	5.03	10/03/1970	3.3 ± 0.2	20° ± 10	0.46	3.3	0.4	[23–25,38,39]
Cadoux, WA	4	6.1	02/06/1979	20 ± 5	60° ± 30	0.54	14	1.4	[28,40–45]
Marryat Creek, SA	5	5.7	30/03/1986	13 ± 1	40° ± 10	0.31	13	0.9	[1,28,46–49]
Tennant Creek 1 (Kunayungku) NT	7	6.27	22/01/1988	9 ± 1	40° ± 5	0.55	10.2	10.9	[1,49–69]
Tennant Creek 2 (Lake Surprise west)	6	6.44	22/01/1988	9 ± 2	60° ± 10	0.84	6.7	1.1	[1,49–69]
Tennant Creek 3 (Lake Surprise east)	3	6.58	22/01/1988	16 ± 0.5	35 ° ± 5	1.23	16	1.8	[1,49–69]
Katanning, WA	10	4.7	10/10/2007	0.5 ± 0.5	40° ± 5	0.2	1.26	0.1	[70,71]
Pukatja, SA	9	5.18	23/03/2012	1.3 ± 0.3	30° ± 10	0.25	1.6	0.5	[9,72]
Petermann, NT	2	6.1	20/05/2016	21 ± 0.5	30° ± 5	0.42	20	1.0	[73–78]
Lake Muir, WA		5.3	08/11/2018				3	0.5	[79]

Other literature with relevant analysis or data regarding historic ruptures: [80–98].

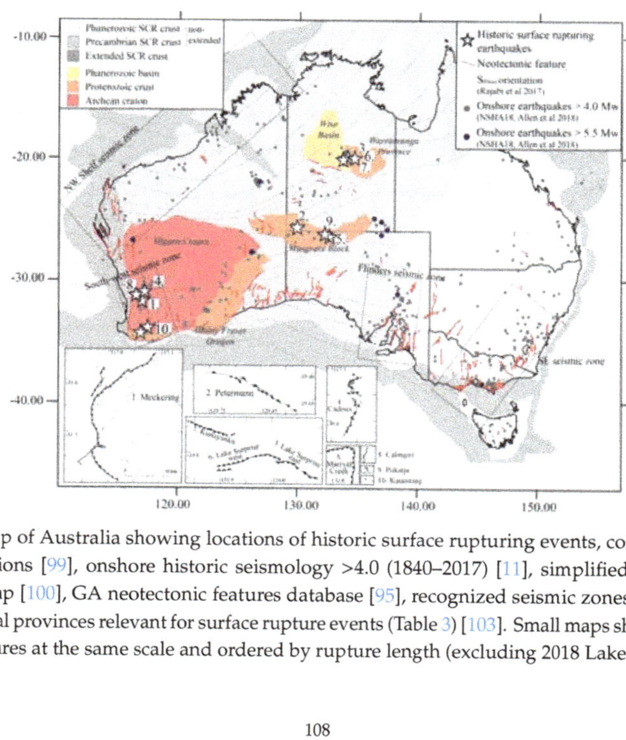

Figure 1. Map of Australia showing locations of historic surface rupturing events, continental scale crustal divisions [99], onshore historic seismology >4.0 (1840–2017) [11], simplified crustal stress trajectory map [100], GA neotectonic features database [95], recognized seismic zones [101,102] and specific crustal provinces relevant for surface rupture events (Table 3) [103]. Small maps show individual surface ruptures at the same scale and ordered by rupture length (excluding 2018 Lake Muir).

Australia is regarded as a stable continental region [104] surrounded by passive margins with an intraplate stress field controlled by plate boundary forces [100,105] (Figure 1). This stress field has been extant throughout much of Australia since the late Miocene, broadly concurrent with a rearrangement of tectonic boundaries in India, New Zealand, New Guinea, and Timor [101]. More than 360 potentially neotectonic features (those showing displacements associated with, or since initiation of, the current stress-field conditions) [5,95] have been recognized in the landscape through field mapping, subsurface geophysical imaging, digital elevation modelling, and palaeoseismic investigation [2,3,5,34,95,106–113] (Figure 1).

Southeast Australia and the Flinders Ranges (Figure 1) have the highest rates of seismicity [101,102] and estimated neotectonic fault slip rates [3,109,111] yet all of the largest onshore historic Australian earthquakes have occurred in Archean to Proterozoic cratonic crust across the central and western parts of the country (Figure 1) [11]. Of four defined zones of high seismicity (Figure 1) [101,102], the South West Seismic Zone (SWSZ) is the only to coincide with historic surface ruptures (Meckering, Calingiri, Cadoux, Katanning, Lake Muir). Other ruptures (Marryat Creek, Tennant Creek, Pukatja, Petermann) have occurred in historically aseismic regions of the cratonic crust (Figure 1, Table 5, Section 3.2).

Table 2. Summary of data sources used in reviewing Australian historic surface ruptures.

Seismological Catalogues	
Primary literature	[9,12,15,18,24,25,38,40,41,43,46,47,57,60,61,63,70,73–75,88]
Geoscience Australia (GA) online catalogue	https://earthquakes.ga.gov.au/
National Seismic Hazard Assessment 2018 (NSHA18)	http://pid.geoscience.gov.au/dataset/ga/123139; [11]
Focal Mechanisms	
Primary literature	[9,23,26,28,41,43,46,55,57,60,70,73,74]
GA compilation	[90]
Global centroid moment tensor catalogue	https://www.globalcmt.org/CMTsearch.html
Surface Rupture Trace	
Primary literature	[9,25,41,48,63,70,73,78]
GA Neotectonic Features Database	[95]; http://pid.geoscience.gov.au/dataset/ga/74056
Google satellite imagery	https://www.google.com/earth/
Bing satellite imagery	https://www.bing.com/maps/aerial
National SRTM DEM	SRTM 1-Sec DEM: http://pid.geoscience.gov.au/dataset/ga/72759
Geological Maps	
Primary literature	[9,16,25,41,65,114–117]
Geological Survey of Western Australia	http://www.dmp.wa.gov.au/Geological-Survey/Geological-Survey-262.aspx
Northern Territory Geological Survey	https://geoscience.nt.gov.au/
Geological Survey of South Australia	http://www.energymining.sa.gov.au/minerals/geoscience/geological_survey
Geoscience Australia	https://ecat.ga.gov.au/geonetwork/srv/eng/catalog.search
Borehole Data	
Northern Territory Government	http://nrmaps.nt.gov.au/nrmaps.html
South Australia Government	https://www.waterconnect.sa.gov.au/Systems/GD/Pages/Default.aspx
Geophysical Maps	
Primary literature	[33,50,51]
Bouguer gravity anomaly	http://pid.geoscience.gov.au/dataset/ga/101104
Total magnetic intensity	http://pid.geoscience.gov.au/dataset/ga/89596
Rupture Offset Data	
Primary literature	[9,25,41,48,62,63,70,73,77]
Historic Photos of Ruptures	
Primary literature	[1,9,12,14,15,25,35,37,40,41,47,48,59,63,64,72,73,87,118]
Websites	https://aees.org.au/; http://fortennantcreekers.com/events/earthquake-friday-22-january-1988/
News articles	https://trove.nla.gov.au/; https://www.abc.net.au/news/

2. Review Data, Methods and Terminology

Publications reviewed for ten of the eleven historic ruptures are provided in Table 1. At the time of writing, no publications are available for the most recent (eleventh) earthquake (8 November 2018 Mw 5.3 Lake Muir earthquake), although one is currently in review [79] and some imagery and data are available online (https://riskfrontiers.com/the-2018-lake-muir-earthquakes/, https://www.abc.net.au/news/2018-11-09/earthquake-hits-lake-muir-western-australia/10480694 (accessed on 21 June 2019)). Available details for this event are included in Tables 1, 3 and 7 but it is otherwise not investigated in this paper. The Tennant Creek event comprises three mainshocks in a 12-hr period on the 22 January 1988, with three separate scarps recognized at the surface. Analysis of available seismological and surface data supports a direct association between each mainshock and an individual rupture (TC1: Kunayungku; TC2: Lake Surprise west; TC3: Lake Surprise east) [57,59,62,69] and they are treated as separate events in this paper.

Relevant papers were identified by reading through either (a) reference lists of recent (2010–2018) publications or (b) the citation history of older publications using Google Scholar. In total N = 67 articles were identified as containing relevant primary data and interpretations for individual or multiple surface rupturing events (Table 1). A further 16 publications were identified containing relevant information on Australian seismicity (drawing on data from the primary publications) or compilations of previously published material (Table 1). Other sources of data used to complement analysis of primary published data are summarized in Table 2.

Epicenter locations and focal mechanisms were collated from primary literature and online databases (Table 2). Geoscience Australia (GA) maintain an online earthquake catalogue that is continuously updated and recently published a national earthquake catalogue (NSHA18) from 1840 to 2017 [11]. The NSHA18 catalogue contains revised magnitude values (Mw) for all surface rupturing events based on a comprehensive reanalysis [11,119], which are used in this study. Epicenters for surface rupturing events are generally located closer to the surface ruptures in the online database than the NSHA18 catalogue.

Published surface rupture maps were previously digitized into GA's publicly available Neotectonic Features Database [95]. In this paper, we sourced the original maps, georeferenced them, and digitized secondary fracturing that was left out of the GA database (Figures 3–10). Some ruptures were relocated up to 200 m from the locations in the Neotectonic Features Database based on infrastructure and visible surface rupture matched on high resolution satellite imagery (Table 2) and primary maps, to correct for datum transformation errors.

For the purposes of this paper we use the terms "surface rupture" and "scarp" to describe the primary zone along which hanging-wall and foot-wall offset is visible at the surface. Fracturing relates to secondary surface features which do not host significant displacement, associated with the primary rupture (e.g., cracking). "Fissures" describe significant extensional cracks often with non-seismic edge collapse extending their width. "Fault" is used to describe the seismologically defined plane of rupture, of which the surface rupture is the observable expression.

3. Results

Detailed summaries of the geology, seismology, surface rupture and palaeoseismology for the eleven considered historical surface ruptures from 1968 to 2016 are available as seven EarthArXiv reports ([120–126]). Figures and data in these reports include available geological maps, geophysical maps, borehole data, surface rupture maps, displacement data, and available palaeoseismic trench logs. In the process of reviewing available literature, a number of inconsistencies in data usage or reproduction were identified. These are summarized in Section 4.1 of this paper, with more detail available in the EarthArXiv reports. Below is a concise summary of the seven reports (the three Tennant Creek ruptures are contained within a single report) with key data presented in Tables 3–9 and Figures 3–10.

3.1. Geology

The Meckering, Calingiri, Cadoux, and Katanning events occurred in the Archean Yilgarn Craton within ~25 km of significant terrane boundaries (Figure 1). The Lake Muir event occurred in the Albany-Fraser Orogen, <15 km south of the south dipping terrane boundary with the Yilgarn Craton (Figure 1). The Marryat Creek, Pukatja and Petermann events occurred within the Mesoproterozoic Musgrave Block (Figure 1) within 0–10 km of major terrane boundaries. The Tennant Creek ruptures extend across the boundary of the Proterozoic Warramunga Province and Neoproterozoic–Cambrian Wiso Basin (Figure 1) (summary of all regional geology in Table 3, comprehensive details in EarthArXiv reports [120–126]).

Table 3. Summary of regional geology for each historic surface rupture.

Rupture	Refs.	Geological Province			Nearby Regional Structure				
		Name	Age	Sub-Division	Name	Age	Geometry	Dist. from Rupture	Approx. Aligned?
Meckering	[33,91,108,127]	Yilgarn Craton	Archean	Jimperding Metamorphic Belt, Lake Grace Terrane	Boundary Lake Grace and Boddington Terranes	Archean	NW-SE, shallow E dipping suture	~25 km on HW	Yes
Calingiri	[91,108,127]	Yilgarn Craton	Archean	Jimperding Metamorphic Belt, Lake Grace Terrane	Boundary Lake Grace and Boddington Terranes	Precambrian	NW-SE, shallow E dipping suture	~10 km on HW	Yes
Cadoux	[91,108,127]	Yilgarn Craton	Archean	Archean greenstone	Boundary Murchison Terrane and Southern Cross Province	Archean	N-S	~10 km?	Yes
Marryat Creek	[128–131]	Musgrave Block	Mesoproterozoic	Fregon Domain	Mann Fault	Neoproterozoic	ENE-WSW, ~1km wide suture	<0.5 km	Yes (part)
Kunayungku ^	[66,68]	Wiso Basin	Neoproterozoic – Cambrian	–	–	–	–	–	–
Lake Surprise west ^	[66,68]	Tennant Creek Region	Paleoproterozoic	Warramunga Province	–	–	–	–	–
Lake Surprise east ^	[66,68]	Tennant Creek Region	Paleoproterozoic	Warramunga Province	–	–	–	–	–
Katanning,	[91,108,127]	Yilgarn Craton	Archean	Boddington Terrane	–	–	–	–	–
Pukatja	[128–131]	Musgrave Block	Mesoproterozoic	Fregon Domain	Woodroffe Thrust	Neoproterozoic	NE-SW, ~30° S, ~3 km wide suture	~10 km on HW	No
Petermann	[116,128–131]	Musgrave Block	Mesoproterozoic	Fregon Domain	Woodroffe Thrust	Neoproterozoic	NW-SE, ~30° S, ~3 km wide suture	~ 10 km on HW	Yes
Lake Muir		Albany Fraser Orogen	Proterozoic	Biranup Zone	Boundary Yilgarn and Albany Fraser Orogen	Mesoproterozoic	E-W, S dipping, ~10–20 km wide shear zone	~ 5–15 km on HW	No

^ Tennant Creek scarp.

Granitic gneiss, migmatite, mylonite, granulite, and/or amphibolite basement rock is observed in trenches or outcrop at <1 m depth at multiple locations along the Petermann (Figure 2), Pukatja, Marryat Creek (Figure 2), Cadoux and Meckering ruptures. Proterozoic basement in the vicinity of the Tennant Creek ruptures is variably overlain by 10 s to 100 s of meters of Phanerozoic basin bedrock. Structural measurements (foliations, intrusive boundaries) for bedrock outcrops within 5 km of surface ruptures are qualitatively well-aligned to surface ruptures in eight of ten events, though dip measurements are only qualitatively well aligned in three cases. This may relate to dip measurement difficulties for heavily weathered bedrock. (Summary of basement/bedrock in Table 4, comprehensive details in EarthArXiv reports [120–126]).

Nine of ten ruptures align with linear magnetic anomalies (Figure 2) and six align with linear gravity anomalies/gradients. The Katanning rupture does not align with either gravity or magnetics at the scale of available geophysical data (Figure 2, Table 2), and the Lake Muir rupture was not studied in this paper (paper in review: [79]). In cases where surface rupture traces are highly curved, arcuate, and/or segmented (Meckering, Marryat Creek, Tennant Creek, Pukatja), the distinctly-oriented rupture traces all align with distinct orientations of linear geophysical anomalies interpreted as faults, dikes, and lithological contacts (e.g., [33]).

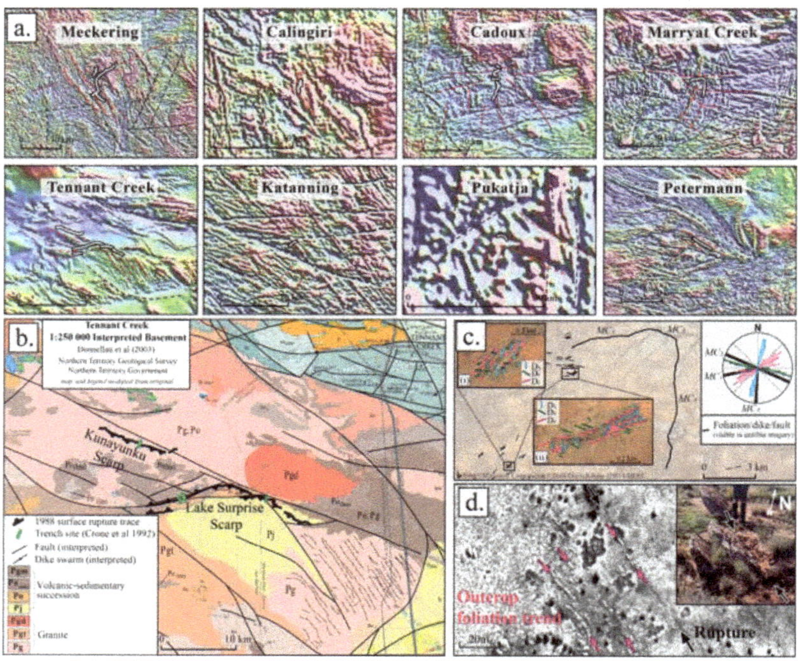

Figure 2. Examples of the relationship between geophysical data and surface outcrop to historic ruptures (**a**) national total magnetic intensity map with ruptures overlaid, and dashed lines indicating linear anomalies (**b**) interpreted basement geology around the three Tennant Creek scarps (no basement outcrops at the surface) demonstrating strong correlation between intrusive/lithological boundaries, basement faults, and historic surface rupture (legend heavily simplified to show lithologies around the ruptures, more details in EarthArXiv report [126] legend and original map. Map used under creative commons NT Gov) (**c**) examples of surface outcrop structures visible in basement around the Marryat Creek rupture including three sets of dike/foliation/fault orientations coincident with the three major orientations of the historic rupture, uninterpreted satellite imagery insets (i) and (ii) available in Marryat Creek EarthArXiv report [123] (**d**) example of mylonite foliation orientation along a section of the Petermann rupture where outcrop occurs within the primary rupture zone.

Table 4. Degree of alignment between rupture, basement structures, and geophysical anomalies.

Rupture	Refs.	Mapped Basement							Geophysics				Evidence of Basement Control on Rupture	
		Description	Depth (m)	Method *	Dominant Strikes	Alignment	Dips	Alignment	Mag.	Alignment	Gravity	Alignment	Y/N	Type #
Meckering	[16,25,33,114]	Granites, gneiss, mafic dikes	0–20	Geol.	NW, SW	Mod.	55°–90°	Low	Yes	Very high	Part	Mod.	Y [1]	FL_{GS}, V_S, D_G, FD_{GS}
Calingiri	[25,114]	Migmatite, gneiss	0–20	Geol.	NE	High	60°–90°	Low	Yes	High	Yes	High	Y	FL/FD_G
Cadoux	[41,115]	Granite, gneiss, mafic dikes	0–20	Geol. + BH	N, NW, NE, W	High	50°–90°	Mod.–High	Yes	Very high	Yes	Mod.	Y [2]	FL_{GS}, V_S, D_G
Marryat Creek	[47,48,117]	Mylonite, gneiss, granite, mafic dikes	0–5	Geol. + BH	W, NNE, NE	Very high	–	–	Yes	Very high	Part	High	Y [3]	FL_{GS}, FT_{GS}, D_{GS}
Kunayungku ^	[50–52,66,68,132]	Schist, granite	50+	BH + Geo.Int.	NE	Very high	–	–	Yes	High	Yes	Very high	Y [4]	FT_G
Lake Surprise west ^	[50,51,66,68,132]	Metavolc., metased., granite	?20+	BH + Geo.Int.	NW	Very high	–	–	Yes	Very high	No		Y [5]	FT_G, L_G
Lake Surprise east ^	[50,51,66,68,132]	Metavolc., metased., granite	50+	BH + Geo.Int.	NE	Very high	–	–	Yes	Very high	Yes	High	Y	FT_G, L_G
Katanning,	[133]	Granite, gneiss	–	Geol.	NW	Very low	–	–	No	–	No	–	N	
Pukatja	[9]	Granite, gneiss	0–10	Geol.	NE	High	30°	Very high	Yes	Very high	No	–	Y [6]	L_{GS}, FL_{GS}
Petermann	[73,128,134]	Mylonite	0–10	Geol. + BH	NW	Very high	20°–50°	Very high	Yes	Very high	Yes	Very high	Y [7]	FL_{GS}

^ Tennant Creek scarp. * 1:250,000 surface geology map (Geol.), ground-water borehole records (BH), geophysical interpretation (Geo.Int.). # Evidence for basement control: foliation (FL); fault (FT); fold (FD); vein (V); dike (D); lithological/batholith (L); geophysical ($_G$)/surface ($_S$) with specific examples detailed below: [1] Rupture along brecciated quartz in cutting / trench [25]. [2] Tank scarp ruptures through surface granite, and along a quartz vein in a hand-dug trench; bedrock within 10 m of the Kalajzic scarp aligned with rupture [41]. [3] Two trenches confirm rupture occurred along a pre-existing Proterozoic fault [118]. [4] Normal fault interpreted below Kunayungku scarp from ground-water boreholes and geophysical data prior to rupture (1981) [52,59]. [5] Lake Surprise west scarp runs along a 1–2 m high quartz ridge, associated with fluid movement along a bedrock fault [59,63,64]. Bedrock does not outcrop at the surface. [6] Rupture aligns with the projected boundary between a granite batholith and granulite facies gneiss; rupture curves around outcrop of granite batholith [9]. [7] Rupture abuts mylonite outcrop with the same strike and dip; mylonite with the same strike and dip outcrops within 1 m of the scarp in multiple locations [73].

3.2. Seismology

The sparse nature of the Australian National Seismograph Network (https://www.fdsn.org/networks/detail/AU/) results in large (i.e., ≥5–10 km) uncertainties in earthquake epicenter and hypocenter location estimates that are difficult to quantify, including those for the earthquakes studied here [102]. Epicentral determinations (Figures 3–10) are typically not sufficiently accurate to unambiguously associate with surface ruptures. Six of ten ruptures have favored epicenter locations that are located on the rupture hanging-wall, within approximated positional uncertainty bounds.

Many publications do not state statistical uncertainties for their epicenter locations. Uncertainties listed in Table 5 include published uncertainties or an assigned value of ± 10 km where no uncertainties are available [102]. Epicenters with lower uncertainties are derived using a variety of relocation methods including extra analysis (e.g., InSAR slip distributions, joint hypocenter determination) or extra data (e.g., surface rupture location, aftershocks from temporary seismometer arrays) (comprehensive details in EarthArXiv reports [120–126]). The epistemic uncertainty relating to the quality of velocity models used to locate epicenters is unconstrainted but appears to be one of the major sources of inaccurate locations where instrumentation is particularly sparse. For instance, epicenters for Pukatja and Marryat Creek are located up to 17 and 30 km from the identified surface rupture respectively, showing large uncertainties still affecting remote earthquake locations between 1986–2012.

Hypocenters derived from mainshock instrumental data do not project onto rupture planes as defined by surface rupture for any of the studied events. Hypocentral depth estimates based on aftershock data and relocated epicenter locations suggest depths of <5 km (for Tennant Creek [59], Petermann [73] and Meckering [37]). Centroid moment tensor depths are <6 km depth, with the authors' preferred best-fits all <4 km depth (Meckering [26–28]; Cadoux [28]; Marryat Creek [28]; Tennant Creek [55]; Katanning [70]; Petermann [74]).

Epicentral location uncertainties limit the study of rupture propagation directions(s) for most events. Model scenarios for the Meckering earthquake support a bilateral rupture [37]. Unilateral upwards propagation has been proposed for the first Tennant Creek mainshock, complex propagation in the second mainshock, and unilateral upwards propagation to the Southeast in the third mainshock (all on separate faults) [57].

Seven of ten events show foreshock activity within six months and 50 km of the mainshock epicenter and six of ten show instrumentally recorded prior seismicity (more than five events within 10 years and 50 km). Precise locations are difficult to obtain due to epistemic and statistical uncertainties, particularly for assessing seismicity prior to 1980 due to sparse instrumentation [102]. Aftershock data are inherently incomplete for most events due to sparse instrumentation. However, temporary seismometers were deployed following most events and magnitude completeness from the national network is >3.0 Mw for all events [102] (though, the locations of these events are generally highly uncertain compared to the temporary arrays, as discussed above). The Musgrave block events (Marryat Creek, Pukatja, Petermann, Figure 1, Table 3) show less aftershock activity in comparison to the Tennant Creek and Western Australia earthquakes (Meckering, Calingiri, Cadoux) which had extended aftershock sequences [5,34].

Table 5. Summary of seismological data and interpretations for each rupture.

Rupture	Refs.	Published Epicenters		Located on Hanging-Wall		Hypocenter Depth [1] (km)		Focal Mechanisms				Rup. Propagation	Foreshocks [4]	Prior Seismicity [5]
		N=	Uncert. [2] (km)	Initial	Relocated	Range	Uncert.	N=	CMT Depth [3] (km)	Dip Range	Strike Corr. with Rupture?			
Meckering	[11,12,15,18,23–26,28,90]	8	~10	1	2	2.5–13	1–10 km	4	1.5–3	29°–45°	Yes	Bilateral	Yes	Yes
Calingiri	[11,23,25,38]	2	~10	0	1	1–15	>5	1		50°	Yes	Unknown	Yes	Yes
Cadoux	[11,28,40,41,43]	6	~10	N/A [6]	N/A [6]	3–15	>5	4	4–6	N/A [6]	Part [6]	Unknown	Yes	Yes
Marryat Creek	[11,28,46,47,88]	7	>10	0	1	5–19	>5	3	0–3	35°–67°	Part	Unknown	No	No
Kunayungku ^	[11,55,57,60,61,63]	4	2–10	?	0	5–6.5	1–4	4	2.7	35°–55°	Yes	Unilateral	Yes	Yes
Lake Surprise west ^	[11,55,57,60,61,63]	4	2–10	?	2	3–4	0.5–3	4	3.0	38°–70°	Part [7]	Unknown	Yes	Yes
Lake Surprise east ^	[11,55,57,60,61,63]	4	2–10	?	4	4.5–5	0.5–3	4	4.2	36°–45°	Yes	Unilateral	Yes	Yes
Katanning,	[70]	3	0.04–5	0	2	<1		1		43°	Yes	Unknown	No	No
Pukatja	[9,11]	6	>10	0	1	4–12	5?	3		45°–72°	Yes	Unknown	No	No
Petermann	[73–75]	6	2.5–8	4	2	1–10	> 5	4	1–2	26°–52°	Yes	Unknown	yes	No

^ Tennant Creek scarp. [1] Includes both initial hypocentral depth estimates, and revised depths based on aftershock depths and locations, uncertainties from source literature. [2] Epicenter uncertainty based on published uncertainties and/or estimate based on published uncertainties for similar events. [3] Centroid moment tensor depth, preferred value from publication. [4] Earthquake within 6 months and 50 km of epicenter (affected by catalogue completeness for very remote events, see EarthArXiv reports ([120–126]) for details). [5] Earthquakes (n > 5) within 10 years and 50 km of the epicenter (affected by catalogue completeness for very remote events, see EarthArXiv reports for details). [6] Geometry of the seismogenic fault is unclear as scarps in the Cadoux rupture dip both east and west. [7] Waveform analysis of the second Tennant Creek mainshock show complicated rupture, potentially related to complex fault interaction

3.3. Surface Ruptures

Methods for the original mapping of individual ruptures are summarized in Table 6 and give some indication of data quality (explored in more detail in EarthArXiv reports [120–126]). Some readjustment of terminology and classification is required when considering the earlier ruptures (e.g., 'fault' may refer to both primary rupture and secondary fractures) and considerable detail of rupture morphology was lost between fine-scale (i.e., 1:500) and whole rupture (1:25,000–1:50,000) for pre-digital maps (Meckering to Tennant Creek). Six of ten ruptures are concave relative to the hanging-wall, three are straight and one is slightly convex (Petermann) (Table 6). All ruptures are reverse, and only two events have surface measurements consistent with secondary lateral movement (Meckering: dextral; Calingiri: sinistral; Table 6, explored in individual EarthArXiv reports [120–126]).

Nine of ten ruptures studied (Katanning was excluded due to lack of field mapping) show a relationship between surface sediments/bedrock depth to rupture morphology. Discrete rupture and duplexing rupture are more common where bedrock is close to the surface or surface sediments are predominately calcrete/ferricrete/silcrete. Where sands dominate in the surface sediments, rupture tends to present as warping and folding, or correspond with breaks in visible surface rupture (e.g., Petermann: morphology explored in individual EarthArXiv report [125])).

Figures 3–10 show digitized versions of published primary ruptures, secondary fracturing, and dip values measured at the surface. Primary sources inconsistently derive published length values to describe their mapped rupture (Tables 1 and 6; explored in detail in EarthArXiv reports [120–126]) which are then used in secondary sources including scaling relationships. This includes simplifying scarps to straight lengths (Calingiri, Cadoux, Marryat Creek), capturing along-rupture complexity to varying degrees (Pukatja, Tennant Creek), excluding segments that have length, offset and morphology characteristics of primary rupture (Meckering, Tennant Creek, Cadoux), and reporting InSAR derived lengths rather than visible rupture (Katanning). (Explored in more detail in individual EarthArXiv reports [120–126]).

Measurements of rupture length in the past have been inconsistent in their approach. Here, we re-classify mapped primary ruptures from original primary sources in order to generate a consistent rupture length dataset (Table 6). We simplify ruptures to straight lines and define new faults where mapped primary rupture has gaps/steps > 1 km and/or where strike changes by > 20° for distances > 1km [135]. The Splinter and Burges scarps (Meckering), Lake Surprise west foot-wall scarp (Tennant Creek), and individual Cadoux scarps were not included in original published lengths. These features show offsets, lengths, and locations consistent with primary slip along basement structures proximal to the main scarps, and therefore we include them in our length values.

Where InSAR is available (Katanning and Petermann) we present fault lengths described by both visible rupture and InSAR (Table 6). Visible rupture in the Petermann event was highly segmented due to ineffective rupture propagation through sand dunes up to six metres high [73]. Due to this we apply a slightly altered set of criteria to this event, faults are defined where strike of visible rupture and InSAR changes by > 20° and/or where steps in InSAR and visible rupture are > 1 km (Figure 10) [75,78].

Under these criteria seven of ten ruptures have more than one source fault defined (i.e., a multi-fault rupture). The total length of faulting is the same as published values for two events, increases by 2%–51% for four events relative to published length, and decreases by 4%–60% for three events (Tables 1 and 6). These lengths describe primary surface ruptures in a consistent way, accounting for all segments of rupture which show evidence of slip along basement structures. Our preferred length for each rupture, including uncertainties, is presented in Tables 1 and 6.

Table 6. Summary of surface measurements for each rupture.

Rupture	Refs.	Method [3]	Shape [4]	Published: Length (km)	Published: Kin.	Published: Dip Range	Simplified faults [1]: N=	Simplified faults [1]: Sum Length (km)	Simplified faults [1]: % Diff. Publ.	Preferred: Length (km)	Preferred: Dip	Displacement [2] (m): Max Vert. Disp.	Displacement [2] (m): Avg. Vert. Disp. [5]	Displacement [2] (m): % Diff.	Displacement [2] (m): Max Net Slip	Displacement [2] (m): Avg. Net Slip [5]	Displacement [2] (m): % Diff	Disp. Profile Shape [6]
Meckering	[25]	FW; A; S_B	CC	37	R(D)	15°–54°	4	44.4	+20%	40 ± 5	35° ± 10	1.98	0.97	51%	3.7	1.78	52%	S. Tg.
(Splinter)	[25]			9	R	24°–42°	3					0.67	0.22	67%	1.34	0.44	67%	AS. Tg.
Calingiri	[25]	FW	S	3.3	R(S)	12°–31°	1	3.3	0%	3.3 ± 0.2	20° ± 10	0.38	0.15	61%	1.26	0.46	63%	AS. Tg.
Cadoux	[41]	FW; A; S_B	CC/S	14	R	20°–80°	6	20.6	+47%	20 ± 5	60° ± 30	1.4	0.35	75%	1.79	0.54	70%	AS. Tg.
Marryat Creek	[48]	FW; A; S_C	CC	13	R	36°–60°	3	13.6	+4%	13 ± 1	40° ± 10	0.9	0.21	77%	1.07	0.31	71%	Avg.
Kunayungku	[63]	FW; A; S_C	S	10.2	R	58°	1	8.6	−15%	9 ± 1	40° ± 5	0.9	0.36	60%	1.41	0.55	61%	S. Tg
Lake Surprise west	[63]	FW; A; S_C	CC	6.7	R	65°–84°	1	10.1	+51%	9 ± 2	60° ± 10	1.13	0.45	60%	2.26	0.84	63%	AS. Sine
(LS west foot-wall)	[63]			3.1	R		1					0.74	0.43	42%	1.16	0.9	22%	Avg.
Lake Surprise east	[63]	FW; A; S_C	CC	16	R	28°–30°	2	15.3	−4%	16 ± 0.5	35° ± 5	1.8	0.61	66%	3.6	1.23	66%	Avg
Katanning (visible) [7]	[70,71]		S	0.3	R		1	0.3	0 %	0.5 ± 0.5	40° ± 5	0.1						
Katanning (InSAR)	[70]	In.		2.5 [8]			1	2.2	−12%			0.2 [8]	0.1	50%	0.32 *	0.2 *	38%	AS. Tg.
Pukatja	[9]	FW	CC	1.6	R	22°–28°	1	1.0	−60%	1.3 ± 0.3	30° ± 10	0.48	0.12	75%	0.96	0.25	74%	AS. Sine
Petermann (visible)	[73,77]	FW; S_C; In; D; SI	CV	20	R	25°–36°	3	21	+5%	21 ± 0.5	30° ± 5	0.96	0.2		1.92	0.42	78%	Avg.
Petermann (InSAR)	[73]	In.		21			2	21.5	+2%									

[1] Where mapped primary rupture has a gap/step > 1 km and/or change in strike > 20° across a length > 1 km (except where InSAR is available to validate rupture continuing along strike across gaps > 1 km). Lengths of individual faults available in EarthArXiv reports [120–126]. [2] Vertical and lateral displacements digitized from original publications. Net slip calculated for this study. [3] Original mapping method: Field work (FW); aerial photographs (A); surveying (levelling, cadastral or GPS) basic (S_B), comprehensive (S_C); InSAR (In); Drone (D); Satellite (SI). [4] Concave (CC) relative to hanging-wall, convex (CV) relative to hanging-wall, straight (minor deviations but overall straight shape). [5] Length weighted average across 0.5 km increments (where rupture length > 5 km) or 0.1 km increments (where rupture length < 5 km). [6] Profile shape based on Wesnousky (2008) [7] from visual fit (e.g., not best-fit regression curves): symmetrical (S); asymmetrical (AS); triangle (Tg); sine; average line (Avg). [7] Katanning visible surface rupture was observed, but no field mapping was conducted [70,71]. Original and subsequent publications describe Katanning length based on best-fit InSAR-derived source parameters (1.26 km) [70], rather than length of InSAR trace (2.5 km). Offset comes from field estimates (0.1 m) and fault modelling from InSAR data [70].

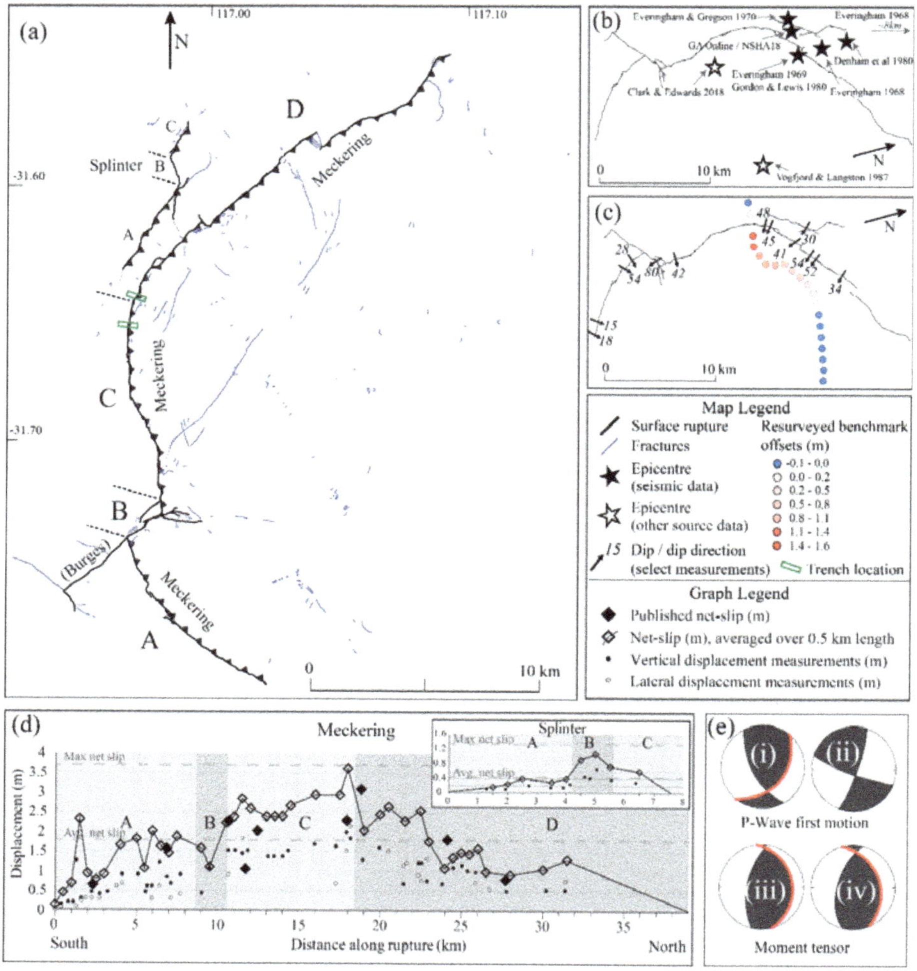

Figure 3. 1968 Mw 6.6 Meckering earthquake (**a**) rupture and fracture map of Meckering and Splinter scarps [25] with faults labelled as per displacement graphs, trench location from [37] (**b**) published epicenter locations, open stars show approximate locations of epicenters without published coordinates (**c**) selected dip measurements of scarp and displacement of resurveyed road bench marks [25] (**d**) graphs of along-rupture vertical and lateral displacement measurements and net slip calculations [25] and net slip calculated from available data averaged over 0.5 km increments (this study) (**e**) focal mechanisms (red line shows preferred plane from original publication) from (i) Fitch et al., 1973, (ii) Fitch et al., 1993 & Leonard et al., 2002, (iii) Fredrich et al., 1988, and (iv) Vogfjord and Langston 1987.

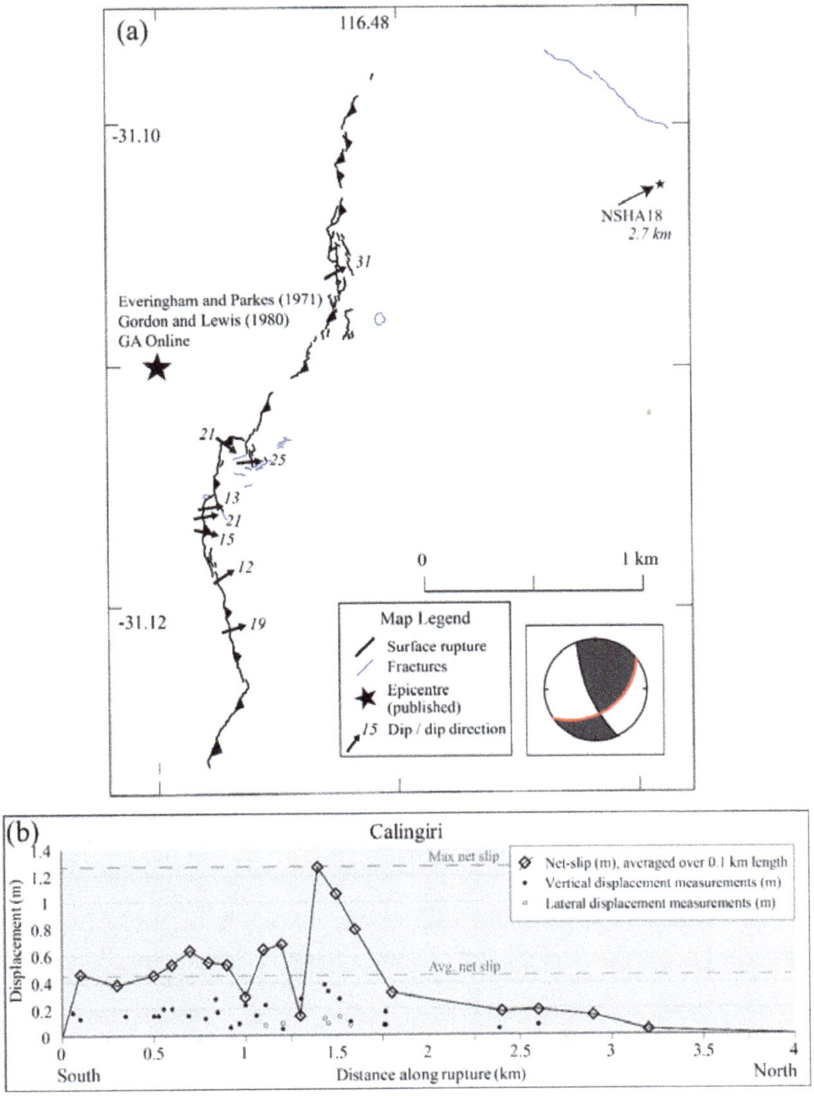

Figure 4. 1970 Mw 5.0 Calingiri earthquake (**a**) rupture and fracture map of Calingiri [25] showing published epicenter locations and dip measurements of scarp [25], focal mechanism (red line shows preferred plane from original publication) from Fitch et al., 1973 (**b**) graph of along-rupture vertical and lateral displacement measurements [25] and net slip calculated from available data averaged over 0.1 km increments (this study).

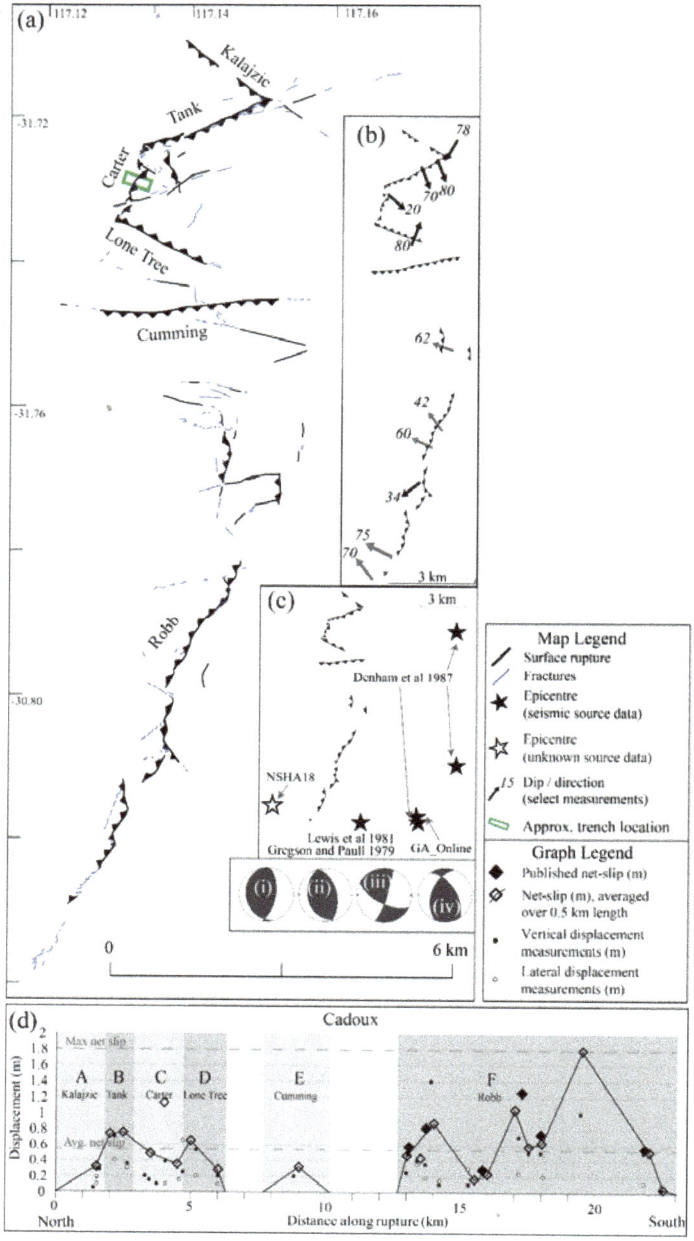

Figure 5. 1979 Mw 6.1 Cadoux earthquake (**a**) rupture scarps and fracturing involved in the Cadoux rupture with named faults [41], focal mechanisms from (i) Denham et al., 1987 (ii) Fredrich et al., 1988 (iii) Everingham and Smith (unpublished, Lewis et al., 1981) (iv) CMT (**b**) available dip measurements, black where directly measured and grey were calculated based on available displacement measurements [41] (**c**) published epicenter locations (**d**) graph along-rupture of vertical and lateral displacement measurements and calculated net slip [41] and net slip calculated from available data averaged over 0.5 km increments (this study).

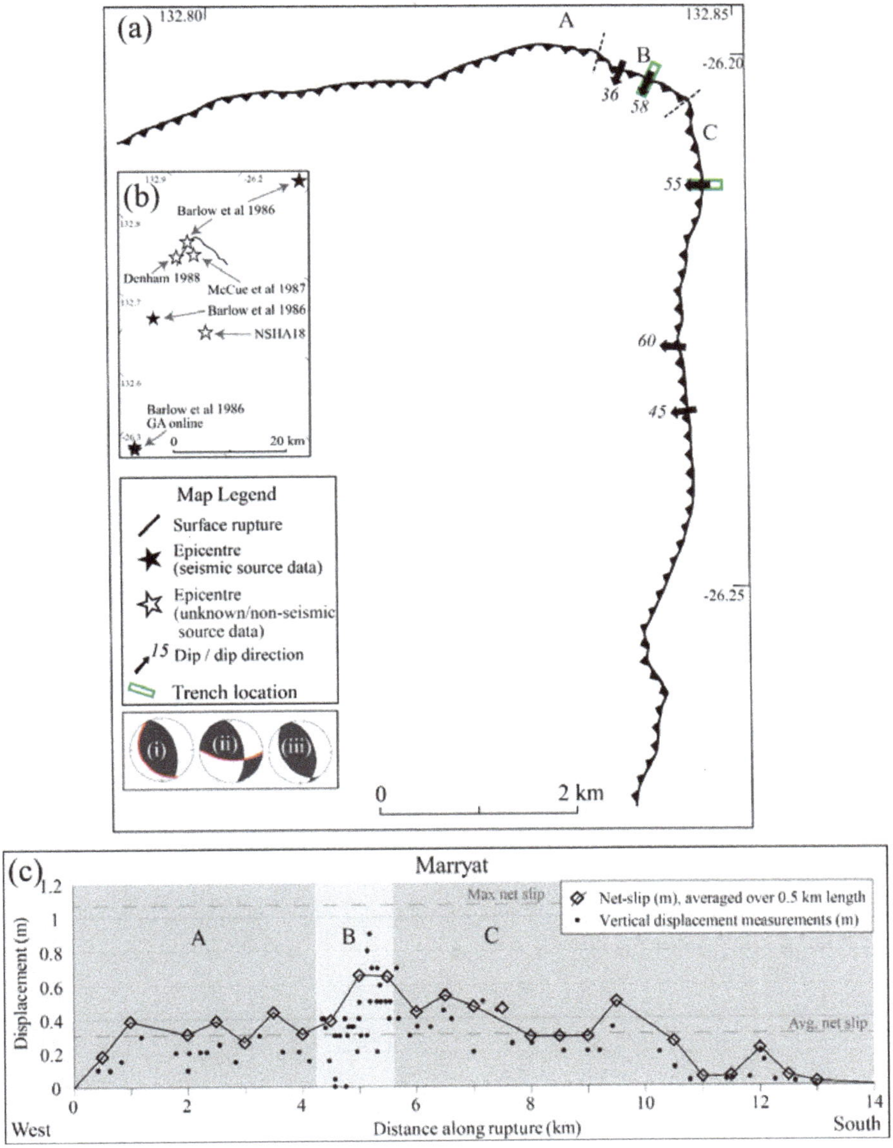

Figure 6. 1986 Mw 5.7 Marryat Creek earthquake (**a**) rupture and fracture map of Marryat Creek scarp and available dip measurements [48,118] with faults labelled as per displacement graphs, focal mechanisms (red line shows preferred plane from original publication) from (i) Fredrich et al., 1968, (ii) Barlow et al., 1986, (iii) CMT, trench location from [118], (**b**) published epicenter locations, and (**c**) graph of along-rupture vertical and lateral displacement measurements [48] and net slip calculated from available data averaged over 0.5 km increments (this study).

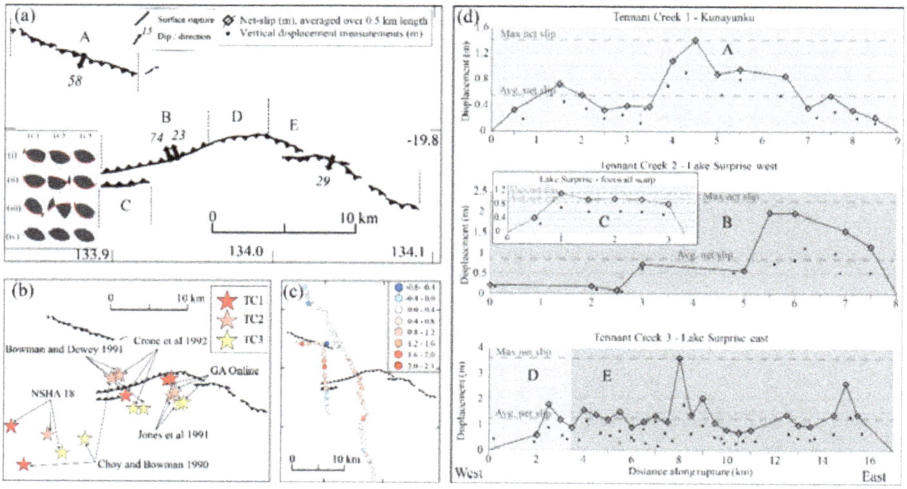

Figure 7. 1988 Mw 6.3 (TC1), 6.4 (TC2) and 6.6 (TC3) Tennant Creek earthquakes (**a**) rupture and cracking map of Kunayungku and Lake Surprise scarps with available dip measurements also the locations of trenches from [63] with faults labelled as per displacement graphs, focal mechanisms (red line shows preferred plane from original publication) from (i) McCaffrey 1989, (ii) Choy and Bowman 1990, (iii) Jones et al., 1991, (iv) CMT, (**b**) published epicenter locations of all three mainshocks (**c**) resurveyed benchmark offsets [63] uncertainties as discussed in text, and (**d**) graphs of along-rupture vertical and lateral displacement measurements [63] and net slip calculated from available data averaged over 0.5 km increments (this study).

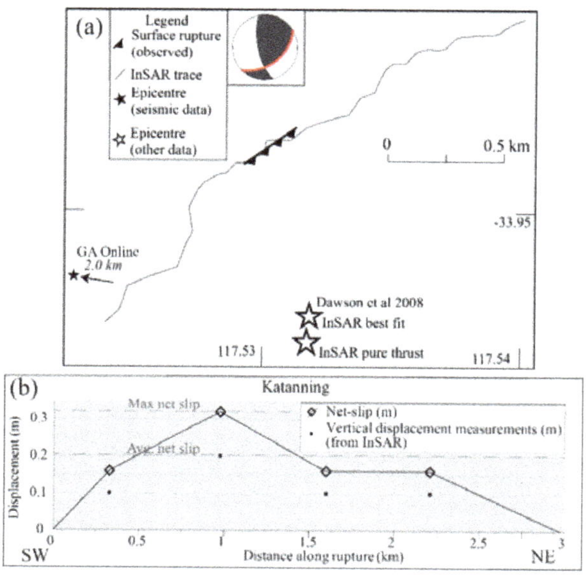

Figure 8. 2008 Mw 4.7 Katanning earthquake (**a**) approximate visible rupture and InSAR trace (digitized from [70]), published epicenter locations and focal mechanism [70] (**b**) graph of along-rupture vertical and displacement taken from InSAR data [70] and net slip calculated from InSAR data (this study).

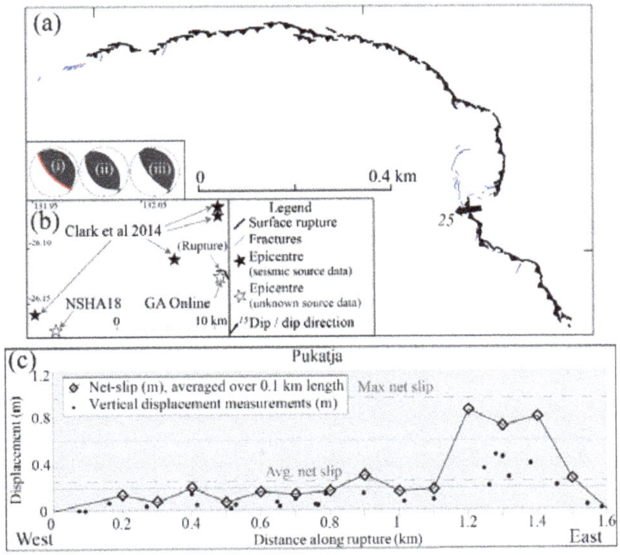

Figure 9. 2012 Mw 5.2 Pukatja/Ernabella earthquake (**a**) rupture and fracture map of Pukatja scarp and available dip measurements also the location of hand-dug trench [9], focal mechanisms as described in [9] from (i) Clark et al., 2014, (ii) GCMT, (iii) St Louis University; (**b**) graph of along-rupture vertical displacement measurements [9] and net slip calculated from available data averaged over 0.1 km increments (this study).

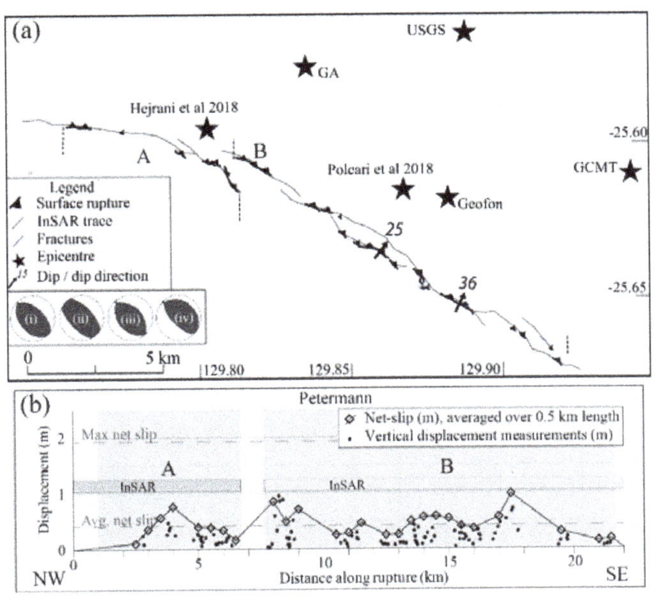

Figure 10. 2016 Mw 6.1 Petermann earthquake (**a**) rupture and fracture map of Petermann scarp [73] showing published epicenter locations and dip measurements of rupture (also the location of hand-dug trenches), focal mechanisms (i–iii) as described in [73], (i) USGS, (ii) GCMT, (iii) Geofon, and (iv) from Hejrani and Tkalcic 2018; (**b**) graph of along-rupture vertical displacement measurements and net slip calculated from available data averaged over 0.5 km increments.

Vertical and lateral offsets for all ruptures were digitized from primary literature (see EarthArXiv reports [120–126] for methods and uncertainties). New net-slip values were calculated for all ruptures from measured offsets, with dips assigned to each field offset measurement based on measured surface dips and/or focal mechanisms (dip measurements from primary literature shown on Figures 3–10 and in Table 6—preferred dip from this paper in Tables 1 and 6). Offset and net-slip data are presented in Figures 3–10 along with length weighted averages to reduce bias towards sections of scarp where high number of measurements were taken (generally where offset is higher). Average offset is between 42%–77% lower than the maximum offset for each rupture (Table 6). Displacement data are visually assigned a displacement profile shape [7] with six of ten ruptures best described by triangle profiles (2= symmetrical, 4= asymmetrical), two assigned an asymmetrical sine profile, and three best represented by a straight average profile (Table 6).

Offset data from resurveyed benchmarks (Tennant Creek [62]) and relevelling along infrastructure (Meckering [25]) were digitized from original publications to visualize distributed deformation across the rupture zone (Figures 1 and 7). No uncertainties are reported for the Meckering data [25], though they are likely in the order of Tennant Creek where original authors report uncertainties of ±9.3–25 cm [62,136]. Despite large uncertainties, the authors of both datasets believe offsets constrain fault geometries and show distributed hanging-wall uplifts (and to a lesser extent, foot-wall depression).

3.4. Environmental Damage

Environmental damage as described in primary literature or visible in published photos for each event were classified under the ESI-07 Scale [10] and summarized in Table 7 (comprehensive details in EarthArXiv reports [120–126]). Seven of eleven historic ruptures (excluding Katanning) can be described as an ESI IX – X despite having a wide range of lengths, magnitudes, and displacements.

Fracturing/cracking is reported for all historic surface ruptures, but generally only in the immediate vicinity of the surface rupture, captured by the rupture ESI value. This may relate to a lack of far-field mapping but is considered to be fairly representative of the true spatial distribution based on described mapping campaigns. The Meckering and Petermann events have the most aerially extensive fracturing, with areas of 580 and 210 km^2 respectively. Of the total area covered by Meckering and Petermann fracturing, approximately 70% and 77% respectively is on the hanging-wall.

Where events occurred close to population centers (Meckering, Cadoux, Calingiri) ground water bores showed evidence of seismic fluctuation (no anomalies were identified in Tennant Creek bore data). The only observed liquefaction for any historic rupture comes from Meckering, where multiple sand blows were observed on the hanging-wall along the Mortlock River. Rockfalls are reported for three ruptures. Concentric or polygonal cracking was reported in the Meckering, Calingiri, Cadoux and Petermann events [25,41,73], and holes (possibly related to collapsible soils e.g., [137]) were reported along the rupture on the hanging-wall for Calingiri and Petermann [41,73]. It is possible that tree damage, hydrological effects, rock falls, polygonal cracking, or holes occurred for other ruptures than those listed but were not observed or described. Until the 2012 Pukatja event, field investigations immediately following the event were conducted by hard-rock geologists or seismologists not necessarily familiar with earthquake mapping techniques.

3.5. Paleoseismology and Slip Rate

In total, 14 trenches are described across the Meckering, Calingiri, Cadoux, Marryat Creek, Tennant Creek, Pukatja and Petermann ruptures (Table 8). Tennant Creek, Marryat Creek and Meckering are the only ruptures where detailed palaeoseismic work is published, including multiple trenches, luminescence dating, and soil descriptions and chemistry [37,63].

Table 7. Summary of environmental effects described for each rupture, assigned ESI level, and approximate area around, or distance from, rupture.

Name	Rupture		Fractures		Displaced Rocks		Sand Blows		Tree Shaking		Slope/Rock fall		Hydrology [1]		Holes [41,73]	Concentric Fracturing [25,41,73]	Vegetation Death/Root Tear along Rupture [9,25,47,48,63,73]
	ESI	Area (km^2)	ESI	Area (km^2)	ESI	Area (km^2)	ESI	Area (km^2)	ESI	Area (km^2)	ESI	Dist. (km)	ESI	Dist. (km)			
Meckering	X	190	VIII	580	–		V	100	●		IV	200	IV	200	–	□	□
Calingiri	IX	2	VII	3	*		–		–		–		–		□	□	□
Cadoux	IX	50	VI	55	–		–		●		–		IV	250	–	□	–
Marryat Creek	X	20	VII	20	–		–		–		–		–		–	–	□
Tennant Creek ^	X	160	VIII	160	–		–		–		–		●		–	–	□
Katanning,	VIII	0.2	–		–		–		–		–		–		–	–	–
Pukatja	IX	1	VII	1	–		–		*		VII	15	–		–	–	□
Petermann	X	12.5	VII	210	IX	290	●		VI	20	VII	20	–		□	□	□
Lake Muir	IX		*		–		–		*		–		*		–	–	–

^ Treating the three Tennant Creek scarps as individual or combined results in the same ESI values. Area given is for a combined scarp. Magnitude is for largest mainshock. [1] Anomalous ground-water levels recorded following earthquake. – No data or observations published. □ Observations of damage outside of ESI-07 descriptions. ● Damage noted as explicitly not present. * Evidence of damage but no detailed description.

Table 8. Summary of available paleoseismic trenching.

Rupture	Ref.	Location	Bedrock		Depth (m) (HW/FW)	Sediments		Pre-Existing:		Possible Rup. Since 100 ka
			Desc.	**Age**		**Desc.**	**Age** *	**Cenozoic Offset**	**Bedrock Offset**	
Meckering 1	[34,36,37]	Tributary of river	Not exposed		>3	Fluvial sands	Late Pleistocene	No	No	N
Meckering 2	[34,36,37]	Farmland	Altered granite	Archean	<1.5	Ferricrete / sand	T?/Holocene	No	Maybe [1]	N
Calingiri	[25]							No	Maybe [1]	
Cadoux/Carter scarp	[41]	Farmland	Altered granite	Archean	<0.5	Sand/soil	Q?	No	No	
Cadoux/Tank scarp	[41]	Farmland	Granite	Archean	0?			No	No	
Marryat Creek west	[118]	10m N of dry creek	Altered gneiss	Proterozoic	<0.3	Ferricrete / eolian sand	T?/*<130 ka*	No	No	N
Marryat Creek south	[118]	'Near' dry creek	Altered greenstone	Proterozoic	<1.25	Ferricrete/gravel, eolian sand	T?/**<130 ka**	No	No	N
Kunayungku	[63]	Max vert. disp.	Altered sedimentary rock	Cambrian?	<2	Ferricrete/eolian sand	T?/**<30 ka**	No	No	N
Lake Surprise east	[63]	Max vert. disp.	Not exposed		>2	Ferricrete/eolian sand	T?/**<46 ka**	No	No	N
Lake Surprise west 1	[63]	Monoclinal rupture (near quartz ridge)	Altered "iron-rich quartzite"	Cambrian?	<0.5/>2	Gravel/eolian sand	Q/*<50 ka*	No	Yes [3]	Y
Lake Surprise west 2	[63]	Discrete rupture (375m E LSW-1)	Altered "hematitic quartzite"	Cambrian?	<0.5/>0.7	Ferricrete/gravel/eolian sand	T?/Q/*<50 ka*	Maybe [2]	Yes [3]	Y
Pukatja	[9]	Max. vert. disp.	Not exposed		>1.5	Sand	$\sim10^4$–10^5	No	No	Y
Petermann 1	Unpub.	Paleovalley	Not exposed		>1.5	Calcrete/eolian sand	T?/Q	No	No	N
Petermann 2	Unpub.	Inter-dune region	Not exposed		>1.5	Calcrete/eolian sand	T?/Q	No	No	N

* ages in bold from direct dating in literature, italics inferred based on nearby dating, underline from estimate in literature, "T?" or "Q?" estimated Tertiary or Quaternary. [1] Thicker soil horizons described on foot-wall relative to hanging-wall [25]. [2] Fissure of potentially seismic origin filled with gravel, overlain by eolian sand; fracture through ferricrete overlain by gravel and eolian sand [63]. [3] Authors propose three possible origins: earthquake rupture bedrock offset; paleochannel along pre-existing bedrock structure; or combination of both; paleotopography greater than twice the height of historic slip [63].

Of seven ruptures with detailed trench data (Table 8), five show evidence of no rupture since the lake Pleistocene (Meckering, Marryat Creek, Kunayungku, Lake Surprise east, Petermann). The only evidence of a pre-existing bedrock scarp exposed in any trench occurs in the second Lake Surprise west trench, and no clear evidence was found to support a seismic-offset origin for this topography (see EarthArXiv report [126] for more detail). Penultimate ruptures since 100 ka are possible for two of seven of these earthquake events, where sediments are estimated to be <100 ka in age, and where either no ferricrete/bedrock is exposed (Pukatja), or a bedrock scarp exists prior to overlying sedimentation (Lake Surprise west) (Table 8).

Maximum slip rates are calculated by applying minimum and maximum erosion rates for bedrock to determine the amount of slip (from average observed historic slip (Table 6)) that could have accumulated and been removed in the past million years. Minimum (0.3 m Myr^{-1}) and maximum (5.7 m Myr^{-1}) cosmogenic nuclide erosion rates from crystalline bedrock inselbergs across the Precambrian crust of central Australia (Figure 1) [138] are applied for ruptures where crystalline basement is exposed in trenches or observed at the surface within two meters of rupture (implying shallow bedrock). Where trenching exposed ferricrete or quartzite, cosmogenic nuclide erosion rates for quartzite exposed on flat bedrock summits in the Flinders Ranges are applied (5–10 m Myr^{-1}) [109].

Applying erosion rates from inselbergs and quartzite bedrock summits to surface bedrock across the different cratonic and erosional environments in which ruptures occurred (e.g., Figure 1) introduces uncertainties that are unavoidable due to a lack of more appropriate erosion data. Based on a lack of evidence of any preceding ruptures for any of the historic events, including topographic or geomorphic, we prefer the minimum erosional estimates, giving maximum slip rates of 0.2–9.1 m Myr^{-1}.

Table 9. Maximum slip rates based on minimum and maximum bedrock erosion rates [109,138] and length-weighted average net-slip values (Table 6).

Name	Rate Applied *	Mw	Pref. Length (km)	Avg. Net-Slip (m)	Maximum Slip Rate (m/Myr)		
					Min.	Max.	Mean
Meckering	CB	6.59	40 ± 5	1.78	0.2	3.2	1.7
Calingiri	CB	5.03	3.3 ± 0.2	0.46	0.7	12.4	6.6
Cadoux	CB	6.1	20 ± 5	0.54	0.6	10.6	5.6
Marryat Creek	CB	5.7	13 ± 1	0.31	1	18.4	9.7
Kunayungku	Q	6.27	9 ± 1	0.55	9.1	18.2	13.7
Lake Surprise west	Q	6.44	9 ± 2	0.84	6	11.9	8.9
Lake Surprise east *	Q	6.58	16 ± 0.5	1.23	4.1	8.1	6.1
Katanning (InSAR)	CB	4.7	0.5 ± 0.5	0.2	1.5	28.5	15
Pukatja	CB	5.18	1.3 ± 0.3	0.25	1.2	22.8	12
Petermann	CB	6.1	21 ± 0.5	0.42	0.7	13.6	7.2

* Erosion rate for crystalline basement (CB) [138]; erosion rate for quartzite (Q) [109].

4. Discussion-Lessons from the Last 50 Years of Australian Surface Ruptures

4.1. Inconsistancies in Data Use

A number of inconsistent uses of data were found while reviewing papers that reference primary sources, as summarized below:

- Some of the original Marryat Creek vertical offset measurements are reproduced incorrectly in subsequent publications [7,118]. We recommend referring to the original source [48] or the data tables from this paper.
- Limbs of the Marryat Creek rupture show sinistral (west limb) and dextral (south limb) components due to SW over NE directed uplift along an arcuate rupture (best described as three faults, Figures 2

and 6). Data tables used in subsequent scaling relationships [6] describe the event as left-lateral based on one of three published focal mechanisms. We recommend a dominantly reverse mechanism for this event based on all available data.

- We recommend referring to the original source of the Calingiri focal mechanism [23] when describing kinematics and preferred rupture plane geometry, as subsequent authors [25] appear to misread the mechanism [121].
- The Tennant Creek rupture has been treated by multiple authors as a single rupture length for fault scaling relationships [7,8,104,139,140] and hazard mapping [119,141,142] as opposed to three separate earthquakes and associated ruptures [6,143–145], a decision which significantly changes the length to magnitude ratio and slip distribution relative to an averaged epicentre location and magnitude.
- An instance of the above decision is seen where a 6 km step over between Tennant Creek scarps is identified as an outlier for reverse fault data [7]. The rupture length and complexity for this event is not anomalous if treated as three separate events.
- Some scaling relationships (e.g., [7]) define only a portion of the Cadoux scarp (the 10km long Robb Scarp), due to "insufficient mapping" of the northern ~6 km. The full rupture includes one step-over that fits the publication analysis criteria (>1 km) but is not represented in the original paper's database [7] and subsequent work [8,139,146]. Mapping of the Cadoux rupture was thorough, but uncertainty regarding which of the mapped scarps (if any) represent the Cadoux mainshock fault complicates the use of this event in scaling relationships.
- A slip rate of 0.005 mm/yr is used to describe the Marryat Creek scarp [147]. This value is likely derived from ~0.5 m measured historic slip in a trench with no evidence of prior rupture between deposition of Quaternary sediments (estimated age from primary source < 130 ka [63]) and formation of the structure along which the modern event ruptured (Proterozoic) [118]. While evidence of prior Quaternary–Tertiary rupture may have been removed by erosion, a slip rate of 0.5 m per 100,000 years is unsupported by available evidence.
- The recently published NSHA18 applies slip rates of 4–8 m/Myr for the historic ruptures discussed in this paper [119,141]. No historic surface ruptures provide convincing evidence of rupture between deposition of Quaternary sediments (50 ka to late Pleistocene) and formation of the host structure (Archean–Cambrian). At least for cratonic areas of Australia (Figure 1), questions arise as to whether historically seismogenic faults are recurrent at all, or if long-term seismic release may be hosted across unique basement structures (e.g., [4]), as well as on recurrent structures (e.g., Lort River, Hyden, Dumbleyung [2,34,108]). We recommend caution in applying these rates in future work.

4.2. Surface rupture Bedrock Controls, Updated Datasets and Environmental Intensity

Analysis of geology and reanalysis of mapped ruptures presented in this paper suggest that in four of the ten events studied (Meckering, Marryat Creek, Cadoux and Peterman) rupture propagated across 2–6 bedrock-controlled faults (e.g., pre-existing fractures and/or foliation planes and/or lithological boundaries and/or intrusive boundaries), and nine of ten events show strong basement controls on rupture location and orientation. Simplistic projection of surface defined faults using our preferred dips results in faults intersecting at depths that are consistent with published centroid depths (e.g., <4 km) in three of the four events with more than one fault defined (Petermann excluded). In all four cases, fault intersections project up-dip to the area of maximum vertical offset (in the case of Petermann, maximum dip occurs where the two faults overlap). It is uncertain with available seismic data whether hypocenters align with these projected fault intersections, and more data would be required to show that surface defined faults can be extended to depth along planar slip zones. However, linear geophysical anomalies in many cases show ruptures associated with basement conjugate fracture/dike orientations underlying rupture, suggesting strong control of the crustal architecture on intraplate earthquake nucleation and/or propagation.

New length, dip, and net-slip data presented for historic ruptures are derived by applying consistent framework and methodology. Past scaling relationships have included and excluded

Australian surface rupturing events inconsistently, generally without clear explanation. They have also relied on vertical offset measurements as most of the original publications do not calculate net-slip. Length-weighted averages of net-slip values calculated in this paper are 32%–67% larger than those for vertical offset data, and maximum net-slip is 68%–89% higher than maximum vertical offset. This suggests that Australian events may be systematically misrepresented in past scaling relationships. Our new data, compiled by thorough analysis of available seismological and field data, and coupled with the recent revision of magnitude values [11], will facilitate more consistent integration of Australian events into earthquake catalogues and displacement-length scaling relationships.

In Table 10 we compare the calculated M_W, area and average displacement from SCR dip-slip scaling relationships of [148] using the surface rupture length used in developing the scaling relationships with the length from this paper. Table 10 then compares the difference between calculated average displacement and magnitude as derived using length of this paper and SCR dip-slip scaling relationship [148] with the average net slip derived from this paper, and update M_W values [11]. Results show differences of over 600% between scaling relationship average displacement and calculated average net slip values of this paper, and up to 18.7% difference in calculated M_W and updated values [11]. This highlights the need to investigate length, magnitude, and net-slip inputs of previous scaling relationships.

Most vertical displacement data for historic surface ruptures are collected as spot-height measurements of foot-wall elevation relative to hanging-wall elevation proximal to the surface trace. Less frequently, scarp perpendicular profiles are captured 5–50 m either side of the rupture. Satellite-based imaging of recent scarps (Petermann [75,77,78], Katanning [70], Lake Muir ([79])) shows permanent distributed displacement of the hanging-wall, and to a lesser degree of the foot-wall that is not captured by these spot-heights and short traverses. Specifically, InSAR imaging shows distributed deformation extending hundreds of metres to kilometres perpendicular to surface scarps [78], and extending along-strike for kilometres beyond the surface rupture detectable in the field [70,78,79]. This is particularly the case for smaller earthquakes (Katanning [70] (Figure 8) and Lake Muir (in review [79])), where the rupture ellipse only partially intersects the surface. Without these satellite derived deformation imaging techniques, the degree to which field observations and spot-height measurements along the visible surface rupture underestimate the length, height and width of surface deformation along a fault cannot be quantified.

Digitized offset data from resurveyed benchmarks across the Tennant Creek (Figure 7), Meckering (Figure 3) and Cadoux (EarthArXiv report [122]) ruptures provide evidence of distributed hanging-wall offset, though published uncertainties are on the order of measured offsets and data should be interpreted with caution. This data cannot be improved upon within the resolution of pre-deformation height data but suggests that the deformation envelope extends beyond the discrete surface rupture, and offset measurements as presented in Figures 3–10 may underestimate the true total vertical displacement values for each event. The ratio of distributed deformation to discrete deformation at a rupture tip might be expected to be larger for surface rupture segments that are more modest in vertical displacement, or cut through relatively more weathered regolith or thicker sedimentary deposits, as much of the initial deformation will be taken up as folding prior to the emergence of the fault tip [149].

This paper reviews primary literature to identify environmental earthquake effects (EEEs) for the purposes of applying the ESI-07 Scale [10,150] to Australian surface rupturing earthquakes. We find that the majority of environmental damage is observed in the immediate rupture zone, with the exception of rare rockfalls in prone-areas (e.g., road cuttings) at distances of ~200 km, and rare ground-water fluctuations up to 250 km away for some but not all events where ground water data was investigated. While this dataset likely does not capture the full range of potential ESI values and affected area due to sparse reporting of EEEs in the literature, it does provide a basis for comparing the maximum ESI and magnitude of reverse earthquakes in intraplate, low-topography, near-surface crystalline bedrock (in most cases), and generally arid settings against events in tectonically and geomorphically diverse regions (e.g., [151–159]).

Table 10. Comparisons between calculated magnitude, area and displacement from previous length scaling relationship [148] using surface rupture length from [148] and length from this paper.

Name	Leonard 2014 [148]		Calculated [148]			This Paper	Calculated (This Paper)			Percent Difference Calculated			Avg. Net Slip [5]	% Diff [6]	M_w [11]	% Diff [6]
	SRL [1] (km)	Source	Mw [2]	A [3] (km^2)	D [4] (m)	L (km)	Mw [2]	A [3] (km^2)	D [4] (m)	Mw	A (km^2)	D (m)				
Meckering	37	[160]	6.93	5546	5.44	40	6.99	6316	5.8	0.9%	12.2%	6.3%	1.78	−225.8%	6.59	−6.1%
Calingiri	3.3	[160]	5.18	99	0.73	3.3	5.18	99	0.73	0.0%	0.0%	0.5%	0.46	−58.7%	5.03	−3.0%
Cadoux	14	[160]	6.23	1098	2.42	20	6.49	1989	3.26	4.0%	44.8%	25.8%	0.54	−503.7%	6.1	−6.4%
Marryat Creek	13	[160]	6.18	970	2.27	13	6.18	970	2.27	0.0%	0.0%	−0.2%	0.31	−632.3%	5.7	−8.4%
Kunayungku	10.2	[160]	6	648	1.86	9	5.91	526	1.67	−1.5%	−23.2%	−11.3%	0.55	−203.6%	6.27	5.7%
Lake Surprise west	6.7	[160]	5.7	321	1.31	9	5.91	526	1.67	3.6%	39.0%	21.7%	0.84	−98.8%	6.44	8.2%
Lake Surprise east *	16	[160]	6.33	1372	2.70	16	6.33	1372	2.7	0.0%	0.0%	−0.1%	1.23	−119.5%	6.58	3.8%
Katanning (InSAR)	1.26	[70]	4.49	20	0.33	0.5	3.82	4	0.15	−17.5%	−400.0%	−117.6%	0.2	25.0%	4.7	18.7%
Pukatja	1.6	[9]	4.66	30	0.40	1.3	4.51	21	0.33	−3.3%	−42.9%	−21.2%	0.25	−32.0%	5.18	12.9%

[1] Surface rupture length. [2] [148]: $M_w = a + b*\log(L)$. [3] [148]: $A = C_1L^{1+\beta}$. [4] [148]: $D = C_2A^{1/2}$. [5] Average net slip calculated in this paper (Table 6). [6] Percent difference between calculated average displacement and M_w using length of this paper [148], and average net slip calculated in Table 6, and M_w of [11].

4.3. Recurrence of Historic Surface Ruptures and Implications for Hazard Modelling

In the fifty years between 1968 and 2018, eleven moderate magnitude reverse earthquakes caused surface ruptures in cratonic Australian. Nine of the ten events analyzed show evidence of rupture along pre-existing structures with little to no evidence of prior Neotectonic movement. While this does not preclude the possibility that evidence of prior rupture was removed prior to the late Pleistocene, the lack of topographic or geomorphic evidence supporting repeated rupture suggests historic surface ruptures may have occurred on faults that could be considered previously inactive in the Neotectonic period (e.g., [4]).

It is unclear whether the historic surface rupturing faults have entered a period of activity and will host future Neotectonic earthquakes, have occurred as isolated events, or have such long recurrence intervals as to obscure all evidence of prior rupture. Paleoseismic work across the Precambrian SCR crust (Figure 1) has shown that faults in similar settings as the historic ruptures have hosted multiple Neotectonic earthquakes [2,34,108], with available dating indicating long recurrence (>30–70 ka [2]), and low topography indicating erosion may outpace seismic slip-rate. In contrast, paleoseismic investigations in the Phanerozoic non-extended crust of eastern Australia identify multiple faults with recurrence frequent enough to maintain topography [3,5,34,101,107,109], despite no historic surface rupturing or large earthquakes in this part of the continent.

Historic surface rupture kinematics are all consistent with S_{Hmax} (as measured from bore-hole breakouts, drilling induce fractures, and focal mechanisms [100]) either directly (e.g., a straight fault perpendicular to S_{Hmax}) or indirectly (e.g., rupture occurred along multiple faults, some of which are aligned oblique to S_{Hmax}, but uplift of the hanging-wall block is perpendicular to S_{Hmax}). The past fifty years of historic surface rupturing events show that in the Precambrian non-extended crust, basement with at least one set of linear structures aligned with S_{Hmax}, or multiple conjugate basement structures, could host a shallow moderate magnitude surface rupturing earthquake along one or multiple (in these cases, previously unrecognized and typically unrecognizable) faults. Eight of eleven surface rupturing earthquakes have occurred in areas of (or proximal to) preceding seismicity, while three (Petermann, Pukatja and Marryat Creek) occurred in areas with low historic seismicity, though instrument density limits the magnitude of completeness and location accuracy and precision of the historic earthquake catalogue in these locations. This suggests that spatially smoothed (distributed) seismicity models may provide the best utility for seismic hazard analyses in the central and western parts of Australia (e.g., [161]). This is also relevant for assessments of earthquake hazard in Precambrian intraplate crust elsewhere (e.g., Canada [162–164]). Further work is required to understand tentative correlations between seismogenic potential and large geophysical anomalies and/or Moho discontinuities (e.g., [165,166]), and whether transient local stress perturbations increase the potential for shallow seismicity (e.g., changes in pore-fluid pressure [76] or surface load variations [4]).

The historic earthquake catalogue for Australia is complete for $M_L > 5.5$ since 1910, and $M_L > 5.0$ since 1960 [102]. The magnitude values of historic earthquakes were recently revised [11]. This new catalogue contains seven $M_W > 5.5$ on-shore earthquakes within the Precambrian non-extended crust that are not related to the historic surface rupturing events, and only one onshore event in the eastern Phanerozoic crust (Figure 1). The Precambrian crustal events include: four events (1941 M_w 5.6, 5.9, 6.5, and 1972 M_w 5.6) in the Simpson Desert NT [80,83,167], one event (1970 M_w 5.9) within the Lake Mackay WA sequence (20 events M_w 4.5–5.5 between 1970–1992) [23,81,83,167], one event 200 km south of Warburton WA (1975 M_w 5.6), and the 1941 M_w 6.8 Meeberrie WA event—Australia's largest recorded onshore earthquake (Figure 1). No surface ruptures have been identified for these events. While depths are poorly constrained due to poor instrumental density, estimates range from 7–33 km [11,83], deeper than the best estimates of depth for surface rupturing events (1–4 km for centroids, <6 km for hypocentral/base of fault depth). This suggests that moderate magnitude and potentially damaging earthquakes (e.g., $M_W > 5.5$) can be generated at depths of up to 33 km within the Precambrian non-extended crust, providing another source of hazard that cannot be effectively captured by active-fault catalogues in seismic hazard analysis.

5. Conclusions

We provide new length, dip, and net-slip data derived using a consistent framework and methodology in order to facilitate more consistent integration of Australian events into earthquake catalogues and displacement-length scaling relationships. Our reanalysis of primary data from 67 publications on ten of eleven historical surface rupturing earthquakes in Australia shows:

- Surface rupture fault orientations aligned with basement structures identified in proximal surface outcrops (foliations ± quartz veins ± intrusive boundaries ± pre-existing faults) and linear geophysical anomalies;
- Rupture involve 1–6 discrete faults based on reanalysis of surface rupture lengths using consistent criteria, with evidence that intersecting basement structures may control rupture initiation and/or propagation;
- Large aleatoric and epistemic uncertainties in seismological data, related to a sparse seismic network, limit determination of hypocenter and fault interaction, rupture propagation, and assessment of whether surface ruptures project to seismogenic depths along planar principle slip zones or whether rupture propagates to multiple basement structures in the near-surface;
- Available analyses of rupture centroids (seven of ten events) show depths of 1–4 km indicating predominately shallow seismic moment release;
- None of the historic surfacing rupturing events have unambiguous geological or geomorphic evidence for preceding earthquakes on the same faults, with five events showing an absence of rupture since at least the late Pleistocene;
- Within the constraints of available basement erosion rates, preferred maximum slip rates are 0.2–9.1 m Myr^{-1} with an estimated minimum epistemic uncertainty of at least one order of magnitude lower. These are considered applicable only within the non-extended Precambrian crust in which all historic surface ruptures have occurred;
- ESI-07 estimates range by ±3 classes in each earthquake and provide new maximum ESI vs. magnitude data for comparison between different tectonic and geomorphic settings;

Author Contributions: Conceptualization, M.Q. and T.R.K.; methodology, T.R.K.; validation T.R.K. and D.C.; formal analysis, T.R.K.; investigation, T.R.K.; data curation T.R.K.; writing—original draft preparation, T.R.K.; writing—review and editing, T.R.K., M.Q. and D.C.; visualization, T.R.K. and M.Q.; supervision, M.Q.; project administration, T.R.K. and M.Q.; funding acquisition, M.Q.

Funding: This research was funded by the Australian Research Council through Discovery Grant #DP170103350. T. King received funding through the Australian Government Research Training Program Scholarship.

Acknowledgments: The authors thank the editors and three reviewers for comments that improved this work. We would like to acknowledge the Noongar people of south-west Western Australia, the Warumungu people of Tennant Creek, and the Antakirinja, Yankunytjatjara, and Pitjantjatjara people of the Western Desert and APY lands in South Australia/Northern Territory, as the traditional custodians of the land on which all historic surface ruptures occurred, and where the data described in this paper were collected. D. Clark publishes with the permission of the Chief Executive Officer of Geoscience Australia.

Conflicts of Interest: The authors declare no conflict of interest.

References

1. Crone, A.J.; Machette, M.N.; Bowman, J.R. Episodic nature of earthquake activity in stable continental regions revealed by palaeoseismicity studies of Australian and North American quaternary faults. *Aust. J. Earth Sci.* **1997**, *44*, 203–214. [CrossRef]
2. Crone, A.J.; De Martini, P.M.; Machette, M.N.; Okumura, K.; Prescott, J.R. Paleoseismicity of Two Historically Quiescent Faults in Australia: Implications for Fault Behavior in Stable Continental Regions. *Bull. Seismol. Soc. Am.* **2003**, *93*, 1913–1934. [CrossRef]
3. Quigley, M.C.; Clark, D.; Sandiford, M. Tectonic geomorphology of Australia. *Geol. Soc. Lond. Spec. Publ.* **2010**, *346*, 243–265. [CrossRef]

4. Calais, E.; Camelbeeck, T.; Stein, S.; Liu, M.; Craig, T.J. A new paradigm for large earthquakes in stable continental plate interiors. *Geophys. Res. Lett.* **2016**, *43*, 10621–10637. [CrossRef]
5. Clark, D.; McPherson, A.; Van Dissen, R.J. Long-term behaviour of Australian stable continental region (SCR) faults. *Tectonophysics* **2012**, *566–567*, 1–30. [CrossRef]
6. Wells, D.L.; Coppersmith, K.J. New Empirical Relationships among Magnitude, Rupture Length, Rupture Width, Rupture Area, and Surface Displacement. *Bull. Seismol. Soc. Am.* **1994**, *84*, 974–1002.
7. Wesnousky, S.G. Displacement and geometrical characteristics of earthquake surface ruptures: Issues and implications for seismic-hazard analysis and the process of earthquake rupture. *Bull. Seismol. Soc. Am.* **2008**, *98*, 1609–1632. [CrossRef]
8. Biasi, G.P.; Wesnousky, S.G. Steps and gaps in ground ruptures: Empirical bounds on rupture propagation. *Bull. Seismol. Soc. Am.* **2016**, *106*, 1110–1124. [CrossRef]
9. Clark, D.; Mcpherson, A.; Allen, T.; De Kool, M. Coseismic surface deformation caused by the 23 March 2012 Mw 5.4 Ernabella (Pukatja) earthquake, central Australia: Implications for fault scaling relations in cratonic settings. *Bull. Seismol. Soc. Am.* **2014**, *104*, 24–39. [CrossRef]
10. Michetti, A.M.; Esposito, E.; Guerrieri, L.; Porfido, S.; Serva, L.; Tatevossian, R.E.; Vittori, E.; Audemard, F.A.; Azuma, T.; Clague, J.; et al. *Intensity Scale ESI 2007*; Guerrieri, L., Vittori, E., Eds.; APAT: Rome, Italy, 2007; Volume 74.
11. Allen, T.; Leonard, M.; Ghasemi, H.; Gibson, G. *The 2018 National Seismic Hazard Assesment: Earthquake Epicentre Catalogue (GA Record 2018/30)*; Geoscience Australia, Commonwealth of Australia: Canberra, Australia, 2018. [CrossRef]
12. Everingham, I.B. *Preliminary Report on the 14 October 1968 Earthquake at Meckering, Western Australia (BMR Record 1968/142)*; 1968/142; Bureau of Mineral Resources, Geology and Geophysics Canberra: Canberra, Australia, 1968. Available online: http://pid.geoscience.gov.au/dataset/ga/12254 (accessed on 2 August 2019).
13. Gordon, F.R. *Reconstruction of Meckering Town, a Geological Appraisal (GSWA Record 1968/14)*; Geological Survey of Western Australia: Perth, Western Australia, 1968.
14. Conacher, A.J.; Murray, I.D. The Meckering earthquake, Western Australia, 14 October 1968. *Aust. Geogr.* **1969**, *11*, 179–184. [CrossRef]
15. Everingham, I.B.; Gregson, P.J.; Doyle, H.A. Thrust Fault Scarp in the Western Australian Shield. *Nature* **1969**, *223*, 701–703. [CrossRef]
16. Lewis, J.D. *The Geology of the Country around Meckering (GSWA Record 1969/18)*; Geological Survey of Western Australia: Perth, Western Australia, 1969.
17. Gordon, F.R. Water level changes preceding the Meckering, Western Australia, earthquake of October 14, 1968. *Bull. Seismol. Soc. Am.* **1970**, *60*, 1739–1740.
18. Everingham, I.B.; Gregson, P.J. *Meckering Earthquake Intensities and Notes on Earthquake Risk for Western Australia (BMR Report 1970/97)*; Bureau of Mineral Resources, Geology and Geophysics: Canberra, ACT, Australia, 1970. Available online: http://pid.geoscience.gov.au/dataset/ga/12510 (accessed on 2 August 2019).
19. Gordon, F.R. Faulting during the earthquake at Meckering, Western Australia: 14 October 1968. *Bull. R. Soc. N. Z.* **1971**, *9*, 85–93.
20. Gordon, F.R.; Wellman, H.W. A mechanism for the Meckering earthquake. *R. Soc. N. Z. Bull.* **1971**, *9*, 95–96.
21. Everingham, I.B.; Gregson, P.J. *Mundaring Geophysical Observatory, Annual Report, 1968 (BMR Record 1971/12)*; Bureau of Mineral Resources, Geology and Geophysics: Canberra, Australia, 1971. Available online: http://pid.geoscience.gov.au/dataset/ga/12549 (accessed on 2 August 2019).
22. Gregson, P.J.; McCue, K.; Smith, R.S. *An Explanation of Water Level Changes Preceding the Meckering Earthquake of 14 October 1968 (BMR Record 1972/101)*; Bureau of Mineral Resources, Geology and Geophysics: Canberra, Australia, 1972. Available online: http://pid.geoscience.gov.au/dataset/ga/12782 (accessed on 2 August 2019).
23. Fitch, T.J.; Worthington, M.H.; Everingham, I.B. Mechanisms of Australian earthquakes and contemporary stress in the Indian ocean plate. *Earth Planet Sci. Lett.* **1973**, *18*, 345–356. [CrossRef]
24. Denham, D.; Alexander, L.G.; Worotnicki, G. The stress field near the sites of the Meckering (1968) and Calingiri (1970) earthquakes, Western Australia. *Tectonophysics* **1980**, *67*, 283–317. [CrossRef]
25. Gordon, F.R.; Lewis, J.D. *The Meckering and Calingiri Earthquakes October 1968 and March 1970*; Geological Survey of Western Australia: Perth, Australia, 1980.
26. Vogfjord, K.S.; Langston, C.A. The Meckering earthquake of 14 October 1968: A possible downward propagating rupture. *Bull. Seismol. Soc. Am.* **1987**, *77*, 1558–1578.

27. Langston, C.A. Depth of Faulting during the 1968 Meckering, Australia, earthquake sequence determined from waveform analysis of local seismogram. *J. Geophys. Res.* **1987**, *92*, 11561–11574. [CrossRef]
28. Fredrich, J.; Mccaffrey, R.; Denham, D. Source parameters of seven large Australian earthquakes determined by body waveform inversion. *Geophys. J.* **1988**, *95*, 1–13. [CrossRef]
29. Gregson, P.J. (Ed.) *Recent intraplate seismicity studies symposium, Perth, Western Australia September 1990 (BMR Record 1990/44)*; Bureau of Mineral Resources, Geology and Geophysics: Canberra, ACT, Australia, 1990. Available online: http://pid.geoscience.gov.au/dataset/ga/14335 (accessed on 2 August 2019).
30. Lewis, J.D. Meckering revisited. Recent Intraplate Seismicity Studies Symposium, Perth Western Australia (BMR Record 1990/44). 1990. Available online: http://pid.geoscience.gov.au/dataset/ga/14335 (accessed on 2 August 2019).
31. Lewis, J.D. *The Meckering Earthquake of 17 January 1990 (GSWA Record 1990/6)*; Geological Survey of Western Australia: Perth, Western Australia, 1990.
32. Dent, V.F. Foreshocks and aftershocks of the 17 Jan 1990 Meckering earthquake. In *Recent Intraplate Seismicity Studies Symposium, Perth Western Australia (BMR Record 1990/44)*; Gregson, P.J., Ed.; Bureau of Mineral Resources, Geology and Geophysics: Canberra, ACT, Australia, 1990. Available online: http://pid.geoscience.gov.au/dataset/ga/14335 (accessed on 2 August 2019).
33. Dentith, M.; Clark, D.; Featherstone, W.E. Aeromagnetic mapping of Precambrian geological structures that controlled the 1968 Meckering earthquake (Ms 6.8): Implications for intraplate seismicity in Western Australia. *Tectonophysics* **2009**, *475*, 544–553. [CrossRef]
34. Clark, D.; Mcpherson, A.; Collins, C. *Australia's Seismogenic Neotectonic Record: A Case for Heterogeneous Intraplate Deformation (GA Record 2011/11)*; Geoscience Australia, Commonwealth of Australia: Canberra, Australia, 2011. Available online: http://pid.geoscience.gov.au/dataset/ga/70288 (accessed on 2 August 2019).
35. Johnston, J.F.; White, S.R. *Understanding the Meckering Earthquake: Western Australia, 14 October 1968*; Geological Survey of Western Australia: Perth, Australia, 2018. [CrossRef]
36. Clark, D. *What Have We Learned in the 50 Years since the 1968 Meckering Earthquake?* Geoscience Australia, Commonwealth of Australia: Canberra, Australia, 2018. Available online: http://pid.geoscience.gov.au/dataset/ga/123342 (accessed on 2 August 2019).
37. Clark, D.; Edwards, M. *50th Anniversary of the 14th October 1968 Mw 6.5 (Ms 6.8) Meckering Earthquake (GA Record 2018/39)*; Geoscience Australia, Commonwealth of Australia: Canberra, ACT, Australia, 2018. Available online: http://dx.doi.org/10.11636/Record.2018.039 (accessed on 2 August 2019).
38. Everingham, I.B.; Parkes, A. *Intensity Data for Earthquakes at Landor (17 June 1969) and Calingiri (10 March 1970) and Their Relationship to Previous Western Australian Observations (BMR Recrod 1971/80)*; 1971/80; Bureau of Mineral Resources, Geology and Geophysics: Canberra, Australia, 1971. Available online: http://pid.geoscience.gov.au/dataset/ga/12617 (accessed on 2 August 2019).
39. Gregson, P.J. *Mundaring Geophysical Observatory Annual Report, 1970 (BMR Record 1971/77)*; Bureau of Mineral Resources, Geology and Geophysics: Canberra, Australia, 1971. Available online: http://pid.geoscience.gov.au/dataset/ga/12614 (accessed on 2 August 2019).
40. Gregson, P.J.; Paull, E.P. *Preliminary Report on the Cadoux Earthquake, Western Australia, 2 June 1979 (BMR Report 1979/215)*; Bureau of Mineral Resources, Geology and Geophysics: Canberra, ACT, Australia, 1979. Available online: http://pid.geoscience.gov.au/dataset/ga/15123 (accessed on 2 August 2019).
41. Lewis, J.D.; Daetwyler, N.A.; Bunting, J.A.; Montcrieff, J.S. *The Cadoux Earthqauke (GSWA Report 11)*; Geological Survey of Western Australia: Perth, Australia, 1981.
42. Dent, V.F.; Gregson, P.J. *Cadoux Microearthquake Survey 1983 (BMR Report 1986/022)*; 1986/22; Bureau of Mineral Resources, Geology and Geophysics: Canberra, ACT, Australia, 1986. Available online: http://pid.geoscience.gov.au/dataset/ga/14114 (accessed on 2 August 2019).
43. Denham, D.; Alexander, L.G.; Everingham, I.B.; Gregson, P.J.; McCaffrey, R.; Enever, J.R. The 1979 Cadoux earthquake and intraplate stress in Western Australia. *Aust. J. Earth Sci.* **1987**, *34*, 507–521. [CrossRef]
44. Dent, V.F. *The Distribution of Cadoux Aftershocks: Additional Results from Temporary Stations near Cadoux, 1983 (BMR Record 1988/51)*; Bureau of Mineral Resources, Geology and Geophysics: Canberra, ACT, Australia, 1988. Available online: http://pid.geoscience.gov.au/dataset/ga/14238 (accessed on 2 August 2019).
45. Dent, V.F. Hypocentre locations from a microearthquake survey, Cadoux, Western Australia, 1983. *BMR J. Aust. Geol. Geophys.* **1991**, *12*, 1–4. Available online: http://pid.geoscience.gov.au/dataset/ga/81278 (accessed on 2 August 2019).

46. Barlow, B.C.; Denham, D.; Jones, T.; McCue, K. The Musgrave Ranges earthquake of March 30, 1986. *Trans. R. Soc. S. Aust.* **1986**, *110*, 187–189. [CrossRef]
47. McCue, K.; Jones, T.; Michael-Leiba, M.; Barlow, B.C.; Denham, D.; Gibson, G. Another chip off the old Australian block. *Eos Trans. Am. Geophys. Union* **1987**, *68*, 609. [CrossRef]
48. Bowman, J.R.; Barlow, B.C. *Surveys of the Fault Scarp of the 1986 Marryat Creek, South Australia, Earthquake (BMR Record 1991/109)*; Australian Seismological Centre, Bureau of Mineral Resources: Canberra, ACT, Australia, 1991. Available online: http://pid.geoscience.gov.au/dataset/ga/14490 (accessed on 2 August 2019).
49. Machette, M.N.; Crone, A.J.; Bowman, J.R.; Prescott, J.R. *Surface ruptures and deformation associated with the 1988 Tennant Creek and 1986 Marryat Creek, Australia, intraplate earthquakes. Abstracts of the U.S. Geological Survey, Central Region; 1991 Poster Review*; U.S. Geological Survey: Reston, VA, USA, 1991; p. 27. Available online: https://doi.org/10.3133/ofr91582 (accessed on 2 August 2019).
50. Bullock, P.W.B. *Tennant Creek Gravity and Magnetic Survey, Northern Territory, 1973 (BMR Record 1977/30)*; Bureau of Mineral Resources, Geology and Geophysics: Canberra, Australia, 1977. Available online: http://pid.geoscience.gov.au/dataset/ga/13559 (accessed on 2 August 2019).
51. Hone, I.G. *Ground Geophysical Survey, Tennant Creek, Northern Territory, 1972 (BMR Record 1974/171)*; Bureau of Mineral Resources, Geology and Geophysics: Canberra, Australia, 1974. Available online: http://pid.geoscience.gov.au/dataset/ga/1321 (accessed on 2 August 2019).
52. Verhoeven, T.J.; Russell, P.W. *Tennant Creek Water Supply 1979 - 1980 Source Investigation [Kelly Well] (Report 27/1981)*; Department of Transport and Works: Alice Springs, Australia, 1981. Available online: http://hdl.handle.net/10070/229202 (accessed on 2 August 2019).
53. Bowman, J.R. Constraints on locations of large intraplate earthquakes in the Northern Territory, Australia from observations at the Warramunga seismic array. *Geophys. Res. Lett.* **1988**, *15*, 1475–1478. [CrossRef]
54. Bowman, J.R.; Gibson, G.; Jones, T. Faulting process of the January 22, 1988 Tennant Creek, Northern Territory, Australia earthquakes. In *Abstracts for the AGU Fall Meeting 1988: EoS Transactions*; American Geophysical Union: Washington, USA, 1988; Volume 69, p. 1301. [CrossRef]
55. McCaffrey, R. Teleseismic investigation of the January 22, 1988 Tennant Creek, Australia, earthquakes. *Geophys. Res. Lett.* **1989**, *16*, 413–416. [CrossRef]
56. Bowman, J.R.; Dewey, J.W.; Peters, N. Recent Results from Tennant Creek. In *Recent Intraplate Seismicity Studies Symposium, Perth Western Australia (BMR Record 1990/44)*; Gregson, P.J., Ed.; Bureau of Mineral Resources, Geology and Geophysics: Canberra, ACT, Australia, 1990. Available online: http://pid.geoscience.gov.au/dataset/ga/14335 (accessed on 2 August 2019).
57. Choy, G.L.; Bowman, J.R. Rupture process of a multiple main shock sequence: Analysis of teleseismic, local and field observations of the Tennant Creek, Australia, earthquakes of January 22, 1988. *J. Geophys. Res.* **1990**, *95*, 6867–6882. [CrossRef]
58. Bouniot, E.; Jones, T.; McCue, K. The pattern of 1987 sequence at Tennant Creek, NT. In *Recent Intraplate Seismicity Studies Symposium, Perth Western Australia (BMR Record 1990/44)*; Gregson, P.J., Ed.; Bureau of Mineral Resources, Geology and Geophysics: Canberra, ACT, Australia, 1990. Available online: http://pid.geoscience.gov.au/dataset/ga/14335 (accessed on 2 August 2019).
59. Bowman, J.R.; Gibson, G.; Jones, T. Aftershocks of the 1988 January 22 Tennant Creek, Australia Intraplate Earthquakes: Evidence For A Complex Thrust-Fault Geometry. *Geophys. J. Int.* **1990**, *100*, 87–97. [CrossRef]
60. Jones, T.; Gibson, G.; McCue, K.; Denham, D.; Gregson, P.J.; Bowman, J.R. Three large intraplate earthquakes near Tennant Creek, Northern Territory, on 22 January 1988. *BMR J. Aust. Geol. Geophys.* **1991**, *12*, 339–343. Available online: http://pid.geoscience.gov.au/dataset/ga/81300 (accessed on 2 August 2019).
61. Bowman, J.R.; Dewey, J.W. Relocation of teleseismically recorded earthquakes near Tennant Creek, Australia: Implications for midplate seismogenesis. *J. Geophys. Res.* **1991**, *96*, 11973–11979. [CrossRef]
62. Bowman, J.R. Geodetic evidence for conjugate faulting during the 1988 Tennant Creek, Australia earthquake sequence. *Geophys. J. Int.* **1991**, *107*, 47–56. [CrossRef]
63. Crone, A.J.; Machette, M.N.; Bowman, J.R. *Geologic Investigations of the 1988 Tennant Creek, Australia, Earthquakes - Implications for Paleoseismicity in the Stable Continental Regions (USGS Bulletin 2032-A)*; U.S. Geological Survey: Washington, DC, USA, 1992.
64. Bowman, J.R. The 1988 Tennant Creek, Northern Territory, earthquakes: A synthesis. *Aust. J. Earth Sci.* **1992**, *39*, 651–669. [CrossRef]

65. Donnelly, K.E.; Morrison, R.S.; Hussey, K.J.; Ferenczi, P.A.; Kruse, P.D. *Tennant Creek 1:250000 Explanatory Notes, Geological Map Series*; Northern Territory Geological Survey: Darwin, Australia, 1999. Available online: https://doi.org/10.1017/CBO9781107415324.004 (accessed on 2 August 2019).
66. Johnstone, A.; Donnellan, N. *Tennant Creek 1:250 000 Integrated Interpretation of Geophysics and Mapped Geology*, 1st ed.; Northern Territory Geological Survey, Alice Springs: Alice Springs, Australia, 2001.
67. Bowman, J.R.; Yong, C. Case 22 A Seismicity Precursor to a Sequence of M 6.3-6.7 Midplate Earthquakes in Australia. *Pure Appl. Geophys.* **1997**, *149*, 61–78. [CrossRef]
68. Donnellan, N. Chapter 9: Warramunga Province. In *Geology and Mineral Resources of the Northern Territory, Special Publication 5*; Ahmad, M., Munson, T.J., Eds.; Special Pu. Northern Territory Geological Survey: Darwin, Australia, 2013.
69. Mohammadi, H.; Quigley, M.; Steacy, S.; Duffy, B. Effects of source model variations on Coulomb stress analyses of a multi-fault intraplate earthquake sequence. *Tectonophysics* **2019**, *766*, 151–166. [CrossRef]
70. Dawson, J.; Cummins, P.R.; Tregoning, P.; Leonard, M. Shallow intraplate earthquakes in Western Australia observed by Interferometric Synthetic Aperture Radar. *J. Geophys. Res. Solid Earth* **2008**, *113*, 1–19. [CrossRef]
71. Dent, V.F. Improved Hypocentral estimates for two recent seismic events in south-western Western Australia, using temporary station data. In Proceedings of the Australian Earthquake Engineering Society Conference 2008, Ballarat, VIC, Australia, 21–23 November 2008.
72. Clark, D.; Mcpherson, A. A tale of two seisms: Ernabella 23/03/2012 (Mw5.4) and Mulga Park 09/06/2013 (Mw 5.6). *Aust. Earthq. Eng. Soc. Newsl.* **2013**, *2013*, 7–11.
73. King, T.R.; Quigley, M.C.; Clark, D. Earthquake environmental effects produced by the Mw 6.1, 20th May 2016 Petermann earthquake, Australia. *Tectonophysics* **2018**, *747–748*, 357–372. [CrossRef]
74. Hejrani, B.; Tkalčić, H. The 20 May 2016 Petermann Ranges earthquake: Centroid location, magnitude and focal mechanism from full waveform modelling. *Aust. J. Earth Sci.* **2018**, *66*, 37–45. [CrossRef]
75. Polcari, M.; Albano, M.; Atzori, S.; Bignami, C.; Stramondo, S.; Polcari, M.; Albano, M.; Atzori, S.; Bignami, C.; Stramondo, S. The Causative Fault of the 2016 Mwp 6.1 Petermann Ranges Intraplate Earthquake (Central Australia) Retrieved by C- and L-Band InSAR Data. *Remote Sens.* **2018**, *10*, 1311. [CrossRef]
76. Wang, S.; Xu, W.; Xu, C.; Yin, Z.; Bürgmann, R.; Liu, L.; Jiang, G. Changes in groundwater level possibly encourage shallow earthquakes in central Australia: The 2016 Petermann Ranges earthquake. *Geophys. Res. Lett.* **2019**, *46*, 3189–3198. [CrossRef]
77. Gold, R.; Clark, D.; King, T.; Quigley, M. *Surface rupture and vertical deformation associated with 20 May 2016 M6 Petermann Ranges earthquake, Northern Territory, Australia*; European Geosciences Union General Assembly: Vienna, Austria, 2017; Volume 19. Available online: http://adsabs.harvard.edu/abs/2017EGUGA..19.8645G (accessed on 2 August 2019).
78. Gold, R.D.; Clark, D.; Barnhart, W.D.; King, T.; Quigley, M.; Briggs, R.W. Surface rupture and distributed deformation revealed by optical satellite imagery: The intraplate 2016 Mw 6.0 Petermann Ranges earthquake, Australia. *Geophys. Res. Lett.* **2019**. [CrossRef]
79. Clark, D.J.; Brennand, S.; Brenn, G.; Allen, T.I.; Garthwaite, M.C.; Standen, S. The 2018 Lake Muir earthquake sequence, southwest Western Australia: Rethinking Australian stable continental region earthquakes. *Solid Earth* **2019**. in review. [CrossRef]
80. Doyle, H.A.; Everingham, I.B.; Sutton, D.J. Seismicity of the Australian continent. *J. Geol. Soc. Aust.* **1968**, *15*, 295–312. [CrossRef]
81. Cleary, J.R.; Simpson, D.W. Seismotectonics of the Australian continent. *Nature* **1971**, *230*, 239–241. [CrossRef]
82. Doyle, H.A. Seismicity and structure in Australia. *Bull. R. Soc. N. Z.* **1971**, *9*, 149–152.
83. Denham, D.; Alexander, L.G.; Worotnicki, G. Stresses in the Australian crust: Evidence from earthquakes and in-situ stress measurements. *BMR J. Aust. Geol. Geophys.* **1979**, *4*, 289–295. Available online: http://pid.geoscience.gov.au/dataset/ga/81007 (accessed on 2 August 2019).
84. Everingham, I.B.; McEwin, A.J.; Denham, D. *Atlas of Isoseismal Maps of Australian Earthquakes*; Bureau of Mineral Resources, Geology and Geophysics: Canberra, Australia, 1982. Available online: http://pid.geoscience.gov.au/dataset/ga/38 (accessed on 2 August 2019).
85. Lambeck, K.; McQueen, H.W.S.; Stephenson, R.A.; Denham, D. The state of stress within the Australian continent. *Ann. Geophys.* **1984**, *2*, 723–742.

86. Rynn, J.M.W.; Denham, D.; Greenhalgh, S.A.; Jones, T.; Gregson, P.J.; McCue, K.; Smith, R.S. *Atlas of Isoseismal Maps of Australian Earthquakes, Part 2*; Bureau of Mineral Resources, Geology and Geophysics: Canberra, ACT, Australia, 1987. Available online: http://pid.geoscience.gov.au/dataset/ga/19 (accessed on 2 August 2019).
87. Johnston, A. Fault Traces Australian Quakes. *Eos Trans. Am. Geophys. Union* **1988**, *69*, 682. [CrossRef]
88. Denham, D. Australian seismicity-the puzzle of the not-so-stable continent. *Seismol. Res. Lett.* **1988**, *59*, 235–240. [CrossRef]
89. McCue, K. Australia's large earthquakes and Recent fault scarps. *J. Struct. Geol.* **1990**, *12*, 761–766. [CrossRef]
90. Leonard, M.; Ripper, I.D.; Yue, L. *Australian Earthquake Fault Plane Solutions (GA Record 2002/019)*; 2002/19; Geoscience Australia: Canberra, ACT, Australia, 2002. Available online: http://pid.geoscience.gov.au/dataset/ga/37302 (accessed on 2 August 2019).
91. Dentith, M.; Featherstone, W.E. Controls on intra-plate seismicity in southwestern Australia. *Tectonophysics* **2003**, *376*, 167–184. [CrossRef]
92. Featherstone, W.E.; Penna T, N.; Leonard, M.; Clark, D.; Dawson, J.; Dentith, M.; Darby, D.; McCarthy, R. GPS-geodetic deformation monitoring of the south-west seismic zone of Western Australia: Review, description of methodology and results from epoch-one. *J. R. Soc. West Aust.* **2004**, *87*, 1–8.
93. Dawson, J.; Tregoning, P. Uncertainty analysis of earthquake source parameters determined from InSAR: A simulation study. *J. Geophys. Res. Solid Earth* **2007**, *112*, 1–13. [CrossRef]
94. Braun, J.; Burbidge, D.R.; Gesto, F.N.; Sandiford, M.; Gleadow, A.J.W.; Kohn, B.P.; Cummins, P.R. Constraints on the current rate of deformation and surface uplift of the Australian continent from a new seismic database and low-T thermochronological data. *Aust. J. Earth Sci.* **2009**, *56*, 99–110. [CrossRef]
95. Clark, D. *Neotectonic Features Database*; Geoscience Australia, Commonwealth of Australia: Canberra, Australia, 2012.
96. Clark, D.; Allen, T. What have we learnt regarding cratonic earthquakes in the fifty years since Meckering? In Proceedings of the Australian Earthquake Engineering Society Conference 2018, Perth, WA, USA, 16–18 November 2018.
97. Tracey, R.M. Analysis of Repeat Levelling Measurements to Give Ground Deformation, Southwest Australia (BMR Record 1982/30). Bureau of Mineral Resources, Geology and Geophysics: Canberra, Australia, 1982.
98. Dent, V.F. *Hypocentre Relocations Using Data from Temporary Seismograph Stations at Burakin and Wyalkatchem, Western Australia (BMR Record 1990/36)*; Bureau of Mineral Resources, Geology and Geophysics: Canberra, Australia, 1990. Available online: http://pid.geoscience.gov.au/dataset/ga/14327 (accessed on 2 August 2019).
99. Leonard, M.; Burbidge, D.R.; Allen, T.; Robinson, D.J.; Mcpherson, A.; Clark, D.; Collins, C. The challenges of probabilistic seismic-hazard assessment in stable continental interiors: An Australian example. *Bull. Seismol. Soc. Am.* **2014**, *104*, 3008–3028. [CrossRef]
100. Rajabi, M.; Tingay, M.; Heidbach, O.; Hillis, R.R.; Reynolds, S.D. The present-day stress field of Australia. *Earth-Sci. Rev.* **2017**, *168*, 165–189. [CrossRef]
101. Hillis, R.R.; Sandiford, M.; Reynolds, S.D.; Quigley, M.C. Present-day stresses, seismicity and Neogene-to-Recent tectonics of Australia's "passive" margins: Intraplate deformation controlled by plate boundary forces. *Geol. Soc. Lond. Spec. Publ.* **2008**, *306*, 71–90. [CrossRef]
102. Leonard, M. One hundred years of earthquake recording in Australia. *Bull. Seismol. Soc. Am.* **2008**, *98*, 1458–1470. [CrossRef]
103. Raymond, O.L.; Totterdell, J.M.; Stewart, A.J.; Woods, M.A. *Australian Geological Provinces: 2018.01 Edition [Digital Dataset]*; Geoscience Australia, Commonwealth of Australia: Canberra, Australia, 2018. Available online: http://pid.geoscience.gov.au/dataset/ga/116823 (accessed on 2 August 2019).
104. Johnston, A.C.; Coppersmith, K.J.; Cornell, C.A. The earthquakes of stable continental regions. In *Electric Power Research Institute Report TR-102261-VI*; EPRI Distribution Centre: Palo Alto, CA, USA, 1994.
105. Rajabi, M.; Heidbach, O.; Tingay, M.; Reiter, K. Prediction of the present-day stress field in the Australian continental crust using 3D geomechanical–numerical models. *Aust. J. Earth Sci.* **2017**, *64*, 435–454. [CrossRef]
106. Clark, D. Identification of quaternary scarps in southwest and central west Western Australia using dem-based hill shading: Application to seismic hazard assessment and neotectonics. *Int. J. Remote Sens.* **2010**, *31*, 6297–6325. [CrossRef]

107. Clark, D.; Cupper, M.; Sandiford, M.; Kiernan, K. Style and timing of late Quaternary faulting on the Lake Edgar fault, southwest Tasmania, Australia: Implications for hazard assessment in intracratonic areas. In *Geological Criteria for Evaluating Seismicity Revisited: Forty Years of Paleoseismic Investigations and the Natural Record of Past Earthquakes: Geological Society of America Special Paper 479*; Audemard, F.A., Michetti, A.M., McCalpin, J.P., Eds.; The Geological Society of America: Boulder, CO, USA, 2011; Volume 2479, pp. 109–131. [CrossRef]
108. Clark, D.; Dentith, M.; Wyrwoll, K.-H.; Yanchou, L.; Dent, V.F.; Featherstone, W.E. The Hyden fault scarp, Western Australia: Paleoseismic evidence for repeated Quaternary displacement in an intracratonic setting. *Aust. J. Earth Sci.* **2008**, *55*, 379–395. [CrossRef]
109. Quigley, M.C.; Sandiford, M.; Fifield, L.K.; Alimanovic, A. Landscape responses to intraplate tectonism: Quantitative constraints from 10Be nuclide abundances. *Earth Planet Sci. Lett.* **2007**, *261*, 120–133. [CrossRef]
110. Quigley, M.C.; Sandiford, M.; Cupper, M. Distinguishing tectonic from climatic controls on range-front sedimentation. *Basin Res.* **2007**, *19*, 491–505. [CrossRef]
111. Quigley, M.C.; Cupper, M.; Sandiford, M. Quaternary faults of south-central Australia: Palaeoseismicity, slip rates and origin. *Aust. J. Earth Sci.* **2006**, *53*, 285–301. [CrossRef]
112. Thom, R. A recent fault scarp in the Lort River area, Ravensthorpe 1:250 000 sheet. In *Geological Survey of Western Australia Annual Report 1971*; Geological Survey of Western Australia: Perth, Australia, 1971; pp. 58–59.
113. Williams, I.R. Recent fault scarps in the Mount Narryer area, byro 1:250 000 sheet. In *Geological Survey of Western Australia Annual Report 1978*; Geological Survey of Western Australia: Perth, Australia, 1978.
114. Wilde, S.A.; Low, G.H.; Lake, R.W. *Perth 1:250 000 Geological Map Sheet*; Geological Survey of Western Australia: Perth, Western Australia, 1978.
115. Blight, D.F.; Chin, R.J.; Smith, R.A.; Bunting, J.A.; Elias, M. *Bencubbin 1:250 000 Geological Map Sheet*; Geological Survey of Western Australia: Perth, Australia, 1983.
116. Scrimgeour, I.R.; Close, D.F.; Edgoose, C.J. *Petermann Ranges SG52-7; Explanatory Notes*; Northern Territory Geological Survey: Darwin, Australia, 1999.
117. Fairclough, M.C.; Sprigg, R.C.; Wilson, B.; Coats, R.P. *Alberga 1:250 000 Geological Map, Digital Edition*; Geological Survey of South Australia: Adelaide, Australia, 2011.
118. Machette, M.N.; Crone, A.J.; Bowman, J.R. *Geologic Investigations of the 1986 Marryat Creek, Australia, Earthquake: Implications for Paleoseismicity in Stable Continental Regions (USGS Bulletin 2032-B)*; U.S. Geological Survey: Washington, DC, USA, 1993. [CrossRef]
119. Allen, T.; Griffin, J.; Leonard, M.; Clark, D.; Ghasemi, H. *The 2018 National Seismic Hazard Assessment: Model Overview (GA Record 2018/27)*; Geoscience Australia, Commonwealth of Australia: Canberra, Australia, 2018.
120. King, T.R.; Quigley, M.; Clark, D. Review paper: The 14th October 1968 Mw 6.6 Meckering surface rupturing earthquake, Australia. *EarthArXiv Prepr.* **2019**, 1–25. [CrossRef]
121. King, T.R.; Quigley, M.C.; Clark, D. Review paper: The 10th March 1970 Mw 5.0 Calingiri surface rupturing earthquake, Australia. *EarthArXiv Prepr.* **2019**. [CrossRef]
122. King, T.R.; Quigley, M.C.; Clark, D. Review paper: The 2nd June 1979 Mw 6.1 Cadoux surface rupturing earthquake, Australia. *EarthArXiv Prepr.* **2019**, 1–19. [CrossRef]
123. King, T.R.; Quigley, M.C.; Clark, D. Review paper: The 30th March 1968 Mw 5.7 Marryat Creek surface rupturing earthquake, Australia. *EarthArXiv Prepr.* **2019**, 1–17. [CrossRef]
124. King, T.R.; Quigley, M.; Clark, D. Review paper: The 23rd March 2012 Mw 5.2 Pukatja surface rupturing earthquake, Australia. *EarthArXiv Prepr.* **2019**, 1–13. [CrossRef]
125. King, T.R.; Quigley, M.; Clark, D. Review paper: The 20th May 2016 Mw 6.1 Petermann surface rupturing earthquake, Australia. *EarthArXiv Prepr.* **2019**, 1–16. [CrossRef]
126. King, T.R.; Quigley, M.C.; Clark, D.; Valkaniotis, S.; Mohammadi, H.; Barnhart, W.D. The 1987 to 2019 Tennant Creek, Australia, earthquake sequence: A protracted intraplate multi-mainshock sequence. *EarthArXiv Prepr.* **2019**. [CrossRef]
127. Wilde, S.A.; Middleton, M.F.; Evans, B.J. Terrane accretion in the southwestern Yilgarn Craton: Evidence from a deep seismic crustal profile. *Precambrian Res.* **1996**, *78*, 179–196. [CrossRef]
128. Edgoose, C.J.; Scrimgeour, I.R.; Close, D.F. *Geology of the Musgrave Block, Northern Territory (NTGS Report 15)*; Munson, T.J., Ed.; Northern Territory Geological Survey: Darwin, Australia, 2004.

129. Raimondo, T.; Collins, A.S.; Hand, M.; Walker-Hallam, A.; Smithies, R.H.; Evins, P.M.; Howard, H.M. The anatomy of a deep intracontinental orogen. *Tectonics* **2010**, *29*. [CrossRef]
130. Neumann, N.L. (Ed.) *Yilgarn Craton – Officer Basin – Musgrave Province Seismic and MT Workshop (GA Record 2013/28)*; Geoscience Australia, Commonwealth of Australia: Canberra, ACT, Australia, 2013. Available online: http://pid.geoscience.gov.au/dataset/ga/76664 (accessed on 2 August 2019).
131. Wade, B.P.; Kelsey, D.E.; Hand, M.; Barovich, K.M. The Musgrave Province: Stitching north, west and south Australia. *Precambrian Res.* **2008**, *166*, 370–386. [CrossRef]
132. Donnellan, N.; Hussey, K.J.; Morrisson, R.S.; Kruse, P.D. *Tennant Creek 1:250 000 Geology*, 2nd ed.; Northern Territory Geological Survey: Darwin, Australia, 1998.
133. Brakel, A.T.; Montcrieff, J.S.; Muhling, P.D.; Chin, R.J. *Dumbleyung 1:250 000 Geological Map*; Geological Survey of Western Australia: Perth, Australia, 1985.
134. Scrimgeour, I.R.; Close, D.F.; Edgoose, C.J. *Petermann Ranges 1:250 000 Geological Map*, 2nd ed.; Northern Territory Geological Survey: Darwin, Australia, 1999.
135. Quigley, M.C.; Mohammadi, H.; Jimenez, A.; Duffy, B.G. Multi-fault earthquakes with kinematic and geometric rupture complexity: How common? INQUA Focus Group Earthquake Geology and Seismic Hazards. In Proceedings of the 8th International INQUA Meeting on Paleoseismology, Active Tectonics and Archeoseismology (PATA), Blenheim, New Zealand, 13–16 November 2017.
136. Bowman, J.R.; Jones, T. *Post-Seismic Surveys of the Epicentral Area of the 1988 Tennant Creek, N.T., Earthquakes (BMR Record 1992/002)*; Bureau of Mineral Resources, Geology and Geophysics: Canberra, Australia, 1991. Available online: http://pid.geoscience.gov.au/dataset/ga/14510 (accessed on 2 August 2019).
137. Rogers, C.D.F. Types and distribution of collapsible soils. In *Genesis and Properties of Collapsible Soils*; Derbyshire, E., Dijkstra, T., Smalley, I.J., Eds.; Springer: Loughborough, UK, 1995; pp. 1–17.
138. Bierman, P.R.; Caffee, M.W. Cosmogenic exposure and erosion history of Australian bedrock landforms. *Bull. Geol. Soc. Am.* **2002**, *114*, 787–803. [CrossRef]
139. Biasi, G.P.; Wesnousky, S.G. Bends and ends of surface ruptures. *Bull. Seismol. Soc. Am.* **2017**, *107*, 2543–2560. [CrossRef]
140. Mcpherson, A.; Clark, D.; Macphail, M.; Cupper, M. Episodic post-rift deformation in the south-eastern Australian passive margin: Evidence from the Lapstone Structural Complex. *Earth Surf. Process. Landf.* **2014**, *39*, 1449–1466. [CrossRef]
141. Allen, T.; Griffin, J.; Clark, D. *The 2018 National Seismic Hazard Assessment: Model Input Files (GA Record 2018/032)*; 2018/32; Geoscience Australia: Canberra, ACT, Australia, 2018. [CrossRef]
142. Clark, D.; Leonard, M.; Griffin, J.; Stirling, M.W.; Volti, T. Incorporating fault sources into the Australian National Seismic Hazard Assessment (NSHA) 2018. In Proceedings of the Australian Earthquake Engineering Society Conference 2016, Melbourne, VIC, Australia, 25–27 November 2016.
143. Boncio, P.; Liberi, F.; Caldarella, M.; Nurminen, F.C. Width of surface rupture zone for thrust earthquakes: Implications for earthquake fault zoning. *Nat. Hazards Earth Syst. Sci.* **2018**, *18*, 241–256. [CrossRef]
144. Leonard, M. Earthquake fault scaling: Self-consistent relating of rupture length, width, average displacement, and moment release. *Bull. Seismol. Soc. Am.* **2010**, *100*, 1971–1988. [CrossRef]
145. Moss, R.E.S.; Ross, Z.E. Probabilistic fault displacement hazard analysis for reverse faults. *Bull. Seismol. Soc. Am.* **2011**, *101*, 1542–1553. [CrossRef]
146. Lavrentiadis, G.; Abrahamson, N. Generation of surface-slip profiles in the wavenumber domain. *Bull. Seismol. Soc. Am.* **2019**, *109*, 888–907. [CrossRef]
147. Anderson, J.G.; Biasi, G.P.; Wesnousky, S.G. Fault-scaling relationships depend on the average fault-slip rate. *Bull. Seismol. Soc. Am.* **2017**, *107*, 2561–2577. [CrossRef]
148. Leonard, M. Self-consistent earthquake fault-scaling relations: Update and extension to stable continental strike-slip faults. *Bull. Seismol. Soc. Am.* **2014**, *104*, 2953–2965. [CrossRef]
149. Finch, E.; Hardy, S.; Gawthorpe, R. Discrete-element modelling of contractional fault-propagation folding above rigid basement fault blocks. *J. Struct. Geol.* **2003**, *25*, 515–528. [CrossRef]
150. Serva, L. History of the Environmental Seismic Intensity Scale ESI-07. *Geosciences* **2019**, *9*, 210. [CrossRef]
151. Ahmad, B.; Sana, H.; Alam, A. Macroseismic intensity assessment of 1885 Baramulla Earthquake of northwestern Kashmir Himalaya, using the Environmental Seismic Intensity scale (ESI 2007). *Quat. Int.* **2014**, *321*, 59–64. [CrossRef]

152. Sanchez, J.J.; Maldonado, R.F. Application of the ESI 2007 scale to two large earthquakes: South Island, New Zealand (2010 Mw 7.1), and Tohoku, Japan (2011 Mw 9.0). *Bull. Seismol. Soc. Am.* **2016**, *106*, 1151–1161. [CrossRef]
153. Nappi, R.; Gaudiosi, G.; Alessio, G.; De Lucia, M.; Porfido, S. The environmental effects of the 1743 Salento earthquake (Apulia, southern Italy): A contribution to seismic hazard assessment of the Salento Peninsula. *Nat. Hazards* **2017**, *86*, S295–S324. [CrossRef]
154. Serva, L.; Vittori, E.; Comerci, V.; Esposito, E.; Guerrieri, L.; Michetti, A.M.; Mohammadioun, B.; Mohammadioun, G.C.; Porfido, S.; Tatevossian, R.E. Earthquake Hazard and the Environmental Seismic Intensity (ESI) Scale. *Pure Appl. Geophys.* **2016**, *173*, 1479–1515. [CrossRef]
155. Mitchell, D.; Paultre, P.; Tinawi, R.; Saatcioglu, M.; Tremblay, R.; Elwood, K.; Adams, J.; DeVall, R. Earthquake Environmental Effect for seismic hazard assessment: The ESI intensity scale and the EEE Catalogue. *Mem Descr della Cart Geol d'Italia* **2015**, *97*, 1–181.
156. Porfido, S.; Esposito, E.; Spiga, E.; Sacchi, M.; Molisso, F.; Mazzola, S. Impact of Ground Effects for an Appropriate Mitigation Strategy in Seismic Area: The Example of Guatemala 1976 Earthquake. In *Engineering Geology for Society and Territory - Volume 2: Landslide Processes*; Lollino, G., Giordan, D., Crosta, G.B., Corominas, J., Azzam, R., Wasowski, J., Sciarra, N., Eds.; Springer International Publishing: Berlin, Germany, 2015; pp. 703–708. [CrossRef]
157. Reicherter, K.; Michetti, A.M.; Silva, P.G.; Silva Barroso, P.G. Palaeoseismology: Historical and prehistorical records of earthquake ground effects for seismic hazard assessment. *Geol. Soc. Lond. Spec. Publ.* **2009**, *316*, 1–10. [CrossRef]
158. Giner-Robles, J.L.; Silva, P.G.; Elez, J.; Rodríguez-Pascua, M.A.; Perez-Lopez, R.; Rodríguez-Escudero, E. Relationships between the ESI-07 scale and expected PGA values from the analysis of two historical earthquakes (≥ VIII EMS) in East Spain: Tavernes 1396 AD and Estubeny 1748 AD events. In Proceedings of the 6th International INQUA Meeting in Paleoseismology, Active Tectonics and Archaeoseismology, Pescina, Fucino Basin, Italy, 19–24 April 2015.
159. Heddar, A.; Beldjoudi, H.; Aurhemayou, C.; SiBachir, R.; Yelles-Chaouche, A.; Boudiaf, A. Use of the ESI-2007 scale to evaluate the 2003 Boumerdès earthquake (North Algeria). *Ann. Geophys.* **2016**, *5*. [CrossRef]
160. Johnston, A.C. Seismotectonic interpretations and conclusions from the stable continental region seismicity database. In *The Earthquakes of Stable Continental Regions—v. 1 Assessment of Large Earthquake Potential*; Johnston, A.C., Coppersmith, K.J., Kanter, L.R., Cornell, C.A., Eds.; Electric Power Research Institute: Palo Alto, CA, USA, 1994.
161. Griffin, J.; Weatherill, G.; Allen, T. Performance of national scale smoothed seismicity estimates of earthquake activity rates. In Proceedings of the Australian Earthquake Engineering Society 2017 Conference, Canberra, ACT, Australia, 24–26 November 2017.
162. Adams, J.; Percival, J.A.; Wetmiller, R.J.; Drysdale, J.A.; Robertson, P.B. Geological controls on the 1989 Ungava surface rupture: A preliminary interpretationn (GSC Paper 92-1C). In *Current Research, Part C*; Geological Survey of Canada: Ottawa, ON, Canada, 1992; pp. 147–155. [CrossRef]
163. Bent, A.L. The 1989 (Ms 6.3) Ungava, Quebec, Earthquake: A Complex Intraplate Event. *Bull. Seismol. Soc. Am.* **1994**, *84*, 1075–1088.
164. Mitchell, D.; Paultre, P.; Tinawi, R.; Saatcioglu, M.; Tremblay, R.; Elwood, K.; Adams, J.; DeVall, R. Evolution of seismic design provisions in the National building code of Canada. *Can. J. Civ. Eng.* **2010**, *37*, 1157–1170. [CrossRef]
165. Beekman, F.; Stephenson, R.A.; Korsch, R.J. Mechanical stability of the Redbank Thrust Zone, Central Australia: Dynamic and rheological implications. *Aust. J. Earth Sci.* **1997**, *44*, 215–226. [CrossRef]
166. Sandiford, M.; Egholm, D.L. Enhanced intraplate seismicity along continental margins: Some causes and consequences. *Tectonophysics* **2008**, *457*, 197–208. [CrossRef]
167. Everingham, I.B.; Smith, R.S. Implications of fault-plane solutions for Australian earthquakes on 4 July 1977, 6 May 1978 and 25 November 1978. *BMR J. Aust. Geol. Geophys.* **1979**, *4*, 297–301. Available online: http://pid.geoscience.gov.au/dataset/ga/81008 (accessed on 2 August 2019).

geosciences

MDPI

Review

Post Seismic Catalog Incompleteness and Aftershock Forecasting

Eugenio Lippiello [1,*], Alessandra Cirillo [1], Cataldo Godano [1], Elefetheria Papadimitriou [2] and Vassilis Karakostas [2]

[1] Department of Mathematics and Physics, University of Campania "L. Vanvitelli", Viale Lincoln 5, 81100 Caserta, Italy
[2] Geophysics Department, Aristotle University of Thessaloniki, GR 541 24 Thessaloniki, Greece
* Correspondence: eugenio.lippiello@unicampania.it

Received: 17 July 2019; Accepted: 6 August 2019; Published: 12 August 2019

Abstract: A growing interest appears among public authorities and society in accurate and nearly real time aftershock forecasting to manage and mitigate post-seismic risk. Existing methods for aftershock forecasting are strongly affected by the incompleteness of the instrumental datasets available soon after the main shock occurrence. The deficit of observed events, in the first part of aftershock sequences, can be naturally attributed to various mechanisms such as the inefficiency of the seismic network and the overlap of earthquake signals in seismic records. In this review, we show that short-term aftershock incompleteness can be explained only in terms of the second mechanism, whereas it is only weakly affected by the quality of the instrumental coverage. We then illustrate how standard models for earthquake forecasting can be modified to take into account this incompleteness. In particular, we focus on forecasting methods based on the data available in real time, in which many events are missing and the uncertainty in hypocenter location is considerable. We present retrospective tests that demonstrate the usefulness of these novel methods compared with traditional ones, which implement average values of parameters obtained from previous sequences.

Keywords: catalog incompleteness; seismic hazard

1. Introduction

Even if a still unanswered question is whether or not the accurate, reliable prediction of individual earthquakes is a realistic scientific goal, the possibility of forecasting future earthquakes exists. The two major examples concern the estimation of the occurrence probability of large shocks over a very long temporal interval (decades up to centuries) and the estimation of the aftershock occurrence rate after a large earthquake. Neither of the two cases is relevant in predicting the occurrence of an impending large earthquake but both examples provide very useful information on mitigating the impact of earthquakes that are likely to occur. As a matter of fact, the first example, usually defined as long-term (LT) seismic forecasting, is probably the most relevant from an engineering point of view, such as urban planning and building constructions: It allows one to address questions such as the maximum magnitude expected in a given area for the next years. Concerning the second example, usually defined as post-seismic Short-Term Aftershock (STA) forecasting, many events (the aftershocks) are always observed soon after the occurrence of a strong shock (the main shock). Aftershocks can attain sizes comparable to their triggering mainshock and can be very dangerous since they impact buildings already damaged by the previous shocks.

This review is focused on STA forecasting that can be potentially very efficient. Indeed the organization in time, space and energy of aftershocks follows well established empirical laws such as the Gutenberg–Richter (GR) and the Omori–Utsu (OU) law [1,2], which can be implemented in forecasting models. The GR law states that the magnitude distribution of earthquakes is an exponential function $P(m) \sim \exp(-\beta m)$, and the OU law characterizes the power law decay of the aftershock rate as function of the time t since the main shock.

Even if the LT and STA forecasting act on two very different time scales, the two problems are intimately related. In the most simple description, seismic occurrence can be viewed as the superposition of two different stochastic processes: background seismicity responsible for mainshocks, which are the target of the LT forecasting, and aftershock occurrence, which is the target of STA. Hence, to achieve an accurate LT forecasting method a so-called declustering procedure is necessary, which allows one to isolate the two processes by means of a detailed knowledge of aftershock features. A clear example is the Epidemic Type Aftershock Sequence (ETAS) model introduced by Ogata [3] and probably representing nowadays the most popular model for STA as well as among the most efficient tools for LT forecasting. Studies of STA forecasting models, such as the ETAS model or more simple models implementing the OU law, have shown [4–14] that the incompleteness of datasets strongly affects the estimation of model parameters. This effect is more relevant in the first part of aftershock sequences when many earthquakes, in particular small ones, are not recorded and therefore not reported in seismic catalogs. This is mainly caused by the overlap of the signal of individual earthquakes in the seismic records. At the same time, incompleteness is also produced by the overload of processing facilities, due to a very large number of events in a narrow temporal window, and the damage caused by the mainshock to the seismic stations. Because of these difficulties, in many cases, operational probability forecasts only start more than 24 h after the mainshock [15].

In this review, we explore the problem of incompleteness of instrumental datasets focusing in particular on the so-called Short Term Aftershock Incompleteness (STAI). This is the main subject of Section 2. In Section 3, we review recent results on the influence of STAI on the estimation of parameters of STA forecasting models. Section 4 is then devoted to show that STAI is an intrinsic property of seismic catalogs which is not related to the efficiency of the seismic network. We conversely show that the main mechanism responsible for STAI is the overlap of aftershock coda waves with the waveforms of other events which obscure small aftershocks that occur close in time after larger ones. In Section 5, we show some approaches recently proposed to take explicitly into account this "obscuration" effect within the ETAS model. These approaches, however, are not simple to be implemented in real-time automatic procedures for aftershock forecasting. This is the topic of Section 6, which presents two different procedures developed to provide accurate STA forecasting, several minutes after the occurrence of a mainshock: the Omi et al. method [7,9,10] and the Lippiello et al. method [16,17]. The test of these two methods in retrospective studies is presented in Section 6 and final conclusions are drawn in the last section.

2. Catalog Incompleteness

Catalog completeness is usually quantified in terms of a magnitude threshold (or lower cut-off) m_c defined as the magnitude above which all events are identified and included in the catalog. An accurate estimate of m_c is fundamental in seismic forecasting. A too high value, discarding usable data, leads to loss information by under-sampling. Conversely, a too low value leads to an unreliable estimation of parameter values and thus to a biased analysis because of the incomplete dataset. A standard way of estimating m_c is to find the minimum magnitude above which the best fit with the GR law is obtained. The value of m_c clearly depends on the ability to filter noise and on the distance between the earthquake epicenter and the seismic stations necessary to trigger an event declaration in a catalog. Instrumental data from Taiwan seismicity, for example, give [18] at a given location $\vec{r}$, $m_c(\vec{r})$

$$m_c(\vec{r}) = 4.83d^{0.09} - 4.36. \tag{1}$$

where $d = |\vec{r} - \vec{r}_3|$ is the distance in kilometers between the epienter and the position $\vec{r}_3$ of the third nearest seismic station. In Figure 1, we present the m_c map for the Southern California obtained in [19] via the method of Amorése [20]. In particular, we observe a region in the central part of Southern California with a higher density of seismic stations, characterized by $m_c \leq 1.4$. This region, defined as Region 1, contains the 36% of $m > 2.5$ events recorded in the entire catalog. The remaining Southern California region (defined as Region 2) has a completeness magnitude starting from $m_c = 1.5$ and becoming as large as $m_c \simeq 3$ near the borders. A similar behavior is found if m_c is evaluated according to the method of Schorlemmer and Woessner [21].

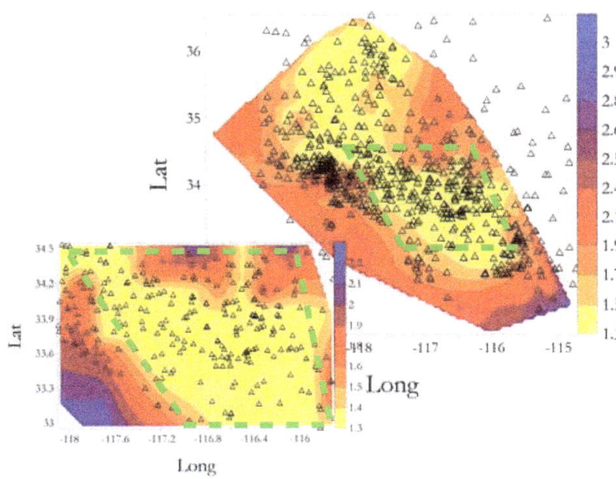

Figure 1. Magnitude completeness in Southern California. The value of m_c can be obtained by the color bar and triangles identify the location of seismic stations. Green dashed lines define Region 1. Region 2 is the complement to Region 1 with respect to the entire Southern California (From [19]).

We stress that m_c estimated from Equation (1) is a static quantity, controlled by the number of seismic stations, and we define it as "the static completeness magnitude". On the other hand, instrumental data show that the m_c value, inside a given region, changes with time reaching much larger values in the first part of the aftershock sequence. As already anticipated in the Introduction, the dependence of the completeness magnitude $m_c(t)$ on the time t since the main shock occurrence is usually termed Short Term Aftershock Incompleteness (STAI). Results in [22–24] give a completeness magnitude $m_c(t)$ which depends logarithmically on the time t since the main shock

$$m_c(t, m_M) = m_M - \frac{1}{d}\left(\log_{10}\left(\frac{t}{C_0}\right)\right), \tag{2}$$

where m_M is the main shock magnitude and d and C_0 are fitting parameters. We refer to Equation (2) as the Kagan–Helmstetter formula with the best fitting parameters $d \simeq 1$ and $C_0 \sim 10^{-4}$ days, when time is measured in days. In Figure 2, we plot the experimental aftershock magnitude distribution evaluated for different temporal intervals after the $m = 7.3$ Landers earthquake, in Southern California. Experimental results show a magnitude distribution with an about flat for values $m < m_c(t)$, whereas curves appear parallel on a semi-logarithmic scale for $m > m_c(t)$ consistently with a GR law with $b \simeq 1$. The crossover magnitude $m_c(t)$ is in agreement with Equation (2).

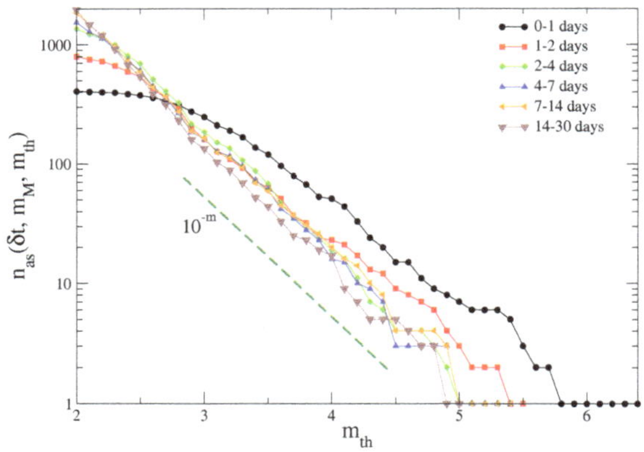

Figure 2. The number of aftershocks with magnitude larger than m_{th} for the $m_M = 7.3$ Landers earthquake in Southern California, evaluated in different temporal windows δt from the main shock. The green dashed line is the exponential behavior expected according to the GR law with $b = 1$.

In a different approach [5,7,9], STAI is taken into account by considering a magnitude distribution

$$P_{\beta,\sigma}(m) \propto e^{-\beta m}\Phi\left(m|\mu(t),\sigma\right) \tag{3}$$

given by the GR law multiplied by the detection rate function Φ, which is represented by an error function

$$\Phi(m|\mu(t),\sigma) = \frac{1}{\sqrt{2\pi\sigma^2}}\int_{-\infty}^{m} e^{-\frac{(x-\mu(t))^2}{2\sigma^2}}dx. \tag{4}$$

In the above equation, the function $\mu(t)$ represents the 50% detection magnitude and σ represents the range of the magnitude of partially detected earthquakes, i.e., at time t, only 50% of the events with $m = \mu(t)$ are expected to be detected whereas more the 98% of events are expected to be detected if $m > \mu(t) + 2\sigma$. A reasonable definition therefore corresponds to assume $m_c(t) = \mu(t) + 2\sigma$. In particular, Ogata and Katsura [5] proposed that $\mu(t)$ obeys the law

$$\mu(t) = \nu_0 + \nu_1 \exp\left(-\nu_2\left(3 + \log_{10}(t)\right)^{\nu_4}\right) \tag{5}$$

where the ν_i are fitting parameters. On the other hand, in a series of papers, Omi et al [7–10,15] developed an elegant method to obtain a non parametric fit of the function $\mu(t)$ and an estimate of σ from the occurrence times and magnitudes of all recorded events in a giving learning period.

In Figure 3, we plot the results by Omi et al. [9] for $\mu(t)$ and $\mu(t) + 2\sigma$ for three aftershock sequences in Japan. These results are compared with the Ogata–Hirata formula (Equation (5)) and the Kagan–Helmstetter formula (Equation (2)). Figure 3 shows that the Omi and the Ogata–Hirata models give similar behavior for $\mu(t)$ and are able to capture the time variation of the detection rate. In contrast with these two models, since the parameters of the Kagan–Helmstetter formula are fixed for all sequences, it cannot reproduce the diverse recovering dynamics of the completeness magnitude that considerably depends on each aftershock sequence. The comparison of the forecasting skill of these three methods, for 38 Japan aftershock sequences, shows that the Omi method performs slightly better than the Ogata–Hirata methods and much better than a Kagan–Helmstetter formula [9].

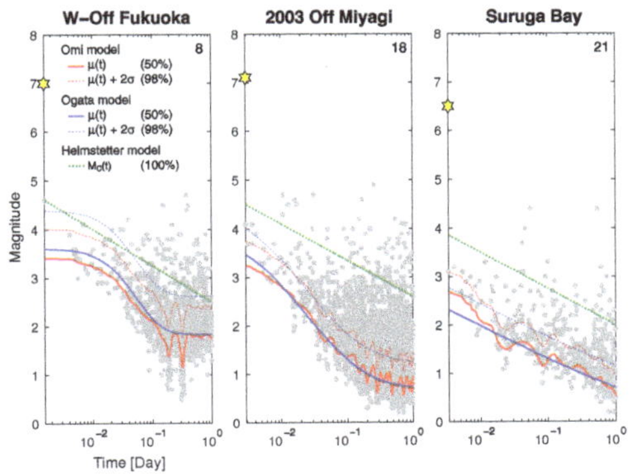

Figure 3. Examples of the estimated time-varying 50% detection rate $\mu(t)$ (solid curve) of magnitudes and estimated various time-varying completeness magnitudes (dotted curve) as indicated in the inset, superimposed on the magnitude-time plot of the observed aftershocks during the first day of the main shock. From [9].

3. The Influence of STAI on Model Parameters

For a complete dataset, one expects that the rate of aftershocks $\rho(t, m_M, m_{th})$ with magnitude larger than a threshold value m_{th} occurring after a time t following a mainshock of magnitude m_M can be obtained by combining the GR law and the OU law

$$\rho(t, m_M, m_{th}) = \frac{K}{(t+c)^p} e^{-\beta m_{th}}. \tag{6}$$

According to the productivity law [25], K depends on the main shock magnitude and Equation (6) can be written as

$$\rho(t, m_M, m_{th}) = \frac{K_0 e^{\alpha m_M - \beta m_{th}}}{(t+c)^p}. \tag{7}$$

As already observed in [2], missing small events in the early stage of the aftershock sequence causes the instability of the estimate of the parameters K_0, α, β, c, p in Equation (6). A problem which becomes particularly relevant at the beginning of aftershock sequences when the completeness magnitude after large earthquakes can temporarily increase by several units [4,22,26,27]. For this reason, long and short term forecasts usually present some corrections which take into account STAI [6,28,29].

Incompleteness, in particular, can make the c-value measured from instrumental catalogs c_{meas} much larger the "true" c-value in the OU law (Equation (6)). Indeed, restricting to aftershocks with magnitudes larger than a reference value m_{th}, if events with magnitudes $m < m_c(t)$ are not recorded, the measured c-value can be obtained from Equation (2) after setting $m_c(c_{meas}) = m_{th}$, which leads to

$$c_{meas} = C_0 10^{d(m_M - m_{th})}. \tag{8}$$

It is evident that this quantity depends on the parameters of Equation (2) but is not related to the c-value of the OU law. Alternatively, an estimate of c_{meas} can be obtained from Equation (5) after setting $\mu(c_{meas}) = m_{th} - 2\sigma$. As a consequence, the incompleteness at short times hides the true value of c that in turn introduces a strong bias in the evaluation of the parameters K_0 and α in Equation (6), strongly affecting routines for short term aftershock forecasting at time $t < c_{meas}$.

3.1. The Influence of STAI on the ETAS Parameters

As anticipated in the Introduction, the ETAS model is probably, nowadays, the most popular one for STA forecasting. The assumptions of the ETAS model include: (1) yhe background seismicity is a stationary Poisson process that depends on the position $\vec{x}$, $\mu(\vec{x})$; (2) every event, whether it is a background or a triggered one by a previous event, triggers its own off-spring independently; (3) the expected number of direct off-springs is an exponential function of the magnitude of the mother event (productivity law); and (4) the time lags between triggered events and the mother event follow the OU law. According to these assumptions, the occurrence rate of events with magnitudes $m \geq m_0$ at the position $\vec{x}$ at time t is given by

$$\Lambda_{ETAS}(m, \vec{x}, t) = \left[\sum_{i=1}^{N} Q\left(|\vec{x}_i - \vec{x}|, t - t_i, m_i\right) + \mu(\vec{x})\right] \beta e^{-\beta(m-m_0)} \tag{9}$$

where the sum extends over all events with magnitude m_i, epicentral coordinate $\vec{x}_i$ and occurrence time $t_i < t$ and

$$Q(\Delta r_i, t - t_i, m_i) = \frac{K_0(p-1)}{c} e^{\alpha(m_i - m_0)} \left(1 + \frac{t - t_i}{c}\right)^{-p} G(\Delta r_i, m_i) \tag{10}$$

with $\Delta r_i = |\vec{x}_i - \vec{x}|$, which is the epicentral distance. The function $G(\Delta r_i, m_i)$ is a spatial kernel that explicitly depends on the triggering magnitude m_i and $\mu(\vec{x})$ is the time independent contribution due to background seismicity.

The influence of STAI on the estimates of the ETAS parameter was addressed by Zhuang et al. [13] in the case of the 15 April 2016, Kumamoto earthquake sequence in Japan. Under the assumption that earthquake magnitudes are independent of their occurrence times, Zhuang et al. [13] replenished the short-term missing data of small earthquakes by using a bi-scale transformation. They then compared the maximum likelihood estimate of the ETAS parameters of the recorded dataset in the JMA catalog with the replenished one, considering only events above a lower magnitude threshold $m_{th} = m_c$. Results plotted in Figure 4, as function of m_c, show that, when the magnitude threshold $m_c \geq 3$, which is approximately the static completeness magnitude of the JMA catalog, the estimated ETAS parameters are about the same for both datasets. Conversely, important differences are found for values of $m_c < 3$. For the replenished dataset, the estimated background rate $\mu(x)$ decreases roughly exponentially when the cut-off magnitude is increased, consistently to what is expected according to the GR law (Figure 4a). The original dataset, conversely, exhibits a flatter behavior, indicating the absence of small magnitude events. Concerning the other parameters, the most striking feature is that in the replenished dataset all parameters only weakly depend on m_c, as expected, whereas we observe a non-trivial dependence on m_c in the JMA catalog.

The results of Zhuang et al. [13] indicate that the estimate of ETAS parameters from the original dataset, when one considers a lower magnitude threshold $m_c < 3$, leads to non-correct results. A similar conclusion was reached by Seif et al. [14] who studied how the ETAS parameters, obtained by the iterative approach of Zhuang et al. [30], depends on the lower magnitude threshold m_{th}. In particular, Seif et al. [14] investigated two simulated ETAS catalogs: a complete one which implements the ETAS parameters estimated from the Southern California catalog and an incomplete one where aftershocks of mainshocks with $m_M > 5$ were removed if their magnitude was smaller than $m_c(t)$ given in Equation (2). Results plotted in Figure 5 show that for sufficiently larger values of m_{th}, the parameter inversion procedure does not give the true values of K_0 and p used to generate synthetic catalogs. Seif et al. [14] attributed the observed discrepancy to the fact that aftershocks triggered by events with $m < m_{th}$ are erroneously identified as direct aftershocks of some previous larger earthquake. This widens the distribution of direct aftershocks leading to a smaller p-value. At the same time, because of the anticorrelation between K_0 and p, K_0 is overestimated. Figure 5, in particular, shows a striking difference between the estimated parameters in the complete and the incomplete catalogs. However, this difference tends to disappear for increasing m_{th} indicating that the influence of aftershock incompleteness is not significant for $m_{th} \gtrsim 3.5$.

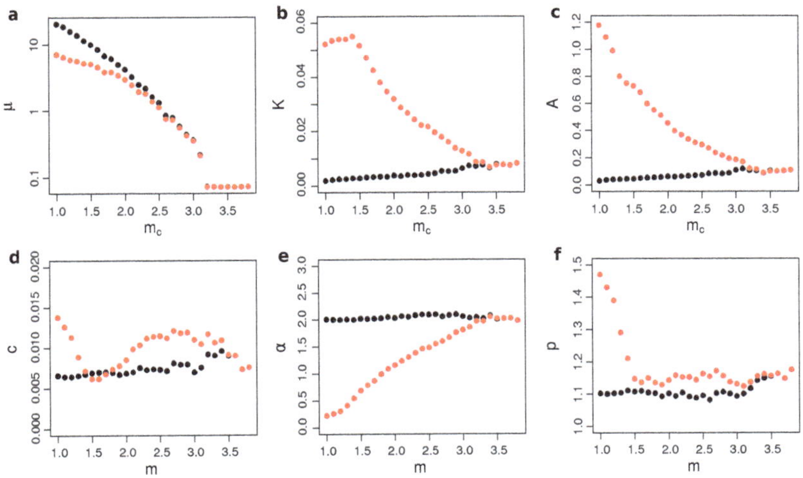

Figure 4. Different panels correspond to the ETAS parameters $\mu, K, A = K\int_0^{\infty}(t+c)^{-p}dt, c, \alpha, p$ (see axis labels) estimated from the Kumamoto aftershock sequence with different magnitude thresholds. The red and black dots are the estimates based on the original and the replenished datasets, respectively. Unit of measures are day^{-1}, day^{p-1}, day for μ, K, c respectively and the other quantities are adimensional except $A = K\int_0^{\infty}(t+c)^{-p}dt$ which represents the productivity from an event of magnitude m_c. From [13].

Results of Figures 4 and 5 indicate that using a lower magnitude threshold m_{th} below the completeness level, especially for some parameters, can lead to incorrect prediction. Unfortunately, it is not simple to establish a strict correspondence between the degree of incompleteness of the catalog and the error expected in the estimate of parameters.

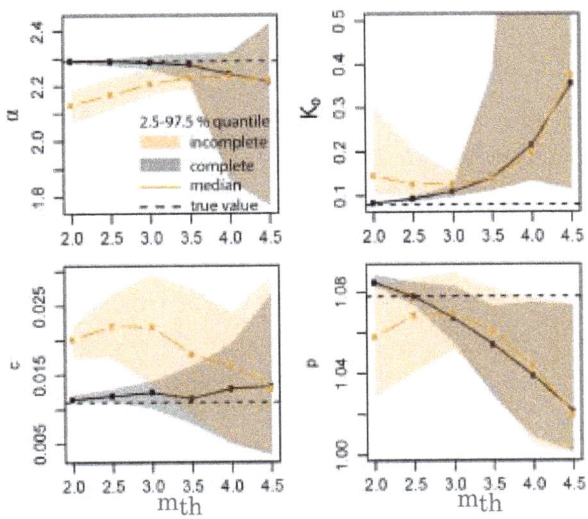

Figure 5. The ETAS parameters are plotted against m_{th} for synthetic catalogs simulated with parameters from Southern California (gray) and compared with the parameter for the incomplete synthetic catalog (orange). The "true" parameter values are plotted with black dashed lines the grey shadowed region represents the 95% quantiles of 30 synthetic ETAS catalogs. The orange shadowed region represents the 95% quantiles of 30 synthetic ETAS incomplete catalogs. From [13].

3.2. Is STAI Related to the Static m_c?

As explained in Section 2 the static m_c is a local quantity which depends on the local density of the seismic network ρ_S, as illustrated by Equation (1) and Figure 1. The influence of the density ρ_S on the STAI was addressed by de Arcangelis et al. [31] by investigating the c_{meas}-value in the two sub-regions of Southern California illustrated in Figure 1. As already explained in Section 2, the inner region (Region 1) comprises a high value of ρ_S and a static $m_c \leq 1.4$. Conversely, a small ρ_S is present in the external region (Region 2) and the static $m_c > 1.5$, with values of $m_c \simeq 3$ close to the borders. To obtain an estimate of c_{meas} in each sub-region, de Arcangelis et al. [31] measured the aftershock daily rate $\rho(t, m_M, m_{th})$ defined as the number of aftershocks with magnitude larger than m_{th} occurring at a temporal distance t after their triggering main shock with magnitude $m \in [m_M, m_M + 1)$, divided by the number of mainshocks with magnitude $m \in [m_M, m_M + 1)$. Three different values of $m_M = (3, 4, 5)$ and $m_{th} = (1.5, 2.5, 3.5)$ were considered. In this study, mainshock–aftershock couples were identified according to the Baiesi–Paczusky (BP) declustering criterion [32–34] using the same parameters adopted by Moradpour et al. [35] and Hainzl [12]. In particular, only aftershocks identified as direct descendants of the mainshock were included in the analysis.

The results (Figure 6) show that the aftershock rate clearly depends on the magnitude difference $m_M - m_{th}$ in both Region 1 and Region 2. In particular, de Arcangelis et al. [31] divided time by $\tau = 10^{d(m_M - m_{th})}$ obtaining that data for different values of m_M and m_{th}, inside each sub-region, exhibits the scaling collapse $\rho(t, m_M, m_{th}) = F(t/\tau)$ (Figure 7a). It is evident from Figure 7a that the Omori decay $\rho \sim t^{-p}$ sets in when t/τ becomes larger than a given value x_0, different between the two regions. Since the c_{meas} can be obtained from the time such that the Omori decay $\rho \sim t^{-p}$ sets in, Figure 7a gives $c_{meas} = x_0\tau$ and one recovers Equation (8) after the identification $x_0 = C_0$. In particular the best fit gives $\log_{10}(C_0) = -3.53 \pm 0.05$ and $d = 1 \pm 0.03$ inside Region 1 and $\log_{10}(C_0) = -3.70 \pm 0.05$ and $d = 0.95 \pm 0.03$ inside Region 2. This leads to a counterintuitive behavior with a c_{meas}-value being larger inside Region 1 even if the static m_c is significantly smaller inside Region 1 than in Region 2. Conversely a smaller c_{meas}-value is found in Region 2 when the static m_c is larger. This result clearly indicates that c_{meas} is not related to ρ_S and that STAI cannot be reduced by increasing the density of the seismic station thus suggesting that STAI originates from a different mechanism (see next section). The same conclusion can be also obtained from the measurement of the correlation between magnitude according to the method proposed in [19,36–38]. This analysis [19,31] has shown significantly larger magnitude correlations in Region 1 than in Region 2.

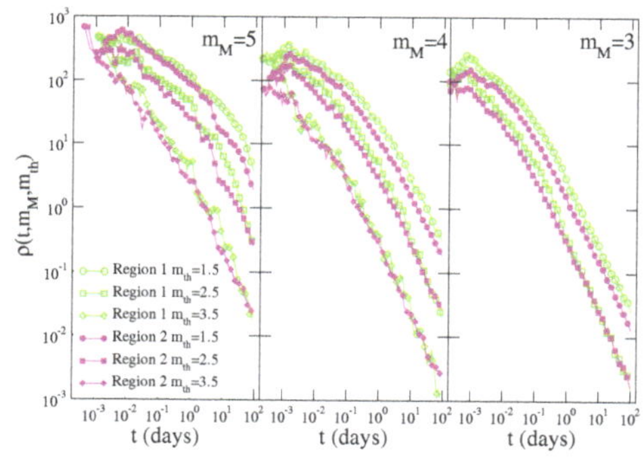

Figure 6. The number of events identified as aftershocks by the BP declustering procedure with magnitude larger than m_{th}, which occurred at a temporal distance t from events identified as mainshocks with magnitude $m \in [m_M, m_M + 1)$, is divided by the number of identified mainshocks and plotted versus t. Different panels correspond to different values of the mainshock magnitude class $m \in [m_M, m_M + 1)$. Different colors correspond to results for different geographic regions: Region 1 (open green symbols) and Region 2 (filled magenta symbols). Different symbols indicate different values of the lower threshold: $m_{th} = 1.5$ (circles), $m_{th} = 2.5$ (squares) and $m_{th} = 3.5$ (diamonds).

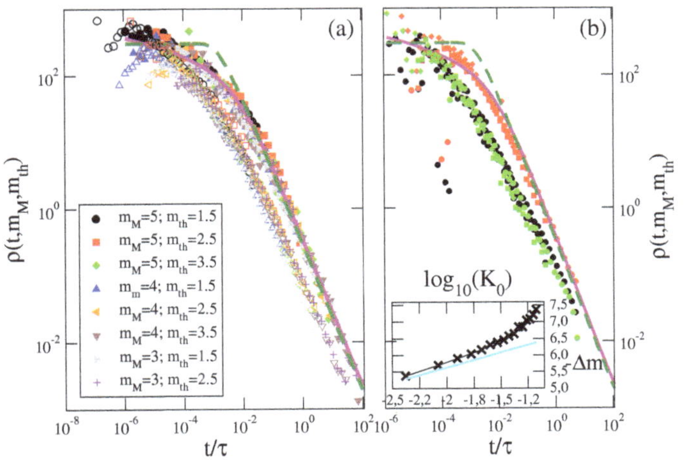

Figure 7. (Color online) (**a**) The same data in Figure 6 are plotted as function of t/τ, with $\tau = 10^{d(m_M - m_{th})}$ proportional to c_{meas} (Equation (8)) with $d = 1$, for different values of m_M and m_{th}. Filled (empty) colored symbols are used for data of Region 1 (Region 2). The magenta continuous line is the scaling function $F(x) = A\log(1 + Bx^{-p})$ with $A = 0.35$, $B = 70$ and $p = 1.1$, whereas the dashed green line is the scaling function $F(x) = A(x/B + 1)^{-p}$ with $A = 300$, $B = 7$ and $p = 1.1$. (**b**) The aftershock density $\rho(t, m_M, m_{th})$ in the ETASI1 catalog, with a blind time $\Delta t = 1$ min, is plotted as a function of t/τ. Different values of m_M and m_{th} are plotted with different symbols: stars for $m_M - m_{th} = 2.5$, crosses for $m_M - m_{th} = 1.5$ and plus for $m_M - m_{th} = 0.5$. Different colors correspond to different values of K_0 and of the average background rate r_B: $K_0 = 0.035$ and $r_B = 4.38$ days^{-1} (black), $K_0 = 0.035$ and $r_B = 8.3$ days^{-1} (green) and $K_0 = 0.068$ and $r_B = 4.38$ days^{-1} (red). Magenta continuous and green dashed lines are the same scaling functions $F(x)$ plotted in (**a**). (Inset) The value of Δm (Equation (8)) as function of $log_{10}(K_o)$ for the ETASI model with a blind time $\Delta t = 1$ min (black crosses). The cyan line is the theoretical prediction (Equation (18)). From [31].

4. The Origin of STAI and the Envelope Function

Results of the previous Section (Section 3.2) suggest that STAI is an intrinsic property of seismic catalogues not related to density of the seismic stations. This conclusion is strongly supported by the study of the envelope function $\mu_e(t)$ after several mainshocks that occurred in Greece and Italy in the last ten years [16]. More precisely, the envelope function $\mu_e(t)$ is obtained from the ground velocity recorded during the first days after the mainshock. The signal of each component is filtered by means of a two-pass Butterworth filter in the range $[1, 10]$ Hz, the envelope of each signal is computed and the signals of the three components are superimposed. $\mu_e(t)$ is finally defined as the logarithm of the resulting signal. This quantity was introduced by Peng et al. [26] to identify aftershocks not reported in the JMA catalog during the first minutes after the main shock. The idea is that the occurrence of an aftershock must produce a double peak in $\mu_e(t)$ corresponding to the coupled pair of P and S arrivals. The local magnitude of the event is given by $m \simeq \mu_{max} + const$, where μ_{max} is the maximum in μ_e and the constant depends on the epicentral distance from the recording station, related to the S-P time difference.

Considering the evolution of $\mu_e(t)$ after a mainshock, occurred at the time t_0, Lippiello et al. [16] found that the envelope function never goes below a given value $\mu_{min}(t)$ which is a logarithmic decreasing function of time (Figure 8)

$$\mu_{min}(t) = \mu_M - \phi \log(t - t_0) - \Delta\mu_{min}. \tag{11}$$

As a consequence, even very accurate analyses of post seismic waveforms, even those which employ sophisticated matched filter detection algorithms [39,40], do not allow one to identify small events which produce peaks smaller than $\mu_{min}(t)$. This reflects a completeness magnitude $m_c(t)$ that depends on the time after the mainshock with a functional dependence similar to $\mu_{min}(t)$ and, therefore, small events cannot found and catalogs are intrinsically incomplete.

To understand the mechanism responsible for the existence of $\mu_{min}(t)$, a closer inspection of the envelope function $\mu(t)$ after all mainshocks reveals the existence of two characteristic times: τ and t_M. The first time τ is of the order of some seconds, whereas t_M is of order of some minutes, and three distinct regimes are observed:

- For $t - t_0 < \tau$, $\mu_e(t)$ increases to a maximum value μ_M.
- For $\tau < t - t_0 < t_M$, $\mu_e(t)$ follows a logarithmic decay as

$$\mu_e(t) \simeq \mu_M - q \log(t - t_0). \tag{12}$$

- For $t - t_0 > t_M$, the average value of the envelope $\langle \mu_e(t) \rangle$ is still logarithmic but with different coefficients:

$$\langle \mu_e(t) \rangle = \mu_M - \phi \log(t - t_0) - \Delta\mu, \tag{13}$$

  with $\phi < q$.

The same three regimes have been found for other mainshocks in Southern California and in Italy [16]. The first two regimes can be easily associated to the mainshock waveform, which can be modeled as $\mu_e(t - t_0) = \mu_M + \log[g(t - t_0)]$, where $g(t - t_0)$ is the mainshock envelope waveform. Experimental results suggest an initial linear increase of $g(t)$ [41] followed by a fast decay consistent with an exponential function $g(t) \sim \exp(-Q^{-1}t)$ [42]. Figure 8 indicates that in the intermediate regime $\tau < t - t_0 < t_M$, with t_M of the order of few minutes, the envelope waveform is more consistent with a power law decay as proposed by Lee et al. [43]. Under these assumptions, the behavior of $g(t)$ up to the time $t - t_0 < t_M$ can be modeled as $g(t) \sim t(t/\tau + 1)^{-1-q}$ with the time τ representing the typical duration of the mainshock signal, leading to

$$\mu_e(t) = \mu_M + \log(t - t_0) - (q + 1) \log\left((t - t_0)/\tau + 1\right). \tag{14}$$

The existence of the third regime, previously enlightened by Sawazaki and Enescu [44], can be interpreted taking into account that not only the main shock but each aftershock of magnitude m_i, occurred at time t_i, produces a signal following the relation $\mu_e(t) = \mu_i + \log[g(t - t_i)]$ and one therefore expects a theoretical envelope of the form

$$\mu_{th}(t) = \log\left\{\max_{t_i < t}\left[10^{\mu_i} g(t - t_i)\right]\right\}, \tag{15}$$

where the maximum must be evaluated for all aftershocks with occurrence times $t_i < t$.

Numerical Generation of the Envelope Function

To verify that Equation (15) reproduces the experimental findings, Lippiello et al. [16] started from a mainshock with magnitude m_M occurring at time t_0 and assumed that the aftershock rate follows the OU law (Equation (6)). Since p-values usually have small fluctuations among different aftershock sequences [45], Lippiello et al. [16] assumed a fixed value of p ($p = 1.1$) and after choosing different values of K and c, they generated an aftershock sequence according to Equation (6) for a temporal window of three days. To each aftershock is then associated a magnitude randomly extracted from the GR law. After fitting the value of τ from the experimental $\mu_e(t)$, the key assumption is that a magnitude m_i aftershock, occurring at time t_i, generates a seismic signal with envelope $A(t) = 10^{m_i} g(t - t_i)$ with $g(t) = t(t/\tau + 1)^{-1-q}$ and $q = 2.5$. The synthetic $\mu_{th}(t)$ is then obtained from Equation (15) and a vertical shift is finally applied in order to have the mainshock peak in $\mu_{th}(t)$ equal to the experimental μ_M. The numerical parameters K, c, implemented in the OU law (Equation (6)) are then tuned in order to reach a good agreement between $\mu_{th}(t)$ and the experimental $\mu_e(t)$, according to the procedure described in Section 6.2. Results of $\mu_{th}(t)$ plotted as orange lines in Figure 8 show that it is possible to generate a synthetic envelope reproducing the experimental one in all the three regimes. The above results indicate that since each aftershock produces its own coda waves which decay as a power law with exponent q, the overlap of coda waves generated by subsequent aftershocks causes the existence of a lower signal $\mu_{min}(t)$ which decays as a power law with an exponent $\phi < q$ (Equation (13)). The same agreement between $\mu_e(t)$ and $\mu_{th}(t)$ is recovered for other mainshocks $m_M > 6$ recorded in Greece, Italy and Southern California [16].

We wish to stress that the mainshock peak μ_M, as well as aftershock peaks μ_i in Equation (15), strongly depends on the distance of the recording station from the mainshock epicenter and on site effects. In addition, the functional form of $g(t)$ can be different at different stations. As a consequence both $\mu_e(t)$ and $\mu_{th}(t)$ are different at different stations but, under the hypothesis that aftershocks occur not too far from the mainshock hypocenter, the values of K and c providing the best agreement between $\mu_e(t)$ and $\mu_{th}(t)$ should be the same for all stations.

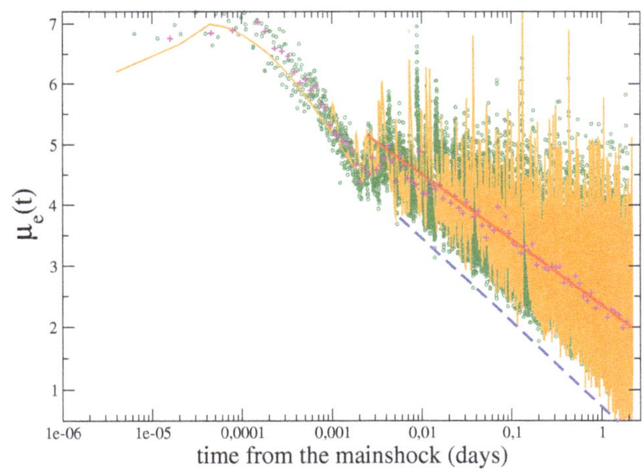

Figure 8. The quantity $\mu_e(t)$ (green circles) after the the Hector Mine earthquake in California recorded at the station CIGSC located at a distance of 92 km from the main shock epicenter. The magenta crosses indicate the (logarithmically binned with bin value 0.1) average value of $\mu(t)$, the red continuous lines represent the results of the logarithmic fit (Equation (13)) for $t - t_0 > t_M$. The dashed blue lines represent the quantity $\mu_{min}(t)$ and orange lines are used for results of numerical simulations for the theoretical envelope $\mu_{th}(t)$, defined in Equation (15). The values of the best-fitting parameters in Equation (6) are $K = 0.95, c = 0.18$ days and $\tau = 8$ s.

5. The ETASI Model

In the previous section, we have shown that STAI is mostly due to the overlap among aftershock coda waves. This ingredient can be incorporated in the ETAS model by multiplying the ETAS occurrence rate Λ_{ETAS} in Equation (9) by a detection function

$$\Lambda_{ETASI}\left(\vec{x}, t, m|\vec{x}_i, t_i, m_i\right) = \Lambda_{ETAS}\left(\vec{x}, t, m|\vec{x}_i, t_i, m_i\right) \times \Phi(m, t, \mu(t)|m_i, t_i). \tag{16}$$

The detection rate can be still described by an error function as in Equation (4) and we define the model described by Equation (16) as the ETAS Incomplete (ETASI) model. The main difference with Equation (3) is that in this approach the detection function $\Phi(m, t, \mu(t)|m_i, t_i)$ depends on the history of all previous earthquakes $\{m_i, t_i\}_{i=1}^{N}$, with $t_i < t$. More precisely, in Equation (3), the 50% detection function $\mu(t)$ depends only on the time and magnitude of the main shock whereas in Equation (16) each event can obscure the recording of subsequent earthquakes.

We observe that the ETASI model differs form the procedure adopted by Seif et al. [14] who generated incomplete ETAS catalogs by removing only aftershocks of mainshock with $m > 5$. In the ETASI model, conversely, any event can obscure subsequent earthquakes independently of its magnitude.

The simplest choice for the detection function $\Phi(m, t|m_i, t_i)$ was proposed by Hainzl [12] and corresponds to an error function with $\sigma \to 0$ and $\mu(t) = m_i$ if $t - t_i \leq \Delta t$, whereas $\mu(t) = 0$ for $t - t_i > \Delta t$, where Δt is a constant blind time. This corresponds to the hypothesis that each earthquake hides all subsequent smaller events occurring at temporal distances smaller than Δt. Notwithstanding the simplicity of this functional form of $\mu(t)$, as already proposed by Hainzl [12], this model, defined as ETASI1 in the following, leads to non-trivial temporal patterns of the aftershock occurrence.

The hypothesis of a constant blind time allows one to achieve an analytical evaluation of c_{meas} [12]. Indeed, the blind time Δt also represents the minimum temporal distance between two subsequent earthquakes reported in a catalog and this leads to a maximum detectable rate $\rho_{max} \simeq 1/\Delta t$. As a consequence, since the "true" aftershock rate is a decreasing function of the time t after the mainshock occurrence (Equation (6)), the measured $\rho(t, m_M, m_{th})$ corresponds to the "true" aftershock rate only if $\rho(t, m_M, m_{th}) < \rho_{max}$, a condition which is always fulfilled at large times. Conversely, at small times, when the "true" aftershock rate is larger than ρ_{max}, the measured ρ exhibits a constant behavior $\rho(t, m_M, m_{th}) \simeq \rho_{max}$. Accordingly, the c_{meas}-value can be identified as the time such as $\rho(c_{meas}, m_M, m_{th}) = \rho_{max}$, and assuming $\alpha \simeq b$ Equation (7) gives

$$\rho(c_{meas}, m_M, m_{th}) = \frac{K_0 e^{b(m_M - m_{th})}}{(c_{meas} + c)^p} = \rho_{max}, \tag{17}$$

giving $c_{meas} = c + (K_0/\rho_{max})^{1/p} \exp{(b/p)}(m_M - m_{th})$, which for $c \ll c_{meas}$ coincides with Equation (8):

$$C_0 = \Delta m = \left(\frac{K_0}{\rho_{max}}\right)^{1/p}, \tag{18}$$

and $d = b/p$.

The ETASI1 model can be implemented numerically via a two step process. At the first step, standard ETAS catalogs are simulated and, at the second step, all events that occurred at a temporal distance smaller than Δt after a larger event are removed from the catalog. de Arcangelis et al. [31] implemented different values of K_0 and analyzed the ETASI1 catalog by the same BP declustering procedure applied to the instrumental catalog. As in Figure 6, the aftershock daily rate $\rho(t, m_M, m_{th})$ for the ETASI1 catalog has been evaluated for different mainshock magnitudes m_M, different thresholds m_{th} and different K_0 values. This study has shown that the c_{meas}-value follows Equation (8) with $d = b/p$ as illustrated in Figure 7b where $\rho(t, m_M, m_{th})$ is plotted as a function of t/τ with $\tau = 10^{d(m_M - m_{th})}$ and $d = b/p$. Data for different m_M and m_{th} and the same K_0 collapse onto the same master curve $F(t/\tau)$, as for the instrumental catalog (Figure 7a). Concerning the value of C_0, de Arcangelis et al. [31] observed that the larger the value of K_0 implemented in ETAS simulations the larger was the value of C_0 fitted from the decay of $\rho(t, m_M, m_{th})$. Results plotted in the inset of Figure 7b show that $-\Delta m = \log_{10} C_0$, becomes more positive for increasing K_0 confirming the strong correlation between C_0 and K_0. In particular, we observe that the dependence of C_0 on K_0 is consistent with Equation (18) only for small values of K_0. Deviations from Equation (18) can be attributed to the cascading process implemented in the ETAS model. Indeed, aftershocks of higher order generation are also followed by a blind time which eventually hides aftershocks of previous generations. This causes a larger total blind time compared to the situation when higher order generation aftershocks are not considered, as in Equation (18).

The comparison between the data collapse observed for the ETASI1 catalog (Figure 7b) with the one observed for the instrumental Southern California catalog (Figure 7a) suggests that the larger value c_{meas} inside Region 1 must be attributed to a larger productivity (larger K_0) of that region. This is in agreement with the behavior of ρ (Figure 6) for times $t > c_{meas}$ when the "true" OU decay $\rho \sim K/t^p$ is expected. Indeed, it is evident that, when $t > c_{meas}$, ρ in Region 1 is systematically larger than in Region 2.

We further observe that the scaling function $F(x)$ presents clear deviations from the OU prediction $F(x) \propto (x+1)^{-p}$ in the intermediate temporal regime. We attribute these deviations to the cascading process which can produce a more gradual decrease of the aftershock number from the initial plateau compared to the situation when higher order generation aftershocks are not taken into account [12]. A better fit for $F(x)$ in numerical and instrumental catalogs is provided by $F(x) = A\log(1 + Bx^{-p})$ obtained by Lippiello et al. [46] under a dynamical scaling assumption [38,45,47–51].

5.1. ETASI2

A more refined expression for $\mu(t)$ within the ETASI model (Equation (16)) is proposed in [31] and corresponds to the so called ETASI2 model. The idea is that the 50% detection function follows the same decay of the envelope function of a single earthquake and according to Equation (12) this corresponds to the assumption that

$$\mu(t) = \max_{i:t_i<t} \left(m_i - q\log(t - t_i) - \delta_0\right), \tag{19}$$

where the maximum is evaluated over all events with magnitude m_i occurred at time $t_i < t$. The model is numerically implemented in [31] taking for the detection rate function Φ an error function as in Equation (4) with $\sigma \to 0$. This corresponds to the two-step procedure illustrated in the previous section with the removal from the original ETAS catalog of all events with magnitude m and occurrence time t such that $m < \mu(t)$. A finite value of σ is considered in [52].

In de Arcangelis et al. [31], the coefficient q in Equation (19) is taken as a model parameter and its value has been tuned in order to achieve the best agreement between the organization of aftershocks in ETASI2 and instrumental catalogs. This study showed that the ETASI2 model provides a more accurate description of aftershock occurrence, with respect to the ETASI1 model, and in particular it better captures the correlation between subsequent magnitudes observed in instrumental catalogs. In particular the agreement between instrumental and ETASI2 catalogs is obtained by setting a K_0 value, in the ETASI2 simulations, significantly larger inside Region 1 of Southern California (Figure 1) than Region 2. As a consequence, de Arcangelis et al. [31] proposed that the value of K_0 which provides the best overlap between ETASI2 and instrumental catalogs can be interpreted as the best estimate for the true productivity coefficient K_0 in each region.

5.2. Dynamical Scaling ETAS Model

A model alternative to the ETASI has been proposed on the basis of a dynamical scaling relation between time and energy [19,36,38,46,47]. Within this hypothesis, different from the general assumption of the ETAS model [3,53,54], time and magnitude are not independent quantities but the magnitude difference fixes a characteristic time scale for aftershock rate relaxation. Deviations from the GR law are a natural consequence of this assumption with a completeness magnitude depending on time in agreement with what is observed in experimental data (Equation (2)). The study of the maximum likelihood [51] has shown that this method provides a more accurate description of the aftershock rate decay than the ETAS model.

6. Automatic Procedures for Short-Term Aftershock Forecasting

In this section, we present two methods which have been developed in order to provide real-time aftershock forecasting: The Omi method [7,9,10] and the Lippiello method [16,17]. The idea of both methods is to extrapolate the parameters of the OU law, or more generally of the ETAS model, by means of an automatic procedure which uses the information available up to a time T_2 after the mainshock. The ETASI model, presented in the previous section, is not suitable for this purpose because it is not possible to apply a maximum likelihood estimation procedure to invert parameters. For the likelihood evaluation, indeed, one should have access to "obscured" events, an information by definition unavailable. In this section, we review the retrospective tests performed with the Omi and the Lippiello methods. Both tests consider the forecasting according to the OU law (Equation (6)) implementing the parameters K and c estimated for each individual mainshock sequences, according to the information available in real time. This forecasting is compared to a generic model where the parameters K and c are taken as average values over many sequences. The results show that the novel methods outperform the generic model.

6.1. The Omi Method

The Omi method, briefly illustrated in Section 2, has been implemented in a real-time system for automatic aftershock forecasting in Japan. A systematic test of the efficiency of the Omi method, using real-time seismic data, was performed by Omi et al. [10] on aftershock sequences of seven inland mainshocks with magnitudes $m \geq 7$ that occurred after the establishment of the Hi-net observation system. The Omi method is based on the evaluation of the parameters K, p, c in Equation (6) using the information from an incomplete dataset, including only the recorded aftershocks. More precisely, Omi et al. [10] considered data in the learning period from two instrumental catalogs: the Hi-net and JMA catalogs. The results of this method are compared to a standard forecasting approach which uses fixed parameter values (the generic model) determined based on many aftershock sequences in Japan. The forecast from the generic model depends only on the main shock magnitude. The performance is compared by means of the log-likelihood ratio score, which is referred to as information gain I. The standard error S_I of the information gain is also numerically evaluated and, under a Gaussian approximation, one forecast performs better than the other one, with a probability larger than the 95%, if $I > 1.64 S_I$. More precisely, Omi et al. [10] considered four learning periods corresponding to the first 3, 6, 12, and 24 h periods of aftershock data to prepare forecasts for the following 3, 6, 12, and 24 h testing periods, respectively. The results of the test, for the seven Japan aftershock sequences, are visually represented in Figure 9 that shows the information gain per aftershock, considering separately data from the Hi-net and JMA catalog, against the generic aftershock model. The error bars correspond to $1.64 S_I$ and, therefore, if their lower bound is greater than zero, the Omi model performs better than the generic model. Omi et al. [10] separately considered two target magnitudes, the smallest one $M_t = M_c$ (Figure 9a) and $M_t = 3.95$ (Figure 9b). Results show that, for the entire forecast period of 3–48 h, both the Hi-net and JMA forecasts significantly outperform the generic model and that the same result is valid in all individual forecast periods for the case of the lowest target magnitude $M_t = M_c$. Conversely, for $M_t = 3.95$, because of the small number of $m > M_t$ aftershocks, the scores tend to have large error bars and, even if the Omi method generally outperforms the generic model, this is not statistically significant for most cases (Figure 9b).

Another interesting item in Figure 9 is the comparison of the performance of the Omi method implementing the JMA catalog against the one implementing the Hi-net automatic catalogs. In general, the results show that the JMA forecast significantly outperforms the Hi-net forecast in the case of the small target magnitude $M_t = M_c$, probably because of the better accuracy of the JMA catalog. On the other hand, the two performances are comparable for $M_t = 3.95$ indicating that, even if the automatic Hi-net catalog is less accurate than the JMA catalog, it provides reasonable results for target magnitudes $M_t \geq 3.95$. This is an important result since it is the only catalog available in real time.

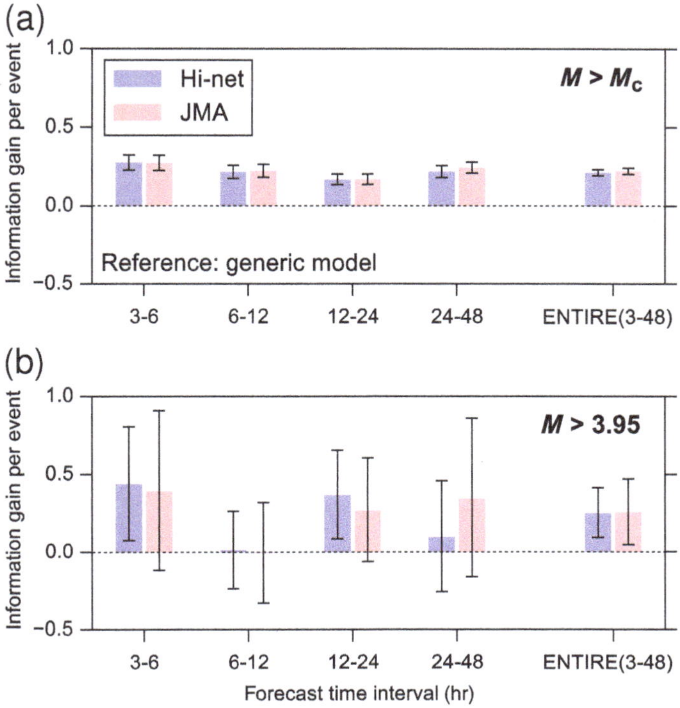

Figure 9. Information gain per aftershock of the forecasts based on the Hi-net and JMA catalogs, respectively, relative to the generic model for the cases with: (**a**) $M_t = M_c$; and (**b**) $M_t = 3.95$. If the lower bound of the error bar is greater than 0, the forecast is significantly better than the generic model with a probability larger than the 95%. From Omi et al. [10].

6.2. The Lippiello Method

Lippiello et al. [16] proposed a method based on the results presented in Section 4, which show that the instrumental envelope $\mu_e(t)$ can be reproduced by the theoretical envelope $\mu_{th}(t)$ given in Equation (15). In particular, $\mu_{th}(t)$ can be tuned to recover Equation (13) with the same parameters ϕ and $\Delta\mu$ of the instrumental $\mu_e(t)$. The central observation is that the value of the coefficients ϕ and $\Delta\mu$ which describe the logarithmic decay of $\mu_{th}(t)$ (Equation (13)) depend on the parameters K and c of the OU law (Equation (6)), implemented in the numerical simulation. This idea has been applied in a procedure which associates the best-fitting parameters $(\phi, \Delta\mu)$ in Equation (13), obtained from the experimental signal, to the pair (K, c) used in numerical simulations of the OU law. The procedure is schematically illustrated in Figure 10. Firstly one evaluates the value of τ which is the best approximation for $\mu_e(t)$ in Equation (14) during the first 60 s. Fixing $p = 1.1$, the estimated value of τ is used to generate many numerical signals $\mu_{th}(t)$ for different choices of K and c according to Equation (15). Then, one compares, in the learning period $t - t_0 \in [T_1, T_2]$, the average value of the numerical signal $\mu_{th}(t)$ with the experimental one $\mu_e(t)$. The slope ϕ of $\mu_{th}(t)$ depends fundamentally on the c-value, whereas K controls its vertical shift $\Delta\mu$. As a consequence, after choosing a given K-value, one varies the c-value until the slopes of $\mu_{th}(t)$ become similar to the experimental $\mu_e(t)$ (Figure 10a). The c-value producing this effect is then defined as $\bar{c}$ and one generates different numerical catalogs with $c = \bar{c}$ and different values of K (Figure 10b). The value of K minimizing the difference between $\mu_e(t)$ and $\mu_{th}(t)$ in the interval $[T_1, T_2]$ is defined as $\overline{K}$. The pair of values $(\overline{K}, \bar{c})$ is considered the best representation of experimental data and is used to forecast aftershock occurrence at times $t - t_0 > T_2$, according to Equation (6).

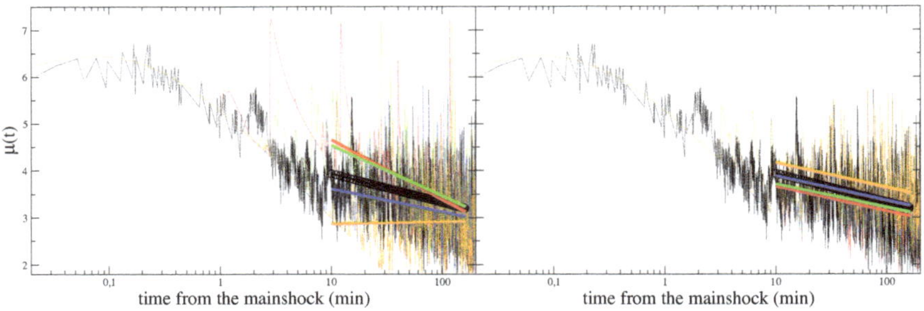

Figure 10. (**Left**) Black dotted-lines represent the envelope function $\mu_e(t)$ of the Lixouri earthquake in Greece recorded at the station LKD2 located at 70 km from the mainshock epicenter. Colored dot-dashed lines are used for $\mu_{th}(t)$ with $\tau = 11$ s, $K = 1.15$ and different values of c ranging in the interval $[0.01, 4.5]$ days. Black circles represent the logarithmic fit (Equation (15)) in the interval $[10, 160]$ min of the experimental envelope function, whereas continuous lines are used for the best fit of the numerical $\mu_{th}(t)$, with c increasing from 0.01 to 4.5 days from top to bottom. (**Right**) The same as in the left panel but plotting numerical data $\mu_{th}(t)$ with $\tau = 11$ s, $c = 1.15$ days and different values of $K \in [0.75, 2.95]$. Continuous lines are the the logarithmic fits of numerical data, in $[10, 160]$ min, with K increasing from bottom to top. From Reference [16].

Test of the Procedure

To test their method, Lippiello et al. [16] considered as target aftershocks all the events producing in the envelope function $\mu_e(t)$ a peak with amplitude larger than $\mu_0 = \mu_M - 3$ and define as $N_3(t)$ their cumulative number in the temporal interval $[T_2, t - t_0]$, after the mainshock. Similarly, the number $N_2(t)$ is the cumulative number of events producing peaks larger than $\mu_0 = \mu_M - 2$.

The quantity $N_2(t)$ and $N_3(t)$ are plotted in Figure 11 for three mainshocks from three different geographic regions, for times $t > 160$ min. Lippiello et al. [16] compared the instrumental number of $N_3(t)$ and $N_2(t)$ with those expected according to the OU law (Equation (6)) after implementing the best values of K and c ($\overline{K}$ and $\overline{c}$) obtained according to the Lippiello procedure. More precisely, Lippiello et al. [16] considered a learning period $[T_1, T_2]$ min with $T_1 = 10$ min and different values of T_2. They found that for values of $T_2 \gtrsim 160$ min the estimate of $\overline{K}$ and $\overline{c}$ became quite stable. Therefore they consider $T_2 = 160$ min and found that at all times $t > T_2$ the Lippiello method predicts with reasonable accuracy the number of occurred aftershocks. Differences between predicted and observed aftershock number are typically smaller than 20% and always within the error bars. For comparison, in the same Figure 11, Lippiello et al. [16] also plotted the expected number $N_3(t)$ and $N_2(t)$ according to a generic model which implements in the OU law Equation (7) the value of K_0, c and α obtained as average over all sequences with $m_M > 5$, recorded in Southern California [55]. We observe that the number of the predicted strong aftershocks according to this generic model is much smaller (approximately ten times) than the observed one. We wish to stress that the estimate of K and c, for each specific sequence, on the basis of the earthquakes recorded in the official catalogs up to the time T_2 leads to unreliable results. As an example, in the case of the Lixouri earthquake only three earthquakes are reported in the Greek catalog in the first thirty minutes after the mainshock. The situation is a little better after the L'Aquila and Hector mine earthquake when 20 events are reported in regional catalogs in the first thirty minutes. These numbers are too small to produce a reasonable estimate of K and c, which, in all cases, would be very biased because of the incompleteness of datasets as confirmed by the absence of earthquakes with magnitude smaller than $m = 3$, in official catalogs in the first thirty minutes.

Summarizing, results of Figure 11 clearly show that the Lippiello method performs much better than the generic model providing a reasonable aftershock forecasting. Very recently, Lippiello et al. [17] proposed a more efficient procedure, still based on the agreement between $\mu_{th}(t)$ and $\mu_e(t)$, which produces even more accurate STA forecasting.

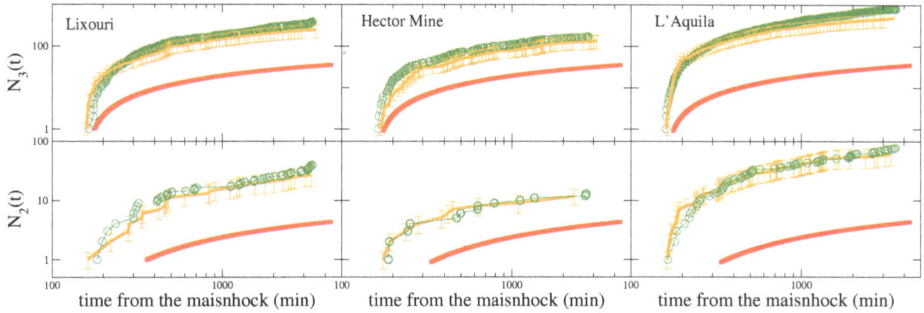

Figure 11. The quantities $N_3(t)$ (**top**) and $N_2(t)$ (**bottom**) are plotted for $t - t_0 > T_2 = 160$ min as green circles for the three main-aftershock sequences of Figure 1: the 26 January 2014 $m = 6.1$ Lixouri earthquake (**left**), the 16 October 1999 $m = 7.1$ Hector Mine earthquake (**middle**) and the 6 April 2009 $m = 5.9$ L'Aquila earthquake (**right**). Red squares are the expected values according to Equation (6) using the average values obtained in Reference [55]. The orange curves are the expected values using in Equation (6) the best parameters K and c inverted from the experimental fit of τ, ϕ and $\Delta\mu$. The error bars in each plot incorporate both the uncertainty in the estimate of $(\Delta\mu, \phi)$ and fluctuations in the aftershock number for given values of $\overline{K}$ and $\overline{c}$. From Reference [16].

6.3. Comparison between the Omi and the Lippiello Methods

The Omi method needs as an earlier stage an automatic routine for real-time automatic detection. The only assumption is that the GR law holds up to the lower considered magnitude m_c, that is a widely accepted idea within the seismological community. Conversely, the key assumption of the Lippiello method is that the theoretical envelope (Equation (15)) reproduces the instrumental one $\mu_e(t)$. This hypothesis is less consolidated but allows one to evaluate seismic hazard directly from the envelope function $\mu_e(t)$ without any information on occurrence times, magnitude and locations of earthquakes producing the observed signal. Overcoming all problems related to event identification and location, the Lippiello method presents some advantages:

(i) It is faster. Indeed, aftershock localization is a non-trivial routine involving the elaboration of at least the seismic signal from three different seismic stations.
(ii) It works when only few events are identified by the automatic detection routine whereas the Omi et al method needs that at least $\sim$30 aftershocks must be identified [15].
(iii) It provides the in-situ occurrence probability by simply installing a seismic station in the site of interest. This could be particularly useful in areas with a very low dense seismic network and where automatic detection routines are not efficient.
(iv) It provides directly in output the probability of peaks of the local ground velocity and therefore it overcomes the large amount of uncertainty [56], which is present in the attenuation relations necessary to convert aftershock occurrence probability to the local ground motion intensity.

Summarizing, the two methods appear as two complementary approaches to the same problem and can be simultaneously adopted.

We finally remark that in the OMI method the spatial dependence of the aftershock occurrence probability can be easily included by multiplying the OU law (Equation (6)) for a decreasing function of the distance from the mainshock epicenter. However, the parameters controlling this decay are very difficult to be inverted from the on-going specific sequence and average quantities must be considered. On the other hand, in the Lippiello method the spatially dependence is, at least in part, implicitly considered. Indeed, the method does not give in output the probability to have a given magnitude aftershock but the occurrence probability of events which produce peaks in the envelope $\mu_e(t)$, in the position where the station is located, larger than a reference value μ_0. This probability, therefore, clearly depends on the distance of the station from the mainshock epicenter.

7. Conclusions

In this review article, we show that, in the first part of aftershock sequence, incompleteness is an intrinsic property of seismic data. Indeed, the overlap of seismic signals makes the envelope function always greater than $\mu_{min}(t)$. This lower threshold $\mu_{min}(t)$ can be related to the minimum aftershock magnitude $m_{min}(t)$ identifiable at time t since the main shock and indicates that it is feasible to obtain more accurate catalogs but it is impossible to reach completeness levels below $m_{min}(t)$. This result also provides an explanation for the dependence of $m_c(t)$ on the time elapsed from the main shock occurrence. We illustrate how the incompleteness affects the estimate of the parameters of STA forecasting models and we present some models which take it explicitly into account. In particular, we present an interpretation of the mechanisms responsible for the existence of $\mu_{min}(t)$ in terms of the overlap of coda-waves generated by each individual aftershock: The combination of the decay of the aftershock rate (OU law) with the power law relaxation of coda waves produces an envelope function $\mu_e(t)$, which, on average, depends logarithmically on the time since the main shock. We illustrate the bias induced in the estimate of model parameters because of the incompleteness of the instrumental catalog. A deeper investigation is necessary to establish a quantitative relationship between the expected error in the estimate of model parameters and the degree of incompleteness of the catalog.

We also show that the parameters of the logarithmic dependence of $\mu_e(t)$ appear strictly related to the parameters of the OU. We then describe a procedure based on this observation and developed in [16] to extract the OU law parameters from a fitting procedure applied to the experimental $\mu_e(t)$. This approach overcomes all problems related to event identification and location since seismic hazard is evaluated directly from the envelope function $\mu_e(t)$ without any information on occurrence times, magnitudes and locations of earthquakes producing the observed signal.

We also illustrate the Omi method [7,9,10,15] proposed to overcome the problems of STA forecasting caused by the incompleteness of instrumental data. We show that the method, based on the detection rate function, provides reliable aftershock forecasting on the basis of incomplete instrumental catalogs.

Summarizing, we review very recent proposals to develop real-time systems for automatic aftershock forecasting. The above procedures have been up to now tested retrospectively but appear already suitable to be implemented in prospective tests. These methods apply the OU law or the ETAS model without taking into account the spatial variability of seismicity. Future developments should correspond to space-time models providing a space dependent forecasting, particularly useful in aftershock sequences with a complex spatial distribution.

Funding: This research received no external funding.

Acknowledgments: E.P. and V.K. acknowledge support of this work by the project HELPOS, Hellenic System for Lithosphere Monitoring (MIS 5002697), which is implemented under the Action Reinforcement of the Research and Innovation Infrastructure, funded by the Operational Programme "Competitiveness, Entrepreneurship and Innovation" (NSRF 2014–2020) and co-financed by Greece and the European Union (European Regional Development Fund). Geophysics Department Contribution 000/2019.

Conflicts of Interest: The authors declare no conflict of interest.

References

1. Omori, F. On the after-shocks of earthquakes. *J. Coll. Sci. Imp. Univ. Tokyo* **1894**, *7*, 111–200.
2. Utsu, T.; Ogata, Y.; Ritsuko, S.; Matsu'ura. The Centenary of the Omori Formula for a Decay Law of Aftershock Activity. *J. Phys. Earth* **1995**, *43*, 1–33. [CrossRef]
3. Ogata, Y. Statistical Models for Earthquake Occurrences and Residual Analysis for Point Processes. *J. Am. Stat. Assoc.* **1988**, *83*, 9–27. [CrossRef]
4. Helmstetter, A.; Kagan, Y.Y.; Jackson, D.D. Comparison of Short-Term and Time-Independent Earthquake Forecast Models for Southern California. *Bull. Seismol. Soc. Am.* **2006**, *96*, 90–106. [CrossRef]
5. Ogata, Y.; Katsura, K. Immediate and updated forecasting of aftershock hazard. *Geophys. Res. Lett.* **2006**, *33*. [CrossRef]
6. Werner, M.J.; Helmstetter, A.; Jackson, D.D.; Kagan, Y.Y. High-Resolution Long-Term and Short-Term Earthquake Forecasts for California. *Bull. Seismol. Soc. Am.* **2011**, *101*, 1630–1648 [CrossRef]
7. Omi, T.; Ogata, Y.; Hirata, Y.; Aihara, K. Forecasting large aftershocks within one day after the main shock. *Sci. Rep.* **2013**, *3*, 2218. [CrossRef]
8. Omi, T.; Ogata, Y.; Hirata, Y.; Aihara, K. Estimating the ETAS model from an early aftershock sequence. *Geophys. Res. Lett.* **2014**, *41*, 850–857. [CrossRef]
9. Omi, T.; Ogata, Y.; Hirata, Y.; Aihara, K. Intermediate-term forecasting of aftershocks from an early aftershock sequence: Bayesian and ensemble forecasting approaches. *J. Geophys. Res. Solid Earth* **2015**, *120*, 2561–2578. [CrossRef]
10. Omi, T.; Ogata, Y.; Shiomi, K.; Enescu, B.; Sawazaki, K.; Aihara, K. Automatic Aftershock Forecasting: A Test Using Real-Time Seismicity Data in Japan. *Bull. Seismol. Soc. Am.* **2016**, *106*, 2450. [CrossRef]
11. Hainzl, S. Rate-Dependent Incompleteness of Earthquake Catalogs. *Seismol. Res. Lett.* **2016**, *87*, 337–344. [CrossRef]
12. Hainzl, S. Apparent triggering function of aftershocks resulting from rate-dependent incompleteness of earthquake catalogs. *J. Geophys. Res. Solid Earth* **2016**, *121*, 6499–6509. [CrossRef]
13. Zhuang, J.; Ogata, Y.; Wang, T. Data completeness of the Kumamoto earthquake sequence in the JMA catalog and its influence on the estimation of the ETAS parameters. *Earth Planets Space* **2017**, *69*, 36. [CrossRef]
14. Seif, S.; Mignan, A.; Zechar, J.D.; Werner, M.J.; Wiemer, S. Estimating ETAS: The effects of truncation, missing data, and model assumptions. *J. Geophys. Res. Solid Earth* **2017**, *122*, 449–469. [CrossRef]
15. Omi, T.; Ogata, Y.; Shiomi, K.; Enescu, B.; Sawazaki, K.; Aihara, K. Implementation of a Real-Time System for Automatic Aftershock Forecasting in Japan. *Seismol. Res. Lett.* **2018**, *90*, 242. [CrossRef]
16. Lippiello, E.; Cirillo, A.; Godano, G.; Papadimitriou, E.; Karakostas, V. Real-time forecast of aftershocks from a single seismic station signal. *Geophys. Res. Lett.* **2016**, *43*, 6252–6258. [CrossRef]
17. Lippiello, E.; Petrillo, C.; Godano, C.; Tramelli, A.; Papadimitriou, E.; Karakostas, V. Forecasting of the first hour aftershocks by means of the perceived magnitude. *Nat. Commun.* **2019**, *10*, 2953. [CrossRef]
18. Mignan, A.; Werner, M.J.; Wiemer, S.; Chen, C.C.; Wu, Y.M. Bayesian Estimation of the Spatially Varying Completeness Magnitude of Earthquake Catalogs. *Bull. Seismol. Soc. Am.* **2011**, *101*, 1371–1385. [CrossRef]
19. Lippiello, E.; Godano, C.; de Arcangelis, L. The earthquake magnitude is influenced by previous seismicity. *Geophys. Res. Lett.* **2012**, *39*, L05309. [CrossRef]
20. Amorése, D. Applying a Change-Point Detection Method on Frequency-Magnitude Distributions. *Bull. Seismol. Soc. Am.* **2007**, *97*, 1742–1749. [CrossRef]
21. Schorlemmer, D.; Woessner, J. Probability of Detecting an Earthquake. *Bull. Seismol. Soc. Am.* **2008**, *98*, 2103–2117. [CrossRef]
22. Kagan, Y.Y. Short-Term Properties of Earthquake Catalogs and Models of Earthquake Source. *Bull. Seismol. Soc. Am.* **2004**, *94*, 1207–1228. [CrossRef]
23. Helmstetter, A.; Kagan, Y.Y.; Jackson, D.D. Importance of small earthquakes for stress transfers and earthquake triggering. *J. Geophys. Res. Solid Earth* **2005**, *110*, B05S08. [CrossRef]
24. Shcherbakov, R.; Turcotte, D.L.; Rundle, J.B. A generalized Omori's law for earthquake aftershock decay. *Geophys. Res. Lett.* **2004**, *31*, L11613. [CrossRef]
25. Reasenberg, P.A.; Jones, L.M. Earthquake Hazard After a Mainshock in California. *Science* **1989**, *243*, 1173–1176. [CrossRef]

26. Peng, Z.; Vidale, J.E.; Ishii, M.; Helmstetter, A. Seismicity rate immediately before and after main shock rupture from high-frequency waveforms in Japan. *J. Geophys. Res. Solid Earth* **2007**, *112*, B03306. [CrossRef]
27. Lennartz, S.; Bunde, A.; Turcotte, D.L. Missing data in aftershock sequences: Explaining the deviations from scaling laws. *Phys. Rev. E* **2008**, *78*, 041115. [CrossRef]
28. Helmstetter, A.; Kagan, Y.Y.; Jackson, D.D. High-resolution Time-independent Grid-based Forecast for $M \geq 5$ Earthquakes in California. *Seismol. Res. Lett.* **2007**, *78*, 78–86. [CrossRef]
29. Marzocchi, W.; Lombardi, A.M. Real-time forecasting following a damaging earthquake. *Geophys. Res. Lett.* **2009**, *36*, L21302. [CrossRef]
30. Zhuang, J.; Ogata, Y.; Vere-Jones, D. Stochastic declustering of space-time earthquake occurrences. *J. Am. Stat. Assoc.* **2002**, *97*, 369–380. [CrossRef]
31. De Arcangelis, L.; Godano, C.; Lippiello, E. The overlap of aftershock coda-waves and short-term post seismic forecasting. *J. Geophys. Res. Solid Earth* **2018**, *123*, 5661–5674. [CrossRef]
32. Baiesi, M.; Paczuski, M. Complex networks of earthquakes and aftershocks. *Nonlinear Process. Geophys.* **2005**, *12*, 1–11. [CrossRef]
33. Zaliapin, I.; Gabrielov, A.; Keilis-Borok, V.; Wong, H. Clustering Analysis of Seismicity and Aftershock Identification. *Phys. Rev. Lett.* **2008**, *101*, 018501. [CrossRef]
34. Zaliapin, I.; Ben-Zion, Y. Earthquake clusters in southern California I: Identification and stability. *J. Geophys. Res. Solid Earth* **2013**, *118*, 2847–2864. [CrossRef]
35. Moradpour, J.; Hainzl, S.; Davidsen, J. Nontrivial decay of aftershock density with distance in Southern California. *J. Geophys. Res. Solid Earth* **2014**, *119*, 5518–5535. [CrossRef]
36. Lippiello, E.; de Arcangelis, L.; Godano, C. Influence of Time and Space Correlations on Earthquake Magnitude. *Phys. Rev. Lett.* **2008**, *100*, 038501. [CrossRef]
37. Lippiello, E.; de Arcangelis, L.; Godano, C. Time, Space and Magnitude Correlations in Earthquake Occurrence. *Int. J. Mod. Phys. B* **2009**, *23*, 5583–5596. [CrossRef]
38. Lippiello, E.; Godano, C.; de Arcangelis, L. Magnitude correlations in the Olami-Feder-Christensen model. *Europhys. Lett.* **2013**, *102*, 59002. [CrossRef]
39. Peng, Z.; Zhao, P. Migration of early aftershocks following the 2004 Parkfield earthquake. *Nat. Geosci.* **2009**, *2*, 877–881. [CrossRef]
40. Enescu, B.; Mori, J.; Miyazawa, M.; Kano, Y. Omori-Utsu Law c-Values Associated with Recent Moderate Earthquakes in JapanShort Note. *Bull. Seismol. Soc. Am.* **2009**, *99*, 884–891. [CrossRef]
41. Odaka, T.; Ashiya, K.; Tsukada, S.; Sato, S.; Ohtake, K.; Nozaka, D. A New Method of Quickly Estimating Epicentral Distance and Magnitude from a Single Seismic Record. *Bull. Seismol. Soc. Am.* **2003**, *93*, 526–532. [CrossRef]
42. Aki, K.; Chouet, B. Origin of coda waves: Source, attenuation, and scattering effects. *J. Geophys. Res.* **1975**, *80*, 3322–3342. [CrossRef]
43. Lee, W.H.K.; Bennett, R.E.; Meagher, K.L. *A Method of Estimating Magnitude of Local Earthquakes from Signal Duration*; US Department of the Interior, Geological Survey: Washington, DC, USA, 1972; Volume 28.
44. Sawazaki, K.; Enescu, B. Imaging the high-frequency energy radiation process of a main shock and its early aftershock sequence: The case of the 2008 Iwate - Miyagi Nairiku earthquake, Japan. *J. Geophys. Res. Solid Earth* **2014**, *119*, 4729–4746. [CrossRef]
45. De Arcangelis, L.; Godano, C.; Grasso, J.R.; Lippiello, E. Statistical physics approach to earthquake occurrence and forecasting. *Phys. Rep.* **2016**, *628*, 1–91. [CrossRef]
46. Lippiello, E.; Bottiglieri, M.; Godano, C.; de Arcangelis, L. Dynamical scaling and generalized Omori law. *Geophys. Res. Lett.* **2007**, *34*, L23301. [CrossRef]
47. Lippiello, E.; Godano, C.; de Arcangelis, L. Dynamical Scaling in Branching Models for Seismicity. *Phys. Rev. Lett.* **2007**, *98*, 098501. [CrossRef]
48. Bottiglieri, M.; Lippiello, E.; Godano, C.; de Arcangelis, L. Identification and spatiotemporal organization of aftershocks. *J. Geophys. Res. Solid Earth* **2009**, *114*, B03303. [CrossRef]
49. Lippiello, E.; de Arcangelis, L.; Godano, C. Role of Static Stress Diffusion in the Spatiotemporal Organization of Aftershocks. *Phys. Rev. Lett.* **2009**, *103*, 038501. [CrossRef]
50. Bottiglieri, M.; de Arcangelis, L.; Godano, C.; Lippiello, E. Multiple-Time Scaling and Universal Behavior of the Earthquake Interevent Time Distribution. *Phys. Rev. Lett.* **2010**, *104*, 158501. [CrossRef]

51. Bottiglieri, M.; Lippiello, E.; Godano, C.; de Arcangelis, L. Comparison of branching models for seismicity and likelihood maximization through simulated annealing. *J. Geophys. Res. Solid Earth* **2011**, *116*, B02303. [CrossRef]
52. Lippiello, E.; Godano, C.; de Arcangelis, L. The Relevance of Foreshocks in Earthquake Triggering: A Statistical Study. *Entropy* **2019**, *21*, 173. [CrossRef]
53. Ogata, Y. Space-time Point-process Models for Earthquake Occurrences. *Ann. Inst. Math. Stat.* **1988**, *50*, 379–402. [CrossRef]
54. Ogata, Y. A Monte Carlo method for high dimensional integration. *Numer. Math.* **1989**, *55*, 137–157. [CrossRef]
55. Reasenberg, P.A.; Jones, L.M. Earthquake Aftershocks: Update. *Science* **1994**, *265*, 1251–1252. [CrossRef]
56. Cornell, C.A. Engineering seismic risk analysis. *Bull. Seismol. Soc. Am.* **1968**, *58*, 1583–1606.

geosciences

MDPI

Article

Landslides Triggered by the 2016 Mw 7.8 Pedernales, Ecuador Earthquake: Correlations with ESI-07 Intensity, Lithology, Slope and PGA-h

Kervin Chunga [1,2,*], Franz A. Livio [3], Carlos Martillo [4], Hernán Lara-Saavedra [5], Maria Francesca Ferrario [3], Ivan Zevallos [1] and Alessandro Maria Michetti [3]

1 Facultad de Ciencias Matemáticas, Físicas y Químicas, Carrera de Ingeniería Civil, Universidad Técnica de Manabí (UTM), Avenue José María Urbina, Portoviejo 130111, Ecuador
2 Facultad de Ciencias de la Ingeniería, Universidad Estatal Península de Santa Elena, Km 1 vía principal, La Libertad 240350, Ecuador
3 Dipartimento di Scienza e Alta Tecnologia, Università degli Studi dell'Insubria, Via Valleggio 11, 22100 Como, Italy
4 Grupo de Investigación en Geociencias Marinas y Costeras (GEMAC-FIMCM-FICT), Escuela Superior Politécnica del Litoral, Campus Gustavo Galindo, Km 30.5 vía Perimetral, Guayaquil 09015863, Ecuador
5 Facultad de Ciencias Naturales, Universidad de Guayaquil, Av. Raúl Gómez Lince s/n y Av. Juan Tanca Marengo, Guayaquil 090150, Ecuador
* Correspondence: kchunga@utm.edu.ec

Received: 9 August 2019; Accepted: 19 August 2019; Published: 26 August 2019

Abstract: We provide a dataset of the landslides induced by the 2016 Pedernales megathrust earthquake, Ecuador (Mw 7.8, focal depth of 20 km) and compare their spatial distribution with mapped bedrock lithology, horizontal peak ground acceleration (PGA-h) and the macroseismic intensity based on earthquake-induced environmental effects (ESI-07). We studied 192 coseismic landslides (classified as coherent, disrupted and lateral spreads) located in the epicentral area, defined by the VII to $IX_{ESI\text{-}07}$ isoseismals. Based on our findings, lahar deposits, tuffs and volcanoclastic units are the most susceptible to landslides occurrence. Alluvial plains with fluvial loose fine sand are the most susceptible setting for lateral spreading, with a maximum intensity of $IX_{ESI\text{-}07}$. The coherent landslides are frequently found in altered shale and siltstone geological units with moderate slopes (8°–16°), with typical intensity ranging between VII and $VIII_{ESI\text{-}07}$. Our analysis draws a typical framework for slope movements triggered by subduction earthquakes in Ecuador. The most dangerous setting is the coastal region, a relatively highly urbanized area located near the epicenter and where liquefaction can trigger massive lateral spreading events. Coherent and disrupted landslides, dominating the more internal hilly region, can be triggered also in moderate slope settings (i.e., less than 10°). Indeed, the regression analysis between seismic intensity, PGA-h and landslide occurrence shows that most of the events occurred at PGA-h values between 0.4 g and 1.2 g, at a distance of 30 to 50 km from the rupture plane. Our database suggests that lithology and hillslope geometry are the main geological/geomorphological factors controlling coseismic landslides occurrence; while the distance from the rupture plane plays a significant role on determining the landslide size. Finally, we underline that coseismically-triggered landslides are among the most common environmental effects occurring during large subduction events that can be effectively used to properly evaluate the earthquake macroseismic field. The landslide inventory we compiled is suitable for assessing the vulnerability of physical environment from subduction earthquakes in Ecuador, and offers a primary data source for future worldwide analysis.

Keywords: coseismic landslides; macroseismic intensity; ESI-07 scale; Pedernales earthquake; Ecuador

1. Introduction

Among all earthquake-induced environmental effects (sensu [1,2]), landslides are the most widespread and surveyed along coastal areas, in intermontane basins and in hilly areas [1,3–9]. Landslide susceptibility is related to lithologic and geomorphic characteristics including highly weathered loose soils, topographic conditions, local hydrogeological and geological setting (e.g., [3,10]). The geological and geotechnical variables play a key role as predisposing factors to earthquake-induced landslides in active tectonic settings such as the Ecuadorian coast region. Large intraslab and megathrust earthquakes due to the convergence between the Nazca, South American and the Caribbean plates (Figure 1) occur in this area. However, there is a lack of information about earthquake-triggered landslides connected to subduction earthquakes in Ecuador (e.g., [11–13]). Some earthquake-triggered landslides due to upper-plate events (in the order of Mw 6–7), in the northern Ecuadorian Andes, have been described in detail, whilst less is known about the distribution of earthquake-triggered landslides due to large subduction events [14–19].

We present a co-seismic landslide inventory for the 2016 Pedernales earthquake sequence. The earthquake sequence started with a foreshock (ML 5.7, at 18:47), some 10 minutes prior the Mw 7.8 mainshock, that occurred at 18:59 on the April 16, 2016 [20], see Data and Resources; Figure 2a). The extent of physical and environmental damage encompassed an area of thousands of square kilometers [21,22]. Locally, building damage and geological ground effects were enhanced by site amplification due to class E and F water saturated soils [23].

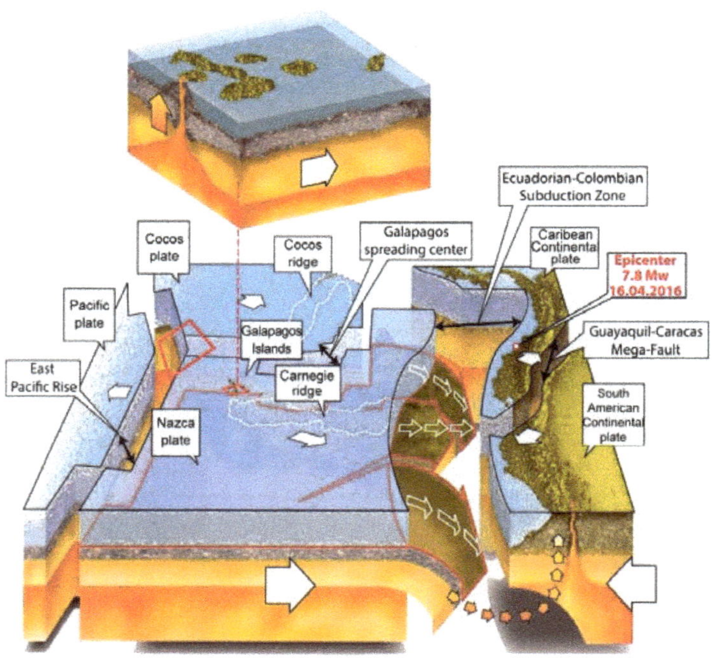

Figure 1. Geodynamic setting of Ecuador, the Galapagos Islands and the Carnegie Ridge; modified after [25].

The epicentral area includes the city of Pedernales and surrounding communities (i.e., Coaque, Jama and Canoa), reaching a maximum intensity of IX_{ESI-07}, assessed from field observation of coastal uplift, soil liquefaction, sinkholes, landslides of natural and stabilized slopes, cracks in natural soils, minor fractures, failure along both the major and minor axis in pavement, particularly asphalt and concrete roads [22]. Coseismic geological effects were observed also across communities near

Cojimíes (north of the epicenter, Figure 2b) and near Portoviejo (south of the epicenter, Figure 2b), resulting in a VIII_{ESI-07} assessment. The area encompassed by isoseismals VII-IX $_{ESI07}$ corresponds to the deformed areas recognized from radar interferometry analysis (e.g., [24]).

Lithology, slope angle, ESI-07 intensity, and PGA-h (horizontal Peak Ground Acceleration, based on information available by the IGEPN; see Data and Resources) are the seismological and geological factors considered in this study to evaluate the hazard of earthquake-triggered landslides. These factors have been analyzed and spatially correlated with the distribution of the coherent, disrupted and lateral spreading landslides in a GIS environment.

We present the outcome of this study together with landslide hazard maps, describing the characteristics of the surveyed landslides and providing new insights into the controlling factors of coseismic landslides during large subduction earthquakes.

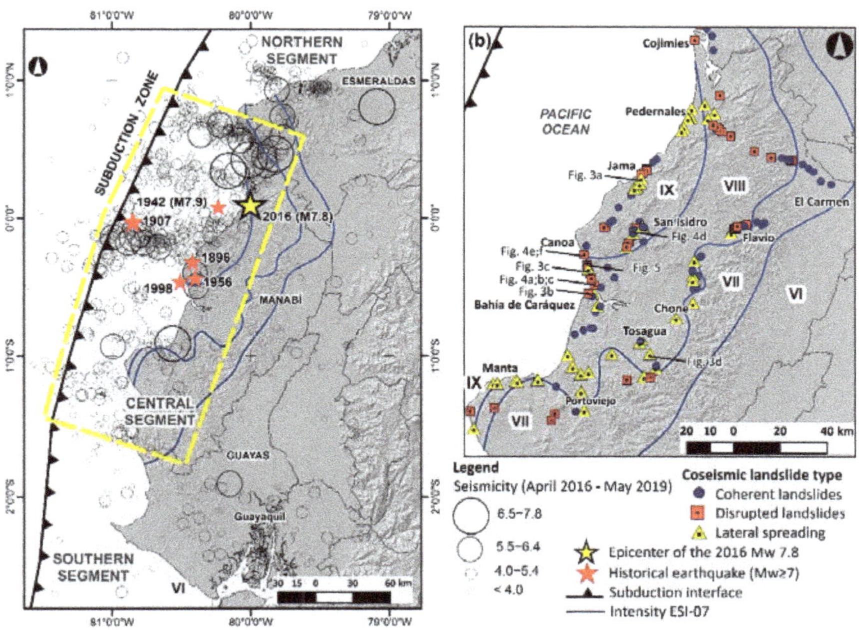

Figure 2. Location maps and seismotectonic setting: **a**) regional view of central tectonic segment of Ecuador coastal region (yellow dashed box): the epicenters of large subduction events occurred offshore the Manabí province (red stars) and the spatial distribution of aftershocks from the 16th April 2016 to 30th May 2019 are reported; **b**) earthquake-triggered landslides accompanying the 2016 Pedernales earthquake, dots mark the types of coseismic landslides and blue lines mark the ESI-07 isoseismals as defined by [22]; the location of the photographic documentation in Figures 3 and 4 is also reported.

2. Description of the Study Area

The epicentral area of the 2016 Pedernales earthquake sequence is located on the central coast of mainland Ecuador. The mapped area covers about 18,000 km^2 of narrow valleys in a hilly landscape, uplifted marine terraces and alluvial plains.

Due to its geomorphological setting, many of the hydrographic basins of the province of Manabí are characterized by alternating periods of erosion and enhanced sedimentation and are typically covered by Quaternary loose sediments. In the eastern sector of the study area, large alluvial fans are the main sedimentary source.

The epicentral area is dominated by altitudes lower than 800 m a.s.l. Based on the Koppen climate classification [26], the study region is characterized by different sub-climates ranging from tropical

Monsoon to dry and tropical Savanna [27]. Annual rainfall for the dry climate zone in southernmost Manabí region is less than 500 mm, measured between January and April, while the summer is very dry and hot. Annual rainfall ranges from 500 to 1000 mm for the tropical Savanna sub-climate where low rainfall occurs between June and November. The tropical Monsoon sub-climate in northernmost Manabí, is as hot as dry, but includes also a dry season from June to November with an annual total rainfall ranging between 1000 and 2000 mm. The predominant vegetation is a dry to very dry tropical forest, followed by mountains with tropical thorn shrub and premontane thorn shrub [26,28,29]. The most populated urban areas (i.e., the cities of Manta, Portoviejo, Chone, Bahía de Caráquez, Jama and Pedernales) have been settled on floodplains, alluvial terraces and colluvium-alluvium, where low-resistance soft soils prevail as well as soft rocks of claystone and siltstones, and sandstones in a lesser proportion.

Figure 3. Example of mapped lateral spread landslides: **a**) deformation in alluvial plain at Jama site, the road is completely displaced for tens of meters, severe damage affected the local kindergarten (IX_{ESI-07}); **b**) deformations in river bank due to lateral spreading: soil cracks 20 cm wide at Bahía de Caráquez city (VII_{ESI-07}); **c**) total collapse of Mejía bridge at Briceño: lateral spreading, river diversion and changes of fluvial dynamics, damage on house sited on the river side (IX_{ESI-07}); **d**) lateral spreading in river bank at Calceta city: soil cracks in natural soil 20 to 25 cm wide, wavy deformations of cobbles, (VII_{ESI-07}); locations on Figure 2.

3. Seismotectonic Setting and the 2016 Pedernales Earthquake Sequence

The Ecuadorian coastal region is within an active plate margin where the oceanic Nazca plate subducts the South American continent (e.g., [30–34]). The Nazca plate carries the Carnegie Ridge, which is a mountain range of seamounts resulting from the Galápagos hotspot volcanism (Figure 1; [35]). The subduction of this submarine ridge, which has an average elevation rise up to 2.5 km, has influenced the geodynamics of the coastal zone of Ecuador [36,37], as well as in its coastal geomorphology [38,39] and seismogenic characteristics [36,40–47].

Historically, the 1942 Jama (Mw 7.9) and the 2016 Pedernales (Mw 7.8) earthquakes are the largest subduction earthquakes recorded in the central segment of the subduction interface zone (Figure 2a; [48,49]). This tectonic segment registered several megathrust earthquakes, including:

i) the May 3, 1896, Mw 7.1; ii) the 1 June 1907, Mw 7.4; iii) the 16 January 1956, Mw 7.4 and iv) the Mw 7.1 on August 4, 1998 earthquakes (Figure 2, [22,50]).

The spatial distribution of the April 16, 2016 Pedernales earthquake sequence indicates a rupture zone ca. 100 to 120 km long [22,24]. The mainshock was located onshore, being less than 10 km northeast from the city of Pedernales, with a hypocentral depth of 21 km (source USGS, see Data and Resources). The focal mechanism obtained by IGEPN indicates a low-angle reverse-fault dipping N119/15. Peak ground acceleration up to 1.4 g was recorded in Pedernales, 0.52 g in Manta, 0.38 g in Portoviejo, 0.23 g in Chone, and 0.19 g in the Esmeraldas (IGEPN, see Data and Resources). The sequence included more than 3500 aftershocks, the most severe ones occurred on May 18, 2016 with Mw 6.7 to 6.9 [22,51] and some 14 months later with a ML 6.3 (IGEPN, see Data and Resources).

Figure 4. Example of coherent landslides: **a**) and **b**) pre- and post-earthquake satellite imagery at San Vicente, provided by the Web-GIS of the Military Geographical Institute of Ecuador (IGM) (VIII_{ESI-07}); **c**) rotational landslide at Briceño site, volume less than 3200 m^3, tree cover fall, active slope (VIII_{ESI-07}); **d**) rotational landslide at San Isidro, soil cracks 10 to 20 meters wide in soil, rotational landslides ca. 60 m long (IX_{ESI-07}); **e**) and **f**) stabilized slope also reported damage in the epicentral area, northern of Canoa site, where part of the hillslope was displaced by landslide of ca. 400 m^3 (drone imagery courtesy of by Global Medic and Servicio Nacional de Gestión de Riesgo y Emergencias de Ecuador), (VIII_{ESI-07}); locations on Figure 2.

This sector of the subduction zone, including the entire province of Manabí, has a short seismic record covering the last 120 years that suggests a recurrence interval of approximately 20 years for Mw >7.0 events, and between 70 and 80 years for Mw >7.5 [22,52]. Such subduction-related earthquakes have a recurrence relatively short in comparison with the subduction zones north and south of it [22].

Figure 5. Example of disrupted landslide at Briceño: rock fall and debris displacement, fractured slopes, completely destroyed road, house damaged by landslide, volume ca 10.000 m^3 ($VIII_{ESI\text{-}07}$); courtesy of drone imagery by Global Medic and Servicio Nacional de Gestión de Riesgo y Emergencias de Ecuador; locations on Figure 2.

4. Methods and Materials

Shortly after the mainshock, we mapped in the field all the observable earthquake-induced environmental effects (sensu, [1]), accessible through the main access roads and towns near Pedernales, which is the location of the instrumental epicenter where the peak ground acceleration of 1.4 g was recorded (IGEPN Accelerograph Ecuadorian Network; see Data and Resources). Government institutions such as MTOP (i.e., Ministerio de Transporte y Obras Pública) and MIDUVI (i.e., Ministerio de Desarrollo Urbano y Vivienda) provided logistics and transportation management due to partial destruction of the main access roads to towns, bridges and buildings by creating fieldwork camp sites for geotechnical engineers and geologists reaching out to the entire Manabí province and the southern part of Esmeraldas province.

Preliminary field mapping evaluated the macroseismic intensities of the primary coseismic geological effects (i.e., coastline uplifting along beaches and cliffs; e.g., [22,53] and secondary effects (i.e., soil liquefaction causing lateral spreading, subsidence, sinkhole and sand boils, as well as cracks in the soil, pavement failure in asphalt roads as well as landslides in stable and unstable slopes), using the ESI-07 intensity scale, with values ranging from VII to $IX_{ESI\text{-}07}$ in a coseismic affected area of about 18,000 km^2 [21,22]. Along with this database, an inventory of 192 seismically induced landslides (in attached pertinent Ⓔ electronic supplement to this article) by the 2016 Mw 7.8 event is herein provided (Figure 2), grouping them as follows:

(a) lateral spreads, landslides which moved in a translational mode in areas of liquefied gravel, sand, or silt, or rarely in sensitive clay transformed into fluid by disturbance [3,54], out of which 58 coseismic landslides were surveyed (Figure 3);

(b) coherent landslides, generally deep-seated, consisting of one or multiple blocks sliding on planar or smoothly curved basal shear surfaces, out of which 78 landslides in the epicentral area were surveyed (Figure 4);

(c) disrupted landslides associated with rock falls, rock slides, soil falls and disrupted soil slide, adding up to 56 surveyed coseismic effects, between earthquake-induced landslides and reactivated older landslides (Figure 5).

In this study, we also manually digitized some landslides on satellite imagery (SPOT-CNES imagery, courtesy of Google Earth) to i) check the extent of landslides at some of the surveyed events, ii) assure that no pre-earthquake landslides were present and iii) check field-observed mapping, in the closeness of some surveyed sites, especially in remote areas where only limited access was possible (Figure 4). Moreover, we checked the coseismic landslide inventory provided by the British Geological Survey BGS and GEER- ATC (see Data and Resources), referred only to the areas of Portoviejo, Bahía de Caráquez, Chone, Muisne and Crucita. This resulted in a statistically significant database of earthquake-induced landslides, though not comprehensive over the entire area interested by coseismic slope movements.

We compared the landslide occurrence with a set of selected variables that can possibly be considered as triggering or predisposing factors: lithology type, horizontal peak ground acceleration (PGA-h), topographic slope (derived from the 1 arc second SRTM digital elevation model) and distance from both the epicenter and the subduction rupture plane.

We derived the statistical correlation of landslide occurrence for each class of variables (i.e., PGA-h and lithology) using a statistical approach, modified after the Information Value Method (InfoVAL). The method allows the quantified prediction of susceptibility by means of a score (*Wi*), calculated according to landslide occurrence on each class, weighed according to the class distribution over the entire study area:

$$Wi = \ln \frac{Densclass_i}{Densmap_i}$$

where *Wi* is the score for the ith class, *Densclass* is the landslide occurrence for the ith class and *Desnmap* is the ith class occurrence on the whole area. This formula normalizes the event occurrences over the spatial distribution of each considered class: positive values indicate positive statistical correlation, negative values a negative one, while values close to zero indicate a random distribution of the data.

Since we did not map the areal extent of landslides and considering that the survey coverage was limited by site accessibility, we adopted simple univariate statistics for the description of topographic slope intervals typically associated with each landslide category.

We calculated the distance of each event from both the mainshock location (i.e., epicentral distance) and from the seismogenetic rupture plane. The latter has been calculated considering a rupture plane orientation on N119/15, as derived from seismological data and the distance have been calculated as the closest distance to the plane (i.e., Rrup distance).

5. Results and Discussion

Most landslides have shallow shear planes, between 2 and 4 m depth. On the contrary, few landslides have deep rupture planes (up to 15–30 m below ground level), as documented in the Briceño and San Isidro sites. A large number of landslides fall in the range of 200 to 3200 m^3 (VIII$_{ESI\text{-}07}$), while only few of them fall in the range of 5000 to 8000 m^3 (marked as intensity IX$_{ESI\text{-}07}$). Isolated rotational landslides and rockfalls smaller than 200 m^3 are regarded as intensity VII$_{ESI\text{-}07}$ (southern part of Esmeraldas province). Landslides located in plains and alluvial terraces are constrained to low resistance cohesive soils and are mainly lateral propagations caused by liquefaction. The earthquake-induced landslides extend as far as to the southern Manabí province, 134 km away from the instrumental epicenter, where lateral-spreading displacements was recorded near the city of Portoviejo in a site within the VIII_{ESI-07} and where 0.35 g of PGA-h was recorded.

According to the geological map at a scale of 1:500,000 of the coasts of Ecuador [55], we divided the study area into 5 lithological units (Figure 6): loose fluvial deposits and/or fine to coarse grained coastal units (AL); mainly sandstones and conglomerates (SS); shales, siltstones and mudstones (SS); tuffs and volcanoclastic deposits (VV); mainly basalts, hyaloclastite and associated greywackes (BB).

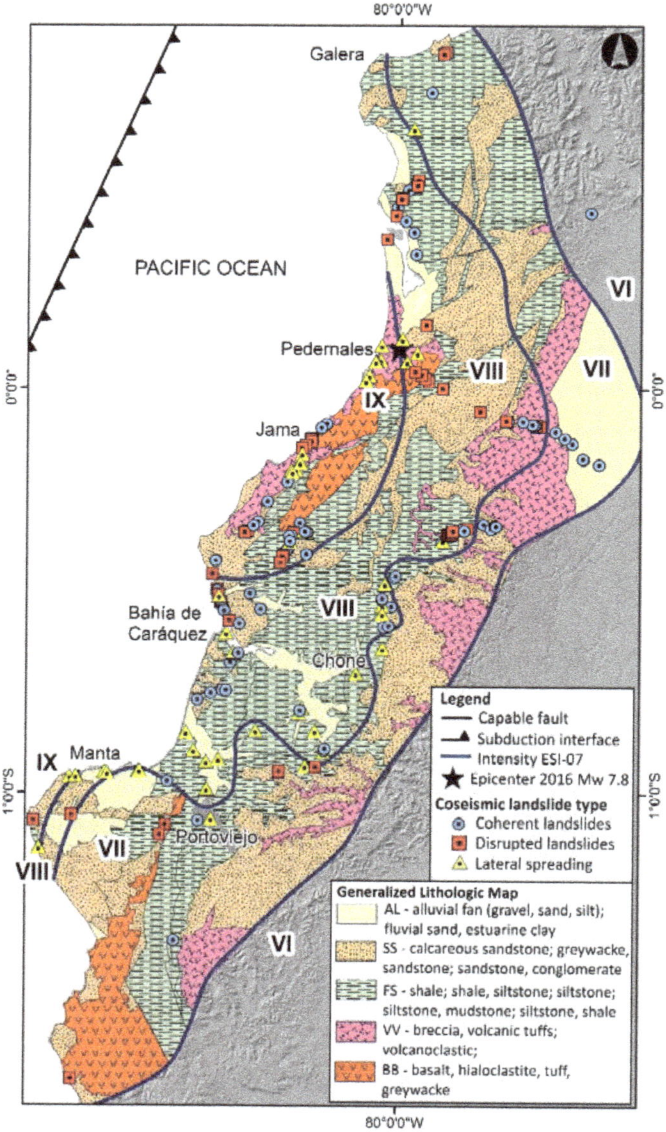

Figure 6. Simplified geologic map (modified from [55]) of the area comprised between isoseismals VII and IX_{ESI07}.

Our analysis shows that tuffs and volcanoclastic units recorded the largest number of coseismic landslides. Shales and siltstones host most coherent landslides while sandstones are mostly associated with disrupted landslides. Fluvial loose deposits and alluvial plains constitute the most susceptible setting for lateral spreading (Figure 7).

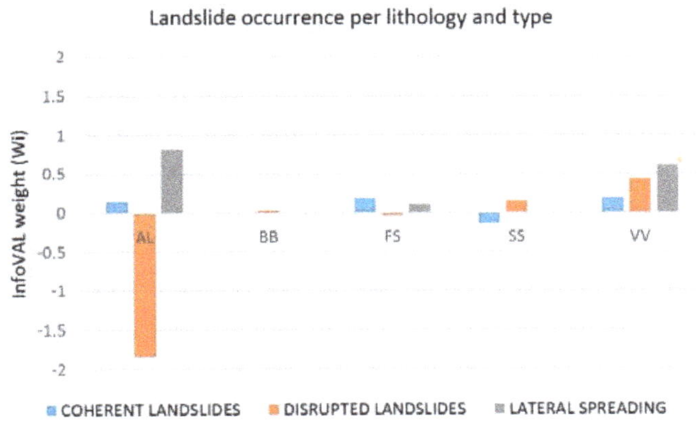

Figure 7. Statistical correlation between landslide occurrence and lithology type; correlation is expressed as an InfoVAL weight (Wi), where positive values indicate a statistically significant correlation (see methods for further details); lithology codes are the same as in Figure 6.

Lateral-spreading landslides are located in plains and alluvial terraces as well as few inclined slopes (Figure 3). Coherent landslides are located on moderate slopes with inclination between 8° and 16° and are mostly developed in Pliocene and Pleistocene geological units, which are composed of volcanic units and highly weathered clastic rocks. Disrupted landslides are developed on escarpments with inclination ranging from 16° to 55°, many of which are formed in high relief within the epicentral area as highlighted in Figure 5, where Miocene sandstone and greywacke lithological units (SS) predominate. These results coincide well with those obtained for the coseismic landslides caused by the 2011 Tohoku (Mw 9.0) and the Maule (Mw 8.8) thrust earthquakes, where most landslides, in the epicentral area, occurred in Neogene geological units [56–58].

The seismic intensity and the PGA-h characterize the degree of destruction and the seismic energy of earthquakes [8]. The ESI-07 intensity map is spatially well-correlated with the PGA-h isolines of the Pedernales earthquake (Figure 8) and shows a distribution of the intensity VII to IX_{ESI07} contours elongated along the with the rupture strike and encompassing the area affected by earthquake-triggered landslides. The isoseismal zone of $VI_{ESI\text{-}07}$ has much fewer landslides, with volumes lower than ca. 10^3 m^3, especially where the slope is unstable and particularly steep.

For the area encompassing the ruptured sector we can estimate a relationship between ESI-07 degree and the PGA-h: $IX_{ESI\text{-}07}$ ranges between 0.73 g and 1.35 g; VII $_{ESI\text{-}07}$ from 0.55 g to 0.73 g and $VII_{ESI\text{-}07}$ from 0.30 g to 0.55 g.

The number of landslide events and their size can be scaled with the epicentral distance up to $VIII_{ESI\text{-}07}$ (Figures 9 and 10, Supplementary Materials). Several landslides ascribable to $IX_{ESI\text{-}07}$ have been mapped in the coastal region but these are mainly lateral spreads and are more influenced by the local geomorphological and geological setting (i.e., coverage lithology and thickness, depth of the aquifer etc.) than by distance or other factors (Figure 11). Collectively, the coseismic landslides caused by the Pedernales earthquake (Mw 7.8) are strongly dependent from the distance to the rupture plane and PGA-h. Most of the landslides are triggered for values that are bounded between 0.4 g and 1.2 g and within a Rrup distance of less than 50 km (Figure 9, Supplementary Materials).

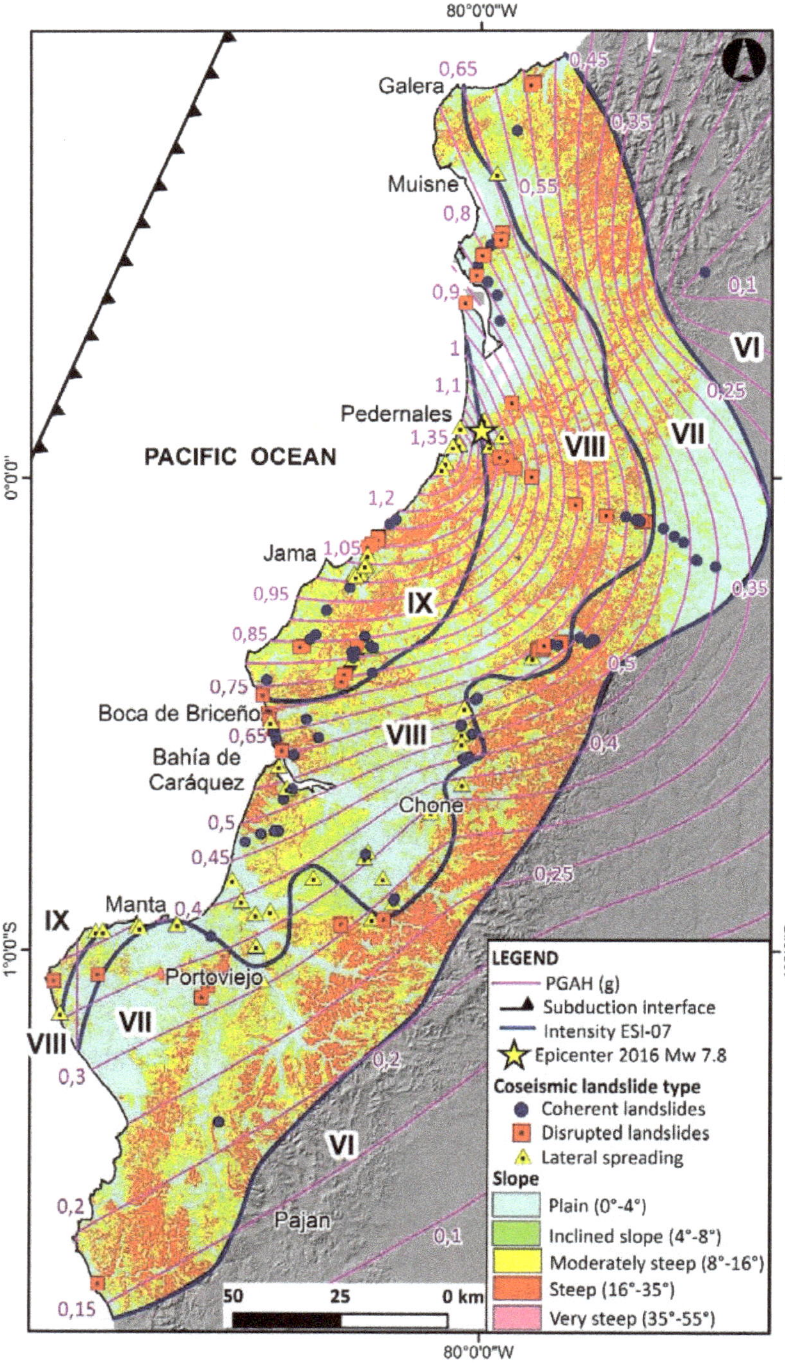

Figure 8. Simplified slope map and PGA-h isolines in the area of 2016 coseismic landslide inventory.

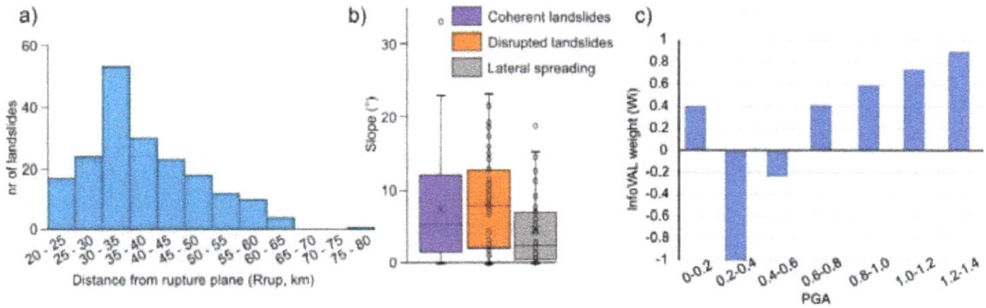

Figure 9. Landslides occurrence per (**a**) distance from the rupture plane (Rrup), (**b**) slope values and **c**) InfoVAL weights for PGA-h values.

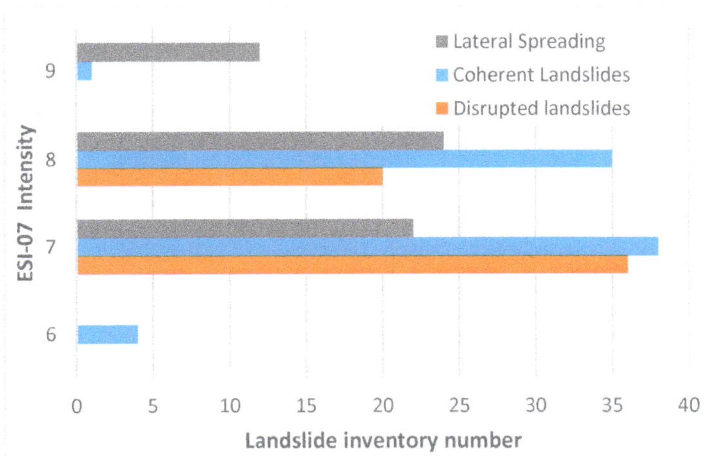

Figure 10. Distribution of coseismic landslides types as a function of assessed ESI-07 macroseismic intensity.

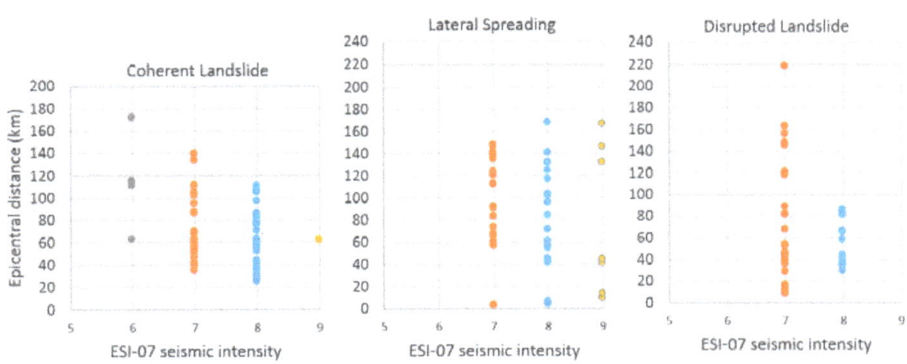

Figure 11. Relationship between epicentral distance, landslide types and ESI-07 seismic intensity.

6. Conclusions

With this first database, we produced an inventory of 192 seismically-induced landslides related to the 2016 Pedernales earthquake (Mw 7.8). Landslides were analyzed through a bivariate approach against: slope, lithological setting, PGA-h and macroseismic intensity data.

The spatial distribution of PGA-h, ranging between 0.4 g and 1.2 g, is consistent with the maximum VII–IX$_{ESI-07}$, recorded over an area of about 18,000 km^2. The disrupted landslide type is preferably located on fractured greywacke/sandstone units and in high relief areas, with slopes ranging between 16° and 55°. Our database suggests that lithology and hillslope geometry are the main geological/geomorphological factors controlling coseismic landslides occurrence; while the distance from the rupture plane plays a significant role on determining the landslide size. Indeed, the regression analysis between seismic intensity, PGA-h and landslide occurrence shows that most of the events occurred at PGA-h values between 0.4 g and 1.2 g, at a distance of 30 to 50 km from the rupture plane.

Our analysis draws a typical framework for slope movements triggered by subduction earthquakes in Ecuador. The most dangerous setting is the coastal region, a relatively highly urbanized area located near the epicenter and where liquefaction can trigger massive lateral spreading events. Coherent and disrupted landslides, dominating the more internal hilly region, can be triggered also in moderate slope settings (i.e., less than 10°).

We highlight that both the dimensions and frequency of landslides scale with distance in accordance with PGA-h and earthquake intensity (ESI-07) and through macroseismic intensity scenarios or strong motion modeling (e.g., USGS SHAKE maps) this information can be effectively used for the prediction of landslides distribution following strong subduction events. Calculated regressions can be considered valid for the South American subduction setting only, still the proposed approach can also be exported in different seismotectonic settings worldwide.

We note that timely field mapping after the mainshock allows to calibrate with unprecedented accuracy the geographic distribution of earthquake environmental effects, specifically landslides; and to understand the factors which control their distribution. We remark that coseismically-triggered landslides are among the most common environmental effects occurring during large subduction events. This effectively augment the relevance of coseismic landslide investigations for properly evaluating the earthquake macroseismic field, as initially suggested by the pioneers in the study of seismic intensity (e.g., [59–62]. We argue that our results represent a clear illustration of this point.

Data and Resources: Seismicity and macroseismic data were collected from IGEPN (http://www.igepn.edu.ec/servicios/eq20160416, last accessed June 2019);

USGS (http://earthquake.usgs.gov/earthquakes/eventpage/us20005j32#dyfi, last accessed May 2019) and the Geological Survey of Colombia (http://studylib.es/doc/6527439/informe-del-sismo-del-16-de-abril-de-2016-en, last accessed June 2019).

The isoseismal map for EMS-98 intensity realized by IGEPN is available at http://www.igepn.edu.ec/servicios/noticias/1324-informe-sismico-especial-n-18-2016 (last accessed May 2019).

Additional information on the Pedernales earthquake made available by BGS (British Geological Survey) and GEER-ATC was obtained at https://www.bgs.ac.uk/research/earthHazards/epom/ecuadorEarthquake.html (last accessed June 2019) and http://www.geerassociation.org/administrator/components/com_geer_reports/geerfiles/ECUADOR_Report_GEER-049-v1b.pdf (last accessed May 2019), respectively.

The DTM used for analysis is the NASA SRTM 3.0 global model, 1 arc second resolution: https://lpdaac.usgs.gov/news/nasa-shuttle-radar-topography-mission-srtm-version-30-srtm-plus-product-release/ (last accessed 6 August, 2019).

Some of the figures were realized using Qgis and Arcmap® software.

Supplementary Materials: The following are available online at http://www.mdpi.com/2076-3263/9/9/371/s1, Table S1: Coseismic landslide inventory for the 2016 Pedernales earthquake sequence.

Author Contributions: K.C. and F.A.L. designed the study. C.M., H.L.-S. and I.Z. analysed the geomorphologic features, landslide types and catalogues of earthquakes, M.F.F. processed and analysed the statistical correlation of landslide occurrence. A.M.M. prepared the manuscript with advice and feedback from all team members.

Funding: This research received no external funding.

Acknowledgments: This work was supported by the Geohazard Research Group of the Universidad Técnica de Manabí (UTM). The authors are grateful with the MTOP (Ministerio de Transporte y Obras Publicas) of Ecuador for the technical support during field-survey activities and data collection. The authors are also grateful to the anonymous reviewers for the constructive comments that helped in improving this paper.

Conflicts of Interest: The authors wish to confirm that there are no known conflicts of interests associated with this publication. This research did not receive any specific grant from funding agencies in the public, commercial, or not-for-profit sectors.

References

1. Michetti, A.M.; Esposito, E.; Guerrieri, L.; Porfido, S.; Serva, L.; Tatevossian, R.; Vittori, E.; Audemard, F.; Azuma, T.; Clague, J.; et al. Environmental Seismic Intensity Scale 2007—ESI 2007. In *Memorie Descrittive della Carta Geologica d'Italia, 74, 7-54*; Servizio Geologico d'Italia—Dipartimento Difesa del Suolo, APAT: Roma, Italy, 2007. Available online: http://www.isprambiente.gov.it/en/publications/technical-periodicals/descriptive-memories-of-the-geological-map-of/intensity-scale-esi-2007?set_language=en (accessed on 15 August 2019).
2. Serva, L. History of the Environmental Seismic Intensity Scale ESI-07. *Geosciences* **2019**, *9*, 210. [CrossRef]
3. Keefer, D.K. Landslides caused by earthquakes. *Geol. Soc. Am. Bull.* **1984**, *95*, 406–421. [CrossRef]
4. Harp, E.L.; Jibson, R.W. Landslides triggered by the 1994 Northridge, California, earthquake. *Bull. Seismol. Soc. Am.* **1996**, *86*, S319–S332.
5. Berzhinskii, Y.A.; Ordynskaya, A.P.; Gladkov, A.S.; Lunina, O.V.; Berzhinskaya, L.P.; Radziminovich, N.A.; Radziminovich, Y.B.; Imayev, V.S.; Chipizubov, A.V.; Smekalin, O.P. Application of the ESI-2007 scale for estimating the intensity of the Kultuk earthquake, August 27, 2008 (south Baikal). *Seism. Instrum.* **2010**, *46*, 307–324. [CrossRef]
6. Delgado, J.; Peláez, J.A.; Tomás, R.; García-Tortosa, F.J.; Alfaro, P.; Casado, C.L. Seismically-induced landslides in the Betic Cordillera (S Spain). *Soil Dynam. Earthq. Eng.* **2011**, *31*, 1203–1211. [CrossRef]
7. Serva, L.; Vittori, E.; Comerci, V.; Esposito, E.; Guerrieri, L.; Michetti, A.M.; Mohammadioun, B.; Mohammadioun, G.C.; Porfido, S.; Tatevossian, R.E. Earthquake Hazard and the Environmental Seismic Intensity (ESI) Scale. *Pure Appl. Geophys.* **2016**, *173*, 1479–1515. [CrossRef]
8. Xu, C.; Ma, S.; Tan, Z.; Xie, C.; Toda, S.; Huang, X. Landslides triggered by the 2016 Mj 7.3 Kumamoto, Japan, earthquake. *Landslides* **2018**, *15*, 551. [CrossRef]
9. Huayong, N.; Hua, G.; Yanchao, G.; Blumetti, A.M.; Comerci, V.; Di Manna, P.; Guerrieri, L.; Vittori, E. Comparison of Earthquake Environmental Effects and ESI intensities for recent seismic events in different tectonic settings: Sichuan (SW China) and Central Apennines (Italy). *Eng. Geol.* **2019**, *258*, 105149. [CrossRef]
10. Di Manna, P.; Guerrieri, L.; Piccardi, L.; Vittori, E.; Castaldini, D.; Berlusconi, A.; Bonadeo, L.; Comerci, V.; Ferrario, F.; Gambillara, R.; et al. Ground effects induced by the 2012 seismic sequence in Emilia: Implications for seismic hazard assessment in the Po Plain. *Ann. Geophys.* **2012**, *55*, 697–703.
11. Egred, J. *Catalogo de Terremotos del Ecuador 1541–2009*; Internal Report; Escuela Politecnica Nacional, Instituto Geofisico: Quito, Ecuador, 2009.
12. Chunga, K. Shallow Crustal Earthquakes and Seismic Zonation for Ecuador through the Integration of Geological, Seismological and Morphostructural Data. Ph.D. Thesis, University of Insubria, Varese, Italy, 2010.
13. Chunga, K.; Toulkeridis, T.; Vera-Grunauer, X.; Gutierrez, M.; Cahuana, N.; Alvarez, A. A review of earthquakes and tsunami records and characterization of capable faults on the northwestern coast of Ecuador. *J. Sci. Tsunami Hazards* **2017**, *36*, 100–127.
14. Tibaldi, A.; Ferrari, L.; Pasquarè, G. Landslides triggered by earthquakes and their relations with faults and mountain slope geometry: An example from Ecuador. *Geomorphology* **1995**, *11*. [CrossRef]
15. Beauval, C.; Yepes, H.; Bakun, W.H.; Egred, J.; Alvarado, A.; Singaucho, J.C. Locations and magnitudes of historical earthquakes in the Sierra of Ecuador (1587-1996). *Geophys. J. Int.* **2010**, *181*, 1613–1633. [CrossRef]
16. Beauval, C.; Yepes, H.; Palacios, P.; Segovia, M.; Alvarado, A.; Font, Y.; Aguilar, J.; Troncoso, L.; Vaca, S. An earthquake catalog for seismic hazard assessment in Ecuador. *Bull. Seism. Soc. Am.* **2013**, *103*, 773–786. [CrossRef]
17. Crespo, E.; O'Rourke, T.D.; Nyman, K.J. Effects on lifelines. In *The March 5, 1987, Ecuador Earthquakes: Mass Wasting & Socioeconomic Effects*; National Research Council: Washington, DC, USA, 1991; Chapter 6.

18. Gómez-Capera, A.A.; Leschiutta, I. Determining the parameters of earthquake sources in South America from macroseismic intensity data (CERESIS database). *Earth Sci. Res. J.* **2004**, *8*, 10.
19. Giesecke, A.; Capera, A.G.; Leschiutta, I.; Migliorini, E.; Valverde, L.R. The CERESIS earthquake catalogue and database of the Andean Region: Background, characteristics and examples of use. *Ann. Geophys.* **2004**, *47*. [CrossRef]
20. IGEPN. Instituto Geofísico Escuela Politécnica Nacional. Quito. Available online: http://www.igepn.edu.ec/solicitud-de-datos (accessed on 10 May 2019).
21. Toulkeridis, T.; Chunga, K.; Rentería, W.; Rodriguez, F.; Mato, F.; Nikolaou, S.; Antonaki, N.; Diaz-Fanas, G.; Besenzon, D. Mw7.8 Muisne, Ecuador 4/16/16 earthquake observations: Geophysical clustering, intensity mapping, tsunami. In Proceedings of the 16th World Conference on Earthquake Engineering, 16WCEE 2017, Santiago, Chile, 9–13 January 2017.
22. Chunga, K.; Livio, F.; Mulas, M.; Ochoa-Cornejo, F.; Besenzon, D.; Ferrario, M.F.; Michetti, A.M. Earthquake ground effects and intensity of the 16 April 2016, Mw 7.8 Pedernales Earthquake (Ecuador): Implications for the source characterization of large subduction earthquakes. *Bull. Seismol. Soc. Am.* **2018**, *108*, 3384–3397. [CrossRef]
23. NEC-11. 2015. Norma Ecuatoriana de la Construcción. Registro Oficial No. 413 del 10 de enero de 2015. Available online: http://www.normaconstruccion.ec/ (accessed on 11 April 2019).
24. Nocquet, J.M.; Jarrin, P.; Vallée, M.; Mothes, P.A.; Grandin, R.; Rolandone, F.; Delouis, B.; Yepes, H.; Font, Y.; Fuentes, D.; et al. Supercycle at the Ecuadorian subduction zone revealed after the 2016 Pedernales earthquake. *Nat. Geosci.* **2016**, *10*, 145–149. [CrossRef]
25. Toulkeridis, T.; Porras, T.; Tierra, A.; Toulkeridis-Estrella, K.; Cisneros, D.; Luna, M.; Carrión, J.L.; Herrera, M.; Murillo, A.; Perez, J.C.; et al. Two independent real-time precursors of the 7.8 Mw earthquake in Ecuador based on radioactive and geodetic processes—Powerful tools for an early warning system. *J. Geodyn.* **2019**, *126*, 12–22. [CrossRef]
26. Pourrut, P. Los Climas del Ecuador: Fundamentos Explicativos. In *Los climas del Ecuador*; CEDIG: Quito, Ecuador; Institut de recherche pour le développement: Marseille, France, 1983; pp. 7–41.
27. UNESCO. *Atlas Pluviométrico del Ecuador*; UNESCO: Quito, Ecuador, 2010.
28. Pourrut, P.; Róvere, O.; Romo, I.; Villacrés, H. Clima del Ecuador. In *El Agua en el Ecuador: Clima, Precipitaciones, Escorrentía*; Institut de recherche pour le développement: Marseille, France, 1995; pp. 13–26.
29. Kottek, M.; Grieser, J.; Beck, C.; Rudolf, B.; Rubel, F. World map of the Köppen-Geiger climate classification updated. *Meteorol. Z.* **2006**, *15*, 259–263. [CrossRef]
30. Hey, R. Tectonic evolution of the Cocos-Nazca spreading center. *Geol. Soc. Am. Bull.* **1977**, *88*, i–vi. [CrossRef]
31. Lonsdale, P. Ecuadorian subduction system. *AAPG Bull.* **1978**, *62*, 2454–2477.
32. Pennington, W.D. Subduction of the eastern Panama Basin and seismotectonics of northwestern South America. *J. Geophys. Res. B Solid Earth Planets* **1981**, *86*, 10753–10770. [CrossRef]
33. Eguez, A.; Alvarado, A.; Yepes, H.; Machette, M.N.; Costa, C.; Dart, R.L.; Bradley, L.A. *Database and Map of Quaternary Faults and Folds of Ecuador and Its Offshore Regions*; Open-File Report 03-289; US Geological Survey: Reston, VA, USA, 2003. Available online: https://pubs.usgs.gov/of/2003/ofr-03-289/ (accessed on 5 March 2019).
34. Veloza, G.; Styron, R.M.; Taylor Mora, A. Open-source archive of active faults for northwest South America. *GSA Today* **2012**, *22*, 4–10. [CrossRef]
35. Sallarès, V.; Charvis, P. Crustal thickness constraints on the geodynamic evolution of the Galapagos Volcanic Province. *Earth Planet. Sci. Lett.* **2003**, *214*, 545–559. [CrossRef]
36. Gutscher, M.A.; Malavieille, J.; Lallemand, S.; Collot, J.Y. Tectonic segmentation of the North Andean margin: Impact of the Carnegie Ridge collision. *Earth Planet. Sci. Lett.* **1999**, *168*, 255–270. [CrossRef]
37. Michaud, F.; Witt, C.; Royer, J.Y. Influence of the subduction of the Carnegie volcanic ridge on Ecuadorian geology: Reality and fiction. In *Backbone of the Americas: Shallow Subduction, Plateau Uplift, and Ridge and Terrane Collision*; Mahlburg Kay, S., Ramos, V., Dickinson, W., Eds.; The Geological Society of America: Boulder, CO, USA, 2009; pp. 217–228.
38. Blanco-Chao, R.; Pedoja, K.; Witt, C.; Martinod, J.; Husson, L.; Regard, V.; Audin, L.; Nexer, M.; Delcaillau, B.; Saillard, M.; et al. The rock coast of South and Central America. In *Memoirs*; Geological Society: London, UK, 2014; Volume 40, pp. 155–191. [CrossRef]

39. Pedoja, K.; Dumont, J.F.; Lamothe, M.; Ortlieb, L.; Collot, J.Y.; Ghaleb, B.; Auclair, M.; Alvarez, V.; Labrousse, B. Plio-Quaternary uplift of the Manta Peninsula and La Plata Island and the subduction of the Carnegie Ridge, central coast of Ecuador. *J. South Am. Earth Sci.* **2006**, *22*, 1–21. [CrossRef]
40. White, S.M.; Trenkamp, R.; Kellogg, J.N. Recent crustal deformation and the earthquake cycle along the Ecuador–Colombia subduction zone. *Earth Planet. Sci. Lett.* **2003**, *216*, 231–242. [CrossRef]
41. Graindorge, D.; Calahorrano, A.; Charvis, P.; Collot, J.Y.; Bethoux, N. Deep structures of the Ecuador convergent margin and the Carnegie Ridge, possible consequence on great earthquakes recurrence interval. *Geophys. Res. Lett.* **2004**, *31*, L04603. [CrossRef]
42. Chlieh, M.; Mothes, P.A.; Nocquet, J.M.; Jarrin, P.; Charvis, P.; Cisneros, D.; Font, Y.; Collot, J.Y.; Villegas-Lanza, J.C.; Rolandone, F.; et al. Distribution of discrete seismic asperities and aseismic slip along the Ecuadorian megathrust. *Earth Planet. Sci. Lett.* **2014**, *400*, 292–301. [CrossRef]
43. Nocquet, J.M.; Villegas-Lanza, J.C.; Chlieh, M.; Mothes, P.A.; Rolandone, F.; Jarrin, P.; Cisneros, D.; Alvarado, A.; Audin, L.; Bondoux, F.; et al. Motion of continental slivers and creeping subduction in the northern Andes. *Nat. Geosci.* **2014**, *7*, 287–291. [CrossRef]
44. Yepes, H.; Audin, L.; Alvarado, A.; Beauval, C.; Aguilar, J.Y.; Font Cotton, F. A new view for the geodynamics of Ecuador: Implication in seismogenic source definition and seismic hazard assessment. *Tectonics* **2016**, *35*, 1249–1279. [CrossRef]
45. Alvarado, A.; Audin, L.; Nocquet, J.M.; Jaillard, E.; Mothes, P.; Jarrín, P.; Segovia, M.; Rolandone, F.; Cisneros, D. Partitioning of oblique convergence in the Northern Andes subduction zone: Migration history and the present-day boundary of the North Andean Sliver in Ecuador. *Tectonics* **2016**, *35*, 1048–1065. [CrossRef]
46. Marcaillou, B.; Collot, J.Y.; Ribodetti, A.; d'Acremont, E.; Mahamat, A.A.; Alvarado, A. Seamount subduction at the North-Ecuadorian convergent margin: Effects on structures, inter-seismic coupling and seismogenesis. *Earth Planet. Sci. Lett.* **2016**, *433*, 146–158. [CrossRef]
47. Collot, J.Y.; Sanclemente, E.; Nocquet, J.M.; Leprêtre, A.; Ribodetti, A.; Jarrin, P.; Chlieh, M.; Graindorges, D.; Charvis, P. Subducted oceanic relief locks the shallow megathrust in central Ecuador. *J. Geophys. Res. B Solid Earth Planets* **2017**, *122*, 3286–3305. [CrossRef]
48. Swenson, J.L.; Beck, S.L. Historical 1942 Ecuador and 1942 Peru subduction earthquakes and earthquake cycles along Colombia-Ecuador and Peru subduction segments. *Pure Appl. Geophys.* **1996**, *146*, 67–101. [CrossRef]
49. Ye, L.; Kanamori, H.; Avouac, J.P.; Li, L.; Cheung, K.F.; Lay, T. The 16 April 2016, Mw 7.8 (Ms 7.5) Ecuador earthquake: A quasi-repeat of the 1942 M S 7.5 earthquake and partial re-rupture of the 1906 Ms 8.6 Colombia–Ecuador earthquake. *Earth Planet. Sci. Lett.* **2016**, *454*, 248–258. [CrossRef]
50. Gómez-Capera, A.A.; Stucchi, M.; Arcila, M.; Bufaliza, M.; Choy, J.; Minaya, E.; Leyton, L.; Pirchiner, M.; Rendón, H.; Rodriguez, L.; et al. Updated Earthquake Catalogue For South America: Time Window Pre-1964. In Proceedings of the 16th World Conference on Earthquake, 16WCEE 2017, Santiago, Chile, 9–13 January 2017.
51. Beauval, C.; Marinière, J.; Yepes, H.; Audin, L.; Nocquet, J.M.; Alvarado, A.; Baize, S.; Aguilar, J.; Singaucho, J.C.; Jomard, H. A New Seismic Hazard Model for Ecuador. *Bull. Seismol. Soc. Am.* **2018**, *108*, 1443–1464. [CrossRef]
52. García-Bustos, S.; Landín, J.; Moreno, R.; Chong, A.S.E.; Mulas, M.; Mite, M.; Cárdenas, N. Statistical analysis of the largest possible earthquake magnitudes on the Ecuadorian coast for selected return periods. *Georisk Assess. Manag. Risk Eng. Syst. Geohazards* **2018**. [CrossRef]
53. He, P.; Hetland, E.A.; Wang, Q.; Ding, K.; Wen, Y.; Zou, R. Coseismic Slip in the 2016 Mw 7.8 Ecuador Earthquake Imaged from Sentinel-1A Radar Interferometry. *Seismol. Res. Lett.* **2017**, *88*, 277–286. [CrossRef]
54. Hancox, G.T.; Perrin, N.D.; Dellow, G.D. Recent studies of historical earthquake-induced landsliding, ground damage, and MM intensity in New Zealand. *Bull. New Zealand Soc. Earthq. Eng.* **2002**, *35*, 59–95.
55. Reyes, P.; Michaud, F. *Mapa Geológico de la Margen Costera Ecuatoriana (1: 500.000)*; Quito, Ecuador; Institut de recherche pour le développement: Marseille, France, 2012.
56. Kelson, K.; Witter, R.C.; Tassara, A.; Ryder, I.; Ledezma, C.; Montalva, G.; Frost, D.; Sitar, N.R.; Moss Johnson, L. Coseismic tectonic surface deformation during the 2010 Maule, Chile, Mw 8.8 earthquake. *Earthq. Spectra* **2012**, *28*, S39–S54. [CrossRef]
57. Wartman, J.; Dunham, L.; Tiwari, B.; Pradel, D. Landslides in Eastern Honshu induced by the 2011 off the Pacific Coast of Tohoku earthquake. *Bull. Seismol. Soc. Am.* **2013**, *103*, 1503–1521. [CrossRef]

58. Serey, A.; Piñero-Feliciangeli, L.; Sepúlveda, S.A.; Iveda, F.; Poblete, D.; Petley, I.; Murphy, W. Landslides induced by the 2010 Chile megathrust earthquake: A comprehensive inventory and correlations with geological and seismic factors. *Landslides* **2019**, *16*, 1153. [CrossRef]
59. Sieberg, A. Geologie der Erdbeden. In *Handbuch der Geophysik*; Harold, J., Ed.; Nature: London, UK, 1930; Volume 2, pp. 550–555.
60. Wood, H.O.; Neumann, F. Modified Mercalli intensity scale of 1931. *Bull. Seismol. Soc. Am.* **1931**, *21*, 277–283.
61. Richter, C.F. *Elementary Seismology*; Freeman: San Francisco, CA, USA, 1958.
62. Medvedev, S.; Sponheuer, W.; Karník, V. *Neue seismische Skala Intensity Scale of Earthquakes, 7. Tagung der Europäischen Seismologischen Kommission vom 24.9. bis 30.9.1962*; Institut für Bodendynamik und Erdbebenforschung in Jena: Jena, Germany; Deutsche Akademie der Wissenschaften zu Berlin: Berlin, Germany, 1964.

geosciences

MDPI

Article

The 1976 Guatemala Earthquake: ESI Scale and Probabilistic/Deterministic Seismic Hazard Analysis Approaches

Mauro Caccavale [1,2,*], Marco Sacchi [1], Efisio Spiga [3] and Sabina Porfido [2,4]

1 CNR-ISMAR, Calata Porta di Massa, Interno Porto, 80133 Napoli, Italy
2 INGV-Osservatorio Vesuviano, 80133 Napoli, Italy
3 Independent Researcher, 83100 Avellino, Italy
4 CNR-ISA-Via Roma 64, 83100 Avellino, Italy
* Correspondence: mauro.caccavale@cnr.it

Received: 30 July 2019; Accepted: 17 September 2019; Published: 19 September 2019

Abstract: A hazard assessment of the 1976 Guatemala earthquake (M = 7.5) was conducted to achieve a better definition of the seismic hazard. The assessment was based on the environmental effects that had effectively contributed to the high destructive impact of that event. An interdisciplinary approach was adopted by integrating: (1) historical data; (2) co-seismic geological effects in terms of Environmental Seismic Intensity (ESI) scale intensity values; and (3) ground shaking data estimated by a probabilistic/deterministic approach. A detailed analysis of primary and secondary effects was conducted for a set of 24 localities, to obtain a better evaluation of seismic intensity. The new intensity values were compared with the Modified Mercalli Intensity (MMI) and Peak Ground Acceleration (PGA) distribution estimated using a probabilistic/deterministic hazard analysis approach for the target area. Our results are evidence that the probabilistic/deterministic hazard analysis procedures may result in very different indications on the PGA distributions. Moreover, PGA values often display significant discrepancy from the macroseismic intensity values calculated with the ESI scale. Therefore, the incorporation of the environmental earth effects into the probabilistic/deterministic hazard analysis appears to be mandatory in order to achieve a more accurate seismic estimation.

Keywords: georisk; probabilistic seismic hazard; ESI scale; ground-shaking map; Guatemala; macroseismic intensity

---

1. Introduction

The last decade has witnessed the development of a novel generation of attenuation models (ground motion prediction relations) [1], accompanied by a significant progress of computation possibilities, implemented GIS software, and the introduction of a new Environmental Seismic Intensity (ESI) scale [2,3]. Particularly, recent research focused on hazard characterization has emphasized the importance of revaluating historical seismic events to improve our ability in developing reliable mitigation strategies [3]. These multidisciplinary approaches have found an ideal application in the case of the Guatemala region, one of the most active seismic zones of Central America. During the last century, several strong earthquakes hit the country with dramatic consequences. These earthquakes were mostly associated with devastating environmental phenomena that caused permanent changes in the landscape [4–6].

This study analyzed the 1976 Guatemala earthquake, one of the most dramatic seismic events to have occurred in the 20th century. Even as destruction was severe both in terms of numbers of dead/injured and geographical extension, the maximum estimated intensity was IX Modified Mercalli Intensity (MMI) in Gualan, the Mixco area and in the center of Guatemala City.

A re-evaluation of the 1976 earthquake was conducted in order to achieve a better definition of the seismic hazard, taking into account the environmental effects that effectively contributed to the destructive impact, both in terms of human life loss and damage, and the effect on houses and infrastructure. The consequences of the earthquake were felt on a large part of the territory and also a few days after the main shock, with a higher concentration in the western part of the Motagua fault. Several phenomena, including building collapses, surface faulting, ground deformation and soil cracking, landslides, and liquefaction of sandy soils afflicted the population and modified the landscape permanently. All effects induced by the earthquake were analyzed to overcome the intrinsic limitations of the macroseismic scale. Ostensibly, the maximum value of IX MMI was not realistic for the degree of destruction generated by the earthquake. In order to face this apparent contradiction, we reviewed the earthquake environmental effects (EEE) recorded for a set of 24 localities [5,7–10]. These effects were analyzed in terms of the ESI scale, with the scope of assigning a new intensity level to each locality. The obtained ESI intensity values were then compared with the MMI and the distribution of Peak Ground Acceleration (PGA) estimates using the seismic hazard analysis approach for the target areas. The aim of this study was achieved by integrating (1) previous earthquakes; (2) ESI values distribution; and (3) hazard maps.

1.1. Geological Framework and Historical Seismicity of Guatemala

The tectonic setting of the Central American region (Figure 1) is characterized by the interaction of three major lithospheric plates: The Cocos, the Caribbean, and the North American plates. In this region the relative plate motion is in the order of 2–9 cm/yr and is accompanied by active volcanism and shallow and intermediate seismicity. The Cocos plate moves towards north-northeast with respect to the Caribbean and North America plates at a velocity of 70–80 mm/yr, and subducts beneath Central America at the Middle America Trench [11]. Where the hypocenters of the largest earthquakes occur [12,13]. Over the past 40 years, Guatemala and neighboring areas have experienced 50 events with $M \geq 6$, out of which two with $M > 7$ (USGS data). The last event occurred in the south of Champerico, near the triple junction of the plates, on 7 November 2012 ($M = 7.4$). Two other strong events occurred in December 1983 ($M = 7.1$), approximately in the same location of the 7 November event, and in September 1993 ($M = 7.2$), about 100 km to the northwest, offshore Mexico [11].

Large earthquakes are also produced along the boundary between the North American and the Caribbean plates, defined by a zone of large left lateral strike–slip faults (the Chixoy–Polochic fault, the Motagua fault, etc.) that run through the Guatemala from the Swan Fracture Zone in the Caribbean Sea. The earthquakes generated along these transcurrent faults have a great importance for seismic hazard in Central America, compared to the subduction-related earthquakes, because of their shallow hypocenters and the proximity of many cities and villages to these active structures.

The most destructive event in this region was the earthquake associated with the Motagua fault, that occurred on 4 February 1976, causing 23,000 deaths and 77,200 injuries. The total number of houses destroyed was 258,000 and 1.2 million people were left homeless [4,6–8]. Following the main earthquake, at least 18 aftershock events ($5.6 \leq M \leq 7.9$) were recorded, that induced ground effects, mostly slope movements, followed by ground cracks, ground collapse, hydrological changes, topographic changes, tsunami, and in some cases surface faults [14,15].

Strong earthquakes also occurred in 1902 ($M = 7.5$) and in 1942 ($M = 7.9$). The event of 19 April 1902 (epicenter in Quetzaltenango) caused landslides and ground fractures within the epicentral area, in Sololà and along the slope of the Agua and Cerro Quemado volcanoes. Significant slope movements dammed the Naranjo and Ixtacapa Rivers, whereas liquefaction and ground settlement phenomena were observed in Cocos. The earthquake of 6 August 1942 (epicentral zone off the southern coast of Guatemala) also induced several landslides and destruction along the west-central highlands in the country. Slope movements affected the Pan-American Highway and secondary roads. Ground settlements also occurred, especially along the western Pacific coast region, [6,16,17].

The historical record of seismicity in the Guatemala region dates back to the 16th century. Destructive historical earthquakes have been accurately described in a series of studies [4,13,16–21] and reported in Figure 1. In some cases, these events were characterized by both high damage and remarkable environmental effects.

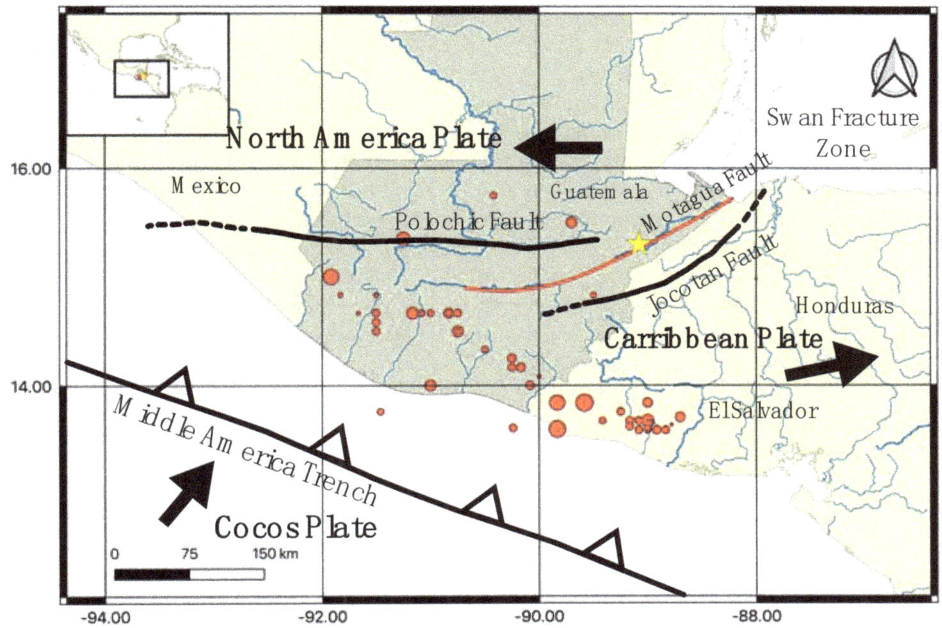

Figure 1. Historical earthquakes in Guatemala included in the Global Historical Earthquake Catalogue (GHEA) for the period 1000–1903 (red points) with Mw = 7 and over; the yellow stars represent the 4th February 1976 Guatemala earthquake epicenter; the red line represents the Motagua fault.

1.2. The 4 February 1976 Earthquake

On 4 February 1976, at 04 09:01 UTC, an earthquake of M = 7.5 hit the Guatemala region with the epicenter located near Los Amates, about 157 Km NE of Guatemala City [3]. The event was generated by a fault segment with a length of ca. 230 km that was activated with a strike slip mechanism [8,15]. The main earthquake was followed by two large aftershocks occurred on 6 and 7 February near Guatemala City, with magnitude equal to 5.8 and 5.7, respectively [14]. As a consequence, ca. 23,000 deaths and more than 77,000 injuries were reported in a vast part of the territory. Severe environmental effects triggered by the ground motions were recorded over an area of about 18,000 km^2, including landslides, liquefaction phenomena, ground cracking, and deformation [4,7,8]. In some districts of the cities of Gualan, Parramos, Patzicia, El Progreso, Rabinal, San Jose Poaquil, and San Martin Jilotepeque, the percentage of houses that collapsed or were seriously damaged even reached 100%.

The main source mechanism was a strike–slip with a mean horizontal displacement of about 1 m and a maximum of 3.4 m [12]. Secondary effects, reported in the Mixo area and other localities, were mainly represented by ground movements of different intensity and volume involved. Harp [5] documented ca. 50,000 landslides, including falls, debris slides and flows, which involved rocks and/or pumiceous pyroclastic deposits over an area of ca. 16,000 km^2. The ground instability reported in the Motagua Valley up to Quebradas and Patzaj cities also involved a thick cover of volcaniclastic deposit, locally producing very large slope failures (V > 100,000 m^3). In the Motagua valley, along Atlantic coast of Guatemala and Honduras, as well as along the shores of Lakes Amatitlan, Atitlan, and Ilopango, in El Salvador, lateral spreads and liquefaction phenomena were also recorded. The geographic distribution of ground effects, such as landslides and ground cracks, indicates a higher concentration of these processes towards the southwestern edge of the Motagua fault. The EEE observation is in a good agreement with the tele-seismic and local data recorded by portable seismic stations during the aftershock sequences. The data confirmed that the highest seismic activity was mostly concentrated in the southwestern edge of the Motagua fault between Guatemala City and Lake Atitlan [14,15].

The intensity level in terms of the ESI scale was evaluated on the basis of both primary effects (e.g., surface faulting) and secondary effects (slope movements, liquefaction, and ground rupture features). The higher ESI intensity level of XI [6] was attributed by taking into account the extent of the area and ground volume involved in slope failures (e.g., landslides, rock-falls, avalanches) [6,22] and was located along the fault zone, as expected. The areal distribution of damages and ground effects reported from localities like Estancia de la Virgen, San Martin Jilotepeque and San Josè Poaquil was helpful in defining the ESI X degree line. The IX level of ESI scale was assigned to some localities in the southwestern part of the fault (Finca San Carlos, Guatemala City, Los Choloyos Mixco Area, Rio Blanco, Rio Los Cubes, Rio Cotzibal) and in Puerto Barrios (eastern edge of the fault).

The total length of the fault zone, together with the extent of the area involved by ground effects represent the criteria used for the definition of the new epicentral intensity (I_0 = XI-ESI) calculated for the Guatemala 1976 earthquake. The new intensity calculated with the ESI scale is a good agreement with an event characterized by a max horizontal offset of 3.40 m and a total length of the surface faulting of 230 km affecting an area of 18,000 km^2. Even if the destruction level was dramatic for both the number of dead/injured and geographical extension, the maximum estimated intensity was IX-MMI in the areas of Gualan, Mixco, and the center of Guatemala City.

2. Methods

2.1. The ESI Scale

Traditional intensity scales are usually based on the evaluation of the damage occurred to buildings and man-made infrastructures of inhabited areas (i.e., the EMS-98 [23]). In 2007, a new approach was proposed [2] to evaluate the intensity of an earthquake, virtually applicable to all areas of the world, even where uninhabited. In the proposed approach, the environmental effects induced by the earthquake become key elements to measure seismic intensity. The new macroseismic scale, named the ESI scale (the Environmental Seismic Intensity 2007 scale) was ratified in 2007 by the International Union for Quaternary Research (INQUA). The ESI scale may be regarded as an extension or evolution of traditional macroseismic scales. The 12-degree subdivision of the ESI scale reflects the need for accuracy in the quantification of the earthquake strength and provides a more realistic measure of the intensity. The assessment of geological, hydrological, geomorphological, and vegetation features, once used only marginally to evaluate the seismic risk, plays a privileged and key rule in the ESI scale approach. The ESI scale has also been applied to modern and paleo-earthquakes [24–30], providing significant input for a better evaluation of seismic hazards in different socio-economic contexts [31–39].

2.2. Seismic Hazard Analysis (SHA)

A seismic hazard analysis (SHA) is carried out according with the most diffused methods used to calculate the hazard maps, namely, the Probabilistic and Deterministic Seismic Hazard Analysis (PSHA and DSHA, respectively). The PSHA methodology is an inclusive approach and can incorporate: (1) Different source zones where the probability of event occurrence is uniformly distributed; (2) continuous magnitude distribution; (3) different ground motion equations; and (4) parameter uncertainties. The intrinsic nature of the probabilistic approach implies the possibility of missing the individual contribution of each source or event.

On the other hand, the DSHA is characterized by a discrete approach. The seismic source is fixed in space and described by a given magnitude value and a GMPE to estimate the ground motion at the site of interest. In this case, information regarding the frequency of magnitude, location, and probability of the event is not taken into account.

Given the characteristics of the two approaches, they should be considered complementary and used simultaneously. In this work, we decided in fact to use both approaches, because of the focus on a specific earthquake.

2.2.1. Probabilistic Seismic Hazard Analysis (PSHA)

The PSHA is the most widely accepted procedure to evaluate and calculate the risk related to possible earthquake events. This approach was defined by Cornel [40] and it is based on the solution of the hazard integral:

$$\sum_{i=1}^{N} E_i(A > A_0) = \sum_{i=1}^{N} \alpha_i \left\{ \int_M \int_R P[A > A_0|m,r] f(m) f(r) dmdr \right\}_i, \quad (1)$$

where E represents the mean annual rate of exceeding of a given threshold value A_0; $f(M)$ and $f(R)$ are the Probability Density Functions (PDF) of magnitude end distance, respectively; The P function represents the conditional probability to overcome the A_0 value for a given magnitude (m) and distance (r) couplet. The α term is an expression of the individual capability to generate an earthquake greater than a fixed lower magnitude bound for each seismic source.

The form of PDFs depends on the specific earthquake recurrence model adopted (e.g., Gutenberg and Richter [41] or McGuire and Arabasaz, [42]) and on the source geometry and Ground Motion Prediction Equation (GMPE) adopted.

By resolving Equation (1) under the hypothesis of a poissonian process, it is possible to demonstrate that the probability to exceed a threshold ground motion value (A_0) in a fixed time interval (t) is

$$P(A \geq A_0, t) = 1 - e^{-\sum_{i=1}^{N} E_i(A > A_0) \cdot t}, \quad (2)$$

and the return period (T_R) of an event strong enough to generate a ground motion higher to the threshold is

$$T_R = -\frac{1}{\ln(1 - P(N \geq 1, t))}, \quad (3)$$

where P represents the poissonian probability to have at least one event ($N \geq 1$) during the time interval t.

The output, in terms of single target site approach of the PSHA, obtained from the solution of Equation (1), is the hazard curve.

In this study we have adopted an ad-hoc Fortran code to resolve the hazard integral Equation (1) and calculated the hazard curve and the hazard maps with the possibility of customizing and controlling each step of the process.

The Fortran code was calibrated and tested using the same seismic zonation, parametrization, and GMPE reported by Benito [43], obtaining compatible results.

In this case, the study area is smaller than the one considered in [32] and is located in Guatemala and connected to the Motagua fault and the secondary effects of the 1976 earthquake.

A regular grid of 756 knots (with vertical and horizontal spacing, Δx and Δy, respectively, of 0.1°), plus 24 points corresponding to target sites, each characterized by an ESI scale value, was generated to cover about 87,000 km^2 (Figures 2 and 3, and Tables 1 and 2)

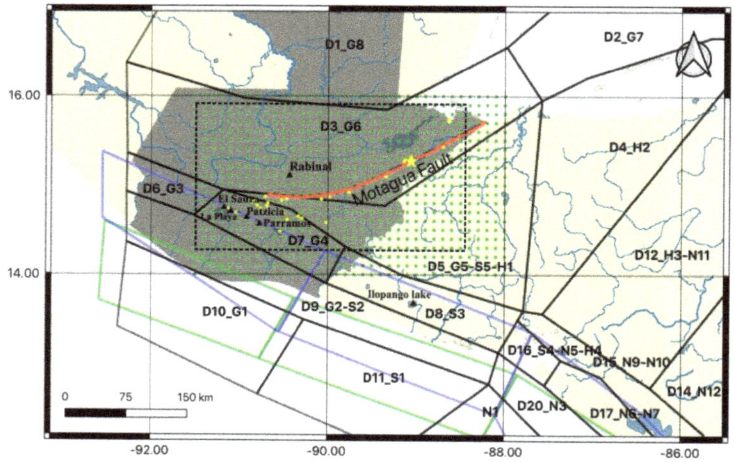

Figure 2. Crustal seismic zones (black polygons); in-slab seismic zones (blue polygons); interplate seismic zones (green polygons); grid points used for the probabilistic approach (green); the red line represents the Motagua fault and the yellow star the 4 February 1976 Guatemala earthquake epicenter. Yellows hexagons represent the target sites and black triangles represent sites of interest. The black dashed box represents the area of interest with the target sites.

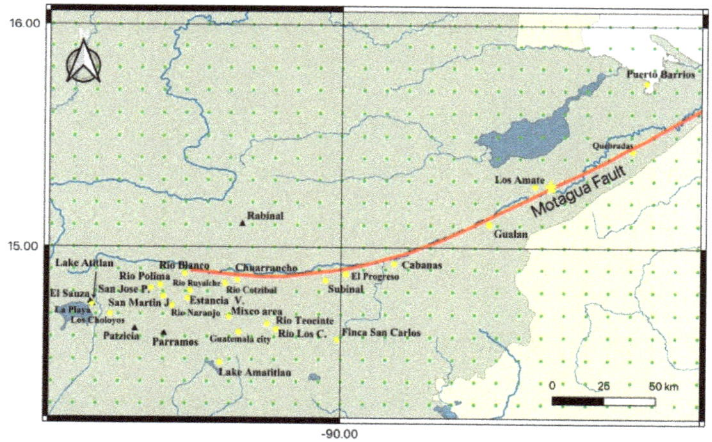

Figure 3. Area of interest with the target sites (yellow hexagons) and other sites of interest (black triangles). In green the grid points used for the probabilistic approach; the red line represents the Motagua fault and the yellow star the 4th February 1976 Guatemala earthquake epicenter.

Table 1. Input parameters of the probabilistic seismic hazard analysis (PSHA) Fortran code.

Grid Configuration		Hazard Integration Parameters	
Horizontal spacing (ΔX)	0.1°	Distance increment (dr) [km]	1.0
Vertical spacing (ΔY)	0.1°	Magnitude increment (dm)	0.5
N° of grid points	780	Integration distance range [km]	0÷300
Grid area (km^2)	~87,000	Return periods [Yrs]	200; 264; 300
Lat. range	14.0, 16.0	Annual probability of exceedance	0.5; 0.4; 0.3; 0.2
Lon. range	−88.0; −91.5	Acceleration threshold range (A_0) [g]	1.0E−4 ÷ 2.4

Table 2. Crustal seismic zones (SZ) parameters used in this study.

Id	Name	$N(M_{min})^{\circ}$	b^{+}	M_{max}^{*}	Depth $(Km)^{\|\|}$	Area (Km^2)
1	D1_G8	0.75	0.78	6.7	10	139,028.293
2	D2_G7	0.93	0.78	7.8	10	76,659.182
3	D3_G6	1.32	0.82	7.8	10	56,656.057
4	D4_H2	0.73	1.21	6.3	10	43,918.544
5	D5_G5-S5-H1	0.97	0.88	6.8	10	38,233.842
6	D6_G3	0.18	0.67	6.7	10	4810.094
7	D7_G4	0.65	0.72	7.0	10	9675.350
8	D8_S3	1.18	0.87	7.0	10	11,967.291
9	D9_G2-S2	1.60	1.03	6.3	20	26,453.291
10	D10_G1	3.51	0.92	7.5	20	22,903.907
11	D11_S1	9.77	1.56	7.5	20	25,079.390
12	D12_H3-N11	0.36	1.09	6.3	10	80,889.087
14	D14_N12	0.44	1.21	6.2	10	61,147.947
15	D15_N9-N10	0.24	0.50	6.8	10	10,023.342
16	D16_S4-N5-H4	0.52	0.97	6.6	10	3005.915
17	D17_N6-N7	1.68	0.95	6.5	10	8024.593
20	D20_N3	1.35	0.88	6.8	20	7934.894
21	N1	13.73	1.18	8	10	26,047.566

$^{\circ}$ $N(M_{min})$ is the number of events with magnitude $M \geq M_{min}$; $M_{min} = 4.5$. + b is the slop of the Gutenberg-Richter relationship [41]. * M_{max} is the maximum magnitude associated to the seismic zone. || Depth is the hypocentral depth of the events.

The solution of Equation (1), for a single target site, allows the calculation of the hazard curves for the site of interest (see the Analysis and Result Section for more details).

The hazard curve represents the annual frequency of exceedance (or exceedance probability, calculated by Equation (2) of each ground motion threshold value (A_0) used in Equation (1). The higher the number of A_0 tested values, the higher will be the hazard curve resolution. This will result in a more accurate evaluation of the ground motion value related to a fixed frequency or probability. In this way, for a set of T_R Equation (3) or exceedance probability, the corresponding values of ground motion parameters (acceleration, velocity, etc.) will be obtained.

By integrating this procedure over a large number of target points (e.g., a regular grid) it is possible to calculate the geographic distribution of the selected ground motion parameters with the same frequency of exceedance or probability of occurrence or T_R (hazard maps).

A set of 200 log-spaced values from 1.0E−4 to 2.4 g (where g is the gravity acceleration = 9.81 m/s^2), has been set-up to better define the shape of the curves. Equation (1) was solved for increments of 1 km (in distance) and 0.5 (in magnitude) (Table 1). These steps were selected to match a reasonable compromise between computation time and precision of results.

Another fundamental parameter in calculating the hazard map is the return period (T_R). This parameter represents the theorical time interval between two earthquakes with the same magnitude, i.e., the earthquake frequency. The Gutenberg and Richter (G–R) relationship parameters [41] associated with the SZ called "D3_G6" was used to calculate the more realistic T_R for the Motagua fault. The fault lies within the SZ reported in Figure 2 and the G–R parameters are also based on the earthquakes that have occurred along this tectonic lineament.

Instead of determining a single T_R value associated with a magnitude of 7.5, we used a range of values ($7.4 \leq M \leq 7.7$) to better analyze the hazard variably. The T_R value was set to 200, 264, and 300 years and reported in Table 2, along with the corresponding annual probability of exceedance calculated by Equation (3) (for sake of simplicity this paper reports the results for 300 years).

Mechanical properties of the ground subsurface are also an important factor when performing a hazard analysis, therefore the GMPEs include a site characterization in terms of the geotechnical properties of rocks. The GMPEs selected in this work follow this approach, by including a term to correct the PGA estimation as a function of the share waves velocity at a depth of 30 m (Vs_{30}).

According to Vs_{30} values, we may distinguish "rock" (basement lithological units) for geological sites with no amplification effects (Vs_{30} > 800 m/s) and "soil" (unconsolidated sediments and/or pedogenetic horizons) for sites that amplify the seismic waves (Vs_{30} < 800 m/s). In this study, we calculated the hazard maps for both "rock" and "soil" ground type. The choice was motivated by the importance of highlighting the likely variability of PGA in association with possible site amplification effects, also in the case of lack of information on Vs_{30} values at the grid points or target site location. A more realistic PGA range estimation is also easier to compare with the earthquake effects.

2.2.2. Deterministic Seismic Hazard Analysis (DSHA)

The DSHA was performed by assuming the Motagua fault (Figure 1) as a linear seismic source of 230 km. The magnitude was fixed to M = 7.5. The PGA at the grid points and at sites of interest was estimated as the linear combination of GMPEs [44,45]. In this case, the hypocentral distance used in the GMPE corresponds to the minimum distance between the fault-segment and the site. The fault depth was fixed at 5.0 km, as reported in literature [14]. Information on the strike–slip focal mechanism of the earthquake was available but could not be included in the deterministic approach due to the intrinsic limitation of the selected GMPEs.

3. Results and Discussion

3.1. ESI Scale

In spite of the destructive impact of the 1976 Guatemala earthquake, the officially estimated intensity, surprisingly, was only IX-MMI for a limited number of localities [4]. To analyze this apparent incongruence, the primary and secondary environmental effects of the mainshock in terms of the ESI scale definition [2,3] were re-examined.

From the collection of all the published data, it was possible to extrapolate the ESI intensity at 24 target sites, allowing the comparison between the different evaluation methodologies (Table 3).

Table 3. Primary and secondary effects triggered by the 1976 Guatemala earthquake.

NO.	Locality	ESI	Type of Effect °	MMI	Note
1	Cabanas	XI	SF, GC	VIII	Main strike slip fault; ground cracks
2	Chuarrancho	XI	SF, GC	VIII	Main strike slip fault; ground cracks
3	El Progreso	XI	SF, GC	VIII	Main strike slip fault; ground cracks
4	Gualan	XI	SF, GC	IX	Main strike slip fault; ground cracks
5	Quebradas	XI	SF, GC, L	VIII	Main strike slip fault; ground cracks; Liquefaction phenomena
6	Subinal	XI	SF, GC	VIII	Main strike slip fault; ground cracks
7	Estancia De La Virgen	X	SM, GC, TL	VIII	Rotational slump/rock-fall avalanche ($V < 10^6$ m^3), Temp. Lake
8	San Martin Jilotepeque	X	SM, GC, L, TL	VIII	Complex rotational slump/eartflow ($V = 10^6$ m^3), lateral spreads; Temp. Lake
9	San Josè Poaquil	X	SM, GC	VIII	Complex block slide/rotational slump/rock-fall avalanche ($V = 3.5\times10^6$ m^3); Temp. Lake
10	Puerto Barrios	IX	SF, GS, GC	IX (VI)	Ground compaction
11	Finca San Carlos	IX	SF, SM, GC		Rotational slump/avalanche ($V < 0.1\times10^6$ m^3)
12	Guatemala City	IX	SF, SM, GC	VII-VIII	Secondary faults = 20 km long
13	Los Choloyos	IX	SM, L, GC	VII-VIII	Block slide/rock-fall avalanche - lateral spreads ($0.75 \times 10^6 < V < 10^6$ m^3)
14	Mixco Area	IX	SF, GC, SM	VII-VIII	Secondary faults = 20 km long
15	Rio Blanco	IX	SM, GC	VII	Complex rock-fall, avalanche ($V < 0.2\times10^6$ m^3)
16	Rio Cotzibal	IX	SM, GC	VIII	Rotational slump ($V < 0.5\times10^6$ m^3)
17	Rio Los Cubes	IX	SM, GC, TL	VIII	Rock-fall avalanche ($V < 0.1\times10^6$ m^3), Temp. Lake
18	Rio Naranjo	IX	SM, GC	VIII	Rotational slump ($V < 0.3\times10^6$ m^3)
19	Rio Polima	IX	SM, GC	VIII	Block slides ($V < 0.2\times10^6$ m^3); Temp. Lake
20	Rio Ruyalchè	IX	SM, GC	VIII	Rotational slump ($V < 0.5\times10^6$ m^3)
21	Rio Teocinte	IV	SM, GC	VIII	Rotational slump, rock-fall avalanche ($0.3 \times 10^6 < V < 0.5\times10^6$ m^3)
22	Lake Amatitlan (La Playa, El Sauza)	VIII	SM, L, GC, GS	V (VII)	Lateral spreads < 1 km; subsidence 1 m; ground cracks > 100 cm
23	Lake Atitlan (Panajachel)	VII	SM, L, GC, GS	V (VI)	Lateral spreads, small subaqueous landslides, subsidence 1 m
24	Los Amates	XI			Epicenter

° Type of effects: SF, Surface Faulting; SM, Slope Movements (V volume in m^3); GC, Ground Cracks; L, Liquefaction phenomena; GS, Ground Settlements; TL, temporary lake.

The data were analyzed and converted into an intensity degree of the ESI scale taking into account the surface faulting, the total length of the fault segment, and the maximum displacement observed, and others secondary co-seismic effects (e.g., landslides, ground deformation) (Figure 4). The maximum intensity value of the ESI scale was attributed to Cabanas, Chuarrancho, El Progreso, Gualan, Quebradas, and Subinal cities, where the damage and the ground effects were extremely intense and could be appropriately described by XI-ESI degree. The secondary effects (mostly landslides) were evaluated in terms of areal extent involved and total volume displaced by slope instability processes [6,22].

Following this assumption, the towns of Estancia de la Virgen, San Martin Jilotepeque, and San Josè Poaquil were characterized by the degree X-ESI, due to the large earthquake-induced landslide that involved a total volume higher than 10E+6 m^3. Other towns, located in the vicinity of the fault (Guatemala City and Mixco, San Martin Jilotepeque, Finca San Carlos, Los Choloyos, Rio Blanco, Rio Cotzibal, Rio Los Cubes, Rio Naranjio, Rio Polima, Rio Ruyalchè, and Rio Teocinte) reached intensity IX-ESI.

A degree VIII-ESI was attributed to the La Playa and El Sauza villages, located approximately 185 km away from the epicenter and 40 km from the Motagua fault, where lateral spreads phenomena affecting a series of houses and settlements were recorded. Outside Guatemala, in El Salvador, liquefaction phenomena were observed in the Ilopango lake (about 158 km from the epicenter) and a VII grade-ESI was considered.

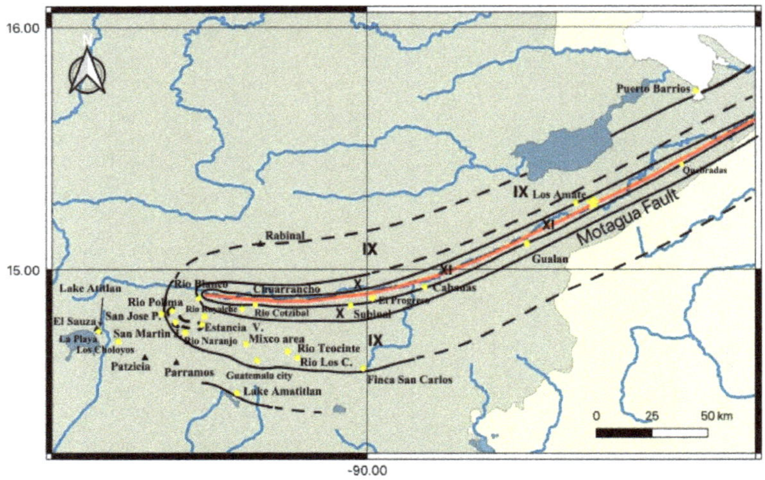

Figure 4. The 4 February 1976 Guatemala earthquake: Isoseismal lines based on the ESI scale [46].

3.2. PSHA and DSHA

A preliminary test for the setup of the Fortran code used in this study was performed by considering all the seismic zones in the area of interest, namely, associated with crustal, interplate and in-slab seismicity. The parameters and GMPEs combination used in this test was selected following the approach of Benito [43]. Hazard curves were calculated using a set of 200 log-spaced acceleration (A_0) values from 1.0E−4 to 2.4 g, to better define the shape of the curves. The results are reported in Figure 5 for the 24 target sites, together with the curves calculated using only the crustal SZ and using only the SZ denoted as "D3_G6".

It can be observed (Figure 5) that the hazard at the 24 sites is controlled by the crustal SZ given the remarkable overlap between the curves obtained by crustal SZ (black lines, Figure 5) and the curves obtained by the crustal, in-slab and intraplate SZs (red lines, Figure 5). The difference in the annual frequency of exceedance for each acceleration value A_0 yielded a maximum residual between the curves of 1.9E−3. This result simplifies the code setup, allowing for the reduction of the input parameters for the hazard computation, by selecting only the crustal SZ. Particularly, only the crustal SZs within a radius of 300 km from each grid knot were considered, consistent with the applicability limits associated with the selected GMPEs [44,45]. The selected crustal SZs is reported in Table 2 with the associated parameters.

The removal of the interplate and in-slab SZ from the hazard computation and the consideration of the GMPE only for the crustal seismicity area proposed by Benito [43–45] also resulted in a simplification of the computation procedure.

In Figures 6 and 7, different maps obtained by PSHA and deterministic approaches at rock and soil sites were reported, respectively. The probabilistic maps were obtained considering only the crustal SZ or the "D3_G6" SZ and the combination of Climent [44] and Zhao [45] GMPEs evaluated at rock and soil site conditions (Figures 6 and 7, respectively).

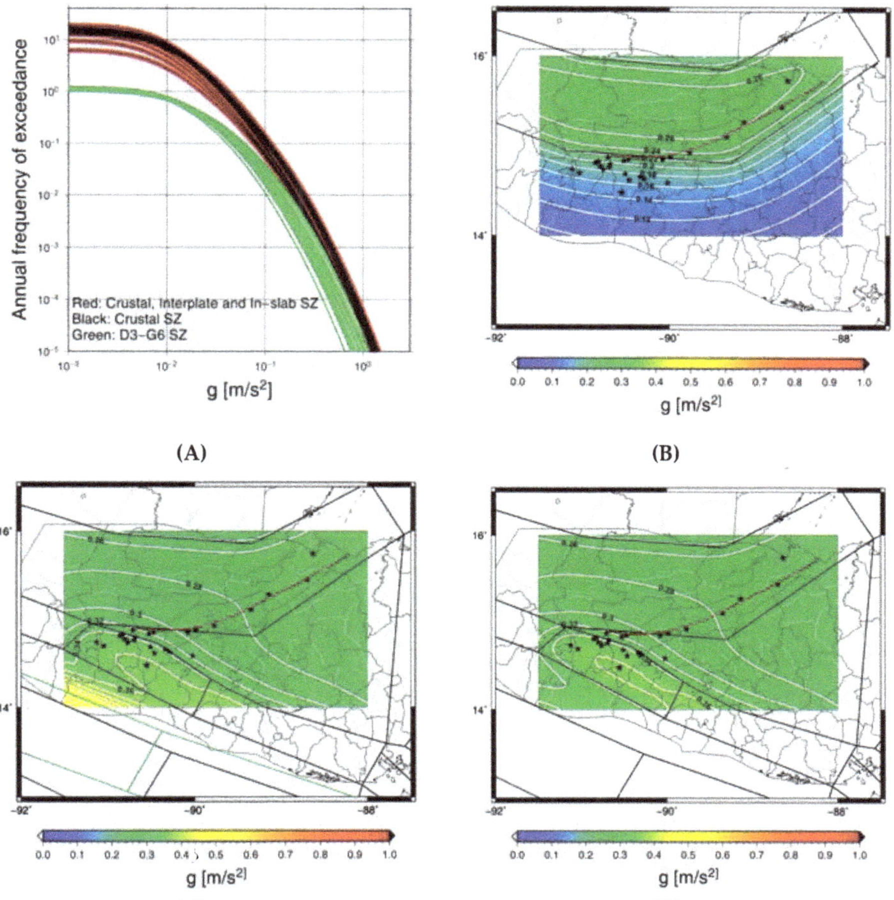

Figure 5. Evaluation of SZ contribution to the hazard. (**A**) Hazard curves for the 24 target sites. Hazard map for T_R = 300 years at rock ground considering only the "D3_G6" SZ (**B**); considering all the SZ (**C**); considering the crustal SZ (**D**). The stars represent the target sites and the red line the Motagua fault.

Figures 6A and 7A report the hazard map for T_R = 300 years obtained for all the crustal seismic zonation in a grid neighborhood of 300 km. In this case the range of PGA values for the 24 targets site is 0.305–0.422 *g* for the soil and 0.229–0.317 *g* at rock. Moreover, as first approach, only the seismogenic zone including the Motagua fault for T_R = 300 years (Figures 6B and 7B) were considered, to better compare the probabilistic and deterministic acceleration values with the intensity values. In this case, the PGA range is 0.106–0.223 *g* for rock and 0.143–0.298 *g* for soil. Finally, the total fault length (Motagua fault—red line in Figure 1, about 230 km), was used as an extended seismic source to compute the acceleration map based on GMPE PGA estimation (Figures 6C and 7C). In the last case, PGA values range from 0.125 to 0.619 *g*, which correspond to values higher than the ones obtained by the probabilistic approach. Such a gap, however, is not uncommon as a first approximation. Every point of the extended seismic source, for a length of 230 km, was considered as a point source of a M = 7.5 event.

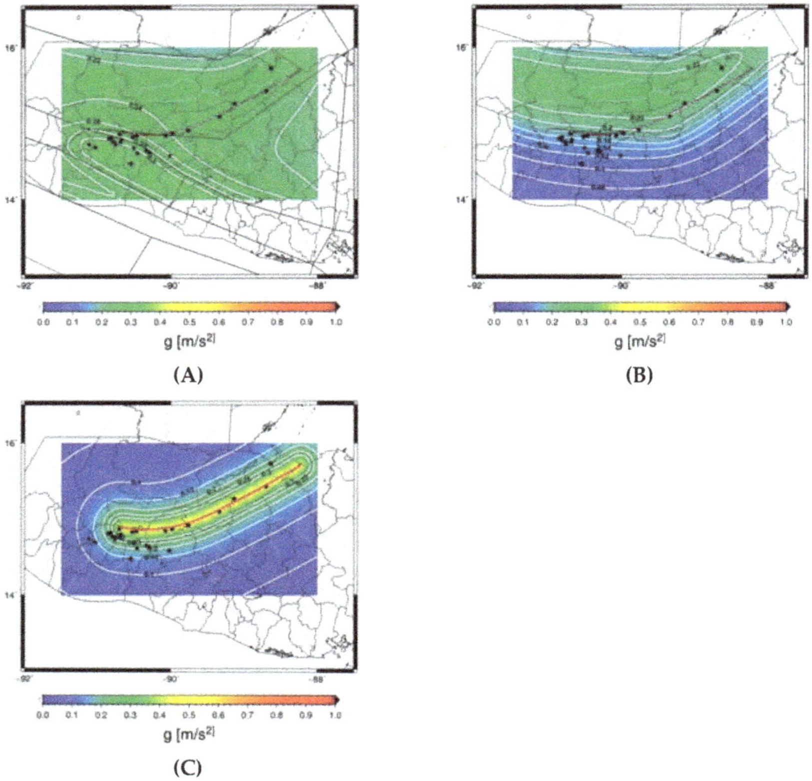

Figure 6. Hazard maps for rock ground and 300 years return period for different configurations: (**A**) Using crustal SZ; (**B**) using only the SZ "D3_G6"; (**C**) deterministic PGA distribution calculated from Motagua fault (red line) for M = 7.5. The black stars represent the target sites.

In Table 4, the intensity (ESI and MMI) and PGA values for 24 different locations in the area for rock and soil site conditions are compared. For the site effects, we reported the PGA interval to better highlight the possible variability of results. PGA values display a significant variability between rock and soil solutions, indicating the complexity of the site effect characterization. For instance, sites yielding values of XI in the ESI scale showed a Deterministic PGA higher than 0.6 *g* whereas the Probabilistic values reach a maximum value of 0.422 *g*. However, other sites, such as Rio Blanco and Rio Cotzipal sites, reached higher Deterministic values (around 0.6 *g*) for lower ESI scale values (IX). In the case of the Puerto Barrios site, there is a general agreement between the ESI scale (IX) and MMI (IX) and both Probabilistic and Deterministic PGA values (0.173–0.365 *g* and 0.224–0.295 *g*, respectively), whereas at Lake Atitlan site (ESI scale and MMI VII and VII, respectively) the Probabilistic and Deterministic PGA values are quite different (0.175–0.352 *g* and 0.125–0.166 *g*, respectively). The apparent discrepancy between the macroseismic and seismic hazard approaches could be reconciled by considering more realistic characterization site effects that may take into account additional factors, including topographic effects, seismic waves amplification as a function of frequency domain, source directivity, and focal mechanism [47].

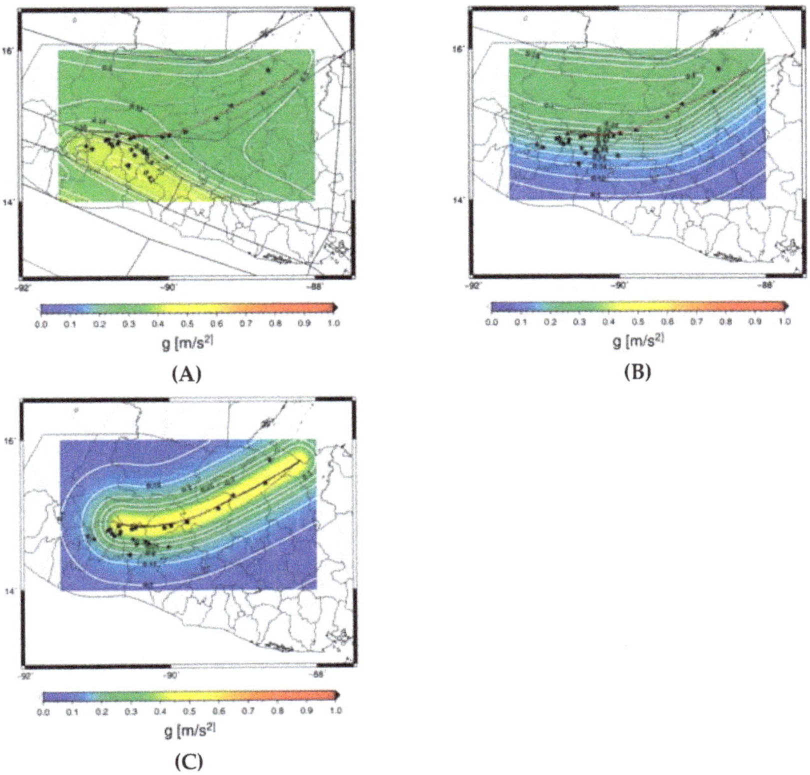

Figure 7. Hazard maps for soil ground and 300 years return period for different configurations: (**A**) Using crustal SZ; (**B**) using only the SZ "D3_G6"; (**C**) deterministic PGA distribution calculated from the Motagua fault (red line) for M=7.5. The black stars represent the target sites.

Table 4. Localities affected by EEE (target sites) with the macroseismic evaluation according to the ESI scale, MMI scales, probabilistic PGA range at rock/soil site for crustal and "D3_G6" SZ and Deterministic Acceleration for rock/soil site.

ID	Town	ESI	MMI	Probabilistic Rock/Soil Crustal SZ [*g*]	Probabilistic Rock/Soil "D3_G6" SZ [*g*]	Deterministic Rock-Soil [*g*]
1	Cabanas	XI	VIII	0.232–0.311	0.223–0.298	0.460–0.608
2	Chuarrancho	XI	VIII	0.308–0.411	0.131–0.176	0.464–0.614
3	El Progreso	XI	VIII	0.317–0.422	0.106–0.143	0.464–0.613
4	Gualan	XI	IX	0.289–0.386	0.131–0.176	0.463–0.612
5	Quebradas	XI	VIII	0.289–0.386	0.154–0.206	0.468–0.618
6	Subinal	XI	VIII	0.292–0.390	0.157–0.210	0.452–0.598
7	Estancia della Virgen	X	VIII	0.306–0.408	0.139–0.186	0.301–0.397
8	San Martin Jilotepeque	X	VIII	0.289–0.385	0.127–0.172	0.255–0.336
9	San Jose Poaquil	X	VIII	0.270–0.361	0.165–0.221	0.235–0.310
10	Puerto Barrios	IX	IX	0.273–0.365	0.173–0.232	0.224–0.295
11	Finca San Carlos	IX	VIII	0.300–0.400	0.134–0.180	0.174–0.230
12	Guatemala City	IX	VIII	0.311–0.415	0.126–0.169	0.198–0.262
13	Los Choloyos	IX	VIII	0.307–0.410	0.123–0.166	0.139–0.184
14	Mixco	X	VIII	0.262–0.350	0.123–0.166	0.241–0.318
15	Rio Blanco	IX	VII	0.229–0.306	0.223–0.298	0.456–0.604
16	Rio Cotzibal	IX	VIII	0.297–0.397	0.152–0.205	0.450–0.596
17	Rio Los Cubes	IX	VIII	0.301–0.402	0.146–0.196	0.210–0.277
18	Rio Naranjo	IX	VIII	0.297–0.396	0.146–0.196	0.248–0.327

Table 4. *Cont.*

ID	Town	ESI	MMI	Probabilistic Rock/Soil Crustal SZ [g]	Probabilistic Rock/Soil "D3_G6" SZ [g]	Deterministic Rock-Soil [g]
19	Rio Polima	IX	VIII	0.243–0.325	0.178–0.238	0.272–0.358
20	Rio Ruyalche	IX	VIII	0.230–0.307	0.215–0.287	0.351–0.463
21	Rio Teocinte	IX	VIII	0.233–0.311	0.219–0.292	0.225–0.297
22	Lake Amatitlan	VIII	VII	0.239–0.320	0.187–0.251	0.143–0.189
23	Lake Atitlan	VII	VII	0.264–0.352	0.171–0.228	0.125–0.166
24	Los Amate	XI	IX	0.235–0.314	0.202–0.270	0.458–0.606

The analysis of the Guatemala 1976 event presented in this study indicates that different approaches used to define and/or estimate an earthquake may lead to different results. For example, the ESI scale, which takes in account the environments effects generated by an earthquake, often provides intensity values in the order of two degrees higher than previous macroseimic intensity evaluation. On the other hand, the probabilistic or deterministic approaches, which utilize mathematical/statistical models to approximate the behavior of complex systems, sometimes yield unrealistic estimations.

4. Conclusions

In this study different methodologies have been used to evaluate the same seismic event by comparing the earthquake environmental effects expressed in terms of ESI scale and the hazard maps calculated with probabilistic and deterministic analyses. Advantages and drawbacks of each individual approach were also highlighted. The 1976 Guatemala earthquake was used as a test case to compare different strategies to define how strong an event could be (SHA approaches) or was (MMI and ESI). A multidisciplinary approach based on the combination of geological, geophysics, seismological, and statistical methods was tested to analyze and evaluate the seismic hazard.

All the available data were re-interpreted in the light of the environmental effects recorded in the area struck by the earthquake. An ad-hoc Fortran code was used to resolve the hazard integral and calculate the hazard curve and maps with the possibility of customizing and controlling each step of the process.

The re-evaluation of the 1976 Guatemala event based on the ESI scale resulted in the definition of a new macroseismic field. The general pattern obtained for the ESI scale isoseismal lines displays an ENE-WSW orientation that is compatible with the fault mechanism and geographic distribution of ground effects. A set of probabilistic hazard maps for rock and soil ground sites were calculated using a selection of GMPEs, each producing a different geographic distribution. Particularly, a basic isotropic geometry is the result from the DSHA and a more complex pattern is provided by the PSHA as a function of the different SZs included in the analysis. The results of the above SHA are also reflected by the PGA values that exhibit a high variability.

This research showed that different approaches may produce significantly diverging results. For example, the ESI scale, which takes in account the ground effects generated by an earthquake, yielded intensity values up to two degree higher than the previous macroseismic classification. On the other hand, the probabilistic or deterministic approaches use mathematical/statistical models that may oversimplify the behavior of complex systems, often underestimating the severity of a seismic event.

Our study confirms the importance of collecting primary and secondary EEE in order to define the nature of an earthquake, and proposes a tool for a more reliable evaluation of the epicentral intensity.

Author Contributions: Conceptualization, investigation, data curation, writing, software—original draft preparation M.C.; investigation, writing—review and editing S.P. and E.S.; writing—review and editing, funding acquisition M.S.

Funding: This research received no external funding.

Conflicts of Interest: The authors declare no conflict of interest. The funders had no role in the design of the study; in the collection, analyses, or interpretation of data; in the writing of the manuscript, or in the decision to publish the results.

References

1. Power, M.; Chiou, C.; Abrahamson, N.; Bozorgnia, Y.; Shantz, T.; Roblee, C. An Overview of the NGA Project. *Earthq. Spectra* **2008**, *24*, 3–21. [CrossRef]
2. Michetti, A.M.; Esposito, E.; Guerrieri, L.; Porfido, S.; Serva, L.; Tatevossian, R.; Vittori, E.; Audemard, F.; Azuma, T.; Clague, J.; et al. Intensity Scale ESI 2007. In *Memorie Descrittive Della Carta Geologica D'Italia*; ISPRA Istituto Superiore per la Protezione e la Ricerca Ambientale: Rome, Italy, 2007; Volume 74, p. 53.
3. Serva, L.; Blumetti, A.M.; Esposito, E.; Guerrieri, L.; Michetti, A.M.; Okumura, K.; Porfido, S.; Reicherter, K.; Silva, P.G.; Vittori, E. Earthquake Environmental Effects, intensity and seismic hazard assessment: The lesson of some recent large earthquakes. In *Memorie Descrittive Della Carta Geologica D'Italia*; ISPRA Istituto Superiore per la Protezione e la Ricerca Ambientale: Rome, Italy, 2015.
4. Espinosa, A.F. (Ed.) *The Guatemalan Earthquake of February 4, 1976, a Preliminary Report*; U.S. Government Printing Office: Washington, DC, USA, 1976.
5. Harp, E.L.; Keefer, D.K.; Sato, H.P.; Yagi, H. Landslide inventories: The essential part of seismic landslide hazard analyses. *Eng. Geol.* **2011**, *122*, 9–21. [CrossRef]
6. Porfido, S.; Esposito, E.; Sacchi, M.; Molisso, F.; Mazzola, S. Impact of Ground Effects for an appropriate mitigation strategy in seismic area: The example of Guatemala 1976 earthquake. *Eng. Geol. Soc. Territ.* **2015**, *2*. [CrossRef]
7. Harp, E.L.; Wilson, R.C.; Wieczorek, G.F. *Landslides from the February 4, 1976, Guatemala Earthquake*; U.S. Geological Survey: Leston, VA, USA, 1981.
8. Plafker, G.; Bonilla, M.G.; Bonis, S.B. *The Guatemalan Earthquake of February 4, 1976*; U.S. Geological Survey: Leston, VA, USA, 1976; pp. 38–51.
9. Hoose, S.N.; Wilson, R.C.; Rosenfeld, J.H. *Liquefaction-Caused Ground Failure during the February 4, 1976, Guatemala Earthquake*; U.S. Geological Survey: Leston, VA, USA, 1978.
10. Plafker, G. Tectonic aspects of the Guatemala 4 February 1976. *Science* **1976**, *193*, 1201–1208. [CrossRef] [PubMed]
11. Hayes, G.P.; Myers, E.K.; Dewey, J.W.; Briggs, R.W.; Earle, P.S.; Benz, H.M.; Smoczyk, G.M.; Flamme, H.E.; Barnhart, W.D.; Gold, R.D.; et al. *Tectonic Summaries of Magnitude 7 and Greater Earthquakes from 2000 to 2015*; U.S. Geological Survey Open-File Report 2016–1192; U.S. Geological Survey: Leston, VA, USA, 2017.
12. Benz, H.M.; Tarr, A.C.; Hayes, G.P.; Villaseñor, A.; Furlong, K.P.; Dart, R.L.; Rhea, S. *Seismicity of the Earth 1900–2010 Caribbean Plate and Vicinity*; U.S. Geological Survey Open-File Report 2010–1083-A, Scale 1:8,000,000; U.S. Geological Survey: Leston, VA, USA, 2011.
13. Benito Oterino, M.B.; Torres Fernández, Y. (Eds.) *Amenaza sísmica en América Central*; Entimema: Madrid, Spain, 2009; p. 371.
14. Person, W.; Spence, W.; Dewey, J.W. Main event and principal aftershocks from teleseismic data. In *The Guatemala Earthquake of February 4, 1976*; Espinosa, A.F., Ed.; U.S. Geological Survey: Leston, VA, USA, 1976.
15. Matumoto, T.; Latham, G.V. Aftershocks of the Guatemalan Earthquake of February 4, 1976. *Geophys. Res. Lett.* **1976**, *3*, 599–602. [CrossRef]
16. Ambraseys, N.N.; Adams, R.D. Large-magnitude Central American earthquakes, 1898–1994. *Geophys. J. Int.* **1996**, *127*, 665–692. [CrossRef]
17. Bommer, J.J.; Rodriguez, C.E. Earthquake-induced landslides in central America. *Eng. Geol.* **2002**, *63*, 189–220. [CrossRef]
18. Peraldo, H.G.; Montero, P.W. *Sismología Histórica de América Central*; IPGH: Mexico City, México, 1999; Volume 513, p. 347.
19. Tanner, J.G.; Shepherd, J.B. *Seismic Hazard in Latin America and the Caribbean, Volume I: Project Catalog and Seismic Hazard Maps*; IRDC: Ottawa, ON, Canada, 1997; p. 143.
20. White, R.A.; Ligorria, J.P.; Cifuentes, I.L. Seismic history of the Middle America subduction zone along El Salvador, Guatemala, and Chiapas, Mexico: 1526–2000. *Spec. Pap. Geol. Soc. Am.* **2004**, *375*, 379–396.
21. Global Historical Earthquake Archive—GHEA. Available online: https://www.emidius.eu/GEH/map.php (accessed on 15 July 2019).
22. Porfido, S.; Esposito, E.; Spiga, E.; Mazzola, S. Application of the ESI Scale: Case study of the February 4, 1976 Guatemala earthquake. In Proceedings of the 3rd INQUA-IGCP-567 International Workshop on Active Tectonics, Paleoseismology and Archaeoseismology, Morelia, Mexico, 19–24 November 2012.

23. Grünthal, G.; Musson, R.; Schwarz, J.; Stucchi, M. *European Macroseismic Scale 1998*; Cahiers du Centre Européen de Geodynamique et de Seismologie: Luxembourg, 1998; Volume 15.
24. Silva, P.G.; Pascua, M.A.R.; Pérez-López, R.; Bardaji, T.; Lario, J.; Alfaro, P.; Martínez-Díaz, J.J.; Reicherter, K.; García, J.G.; Giner, J.; et al. Catalogacion de los efectos geologicos y ambientales de los terremotos en Espana en la Escala ESI 2007 y su aplicacion a los estudions paleosismologicos. *Geotemas* **2008**, *6*, 1063–1066.
25. Reicherter, K.; Michetti, A.M.; Silva, P.G. (Eds.) Paleoseismology: Historical and prehistorical records of earthquake ground effects for seismic hazard as-sessment. *Geol. Soc.* **2009**, *316*, 1–10. [CrossRef]
26. Espinosa, A.F.; Husid, R.; Quesada, A. *Intensity and Source Parameters from Filed Observations. The Guatemalan Earthquake of February 4, 1976*; U.S. Geological Survey: Leston, VA, USA, 2013; pp. 52–66.
27. Papanikolaou, I.D. Uncertainty in intensity assignment and attenuation rela-tionships: How seismic hazard maps can benefit from the implementation of the Environmental Seismic Intensity scale (ESI 2007). *Quat. Int.* **2011**. [CrossRef]
28. Bobrowsky, P.T. (Ed.) *Encyclopedia of Natural Hazards*; Springer: Berlin/Heidelberg, Germany, 2013; ISBN 978-90-481-8699-0.
29. Serva, L. History of the Environmental Seismic Intensity Scale ESI-07. *Geosciences* **2019**, *9*, 210. [CrossRef]
30. Porfido, S.; Nappi, R.; de Lucia, M.; Gaudiosi, G.; Alessio, G.; Guerrieri, L. The ESI scale, an ethical approach to the evaluation of seismic hazards. In Proceedings of the EGU General Assembly 2015, Vienna, Austria, 12–17 April 2015.
31. Chunga, K.; Livio, F.; Mulas, M.; OchoaCornejo, F.; Besenzon, D.; Ferrario, M.F.; Michetti, A.M. Earthquake Ground Effects and Intensity of the 16 April 2016 Mw 7.8 Pedernales, Ecuador, Earthquake: Implications for the Source Characterization of Large Subduction Earthquakes. *Bull. Seism. Soc. Am.* **2018**, *108*, 3384–3397. [CrossRef]
32. Serva, L.; Vittori, E.; Comerci, V.; Esposito, E.; Guerrieri, L.; Michetti, A.M.; Mohammadioun, B.; Mohammadioun, G.C.; Porfido, S.; Tatevossian, R.E. Earthquake Hazard and the Environmental Seismic Intensity (ESI) Scale. *Pure Appl. Geophys.* **2015**, *173*, 1479–1515. [CrossRef]
33. Chunga, K.; Livio, F.A.; Martillo, C.; Lara-Saavedra, H.; Ferrario, M.F.; Zevallos, I.; Michetti, A.M. Landslides Triggered by the 2016 Mw 7.8 Pedernales, Ecuador Earthquake: Correlations with ESI-07 Intensity, Lithology, Slope and PGA-h. *Geosciences* **2019**, *9*, 371. [CrossRef]
34. Silva, P.G.; Rodríguez-Pascua, M.A.; Giner Robles, J.L.; Élez, J.; Pérez-López, R.; Davila, M.B.B. Catalogue of the Geological Effects of Earthquakes in Spain Based on the ESI-07 Macroseismic Scale: A New Database for Seismic Hazard Analysis. *Geosciences* **2019**, *9*, 334. [CrossRef]
35. Nappi, R.; Gaudiosi, G.; Alessio, G.; De Lucia, M.; Porfido, S. The environmental effects of the 1743 Salento earthquake (Apulia, southern Italy): A contribution to seismic hazard assessment of the Salento Peninsula. *Nat. Hazards* **2017**, *86*, 295–324. [CrossRef]
36. Audemard, F.; Azuma, T.; Baiocco, F.; Baize, S.; Blumetti, A.M.; Brustia, E.; Clague, J.; Comerci, V.; Esposito, E.; Guerrieri, L.; et al. Earthquake Environmental Effect for seismic hazard assessment: the ESI intensity scale and the EEE Catalogue. *Mem. Descr. Carta Geol. D'Italia* **2015**, *97*, 1–181.
37. Mavroulis, S.D.; Fountoulis, I.G.; Skourtsos, E.N.; Lekkas, E.; Papanikolaou, I.D. Seismic intensity assignments for the 2008 Andravida (NW Peloponnese, Greece) strike-slip event (June 8, Mw = 6.4) based on the application of the Environmental Seismic Intensity scale (ESI 2007) and the European Macroseismic scale (EMS-98), Geological structure, active tectonics, earthquake environmental effects and damage pattern. *Ann. Geophys.* **2013**, *56*, S0681.
38. Mosquera-Machado, S.; Lalinde-Pulido, C.; Salcedo-Huarto, E.; Michetti, A.M. Ground effects of the October 18, 1992, Murindo Earthquake (NW Colombia), using the Environmental Seismic Intensity Scale (ESI 2007) for the assessment of the intensity. *Geol. Soc.* **2009**, *316*, 123–144. [CrossRef]
39. Lekkas, E.L. The 12 May 2008 Mw 7.9 Wenchuan, China, earthquake: Macroseismic intensity assessment using the EMS-98 and ESI 2007 Scales and their correlation with the geological structure. *Bull. Seism. Soc. Am.* **2010**, *100*, 2791–2804. [CrossRef]
40. Cornell, C.A. Engineering seismic risk analysis. *Bull. Seism. Soc. Am.* **1968**, *58*, 1583–1606.
41. Gutenberg, B.; Richter, C.R. Frequency of earthquakes in California. *Bull. Seism. Soc. Am.* **1944**, *34*, 185–188.
42. McGuire, R.K.; Arabasz, W.J. An introduction to probabilistic seismic hazard analysis. In *Geotechnical and Environmental Geophysics, Society of Exploration Geophysicists*; Ward, S.H., Ed.; Society of Exploration Geophysics: Tulsa, OK, USA, 1990; Volume 1, pp. 333–353.

43. Benito, M.B.; Lindholm, C.; Camacho, E.; Climent, A.; Marroquin, G.; Molina, E.; Rojas, W.; Escobar, J.J.; Talavera, E.; Alvarado, G.E.; et al. A New Evaluation of Seismic Hazard for the Central America Region. *Bull. Seism. Soc. Am.* **2012**, *102*, 504–523. [CrossRef]
44. Climent, Á.; Taylor, W.; Real, M.C.; Strauch, W.; Villagran, M.; Dahle, A.; Bungum, H. *Spectral Strong Motion Attenuation in Central America*; Internal NORSAR Technical Report: Kjeller, Norway, No. 2–17; 1994; p. 46.
45. Zhao, J.X.; Zhang, J.; Asano, A.; Ohno, Y.; Oouchi, T.; Takahashi, T.; Ogawa, H.; Irikura, K.; Thio, H.K.; Somerville, P.G. Attenuation relations of strong ground motion in Japan using site classification based on predominant period. *Bull. Seism. Soc. Am.* **2006**, *96*, 898–913. [CrossRef]
46. Porfido, S.; Caccavale, M.; Spiga, E.; Sacchi, M. The 1976 Guatemala earthquake revised. Macroseismic data for an appropriate seismic hazard assessment. In Proceedings of the Geo-Risks in the Mediterranean and their Mitigation, Msida, Malta, 20–21 July 2015.
47. Steven, K.L. *Geotechinical Earthquake Engineering*; Prentice-Hall International Series; Prentice-Hall: Upper Saddle River, NJ, USA, 1996.

geosciences

MDPI

Article

Geological and Structural Control on Localized Ground Effects within the Heunghae Basin during the Pohang Earthquake (M_W 5.4, 15th November 2017), South Korea

Sambit Prasanajit Naik [1,2], Young-Seog Kim [1,2,*], Taehyung Kim [2] and Jeong Su-Ho [2]

[1] Active Fault and Earthquake Hazard Mitigation Research Center, Pukyong National University, Busan 48513, Korea; sambitnaik@gmail.com

[2] Geological Structure and Geohazard Research Laboratory, Department of Earth and Environmental Sciences, Pukyong National University, Busan 48513, Korea; goth4453@naver.com (T.K.); jsh1212@pknu.ac.kr (J.S.-H.)

* Correspondence: ysk7909@pknu.ac.kr; Tel.: +82-051-629-7909

Received: 29 January 2019; Accepted: 8 April 2019; Published: 16 April 2019

Abstract: On 15th November 2017, the Pohang earthquake (M_w 5.4) had strong ground shaking that caused severe liquefaction and lateral spreading across the Heunghae Basin, around Pohang city, South Korea. Such liquefaction is a rare phenomenon during small or moderate earthquakes (M_W < 5.5). There are only a few examples around the globe, but more so in the Korean Peninsula. In this paper, we present the results of a systematic survey of the secondary ground effects—i.e., soil liquefaction and ground cracks—developed during the earthquake. Most of the liquefaction sites are clustered near the epicenter and close to the Heunghae fault. Based on the geology, tectonic setting, distribution, and clustering of the sand boils along the southern part of the Heunghae Basin, we propose a geological model, suggesting that the Heunghae fault may have acted as a barrier to the propagation of seismic waves. Other factors like the mountain basin effect and/or amplification of seismic waves by a blind thrust fault could play an important role. Liquefaction phenomenon associated with the 2017 Pohang earthquake emphasizes that there is an urgent need of liquefaction potential mapping for the Pohang city and other areas with a similar geological setting. In areas underlain by extensive unconsolidated basin fill sediments—where the records of past earthquakes are exiguous or indistinct and there is poor implementation of building codes—future earthquakes of similar or larger magnitude as the Pohang earthquake are likely to occur again. Therefore, this represents a hazard that may cause significant societal and economic threats in the future.

Keywords: Pohang earthquake; South Korea; ground effects; liquefaction; geological control; fault barrier; seismic hazard

1. Introduction

Most of the damages that occurred during an earthquake across the sedimentary basins around the world are mainly due to strong shaking and amplification of soft sediments, which are commonly associated with earthquakes of magnitudes (M) more than 6 [1–9]. The more recent 2015 Nepal earthquake or the 2018 Palu earthquake in Indonesia are good examples of such phenomenon. Post-earthquake field survey and its documentation can help us understand the cause and mechanism involved, as well as future hazard predictions [6]. In the past and in recent history, several studies [10–23] have been conducted to understand the mechanism and factors controlling the liquefaction of related hazards. Almost all of these studies are associated with large magnitude earthquakes. However, the liquefaction and associated damages due to moderate to small magnitude earthquakes like the 2009 Olancha earthquake in the United States (M_w 5.2), 2011 Yunnan earthquake in China (M_w 5.4),

2011 Mineral Virginia earthquake in Canada (M_w 5.8), 2012 Emilia earthquake in Italy (M_w 6.1), 2011 Sikkim earthquake in India (M_w 6.9), 2016 Manipur earthquake in India (M_w 6.7), and the recent 2017 Pohang earthquake in South Korea (M_w 5.5) indicate that moderate earthquakes need a similar kind of attention as large magnitude earthquakes in terms of seismic hazard.

Actually, reporting on the phenomena and processes involved for the liquefactions associated with earthquakes of $M \leq 5.5$ is only four to five cases so far [24–27]. Thus, the ability and exposure towards understanding of the causes of liquefaction during $M_w \leq 5.5$ earthquake from a geological perspective still remains poor, especially in the Korean Peninsula, where this is the first ever recorded liquefaction phenomena in the history of instrumental seismicity in South Korea.

The Korean Peninsula has been considered to be seismically stable in comparison to neighboring countries such as Japan and Taiwan [28]. Thus, the earthquake was a shock even to the Korean geologists as well as to the public. During the M_w 5.4 Pohang earthquake on 15th November 2017, a widespread liquefaction was observed in and around the Pohang area, especially around the Heunghae Basin composed of recent alluvial sediments. The liquefaction caused ground failures such as sand boils and cracks; it also damaged the engineering structures like school buildings and residential apartments. The documentation of liquefaction features may be useful from both geological and geotechnical point of view in terms of seismic hazards [6].

Internationally adopted empirical procedures for the prediction of liquefaction potential are completely based on field observations from large magnitude earthquakes, except in one case of less than $M_w \leq 6$ [2,4,29,30]. This approach may overestimate or underestimate the hazards posed by a potential liquefaction. Therefore, proper liquefaction analyses associated with moderate earthquakes like the Pohang or Olancha earthquakes may help to properly estimate the potential liquefaction hazard in areas with similar geological settings. Most of the liquefaction potential analyses that have been conducted in Korea [31–33] are mainly based on these empirical relationships and considering earthquake scenario of M_W 6.5. Thus, it is necessary to check the reliability of the procedures used so far and to propose a better liquefaction boundary curve for medium size earthquakes in Korea.

In this paper, we are presenting field evidences of secondary ground effects [34] i.e., mainly liquefaction and its localized distribution within the Heunghae basin. Based on the analyzed data, we propose a possible mechanism from the geological perspective and a conceptual model with geological structures to explain the localized sand boil occurrences. In addition, we have discussed the role of geological factors such as the fault barrier zone and mountain basin effect, which is helpful for future liquefaction hazard mapping in the study area and similar geological settings around the world.

2. General Characteristics of the Pohang Earthquake on 15th November 2017

On 15th November 2017, a moderate earthquake (M_W 5.4) occurred in Pohang (36.065° N, 129.269° E) at 2:29 p.m. (Local time). The focal depth was about 4–6 km and associated with a NE-SW striking reverse dominant oblique-slip fault (Figure 1). The main shock was followed by more than 65 aftershocks of magnitude ranging from M_W 2 to M_W 4.6. The Peak Ground Acceleration (PGA) measured by the nearby seismic station was about 0.58 g, which is very large and almost equivalent to Modified Mercalli intensity (MMI) of VIII to IX (Korean Meteorological Administration (KMA) report [35]). Due to Pohang City's poor subsoil condition, the seismic wave was amplified while passing through the alluvial soil, making the damage somewhat heavier than the previous Gyeongju earthquakes (M_L 5.8; 12th September 2016). There is also no permanent seismic station within the Heunghae basin. Other seismic stations show the following PGA values: Pohang 0.29 g, 10 km from the epicenter; Deokjeong-ri 0.035 g, 29.3 km from the epicenter; and Cheongsong 0.042 g, 49 km away from the epicenter. Because there are no permanent stations close to the epicenter, we have analyzed the temporary stations installed by KMA around the epicenter area after the Main shock to analyze the Horizontal and Vertical Spectrum ratio (HVSR) and the subsoil characteristics. The lower shear wave velocity (120 m/S to 275 m/S, [35]) and higher frequency indicates higher amplification within the Heunghae basin (See Figure S1, Table S1; modified from [35]).

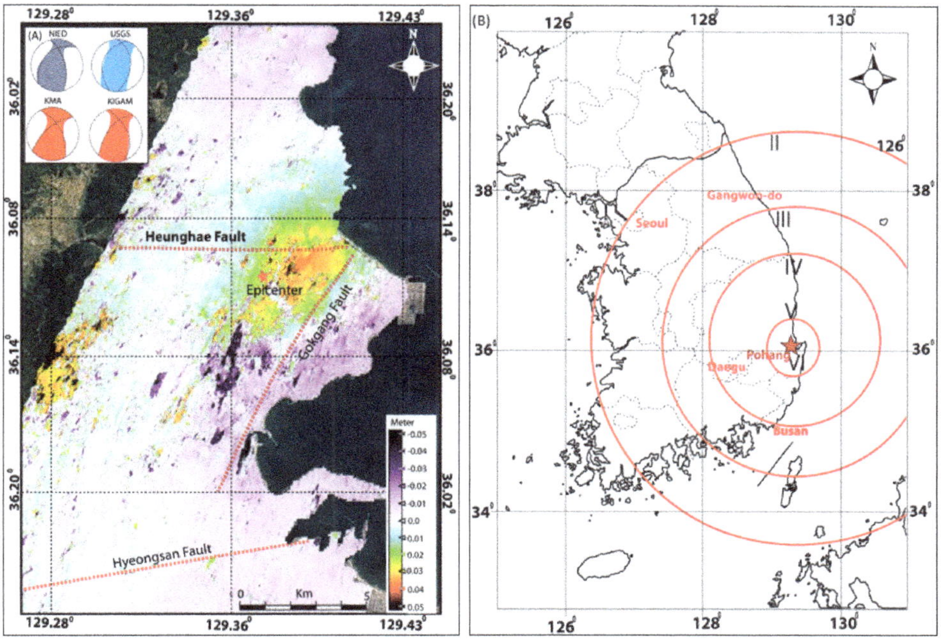

Figure 1. (**A**) InSAR image showing the deformation (~3–4 cm uplift) in the epicenter area, previously reported faults and the regional stress condition with focal mechanism (National Research Institute for Earth Science and Disaster Resilience (NIED), United States Geological Survey (USGS), Korean Meteorological Administration (KMA), and Korea Institute of Geoscience and Mineral Resources (KIGAM). (**B**) General Modified Mercalli intensity (MMI) distribution of Pohang earthquake, according to USGS. KMA suggested intensity of VIII to IX in the epicenter area.

Due to its epicenter within the basin, shallow depth (5 km) and the presence of thick alluvial soil, the earthquake caused severe amplification of the seismic wave and caused extensive damages around the epicenter area. According to the Ministry of the Interior and Safety of South Korea, the Pohang earthquake is the most damaging earthquake in the Korean Peninsula since 1978, which caused injuries to more than 90 people and estimated property damage was about USD 52 million [35]. The earthquake caused damages to 2165 private houses, 227 school buildings, many roads, and 11 bridges.

Although no primary surface rupture was reported during the Pohang earthquake [36], numerous liquefactions and related phenomena were witnessed by local people such as differential settlement of buildings and lateral spreading [36]. Because the liquefaction phenomenon occurred during this earthquake is the first reporting since 1978, it can provide useful information of liquefaction associated with paleo-earthquakes and future earthquakes in the areas of similar geological settings around the Korean Peninsula. It is also a good opportunity for the geological community and geotechnical engineers to consider liquefaction in the seismic hazard assessment for such basins in the near future. Since liquefaction is a rare phenomenon in the Korean Peninsula, understanding the factors (like geological structures) controlling the distribution of liquefaction zone should be taken into consideration for liquefaction zonation mapping.

3. Geological, Geomorphological, and Tectonic Setting of the Area

The Korean peninsula lies on the margin of the Eurasian Plate. During the opening of the East Sea (Japan Sea) (30–15 million years ago), several back-arc basins including Pohang Basin were developed in southeast Korea and adjacent offshore areas [28,37,38]. These basins are bounded by several NNE~NNW-striking strike-slip faults and NNE-NE-striking normal faults. Some of these faults have been reactivated as strike-slip or thrust faults due to the ongoing compression of the Korean mainland [28,37,38].

The Pohang basin is composed of middle Miocene non-marine to deep marine sedimentary deposits of (~20 million years ago) up to ~200–400 m deep. The middle Miocene sedimentary deposits is covered at surface by the recent alluvial deposits (<10 m). The basement of the Pohang basin is mainly composed of Cretaceous to Eocene sedimentary and volcanic rocks ~1000 m thick (Figure 2) [39–41]. The previous studies [42–45] suggested that Pohang basin is bounded by several normal faults and transfer faults (Figure 2) [38,39]. These normal faults divided the Pahang basin into several small sub basins. The epicentral area which is known as Heunghae basin is a small sub-basin within the Pohang Basin [42–45] and mainly composed of fluvial deposits.

Recent paleoseismic studies suggested Pohang basin is having seismic threats from the Yangsan fault which one of the prominent dextral strike slip fault around the study area [36] (Figure 2). A recent study reported Paleo-liquefaction features in epicenter area of the Pohang earthquake [40], which indicates that the area has undergone repeated liquefactions. Therefore, it is necessary to explore the role of regional structures in liquefaction phenomenon and distribution characteristics.

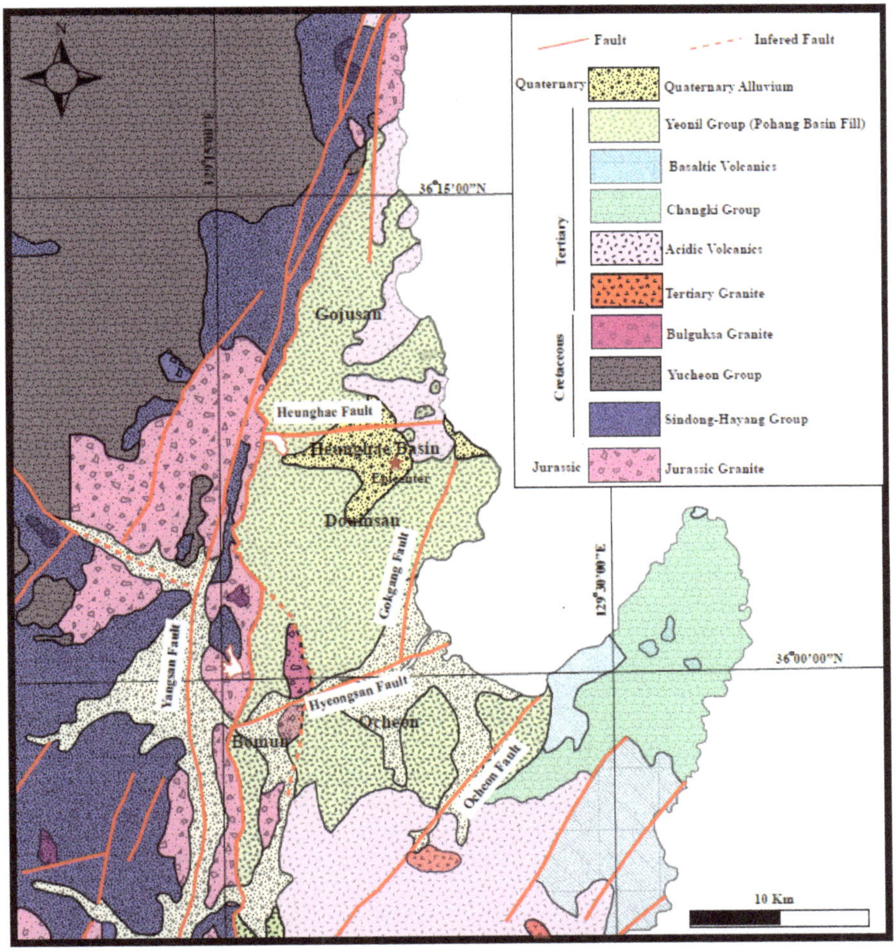

Figure 2. Geological and tectonic map of the study area showing major geological structures and lithological units (Modified from [37]).

4. Coseismic Damages Within the Meizoseismal Area

4.1. Mapping of Liquefaction and Related Features

Major striking features associated with the Pohang earthquake were the occurrence of severe liquefaction, structural damages including residential and academic buildings, as well as lateral spreading/ground cracks. Because this is the first earthquake related liquefaction in the modern seismic history of Korea, the field survey and mechanism involved will help to guide us for the future liquefaction hazard analysis and assessment [36]. For the detailed documentation of ground deformation features (sand boils, cracks, damaged buildings, etc.), we have used Pleiades high resolution satellite images with a spatial resolution of 0.5 m, which was taken one day after the earthquake and provided by the International Charter [36]. The data collected from the Pleiades satellite were complemented by a five-week field survey and Unmanned Air Vehicle (UAV) images taken by Phantom 4 UAV images. The high-resolution UAV images taken from the Heunghae basin were merged using Pix4D software, and high-resolution Digital Surface Model (DSM) of ~2.5 cm resolution has been constructed using ArcGIS 10.0.

During the field work, we performed a detailed field survey that catalogued sand boils, lateral spreading, and damaged buildings. In addition, an interesting phenomenon i.e., waterlogging from the rice fields around the epicenter area was observed immediately after the earthquake. To observe the ground water level fluctuation associated with the earthquake, we have collected the hourly basis ground water data for a span of 11 days (before and after the earthquake) from the ground water monitoring wells installed by the Korea Rural Community Corporation, which monitors the quantity and quality of ground water and provides the information to the general public. In the following section, we have provided an overview of the coseismic ground deformations observed during the Pohang earthquake.

4.2. Liquefaction Features, Building Damages, and Ground Cracks

Our field survey area covers a zone of 8 km length in E-W direction and 3 km wide in N-S direction around the epicenter. We have incorporated an integrated survey for the present study, which includes the use of high-resolution satellite images, UAV images, and detailed field survey.

The Pleiades image (0.5 m resolution) and UAV images were used for the mapping of sand boils/ground cracks that are larger than the spatial/ground resolution of the images. It was confirmed by the field survey and additional small features during the five-week detailed field survey. Within the epicenter area, most of the sand boils and lateral spreads were observed in the agricultural land or along the cracks between the concrete canal walls and agricultural lands having a gentle slope (≤5%) (Figure 3). Most of the sand boils are confined between the Gokgang River and Chogok River where the area is generally composed of Quaternary deposits. Eyewitnesses reported water and sand spouting from those cracks, isolated sand boils within the agricultural lands, and some of the riverbeds.

Field observations and DSM images allow us to identify several kind of liquefaction features such as liquefaction in gravelly soil, isolated circular to semicircular sand boils, aligned sand boils, en-echelon patterned sand boils, and linear sand boils along artifacts (Figure 4). We have collected data sets for more than 600 liquefaction related features and phenomena, out of which more than 70% data sets were assigned to liquefaction, 23% were assigned to cracks/liquefaction, and 7% were assigned to building damages due to liquefaction or shaking.

The individual sand boil diameter ranges from few centimeters to more than 2 m. The orientation of the linear chains of the sand boils measured in the field are plotted on the rose-diagram, which shows three main directions of ejection (Figure 4). From the rose diagram of liquefaction features, two preferential trends can be inferred, i.e., NNE-SSW and NE-SW (our datasets combined with [40], Table S2; [40]). Based on the surface expression of the sand boils, the liquefaction features can be classified into two categories: (a) solitary or clustered very flat in nature (diameter from a few cms to 1–2 m) or (b) water and sand/gravel mixtures ejected using preexisting cracks as their pathways to the surface induced by seismic shaking (Figure 5). Similar types of sand boils have been reported during 2005 Kashmir earthquake in India (M_W 7.6), 2008 Wenchuan earthquake in China (M_W 7.9), 2011 Yingjiang earthquake in China (M_W 5.4), and 2011 Tohoku in Japan (M_W 9.0) earthquake [16,46–48], respectively.

The farthest liquefaction feature observed during the Pohang earthquake is 15 km away from the epicenter, which is relatively far away, but is probably related to the shallow focal depth. Considering this point, it is more or less consistent with the empirical relationship developed on the basis of recorded liquefactions from global earthquakes, including both large and some moderate earthquakes.

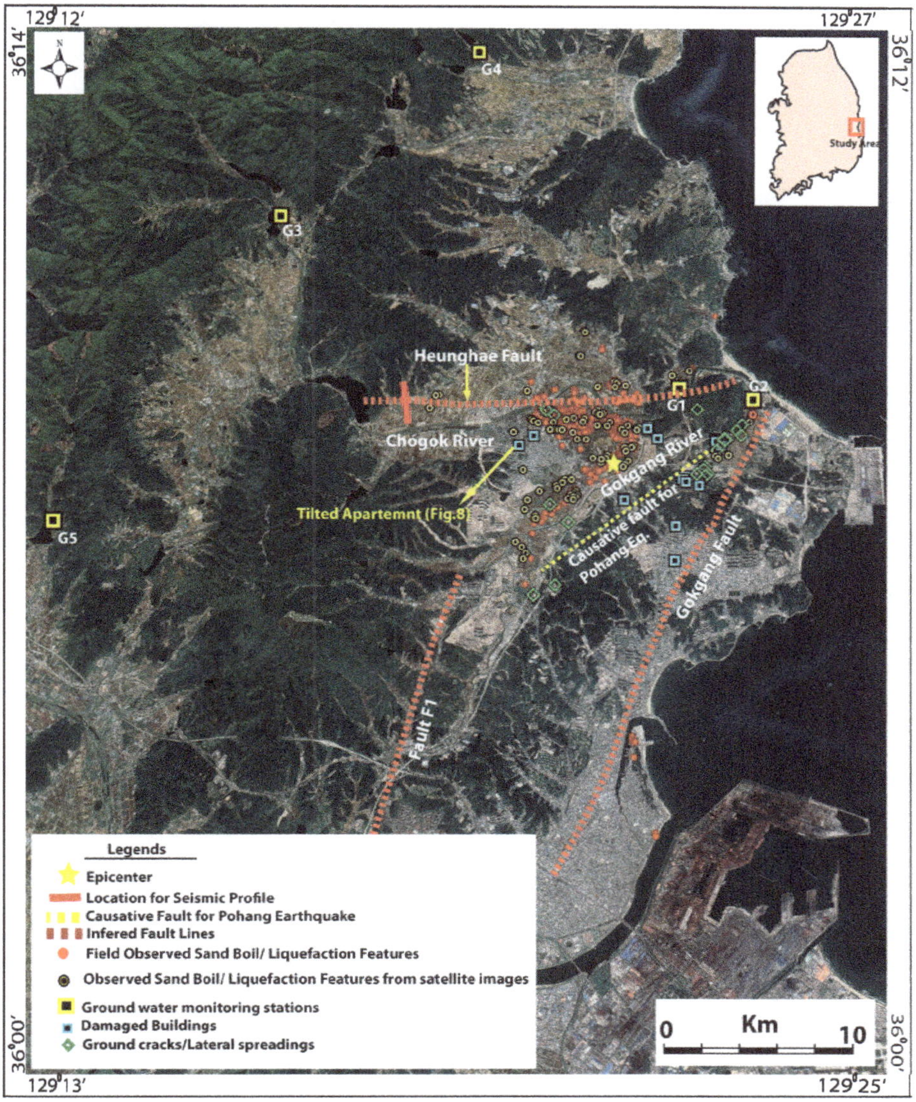

Figure 3. Google Earth image showing the major faults around the study area with the collected liquefaction features during the M_W 5.4 Pohang earthquake. Most of the liquefaction features were clustered in the southern part of the Heunghae Basin and to the south of the Heunghae Fault (Sand boil data sets used in this diagram were collected ourselves and also taken from [40]). The red line indicates the location of geophysical profile taken across the Heunghae Fault.

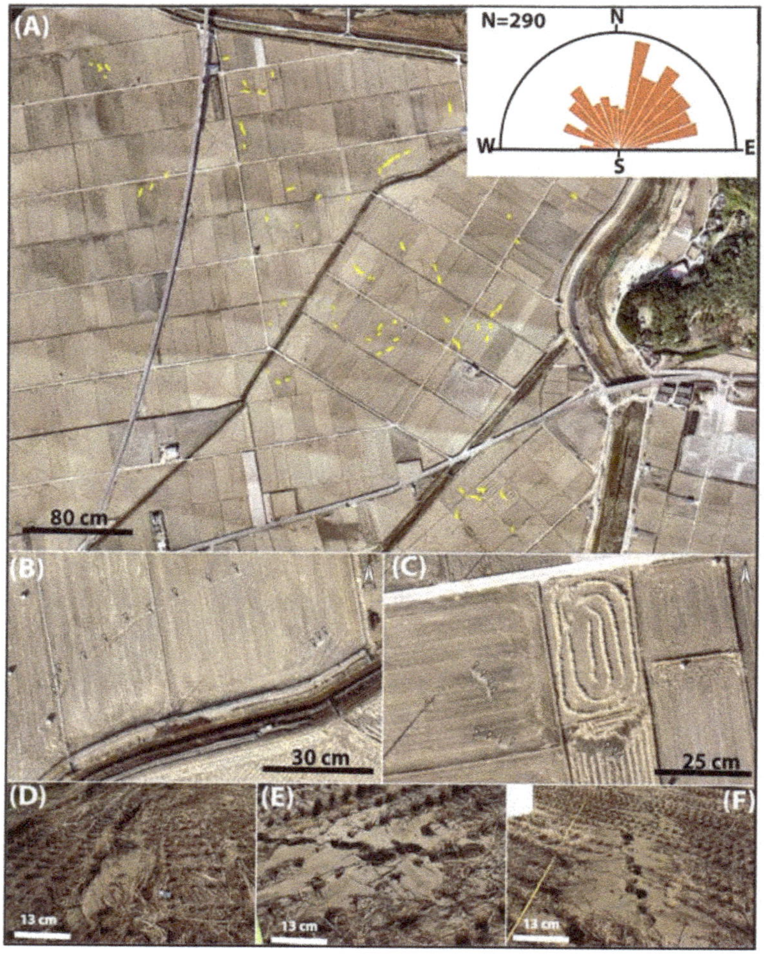

Figure 4. Unmanned Air Vehicle (UAV) images and field photographs showing typical liquefaction features observed during the Pohang earthquake. (**A–C**) UAV images showing the distribution of sand boils and rose diagram showing two preferential trends: NNE ~SSW and NE-SW. (**D–F**) Field photographs showing typical isolated and series of sand boils observed in the agricultural field.

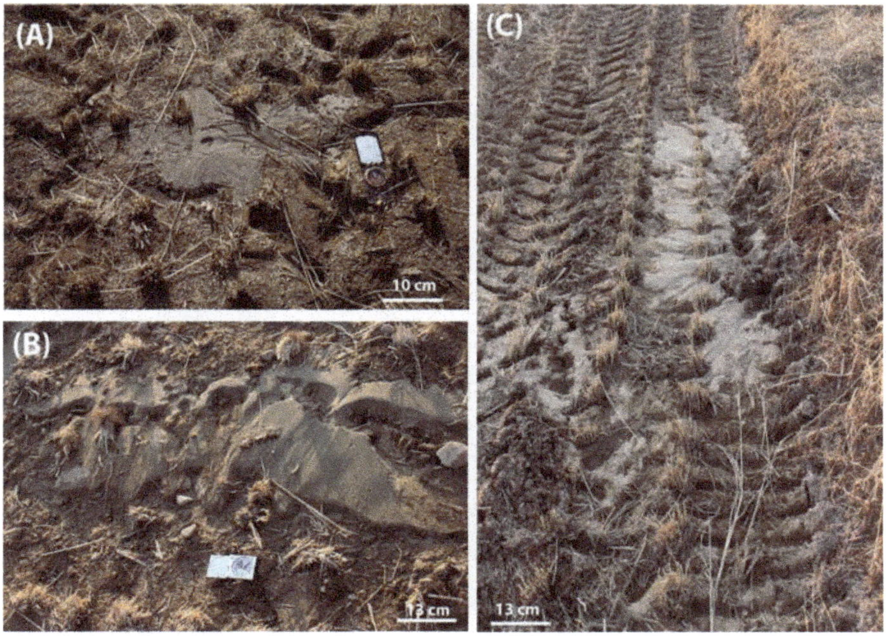

Figure 5. Typical field photos showing different types of liquefaction feature observed during the Pohang earthquake. (**A**) Isolated sand boil in the rice field. (**B**) Chain/Clustered sand boils in the rice field. (**C**) Sand/gravel mixtures ejected using preexisting cracks.

The presence of more than 600 liquefaction features within a radius of 3 km from the epicenter during a moderate magnitude earthquake (M_W 5.4) indicates that the Heunghae basin and the Pohang area are susceptible to liquefaction. More detailed liquefaction hazard mapping is required for this area [25,27,47,49–51].

During the field work after the earthquake, we collected ejected sand samples from the sand boils and from the drilled borehole conducted by Pusan National University, in order to analyze and classify grain sizes. We have taken the representative sand samples from one of the biggest sand boils and from the bore hole, which was drilled within the sand boil observed at the center of the Heunghae basin. Both the sand samples were collected within the 500–700 m radius of the epicenter area.

The grain size analysis of the ejected soil samples and sand samples collected from the borehole has been carried out as standard ASTM D6913. The analyzed grain sizes were plotted on the curve proposed by Tsuchida [52] for the possibility of liquefaction (Figure 6). The grain size analysis suggests that sand samples from the ejected sand boils and borehole have an identical nature with more than 90% of grain size between 4.75–0.075 mm and fall within the zone of most liquefiable soil in the potentially liquefiable soil range. Thus, liquefaction hazard zonation study is required for the soils of the Heunghae Basin.

Differential settlement and lateral spreading were most widely observed phenomena during this Pohang earthquake like other earthquakes around the world. The lateral spreading was observed around the levees, shore lines and reclaimed lands for construction purposes, road embankments with soft soil and some portions of agricultural lands (Figure 7). The dimension of the cracks varied from a few meters to 10 s of meters in length. Most of the cracks developed in the N-S direction.

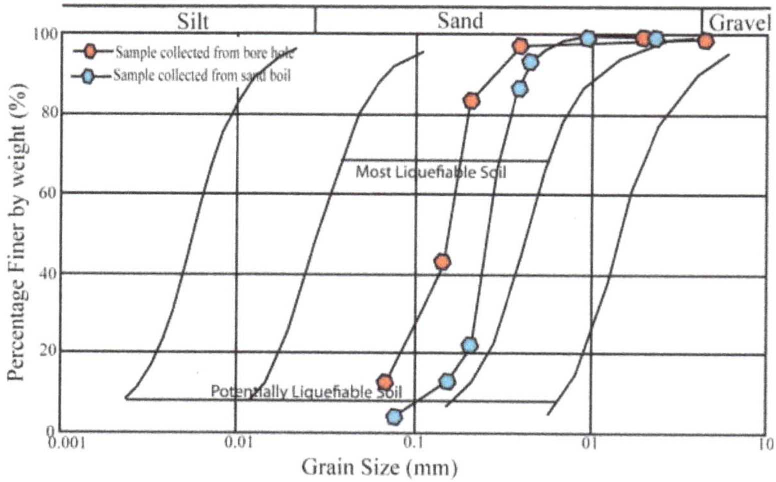

Figure 6. Grain size distribution graph for the samples collected from the ejected sand boils and drilled borehole. The grain size distributions were compared with the range of grain size distribution for liquefiable soil (Modified from [52]).

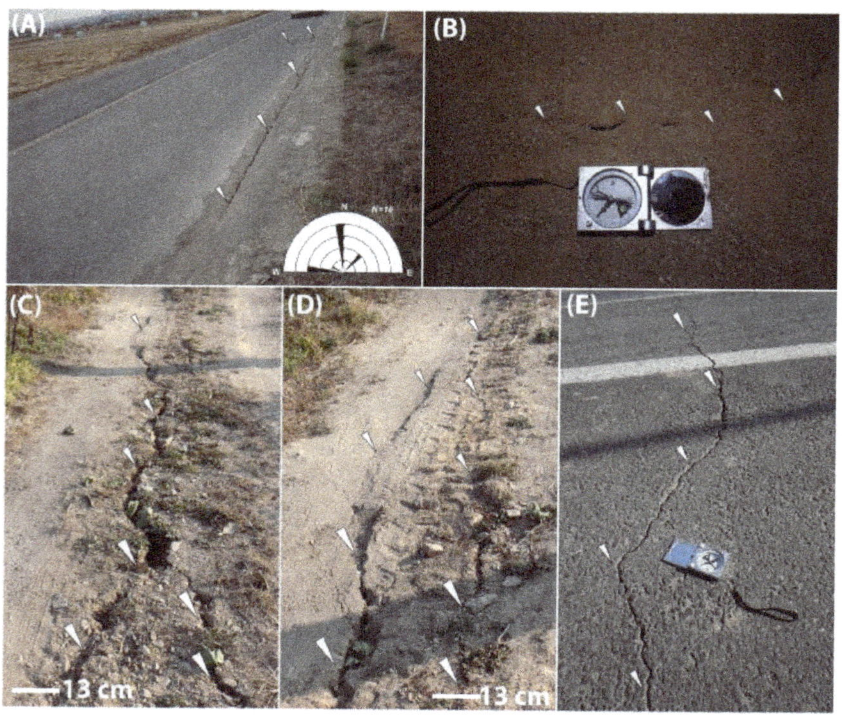

Figure 7. Field photographs showing (**A**) The cracks on a paved road due to differential settlement; (**B–D**). Ground cracks and lateral spreading along the canal embankments during the Pohang earthquake. Most of the cracks showed the N-S trend; (**E**) En-echelon cracks along a paved road.

The Pohang earthquake caused damages to more than 2000 individual houses and school buildings. Some of the houses close to the epicenter were tilted about 15 cm towards the north and the balcony of the ground floor got subsided more than 30 cm. Referable to the tilting of the building, some of the doors in the ground floor got buckled (Figure 8). Additionally, a church building near the epicenter area got tilted about 10 cm due to ground subsidence, accompanied by liquefaction.

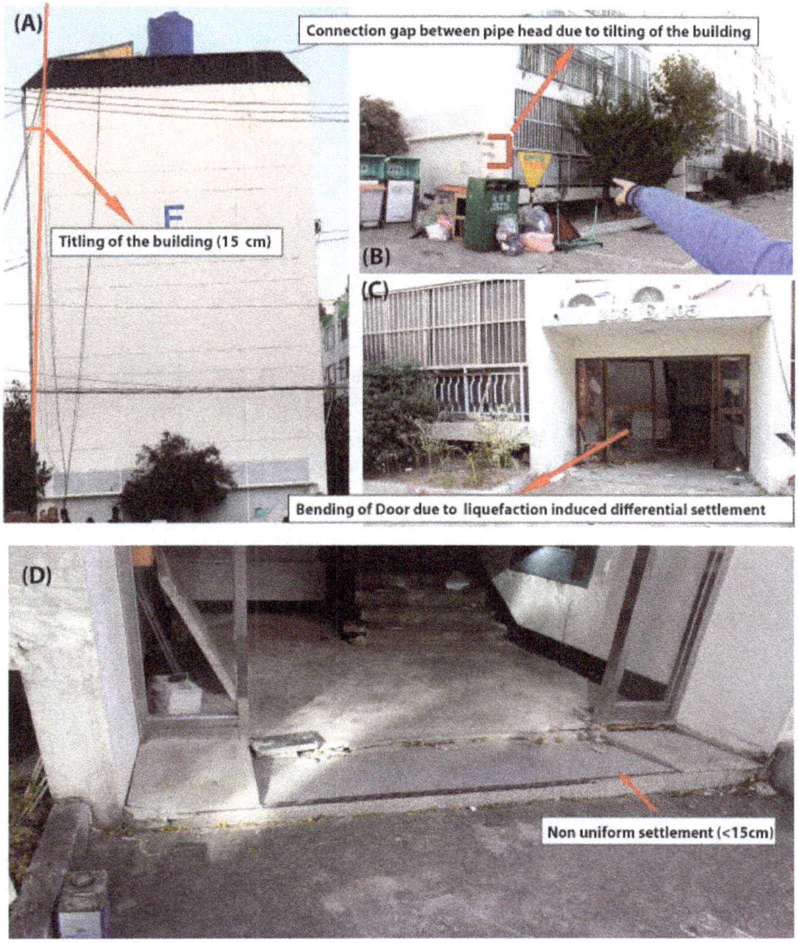

Figure 8. Photographs showing (**A**) a titled residential building due to differential settlement (the building was titled about 15 cm); (**B**) separation of water pipeline due to titling of the building; (**C**,**D**) differential settlement observed in the basement of the residence buildings in the epicentral area (he basement of the building settled almost 15–30 cm).

One of the field photographs of the titled apartment building shows that the two ends of the gas pipeline connected to the apartment got separated about more than 15 cm due to the titling (Figure 8B). The liquefaction and lateral spreading damages to residential buildings and schools during this earthquake indicate that the Korean building code needs to be revised and should include the factor of safety against liquefaction (FSL) as an important parameters of seismic resistant building design.

4.3. Liquefaction of Gravelly Soil

During the recent M_W 5.4 Pohang earthquake, we have observed evidences from the field of gravelly soil liquefaction in the river bed and in the agricultural land (Figure 9). The diameter of ejected gravel varies from 1 cm to < 4 cm. Liquefaction of sand or silt during an earthquake is common, but liquefaction of gravelly sand is very rare, specifically during an earthquake of magnitude $M_W < 5.5$, such as the Pohang earthquake. This phenomenon has been reported by a few earthquakes in recent history, such as the 1976 Friuli earthquake in Italy (M_W 6.5), 1983 Borah Peak earthquake in the USA (M_W 6.9), 1994 Hokkaido earthquake in Japan (M_W 7.7), 1995 Kobe earthquake in Japan (M_W 6.9), 1999 Chi–Chi earthquake in Taiwan (M_W 7.7), 2008 Wenchuan earthquake in China (M_W 7.9), and 2011 Tohoku earthquake in Japan (M_W 9.0) [7,53–59].

Figure 9. Field photos showing evidences of the gravelly soil liquefaction. (**A**) Photographs showing ejected gravels due to liquefaction through ground cracks; (**B**) sand boil showing evidence of gravelly soil liquefaction along a dry river bed; (**C**) sand boil showing ejected pebbles with fine sand in rice farm during the earthquake.

4.4. Water Logging and Ground Water Fluctuations

The liquefaction features associated with the Pohang earthquake were mostly confined between the Gokgang River and Chogok River (Figure 3). It may indicate that the shallow ground water level and soft sediments around the rivers are the main controlling factors to this liquefaction. The physical appearances of the ejected soils from the sand boils and borehole (Figure 6) suggest that the soils in the study area are prone to liquefaction.

The amount and consistency of water spring and logging in the agricultural fields, which lasted several days after the earthquake [60], indicate the significant rise of the water table during the earthquake. It is a rare phenomenon during medium scale earthquakes, although a similar water ejection phenomenon was reported during the Wenchuan earthquake in China, which was a large magnitude earthquake of M_W 7.9. It is likely that the Pohang earthquake must be an unusual case to raise the water table seriously and cause extensive water logging and liquefaction associated with a medium scale earthquake. It indicates that the ground water within this area is geologically highly confined.

To confirm the reason for the water table variation, we have collected and plotted hourly based ground water data for 11 days from three respective monitoring wells around the epicenter area of the Pohang earthquake. The variation of the water level from the ground water monitoring stations (Figure 10) is evident on the date of the earthquake. From the figure, it is evident that the station closest to the epicenter and situated within the Heunghae basin (G1) shows a sudden rise in water level (~0.5 m) during the time of earthquake followed by a decrease in water level and taken several days to be normalized to the regular level. Other monitoring wells (G3, G5, located at higher elevation than the G1) shows a sudden drop in water level (~0.3 m) at the time of the earthquake and takes two days to revert back to the original level. Though there is an interesting ground water response to the earthquake, it is necessary to do more detailed analysis which is beyond the scope of the present work.

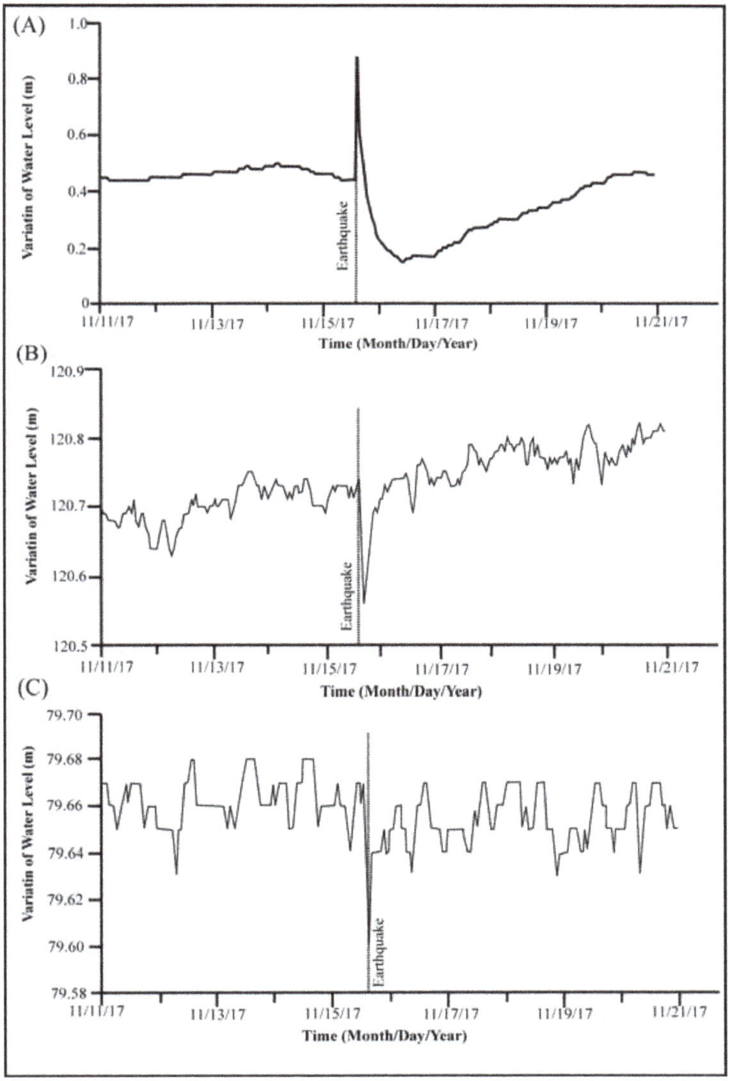

Figure 10. Changes of ground water levels associated with the Pohang earthquake. The ground water monitoring stations show the variation of the water levels before and after the 15th November 2017 Pohang earthquake (**A**) G1, (**B**) G3, and (**C**) G5. For the station locations: see Figure 3 (Data taken from https://www.groundwater.or.kr).

On the basis of preliminary observation from the water level fluctuations of the monitoring wells and local residents' reports, the observed hydrologic phenomenon might be relevance to the earthquake.

It is well known that water wells respond to the seismic activity near field as well as far field [13,60–62], which depends upon different geological conditions and the location of the monitoring wells within the epicenter area. Several studies suggest [13,60–65] that the rise or fall in water level in the monitoring wells is related to the elastic volumetric compression or expansion of the well aquifer system caused by seismic waves, but steep rise or fall in the water level may also be related to coseismic changes in the well aquifer media.

Though we have limited ground water data, we think the difference in behavior of the monitoring wells was controlled by two factors: (1) change in the static stress field and (2) local geological changes triggered by earthquakes. The ground shaking may cause the rearrangement of unconsolidated sediments favoring compaction, which has led the rise in the water table (case of G1), as well as the opening or closing of the deeper aquifer system due to seismic shaking that might have caused the water level drop or rise (case of the G3 and G5) [63–65]. A similar phenomenon was also observed around the epicenter area of the 1980 Irpinia earthquake in Italy (M_W 6.9), 1989 Loma Prieta earthquake in the USA (M_W 7.1), 1998 Pymatuning earthquake in Pennsylvania, USA (M_W 5.2), and 2009 L'Aquilla in Italy (M_W 6.3) earthquake [63–65].

5. Discussion on Geological Aspects of the Mechanism Involved in Liquefaction

The distribution of sand boils, lateral spreading, and cracks allow us to relate the damaging features to the geological structures in this area. For instance, most of the liquefaction sites were clustered between the Gokgang River and Chogok River, and towards the south of the Heunghae fault. The higher concentrations of liquefaction in the southern part of the Heunghae basin suggest that the ground motion duration or amplification was higher in that area. Liquefaction is one of the main responses to seismic waves. Furthermore, the clustering of the liquefaction features in this area may indicate that the seismic waves were also controlled by geological structures. Based on previous studies and our fieldwork, the Heunghae fault and an inferred blind fault (Figures 1, 3 and 11) generating this earthquake are the major structures in this basin, which might strongly contribute to releasing the generated seismic waves.

To examine the role of geological structures, especially related to the Heunghae fault [38,62] in the liquefaction feature distribution, detailed literature review and field studies have been carried out. Previous studies [38,66] reported that the Pohang basin is bounded by several E-W trending faults, which offset the tertiary formations and merged to the Yangsan fault. These faults are named as Hyongsan fault and Heunghae fault, which divide the Pohang basin into several sub-basins. During the post-earthquake survey, we were able to find several evidences of NE-SW striking small to medium scale normal faults in the northern part of the Heunghae basin with a fault gouge of 1-5 cm thick (see Figure 3 for the location and Figure 11 for the evidences of faults in seismic profile). These faults probably are subsidiary faults of the E-W trending Heunghae fault, because the structural parameters of these faults are well matched with the previously mapped Heunghae fault. The main Heunghae fault might have been covered by the basin fill deposits and was not clearly traced.

To confirm the existence of the main Heunghae fault within the basin, a shallow subsurface seismic refraction survey has been carried out using OYO McSeis SX 1125 instrument (Tsukuba, Japan) with 24 channel (28 Hz) seismographs. A Sledge Hammer was used for generating seismic waves at the surface and after recording the data it was processed using SeisImager software. The 1 km long seismic survey was taken perpendicular to the Heunghae fault towards the western margin of the Heunghae basin (see Figure 3 for location of seismic survey). The same 1 km long profile has been divided into six divisions (SP-1 to SP-6). The geophone spacing was taken at 5 m for SP-1 and SP-6, and 7.5 m for SP-2 to SP-5. The data processing of the seismic survey is mainly based on seismic refraction tomography techniques, as well as inverse travel time modelling of the refracted seismic waves. Using the estimated velocity (1.6–3.6 km/S), we have detected unconsolidated basin fill deposits lies from surface to 10 m deep, and consolidated basin deposits or weathered rock has been detected from 10 m to 80 m deep (Figure 11).

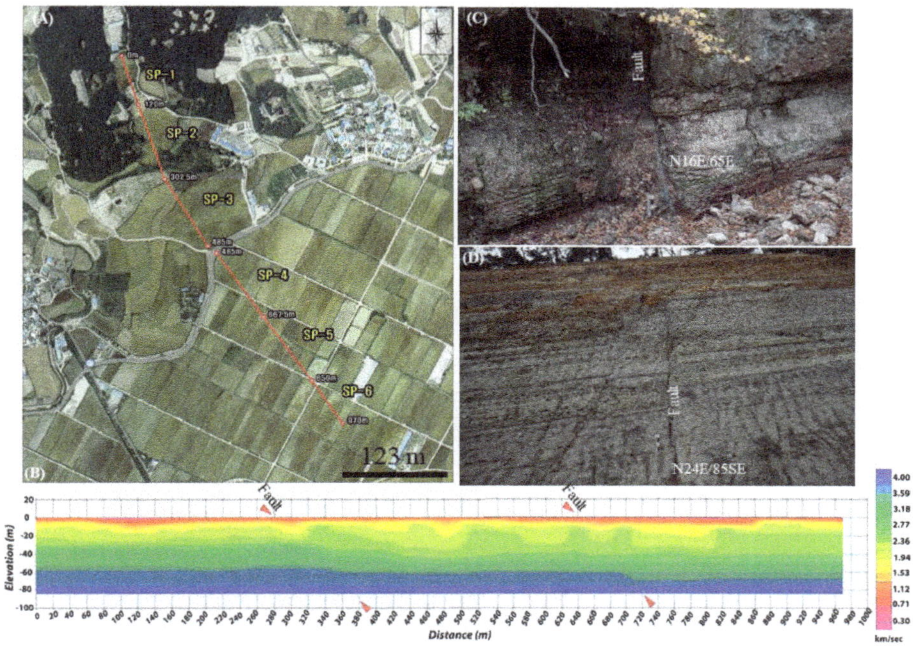

Figure 11. (**A**,**B**) Seismic refraction profile showing existences of the south dipping normal fault within Heunghae basin, which could be the traces of Heunghae Fault (red arrow shows the traces of faults). (**C**,**D**) Field photographs showing evidence of normal faulting. The faults offset the tertiary deposits within the Heunghae basin.

On the basis of seismic profile and velocity difference, we have detected two south dipping normal faults which could be the traces of the Heunghae fault, which is well matched with the field data collected. The seismic profile and field evidences of normal faulting along the Heunghae fault is shown in Figure 11. We argue that this fault played a role in the passage of seismic waves and amplifications within the Heunghae basin and the distribution of liquefaction features. Moreover, the mountain basin effect may do play an important role in extensive liquefaction within the basin and the distribution of sand boils (Figures 3 and 11) [11]. On the basis of these observations, we will discuss the possible mechanism involved in the liquefaction and its distribution within this study area in the following section.

Possible Mechanisms Involved in Liquefaction Clustering in the South Part of the Heunghae Basin

The cause and distribution characteristics of liquefaction and related damages are associated with the combined effects of several factors such as earthquake magnitude, duration of shaking, distance from the epicenter, type of soil content, relative density, drainage condition, degree of consolidation, thickness of liquefiable sand/silt layer, and depth of groundwater table [5,67]. Sometimes anthropogenic structures such as clay lining in rice fields and reclaimed land also influence to the severity of the hazard [5,40].

By analyzing the source of the earthquake and the geological setting of the Heunghae basin area underwent liquefaction, we argue that the major structural factors for the liquefaction clustering during the Pohang earthquake might be the combination of mountain basin effect and trapping of seismic waves within fault zones. During the 2012 Emilia earthquake (M_W 5.9) in Italy, clustering of liquefaction features was observed with in the Po-Plain. This indicates that even if the affected area appears to be homogeneous from a geological point of view there are other local geological factors that controls the liquefaction susceptibility of the area within a basin or an alluvial plain [68–70]. Furthermore, previous studies [70,71] observed the mechanism involved in severe liquefaction within basins and suggested that the wedge shaped basement-to-sediment basin interface, which acted as acoustic lens, caused localized seismic wave amplification and extensive damage within the basin [7,71–75]. Though the basin effect is poorly understood and included in the routine seismic hazard assessment, it has been well evidenced that several large and small magnitude earthquakes (1985 Mexico City earthquake, Mexico, M_W 8.0; 1994 North Ridge earthquake, USA, M_W 6.7; 1999 Izmit earthquake, Turkey M_W 7.6; 2008 Wenchuan earthquake, China, M_W 7.9; 2009 Olancha earthquake, USA, M_W 5.2; 2011 Tohoku earthquake, Japan (M_W-9.0); 2012 Emilia earthquake, Italy, M_W 5.9; and 2017 Tripura earthquake, India, M_W 5.7) [7,75–79].

Other factor which controls the severity and distribution of liquefaction and seismic ground deformation is the trapping of seismic waves by the major fault zones within the basin. It was suggested that large faults within the sedimentary basin with fault gouges, fractured rocks and fluids can trap the seismic waves within the block bounded by fault zones [5], which amplifies the upper bound in soft sediments of the basin. This amplification could be stronger within the basin surrounded by fault zones covered by unconsolidated Holocene alluvial deposits (Figures 11 and 12) [7,68,80–84]. A similar observation was reported during the 2008 Wenchuan earthquake, where most of the liquefaction features were confined to the recent alluvial deposits close to the range front blind fault, and damaged buildings were clustered near or top of the Qingchuan blind fault in Sichuan province in China [7]. During the 1994 Northridge earthquake in the USA, (M_W 6.7) [11], the basin structure was an important factor for the enhancement of liquefaction hazard. In the 2001 Bhuj earthquake in India (M_W 7.7), most of the liquefaction features were distributed close to the fault [83]. This clustering of sand boils indicates a fault barrier mechanism for passage of seismic waves within a basin.

The borehole log drilled across the Heunghae basin for the pilot project of the potential CO_2 storage site [84] suggests that the Pohang basin has a typical wedge-shaped structure bounded and dissected by several faults and covered by soft sediments. The previous study [79] for a CO_2 storage project suggests that the Heunghae basin is bounded by the east dipping Gokgang fault to the east and the south dipping Heunghae fault to the north. The seismic refraction profile and field evidences about the presence of the two NE-SW/E-W striking and S-SE dipping subsidiary normal faults within the Heunghae basin help us to suggest the presence of the E-W striking Heunghae fault.

The potential seismogenic fault for the Pohang earthquake is the west dipping thrust fault, which might be an antithetic fault of the Gokgang fault (Figure 12). On the basis of the geometry and location of the Heunghae fault and the seismogenic fault, the seismic waves generated during the earthquake were trapped and caused more amplification in the southern part of the basin than the northern part. The geophysical and field results suggesting the presence of the E-W trending Heunghae fault and the field evidence of clustered sand boils along the NE-SW causative fault for the Pohang earthquake proved the trapping of the seismic waves by fault zones (Figure 3) [40]. The distribution of sand boils within the Heunghae basin (Figure 3) shows clustered sand boils indicating two preferred orientations. One set of sand boils shows E-W trend, whereas another set shows NE-SW trend, which are similar to the trend of the Heunghae fault and the antithetic fault caused the Pohang earthquake, respectively.

Based on the field observations, geological structures presented in the study area—i.e., spatial distribution plot of sand boils around the epicenter—it can be inferred that the distribution of liquefaction features is mostly controlled by the geological structures within the Heunghae Basin (Figures 3 and 12). Thus, the presence of the Heunghae fault and the antithetic blind fault led the differential amplification due to trapping of seismic waves within the same basin and the differential distribution of liquefaction features. Using this, we have proposed a conceptual model (Figure 12) to explain the local clustering of sand boils within the Heunghae basin, which is well matched with the previous observations in Sichuan province, China [7] and San Fernando Valley, USA [11].

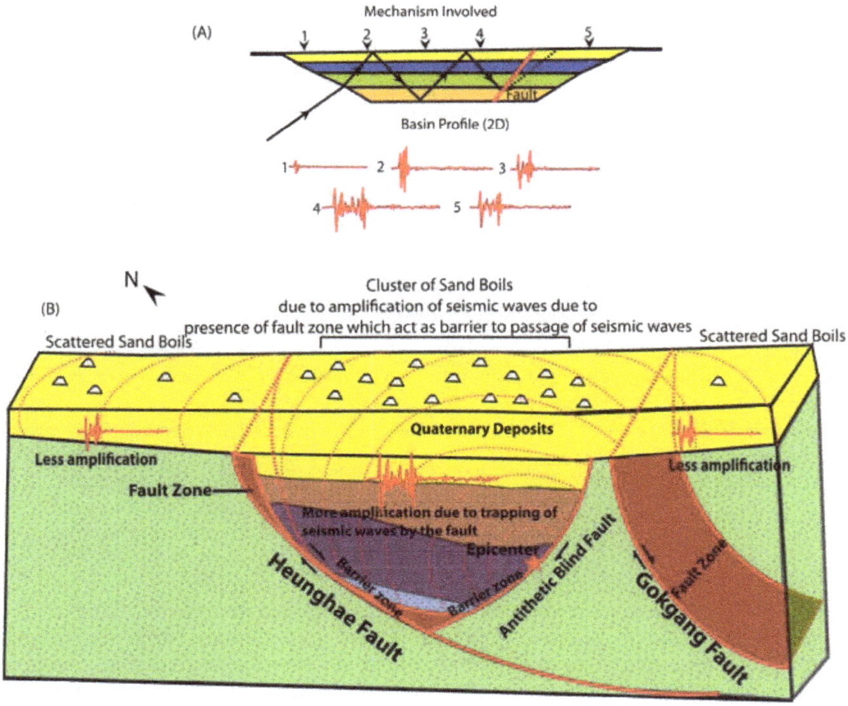

Figure 12. Conceptual models for the clustering of liquefaction during the Pohang earthquake. (**A**) Model for the mechanism involved in the localized amplification of seismic waves within a basin due to mountain basin effect. (**B**) Model for the mechanism explaining the role of the Heunghae fault in clusters of sand boils in the southern part of the fault. From the field survey and UAV image (Figure 3), it was clear that most of the sand boils were clustered densely in the southern part of the fault (model is not up to the scale).

Many of the liquefactions that have caused damages within sedimentary basins have been reported in recent earthquakes such as the Sichuan basin during the 2008 Wenchuan earthquake in China (M_W 7.9), the Kanto basin near Tokyo during the 2011 Tohoku earthquake in Japan (M_W 9.0), Po-Plain in Italy during the 2012 Emilia earthquake (M_W 5.8), and the Kathmandu valley during the 2015 Nepal earthquake (M_W 7.8). The observed damages during these earthquakes poses a real seismic threat to the areas with similar geological settings around the globe. However, the involved mechanisms of geological aspect have received little to no attention, especially during small to moderate earthquakes. Although several numerical studies suggested the trapping and amplification of seismic waves within the basins, geological evidences were very rare and difficult to prove it during a small or a moderate earthquake.

Currently, the liquefaction during small to moderate earthquakes without any surface ruptures is another issue (e.g., such as in the 2009 Olancha earthquake in the USA and the 2017 Pohang earthquake in South Korea), because it can cause serious damages compared to its magnitude. This is especially true of the effect of geological structures within the basin, which is an important concern in earthquake hazard assessment. The present study and proposed inferences will help in understanding the geological phenomenon involved in more localized seismic damages, especially where serious liquefaction and related damages compared with its magnitude are reported in a similar geological and depositional setting. However, it needs more geophysical or seismic data for a conclusive interpretation.

6. Conclusions

The 15th November 2017 Pohang earthquake was the most devastating earthquake with extensive liquefaction in the modern seismic history of the Korean Peninsula. The MMI Intensity of the Pohang earthquake was VIII to IX and caused injury to more than 90 people and estimated property damage of 52 million (USD) with many structural damages. The liquefaction and lateral spreading during the Pohang earthquake provide a very good opportunity to understand the factors involved in liquefaction during a small to moderate earthquakes, which is a very rare phenomenon in Korea and around the world. Based on the results of a systematic field survey, we identified several areas showing unusually extensive liquefaction around the Heunghae basin.

Several kinds of liquefaction features—i.e., liquefaction in gravelly soil, isolated circular to semicircular sand boil, aligned sand boils, en-echelon patterned sand boils, and linear sand boils along artifacts—were identified during the field survey. Most of the sand boils were concentrated along the major faults and especially in the southern part of the Heunghae fault between the Gokgang River and Chogok River. The grain size analysis suggests that most of the ejected materials can be categorized as most favorable liquefiable soil to potentially liquefiable soil. The liquefaction associated with a small to moderate earthquake like the Pohang earthquake could be related to a favorable site condition and strong ground motion. The borehole data drilled in one of the sand boils suggests that the presence of shallow water table and sandy layer was a prime factor for the liquefaction.

The PGA recorded from the Pohang earthquake (i.e., 0.58 g) was very high as a large magnitude earthquake, which may affect to cause a strong ground motion and related to the shallow focus of the earthquake. Furthermore, we interpreted that the amplification of seismic waves within the basin bounded by the major fault zones and mountain basin effects have contributed to the intense shaking and the vast occurrence of liquefaction within the basin. Based on this interpretation, we proposed a conceptual model showing the Heunghae fault zone acting as a barrier for the passage of seismic waves, which produced clusters of sand boils along the southern part of the Heunghae fault. This study suggests that immediate and careful coseismic geological investigation can be effective for proper earthquake parameter estimation and for seismic hazard evaluation on vulnerability of the particular area. Moreover, this kind of study can significantly contribute to engineering implications for a realistic seismic hazard assessment particularly in liquefaction zonation. These aspects need to be considered in liquefaction hazard mapping for similar geological settings on a local as well as regional scale.

Supplementary Materials: The following are available online at http://www.mdpi.com/2076-3263/9/4/173/s1, Figure S1: Google Earth Image showing temporary stations installed by KMA around the epicentral area immediately after the earthquake to measure HVSR ratio and understand the subsoil conditions. The yellow square without number indicates temporary stations without any data (modified from [35]); Table S1: Recorded PGA at different seismic stations around epicentral area of M_w 5.4 Pohang earthquake (KMA Report (modified from [35]); Table S2: Detailed liquefaction features mapped during M_w 5.4 Pohang post-earthquake field survey (Our data from the present study combined with [40]).

Author Contributions: In the present research, S.P.N. conducted the field survey and analyzed the field data, developed the idea, and prepared the manuscript. T.K. and J.S.-H. helped in the field work and geophysical survey. Y.-S.K. encouraged the team and supervised the findings of this work with continuous technical inputs during the preparation of the manuscript.

Funding: This research was supported by a grant (2017-MOIS31-006) of the Fundamental Technology Development Program for Extreme Disaster Response funded by Ministry of Interior and Safety (MOIS, Korea).

Acknowledgments: Authors are thankful to Young Sik Gihm (KIGAM) for providing the additional datasets for the sand boil locations. Also, we are thankful to Master and Bachelor degree students of GSGR Lab for providing support during post-earthquake field survey.

Conflicts of Interest: The authors declare no conflict of interest.

References

1. Ambraseys, N.; Sarma, S. Liquefaction of soils induced by earthquakes. *Bull. Seismol. Soc. Am.* **1969**, *59*, 651–664.
2. Seed, H.B.; Idriss, I.M. Simplified procedure for evaluating soil liquefaction potential. *J. Soil Mech. Found.* **1971**, *97*, 1249–1273.
3. Ishihara, K. Liquefaction and flow failure during earthquakes. *Geotechnique* **1993**, *43*, 351–451. [CrossRef]
4. Youd, T.L.; Idriss, I.M. Liquefaction resistance of soils: Summary report from the 1996 NCEER and 1998 NCEER/NSF workshops on evaluation of liquefaction resistance of soils. *J. Geotech. Geoenviron. Eng.* **2001**, *127*, 297–313. [CrossRef]
5. Wang, C.Y. Liquefaction beyond the near field. *Seismol. Res. Lett.* **2007**, *78*, 512–517. [CrossRef]
6. Holzer, T.L.; Jayko, A.S.; Hauksson, E.; Fletcher, J.P.; Noce, T.E.; Bennett, M.J.; Dietel, C.M.; Hudnut, K.W. Liquefaction caused by the 2009 Olancha, California (USA), M5. 2 nearthquakes. *Eng. Geol.* **2010**, *116*, 184–188. [CrossRef]
7. Liu-Zeng, J.; Wang, P.; Zhang, Z.; Li, Z.; Cao, Z.; Zhang, J.; Yuan, X.; Wang, W.; Xing, X. Liquefaction in western Sichuan Basin during the 2008 M_W 7.9 Wenchuan earthquake, China. *Tectonophysics* **2017**, *694*, 214–238. [CrossRef]
8. Naik, S.P.; Patra, N.R. Generation of Liquefaction Potential Map for Kanpur City and Allahabad City of Northern India: An Attempt for Liquefaction Hazard Assessment. *Geotech. Geol. Eng.* **2018**, *36*, 293–305. [CrossRef]
9. Naik, S.P.; Patra, N.R.; Malik, J.N. Spatial distribution of Shear wave velocity for late Quaternary Alluvial soil of Kanpur city, Northern India. *Geotech. Geol. Eng.* **2014**, *32*, 131–149. [CrossRef]
10. Seed, H.B. Landslides during earthquakes due to liquefaction. *J. Soil Mech. Found.* **1968**, *94*, 1055–1122.
11. Holzer, T.L.; Bennett, M.J.; Ponti, D.J.; Tinsley, J.C., III. Liquefaction and soil failure during 1994 Northridge earthquake. *J. Geotech. Geoenviron. Eng.* **1999**, *125*, 438–452. [CrossRef]
12. Rajendran, K.; Rajendran, C.P.; Thakkar, M.; Tuttle, M.P. The 2001 Kutch (Bhuj) earthquake: Coseismic surface features and their significance. *Curr. Sci.* **2001**, *80*, 1397–1405.
13. Wang, C.Y.; Wang, C.H.; Manga, M. Coseismic release of water from mountains: Evidence from the 1999 (M_W = 7.5) Chi-Chi, Taiwan, earthquake. *Geology* **2004**, *32*, 769–772. [CrossRef]
14. Gómez, J.C.; Tavera, H.J.; Orihuela, N. Soil liquefaction during the Arequipa M_W 8.4, June 23, 2001 earthquake, Southern Coastal Peru. *Eng. Geol.* **2005**, *78*, 237–255.
15. Yin, R.Y.; Liu, Y.M.; Li, Y.L.; Zhang, S.M. The relation between earthquake liquefaction and landforms in Tangshan region. *Res. Soil Water Conserv.* **2005**, *12*, 110–112.
16. Bhattacharya, S.; Hyodo, M.; Gouda, K.; Tazoh, T.; Taylor, C.A. Liquefaction of soil in the Tokyo Bay area from the 2011 Tohoku (Japan) earthquake. *Soil Dyn. Earth Eng.* **2011**, *31*, 1618–1628. [CrossRef]
17. Atzori, S.; Tolomei, C.; Antonioli, A.; Merryman Boncori, J.P.; Bannister, S.; Trasatti, E.; Pasquali, P.; Salvi, S. The 2010–2011 Canterbury, New Zealand, seismic sequence: Multiple source analysis from InSAR data and modeling. *J. Geophyl. Res. Solid Earth* **2012**, *117*, 1–16. [CrossRef]
18. Ishitsuka, K.; Tsuji, T.; Matsuoka, T. Detection and mapping of soil liquefaction in the 2011 Tohoku earthquake using SAR interferometry. *Earth Planets Space* **2012**, *64*, 1267–1276. [CrossRef]
19. Papathanassiou, G.; Caputo, R.; Rapti-Caputo, D. Liquefaction phenomena along the paleo-Reno River caused by the May 20, 2012, Emilia (northern Italy) earthquake. *Ann. Geophys.* **2012**, *55*, 735–742.
20. Quigley, M.C.; Bastin, S.; Bradley, B.A. Recurrent liquefaction in Christchurch, New Zealand, during the Canterbury earthquake sequence. *Geology* **2013**, *41*, 419–422. [CrossRef]
21. De la Maza, G.; Williams, N.; Sáez, E.; Rollins, K.; Ledezma, C. Liquefaction-Induced Lateral Spread in Lo Rojas, Coronel, Chile: Field Study and Numerical Modeling. *Earth Spectra* **2017**, *33*, 219–240. [CrossRef]
22. Sharma, K.; Deng, L.; Khadka, D. Reconnaissance of liquefaction case studies in 2015 Gorkha (Nepal) earthquake and assessment of liquefaction susceptibility. *Int. J. Geotech. Eng.* **2017**, 1–13. [CrossRef]
23. Gautam, D.; de Magistris, F.S.; Fabbrocino, G. Soil liquefaction in Kathmandu valley due to 25 April 2015 Gorkha, Nepal earthquake. *Soil Dyn. Earth Eng.* **2017**, *97*, 37–47. [CrossRef]
24. Kuribayashi, E.; Tatsuoka, F. Brief review of liquefaction during earthquakes in Japan. *Soils Found.* **1975**, *15*, 81–92. [CrossRef]
25. Ambraseys, N.N. Engineering seismology: Part II. *Earth Eng. Struct. Dyn.* **1988**, *17*, 51–105. [CrossRef]

26. Sims, J.D.; Garvin, C.D. Recurrent liquefaction induced by the 1989 Loma Prieta earthquake and 1990 and 1991 aftershocks: Implications for paleoseismicity studies. *Bull. Seismol. Soc. Am.* **1995**, *85*, 51–65.
27. Galli, P. New empirical relationships between magnitude and distance for liquefaction. *Tectonophysics* **2000**, *324*, 169–187. [CrossRef]
28. Choi, J.H.; Kim, Y.S.; Choi, S.J. Identification of a suspected Quaternary fault in eastern Korea: Proposal for a paleoseismic research procedure for the mapping of active faults in Korea. *J. Asia Earth Sci.* **2015**, *113*, 897–908. [CrossRef]
29. Cetin, K.O.; Seed, R.B.; Kayen, R.E.; Moss, R.E.; Bilge, H.T.; Ilgac, M.; Chowdhury, K. *Summary of SPT Based Field Case History Data of CETIN (2016) Database (No. METU/GTENG 08/16-01)*; Middle East Technical University: Ankara, Turkey, 2016.
30. Moss, R.E.S.; Seed, R.B.; Kayen, R.E.; Stewart, J.P.; Der Kiureghian, A.; Cetin, K.O. CPT-based probabilistic and deterministic assessment of in situ seismic soil liquefaction potential. *J. Geotech. Geoenviron. Eng.* **2006**, *132*, 1032–1051. [CrossRef]
31. Kim, S.I.; Park, I.J.; Choi, J.S. A Study on the Assessment of Liquefaction Potential in Korea. *J. Korean Soc. Civ. Eng.* **2000**, *20*, 129.
32. Park, D.; Kwak, D.Y.; Cho, C.K.; Chun, B.S. Evaluation of liquefaction potential of port structures with earthquake magnitude adjustment. *J. Coast. Res.* **2009**, *2*, 1035–1039.
33. Seo, M.W.; Olson, S.M.; Sun, C.G.; Oh, M.H. Evaluation of liquefaction potential index along western coast of South Korea using SPT and CPT. *Mar. Geores. Geotech.* **2012**, *30*, 234–260. [CrossRef]
34. Michetti, A.M.; Esposito, E.; Guerrieri, L.; Porfido, S.; Serva, L.; Tatevossian, R.; Vittori, E.; Audemard, F.; Azuma, T.; Clague, J.; et al. Intensita' scale ESI 2007. In *Memorie Descrittive della. Carta Geologica d'Italia, Servizio Geologico d'Italia*; Guerrieri, L., Vittori, E., Eds.; Dipartimento Difesa del Suolo, APAT: Rome, Italy, 2007; pp. 1–54.
35. Korean Meteorological Administration. *KMA Report on Pohang Earthquake (Korean)*; Korean Meteorological Administration: Seoul, Korea, 2018; pp. 1–41.
36. Choi, J.H.; Ko, K.; Gihm, Y.S.; Cho, C.S.; Lee, H.; Song, S.G.; Bang, E.S.; Lee, H.J.; Bae, H.K.; Kim, S.W.; et al. Surface Deformations and Rupture Processes Associated with the 2017 Mw 5.4 Pohang, Korea, Earthquake. *Bul. Seism. Soc. Am.* **2019**, *109*, 756–769. [CrossRef]
37. Sohn, Y.K.; Son, M. Synrift stratigraphic geometry in a transfer zone coarse-grained delta complex, Miocene Pohang Basin, SE Korea. *Sedimentology* **2004**, *51*, 1387–1408. [CrossRef]
38. Son, M.; Kim, J.S.; Chong, H.Y.; Lee, Y.H.; Kim, I.S. Characteristics of the Cenozoic crustal deformation in SE Korea and their tectonic implications. *Korean J. Petrol. Geol.* **2007**, *13*, 1–16.
39. Kim, K.H.; Ree, J.H.; Kim, Y.; Kim, S.; Kang, S.Y.; Seo, W. Assessing whether the 2017 M_W 5.4 Pohang earthquake in South Korea was an induced event. *Science* **2018**, *26*, 1007–1009. [CrossRef]
40. Gihm, Y.S.; Kim, S.W.; Ko, K.; Choi, J.H.; Bae, H.; Hong, P.S.; Lee, Y.; Lee, H.; Jin, K.; Choi, S.J.; et al. Paleoseismological implications of liquefaction-induced structures caused by the 2017 Pohang Earthquake. *Geosci. J.* **2018**, *22*, 1–10. [CrossRef]
41. Grigoli, F.; Cesca, S.; Rinaldi, A.P.; Manconi, A.; López-Comino, J.A.; Clinton, J.F.; Westaway, R.; Cauzzi, C.; Dahm, T.; Wiemer, S. The November 2017 M_W 5.5 Pohang earthquake: A possible case of induced seismicity in South Korea. *Science* **2018**, *360*, 1003–1006. [CrossRef] [PubMed]
42. Han, S.J.; Kim, H.J.; Huh, S.; Park, C.H.; Kim, S.R.; Lee, Y.K.; Yoo, H.S.; Choi, D.L.; Park, B.K. Basin structure of the northeastern Ulleung basin (Ulleung and Dok island areas), East Sea of Korea. *J. Geol. Soc. Korea* **1997**, *33*, 127–138.
43. Kim, I.S. Origin and Tectonic Evolution of the East Sea (Sea of Japan) and the Yangsan Fault System: A new synthetic Interpretation. *J. Geol. Soc. Korea* **1992**, *28*, 84–109.
44. Hwang, I.G.; Chough, S.K.; Hong, S.W.; Choe, M.Y. Controls and evolution of fan delta systems in the Miocene Pohang Basin, SE Korea. *Sediment. Geol.* **1995**, *98*, 147–179. [CrossRef]
45. Chough, S.K.; Kwon, S.T.; Ree, J.H.; Choi, D.K. Tectonic and sedimentary evolution of the Korean peninsula: A review and new view. *Earth Sci. Rev.* **2000**, *52*, 175–235. [CrossRef]
46. Sahoo, R.N.; Reddy, D.V.; Sukhija, B.S. Evidence of liquefaction near Baramulla (Jammu and Kashmir, India) due to the 2005 Kashmir earthquake. *Curr. Sci.* **2007**, *92*, 293–295.
47. Huang, Y.; Jiang, X. Field-observed phenomena of seismic liquefaction and subsidence during the 2008 Wenchuan earthquake in China. *Nat. Hazards* **2010**, *54*, 839–850. [CrossRef]

48. Yao, X.; Zhang, J.G.; Zhang, Y.S.; Yang, B.; Yu, K. Study of sand liquefaction hazard features induced by Yingjiang Ms 5.8 earthquake on March 10, 2011. *J. Eng. Geol.* **2011**, *19*, 152–161.
49. Castilla, R.A.; Audemard, F.A. Sand blows as a potential tool for magnitude estimation of pre-instrumental earthquakes. *J. Seismol.* **2007**, *11*, 473–487. [CrossRef]
50. Kramer, S.L. *Geotechnical Earthquake Engineering. International Series in Civil Engineering and Engineering Mechanics*; Prentice-Hall: Englewood Cliffs, NJ, USA, 1996.
51. Kumar, A.; Borah, N.; Naik, S.P.; Olympa, B. Detailed review on methodologies available to find preinstrumental missing earthquakes of the present catalogue with the relevance to seismicity assessment of the Northeast India. *Ind. Geotech. J.* **2018**, 1–15. [CrossRef]
52. Tsuchida, H.; Hayashi, S. *Estimation of Liquefaction Potential of Sandy Soils*; Publication of Mcgraw Hill Book Company: New York, NY, USA, 1972.
53. Andrus, R.D. In-Situ Characterization of Gravelly Soils That Liquefied in the 1983 Borah Peak Earthquake. Ph.D. Thesis, University of Texas, Austin, TX, USA, 1994.
54. Sirovich, L. Repetitive liquefaction at a gravelly site and liquefaction in overconsolidated sands. *Soils Found.* **1996**, *36*, 23–34. [CrossRef]
55. Hatanaka, M.; Uchida, A.; Ohara, J. Liquefaction characteristics of a gravelly fill liquefied during the 1995 Hyogo-Ken Nanbu earthquake. *Soils Found.* **1997**, *37*, 107–115. [CrossRef]
56. Kokusho, T.; Matsumoto, M. Nonlinearity in site amplification and soil properties during the 1995 Hyogoken-Nambu earthquake. *Soils Found.* **1998**, *38*, 1–9. [CrossRef]
57. Lin, P.S.; Chang, C.W.; Chang, W.J. Characterization of liquefaction resistance in gravelly soil: Large hammer penetration test and shear wave velocity approach. *Soil Dyn. Earthq. Eng.* **2004**, *24*, 675–687. [CrossRef]
58. Cao, Z.; Hou, L.; Xu, H.; Yuan, X. Distribution and characteristics of gravelly soil liquefaction in the Wenchuan M s 8.0 earthquake. *Earthq. Eng. Eng. Vibr.* **2010**, *9*, 167–175. [CrossRef]
59. Chen, L.; Yuan, X.; Cao, Z.; Hou, L.; Sun, R.; Dong, L.; Wang, W.; Meng, F.; Chen, H. Liquefaction macro phenomena in the great Wenchuan earthquake. *Earthq. Eng. Eng. Eng Vibra.* **2009**, *8*, 219–229. [CrossRef]
60. Liu, C.Y.; Chia, Y.; Chuang, P.Y.; Chiu, Y.C.; Tseng, T.L. Impacts of hydrogeological characteristics on groundwater-level changes induced by earthquakes. *Hydrogeol. J.* **2018**, *26*, 451–465. [CrossRef]
61. Nespoli, M.; Micol, T.; Enrico, S.; Maria, E.B.; Maurizio, B.; Marco, M.; Antonio, P.R.; Letizia, A.; Adriano, G. Modeling earthquake effects on groundwater levels: Evidences from the 2012 Emilia earthquake (Italy). *Geofluids* **2016**, *16*, 452–463. [CrossRef]
62. Yun, S.M.; Hamm, S.Y.; Cheong, J.Y.; Lee, C.M.; Seo, W.S.; Woo, N.C. Analyzing groundwater level anomalies in a fault zone in Korea caused by local and offshore earthquakes. *Geosci. J.* **2019**, *23*, 137–148. [CrossRef]
63. Porfido, S.; Esposito, E.; Vittori, E.; Tranfaglia, G.; Guarrieri, L.; Pece, R. Seismically induced ground effects of the 1805, 1930 and 1980 earthquakes in the Southern Apennines, Italy. *Ital. J. Geosci.* **2007**, *126*, 333–346.
64. Amoruso, A.; Crescentini, L.; Petitta, M.; Rusi, S.; Tallini, M. Impact of the 6 April 2009 L'Aquila earthquake on groundwater flow in the Gran Sasso carbonate aquifer, Central Italy. *Hydrol. Process.* **2011**, *25*, 1754–1764. [CrossRef]
65. Fleeger, G.M.; Goode, D.J.; Buckwalter, T.F.; Risser, D.W. *Hydrologic Effects of the Pymatuning Earthquake of September 25, 1998, in Northwestern Pennsylvania*; US Department of the Interior, US Geological Survey: Reston, VA, USA, 1999; No. 99-4170.
66. Yun, H.; Min, K.D.; Moon, H.S.; Lee, H.K.; Yi, S.S. Biostratigraphic, Chemostratigraphic, Paleomagne Tostratigraphic, and Tephrochronological Study for the Correlation of Tertiary Formations in Southern Part of Korea. *Paleontology* **1991**, *7*, 1–12.
67. Yan, R.; Woith, H.; Wang, R.J. Groundwater level changes induced by the 2011 Tohoku earthquake in China mainland. *Geophys. J. Int.* **2014**, *199*, 533–548. [CrossRef]
68. Alessio, G.; Alfonsi, L.; Brunori, C.A.; Burrato, P.; Casula, G.; Cinti, F.R.; Civico, R.; Colini, L.; Cucci, L.; De Martini, P.M.; et al. Liquefaction phenomena associated with the Emilia earthquake sequence of May-June 2012 (Northern Italy). *Nat. Hazards Earth Syst. Sci.* **2013**, *13*, 935–947. [CrossRef]
69. Alessio, G.; Alfonsi, L.; Brunori, C.A.; Burrato, P.; Casula, G.; Cinti, R.F.; Civico, R.; Colini, L.; Cucci, L.; De Martini, P.M.; et al. A photographic dataset of the coseismic geological effects induced on the environment by the 2012 Emilia (Northern Italy) earthquake sequence. *J. Jpn. Geotech. Soc. Soils Found.* **1997**, *37*, 107–115.
70. Davis, P.M.; Rubinstein, J.L.; Liu, K.H.; Gao, S.S.; Knopoff, L. Northridge earthquake damage caused by geological focusing of seismic waves. *Science* **2000**, *289*, 1746–1750. [CrossRef]

71. Lee, S.J.; Komatitsch, D.; Huang, B.S.; Tromp, J. Effects of topography on seismic-wave propagation: An example from northern Taiwan. *Bull. Seismol. Soc. Am.* **2009**, *99*, 314–325. [CrossRef]
72. Graves, R.W.; Pitarka, A.; Somerville, P.G. Ground-motion amplification in the Santa Monica area: Effects of shallow basin-edge structure. *Bull. Seismol. Soc. Am.* **1998**, *88*, 1224–1242.
73. Stewart, J.P.; Bray, J.D.; McMahon, D.J.; Smith, P.M.; Kropp, A.L. Seismic performance of hillside fills. *J. Geotech. Geoenviron. Eng.* **2001**, *127*, 905–919. [CrossRef]
74. Olsen, K.B.; Day, S.M.; Minster, J.B.; Cui, Y.; Chourasia, A.; Faerman, M.; Moore, R.; Maechling, P.; Jordan, T. Strong shaking in Los Angeles expected from the southern San Andreas earthquake. *Geophy. Res. Lett.* **2006**, *33*, L073054. [CrossRef]
75. Pitarka, A.; Irikura, K.; Iwata, T.; Sekiguchi, H. Three-dimensional simulation of the near-fault ground motion for the 1995 Hyogo-ken Nanbu (Kobe), Japan, earthquake. *Bull. Seismol. Soc. Am.* **1998**, *88*, 428–440.
76. Ergin, M.; Özalaybey, S.; Aktar, M.; Yalcin, M.N. Site amplification at Avcılar, Istanbul. *Tectonophysics* **2004**, *391*, 335–346. [CrossRef]
77. Debbarma, J.; Martin, S.S.; Suresh, G.; Ahsan, A.; Gahalaut, V.K. Preliminary observations from the 3 January 2017, M_W 5.6 Manu, Tripura (India) earthquake. *J. Asian Earth Sci.* **2017**, *148*, 173–180. [CrossRef]
78. Pratt, T.L.; Brocher, T.M.; Weaver, C.S.; Creager, K.C.; Snelson, C.M.; Crosson, R.S.; Miller, K.C.; Tréhu, A.M. Amplification of seismic waves by the Seattle basin, Washington State. *Bull. Seismol. Soc. Am.* **2003**, *93*, 533–545. [CrossRef]
79. Lee, T.J.; Song, Y.; Uchida, T. Three dimensional magnetotelluric surveys for geothermal development in Pohang, Korea. *Explor. Geophy.* **2007**, *38*, 89–97. [CrossRef]
80. Donati, S.; Marra, F.; Rovelli, A. Damage and ground shaking in the town of Nocera Umbra during Umbria-Marche, central Italy, earthquakes: The special effect of a fault zone. *Bull. Seismol. Soc. Am.* **2001**, *91*, 511–519. [CrossRef]
81. Li, Y.G.; Vidale, J.E.; Cochran, E.S. Low-velocity damaged structure of the San Andreas Fault at Parkfield from fault zone trapped waves. *Geophy. Res. Lett.* **2004**, *31*, L12S06. [CrossRef]
82. Lombardi, D.; Bhattacharya, S. Liquefaction of soil in the Emilia-Romagna region after the 2012 Northern Italy earthquake sequence. *Nat. Hazards* **2014**, *73*, 1749–1770. [CrossRef]
83. Thakkar, M.G.; Goyal, B. On the relation between magnitude and liquefaction dimension at the epicentral zone of 2001 Bhuj earthquake. *Curr. Sci.* **2004**, *87*, 811–817.
84. Lee, T.J.; Yoonho, S.; Deok-Won, P.; Jaesoo, J.; Woon, S.Y. Three dimensional geological model of Pohang EGS pilot site, Korea. In Proceedings of the World Geothermal Congress, Melbourne, Australia, 19–25 April 2015; Volume 19.

geosciences

MDPI

Article

Earthquake Environmental Effects of the 1992 M_S7.3 Suusamyr Earthquake, Kyrgyzstan, and Their Implications for Paleo-Earthquake Studies

Christoph Grützner [1,2,*], Richard Walker [3], Eleanor Ainscoe [3], Austin Elliott [3] and Kanatbek Abdrakhmatov [4]

1 Department of Earth Sciences, University of Cambridge, Cambridge CB3 0EZ, UK
2 Institute of Geological Sciences, Friedrich Schiller University Jena, 07749 Jena, Germany
3 Department of Earth Sciences, University of Oxford, Oxford OX1 3AN, UK; Richard.Walker@earth.ox.ac.uk (R.W.); eleanorainscoe@hotmail.co.uk (E.A.); austin.elliott@earth.ox.ac.uk (A.E.)
4 Kyrgyz Seismological Institute, Academy of Sciences of the Kyrgyz Republic, 720060 Bishkek, Kyrgyz Republic; kanab53@yandex.ru
* Correspondence: christoph.gruetzner@uni-jena.de

Received: 20 May 2019; Accepted: 19 June 2019; Published: 21 June 2019

Abstract: Large pre-historical earthquakes leave traces in the geological and geomorphological record, such as primary and secondary surface ruptures and mass movements, which are the only means to estimate their magnitudes. These environmental earthquake effects (EEEs) can be calibrated using recent seismic events and the Environmental Seismic Intensity Scale (ESI2007). We apply the ESI2007 scale to the 1992 M_S7.3 Suusamyr Earthquake in the Kyrgyz Tien Shan, because similar studies are sparse in that area and geological setting, and because this earthquake was very peculiar in its primary surface rupture pattern. We analyze literature data on primary and secondary earthquake effects and add our own observations from fieldwork. We show that the ESI2007 distribution differs somewhat from traditional intensity assessments (MSK (Medvedev-Sponheuer-Karnik) and MM (Modified Mercalli)), because of the sparse population in the epicentral area and the spatial distribution of primary and secondary EEEs. However, the ESI2007 scale captures a similar overall pattern of the intensity distribution. We then explore how uncertainties in the identification of primary surface ruptures influence the results of the ESI2007 assignment. Our results highlight the applicability of the ESI2007 scale, even in earthquakes with complex and unusual primary surface rupture patterns.

Keywords: earthquake environmental effects; Suusamyr earthquake; Kyrgyzstan; Tien Shan; surface rupture; landslide; digital elevation model (DEM); Structure-from-Motion

1. Introduction

Paleo-earthquake magnitudes are usually estimated from the lengths of mapped primary surface ruptures and the single-event offsets of geomorphological markers. Empirical relationships between magnitude, primary surface rupture length, and average/maximum offset then allow calculating magnitudes (e.g., References [1,2]). Those relationships are derived from recent surface-rupturing earthquakes with well-constrained magnitudes. Using this approach in paleoseismological studies comes with several limitations: (i) It can only be applied to earthquakes with primary surface ruptures. Large events without significant surface offset like the 2015 Gorkha M_W7.8 Earthquake [3] will go unnoticed in the paleoseismological record; (ii) if earthquake recurrence intervals on a fault are long, erosion or sedimentation may have intensely altered the primary surface ruptures or even completely eradicated them (e.g., Reference [4]); and (iii) complex primary surface ruptures on several faults may not be interpreted as a single earthquake. For example, the 2016 Kaikoura Earthquake in New Zealand with a magnitude of M_W7.8 ruptured at least twelve major crustal faults and produced highly variable primary surface ruptures [5]. It is unlikely that such complexity can be understood with paleoseismological methods, and the magnitude of the paleo-earthquake will probably be under-estimated. (iv) Incoherent and anomalous primary surface ruptures can not only be hard to detect, but may also cause problems when it comes to applying the empirical relationships. For example, the 1992 M_S7.3 Suusamyr Earthquake, Kyrgyzstan, broke the surface in two short sets of primary ruptures with a 25 km gap in between [6,7].

One approach to overcome the problem of estimating paleo-earthquake magnitudes based solely on primary surface ruptures is the application of the Environmental Seismic Intensity Scale, ESI2007 [8–10]. In contrast to classical intensity scales, such as MSK (Medvedev-Sponheuer-Karnik) or MM (Modified Mercalli), the ESI2007 only uses effects on the environment to assign earthquake intensities. It, therefore, avoids the influence of building styles and the problem of saturation at high intensities, and allows applying the scale to paleo-earthquakes. The conversion of ESI2007 intensities to magnitudes, however, can only be based on a large set of modern case studies, which are currently being collected in the Earthquake Environmental Effects (EEE) catalogue hosted by ISPRA [11]. Preferably, this set of case studies should include earthquakes with different mechanisms, magnitudes, depths, tectonic and geological settings, distributed all across the globe. A glance at the events included in the catalogue so far makes clear that this is not yet the case.

In this paper, we apply the ESI2007 scale to the 1992 Suusamyr Earthquake. There are several reasons for doing this: (i) The majority of entries in the EEE catalogue are currently from Europe and NW South America. Only a few Central Asian earthquakes are included. Thus, this study contributes to extending the entries for central Asia and compressional tectonic settings; (ii) the Suusamyr Earthquake was special in its primary surface rupture pattern, and the exercise of applying the ESI2007 scale to it thereby points out a general problem; (iii) the earthquake occurred in a sparsely populated area, which hampers the application of traditional intensity scales that mainly focus on damage to human-made infrastructure; and (iv) we document secondary cracks of the Suusamyr Earthquake with a high-resolution digital elevation model (DEM) computed from drone aerial imagery and the Structure-from-Motion (SfM) technique [7,12,13].

Here, we first provide an overview of the 1992 Suusamyr Earthquake and review the existing intensity assessments. Then we evaluate the published EEEs from the literature, we include our field observations, and we present an ESI2007 intensity map. Finally, we discuss ESI2007 intensities vs. MM and MSK intensities and the implications of our study for the application of the ESI2007 scale on paleo-earthquake in general.

2. The 1992 Suusamyr Earthquake

An earthquake with a magnitude of M_S7.3 hit the Suusamyr Basin in Kyrgyzstan on 19 August 1992, 02:04 GMT (Figure 1). The Suusamyr Basin is an east-west elongated intramontane basin in the Kyrgyz Tien Shan, surrounded by Paleozoic bedrock. The thickness of the Cenozoic basin fill reaches several hundreds of meters in its widest part near the town of Suusamyr ([14,15] and references therein). A thick Quaternary cover blankets most of the Neogene rocks in the basin center. The basin is bound by E-W striking thrust faults to the north and the south (Figure 1c). North–South shortening is accommodated by the oppositely-vergent thrust. Reference [7] showed that several surface-rupturing earthquakes occurred in the Late Quaternary on these faults, testifying to the tectonic activity of the area. Several ~E-W trending anticlinal ridges in the basin also record ongoing shortening. The Chet Korumdy ridge, which will be discussed in this paper, is a 250 m high, 7 km long ridge paralleling the Suusamyr River (Figure 1). It is made up of steeply north-dipping sediments of Pliocene-Quaternary age [6]. At its western tip, prominent wind gaps record Late Quaternary uplift [7,15,16].

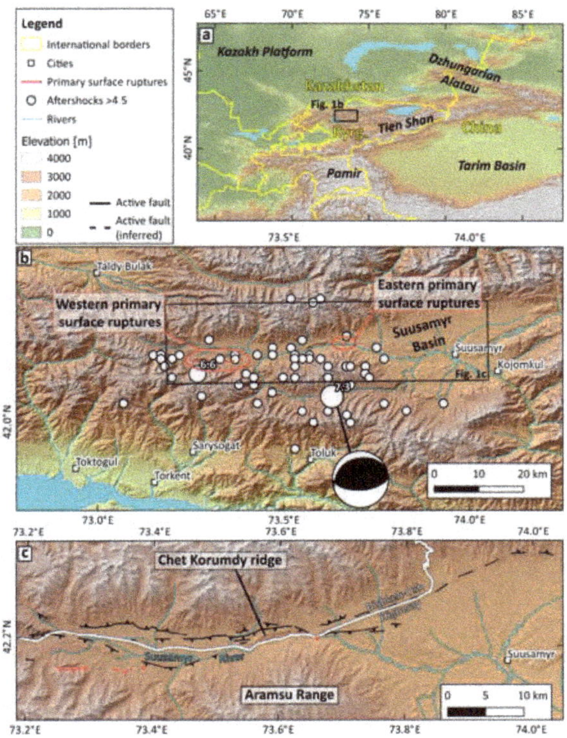

Figure 1. (**a**) The Suusamyr Basin in the Kyrgyz Tien Shan. Kyrg.: Kyrgyzstan. (**b**) Overview of the Suusamyr Basin and the 1992 earthquakes. Large white circles: M_S7.3 main shock with the Harvard CMT solution [17] and the strongest aftershock (M6.6) located with data from the Kyrgyz regional network [18]. All other aftershocks are from the compilation of Reference [19]. Elevation data: SRTM1. Note the >20 km gap in between the two sets of primary surface ruptures. (**c**) The Suusamyr fault system (from Reference [7]) and location of the Bishkek-Osh Highway.

The Suusamyr Earthquake ruptured an east-west-trending, south-dipping thrust fault in the Suusamyr Basin with the epicenter beneath the Aramsu Range. Reference [18] reports a centroid depth between 5–21 km and a fault dip of 50° ± 13° based on the teleseismic body-wave inversion of the main shock and an aftershock study using broadband stations and temporal and regional networks. Reference [20] analyzed body waves and surface waves, and reported a hypocenter at a depth of 14 ± 2 km, a fault dip of 49° ± 6° to the south, and a rake of 105° ± 3°. Within two hours of the main shock, three large aftershocks with magnitudes of M_S6.6, M_S6.6, and mb6.0 occurred west of the Aramsu Range, but their mechanisms and depths are unknown [18]. The more than 900 aftershocks that were registered in the months after the Suusamyr Earthquake occurred at depths shallower than 15 km. They occurred within a rather narrow, south-dipping zone at depths between 5 and 15 km and showed a more diffuse pattern near the surface [18].

The most surprising effect of the earthquake was its primary surface rupture pattern. Two sets of relatively short primary ruptures were identified. At the eastern tip of Chet Korumdy ridge, a short primary rupture occurred in the Suusamyr River bed [6]. This site is referred to as the 'eastern primary surface ruptures'. The second set of primary surface ruptures formed more than 20 km further west (the 'western primary surface ruptures'), with no hints for surface breaks in between these two (Figure 1). This peculiarity and its implications for ESI2007 intensity assignment will be discussed later in this paper. Apart from these primary surface ruptures, the Suusamyr Earthquake also caused secondary cracks, often referred to as secondary ruptures in the literature. Here we will use the term secondary ruptures to summarize all types of cracks and ruptures that cannot be attributed to primary faulting. The 1992 Suusamyr Earthquake also caused widespread mass movements, mud eruptions, and jumping stones.

3. Macroseismic Effects of the Suusamyr Earthquake

The Suusamyr Earthquake resulted in more than 50 deaths, most of which were due to collapsed buildings throughout the valley and mass movements that occurred at Belaldy River [6,21,22]. The epicentral area was surveyed by field crews and from a helicopter during the months following the earthquake. The results of these surveys were reported in great detail in References [6,21–23]. Reference [7] revisited the Suusamyr Basin and reported additional information on the earthquake effects. In the following, we briefly report on the published intensity assessments and review the earthquake environmental effects of the Suusamyr Earthquake in light of the ESI2007 scale.

Reference [6] published a Modified Mercalli (MM) intensity map of the earthquake (Figure 2). They used data from 41 individual sites, assessing the damage to human-made structures and also environmental effects. They report intensities of MM ≥ X for two sites. The first one is an E-W elongated zone around the eastern primary surface ruptures and the Chet Korumdy ridge, the second one is an NNE-SSW elongated zone in the Belaldy River valley region. The first site is characterized by primary and secondary surface ruptures, large landslide volumes, and strong peak ground accelerations (PGA). The second site, however, does not encompass primary or secondary ruptures, but a very large mass movement in the Belaldy Valley. The combined area of MM ≥ X is 62 km^2. Intensities of MM = IX were assigned to an ENE-WSW elongated area of 835 km^2, surrounding the primary surface ruptures and encompassing the northern part of the area where numerous mass movements occurred. An E-W elongated area of 3000 km^2 was assigned intensity MM = VIII. This isoseismal includes nearly all mass movements. The area of intensity MM = VII is NNW-SSE elongated (roughly parallel to the orogenic trend) and encloses 13,000 km^2 (Table 1).

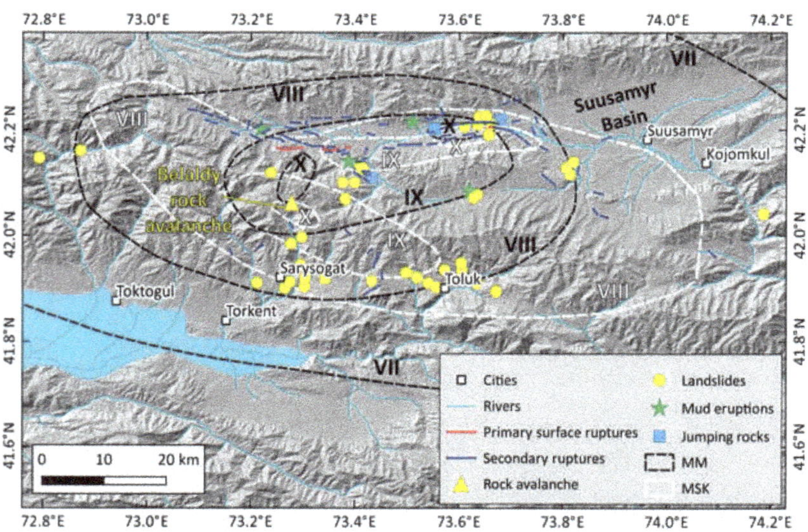

Figure 2. Earthquake environmental effects (EEEs) of the 1992 Suusamyr Earthquake. Primary surface ruptures are from References [6,7,21]. Secondary ruptures are from References [21,23]. Landslides, the location of the Belaldy rock avalanche, mud eruptions, and jumping rocks are from References [19,23]. Modified Mercalli (MM) intensities are from Reference [6]; Medvedev-Sponheuer-Karnik (MSK) intensities are from Reference [19]. Hillshade is based on SRTM1 elevation data.

Table 1. Comparison of the intensity distribution of the 1992 Suusamyr Earthquake.

Intensity	Area MM in km^2 [6]	Area MSK in km^2 [19]	Area ESI2007 in km^2 (this Study)
X	62	51	317
IX	835	600	1735
VIII	3000	3200	2700
VII	13,000	-	6500

Intensity data were also compiled within the framework of the SNSF Project No IB7320-110694 (http://www.kyrgyzstan.ethz.ch) by Reference [19], who published Medvedev-Sponheuer-Karnik (MSK) intensities (Figure 2). The shape of the isoseismals differs from the MM intensities reported by Reference [6], but the affected locations and extent are similar. Highest intensities of MSK = X were also assigned to two separate locations, covering 51 km^2. The one near the eastern primary surface ruptures is located south of the Suusamyr River and has a shape similar to the MM ≥ X intensities, although shifted to the south by ~1 km. The other one near the Belaldy River valley is elongated in NW-SE direction and also located further south compared to MM ≥ X. Intensities of MSK = IX were assigned to two individual patches with a total of 600 km^2. The first one runs parallel to the Suusamyr River and is stretched in E-W direction, the second one is elongated NW-SE (parallel to the orogenic trend), covering the high mountains that border the Suusamyr Valley to the southwest. The zone of intensity MSK = VIII overlaps with the MM = VIII intensities, but is more E-W elongated and also followed the orogenic trend. The total area of MSK = VIII is 3200 km^2. No MSK = VII intensities were reported by Reference [19] (Table 1).

In the following, we discuss all earthquake environmental effects described in the literature and present new data from our fieldwork. We assign ESI2007 intensities for all EEEs, which are the basis for our intensity map.

3.1. *Primary Surface Ruptures*

The eastern primary surface ruptures occurred at the eastern tip of Chet Korumdy ridge next to the Bishkek-Osh highway (162 km marker; Figures 1–3). As a result of the main primary surface rupture, the Suusamyr River temporarily changed its course. The ruptures were trenched by Reference [24] who confirmed that the fault reached the surface and that the surface deformation is due to thrust motion on an S-dipping fault plane. Reference [21] report a 400 m long, N-facing scarp of up to 2.7 m height and up to 0.3 m of lateral slip. Here, Reference [6] report a fold-scarp geometry and 500 m of primary ruptures with a height of 2.7 m. They estimate a total slip of 4.2 m assuming no horizontal component of motion. Both authors mention numerous rupture segments that also affected the nearby highway, but state that the scarp in the river bed is by far the dominating feature. Furthermore, the authors state that the other segments are likely secondary ruptures. Reference [7] surveyed the eastern primary surface ruptures again in 2015/16 with a drone and produced a high-resolution DEM with the SfM technique. They document a 600 m-long primary surface rupture with up to 3.1 m vertical displacement and a total slip of ~3.6 m that occurred in the Suusamyr River bed. No evidence for the other, smaller scarps on the highway was preserved in 2015/16, due to road repairs, but small secondary scarps were still visible south of the road. Reference [21] reported additional ~4 km of N-facing scarps on the hills further to the east, which could be a continuation of the large main scarp in the river valley. A 300 m wide deformation zone encompasses numerous fault strands with vertical offsets of 0.1–1.05 m height. These scarps face north and are at least in one location formed by a steeply S-dipping fault. References [6,23] do not mention these additional ruptures.

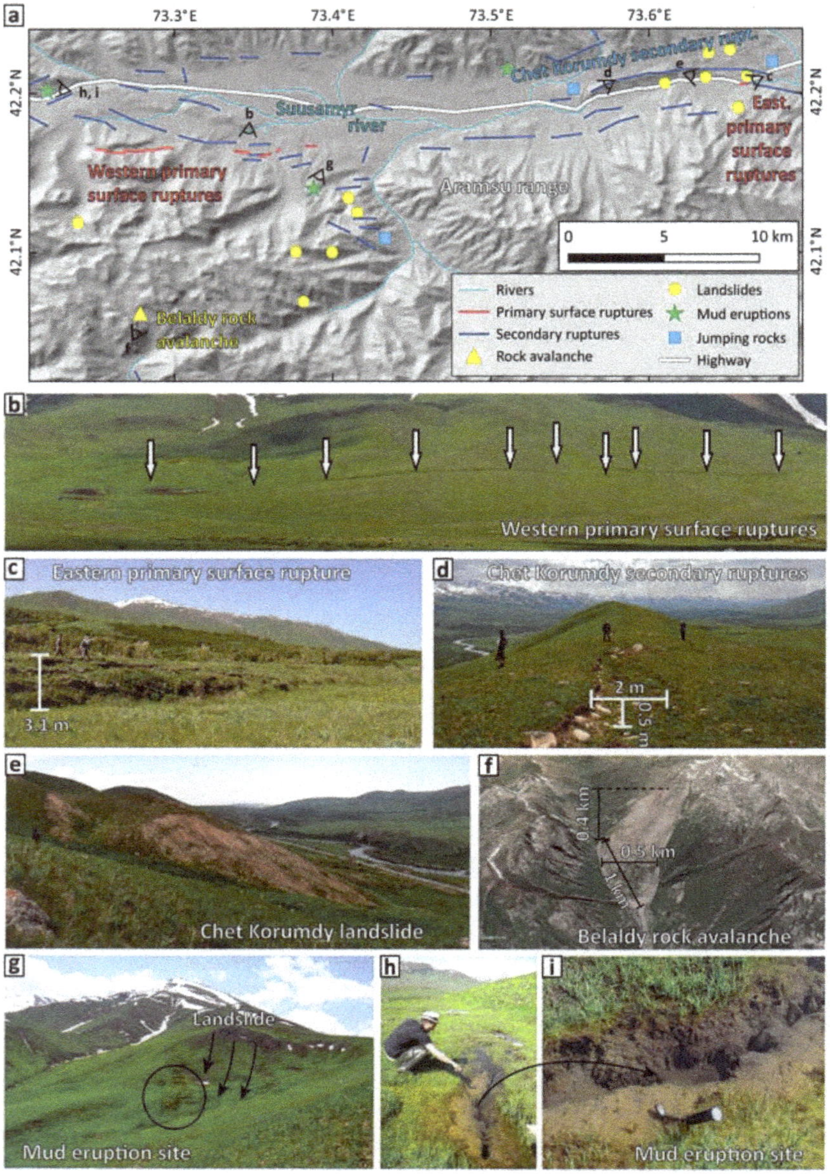

Figure 3. Earthquake environmental effects of the 1992 Suusamyr Earthquake. (**a**) Map with the photo locations. (**b**) The central section of the western primary surface ruptures in summer 2015. View to the south. (**c**) The highest section of the eastern primary surface ruptures in summer 2016. View to the west. (**d**) Secondary ruptures on top of the Chet Korumdy ridge in summer 2015. View to the west. (**e**) The 1992 Chet Korumdy landslide that damaged the Bishkek-Osh highway (visible in the background) in summer 2015. View to the west. (**f**) Oblique satellite image of the Belaldy rock avalanche. Source: GoogleEarth/CNES/Airbus (2019), imagery from16 August 2013. View to the northeast. (**g**) Mud eruption site W of the Aramsu range. Note that the eruption site sits at the landslide toe. View to the southwest. (**h**) Mud eruption site in the W of the valley. This site also sits on top of a landslide. View to the west. (**i**) Close-up of the cracks from which the mud erupted.

The western primary surface ruptures occurred on six segments with a total length of 7 km, spanning a 12 km long rupture zone [6,7,21] (Figures 2 and 3). N-facing scarps are typically 0.7–1.2 m high, but locally reach up to 2 m. Reference [7] also report that some scarps show up on pre-1992 satellite imagery, but their pre-1992 height in unknown. This is why the total vertical offset measured from high-resolution DEMs of marker surfaces has to be taken as a maximum value. However, References [6,21] report 0.9–1.8 m high fresh ruptures scarps that formed in 1992, and we take this information as the best available data on their height.

The two sets of primary surface ruptures with their >20 km separation and the unusual scarp length-to-height ratio of the eastern primary surface ruptures can be treated in three different ways in an ESI2007 intensity assignment [8,9]. (i) If we consider the ruptures separately, the western primary surface ruptures correspond to intensity ESI = IX because of their length of 7 km. Using the 12 km long rupture zone reported by Reference [7] would correspond to ESI = X. The surface displacement also results in ESI = X. The eastern primary surface ruptures are classified as ESI = VIII if we use the length of the main scarp only (600 m). The additional ~4 km of scarps reported by Reference [21] would lead to ESI = IX. The maximum surface displacement of 3.6–4.2 m results in ESI = XI, since 3 m offset is recommended as the threshold between ESI = X and ESI = XI. (ii) The second option is to add up the primary surface rupture lengths of the western and the eastern sites and to ignore the spatial separation. Using the 7 km primary rupture length in the west and the 600 m length in the east gives ESI = IX. Using the 12 km long primary rupture zone in the west leads to ESI = X, regardless whether the additional 4 km in Reference [21] from the east are taken into account or not. Similarly, using the additional 4 km from Reference [21] leads to ESI = X, using either the 7 km primary rupture length or the 12 km primary rupture zone in the west. (iii) The third option is to measure a total primary rupture zone length from the western tip of the western primary surface ruptures to the eastern tip of the eastern primary surface ruptures. This leads to a length of 35 km without and 40 km with the additional 4 km of primary surface ruptures reported by Reference [21], corresponding to ESI = X.

3.2. Secondary Ruptures

Extensive ground cracks were caused by the Suusamyr Earthquake throughout the Suusamyr Basin and along with the surrounding mountain ranges (Figures 2 and 3). These effects were mapped in detail after the earthquake and published in References [6,21,22]. Reference [21] report that secondary ruptures covered an area of more than 4000 km^2. Reference [6], however, estimates that the total area affected by secondary earthquake effects (including secondary ruptures, mass movements, and mud eruptions) is only ~2500 km^2, based on the report in Reference [25]. We used the map from Reference [22] and calculated a total secondary rupture length of 114 km, covering an area of 2520 km^2.

Most secondary cracks occurred in the E-W running Suusamyr River valley between the western and the eastern primary ruptures, but secondary cracks were also reported from SE of the Aramsu Range and the area between Toluk, Sarysogat, and the Belaldy rock avalanche [21,22] (Figure 2). Along the flanks of the mountain ranges, seismic shaking led to gravitational cracks that mostly paralleled the valley. These reached individual lengths of several tens of meters, spanning fissure zones of several kilometers length. The Chet Korumdy ridge was particularly affected by secondary cracks, which mainly formed as E-W elongated grabens on top of the ridge as a result of extension (Figures 3 and 4) and S-facing scarps on its southern side. We surveyed parts of the secondary ruptures on the Chet Korumdy ridge with a drone and created a high-resolution DEM, shown in Figure 4.

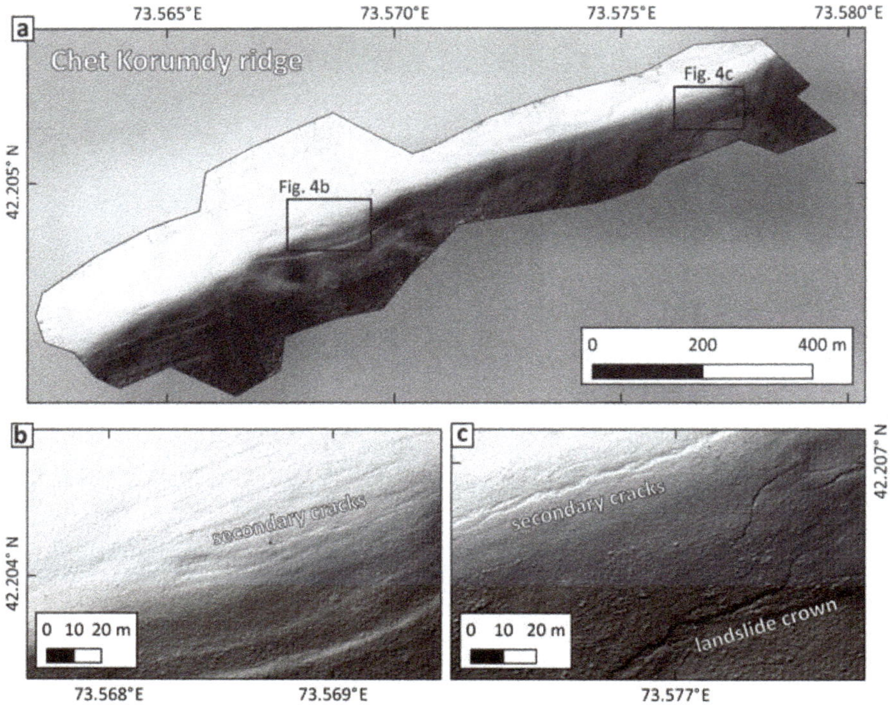

Figure 4. Secondary ruptures at Chet Korumdy ridge included small grabens that formed at the top of the ridge and mainly south-facing scarps on its southern side. (**a**) Overview of the western part of Chet Korumdy ridge. (**b**)Detail of the small grabens and secondary cracks that formed on top of the ridge. Also see Figure 3d. (**c**) Secondary cracks on top of the ridge and south-facing crown of a landslide that formed in the 1992 earthquake. The hillshade is from drone survey imagery and has a resolution of 2.91 cm/pixel.

Secondary ruptures were—to a lower extent—also reported from some flat areas in the Suusamyr basin and west of Chet Korumdy ridge. Most of them occurred close to rivers and are likely related to the presence of unconsolidated, saturated sediments subject to lateral spread.

We assign intensities of ESI = X to the Chet Korumdy area and the eastern primary surface ruptures, based on the clustering and the dimensions of the secondary ruptures. Near the western primary surface ruptures, we assign ESI = IX also because of the number, the size, and the extent of the phenomena. Intensities of ESI = VIII were assigned to the remaining, more isolated occurrences of secondary cracks.

3.3. Mass Movements

The 1992 Suusamyr Earthquake caused widespread mass movements, including landslides, rock falls, and rock avalanches. These mass movements have been intensively studied after the earthquake [6,19,21,22,26–31], and we compiled the intensity assignment from these studies.

The largest mass movement was the Belaldy rock avalanche (Figures 2 and 3) with an estimated volume of 10^7–10^8 m^3 [6,22,30]. Additional large landslides were reported from the downstream section of the Belaldy river, with volumes of a few thousand to a few tens of thousand cubic kilometers [21]. Those deposits were re-activated in the year after the earthquake and caused a catastrophic mud flow that affected several villages [22]. At the southern flank of Chet Korumdy ridge, landslides with up to 0.5–1 × 10^6 m^3 volume were mobilized and damaged the Bishkek-Osh highway [26]. A large cluster of mass movements occurred in the Toluk-Sarysogat area, where also intense secondary ruptures were observed (Figure 2). Based on the maps in References [6,19,21,22], we compute a total affected area of 2336 km^2.

The volume of the largest mass movement corresponds to intensities of ESI = X-IX, smaller slides are associated with ESI = VIII-IX. Little information is available on the volume of individual mass movements apart from the examples discussed above. For ESI2007 intensity assessments, the total affected area is also an input parameter. The southern cluster of landslides in the Toluk-Sarysogat area covers ~500 km^2, which is much higher than the threshold of ~100 km^2 recommended for ESI = VIII zoning. We, therefore, assigned intensities of ESI = IX to areas where mass movements cluster and intensities of ESI = VIII to more isolated mass movements. The isolated mass movements in the far west and the far east of the study area were assigned intensity ESI = VII. Using the total affected area (2636 km^3) as an input value leads to a maximum intensity of ESI = X.

3.4. Mud Eruptions

No liquefaction was reported from the Suusamyr Earthquake, bud mud eruptions in the days following the earthquake were described from four locations [19,21]. Reference [21] report that the explosive ejections occurred at 'boggy sites on young active faults' and that they were even accompanied by gas combustion. Reference [6] cites eye witnesses who heard 'loud noise' and saw 'smoke that bore a pungent aroma'. They also report radially scattered rocks and mention the connection underlying active faults. Reference [21] mention that one of the eruptions resulted in a 400 m long and 100 m wide mud flow. We visited two of the sites in 2015 (Figure 3) and found muddy bogs related to landslides. At the central site near the western primary surface ruptures a muddy layer was found at a landslide toe (Figure 3g). We excavated a 25 cm deep pit into the deposits and found them to be dark, uniform, saturated soils without any hints of burned layers. We did not find any radially scattered rocks or similar signs of an explosive event. Instead, the site appeared like an over-pressured landslide mass from which water was seeping. At the westernmost site near the highway (Figure 3h,i), we encountered swampy areas on top of a landslide deposit. Puddles of several meters length contained fissures with seeping water (Figure 3i). No hints for explosive eruptions could be identified in 2015/16. If those mud eruptions happened as reported by the eyewitnesses, their traces were entirely gone 23 years after the earthquake and the sites looked like ordinary, yet very wet landslides.

These effects are hard to evaluate with the ESI2007 scale, since they are rather unusual. We decided to assign them intensity ESI = IX to account for the fact that they are unusual phenomena related to the seismic shaking, and that they had a considerable impact on the natural environment. However, the observation that they occurred after the earthquake and not co-seismically complicates the assessment. With the ESI = IX assignment, they fall well within the range of values of the neighboring EEEs.

3.5. Jumping Rocks

At several sites, jumping rocks were found during the post-earthquake surveys [21]. These are boulders of considerable size sitting on flat or nearly flat surfaces that were displaced as a result of the strong ground motion, not to be confused with rock falls or other types of gravitational effects. These effects require >1 g peak ground acceleration. Reference [21] describe rocks with a weight of 40–70 kg that moved 2–5 m on a flat surface west of the Aramsu range and an ~8 ton boulder that moved 30 cm uphill at the same site (Figure 2). Near the eastern primary surface ruptures, a ~1.3 ton block reportedly moved 2 m and two boulders of ~10 tons sitting on flat ground underwent displacement of 25–30 cm and 10° rotation. In the same area, an undisclosed number of rocks were moved laterally 85 cm [21].

We visited the Chet Korumdy ridge in 2015/16 and looked for the jumping rocks, but we failed to identify the boulders that moved in 1992. Most likely the footprints of the pre-earthquake position were already completely altered in the 23 years after the earthquake, so that there is no way to distinguish between a boulder that had moved and ones that did not jump.

The ESI2007 guidelines assign intensities of ESI = IX-XII to the phenomenon of jumping rocks, depending on their size, roundness, travel distance, and the local slope. Based on the values reported by Reference [21], we assign ESI = X to all the sites with large jumping boulders.

3.6. Summary of the ESI2007 Assignment

We used 132 data points to assign ESI2007 intensities (Figure 5; Table 1; see Supplementary Materials for a list with the details of each point).

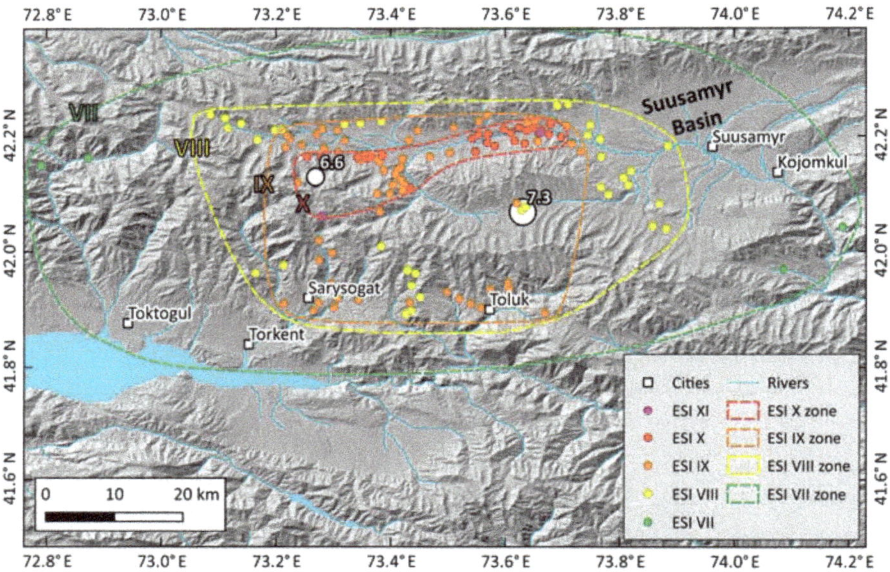

Figure 5. Map of ESI2007 intensities compiled in this study. The location of the main shock and the strongest well-located aftershock are shown. See the Supplementary Data for details on each data point.

ESI2007 intensity ESI = XI was assigned to two sites only: To the eastern primary surface ruptures because of the amount of offset, and to the Belaldy rock avalanche because of the large volume of this mass movement.

Intensities of ESI = X cover an area of 317 km^2 and are mainly based on the evaluation of primary surface ruptures, clusters of secondary ruptures, jumping rocks, and large mass movements. This isoseismal has an E-W elongated shape paralleling the fault strike and includes the Suusamyr river valley and the slopes south of it.

ESI = IX intensities cover an area of 1735 km^2 and include secondary ruptures, clusters of mass movements, and mud eruption sites. This isoseismal has an E-W elongated, rectangular shape.

Intensities of ESI = VIII were found in an E-W elongated ellipse with an area of 2700 km^2. This intensity is based on rather isolated secondary ruptures and individual mass movements.

For intensity ESI = VII, only a few data points were available in the far east and the far west of the study area. These intensity values are based on four data points only, two mass movements and two secondary cracks. Thus, they come with high uncertainty, and the ESI = VII isoseismal could only be roughly estimated to cover ~6500 km^2. This value is highly uncertain. In contrast, drawing a concave hull around all secondary effects combined (cracks, mud eruptions, landslides, rock falls, jumping stones) results in an area of 3254 km^2 only, which is little larger than the ESI = VIII isoseismal.

4. Discussion

We combined our intensity assessment with the macroseismic effects on the environment reported by Reference [6], who list 41 individual sites with macroseismic observations. We followed the procedures described in detail in References [8–10].

4.1. The ESI2007 Map of the Suusamyr Earthquake and Comparison with Other Intensity Scales

The assignment of ESI2007 values for the primary surface ruptures is not straight forward as pointed out in Section 3.1. Depending on how the spatial separation between the two sets of primary surface ruptures is treated, one option is to plot two areas with different intensities around either location (ESI = X in the west and up to ESI = XI in the east). The other option is to average the observations and to plot one or two zones with ESI = X. Taking into account the unusual scarp length-to-height ratio of the eastern primary surface ruptures, for which the underlying mechanism is not yet well understood, we decide to treat them as an exception that does not justify assigning ESI = XI. Based on the scarp length, there we assign ESI = X.

Secondary effects allow drawing a more coherent image of the intensity distribution. Although the Belaldy rock avalanche stands out, due to its huge volume that would justify intensity ESI = XI, we treat it as an isolated event that fits more into the general pattern of ESI = X mass movements in the epicentral area. In general, we found that the highest ESI intensities can be assigned, due to the size of individual effects and their spatial distribution. Intensity values of ESI = IX and ESI = VIII, on the contrary, can be distinguished based on the spatial clustering of secondary effects only. This is mainly due to the lack of detailed descriptions for each locality, such as landslide volume or length of individual secondary ruptures and cracks.

The highest ESI2007 intensity X covers an E-W elongated area that roughly includes MM = X and MSK = X intensities from previous studies (Figures 5 and 6). In contrast to the ESI2007 intensities, MM and MSK intensities were assigned to two isolated patches surrounding the eastern and western primary surface ruptures, respectively, and to the Belaldy rock avalanche area (Figure 6). Intensity MSK = IX was also attributed to two isolated patches by Reference [19], whereas MM = IX was assigned to an E-W elongated patch [6]. Our intensity assessment results in a larger patch of ESI = IX, which reaches further south than MM and MSK intensities. This is due to a large cluster of mass movements in the Sarysogat-Toluk area and probably reflects the fact that this area is sparsely populated, but sits on the hanging wall above the actual fault plane. The mud eruption sites were also assigned ESI = IX, although they occurred after the earthquake [21]. However, they are related to the Suusamyr event. In any case, they do not have a strong influence on the overall ESI2007 map because they do fall in areas where other effects were mapped with ESI = IX, too. Further examples of similar phenomena are needed to understand these effects and their potential for ESI assignments better. Intensity ESI = VIII was found in an area smaller than that assigned MM = VIII and MSK = VIII, and the same is true for intensity ESI = VII. This can be explained by the fact that minor shaking can still cause damage to houses and infrastructure, while it is not sufficient anymore to cause significant earthquake environmental effects, such as landslides or secondary ruptures.

The observations on the ESI2007 intensity distribution and the comparison with MM and MSK intensity assessments highlight three peculiarities of the ESI2007 scale. First, it allows distinguishing high intensities in places were traditional scales saturate because of the total collapse of human-made structures that are not reinforced. Second, it allows assigning intensities in sparsely populated places, where traditional scales cannot be applied, due to the lack of observations. Third, it rather captures high intensities (>VII), which cause significant environmental effects. Thus, for modern earthquakes, it is best applied in combination with the traditional scales to get the full picture of earthquake effects. For paleo-earthquakes, however, the insensitivity of small to moderate events is useful because it makes it less likely to confuse moderate earthquakes with significant events that release a large seismic moment.

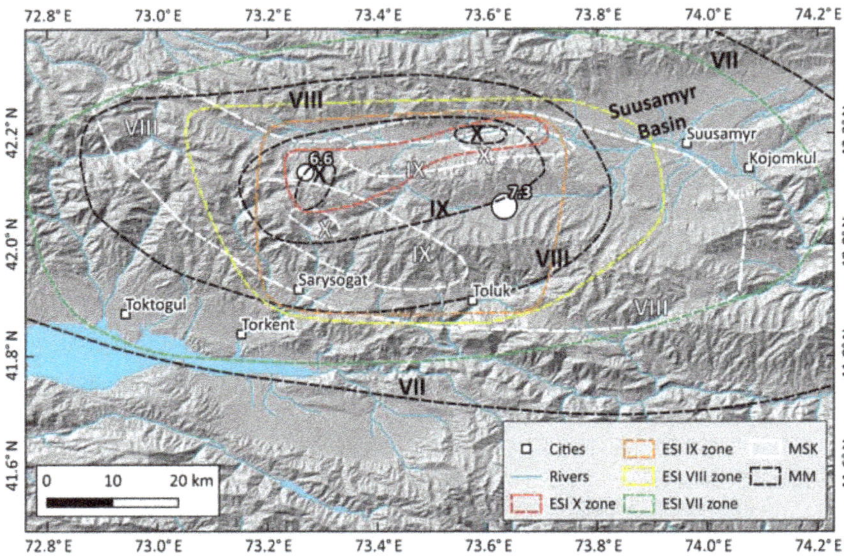

Figure 6. Comparison of ESI2007 intensities with MM isoseismals [6] and MSK intensities from Reference [19]. Note the shift to the south of the ESI = IX, which we attribute to the sparse population in the high mountains of the Aramsu range, which hampers traditional MM and MSK intensity assessments.

4.2. Treating Uncertainties in the Application of the ESI2007 Scale

Uncertainties in intensity assignment mainly arise from the unknown volumes of most landslides. However, the maximum ESI2007 intensity is well-constrained by the primary surface ruptures and the Belaldy rock avalanche. No mass movements of similar size have been reported, so maximum intensities are unlikely to be higher. The lowest intensities are defined by the extent of documented environmental effects, which is also well-known. Therefore, the shape of the isoseismals between maximum and minimum may somewhat vary, but the overall picture and the maximum intensity are unlikely to be affected by these uncertainties.

Another source of uncertainty is the unknown height of the pre-1992 scarps at the western primary surface ruptures that [7] identified on satellite imagery from 1968. The post-earthquake surveys report a height of 0.9–1.8 m [6,21], while the survey in 2015/16 found heights of 0.7–1.2 m with some outliers reaching 2 m height [7]. The imagery of the pre-1992 scarps does not reveal a free-face (Figure 5 in Reference [7]), which indicates that the height of the fresh ruptures reported by References [6,21] should be reliable because the height of the free-face is easy to measure. These heights would correspond to ESI = X. If the pre-1992 vertical offset was larger than 0.5–1.4 m, a lower intensity of ESI = IX would have to be assigned [8,10]. Given the state of preservation of the free-face in 2015/16, we consider this very unlikely (Figure 3b, Reference [7]). If the length of the western primary surface ruptures is taken into account, ESI2007 intensities do not change of course. Therefore, the height of the 1992 western primary surface ruptures does not have a large effect on the overall ESI2007 assignment. While the problem of pre-existing scarps is usually a source of uncertainty hard to deal with in paleoseismology studies, this effect does not pose a significant problem in the case of the 1992 Suusamyr earthquake. Extending the use of scarp-diffusion modelling to multi-event scarps [32–34] may help to improve the interpretation of multi-event scarps.

From a paleoseismology perspective, the peculiar offset-to-length ratio of the eastern primary surface ruptures is interesting. Although the nature of the eastern primary surface ruptures of the 1992 Suusamyr Earthquake is well understood [6,7,21,23–25], such unusual primary effects are likely hard to interpret when attributed to a paleo-earthquake. One obvious interpretation of such an unusal feature would be that a longer primary scarp once existed, and that erosion has eradicated the prolongation of the scarp along strike. In that case, using the vertical offset only would result in ESI = X-XI. Using the mapped length to assign ESI2007 intensities would result in ESI = VIII only. In that case, the earthquake intensity, and thus, also the magnitude, would be dramatically underestimated.

If in case of a paleo-earthquake, the eastern and western primary ruptures would be assigned to the same event depends on the precision of the dating methods and the interpretation of the individual researcher. Reference [7] have shown in the Suusamyr case, this may indeed be challenging. Another possibility would be that, due to the unusual offset-to-length ratio of the eastern primary surface ruptures, they are not even identified as such, but attributed to other effects, such as being a landslide toe or being a terrace riser. Paleoseismological trenching could help to solve this problem. However, in the case of false attribution, there would be no intensity ESI = XI in this area. The clustering of landslides and secondary cracks could probably be ascribed to topography and preservation effects, and the epicenter of the earthquake would be located further to the west near the western surface ruptures.

4.3. The Problem of Strong Aftershocks

A significant source of uncertainty in the ESI2007 assignment is related to the three strong aftershocks that occurred within two hours of the main event. These earthquakes had magnitudes of M_S6.6, M_S6.6, and mb6.0, and they occurred west of the Aramsu Range [18], although their mechanism and location are not well constrained. Given their magnitudes, they may have even caused the western primary surface ruptures [7] or at least contributed to their total lengths. The same holds true for the secondary effects. There is virtually no possibility to distinguish between the effects of the main shock and those of the three strong aftershocks. The earthquakes occurred so close in time, and in such a sparsely populated region, that there are no eyewitness accounts that could help to solve this problem. Perhaps the shape of the isoseismals with the highest intensities could support the interpretation of a combined effect of the main shock and the aftershocks, but we will most likely never know.

This problem usually arises in paleoseismology, when due to the limits of Quaternary dating methods, it is impossible to distinguish between one strong earthquake and two smaller ones, or similar combinations. However, the Suusamyr case illustrates that even for modern events this may be an issue, especially in sparsely populated regions and when significant aftershocks occur right after the main shock before any documentation of the earthquake effects has been undertaken.

4.4. Lessons for ESI2007 Assignments

The peculiar primary surface rupture pattern of the Suusamyr Earthquake poses a problem for ESI2007 assignments as discussed above. Thanks to the widespread secondary effects, we were able to produce an intensity map that relies on a large number of observations. If only information on the primary surface ruptures were available, for example from a similar paleo-earthquake, the picture would look different. This is important from a paleoseismological point of view. For many, if not most pre-historic earthquakes, only data on the primary surface ruptures are derived from field mapping and/or trenching studies. Reference [7] have shown that this may be misleading in the Suusamyr Basin. Another example is the M_W6.6 1998 Fandoqa Earthquake that occurred on the Gowk Fault in Iran [35]. This earthquake re-ruptured a section of the fault that had already ruptured in an M_W7.1 earthquake in 1981. Primary surface ruptures were observed in both cases, but the offsets caused by the smaller earthquake were several times larger than those of the stronger event. Reference [35] report that the two earthquakes ruptured the fault at different depths and rightly emphasized the problem this causes for paleoseismological studies in similar settings.

Similarly, complex primary surface ruptures from paleo-earthquakes may not be entirely understood and, therefore, lead to a wrong ESI2007 assignment [36]. Modern examples of such complex primary surface ruptures are among others the 1889 Chilik Earthquake [37], the 1911

Chon Kemin Earthquake [38], the 2010 El Mayor Cucapah Earthquake [39], and the 2016 Kaikoura Earthquake [5].

Modern earthquake ruptures can be studied extensively in the field [40], using seismological methods, and with (space) geodesy [41]. Their secondary effects can also be documented in great detail. Most studies of paleo-earthquakes, however, mainly focus on the primary surface ruptures, because this is often the only definitively seismogenic evidence left in the landscape. Notwithstanding, with the Suusamyr example in mind, we emphasize the importance of collecting as much information on the environmental effects of such earthquakes as possible.

Unfortunately, secondary effects may also be problematic for intensity assessment. The Belaldy rock avalanche formed a landslide lake which burst ten months after the Suusamyr Earthquake in June 1993. Triggered by snowmelt, a debris flow mainly incorporating material from the 1992 landslides reached and damaged the village of Torkent 30 km downstream [22,42]. Such phenomena may be easily confused with coseismic effects. In case of the Suusamyr Earthquake, the mud eruption sites that we visited a few years after the earthquake were virtually indistinguishable from ordinary landslides. Not only will they be interpreted as such in the case of a paleo-earthquake, but mud eruptions may also happen in any future landslide without a seismic trigger. From a paleoseismological perspective, temporal clustering of earthquake environmental effects can help to identify seismic events, but as can be seen from the 1998 Fandoqa Earthquake [35] or the Central Italy earthquake series [43], earthquakes can occur very close in time on the same fault, which is beyond the resolution limit of any Quaternary dating method (perhaps excluding varves).

5. Conclusions

We have compiled an ESI2007 intensity map of the 1992 Suusamyr Earthquake based on earthquake environmental effects only. This assignment was based on published data on primary and secondary earthquake effects, and in our field mapping. The ESI2007 intensity distribution is roughly comparable to the MM and MSK intensities reported, although we note important differences mainly in the shape of the isoseismals (Figure 6), but also in the area they encompass (Table 1). One major advantage of the ESI2007 scale is that it allows incorporating data from uninhabited areas, which led to a shift to the south of the overall intensity pattern.

Due to the peculiar pattern of 1992 primary surface ruptures, the assignment of ESI2007 intensities to these effects is not straight forward and allows different approaches. Depending on the input data used, the resulting ESI2007 intensities vary significantly. When combined with the secondary effects of the earthquake, however, a coherent picture emerges.

We show that this may not be the case when dealing with a paleo-earthquake of similar characteristics. The unusual offset-to-length ratio of the eastern primary surface ruptures and the gap between the western and eastern primary surface ruptures are prone to false or simplified interpretations which may hamper the correct assignment of ESI2007 intensities. This is especially important as the secondary effects, such as gravitational cracks and mass movements are likely to be short-living features that could also happen without a seismic trigger. The eastern primary surface ruptures occurred in a river bed and are also likely to be eroded within a relatively short time span. Thus, a paleo-earthquake that resembles the 1992 event will likely be described incompletely with the ESI2007 scale. This is an important lesson for the application of this scale to paleo-earthquakes in similar environments, and to events with a complex or anomalous primary surface rupture pattern.

Supplementary Materials: The following are available online at http://www.mdpi.com/2076-3263/9/6/271/s1.

Author Contributions: Conceptualization, C.G.; methodology, C.G.; validation, C.G., R.W., E.A., A.E. and K.A.; formal analysis, C.G.; investigation, C.G., R.W., E.A., A.E. and K.A.; resources, C.G., R.W., E.A., A.E. and K.A.; data curation, C.G., E.A. and A.E.; writing—original draft preparation, C.G.; writing—review and editing, E.A.; project administration, R.W.; funding acquisition, R.W.

Funding: This study was funded by the Earthquakes without Frontiers project, NERC and ESRC grant code: EwF_NE/J02001X/1_1), by the Center for Observation and Modelling of Earthquakes, Volcanoes and Tectonics (COMET, GA/13/M/031), and by Looking inside the Continents from Space (LiCS) (NE/K011006/1).

Acknowledgments: We thank three anonymous reviewers for their constructive comments that helped to improve the manuscript. We thank the editor Alicia Wang for handling our paper.

Conflicts of Interest: The authors declare no conflict of interest.

References

1. Wells, D.L.; Coppersmith, K.J. New empirical relationships among magnitude, rupture length, rupture width, rupture area, and surface displacement. *BSSA* **1994**, *84*, 974–1002.
2. Stirling, M.; Goded, T.; Berryman, K.; Litchfield, N. Selection of earthquake scaling relationships for seismic-hazard analysis. *BSSA* **2013**, *103*, 2993–3011. [CrossRef]
3. Elliott, J.R.; Jolivet, R.; González, P.J.; Avouac, J.P.; Hollingsworth, J.; Searle, M.P.; Stevens, V.L. Himalayan megathrust geometry and relation to topography revealed by the Gorkha earthquake. *Nat. Geosci.* **2016**, *9*, 174–180. [CrossRef]
4. Grützner, C.; Carson, E.; Walker, R.T.; Rhodes, E.J.; Mukambayev, A.; Mackenzie, D.; Elliott, J.R.; Campbell, G.E.; Abdrakhmatov, K. Assessing the activity of faults in continental interiors: Palaeoseismic insights from SE Kazakhstan. *EPSL* **2017**, *459*, 93–104. [CrossRef]
5. Hamling, I.J.; Hreinsdóttir, S.; Clark, K.; Elliott, J.; Liang, C.; Fielding, E.; Litchfield, N.; Villamor, P.; Wallace, L.; Wright, T.J.; et al. Complex multifault rupture during the 2016 Mw 7.8 Kaikōura earthquake, New Zealand. *Science* **2017**, *356*, eaam7194. [CrossRef] [PubMed]
6. Ghose, S.; Mellors, R.J.; Korjenkov, A.M.; Hamburger, M.W.; Pavlis, T.L.; Pavlis, G.L.; Omuraliev, M.; Mamyrov, E.; Muraliev, A.R. The M_S = 7.3 1992 Suusamyr, Kyrgyzstan, earthquake in the Tien Shan: 2. Aftershock focal mechanisms and surface deformation. *BSSA* **1997**, *87*, 23–38.
7. Ainscoe, E.A.; Abdrakhmatov, K.E.; Baikulov, S.; Carr, A.S.; Elliott, A.J.; Grützner, C.; Walker, R.T. Variability in surface rupture between successive earthquakes on the Suusamyr Fault, Kyrgyz Tien Shan: Implications for palaeoseismology. *GJI* **2019**, *216*, 703–725. [CrossRef]
8. Michetti, A.M.; Esposito, E.; Guerrieri, L.; Porfido, S.; Serva, L.; Tatevossian, R.; Vittori, E.; Audemard, F.; Azuma, T.; Clague, J.; et al. Intensity scale ESI 2007. In *Memorie Descrittive Carta Geologica d'Italia. Servizio Geologico d'Italia*; Guerrieri, L., Vittori, E., Eds.; 74. Dipartimento Difesa del Suolo, APAT: Rome, Italy; Volume 74, p. 53.
9. Audemard, F.; Azuma, T.; Baiocco, F.; Baize, S. *Earthquake Environmental Effect for seismic hazard assessment: The ESI intensity scale and the EEE Catalogue*; ISPRA: Rome, Italy, 2015; Volume 97.
10. Serva, L.; Vittori, E.; Comerci, V.; Esposito, E.; Guerrieri, L.; Michetti, A.M.; Mohammadioun, B.; Mohammadioun, G.C.; Porfido, S.; Tatevossian, R.E. Earthquake Hazard and the Environmental Seismic Intensity (ESI) Scale. *Pure Appl. Geophys.* **2015**, *173*, 1479–1515. [CrossRef]
11. ISPRA. The Earthquake Enviromental Effects (EEE) Catalogue. 2019. Available online: http://193.206.192.211/wfd/eee_catalog/viewer.php (accessed on 28 February 2019).
12. Bemis, S.P.; Micklethwaite, S.; Turner, D.; James, M.R.; Akciz, S.; Thiele, S.T.; Bangash, H.A. Ground-based and UAV-based photogrammetry: A multi-scale, high-resolution mapping tool for structural geology and paleoseismology. *J. Struct. Geol.* **2014**, *69*, 163–178. [CrossRef]
13. Johnson, K.; Nissen, E.; Saripalli, S.; Arrowsmith, J.R.; McGarey, P.; Scharer, K.; Williams, P.; Blisniuk, K. Rapid mapping of ultrafine fault zone topography with structure from motion. *Geosphere* **2014**, *10*, 969–986. [CrossRef]
14. Sadybakasov, I. *Neotectonics of High Asia*; Nauka: Moscow, Russia, 1990; 181p. (In Russian)
15. Korjenkov, A.; Bowman, D.; Haselton, K.; Porat, N. Recent drainage diversions under thrusting conditions in the Suusamyr valley, the Tien Shan Range, Kyrgyzstan. *Isr. J. Earth Sci.* **1999**, *48*, 63–79.
16. Chediya, O.K.; Mikhel, G.; Mikhaylyov, V. On rates of Holocene tectonic movements and an attempt to determine the duration of seismic calm in the central Tien Shan. *Echo Sci. Proc. Natl. Acad. Sci. Kyrgyz Repub.* **1997**, *1*, 13–17. (In Russian)
17. Dziewonski, A.M.; Ekström, G.; Salganik, M.P. Centroid moment tensor solutions for July–September 1992. *Phys. Earth Planet. Interiors* **1993**, *79*, 287–297. [CrossRef]

18. Mellors, R.J.; Vernon, F.L.; Pavlis, G.L.; Abers, G.A.; Hamburger, M.W.; Ghose, S.; Iliasov, B. The Ms = 7.3 1992 Suusamyr, Kyrgyzstan, earthquake: 1. Constraints on fault geometry and source parameters based on aftershocks and body-wave modeling. *BSSA* **1997**, *87*, 11–22.
19. Frolova, A.G.; Kalmetyeva, Z.A. Peak Ground Acceleration Map, (Inner Tien-Shan), Kyrgyzstan. In *SNSF Project No IB7320-110694*; Burg, J.-P., Mikolaichuk, A.V., Eds.; ETH Zürich: Zürich, Switzerland; Available online: http://www.kyrgyzstan.ethz.ch (accessed on 20 June 2019).
20. Gomez, J.M.; Bukchin, B.; Madariaga, R.; Rogozhin, E.A.; Bogachkin, B. Rupture process of the 19 August 1992 Susamyr, Kyrgyzstan, earthquake. *J. Seismol.* **1997**, *1*, 219–235. [CrossRef]
21. Bogachkin, B.M.; Korzhenkov, A.M.; Mamyrov, E.; Nechaev, Y.V.; Omuraliev, M.; Petrosyan, A.E.; Pletnyov, K.G.; Rogozhin, E.A.; Charimov, T.A. The structure of 1992 Susamyr earthquake source based on its geological and seismological manifestations. *Izv. Phys. Solid Earth* **1997**, *33*, 867–882.
22. Korjenkov, A.M.; Mamyrov, E.; Omuraliev, M.; Kovalenko, V.A.; Usmanov, S.F. Rock avalanches and landslides formed in result of strong Suusamyr (1992, M = 7,4) earthquake in the Northern Tien Shan-test structures for mapping of paleoseismic deformations by satellite images. In *High Mountain Remote Sensing Cartography VII (HMRSC VII)*; Buchroithner, M.F., Ed.; Kartographische Bausteine: Dresden, Germany, 2004; Volume 28, pp. 117–136.
23. Korzhenkov, A.M. *Seismogeology in the Tian Shan (within the Territory of Kyrgyzstan and Surrounding Areas)*; Ilim: Bishkek, Kyrgyz Republic, 2006; p. 290. (In Russian)
24. Zongzheng, S.; Abdrakhmatov, K.; Xinyuan, C.; Lemzin, I.N.; Zhengming, Y. Seismotectonics and Seismic Rupture of Susamyr Earthquake with M= 7.3, Kirghizistan, 1992. *Earthq. Res. Shanxi* **1996**, *85*, 3. (In Chinese)
25. Korjenkov, A.M.; Omuraliev, M. Features on relief (surface) formed during the strong 1992 Suusamyr earthquake in the northern Tien Shan. In *Proceedings of the Geomorphological Seminar, Irkutsk*; Geomorphological Commission, Russian Academy of Sciences: Moscow, Russia; 15–17 December 1993, pp. 105–106. (Abstract in Russian)
26. Havenith, H.B.; Jongmans, D.; Abdrakhmatov, K.; Trefois, P.; Delvaux, D.; Torgoev, I.A. Geophysical investigations of seismically induced surface effects: Case study of a landslide in the Suusamyr valley, Kyrgyzstan. *Surv. Geophys.* **2000**, *21*, 351–370. [CrossRef]
27. Havenith, H.B.; Strom, A.; Jongmans, D.; Abdrakhmatov, A.; Delvaux, D.; Tréfois, P. Seismic triggering of landslides, Part A: Field evidence from the Northern Tien Shan. *Nat. Hazard Earth Syst. Sci.* **2003**, *3*, 135–149. [CrossRef]
28. Havenith, H.B.; Strom, A.; Caceres, F.; Pirard, E. Analysis of landslide susceptibility in the Suusamyr region, Tien Shan: Statistical and geotechnical approach. *Landslides* **2006**, *3*, 39–50. [CrossRef]
29. Havenith, H.B.; Torgoev, A.; Schlögel, R.; Braun, A.; Torgoev, A.; Ischuk, A. Tien Shan geohazards database: Landslide susceptibility analysis. *Geomorphology* **2015**, *249*, 32–43. [CrossRef]
30. Havenith, H.B.; Strom, A.; Torgoev, I.; Torgoev, A.; Lamair, L.; Ischuk, A.; Abdrakhmatov, K. Tien Shan geohazards database: Earthquakes and landslides. *Geomorphology* **2015**, *249*, 16–31. [CrossRef]
31. Bourdeau, C.; Havenith, H.B. Site effects modelling applied to the slope affected by the Suusamyr earthquake (Kyrgyzstan, 1992). *Eng. Geol.* **2008**, *97*, 126–145. [CrossRef]
32. Andrews, D.J.; Hanks, T.C. Scarp degraded by linear diffusion: Inverse solution for age. *J. Geophys. Res. Solid Earth* **1985**, *90*, 10193–10208. [CrossRef]
33. Avouac, J. Analysis of scarp profiles: Evaluation of errors in morphologic dating. *J. Geol. Res.* **1993**, *98*, 6745–6754. [CrossRef]
34. Hanks, T.C. The age of scarplike landforms from diffusion-equation analysis. In *Quaternary Geochronology: Methods and Applications*; Noller, J.S., Sowers, J.M., Lettis, W.R., Eds.; American Geophysical Union: Washington, DC, USA, 2000.
35. Berberian, M.; Jackson, J.A.; Fielding, E.; Parsons, B.E.; Priestley, K.; Qorashi, M.; Talebian, M.; Walker, R.; Wright, T.J.; Baker, C. The 1998 March 14 Fandoqa earthquake (Mw 6.6) in Kerman province, southeast Iran: re-rupture of the 1981 Sirch earthquake fault, triggering of slip on adjacent thrusts and the active tectonics of the Gowk fault zone. *GJI* **2001**, *146*, 371–398.
36. Tatevossian, R.E. On the problem of magnitude calibration of palaeoearthquakes. *Quat. Int.* **2011**, *242*, 11–19. [CrossRef]

37. Abdrakhmatov, K.E.; Walker, R.T.; Campbell, G.E.; Carr, A.S.; Elliott, A.; Hillemann, C.; Hollingsworth, J.; Landgraf, A.; Mackenzie, D.; Mukambayev, A.; et al. Multisegment rupture in the 11 July 1889 Chilik earthquake (Mw 8.0–8.3), Kazakh Tien Shan, interpreted from remote sensing, field survey, and paleoseismic trenching. *J. Geophys. Res. Solid Earth* **2016**, *121*, 4615–4640. [CrossRef]
38. Arrowsmith, J.R.; Crosby, C.J.; Korzhenkov, A.M.; Mamyrov, E.; Povolotskaya, I.; Guralnik, B.; Landgraf, A. Surface rupture of the 1911 Kebin (Chon–Kemin) earthquake, Northern Tien Shan, Kyrgyzstan. *Geol. Soc. Lond. Spec. Publ.* **2016**, *432*, 233–253. [CrossRef]
39. Fletcher, J.M.; Teran, O.J.; Rockwell, T.K.; Oskin, M.E.; Hudnut, K.W.; Mueller, K.J.; Spelz, R.M.; Akciz, S.O.; Masana, E.; Faneros, G.; et al. Assembly of a large earthquake from a complex fault system: Surface rupture kinematics of the 4 April 2010 El Mayor–Cucapah (Mexico) Mw 7.2 earthquake. *Geosphere* **2014**, *10*, 797–827. [CrossRef]
40. Villani, F.; Civico, R.; Pucci, S.; Pizzimenti, L.; Nappi, R.; De Martini, P.M.; the Open EMERGEO Working Group. A database of the coseismic effects following the 30 October 2016 Norcia earthquake in Central Italy. *Sci. Data* **2018**, *5*, 180049. [CrossRef] [PubMed]
41. Elliott, J.R.; Walters, R.J.; Wright, T.J. The role of space-based observation in understanding and responding to active tectonics and earthquakes. *Nat. Commun.* **2016**, *7*, 13844. [CrossRef] [PubMed]
42. Strom, A. Geological prerequisites for landslide dams' disaster assessment and mitigation in Central Asia. In *Progress of Geo-Disaster Mitigation Technology in Asia*; Wang, F., Miyajima, M., Li, T., Shan, W., Fathani, T.F., Eds.; Springer: Berlin/Heidelberg, Germany, 2013; pp. 17–53.
43. Blumetti, A.M.; Grützner, C.; Guerrieri, L.; Livio, F. Quaternary earthquakes: Geology and palaeoseismology for seismic hazard assessment. *Quat. Int.* **2017**, *451*, 1–10. [CrossRef]

geosciences

MDPI

Article

Earthquake-Induced Landslide Risk Assessment: An Example from Sakhalin Island, Russia

Alexey Konovalov *, Yuriy Gensiorovskiy, Valentina Lobkina, Alexandra Muzychenko, Yuliya Stepnova, Leonid Muzychenko, Andrey Stepnov and Mikhail Mikhalyov

Sakhalin Department of Far East Geological Institute, Far Eastern Branch, Russian Academy of Sciences, Yuzhno-Sakhalinsk 693023, Russia
* Correspondence: a.konovalov@geophystech.ru

Received: 30 April 2019; Accepted: 8 July 2019; Published: 11 July 2019

Abstract: Damages caused by earthquake-induced ground effects can be of the order or significantly exceed the expected damages from ground shaking. A new probabilistic technique is considered in this study for earthquake-induced landslide risk assessment. A fully probabilistic technique suggests a multi-stage hazard assessment. These stages include the determination of seismic hazard curves and landslide probabilistic models, a vulnerability assessment, and geotechnical investigations. At each of the stages, the uncertainties should be carefully analyzed. A logic tree technique, which handles all available models and parameters, was used in the study. The method was applied considering child education facilities located at the foot of a natural slope in the south of Sakhalin Island which is known as an active seismic and land sliding area. The significant differences in the ground motion scenario in terms of the 475-year seismic hazard map and the fully probabilistic approach considered suggests that seismic landslide risk could be underestimated or overestimated when using the 475-year seismic hazard map for risk assessment. The given approach follows the rational risk management idea that handles well all possible ground motion scenarios, slope models, and parameters. The authors suggest that the given approach can improve geotechnical studies of slope stability.

Keywords: earthquake-induced landslide; fully probabilistic technique; Newmark's method; Sakhalin Island; risk

---

1. Introduction

Large earthquakes affecting urban areas are one of the most destructive natural hazards and can lead to significant impacts on the built and human environment. Generally, earthquake loss models consider ground shaking and ground failure (such as landslides, liquefaction, and faulting) hazards. Damages caused by earthquake-induced ground effects, in some cases, significantly exceed the damages from direct ground shaking [1,2]. Damages related to seismically-induced landslides can be considerable due to the full collapse or loss in functionality of facilities, roads, pipelines, and other lifelines [3–7].

There are numerous causative factors for seismically-induced gravity-slope processes on Sakhalin Island, which is recognized as an area with a high level of geohazards. A total of 70% of the South Sakhalin territory is susceptible to landslide activity [8]. According to general seismic hazard maps, Sakhalin Island is a seismically active area, with an 8–9 MSK-64 macroseismic intensity for the 475-year return period [9].

As a recent example of earthquake-induced landslides, the Mw = 6.2 2 August 2007 Nevelsk earthquake should be noted. The Nevelsk earthquake was followed by aftershock sequences with a relatively high productivity level [10]. Focal mechanisms of the mainshock and aftershocks indicates the west dipping (38–40°) fault planes.

The largest aftershock with a magnitude of Mw = 5.9 followed about two hours after the mainshock origin time. They generated tsunami waves up to 3.2 m high. The ground shaking effects in Nevelsk caused by the mainshock and largest aftershock corresponded to a 7–8 MSK-64 macroseismic intensity (Figure 1). Massive damages of buildings (>200), bridges, railway and roads were found during macroseismic inspection [10].

The 2007 Nevelsk earthquakes caused massive release of methane from the coal beds in the coastal zone of 40 km in length and an uplift of benches in the area of the Nevelsk sea port [10].

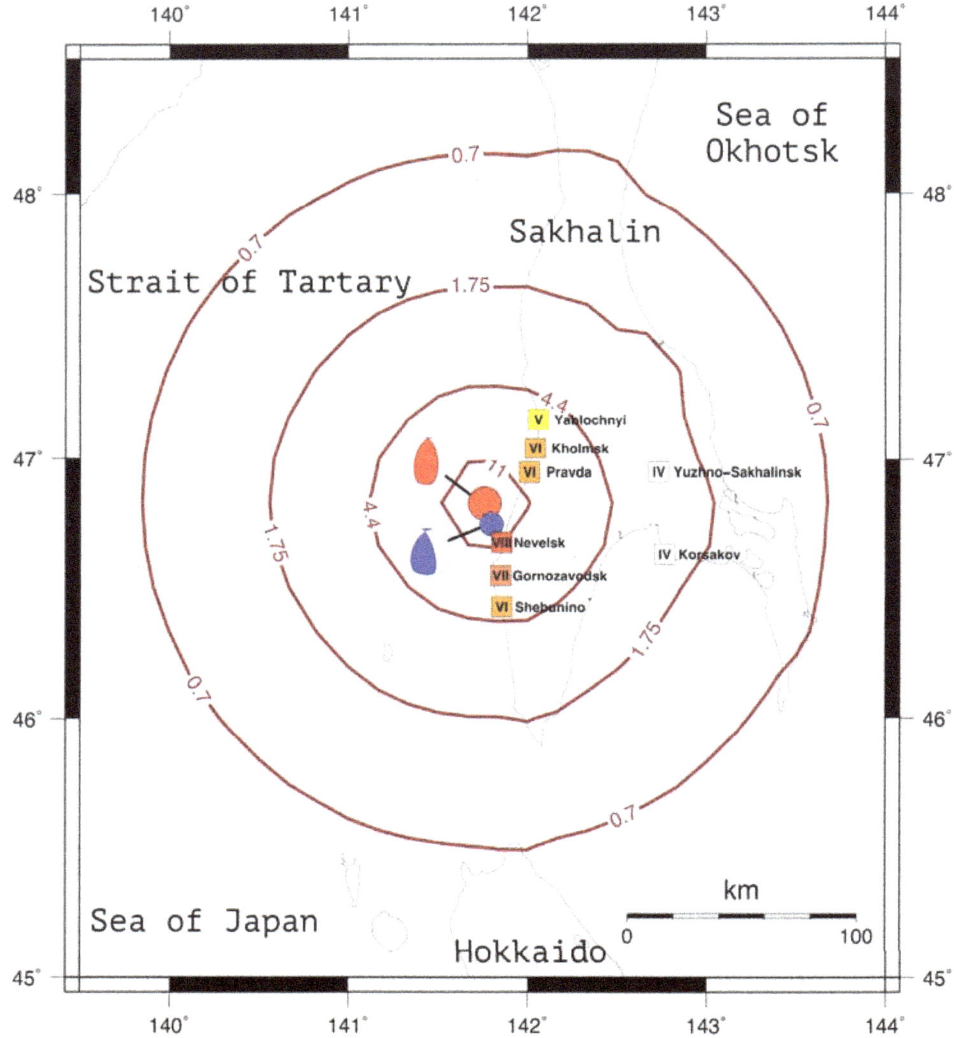

Figure 1. Contour map of peak ground acceleration (% g) for the 2 August 2007 Nevelsk earthquake (Mw = 6.2). Filled boxes indicate the settlements with MSK-64 felt reports. The mainshock (Mw = 6.2) is shown by the red filled circle, the largest aftershock (Mw = 5.9) by the blue filled circle.

As a result of significant ground shaking, subsidence cracks and shallow landslides (up to 200–300 m^3) were widely recorded (Figure 2) within the Nevelsk urban area (16–21 km from the epicenter). The 2007 Nevelsk earthquake occurred in a relatively dry period. There was a recorded

69 mm of rain precipitation, representing 45% of the mean annual value, for the two months before the mainshock [11]. Therefore, seismically-induced landslides remain a major natural hazard on Sakhalin Island that should be considered in the risk assessment strategy.

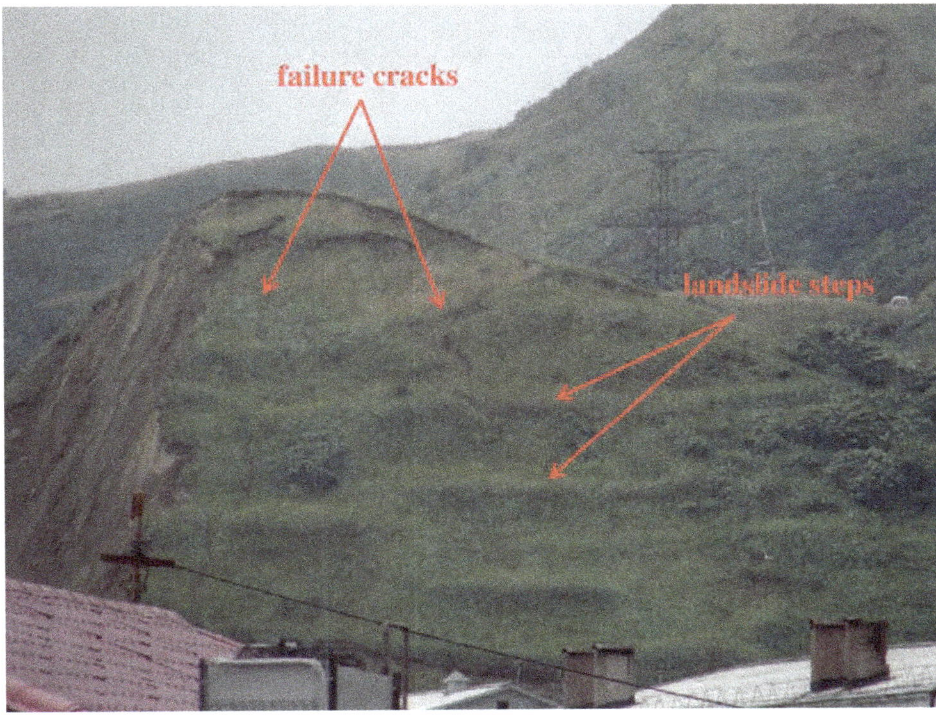

Figure 2. An example of shallow landslides caused by the 2007 Nevelsk earthquake.

Most hazard assessment techniques are designed for producing susceptibility or likelihood maps on large scales (regional or global) [2,12,13]. Because material parameters are difficult to identify in detail for large areas, slope parameters are estimated from topographic, geologic, and other geospatial information. These maps are most commonly used for an estimation of the general hazard level and specifying sites where detailed geotechnical investigations are needed.

The aim of this study is to apply the earthquake-induced landslide risk assessment technique at a local scale. This paper proposes a fully probabilistic approach that handles all available ground motion scenarios and geomechanical slope models well. The important issue is the estimation of the uncertainties from the spatial and temporal variability of soil parameters.

The child education facilities (Figure 3) located under the natural slope in Nevelsk (Sakhalin Island, Russia) were considered in this study as an example of the given approach. This site was proposed for application of the methodology due to available slope material parameters previously obtained from geotechnical investigations.

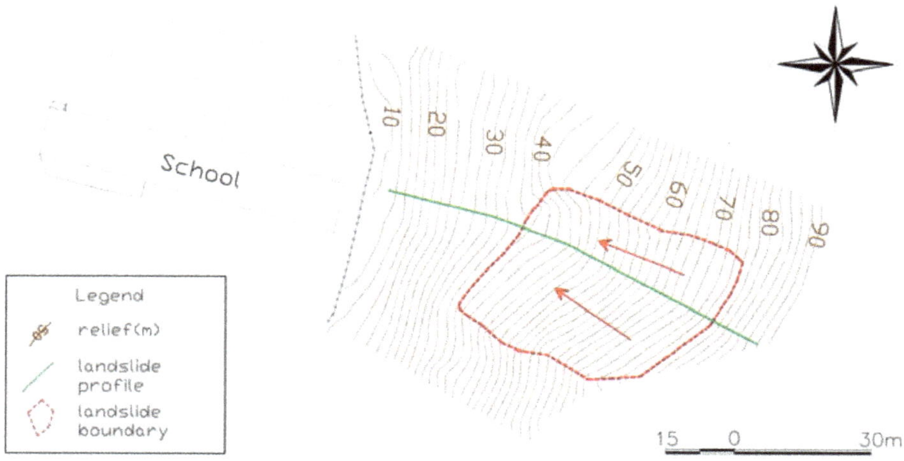

Figure 3. Map of the slope fragment with an expected landslide boundary.

2. Materials and Methods

2.1. Fully Probabilistic Risk Assessment Technique

The seismic hazard maps in terms of the 475-year ground shaking intensity are most commonly used as a triggering condition for analyzing the slope stability under seismic loading [2,13,14]. Uncertainties about the 'best' ground motion scenario for land sliding prediction are still being discussed [15].

A fully probabilistic risk assessment technique that handles all available ground motion scenarios well is considered in this study [16]. The given approach follows a rational risk management idea that deals with all possible models, scenarios, and uncertainties.

The total risk of an earthquake-induced landslide hazard to the facility located at the foot of the slope in the next T years can be expressed in terms of probability and vulnerability. Formally, the probability of the occurrence of N seismically-induced landslides within a certain time interval should be considered. In practice, the probability of the triggering of two or more slope failures within 50 years is extremally small.

A fully probabilistic approach suggests determination of the occurrence probabilities of various ground shaking levels and probabilities of landslide triggering for these ground motion scenarios. That gives the total probability of earthquake-induced slope failure in a certain time interval considering all possible ground motion scenarios:

$$P_{sf}(T) = \sum_j \sum_i w_j P(\mathrm{PGA} = a_i | T) \cdot Pr_j(a_i) = \sum_j \sum_i w_j p_{ij}, \quad (1)$$

where $P(\mathrm{PGA} = a_i|T)$ is the probability of the occurrence of ground shaking level a_i in the next T years (ground motion scenario), $Pr_j(a_i)$ is the probability that seismic loading a_i will trigger a landslide in the frame of the geomechanical slope model j, p_{ij} is the probability of slope failure in the next T years under ground motion scenario a_i in the frame of the geomechanical slope model j, and PGA is the peak ground acceleration at the given site.

The summation in (1) is carried out for all available geomechanical slope models ranked by weights w_j, where:

$$\sum_j w_j = 1. \quad (2)$$

The resistance of the facility to prevent the land sliding can be expressed in terms of the structural damage risk (physical vulnerability). The vulnerability is defined in this study as the degree of loss of a given element at risk and ranging from 0 (no loss) to 1 (total loss).

Vulnerability depends on a number of factors [17,18], such as the sliding mass volume and its velocity, position of landslide initiation, etc. For an accurate risk assessment, all of these factors should be considered. In many case studies, the vulnerability assessment is still characterized by large uncertainties.

In a simple case, vulnerability depends on the intensity of the landslide that can hit the facility located at some distance from the landslide source [18,19]. It means that a certain geomechanical slope model can be associated with the expected intensity level of the landslide.

Then, the total risk $R(T)$ of facility damage caused by seismic land sliding within T years can be written as:

$$R(T) = \sum_j \sum_i w_j v_j p_{ij}, \tag{3}$$

where v_j is the vulnerability of the facility in the frame of the geomechanical slope model j.

Basically, Equations (1) and (3) suggest a multi-stage and multi-hazard approach. These stages include site-specific probabilistic seismic hazard analysis, vulnerability assessment, geotechnical investigations, and landslide probability calibration.

2.2. *Causative Factors of Studied Area*

Seismically-induced landslide risk assessment is complicated by the numerous factors and conditions contributing to slope failure. This section discusses the major causative factors of the studied area, helping to constrain the slope models and parameters for the probabilistic hazard analysis (Figure 4).

Figure 4. Side view of the location of the natural slope and school considered in this study. The red line indicates the expected landslide breakout wall. The red arrow shows the path direction of the expected landslide.

2.2.1. Geology

From the basis of the tectonic terrains scheme, the studied area is related to the southern part of the West Sakhalin Terrane located to the north of the Lopatino Cape. The target area consists of Late Cretaceous and Cenozoic outcropping rocks (Figure 5).

Quaternary and Neogene age deposits are widely presented in the studied area. The Quaternary soils of low-thickness are presented by deluvial-eluvial deposits. The Neogene soils of the Nevelskaya formation are presented by volcanic sandstones, siltstones, tuffs, and tuffits. A geologic map with detailed formation lithology is given in Figure 5 [20].

From the view of general engineering geology, the following layers are highlighted in the lithological slope section: 1—topsoil (0.2 m); 2—fine-gravelly with sandy filler up to 50%, dense (deluvial-eluvial) (0.2 m); 3—tuff-sandstone on clayey cement with low-strength siltstone interlayers (3.6 m).

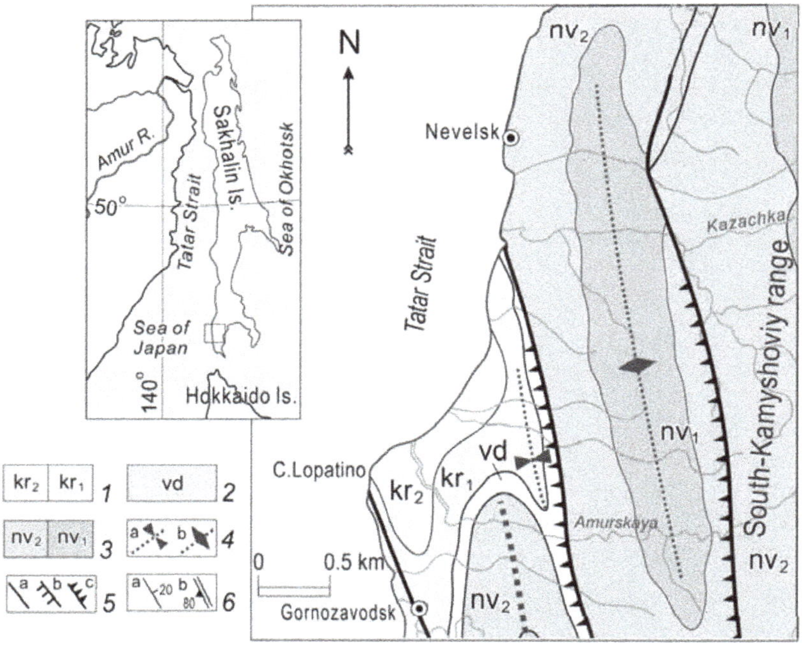

Figure 5. Geological structural map of studied area compiled according to [20]: 1—Upper Miocene Kurasiiskay formation: lower subformation (kr1)—siliceous claystones and upper subformation (kr2)—alternation of sandstones and siltstones; 2—Middle Miocene Verhneduiskaya formation (vd)—alternation of sandstones and siltstones and claystones, hards coal; 3—Lower Miocene Nevenlskaya formation: lower subformation (nv1)—alternation of sandstones and siltstones and upper subformation (nv2)—alternation of sandstones and siltstones; 4—synclineaxis (a) anticlineaxis (b); 5—fault zone; 6—studied area.

2.2.2. Geomorphology

Nevelsk is located on the western flanks and sea terrace's foot, which are western branches of the South Kamyshoviy range. Terraces have an absolute elevation till 200 m and are dissected by narrow, deeply incised V-shaped valleys of rivers and rills [21].

Landscape dissection depth reaches 200 m, and the slope steepness is 45° and more. Terrace surfaces are inclined to the sea side.

Abrasion sea terrace scarps, rivers, and rill valley slopes are well-sodded and covered by grassy-shrub vegetation, and in some places are forested. Vegetation does not prevent landslides.

River valley and sea terrace slopes are complicated by old block slides, which make slopes stepped.

The child education facilities are located close to the foot of the high sea terrace (Figures 3 and 4).

2.2.3. Climatic Settings

The Nevelsk district is located along the Tatar Strait of the Japanese Sea, and is strongly affected by a warm Tsushima current, so it witnesses the warmest summer and winter of the entire island. The territory is characterized by a large precipitation quantity during the second part of the summer.

The mean annual precipitation quantity in Nevelsk is 911 mm [22]. Precipitation falls inhomogenously. The major precipitation period is related to the second half of the summer and early autumn, and the minimum period is the second half of winter.

Two landslide activation periods related to soil moisture conditions are typical for the studied area.

The first period (end of May to early June) is related to melting snow which leads to rapid soil humidification and often, to shallow (1 m deep) landslides. Rainfall can trigger landslides during this period.

The second period (August to October) is related to high cyclone activity, leading to a large rain precipitation quantity.

The long-term annual average precipitation quantity during the warm period is 579 mm. The rainiest month is September, with a precipitation quantity of ~111 mm. The precipitation quantity during a cyclone event (in August–September) can exceed the monthly average level. The maximum precipitation quantity (211 mm) was registered during the Phillis cyclone event (2–7 August 1981).

2.2.4. Soil Moisture Conditions

The soil moisture conditions of a potentially sliding mass represent one of the most important factors for landslide prediction.

Three soil moisture models are assumed. The models are constrained by generalizing the climatic settings in the studied area. These simplified models are characterized by the following properties:

1. Slightly wet—Shallow invasion of the water into the soil mass. Dry soil conditions are typical for drought periods and for periods with stable snow cover. The overall period duration is about five months;
2. Moist—Invasion of the water into the soil mass to a 1 m depth. The overall period duration is three months;
3. Water saturated—Soil mass is waterlogged to a ~2 m depth. This model is typical for the rain precipitation and rapid snow melting coincidence period and for cyclone/typhoon occurrences. The overall period duration is about four months.

2.3. Materials

2.3.1. Ground Motion Scenarios

The main goal of probabilistic seismic hazard analysis (PSHA) is to determine the probability of exceeding a certain ground shaking level at a given site within the time interval of interest [23]. The result of such analysis is a seismic hazard curve, which demonstrates the relationship between the probability of exceeding $P(\text{PGA} > a|T)$ and ground shaking level a. Under ground shaking conditions, we consider PGA.

The occurrence probability $P(\text{PGA} = a_i|T)$ is recognized in this study as a ground motion scenario. The transition from the probability of exceeding to the discrete probability is given as:

$$P(\text{PGA} = a_i|T) = P(\text{PGA} > a_i|T) - P(\text{PGA} > a_{i+1}\big|T). \quad (4)$$

Therefore, Equation (4) defines the probability of the occurrence of a certain ground shaking level in the next T years. It was substituted into Equations (1) and (3).

The seismic hazard curve for a given site was computed using CRISIS 2015 software [24]. The authors used the regional seismic source models and ground motion prediction equations tested in the previous PSHA studies [25].

The average 30-m ground layer shear-wave velocity V_{S30} at the site is of the order of 300 m/s. It was used for site correction within the PSHA stage considering the site-correction term in the ground motion prediction equations.

The seismic hazard curve used in this study is shown in Figure 6. The 10% exceeding probability (or 475-year return period probability) corresponds to the PGA value of 0.38 g (Figure 6). Peak ground acceleration of 0.38 g corresponds to a ground shaking intensity of 9 MSK64. It is in good agreement with the estimates based on the maps of general seismic zonation [9].

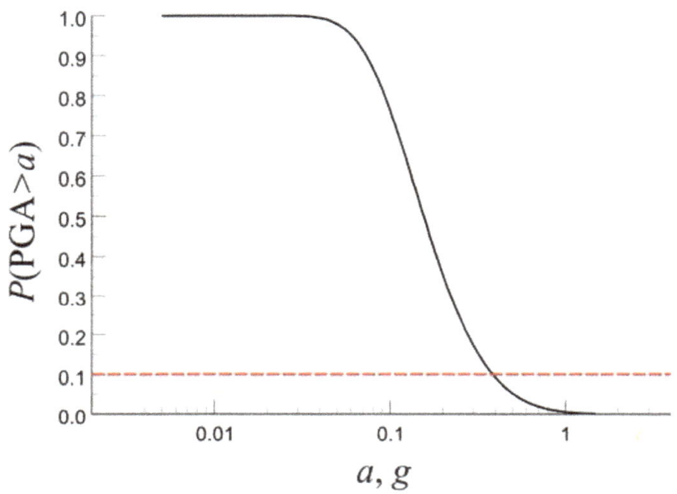

Figure 6. The probability of exceeding the given peak ground acceleration. Red dotted line indicates the 10% probability level.

2.3.2. Probability of Landslide Triggering

Jibson et al. [12] calibrated the slope failure probability by analyzing landslide field data after the 1994 Northridge earthquake and predicted Newmark displacement (D_N). The authors used the Newmark sliding block model [26] to quantify the probability of landslide triggering given the PGA value (a) and critical slope acceleration (a_c). The Jibson probability model follows the Weibull distribution:

$$Pr(D_N) = 0.335\left[1 - \exp\left(-0.048D_N^{1.565}\right)\right], \tag{5}$$

where

$$\log D_N(a) = -0.215 + \log\left[\left(1 - \frac{a_c}{a}\right)^{2.341}\left(\frac{a_c}{a}\right)^{-1.438}\right] \pm 0.51. \tag{6}$$

The critical slope acceleration a_c of a sliding mass is considered as a simple function of the static factor of safety F_S and the slope geometry:

$$a_c = (F_S - 1)g\sin\alpha, \tag{7}$$

where g is the gravity factor and α is the dip angle of the potential sliding surface.

According to the limit equilibrium theory, the static factor of safety F_S is defined as the relationship between the force keeping the sliding mass on the slope and the force moving the sliding mass down the slope [12]:

$$F_S = \frac{c'}{\gamma z \sin\alpha} + \frac{\tan\varphi'}{\tan\alpha} - \frac{m\gamma_w \tan\varphi'}{\gamma \tan\alpha}, \quad (8)$$

where c' is the effective cohesion, z is the slope-normal thickness of the potential sliding mass, γ is the material unit weight, γ_w is the unit weight of ground water, φ' is the effective friction angle, and m represents the fractional depth of the water table with respect to the total slide depth. The sliding mass is stable if $F_S > 1$, and unstable when $F_S < 1$.

Soil moisture conditions in the studied area and southern California are significantly different. Virtually no rain had fallen prior to the 1994 Northridge earthquake, which was used for calibrating the seismically-induced landslide probability [12]. The pore-water pressure term was dropped from Equation (8) in the Jibson probability model. The soils in the studied area are expected to be saturated most of the year, so the authors paid great attention to the third term of Equation (8). Several models with a non-zero pore-water pressure term are hypothesized in this study.

Therefore, the Jibson probabilistic model for California (5) was imported into the Equations (1) and (3).

2.3.3. Geomechanical Slope Models and Logic Tree

The physico-mechanical parameters of the slope soils were obtained during geotechnical investigations at the site (see Acknowledgments Section). Geotechnical studies included sampling undisturbed cores and laboratory tests. The material parameters of the soil mass are given in Table 1.

Table 1. Physico-mechanical parameters of the soil mass.

Soil Type	c', Kpa	α, deg.	γ, kN/m^3	γ_w, kN/m^3	φ', deg.
Tuffaceous sandstone	24	40	26.8	9.8	30

Identifying the slab thickness z of potential sliding mass requires high-quality seismic profile, geological, and geophysical data. These can give additional information about the layer structure that helps to fix the slope-normal thickness or to identify the preexisting landslides. Such data is not available for the studied area. Therefore, the slab thickness variability was hypothesized through the logic tree approach, which is commonly used in PSHA. Four models with varied slab thicknesses (1, 2, 4, and 8 m) were considered. The corresponding weights are equal to 0.3, 0.3, 0.3, and 0.1 (Figure 7). The weights were determined by an expert's view.

The given difference in weighting reflects our view that the landslide initiation depth of an order of 8 m seems to be unrealistic. However, the authors are aware of the uncertainties connected with this choice in weighting scheme.

The accurate weighting of models with slab thickness uncertainties can be realized through the corresponding probability density distribution. The probability that the landslide has a certain slip surface depth value is defined by its statistical distribution. A significant amount of observations in the studied area are required for developing such a probability model.

The second variable parameter is the proportion of the sliding mass thickness that is saturated (m in Equation (8)). The water saturation parameter was defined as a simplified relation given by soil moisture conditions (see Section 2.2.4):

$$m = \frac{\text{depth of water invasion}}{\text{slab} - \text{normal thickness}}. \quad (9)$$

The proportion between the period duration of water invasion into the corresponding soil depth and the annual period was used for the weighting. For instance, for dry soil conditions, the corresponding weight was defined as 5/12~0.4 (Figure 7).

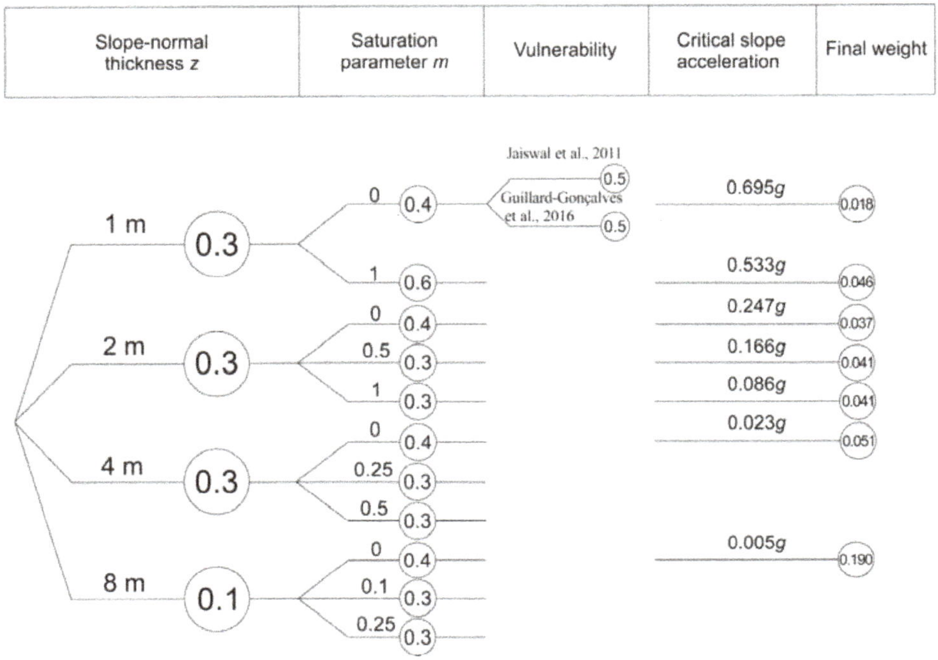

Figure 7. Logic tree for handling the epistemic uncertainties in the considered models and parameters. The weights of the branches are given in circles.

A full logic tree that handles uncertainties in the considered models and parameters is shown in Figure 6. The resulting critical slope acceleration for each branch was calculated according to (7). For the unstable case ($F_S < 1$), the critical slope acceleration was defined as the minimum acceleration in the given seismic hazard curve (0.005 g).

2.3.4. Vulnerability Assessment

The physical vulnerability term is widely used in many scientific and engineering fields [27]. It defines the probability that a given element at risk will be damaged under a certain impact. Recent seismic risk analyses have usually dealt with seismic fragility curves [28]. The fragility curve defines the conditional probability of exceeding the certain damage state of the building for a given level of ground shaking intensity. Increasing the research volume of earthquake damage data allows researchers to calibrate seismic fragility curves in many seismically-active regions.

Unfortunately, accurate calibrating of the fragility curves is still complicated by limited damage data for most types of landslides. Moreover, different physical mechanisms are associated with different types of landslides (rockfalls, block slides, debris flow etc.). Vulnerability depends on a number of factors [17,18], such as the sliding mass volume, landslide velocity, landslide source, etc.

A physical vulnerability assessment is characterized by uncertainties that can be either epistemic or aleatory [29]. Epistemic uncertainties can be associated with the simplification of the landslide intensity estimation, the characterization of elements at risk, the vulnerability models, expert view, etc. Aleatory uncertainties can be associated with the spatial variability of parameters [29], such as landslide source location. For this type of uncertainty, the expected structural damage differs, depending on whether the facility is located on the foot or landslide body [19].

Russian regulation norms have a specific definition of landslide hazard, depending on its type and volume, but quantitative estimations of the structural response are not available.

Due to the reasons mentioned above, vulnerability assessment still remains somewhat subjective and is usually performed in small study areas (local scale). The main issue of the vulnerability assessment in our study is to give a suitable estimation of expected damage using different sliding mass models.

Available data that defines the physical loss of buildings for the wide range of sliding processes were used in this study [18,19]. Most available data for physical damage estimates come from debris flow studies. However, the authors are aware of the uncertainties connected with the choice of vulnerability values.

Different vulnerability values in connection to the expected intensity level of a landslide and facility type are given in [18]. Jaiswal et al. [18] used the landslide volume for intensity estimation (M-I, M-II and M-III). The M-I intensity class corresponds to shallow landslides with a volume of less than 1000 m^3. If a landslide becomes bigger, with a volume ranging from 1000 m^3 to 10,000 m^3, the intensity follows the M-II class. For landslides with a volume greater than 10,000 m^3, the intensity reaches the M-III class.

For similarity, the landslide volume was estimated as $V = AL \times z$, where AL is the sliding area. The equation $AL = \min(706 \times z, AL_{max})$ links the normal slope thickness and sliding area [19]. According to field studies, the maximum expected sliding area AL_{max} is of the order of 2335 m^2 (Figure 3). Despite the fact that the given relationships have significant uncertainties, they help to constrain the landslide intensity class in the studied area.

According to [19], the landslide intensity is associated with the slip surface depth, which is directly related to the slab thickness of the sliding mass. There are [12] five different landslide intensity classes that correspond to the given slip surface depths of 1 m, 2 m, 6 m, 10 m, and 20 m. The estimated vulnerabilities for slip surface depths of 1 m, 2 m, 6 m, and 10 m were imported to our model with corresponding slab thicknesses of 1 m, 2 m, 4 m, and 8 m.

The reinforced concrete building type (Type-4 or SBT4) was used in this study for the vulnerability assessment. This building type is closely related to the child education facilities considered by its material strength properties.

The vulnerability exhibits average, minimum, and maximum values. For dry soil conditions, we use average vulnerability values, and for water-saturated soil conditions, we use maximum vulnerability values, as we expect that a water-saturated sliding mass has a relatively higher landslide velocity.

We used the estimated vulnerability of elements at risk located within run-out paths of a landslide. Table 2 contains the vulnerability estimations for the considered facilitates and slope models in full accordance with [18,19]. The final weights for each of the considered slope models in Figure 6 are given with respect to the vulnerability values from Table 2.

Table 2. Simplified vulnerability model for the facilities located within landslide run-out paths.

Thickness of Sliding Mass, m	Estimated Landslide Volume, m^3	Landslide Intensity Class	Vulnerability According to [11]	Vulnerability According to [8]
1	<1000	M-I	0.05 ± 0.05	0.25 ± 0.16
2	2824	M-II	0.30 ± 0.10	0.31 ± 0.19
4	9340	M-II	0.30 ± 0.10	0.54 ± 0.19
8	18,680	M-III	0.80 ± 0.20	0.72 ± 0.20

3. Results and Discussion

Since the critical slope acceleration becomes available for each of the considered slope models, it is substituted into the Jibson probabilistic model (4). Next, the probability of occurrence of a landslide ($Pr(D_N(a_i))$) under the given seismic loading is multiplied by the probability of the occurrence a certain ground shaking level in the next 50 years ($P(\mathrm{PGA} = a_i|T)$). This gives the probability (p_{ij}) of slope failure in the next 50 years under ground motion scenario a_i in the frame of the geomechanical slope model j. Then, the summation of discrete probabilities p_{ij} is carried out for all available ground motion scenarios a_i and geomechanical slope models ranked by weights w_j. Since the final weights for each of the considered slope model are calculated with respect to the vulnerability (Figure 7), the summation of p_{ij} multiplied by its final weights w_j gives the total risk value.

The total risk of an earthquake-induced landslide hazard for the child education facility located at the foot of the slope within the next 50 years was computed according to Equation (3) and the weighting scheme illustrated in Figure 7. The corresponding risk value appeared to be 7.4%. It reflects the high level of seismic and landslide hazards and high vulnerability value for the facilities located at the foot of the natural slope. Therefore, the high risk value shows a considerable hazard of seismically-induced landslides in terms of civil engineering.

In order to estimate the most probable ground motion scenario that will cause damage to the facility by an earthquake-induced landslide, the discrete probabilities p_{ij} from Equation (3) were plotted against the corresponding ground motion level. This gives the contribution to the total risk from each slope model and each ground shaking level (Figure 8). As can be seen from Figure 8, the distributions have a modal form.

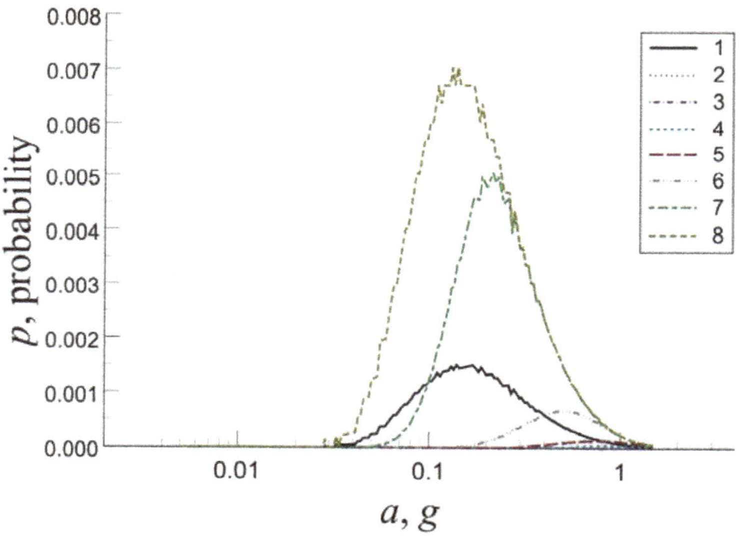

Figure 8. Average weighted (1) and unweighted (2–8) curves of probability of slope failure in the next 50 years as a function of peak ground acceleration: 2—$a_c = 0.695\ g$; 3—$a_c = 0.533\ g$; 4—$a_c = 0.247\ g$; 5—$a_c = 0.166\ g$; 6—$a_c = 0.086\ g$; 7—$a_c = 0.023\ g$; 8—$a_c = 0.005\ g$.

The average weighted probability density function computed according to the logic tree scheme (Figure 7) has a peak at PGA = 0.15 g (7–8 MSK64 intensity level). The PGA value of the order of 0.15 g can be recognized as the 'best' ground motion scenario for the considered slope models and parameters. In terms of PSHA, the given ground shaking level is closely related to the 100-year return period (according to Figure 6).

At the same time, Figure 7 shows the contribution to the total risk from each slope model. The saturated soil's model (a_c = 0.005 plot in Figure 8) should be considered as the most probable scenario for the considered area. Generally, it makes the slope unstable, regardless of the triggering conditions.

The slope model corresponding to the slip surface depth of $z = 4$ m and dry soil conditions ($m = 0$) has a clear peak around PGA = 0.2–0.3 g ($a_c = 0.023\ g$ plot in Figure 8). This ground shaking value is closely related to the 475-year probability.

It should be noted that a significant ground shaking level corresponding to a 7–8 MSK-64 intensity caused shallow landslides within the Nevelsk urban area in 2007. The PGA value corresponding to the 'best' ground motion scenario is the same order as the ground shaking intensity in the studied area generated by the 2007 Nevelsk earthquake. No facilities were damaged by the earthquake-induced landslide. This may be explained in terms of soil moisture conditions. The point is that several months prior to the Mw = 6.2 2 August 2007 Nevelsk earthquake, no rain had fallen. As a result, the 2007 Nevelsk earthquakes occurred in a relatively dry period. For such untypical conditions, the corresponding ground motion scenario is the order of PGA = 0.2–0.3 g or an 8–9 MSK64 intensity.

When generalizing results of the study, several assumptions and models should be mentioned.

Constants of the Jibson landslide probability model in the target area are generally not the same as those in southern California. The shape of the probability curve depends on the geological, geomorphological, and soil moisture conditions. Jibson et al. [12] argue that the variability of constants in Equation (4) should be proposed if a significant difference between the target area and southern California is found. This means that the logic tree will be complicated by uncertainties from the landslide probability model.

One more source of epistemic uncertainty comes from the spatial and seasonal variability of slope material parameters. Geotechnical studies of rocks and soils within the large area of the natural slope within different seasons should significantly reduce the uncertainties associated with the variability of geomechanical slope models (Figure 8).

4. Conclusions

A fully probabilistic technique is considered in this study for an earthquake-induced landslide risk assessment in a relatively small area. The given method suggests a multi-stage and multi-hazard approach. These stages include a site-specific probabilistic seismic hazard analysis, vulnerability assessment, geotechnical investigations, and landslide probability calibration.

As a case study, the child education facility located under the natural slope was considered in this study. The total risk of earthquake-induced landslide hazard to the child education facility located at the foot of the slope within the next 50 years was of the order of 7.4%.

A significant difference between the ground motion scenario in terms of the 475-year seismic hazard map and considered fully probabilistic approach suggests that seismic landslide risk could be underestimated or overestimated when using the 475-year seismic hazard map for landslide risk assessment. The given approach follows the rational risk management idea that handles well all possible ground motion scenarios and geomechanical slope models.

An important factor that leads to an increase of the total risk is the saturated soil mass. Geotechnical studies of rocks and soils within large areas of a natural slope within different seasons should significantly reduce the uncertainties associated with the variability of geomechanical slope models.

The aim of future research is to produce regional seismically-induced landslide hazard maps using the fully probabilistic approach.

Author Contributions: Conceptualization, A.K.; methodology, A.K.; software, A.S.; validation, A.S.; formal analysis, Y.G.; investigation, Y.G., V.L., A.M., Y.S., L.M., A.S. and M.M.; resources, A.K. and Y.S.; writing—review and editing, A.K. and L.M.; visualization, Y.S. and A.M.; project administration, A.K. and Y.G.

Acknowledgments: This study was supported through computational resources provided by the Shared Facility Center "Data Center of FEB RAS" (Khabarovsk) [30]. Material parameters of the studied area were supported by Biolit LLC (Yuzhno-Sakhalinsk).

Conflicts of Interest: The authors declare no conflict of interest.

References

1. Mohammad, E.; Babak, O.; Mahdi, M.; Mohammad, A.N.; Ali, A.A. Sensitivity analysis in seismic loss estimation of urban infrastructures. *Geomat. Nat. Hazards Risk* **2018**, *9*, 624–644.
2. Martino, S.; Battaglia, S.; Delgado, J.; Esposito, C.; Martini, G.; Missori, C. Probabilistic Approach to Provide Scenarios of Earthquake-Induced Slope Failures (PARSIFAL) Applied to the Alcoy Basin (South Spain). *Geosciences* **2018**, *8*, 57. [CrossRef]
3. Forte, G.; Fabbrocino, S.; de Magistris, F.S.; Silvestri, F.; Fabbrocino, G. Earthquake Triggered Landslides: The Case Study of a Roadway Network in Molise Region (Italy). In *Engineering Geology for Society and Territory*; Springer: Cham, Switzerland, 2015; Volume 2, pp. 765–768.
4. Keefer, D.K. Statistical analysis of an earthquake-induced landslide distribution—The 1989 Loma Prieta, California event. *Eng. Geol.* **2000**, *58*, 231–249. [CrossRef]
5. Xu, C.; Xu, X.; Yao, X.; Dai, F. Three (nearly) complete inventories of landslides triggered by the May 12, 2008 Wenchuan Mw 7.9 earthquake of China and their spatial distribution statistical analysis. *Landslides* **2014**, *11*, 441–461. [CrossRef]
6. Esposito, E.; Guerrieri, L.; Porfido, S.; Vittori, E.; Blumetti, A.M.; Comerci, V.; Michetti, A.M.; Serva, L. Landslides induced by historical and recent earthquakes in Central-Southern Apennines (Italy): A tool for intensity assessment and seismic hazard. *Landslide Sci. Pract.* **2013**, 295–303.
7. Porfido, S.; Esposito, E.; Spiga, E.; Sacchi, M.; Molisso, F.; Mazzola, S. Impact of ground effects for an appropriate mitigation strategy in seismic area: The example of Guatemala 1976 earthquake. In *Engineering Geology for Society and Territory*; Springer: Cham, Switzerland, 2015; Volume 2, pp. 703–708.
8. Lobkina, V.A.; Kazakova, E.N.; Zhiruev, S.P.; Kazakov, N.A. Methods of landslide risk assessment for territory of settlements of Sakhalin Region (Makarov city, Sakhalin). *Russ. J. Pac. Geol.* **2013**, *32*, 100–109. (In Russian)
9. Ulomov, V.I.; Bogdanov, M.I. Explanatory note on the GSZ-2016 maps set of general seismic zoning of the Russian Federation territory. *Eng. Surv.* **2016**, *7*, 49–122. (In Russian)
10. Konovalov, A.V.; Nagornykh, T.V.; Safonov, D.A.; Lomtev, V.L. Nevelsk earthquakes of August 2, 2007 and seismic setting in the southeastern margin of Sakhalin Island. *Russ. J. Pac. Geol.* **2015**, *9*, 451–466. [CrossRef]
11. Reliable Prognosis. Available online: https://rp5.ru/%D0%90%D1%80%D1%85%D0%B8%D0%B2_%D0%BF%D0%BE%D0%B3%D0%BE%D0%B4%D1%8B_%D0%B2_%D0%9D%D0%B5%D0%B2%D0%B5%D0%BB%D1%8C%D1%81%D0%BA%D0%B5 (accessed on 10 April 2019).
12. Jibson, R.W.; Harp, E.L.; Michael, J.A. A method for producing digital probabilistic seismic landslide hazard maps. *Eng. Geol.* **2000**, *58*, 271–289. [CrossRef]
13. Lee, C.T. Statistical seismic landslide hazard analysis: An example from Taiwan. *Eng. Geol.* **2014**, *182*, 201–212. [CrossRef]
14. Jibson, R.W. Mapping seismic landslide hazards in Anchorage, Alaska. In Proceedings of the 10th National Conference in Earthquake Engineering, Earthquake Engineering Research Institute, Anchorage, AK, USA, 21–25 July 2014.
15. Martino, S.; Battaglia, S.; D'Alessandro, F.; Della Seta, M.; Esposito, C.; Martini, G.; Pallone, F.; Troiani, F. Earthquake-induced landslide scenarios for seismic microzonation: Application to the Accumoli area (Rieti, Italy). *Bull. Earthq. Eng.* **2019**, 1–19. [CrossRef]
16. Del Gaudio, V.; Wasowski, J.; Pierri, P. An Approach to Time-Probabilistic Evaluation of Seismically Induced Landslide Hazard. *Bull. Seismol. Soc. Am.* **2003**, *93*, 557–569. [CrossRef]
17. Galli, M.; Guzzetti, F. Landslide vulnerability criteria: A case study from Umbria, Central Italy. *Environ. Manag.* **2007**, *40*, 649–664. [CrossRef] [PubMed]
18. Jaiswal, P.; van Westen, C.J.; Jetten, V. Quantitative estimation of landslide risk from rapid debris slides on natural slopes in the Nilgiri hills, India. *Nat. Hazards Earth Syst. Sci.* **2011**, *11*, 1723–1743. [CrossRef]
19. Guillard-Gonçalves, C.; Zêzere, J.L.; Pereira, S.; Garcia, R.A.C. Assessment of physical vulnerability of buildings and analysis of landslide risk at the municipal scale: Application to the Loures municipality, Portugal. *Nat. Hazards Earth Syst. Sci.* **2016**, *16*, 311–331. [CrossRef]
20. Golozubov, V.V.; Kasatkin, S.A.; Grannik, V.M.; Nechayuk, A.E. Deformation of the Upper Cretaceous and Cenozoic complexes of the West Sakhalin terrane. *Geotectonics* **2012**, *46*, 333–351. [CrossRef]
21. *Atlas of the Sakhalin Region*; Komsomolskiy, G.V.; Siryk, I.M., Eds.; GUGK Sovmin USSR: Moscow, Russia, 1967; 137p. (In Russian)

22. *Handbook of Climate of the USSR*; Pilnikova, Z.N., Ed.; Sakhalin Region, Gidrometeoizdat: Saint Petersburg, Russia, 1990; pp. 48–64. (In Russian)
23. Cornell, C.A. Engineering seismic risk analysis. *Bull. Seismol. Soc. Am.* **1968**, *58*, 1583–1606.
24. Aguilar-Meléndez, A.; Ordaz Schroeder, M.G.; De la Puente, J.; Gonzalez Rocha, S.N.; Rodrigez Lozoya, H.E.; Cordova Ceballos, A.; Garcia Elias, A.; Calderon Ramon, C.M.; Escalante Martinez, J.E.; Laguna Camacho, J.R.; et al. Development and Validation of Software CRISIS to Perform Probabilistic Seismic Hazard Assessment with Emphasis on the Recent CRISIS2015. *Computacion y Sistemas* **2017**, *21*, 67–90. [CrossRef]
25. Konovalov, A.V.; Sychov, A.S.; Manaychev, K.A.; Stepnov, A.A.; Gavrilov, A.V. Testing of a New GMPE Model in Probabilistic Seismic Hazard Analysis for the Sakhalin Region. *Seismic Instr.* **2019**, *55*, 283–290. [CrossRef]
26. Newmark, N.M. Effects of earthquakes on dams and embankments. *Geotechnique* **1965**, *15*, 139–159. [CrossRef]
27. Fabbrocino, S.; Paduano, P.; Lanzano, G.; Forte, G.; de Magistris, F.S.; Fabbrocino, G. *Engineering Geology Model for Seismic Vulnerability Assessment of Critical Infrastructures*; Engineering Geology Special Publications; Geological Society: London, UK, 2016.
28. Rossetto, T.; Elnashai, A. Derivation of vulnerability functions for European-type RC structures based on observational data. *Eng. Struct.* **2003**, *25*, 1241–1263. [CrossRef]
29. Ciurean, R.L.; Schröter, D.; Glade, T. Conceptual Frameworks of Vulnerability Assessments for Natural Disasters Reduction. In *Approaches to Disaster Management—Examining the Implications of Hazards, Emergencies and Disasters*; Tiefenbacher, J., Ed.; InTech: Rijeka, Croatia, 2013; pp. 3–32.
30. Sorokin, A.A.; Makogonov, S.I.; Korolev, S.P. The Information Infrastructure for Collective Scientific Work in the Far East of Russia. *Sci. Tech. Inf. Process.* **2017**, *44*, 302–304. [CrossRef]

geosciences

MDPI

Article

Site Effect Assessment in Ulaanbaatar, Mongolia through Inversion Analysis of Microtremor H/V Spectral Ratios

Zorigt Tumurbaatar [1,*], Hiroyuki Miura [1] and Tsoggerel Tsamba [2]

1 Department of Architecture, Hiroshima University, 1-4-1 Kagamiyama, Higashi-Hiroshima, Hiroshima 739-8527, Japan; hmiura@hiroshima-u.ac.jp
2 Department of Civil Engineering, Mongolian University of Science and Technology, 8th khoroo, Baga toiruu 34, Ulaanbaatar 14191, Mongolia; tsoggerelt@must.edu.mn
* Correspondence: zorigt.ces@gmail.com; Tel.: +81-82-424-7798

Received: 11 April 2019; Accepted: 14 May 2019; Published: 17 May 2019

Abstract: Due to the population growth and urban sprawl in Ulaanbaatar city (UB), Mongolia, hazard and risk analysis for future earthquakes have become an important issue for disaster mitigation planning. Evaluation of a site effect is one of the essential parts of the earthquake hazard estimation in this area. The site effect can be evaluated by site amplifications calculated from shear-wave velocity (V_S) models including from bedrock to surface layers. However, it is difficult to assess the pattern of the site effects in UB because shallow mostly up to 15 m and a small number of investigated V_S models are available in previous studies. In this study, the V_S models are estimated using microtremor data at 50 sites and inversion analysis is applied to the observed data in order to evaluate site amplifications in UB. In particular, the joint inversion technique based on a diffuse field approach is applied to estimate the V_S structures at three sites using the observed horizontal-to-vertical (H/V) spectral ratios and surface wave phase velocities obtained by Odonbaatar (2011). The rest of the sites are estimated by the single inversion technique using the observed microtremor H/V spectral ratios considering the results of the joint inversions. The seismic microzoning in UB is performed based on the site amplifications computed from the inverted V_S models to characterize the pattern of seismic hazard. The result shows the largest site amplification zone is computed along the Tuul river in the southeastern part of UB.

Keywords: microtremor H/V spectral ratio; inversion; shear-wave velocity structure; site amplification; Ulaanbaatar

1. Introduction

Mongolia has been one of the most seismically active intracontinental regions in the world with four large earthquakes with the magnitudes of around 8 along with their active faults in the western part of the country during the last century [1]. Compared with western Mongolia, central Mongolia including Ulaanbaatar's region (UB) is considered a relatively low seismically active zone. However, according to the previous investigation [1], there are several active faults such as Hustai, Gunj, and Emeelt faults. In particular, the Emeelt fault earthquake with the estimated maximum magnitude of 7 [2] has been expected to produce great impact on the UB region because the fault is located at only around 20 km from UB to the west–southwest direction (see Figure 1).

In order to consider the disaster mitigation planning for such large earthquakes, the assessments of ground shaking intensities and building damage for scenario earthquakes are indispensable. Generally, the ground shaking intensity is determined by evaluating the source effect, path effect and site effect. The evaluation of site effects is especially essential because the effect of surface soil can generate large ground motion amplification during the earthquakes. The increase of the amplitude of seismic shaking can produce severe damage to buildings and infrastructures accompanying a loss of lives. The UB basin located at the valley of Tuul river has 30 km length and 4 to 10 km width and is filled with alluvial deposits. According to borehole data and the geological map, the deposit thicknesses are around 10 to 80 m in most of the area, and the maximum depth is 120 m expected around the Tuul riverside [3,4]. Since the population of UB city has grown around three times in the last two decades, the number of building constructions and densities are highly increased, and new construction districts have been expanded to everywhere in the city area. Especially in the last six years, moderate and high-rise buildings have been constructed near the Tuul Riverside such as the Zaisan, Yarmagiin Guur, and Shine Yarmag areas (see Figure 1).

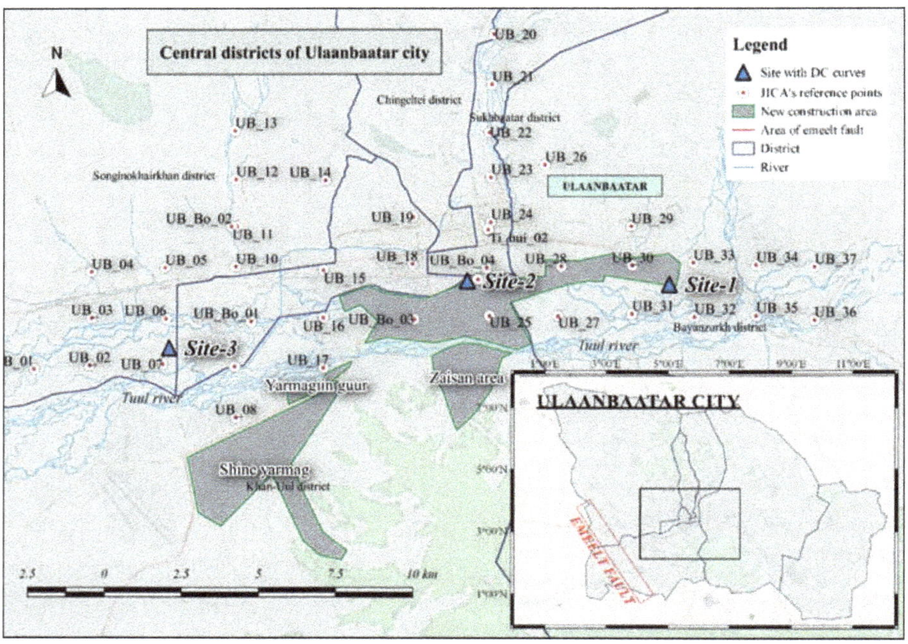

Figure 1. Location map of the study area with Japan International Cooperation Agency's (JICA) report reference sites and the location of the Emeelt fault.

Odonbaatar [3] performed microtremor array observations at three sites in UB in order to identify the deep underground structures and estimated the three-dimensional (3D) UB basin model. He also simulated ground motions using the basin model to discuss the 3D effect on the surface ground motions. However, the detailed V_S structures in UB have not been validated since he developed two layer models with surface and bedrock layers. Due to the recent project of the Japan International Cooperation Agency (JICA) in UB city [5], the results of downhole seismic method (PS-logging) investigations and Spectral Analysis of Surface Wave (SASW) surveys at around 40 sites were reported. Since the explorations were up to 30 m depth at the maximum, the intermediate and deeper V_S structures have not been elucidated.

Microtremors (ambient noise of ground) analysis are a low-cost and convenient method to estimate site characterizations. For example, while horizontal motions are larger than vertical motion on the soft ground, both horizontal and vertical motions are similar to each other on the stiff ground. In particular, the Horizontal-to-Vertical (H/V) spectral ratio technique firstly proposed by Nakamura [6] has been utilized to evaluate the fundamental resonant frequency of grounds [7,8]. The microtremor H/V spectral ratio (MHVR) has also been used to estimate V_S profiles by assuming that the microtremors are mainly composed of surface waves [9,10]. Recently, diffuse field assumption (DFA) has been proposed to understand MHVR by assuming a diffuse wavefield containing all types of elastic waves [11,12]. In this theory, MHVR can be numerically computed as the imaginary parts of Green's functions, and the theoretical H/V ratio depends on underground layer thickness and elastic properties of soil. Hence the theory allowed for inverting V_S profiles from MHVR [13–16]. One of the authors also examined the applicability of the DFA for the MHVR-based V_S profiling and presented the validity of the technique for the site characterizations [17].

In this paper, we observe the microtremors at multiple sites in UB areas and apply the inversion technique to the MHVR in order to reveal the patterns of the V_S profiles considering the observations in the previous studies such as Odonbaatar [3] and JICA's report [5]. The estimated V_S profiles and the site characteristics are discussed by comparing the existing borehole data, geological and terrain maps. Finally, the site amplification characteristics obtained from the estimated V_S profiles are assessed by zoning the UB area.

2. Previous Studies for Site Characterizations in UB

In the UB area, around 5000 boreholes were explored by Russian and Mongolian research expeditions for geotechnical purpose before 1990. Although these data include some geotechnical information with laboratory tests, those data were surveyed at less than 10 m depth in most of the sites [18], and they did not include seismic velocity information.

In the previous study by Odonbaatar [3], this study hereafter is referred to as OD, the microtremor array observations were performed at three sites in UB (Site-1, -2 and -3 in Figure 1) to estimate the deep underground structures. Since the estimated profiles were developed to create the UB basin model, the shallow V_S structures in the basin were not discussed in detail. Bedrock depth with the V_S of 1600 to 1800 m/s was determined at 40 m to more than 100 m depending on the location.

In the JICA report, on the other hand, the V_S profiles were determined by PS-loggings at four sites (UB_Bo_01 to 04 in Figure 1) and the SASW method at 37 sites (UB_01 to 36 in Figure 1) in the central UB area [5]. Since the depths of the V_S profiles are less than 30 m and around 10 m at most sites, deeper V_S structures have not been explored. They were never validated by other data sources such as MHVR. Furthermore, significant discrepancies between the V_S profiles derived by these two techniques were found even at the same site, indicating the uncertainty of the techniques. For those reasons, we conduct the microtremor observations at UB and apply the inversion technique to the observed MHVRs to derive the V_S profiles and evaluate the site amplification characteristics.

3. Microtremor Observations

The microtremors at more than 50 sites are measured by single-site observations in the city from 25 March 2018 to 15 April 2018. The locations of the observation sites are shown in Figure 2. The coordinates of the array observation sites in OD's study and the JICA report sites are used for the observations, and we measure at the same locations in the most sites. Only a few sites are located in a special prohibited zone such as a pure water collecting zone of UB city. It was forbidden to enter that zone, so the microtremors are measured as close to the target sites as possible.

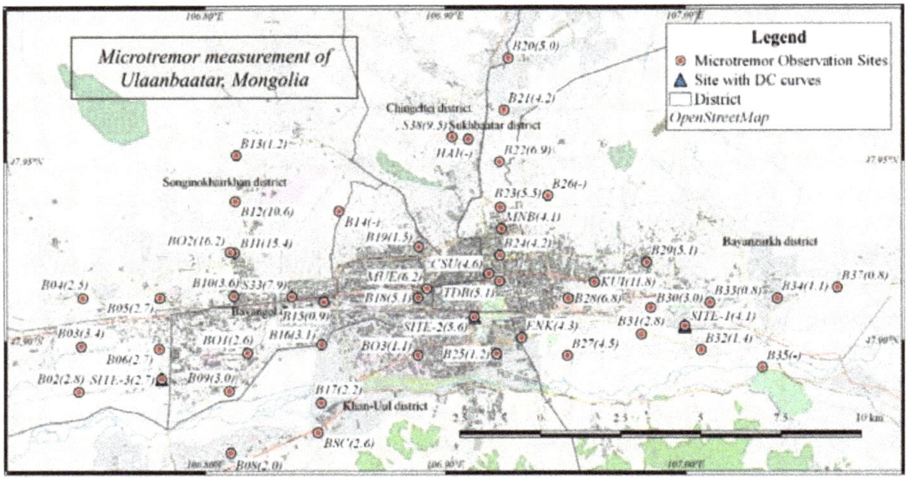

Figure 2. Microtremor observation sites in Ulaanbaatar city. Values in brackets indicate peak frequencies identified from microtremor horizontal-to-vertical ratios (MHVRs).

In the observations, we use the portable handheld microtremor sensor of GEODAQS-2S3D developed by ANET Inc., Ueda-shi, Japan. The sensor can measure the ambient ground motions in velocity with the lower limit of the frequency of 0.5 Hz (period of 2.0 s). The sampling of 100 Hz is applied during the observation, indicating the frequency coverage from 0.5 Hz to 50 Hz. We have measured the microtremor at the surface of the ground and recorded for three to ten minutes. Due to the heavy traffic, the measurements are basically conducted in the early morning.

The MHVRs (HV_{obs}) are calculated from Equation (1) shown below:

$$HV_{obs}(f) = \frac{\sqrt{NS(f)^2 + EW(f)^2}}{UD(f)} \quad (1)$$

where the scripts *NS* and *EW* represent the Fourier spectra for north–south (NS) and east–west (EW) components, respectively, while the *UD* corresponds to the up-down (UD) component in frequency f in Hz. The Parzen window with a bandwidth of 0.3 Hz is applied for the Fourier spectrum. Three 20.48 s duration parts with less traffic and artificial noises are extracted from the records, and mean values of the three MHVRs are calculated.

The observation conditions and the observed MHVRs at the typical three sites are shown in Figure 3. The red number with an arrow indicates peak frequencies of the MVHRs. The peak frequencies vary from 0.8 to 16.2 Hz, showing different site conditions in the city. The identified peak frequencies of the MHVRs are shown in the values in brackets in Figure 2. The peak frequencies are different site by site. The northern locations such as BO2, B12, S38, KUI and B26 near the mountain have a higher frequency (9.5 to 16.2 Hz) or no peak. These results indicate shallow sedimentary depth or rock sites. On the other hand, lowest peak frequencies (0.8 to 5.0 Hz) are observed in the most sites around the Tuul river basin such as BO3, B25, B32, and B37 probably due to the thick sediment.

In addition, the peak amplitudes of MHVRs near the mountain sites are much lower than those at the Tuul riverside area. This result suggests that large site amplifications are expected in the Tuul riverside area.

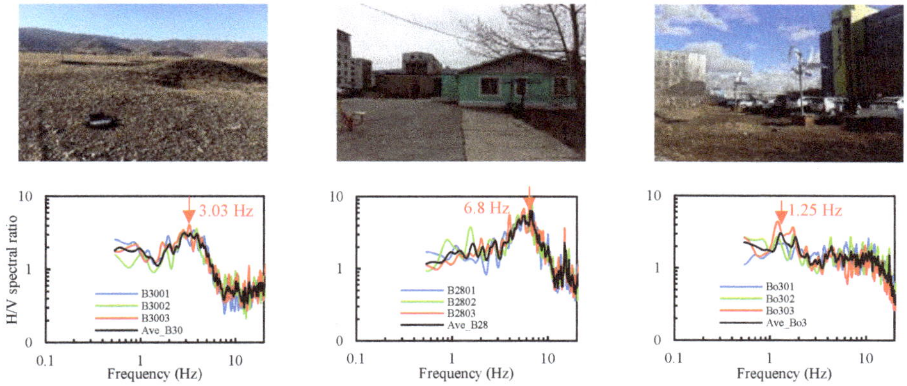

Figure 3. Typical site observation photos and observed H/V spectral ratios at B30, B28, and BO3.

4. Inversion Analysis of Microtremor Data

4.1. Joint Inversion of H/V Spectral Ratio and Surface Wave Dispersion Curve

According to the previous inversion studies of MHVR [10,15], joint inversion of MHVR and surface wave dispersion curve is more effective than a single inversion of MHVR to avoid the non-uniqueness of V_S and thickness during the V_S profile estimation. Therefore, the joint inversion of the MHVRs with dispersion curve obtained in the OD's microtremor array observations [3] are applied at Sites 1, 2 and 3.

First, we evaluate the estimated V_S profiles in the OD's study (OD models) by comparing theoretical H/V spectral ratio (HVR) with the observed MHVRs. The circles in Figure 4a indicate the observed MHVRs. The data is resampled at the equal intervals in the common logarithm scale for the following inversion analysis. The dotted lines in Figure 4a show the theoretical HVRs computed from the OD models by the diffuse field theory at Site-1, 2 and 3. The observed phase velocities and V_S profiles of the OD models [3] are shown by circles in Figure 4b and dotted lines in Figure 4c, respectively. The dotted lines in Figure 4b show the standard deviations of the phase velocities obtained from the array observations. As shown in Figure 4a, the theoretical values by the OD models do not match the observed MHVRs especially at Site-2 and Site-3, suggesting that the OD models need to be modified using not only the dispersion curves but also the MHVRs.

The inversion technique requires search limits for the thickness and V_S for each soil layer generally based on geology, borehole, and geophysical test results. In this study, search limits are assumed for each site considering the OD model for deeper structures and JICA's report at the closest sites for shallower structures. The search limits for Site-1, 2, and 3 are shown in Tables 1–3. The search limits are set to cover the existing models. The V_S of the lowest layers are fixed at 1800 m/s because the V_S of the bedrock was defined at around 1800 m/s in the OD models [3], and the phase velocities were obtained up to about 1500 m/s in the dispersion curves as shown in Figure 4b.

In addition, V_P and soil density for each layer requires calculating theoretical HVR. Based on the previous PS-loggings in the JICA's report [5], the relationship between V_S and V_P is modeled by linear regression analysis as shown in Equation (2):

$$V_P = 0.00162V_S^2 + 1.403V_S + 14.9. \tag{2}$$

Equation (2) is used to set the search limit for V_P in the inversion analysis for the layer of V_S of less than 800 m/s because the shallow V_S structures up to the V_S of around 800 m/s were obtained in JICA's report. If the V_S is higher than 800 m/s, the relationship developed by Kitsunezaki et al. [19] shown in Equation (3) is applied:

$$V_p = 1.11V_S + 1290 \tag{3}$$

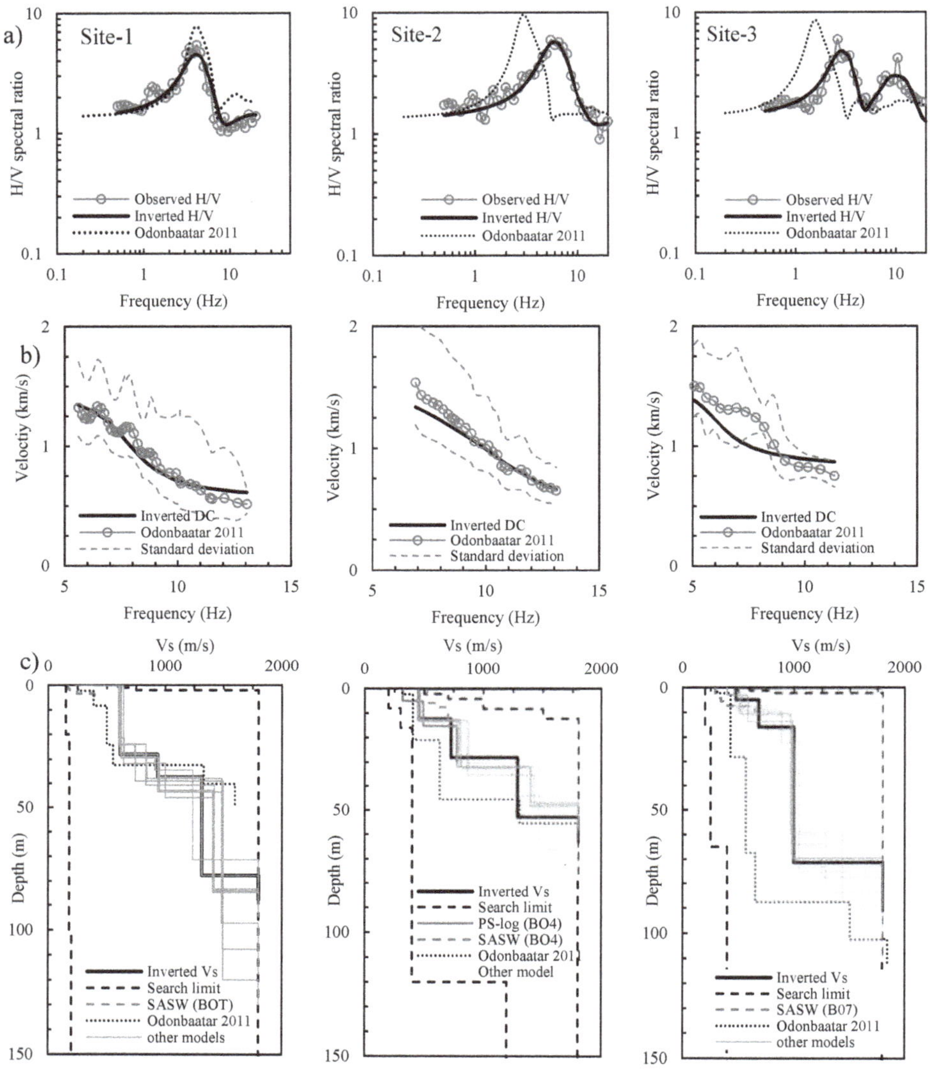

Figure 4. (**a**) Comparison of observed and theoretical and observed H/V spectral ratios; (**b**) Comparison of observed and theoretical surface wave dispersion curves and (**c**) Shear-wave velocity profiles obtained by joint inversion at Site-1, 2, and 3.

Soil density for each layer is assumed at general values as shown in Tables 1–3.

In this study, the Simulated Annealing method [20] is applied during the inversion analysis because it was already successfully used in the inversion of dispersion curves and MHVR [15]. We performed 10 inversions per site with different initial sets of randomly-generated models and selected the best structure with the smallest misfit (best fit model).

Misfit between the theoretical values and observed values in MHVR and dispersion curve is minimized in the joint inversion process. Misfit Γ is defined in Equation (4) as shown below:

$$\Gamma = \frac{2(1-\xi)}{n}\sum_{i=1}^{n}\frac{\left(HV_{obs}(f_i) - HV_{th}(f_i)\right)^2}{HV_{obs}^2(f_i)} + \frac{2\xi}{m}\sum_{i=1}^{m}\frac{\left(DC_{obs}(f_i) - DC_{th}(f_i)\right)^2}{{DC_{obs}}^2(f_i)} \tag{4}$$

where $\xi = n/(n + m)$ is cost function, n number of data with H/V spectral ratio is 41 at each site, and m number of data with dispersion curves is different for each location. Because the amount of data in the obtained dispersion curves varied, m = 42 at Site-1, m = 27 at Site-2 and m = 16 at Site-3 are used. Even if the amount of data of both observables is not the same, the cost function does not change [13]. H/V_{obs} and H/V_{th} represent the target measurements and calculated MHVRs for the current model in a frequency f_i. DC_{obs} and DC_{th} are the experimental measurements and the estimated dispersion curve for the current model. In the preliminary analysis with the cost function of 0.5, the observed and inverted MHVRs showed a significant difference. Therefore, the cost is changed to 0.8 to increase the weight for the MHVR than for the dispersion curve. The results of the inversions are shown in solid thick lines in Figure 4. The solid lines in Figure 4a,b indicate the theoretical values obtained from the inverted V_S profiles shown in Figure 4c. Gray lines in Figure 4c indicate variations of the inverted models, Site-2 and Site-3's other models are almost identical to each other. However, a shallow part of the models is the same; a deeper part has around 50-meter difference in Site-1. It means that variabilities of the deeper structure are much smaller in the joint inversion. The theoretical values show good agreement with the observed values, especially better agreement for the MHVRs than those by the OD models. On the other hand, the theoretical dispersion curves show less agreement with the observed values than those by the OD models, especially at Site-3. This difference indicates that it is difficult to find a model that accurately reproduces both types of observation data. Considering the variability of the dispersion curve at Site-3, we accepted the inversion result since the theoretical values are within the standard error (mean plus minus one standard deviation).

Table 1. Initial inversion parameter ranges for Site-1.

Layer	Thickness (m)	V_S (m/s)	V_P (m/s)	Density (kg/m^3)
1	0–20	200–600	350–1400	1600–1800
2	0–80	250–800	450–2400	1700–1900
3	0–100	400–1200	830–2600	1800–2000
4	0–100	1200–1500	2600–2950	1900–2100
5	-	1800	3280	2000–2200

Table 2. Initial inversion parameter ranges for Site-2.

Layer	Thickness (m)	V_S (m/s)	V_P (m/s)	Density (kg/m^3)
1	2–8	200–500	360v1120	1600–1800
2	2–8	300–700	580–1790	1700–1900
3	4–104	400–1000	830–2400	1800–2000
4	4–100	1200–1500	2620–2950	1900–2100
5	-	1800	3280	2000-2200

Table 3. Initial inversion parameter ranges for Site-3.

Layer	Thickness (m)	V_S (m/s)	V_P (m/s)	Density (kg/m^3)
1	2–8	160–600	280–1400	1600–1800
2	2–8	300–700	580–1790	1700–1900
3	4–104	400–1000	830–2400	1800–2000
4	4–100	1200–1500	2600–2950	1900–2100
5	-	1800	3280	2000–2200

The estimated bedrock depths vary 50 to 80 m showing larger depths at Site-1 and 3 than at Site-2. Since the Site-1 and Site-3 are located near the Tuul river as shown in Figure 1, thicker sediments are expected in the riverside area than in the central UB such as Site-2.

4.2. Single Inversion of H/V Spectral Ratio

In this part, single inversion analysis is applied to the observed MHVRs to estimate V_S profiles at the observation sites except for Site-1, 2 and 3. Blind single inversion can be quite costly because it may imply a huge search within the parameter space and the convergence can require a long time or even not be reached. In order to sharpen the search of solutions, we define search limits of target values using existing shallow V_S profiles [5] with the previous joint inverted results for deep underground structures. For example, a typical site search limit comparing with JICA's report is shown in Table 4.

In the single inversion, the misfit Γ_{HV} was defined in Equation (5) as below:

$$\Gamma_{HV} = \sum_{i=1}^{n} \frac{(HV_{obs}(f_i) - HV_{th}(f_i))^2}{HV_{obs}{}^2(f_i)} \tag{5}$$

where HV_{obs} and HV_{th} represent target measurements and calculated MHVR for a current model at frequencies f_i [15].

Table 4. Initial inversion parameter of the typical site (UB25).

	JICA's Report		This Study	
Layer	**Thickness (m)**	**V_S (m/s)**	**Thickness (m)**	**V_S (m/s)**
1	0–4	207–323	0–20	150–500
2	0–6	436–703	0–50	180–1000
3	-	-	0–100	600–1500
4	-	-	-	1800

The H/V spectral amplitudes have no peak or are smaller than a factor of 2 at a few sites located in the mountain areas. These sites are eliminated in the inversion process because it would be difficult to accurately estimate V_S profiles due to the uncertainties during the inversion analysis. Figure 5 shows the observed and inverted MHVRs of 40 sites. The black line indicates the observed MHVRs, the red line indicates inverted best results of MHVRs, and gray lines show the results of other inverted models. At most of the sites, the misfits of the inversion process are small. It is considered that the difference between the observed and inverted MHVRs are similar in shape and amplitude. Figure 6 shows the inverted V_S profiles of the sites in UB. The V_S profiles are up to 160 m depth in most of the sites. In some sites such as UB_08, UB_33, UB_34, and UB_Bo_03, the bedrock depths are estimated at around 200 m depth, which is slightly deeper than those expected in the previous research [3,4] because the predominant frequencies are found at approximately 1 Hz or less. Since the information for the deeper underground structures in UB is very limited, further detailed observations such as much larger-scale microtremor array observations would be required to justify the deeper parts of the estimated V_S profiles. The main objective of this study is to evaluate the site effects in UB area. In order to discuss the effect of the bedrock depth on site amplifications, sensitivity analysis is performed by using the estimated V_S profile in UB33 where the estimated bedrock depth is 203 m. Table 5 shows the V_S profile calculated in UB33. We prepare three other models (Model-1, 2 and 3) by changing the thickness of the third layer from 98 m to 80 m, 60 m, and 40 m. The bedrock depths of the three models correspond to 185 m, 165 m, and 145 m, respectively. The site amplifications are calculated from the four models by one-dimensional elastic wave propagation theory of SH-wave [21]. The damping factor of 0.03 is given to all the layers. Figure 7 shows the calculated site amplifications for the models. The amplification factors around 1.0 Hz are slightly shifted to a higher frequency by reducing the thickness of the third layer. However, the change of the site amplification factors among the models are quite small and negligible. This indicates that the thickness of the deep layers does not largely influence the site amplifications.

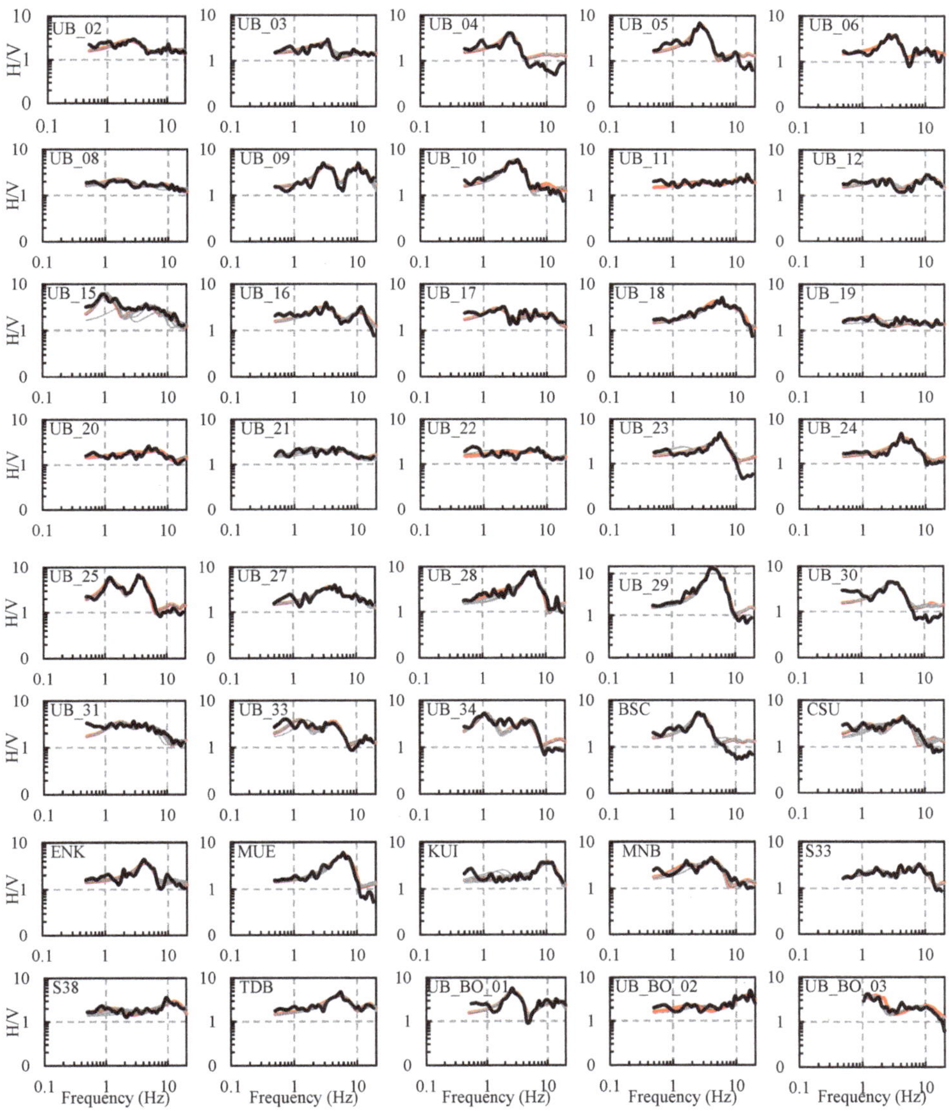

Figure 5. Observed and theoretical MHVRs obtained by single inversion; the black line indicates observed MHVR; the red line indicates inverted best MHVR; the gray lines indicate inverted models.

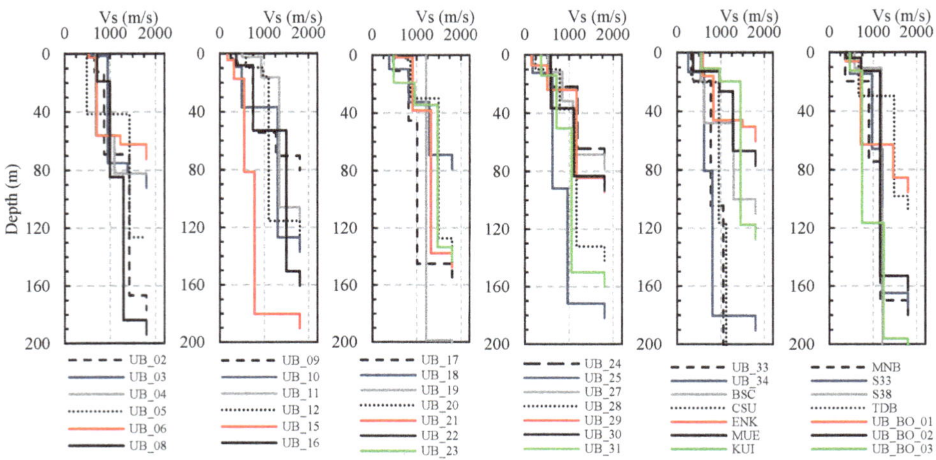

Figure 6. Estimated shear-wave velocity profiles of all sites obtained from microtremor single inversions.

4.3. Comparison with Borehole Data

In order to validate the shallow V_S profiles, we compare the estimated profiles with the existing borehole data. In the geotechnical map in UB [18], cross-sections of geological profiles were included. Based on the map, the profiles are compared at five sites located close to the cross-sections with the borehole information. Figure 8 shows the comparison of the borehole data and the inverted V_S profiles. Depths of the cross-sections are up to 60 m in the sites. According to the borehole data, the soil consists of gravel with clay or sand in upper layers and weathered shale or rocks in lower layers at most of the sites. The thickness of the inverted layers almost corresponds to the boundaries of the soil types in the borehole.

The V_S corresponding to the gravel with clay and sand layers are 400 to 1000 m/s while the V_S corresponding to the weathered shale and rock are 1000 to 1500 m/s. The results show some justification for the inverted shallow V_S structures.

Table 5. The parameters for sensitivity analysis (UB33).

№	Estimated		Model-1	Model-2	Model-3
	V_S (m/s)	Thickness (m)	Thickness (m)	Thickness (m)	Thickness (m)
1	327.7	19.4	19.4	19.4	19.4
2	769.2	85.2	85.2	85.2	85.2
3	1054.02	98.5	80.0	60.0	40.0
4	1800.0	-	-	-	-

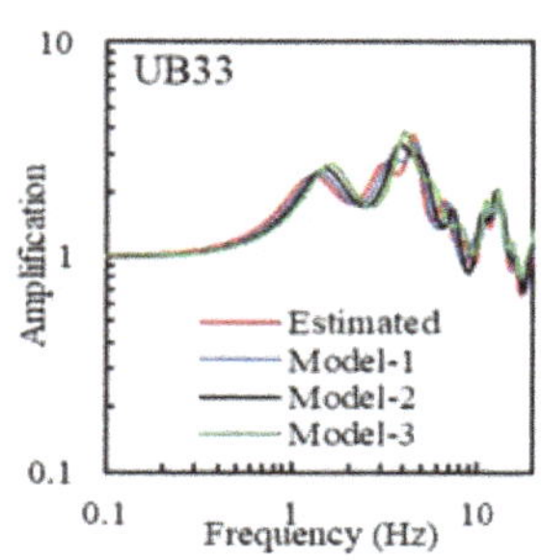

Figure 7. Variations of theoretical site amplification analysis.

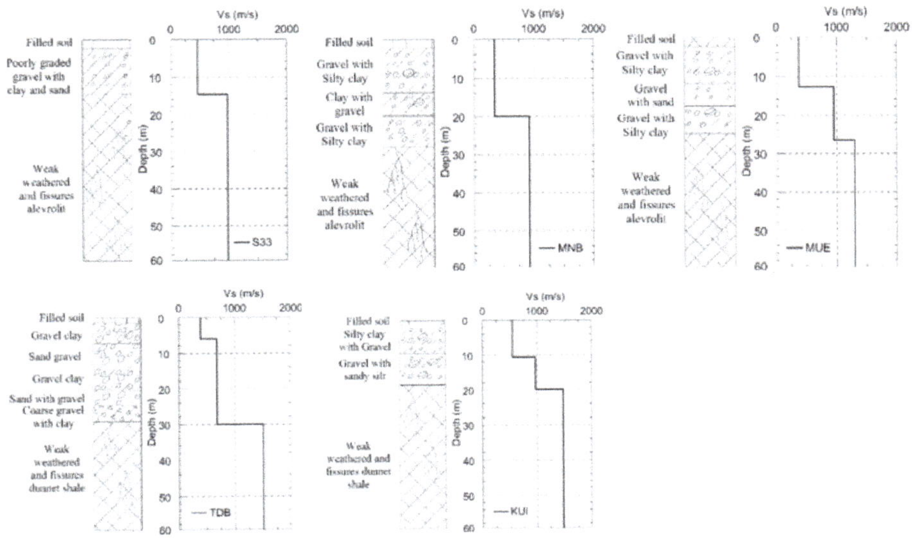

Figure 8. Comparison of estimated shear-wave velocity profiles and existing borehole data.

4.4. V_{S30} Distribution in UB.

Time-averaged shear-wave velocity in upper 30 m depth (V_{S30}) has been used as an index of site characteristics in many applications such as ground motion prediction equations, building codes and earthquake hazard maps. In order to evaluate the pattern of V_{S30}s in UB, the V_{S30}s are calculated from the V_S profiles inverted in the previous section.

Figure 9a shows V_{S30} distribution with the geotechnical map of UB [18]. According to this map, the UB basin has consisted mostly of Alluvium, Alluvium-Diluvium, Fluvial deposits, and the northern part has included of Neogene. The geological period of deposits is mostly Quaternary; an epoch is Holocene to Pleistocene. Bedrocks are mainly comprised of carboniferous shallow marine sedimentary rocks and Devonian to Carboniferous accretionary complex. Along the Tuul river, it is covered by sediments of sand, gravel, and mud, which is transported by the rivers. In addition, Selbe, Uliastai, Bilgekh, and Tahilt rivers are located around Ulaanbaatar city. All of these rivers are affected by the boundary of deposits and geotechnical conditions of the UB basin.

Figure 9b shows the map of V_{S30} distribution in UB paired with the 7.5-arcsecond resolution digital elevation model (DEM) of the Global multi-resolution terrain elevation data 2010 [22]. The elevation in Ulaanbaatar's central area is 1280 to 1350 m above the sea level, and those in northern and southern mountains are higher than 2100 m.

Larger V_{S30} sites (larger than 750 m/s) are located near the mountain area such as UB20, UB22, KUI, and UB11. Smaller V_{S30} sites (less than 450 m/s) are distributed in the lower elevation area. In addition, the V_{S30}s smaller than 450 m/s are expected near the Uliastai river such as UB34, UB33 and UB29 located in the eastern part of UB. In the central part of UB, the V_{S30} with 450 to 600 m/s are distributed such as CSU and TDB sites. Most of the sites in UB are classified to class C or B in the V_{S30} categories of the National Earthquake Hazards Reduction Program (NEHRP) [23], indicating that the site conditions are basically stiff and almost no very soft soils are expected in UB.

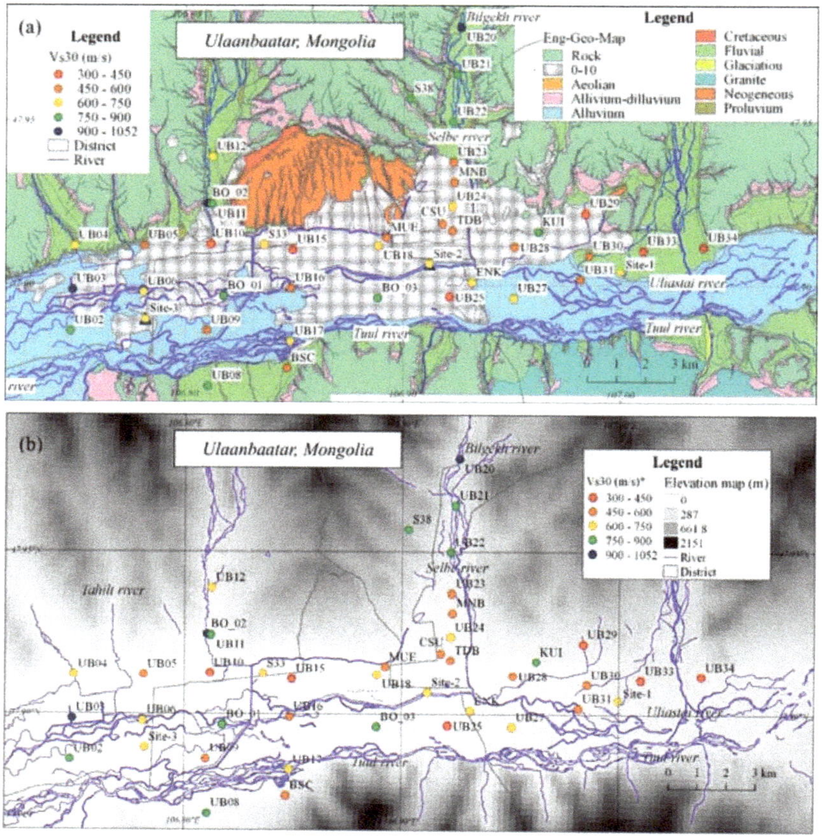

Figure 9. (**a**) V_{S30} distribution map with geotechnical map [18]; (**b**) V_S30 distribution map with a digital elevation model (DEM).

5. Site Effect Assessment

Concerning seismic hazards, key questions are whether high seismicity rate is representative of what we may expect in the future and what the hazard assessment is with smaller, but more frequent events. Site effects characterized by site amplifications would be necessary for seismic hazard mapping and microzoning would also convenient for understanding the patterns of the site amplification. In this study, the microzonation in UB is evaluated based on the site amplifications calculated from the inverted V_S profiles.

At first, when we consider the geotechnical map [18] and the terrain map, the total area of UB is divided into three zones, such as the northern part mountain area, the central part along the Tuul river and the southern part. Then, we consider the site amplification characteristic at each site; almost the same amplitude and amplified frequency (around 2 to 3 Hz) are observed at sites in the western side of the central part. Finally, if we carefully look at the site amplifications at the rest of sites in the central part, largest amplitude and amplified frequency (around 5 to 6 Hz) sites are located in the southeastern zone along the Tuul river. Finally, we can divide all the sites into five amplification zones as shown in Figure 10. The distribution of the average site amplifications for 1.0–1.25 Hz, 3.33–5.0 Hz, and 6.67–10.0 Hz in the sites are displayed in Figure 10. The H/V spectral ratios for each zone and theoretically calculated site amplification in the zones are shown in Figures 11 and 12. In addition, if we look at the H/V spectral ratios, predominant frequency and amplitude of each site's H/V spectral ratios are also similar to each other in the same zone. Zone-I is located in the western part of the UB basin and north of the Tuul river, showing larger amplifications in around 2 to 3 Hz. Zone-II is in the southwestern part of UB and south of the Tuul river. Recently, newly developed built-up areas have been expanded to this zone. Moderate to small amplifications are expected in the zone. Zone-III is in the northern part of UB along the Bilgekh and Selbe rivers. As described above, these sites are located in higher elevation zones, and the V_{S30}s are also larger than other areas. Small amplifications are expected for all frequency ranges in the zone. Zone-IV includes the central part of UB located in north of the Selbe river. Intermediate to large amplifications are calculated. Zone-V covers the eastern and southern parts of UB. This zone contains the major rivers such as Uliastai and Tuul rivers. Since the smaller V_{S30}s are estimated in the region, the highest amplifications are obtained in UB, especially for the frequencies higher than 3 Hz as shown in Figure 12. Since thick sediments are determined in the zone as discussed above, stronger ground shaking would be excited in future earthquakes than other zones. In particular, newly urbanized areas have been expanded to the area of Zone-V as shown in Figure 13, indicating that it might be a higher risk in such new construction districts.

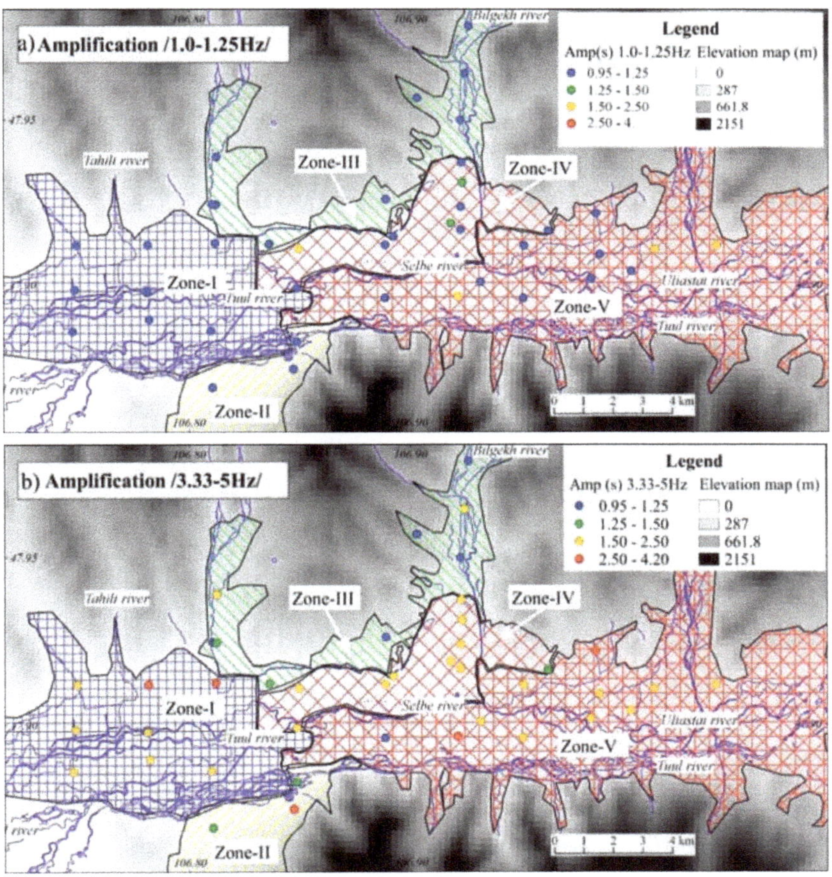

Figure 10. *Cont.*

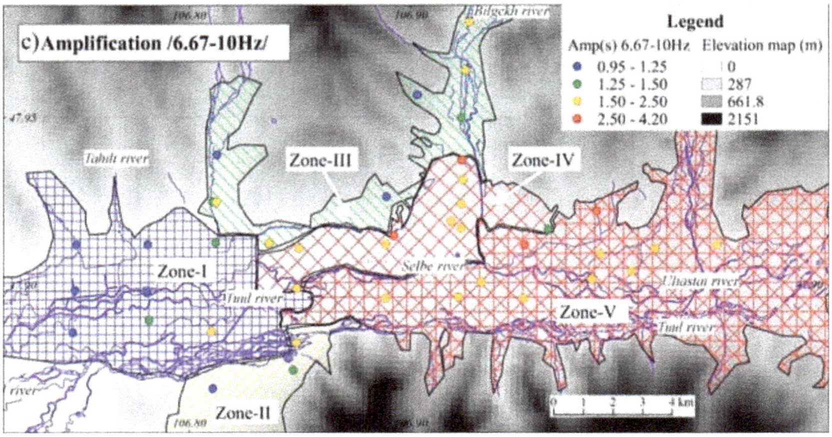

Figure 10. Distribution of site amplifications with the zoning map for (**a**) 1.0–1.25 Hz, (**b**) 3.33–5.0 Hz, and (**c**) 6.67–10 Hz.

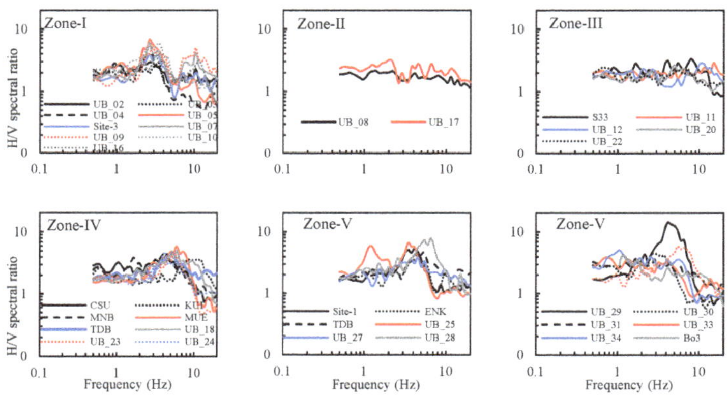

Figure 11. Observed H/V spectral ratios for each zone in Ulaanbaatar city (UB).

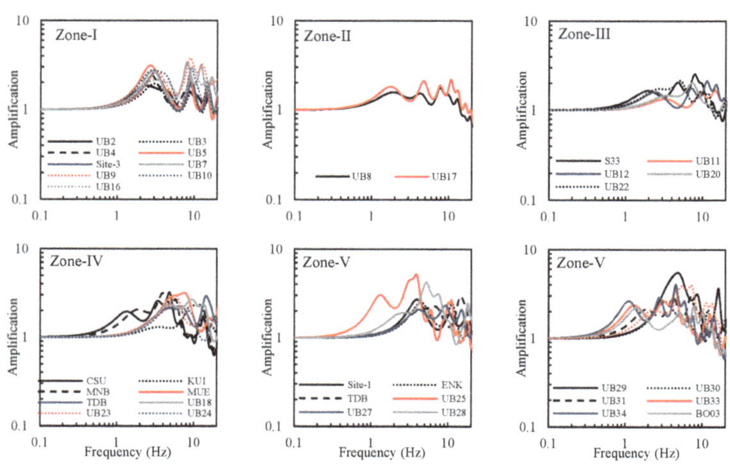

Figure 12. Theoretical site amplifications for each zone in Ulaanbaatar (UB).

Figure 13. Photographs of newly constructed high-rise buildings in Zone-V. (**Left**: taken by authors on 20 March 2018, **Right**: taken on 14 March 2019).

6. Conclusions

In this research, the shear-wave velocity model of UB is computed using microtremor data and the previous investigations for evaluating site amplifications and assessing seismic zoning in Ulaanbaatar city (UB). At first, the shear-wave velocity (V_S) profiles in UB are estimated from the inversion analysis of the observed microtremor H/V spectral ratios (MHVRs) based on the diffuse field assumption. The joint inversions of the MHVRs and the surface wave dispersion curves obtained in the previous study are applied to the three sites in UB. In addition, the single inversion of the MHVRs estimates the V_S profiles for the rest of the sites. A comparison between the inverted shallow V_S profiles and the geological borehole information has shown that reliable results within the capabilities and the resolution of the methods could be achieved.

Finally, the seismic microzoning is performed based on the site amplifications obtained from the inverted V_S profiles to characterize the patterns of seismic hazards. The UB areas are divided into five zones considering the site characterizations. Most significant site amplifications are expected around the Tuul river area, which consists of the deepest sedimentary deposit. In particular, Zone-V is the largest site amplified area, which means considerable risk for a newly constructed building district. Strong ground motion predictions and building damage estimation for scenario earthquakes based on the site characterization of this study will be performed in future works.

Author Contributions: Z.T. and H.M. conceived, designed the experiments and performed the field measurements. Z.T. analyzed the data and wrote the manuscript. T.T. supervised the field measurements. H.M. and T.T. also provided the corrections of the manuscript.

Funding: This research received no external funding.

Acknowledgments: The authors thank MJEED (Mongol-Japan Higher Engineering Education Development Project) for supporting this study and the Environmental and Earthquake Engineering project unit for helping the microtremor site observations.

Conflicts of Interest: The authors declare no conflict of interest.

References

1. Adiya, M. Seismic Activity Near Ulaanbaatar: Implication for Seismic Hazard Assessment. Ph.D. Thesis, University of Strasbourg, Strasbourg, France, 2016.
2. Oyun-Erdene, M.; Odonbaatar, C.; Ganzorig, D. Predicting strong ground motion of Emeelt fault. *Sci. Acad. News* **2014**, 2, 48–56. (In Mongolian)
3. Odonbaatar, C. Site effects characterization in the basin of Ulaanbaatar. Ph.D. Thesis, University of Strasbourg, Strasbourg, France, 2011.
4. Dzurick, V.I.; Kluchevskii, A.I.; Serebnikov, S.P.; Demiyanovich, B.M.; Batsaikhan, T.; Bayaraa, G. *Seismicity and Seismic Hazard Zoning of Mongolian Region*; Earth's Crust Institute: Irkutsk, Russia, 2009.

5. The Government of Ulaanbaatar City, Japan International Cooperation Agency (JICA) and Emergency Management Department of the Capital City. Report of Mongolia "The Project for Strengthening the Capacity of Seismic Disaster Risk Management in Ulaanbaatar City". Available online: http://open_jicareport.jica.go.jp/pdf/12129110_01.pdf (accessed on 11 April 2019).
6. Nakamura, Y. A method for dynamic characteristics estimation of the subsurface using microtremor on the ground surface. *Q. Rep. RTRI* **1989**, *30*, 25–33.
7. Lermo, J.; Chavez-Garcia, F.J. Are microtremors useful in site response evaluation? *Bull. Seismol. Soc. Am.* **1994**, *84*, 1350–1364.
8. Lachet, C.; Bard, P.-Y. Numerical and theoretical investigations of the possibilities and limitations of Nakamura's technique. *J. Phys. Earth* **1994**, *42*, 377–397. [CrossRef]
9. Arai, H.; Tokimatsu, K. S-wave velocity profiling by inversion of microtremor H/V spectrum. *Bull. Seismol. Soc. Am.* **2004**, *94*, 53–63. [CrossRef]
10. Arai, H.; Tokimatsu, K. S-wave velocity profiling by joint inversion of microtremor dispersion curve and horizontal-to-vertical (H/V) spectrum. *Bull. Seismol. Soc. Am.* **2005**, *95*, 1766–1778. [CrossRef]
11. Sánchez-Sesma, F.J.; Weaver, R.L.; Kawase, H.; Matsushima, S.; Luzon, F.; Campillo, M. Energy partitions among elastic waves for dynamic surface loads in a semi-infinite solid. *Bull. Seismol. Soc. Am.* **2011**, *101*, 1704–1709. [CrossRef]
12. Sánchez-Sesma, F.J.; Rodríguez, M.; Iturrarán-Viveros, U.; Luzón, F.; Campillo, M.; Margerin, L.; Rodríguez-Castellanos, A. A theory for microtremor H/V spectral ratio: Application for a layered medium. *Geophys. J. Int.* **2011**, *186*, 221–225. [CrossRef]
13. García-Jerez, A.; Piña-Flores, J.; Sánchez-Sesma, F.J.; Luzón, F.; Perton, M. A computer code for forward calculation and inversion of the H/V spectral ratio under the diffuse field assumption. *Comput. Geosci.* **2016**, *97*, 67–78. [CrossRef]
14. Piña-Flores, J.; Perton, M.; García-Jerez, A.; Carmona, E.; Luzón, F.; Molina-Villegas, J.C.; Sánchez-Sesma, F.J. The inversion of spectral ratio H/V in a layered system using the diffuse field assumption (DFA). *Geophys. J. Int.* **2017**, *208*, 577–588. [CrossRef]
15. Sánchez-Sesma, F.J. Modeling and inversion of the microtremor H/V spectral ratio: Physical basis behind the diffuse field approach. *Earth Planets Space* **2017**, *69*, 92. [CrossRef]
16. Kawase, H.; Mori, Y.; Nagashima, F. Difference of horizontal-to-vertical spectral ratios of observed earthquakes and microtremors and its application to S-wave velocity inversion based on the diffuse field concept. *Earth Planet Space* **2018**, *70*, 1. [CrossRef]
17. Fujishima, D.; Miura, H.; Kanno, T.; Shigefuji, M.; Abiru, T.; Nakamura, H.; Aoi, S. Estimation accuracy of shear wave velocity profiles inverted from earthquake and microtremor H/V spectral ratios. In Proceedings of the 15th Japan Earthquake Engineering Symposium, Sendai, Japan, 6–8 December 2018; pp. 2141–2149. (In Japanese with English abstract)
18. Orkhon Hydrogeology Co. Ltd. *Engineering geological map of Ulaanbaatar*; Ministry of Construction of Mongolia: Ulaanbaatar, Mongolia, 2014.
19. Kitsunezaki, C.; Goto, N.; Kobayashi, Y.; Ikawa, T.; Horike, M.; Saito, T.; Kurota, T.; Yamane, K.; Okuzumi, K. Estimation of P and S-wave velocities in deep soil deposits for evaluating ground vibrations in earthquake. *J. Nat. Dis. Sci.* **1990**, *9*, 1–17, (In Japanese with English abstract).
20. Kirkpatricks, S.; Gelatt, C.D., Jr.; Vecchi, M.P. Optimization by simulated annealing. *Science* **1983**, *220*, 671–680. [CrossRef] [PubMed]
21. Haskell, N.A. Crustal reflection of plane SH-waves. *J. Geophys. Res.* **1960**, *65*, 4147–4150. [CrossRef]
22. Danielson, J.J.; Gesch, D.B. Global Multi-Resolution Terrain Elevation Data 2010 (GMTED2010). Available online: https://pubs.usgs.gov/of/2011/1073/pdf/of2011-1073.pdf (accessed on 11 April 2019).
23. Building Seismic Safety Council (BSSC). *NEHRP Recommended Seismic Provisions for New Buildings and other Structures, Part 1 (Provisions) and Part 2 (Commentary to ASCE/SEI 7)*; Federal Emergency Management Agency: Washington, DC, USA, 2009.

geosciences

Article

Bayesian Variable Selection for Pareto Regression Models with Latent Multivariate Log Gamma Process with Applications to Earthquake Magnitudes

Hou-Cheng Yang [1], Guanyu Hu [2,*] and Ming-Hui Chen [2]

1 Department of Statistics, Florida State University, Tallahassee, FL 32306, USA; hou-cheng.yang@stat.fsu.edu
2 Department of Statistics, University of Connecticut, Storrs, CT 06269, USA; ming-hui.chen@uconn.edu
* Correspondence: guanyu.hu@uconn.edu

Received: 10 January 2019; Accepted: 9 April 2019; Published: 12 April 2019

Abstract: Generalized linear models are routinely used in many environment statistics problems such as earthquake magnitudes prediction. Hu et al. proposed Pareto regression with spatial random effects for earthquake magnitudes. In this paper, we propose Bayesian spatial variable selection for Pareto regression based on Bradley et al. and Hu et al. to tackle variable selection issue in generalized linear regression models with spatial random effects. A Bayesian hierarchical latent multivariate log gamma model framework is applied to account for spatial random effects to capture spatial dependence. We use two Bayesian model assessment criteria for variable selection including Conditional Predictive Ordinate (CPO) and Deviance Information Criterion (DIC). Furthermore, we show that these two Bayesian criteria have analytic connections with conditional AIC under the linear mixed model setting. We examine empirical performance of the proposed method via a simulation study and further demonstrate the applicability of the proposed method in an analysis of the earthquake data obtained from the United States Geological Survey (USGS).

Keywords: earthquake hazard; DIC; CPO; model selection

1. Introduction

The earthquake magnitude data has become increasingly popular over the last decade. Statistical models for earthquake have been proposed since 1800s. Since large earthquakes are rare, it is difficult to fit simple linear models. Many different parametric models (Gamma model, Weibull model) have been considered to analyze earthquake magnitudes, but some earthquakes with very small magnitudes are not reported by seismic centers. The Pareto-type distribution is a popular choice for analyzing earthquake magnitudes data (e.g., [1–3]), as the Pareto distribution is a heavy-tailed distribution with a lower threshold. In statistical analysis, a regression model is used to connect dependent covariates of earthquakes to the magnitude of the earthquake. A generalized linear model strategy can be used for the Pareto regression. Existing seismology literatures pay less attention to spatially dependent structure on earthquake magnitudes. They just built simple linear regression models or generalized linear models to explore covariates effects on earthquake magnitudes [4]. Hu and Bradley [5] proposed using the Pareto regression with spatial random effects for earthquake magnitudes, but they did not consider the model selection problems. In order to have more explicit understanding of dependent covariates of earthquake magnitudes, variable selection approaches should be considered in a Pareto regression model with spatial random effects.

Variable selection and Bayesian statistics have received widespread attention and become increasingly important tools in the field of environment and ecology [6,7]. For hierarchical spatial model, it is difficult to do inference for latent variables. Bayesian approach provides a convenient way for estimating latent variables in hierarchical models. Compared with the frequentist approach, a Bayesian approach can bring some prior information on parameters of the model. It is an important part of a statistical analysis. In practice, we may want to measure how good a model is for answering a certain question or comparing different models to see which model is best suited. There are many popular variable selection criteria, including Akaike's information criterion (AIC) [8] and Bayesian information criterion (BIC) [9], Bayes factor, conditional predictive ordinate (CPO) [10,11], L measure [12], and the deviance information criterion (DIC) [13]. Chen et al. [14] provide the connections between these popular criteria for variable subset selection under generalized linear models. However, there are some difficulties for Bayesian variable selection to carry out because of the challenge in assigning prior distributions for the parameters. In order to tackle this issue, we consider the multivariate log-Gamma distribution (MLG) based on Bradley et al. [15], which is conjugate with the Pareto distribution [5]. Hence, the Bayesian approach to variable selection is straightforward for our model. Consequently, we use CPO and DIC criteria to carry out Bayesian variable selection for Pareto regression models due to the performance of the conjugate priors (see [16], for a discussion).

Both CPO and DIC are criteria-based methods and they have some advantage over other criteria. Compared with regularized estimation approach, these two criteria consider goodness of fit of the candidate models. Furthermore, compared with negative log probability density or RMSE for predictions, these two criteria consider the model complexity. Like the AIC or BIC, these two criteria compromise the tradeoff between the goodness of fit and model complexity. The CPO provides a site-specific model fit metric that can be used for exploratory analysis and can be combined at the site to generate a logarithm pseudo marginal likelihood (LPML) as an overall model fit measure. The CPO is based on leave-one-out-cross-validation. It estimates the probability of observing data on one particular location in the future if after having already observed data. The LPML is a leave-one-out cross-validation with log likelihood as the criteria which can be easily obtained from an Markov chain Monte Carlo (MCMC) output (see [17]). More details about two criteria will be discussed in Section 2.2. The major contribution of this paper is that we introduce two Bayesian model selection criteria in generalized linear model with spatial random effects. Furthermore, we exam the relationship between the two criteria with conditional AIC (cAIC) in random effects model. Other than the variable selection problem in regression model, our criteria can also be used in model selection in the presence of spatial random effects. In general, our proposed criteria can select important covariates and random effects model simultaneously.

The remaining sections of this article are organized as follows. Section 2 introduces our proposed statistical model, and review two Bayesian model assessment Criteria including LPML and DIC [13]. In Sections 3 and 4, we present MCMC scheme and a simulation study for two scenarios, and use two criteria to select true model. In Section 5, we carry out a detailed analysis of the US earthquake dataset from United States Geological Survey (USGS) and use two criteria to select the best model(s). Finally, Section 6 contains a brief summary of this paper. For ease of exposition all proofs are given Appendix A.

2. Methodology

2.1. Pareto Regression with Spatial Random Effects

In many regression problems, normality may not be always held. Generalized linear models allow a linear regression model to connect the response variable with a proper link function. For some heavy tailed data with minimum value, it is common to use the Pareto model to fit these data. From the expression of Gutenberg–Richter law, it is possible to derive a relationship for the logarithm of the probability to exceed some given magnitude. The standard distribution used for seismic moment is the Pareto distribution. The Pareto distribution has a natural threshold. In practice, people do not take more consideration on "micro" (magnitude from 1–1.9) or "minor" (magnitude from 2–2.9) earthquakes. Compared with exponential distribution, Pareto distribution is a heavy tailed distribution. Heavy tailed distributions tend to have many outliers with very high values. The heavier the tail, the larger the probability that you will get one or more disproportionate values in a sample. In earthquake data, most recorded earthquakes have a magnitude around 3–5, but sometime there will have some significant earthquakes with large magnitude. Hu [5] used Pareto regression to model earthquake magnitudes, since the Pareto distribution is a heavy tailed distribution with a threshold. Earthquake magnitude data also has a threshold, since people consider earthquake only over a certain magnitude. Based on the generalized linear model setting, we can build Pareto regression model as

$$f(z) = \exp(\mu(s)) z_m^{\exp(\mu(s))} z^{-1-\exp(\mu(s))} \quad z \geq z_m, \tag{1}$$

where $s \in D \subset \mathcal{R}^2$ is a spatial location, $\mu(s) = \beta_0 + \beta_1 X_1(s) + ... + \beta_p X_p(s)$, $X_i(s)$ is i-th covariate on location s and z_m is the minimum value of the response variable. Under this model, the log shape parameter is modeled with a fixed effects term.

The model in Equation (1) does not include spatial random effects. Consequently, it is implicitly assumed that $\alpha(s)$ and $\alpha(w)$ are independent for $s \neq w$. But for many spatial data, it is not realistic to assume that $\alpha(s)$ and $\alpha(w)$ are independent. We can add the latent Gaussian process in the log-linear model so that the generalized linear model becomes a generalized linear mixed model (GLMM). Specifically, we assumed

$$\log\{\alpha(s)\} = \boldsymbol{\beta}' \boldsymbol{X}(s) + w(s) \quad s \in D, \tag{2}$$

$$\boldsymbol{W} \sim N\{\boldsymbol{0}, \sigma_w^2 \boldsymbol{H}(\phi)\}, \tag{3}$$

where $\boldsymbol{W}$ is an n-dimensional vector of $(w(s_1), \ldots, w(s_n))'$, $\boldsymbol{H}(\phi)$ is a $n \times n$ spatial correlation matrix, and $\{s_1, \ldots, s_n\} \in D$ are the observed spatial locations. The natural strategy to consider spatial correlation is to use in light of Tobler's first law that "near things are more related than distant things" [18]. Spatial random effects allow one to leverage information from nearby locations. Latent Gaussian process models have become a standard method for modeling spatial random effects [19]. Based on Gaussian process structure, the nearby observations will have higher correlation.

For the latent Gaussian process GLMM, we can build the following hierarchical model:

$$\begin{aligned}
&\text{Data Model}: Z(s_i)|\boldsymbol{W}, \boldsymbol{\beta}, \sigma^2, \phi \overset{\text{ind}}{\sim} \text{Pareto}(Z_m, e^{\mu(s_i)}); \; i = 1, \ldots, n \\
&\text{Process Model}: \boldsymbol{W}|\phi, \sigma_w^2 \sim \text{MVN}\{\boldsymbol{0}, \sigma_w^2 \boldsymbol{H}(\phi)\} \\
&\text{Parameter Model 1}: \beta_j \sim N(0, \sigma_j^2); \; j = 1, ..., p \\
&\text{Parameter Model 2}: \sigma_j^2 \sim \text{IG}(a_1, b_1); \; j = 1, ..., p \\
&\text{Parameter Model 3}: \sigma_w^2 \sim \text{IG}(a_2, b_2) \\
&\text{Parameter Model 4}: \phi \sim \text{IG}(a_3, b_3),
\end{aligned} \tag{4}$$

where "IG" is a shorthand for inverse gamma, "MVN" is a shorthand for multivariate normal, and "N" is a shorthand for a univariate normal distribution. For the Pareto regression model, the normal prior is not conjugate. A proper conjugate prior for the Pareto regression will facilitate the development of an efficient computational algorithm. Chen and Ibrahim [16] proposed a novel class of conjugate priors for the family of generalized linear model. But they did not show the connection between their conjugate prior and gaussian prior. Bradley et al. [20] proposed the multivariate log-gamma distribution as a conjugate prior for Poisson spatial regression model and established a connection between a multivariate log-gamma distribution and a multivariate normal distribution. The multivariate log-gamma distribution is an attractive alternative prior for the Pareto regression model due to its conjugacy.

We now present the multivariate log-gamma distribution from Bradley et al. [20]. We define the n-dimensional random vector $\boldsymbol{\gamma} = (\gamma_1, ..., \gamma_n)'$, which consists of n mutually independent log-gamma random variables with shape and scale parameters organized into the n-dimensional vectors $\boldsymbol{\alpha} \equiv (\alpha_1, ..., \alpha_n)'$, and $\boldsymbol{\kappa} \equiv (\kappa_1, ..., \kappa_n)'$, respectively. Then define the n-dimensional random vector $\boldsymbol{q}$ as follows:

$$\boldsymbol{q} = \boldsymbol{\mu} + \boldsymbol{V}\gamma, \tag{5}$$

where $\boldsymbol{V} \in \mathcal{R}^n \times \mathcal{R}^n$ and $\boldsymbol{\mu} \in \mathcal{R}^n$. Bradley et al. [20] called $\boldsymbol{q}$ the multivariate log-gamma random vector. The random vector $\boldsymbol{q}$ has the following probability density function:

$$f(\boldsymbol{q}|\boldsymbol{c}, \boldsymbol{V}, \boldsymbol{\alpha}, \boldsymbol{\kappa}) = \frac{1}{\det(\boldsymbol{V})} \left(\prod_{i=1}^{m} \frac{1}{\Gamma(\alpha_i)\kappa_i^{\alpha_i}} \right) \exp[\boldsymbol{\alpha}'\boldsymbol{V}^{-1}(\boldsymbol{q} - \boldsymbol{\mu}) - \boldsymbol{\kappa}^{(-1)'} \exp\{\boldsymbol{V}^{-1}(\boldsymbol{q} - \boldsymbol{\mu})\}]; \; \boldsymbol{q} \in \mathcal{R}^n, \tag{6}$$

where "det" represents the determinant function. We use "MLG $(\boldsymbol{\mu}, \boldsymbol{V}, \boldsymbol{\alpha}, \boldsymbol{\kappa})$" as a shorthand for the probability density function in Equation (6).

According to Bradley et al. [20], the latent Gaussian process is a special case of the latent multivariate log-gamma process. If $\boldsymbol{\beta}$ has a multivariate log-gamma distribution $\text{MLG}(\boldsymbol{0}, \alpha^{1/2}\boldsymbol{V}, \alpha\boldsymbol{1}, 1/\alpha\boldsymbol{1})$. When $\alpha \to \infty$, $\boldsymbol{\beta}$ will converge in distribution to the multivariate normal distribution vector with mean $\boldsymbol{0}$ and covariance matrix $\boldsymbol{V}\boldsymbol{V}'$. $\alpha = 10{,}000$ is sufficiently large for this approximation. MLG model is a more saturated model than Gaussian process model. For the Pareto regression model, the MLG process is more computationally efficient than the Gaussian process. In following hierarchical model, we refer to $\boldsymbol{\beta}$ and $\boldsymbol{W}$ as following an MLG distribution with $\boldsymbol{q}$, $\boldsymbol{0}_p$ and $\boldsymbol{0}_n$ being the first parameter of MLG corresponding to $\boldsymbol{\mu}$, and $\boldsymbol{\Sigma}_W^{1/2}$ and $\boldsymbol{\Sigma}_\beta^{1/2}$ are the second parameter of MLG like $\boldsymbol{V}$.

In order to establish conjugacy, we build a spatial GLM with latent multivariate log gamma process as follows:

$$\begin{aligned}
&\textbf{Data Model}: Z(\boldsymbol{s}_i)|\boldsymbol{W}, \boldsymbol{\beta}, \sigma^2, \phi \overset{\text{ind}}{\sim} \text{Pareto}(Z_m, e^{\mu(\boldsymbol{s}_i)}); \; i = 1, \ldots, n \\
&\textbf{Process Model}: \boldsymbol{W}|\phi, \sigma_w \sim \text{MLG}(\boldsymbol{0}_n, \boldsymbol{\Sigma}_W^{1/2}, \alpha_W\boldsymbol{1}_n, \kappa_W\boldsymbol{1}_n) \\
&\textbf{Parameter Model 1}: \boldsymbol{\beta} \sim \text{MLG}(\boldsymbol{0}_p, \boldsymbol{\Sigma}_\beta^{1/2}, \alpha_\beta\boldsymbol{1}_p, \kappa_\beta\boldsymbol{1}_p) \\
&\textbf{Parameter Model 2}: \sigma^2 \sim \text{IG}(a_1, b_1); \\
&\textbf{Parameter Model 3}: \sigma_w^2 \sim \text{IG}(a_2, b_2) \\
&\textbf{Parameter Model 4}: \phi \sim \text{IG}(a_3, b_3),
\end{aligned} \tag{7}$$

where Z_m defined baseline, $\mu(\boldsymbol{s}_i) = \boldsymbol{X}(s_i)\boldsymbol{\beta} + \boldsymbol{W}$, $\boldsymbol{\Sigma}_W = \sigma_w^2\boldsymbol{H}(\phi)$, $\boldsymbol{\Sigma}_\beta = \sigma^2\text{diag}(p)$, $\alpha_W > 0$, $\alpha_\beta > 0, \kappa_W > 0$, and $\kappa_\beta > 0$.

2.2. Bayesian Model Assessment Criteria

In this section, we consider two Bayesian model assessment criteria, DIC and LPML. In addition, we introduce the procedure to calculate DIC and LMPL for the Pareto regression model with spatial random effects. Let $\boldsymbol{\beta}^{(M)}$ denote the vector of regression coefficient under the full model M. Also let $\boldsymbol{\beta}^{(m)}$ and $\boldsymbol{\beta}^{(-m)}$ denote the corresponding vectors of regression parameters included and excluded in the subset model m. Then, $\boldsymbol{\beta}^{(M)} = \boldsymbol{\beta} = ((\boldsymbol{\beta}^{(m)})', (\boldsymbol{\beta}^{(-m)})')'$ holds for all m, and $\boldsymbol{\beta}^{(-M)} = \emptyset$.

2.2.1. DIC

The deviance information criterion is defined as

$$\text{DIC} = \text{Dev}(\bar{\theta}) + 2p_D, \tag{8}$$

where $\text{Dev}(\bar{\theta})$ is the deviance function, $p_D = \overline{\text{Dev}}(\theta) - \text{Dev}(\bar{\theta})$ is the effective number of model parameters, and $\bar{\theta}$ is the posterior mean of parameters θ, and $\bar{\text{Dev}}(\theta)$ is the posterior mean of $\text{Dev}(\theta)$. To carry out variable selection, we specify the deviance function as

$$\text{Dev}(\boldsymbol{\beta}^{(m)}) = -2\sum_{i=1}^{n} \log(f(\boldsymbol{\beta}^{(m)})|D_i), \tag{9}$$

where $D_i = (Y_i, \mathbf{X}_i, \hat{W}_i)$, $f(.)$ is the likelihood function in Equation (7), $\hat{W}_i$ is the posterior mean of the spatial random effects on location s_i, $\boldsymbol{\beta}^{(m)}$ is the vector of regression coefficient under the m-th model. In this way, the DIC criterion is given by

$$\text{DIC}_m = \text{Dev}(\bar{\boldsymbol{\beta}}^{(m)}) + 2p_D^{(m)}, \tag{10}$$

where

$$2p_D^{(m)} = \text{Dev}(\bar{\boldsymbol{\beta}}^{(m)}) - \text{Dev}(\bar{\boldsymbol{\beta}}^{(m)}), \tag{11}$$

where $\bar{\boldsymbol{\beta}}^{(m)} = E[\boldsymbol{\beta}^{(m)}|D]$, and $\text{Dev}(\bar{\boldsymbol{\beta}}^{m}) = E[\text{Dev}(\boldsymbol{\beta}^{(m)})]$.

2.2.2. LPML

In order to calculate the LPML, we need to calculate CPO first [14]. Then LPML can be obtained as

$$\text{LPML} = \sum_{i=1}^{n} \log(\text{CPO}_i), \tag{12}$$

where CPO_i is the CPO for the i-th subject.

Let $D_{(-i)}$ denote the observation data with the i-th observation deleted. The CPO for the i-th subject is defined as

$$\text{CPO}_i = f(Y_i|\mathbf{X}_i, D_{(-i)}) = \int f(Y_i|\mathbf{X}_i, \boldsymbol{\beta})\pi(\boldsymbol{\beta}|D_{(-i)})d\boldsymbol{\beta}, \tag{13}$$

where $\pi(\boldsymbol{\beta}|D_{(-i)})$ is the posterior distribution based on the data $D_{(-i)}$.

From Chapter 10 of Chen et al. [21], CPO in (13) can be rewritten as

$$\text{CPO}_i = \frac{1}{\int \frac{1}{f(y_i|\boldsymbol{\beta}, \mathbf{W}, \mathbf{X}_i)} \pi(\boldsymbol{\beta}|D)d\boldsymbol{\beta}}. \tag{14}$$

A popular Monte Carlo estimate of CPO using Gibbs samples form the posterior distribution is given as D instead of $D_{(-i)}$. Letting $\{\boldsymbol{\beta}_b, b = 1, \cdots, B\}$ denote a Gibss sample of $\boldsymbol{\beta}$ from $\pi(\boldsymbol{\beta}|D)$ and using (14), a Monte Carlo estimate of CPO_i^{-1} is given by

$$\widehat{\text{CPO}}_i^{-1} = \frac{1}{B}\sum_{b=1}^{B} \frac{1}{f(Y_i|\boldsymbol{\beta}_b, X_i, \hat{W}_i)}. \tag{15}$$

So the LPML defined as

$$\text{LPML}_m = \sum_{i=1}^{n} \log(\widehat{\text{CPO}}_i). \tag{16}$$

In the context of variable selection, we select a subset model, which has the largest LPML value and/or the smallest DIC value. In practice, if we have two different results based on two criteria, we will choose both models which were selected by two criteria as the best models. In addition, we can do more diagnostics for the two candidate models. DIC compromises the goodness of fit and the complexity of the model. The CPO is based on leave-one-out-cross-validation. The LPML, the sum of the log CPO's, is an estimator for the log marginal likelihood.

2.3. Analytic Connections between Bayesian Variable Selection Criteria with Conditional AIC for the Normal Linear Regression with Spatial Random Effects

The Akaike information criterion (AIC) has been applied to choose candidate models in the mixed-effects model by integrating out the random effects. A conditional AIC was proposed to be used for the linear mixed-effects model [22] under the assumption that the variance-covariance matrix of random effects is known. Under the this assumption, we establish analytic connections of DIC and LPML we proposed in Section 2.3 with cAIC. We have the following linear regression model with spatial random effects:

$$y_i = \boldsymbol{X}_i\boldsymbol{\beta} + w_i + \epsilon_i, \tag{17}$$

where $\boldsymbol{\beta}$ is a $p \times 1$ vector of fixed effects, w_i is spatial random effects for individual i. The cAIC is defined as:

$$\text{cAIC} = -2\log(\hat{\boldsymbol{\beta}}|\boldsymbol{X}, \boldsymbol{y}, \hat{\boldsymbol{w}}) + 2k, \tag{18}$$

where $\boldsymbol{X}$ is with full rank k. Having the MLE of $\boldsymbol{\beta}$, we can have

$$\text{cAIC} = -n\log(\frac{1}{2\pi\sigma^2}) + \frac{1}{\sigma^2}\text{SSE} + 2k, \tag{19}$$

where $\text{SSE} = (\boldsymbol{y} - \hat{\boldsymbol{y}})'(\boldsymbol{y} - \hat{\boldsymbol{y}}), \hat{\boldsymbol{y}} = (\hat{y}_1, ..., \hat{y}_n)', \hat{y}_i = \boldsymbol{X}_i\hat{\boldsymbol{\beta}} + \hat{w}_i$.

From [14], we can have DIC and LPML for the linear regression model with spatial random effects as follows

$$\text{DIC}(a_0) = -n\log(\frac{1}{2\pi\sigma^2}) + \frac{1}{\sigma^2}\text{SSE}^* + \frac{2(1+a_0)}{1+2a_0}2k, \tag{20}$$

and

$$\text{LPML}(a_0) = -n\log(\frac{1}{2\pi\sigma^2}) + \frac{1}{\sigma^2}\text{SSE}^* + \frac{(1+a_0)}{1+2a_0}k + R, \tag{21}$$

where SSE^* is calculated by posterior mean, $a_0 = 0$ with conjugate prior for likelihood model, $R = -\frac{2(1+a_0)^2}{1+2a_0}R^*$, R^* is the remainder of Taylor expansion. So in the conjugate prior condition, our proposed Bayesian variable selection criterion is similar with cAIC for the linear regression model with spatial random effects.

3. MCMC Scheme

The algorithm requires sampling the all parameters in turn from their respective full conditional distributions. We assume that $\boldsymbol{\beta}$, $\boldsymbol{W}$ are independent a priori. We further assume $\boldsymbol{\beta} \sim \text{MLG}(\boldsymbol{0}_p, \boldsymbol{\Sigma}_\beta^{1/2}, \alpha_\beta\boldsymbol{1}_p, \kappa_\beta\boldsymbol{1}_p)$ and $\boldsymbol{W} \sim \text{MLG}(\boldsymbol{0}_n, \boldsymbol{\Sigma}_W^{1/2}, \alpha_W\boldsymbol{1}_n, \kappa_W\boldsymbol{1}_n)$. Thus, sampling from $p(\boldsymbol{\beta}|\cdot) \propto \exp\left\{\boldsymbol{\alpha}_\beta'\boldsymbol{H}_\beta\boldsymbol{\beta} - \boldsymbol{\kappa}_\beta'\exp(\boldsymbol{H}_\beta\boldsymbol{\beta})\right\}$ and $p(\mathbf{W}|\cdot) \propto \exp\left\{\boldsymbol{\alpha}_W'\boldsymbol{H}_W\boldsymbol{W} - \boldsymbol{\kappa}_W'\exp(\boldsymbol{H}_W\boldsymbol{W})\right\}$ is straightforward. For $\boldsymbol{\Sigma}_W = \sigma_w^2\boldsymbol{H}(\phi)$ and $\boldsymbol{\Sigma}_\beta = \sigma^2\text{diag}(p)$, we assume that $\sigma^2 \sim \text{IG}(a_1, b_1)$, $\sigma_w^2 \sim \text{IG}(a_2, b_2)$ and $\phi \sim \text{IG}(a_3, b_3)$, that is, $p(\sigma^2|a_1, b_1) \propto \text{MLG}(\boldsymbol{0}, \boldsymbol{\Sigma}_\beta^{1/2}, \alpha_\beta\boldsymbol{1}_p, \kappa_\beta\boldsymbol{1}_p) \times \text{IG}(a_1, b_1)$, $p(\sigma_w^2|a_2, b_2) \propto \text{MLG}(\boldsymbol{0}, \boldsymbol{\Sigma}_W^{1/2}, \alpha_w\boldsymbol{1}_n, \kappa_w\boldsymbol{1}_n) \times \text{IG}(a_2, b_2)$, and $p(\phi|a_3, b_3) \propto \text{MLG}(\boldsymbol{0}_n, \boldsymbol{\Sigma}_W^{1/2}, \alpha_w\boldsymbol{1}_n, \kappa_w\boldsymbol{1}_n) \times \text{IG}(a_3, b_3)$. The sampling scheme for these three parameters is not straightforward. We use a Metropolis–Hasting algorithm to sampling from three parameters. The other difficulty is how to compute the log-determinant of a matrix. Because we are using a log-likelihood function, the formula for the log-likelihood involves the expression $\log(\det(\Sigma_\beta))$ or $\log(\det(\Sigma_W))$. To compute the logarithm of a determinant, we encourage not try to compute the determinant itself. Instead, computing the log-determinant directly. For a matrix with a large determinant, the computation of the log-determinant will usually be achieved, however, the computation of the determinant might cause a numerical error. The method is given by

$$\log(\det(\Sigma_\beta)) = 2 * \sum(\log(\text{diag}(L_\beta))),$$
$$\log(\det(\Sigma_W)) = 2 * \sum(\log(\text{diag}(L_W))), \tag{22}$$

where the L_β is the Cholesky root of matrix Σ_β, L_W is the Cholesky root of matrix Σ_W, and "diag" denotes a column vector whose elements are the elements on the diagonal of matrix. The derivative details for the full conditional distributions given in Appendix A.

Note that $\alpha_\beta, \kappa_\beta, \alpha_W, \kappa_W, a_1, b_1, a_2, b_2, a_3$, and b_3 are prespecified hyperparameters. In this article, we use $\alpha_\beta = 10{,}000, \kappa_\beta = 0.0001, \alpha_W = 1, \kappa_W = 1, a_1 = 1, b_1 = 1, a_2 = 1, b_2 = 1, a_3 = 1$ and $b_3 = 1$. For more flexibility, we can also assume $\alpha_\beta, \kappa_\beta, \alpha_W$ and κ_W each following a Gamma distribution with suitable hyperparameters.

4. Simulation Study

The spatial domain for the two simulation studies are chosen to be $D \in [0, 50] \times [0, 50]$. The locations s_i is selected uniformly over $D\ \forall i = 1 \ldots 50$. We present the two different simulation settings and generate 100 replicate data sets for each scenario. We assume $\boldsymbol{\beta} = (\beta_1, \beta_2, \beta_3)$ so that we have seven candidate models. We generate $\boldsymbol{W}$ from a multivariate normal distribution with mean zero and covariance $\boldsymbol{\Sigma}_W = \boldsymbol{H}(\phi)$. We set $\phi = 1$ and fix $\sigma^2 = 1$ in both Simulations 1 and 2. We generate the elements of $X(s_i)$ independently from the uniform distribution U(0,1). We define the baseline threshold (scale parameter) equal to three in both simulations.

4.1. Simulation for the Connection between Multivariate Log Gamma and Multivariate Normal Distribution

In this section, we examine the connection between the multivariate log-gamma distribution and the multivariate normal distribution. First, we draw the quantile-quantile (QQ)-plot in Figure 1 to show the normality of $\boldsymbol{q}$ generated from $\text{MLG}(\boldsymbol{0}, \alpha^{1/2}\boldsymbol{V}, \alpha\boldsymbol{1}, 1/\alpha\boldsymbol{1})$, when $\alpha = 10{,}000$. In addition, we use the Kolmogorov–Smirnov test to examine the connection for one dimensional data. We use a multivariate two-sample test [23] for multivariate dimensional data. We generated one data set of size 100 from the multivariate log-gamma distribution and another data set of size 100 from the multivariate normal distribution and then calculated the p-value from the multivariate two-sample test for comparing these two data sets. Then, we repeated this process 1000 times. We found that 992 out of these 1000 p-values were larger than the significance level of 0.05. That is, in 992 of 1000 times, we did not reject the null hypothesis that the two samples were drawn from the same distribution.

4.2. Simulation for Estimation Performance

In this simulation study, our goal was to examine the estimation performance of the hierarchical model. We set $\boldsymbol{\beta} = (1, 1, 1)$. We estimated the parameters in this simulation and report the bias (bias $= \frac{1}{m}\sum_{j=1}^{m}(\beta_i^{(j)} - \beta_i^*)$), the standard error (SE) (SE $= \left[\frac{1}{m}\sum_{j=1}^{m}(\beta_i^{(j)} - \bar{\beta}_i)^2\right]^{1/2}$, where $\bar{\beta}_i = \frac{1}{m}\sum_{j=1}^{m}\beta_i^{(j)}$), and the mean square error (MSE) (MSE $= \frac{1}{m}\sum_{j=1}^{m}(\beta_i^{(j)} - \beta_i^*)^2$) in Table 1, where β_i^* is the true value of β_i.

Table 1. Estimation performance.

Parameter	True Value	Bias	SE	MSE	Coverage Probability
β_1	1	−0.0272	0.2903	0.085	0.94
β_2	1	−0.0024	0.2939	0.0863	0.94
β_3	1	−0.0102	0.3369	0.1135	0.94

We try to predict the parameters close to true mean value of our target random variable and the variance is how scattered for our predictions. From Table 1, using the MLG prior for $\boldsymbol{\beta}$, we got a reasonable estimation result because it achieves low bias and low variance simultaneously. Besides, we calculated the coverage probability for each variable, it indicates the 94% coverage probability for each parameter.

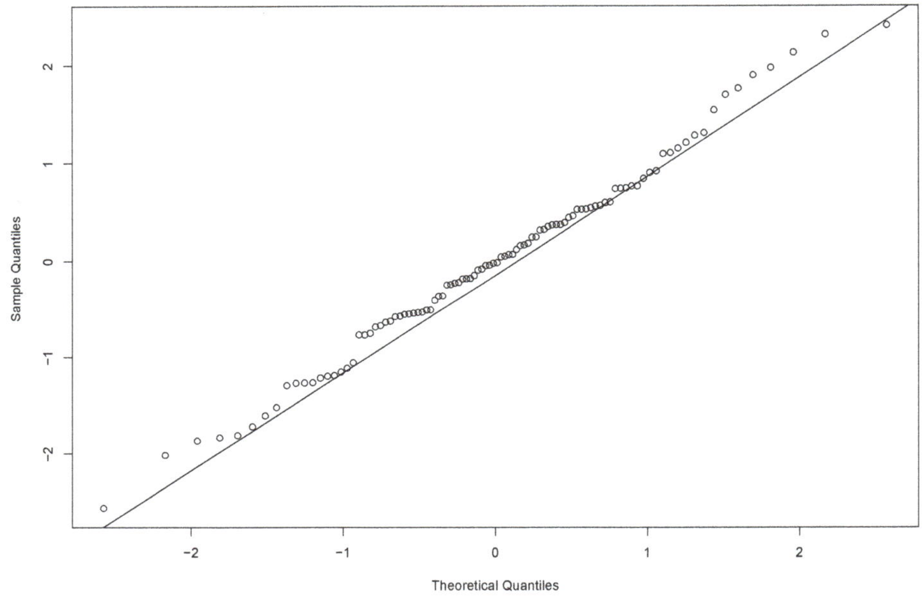

Figure 1. QQ-plot.

4.3. *Simulation for Model Selection*

In this simulation study, our goal was to study the accuracy of our model selection criteria. We have two different simulations in this section. Simulation 1: we set true $\boldsymbol{\beta} = (3, 0, 0)$ and calculated the difference between the true model and other candidate models for both criteria. In Figure 2, a difference beyond zero means that the true model had smaller DIC than the candidate model and the difference below zero means that the true model had higher LPML than the candidate model in Figure 2. The true model had the smallest DIC and the largest LPML in 99 of 100 simulated data sets. Simulation 2: we set true $\boldsymbol{\beta} = (1, 0, 0)$ and the results are shown in Figure 3. In each simulation, we have seven candidate models and one of them is true model and denote the true model as model 5. In Figures 2 and 3, the y-axis is the difference between "candidate model i" with true model. The true model had the smallest DIC in 81 of 100 simulated data sets and the largest LPML in 80 out of 100 simulated data sets. For each replicate dataset, we fit our model with 5000 Markov chain Monte Carlo iterations and treated the first 2000 iterations as burn-in. From Figures 2 and 3, in both simulation studies, we find that DIC and LPML yielded relatively consistent model selection results.

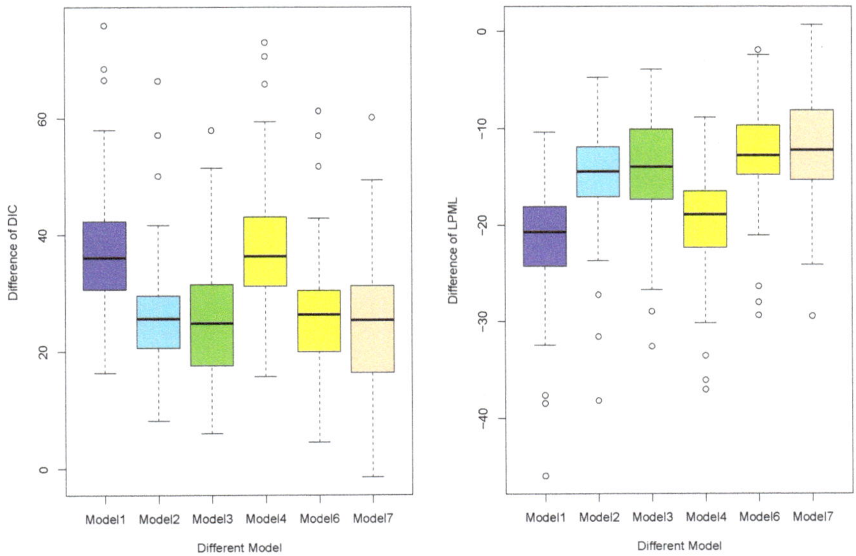

Figure 2. Deviance information criterion (DIC) and logarithm pseudo marginal likelihood (LPML) difference between candidate models and true model (model 5) of Simulation 1 ((**left**) DIC, (**right**) LPML).

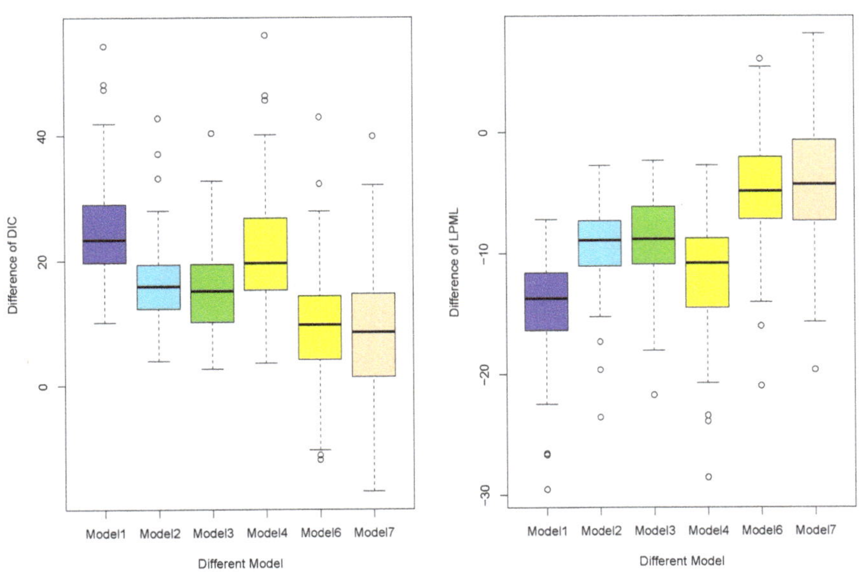

Figure 3. DIC and LPML difference between candidate models and true model (model 5) of Simulation 2 ((**left**) DIC, (**right**) LPML).

4.4. Simulation for Model Comparison

In this simulation study, our goal is to evaluate the accuracy of our model selection criteria for different spatial random effects model. In this section, we generate the spatial random effects from MLG, $\mathbf{W} \sim \text{MLG}(\mathbf{0}_n, \boldsymbol{\Sigma}_W^{1/2}, \alpha_W \mathbf{1}_n, \kappa_W \mathbf{1}_n)$, where $\alpha_W = \kappa_W = 1$. Other settings are same with previous simulations. We generated 100 data sets in these settings. Then, we compared the model fitness based on two following priors:

$$\begin{aligned} &\textbf{Prior 1}: W|\phi,\sigma_w \sim \text{MLG}(\mathbf{0}_n, \mathbf{\Sigma}_W^{1/2}, \alpha_W\mathbf{1}_n, \kappa_W\mathbf{1}_n), \\ &\textbf{Prior 2}: W|\phi,\sigma_w \sim \text{N}(\mathbf{0}_n, \mathbf{\Sigma}_W). \end{aligned} \tag{23}$$

For each replicate dataset, we fit our model with 5000 Markov chain Monte Carlo iterations and treat first 2000 iterations as burn-in. Then, we calculated the difference of DICs and the difference of LPMLs between these two priors. In Figure 4, the values below zero in the left plot imply that prior 1 has smaller DIC than prior 2. Also, the values above zero in the right plot in Figure 4 indicate that prior 1 has higher LPML than prior 2. The results shown in Figure 4 that we have a better result when we use the MLG prior than the Gaussian prior.

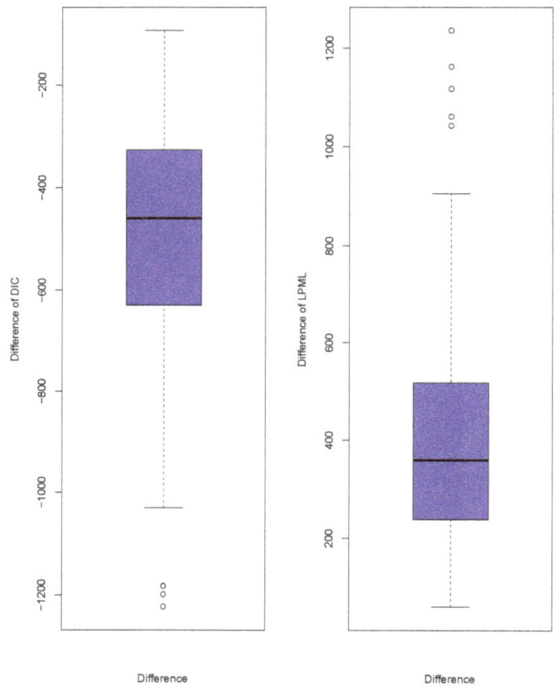

Figure 4. DIC and LPML difference ((**left**) DIC, (**right**) LPML).

5. A Real Data Example

5.1. Data Description

We analyzed seven days of US earthquake data collected in 2018, which includes $n = 228$ earthquakes that have magnitudes over $Z_m = 2.44$ (https://earthquake.usgs.gov/). We present the earthquake data in Figure 5. We find the data most lie in seismic belts. In Figures 6 and 7, we present the histogram of this data and the scatter plot of this data set. In this analysis we have three variables (depth, gap, rms). The depth is where the earthquake begins to rupture. The gap is the largest azimuthal gap between azimuthally adjacent stations (in degrees). RMS is the root-mean-square (RMS) travel time residual, in sec, using all weights.

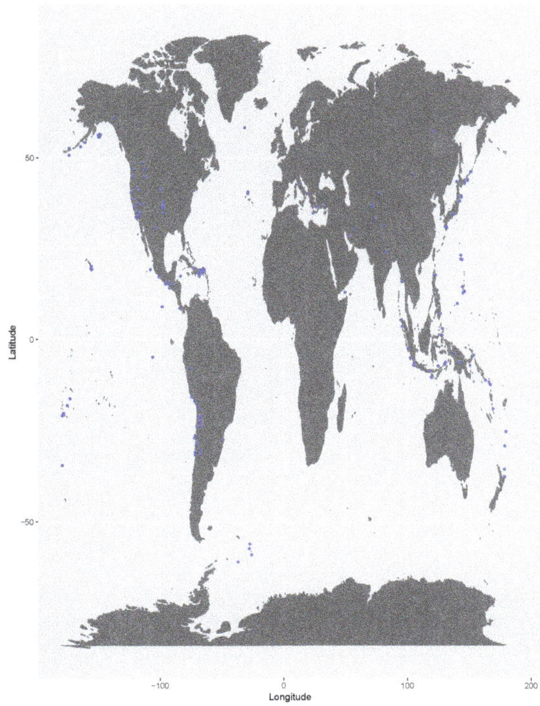

Figure 5. Map of US earthquake data.

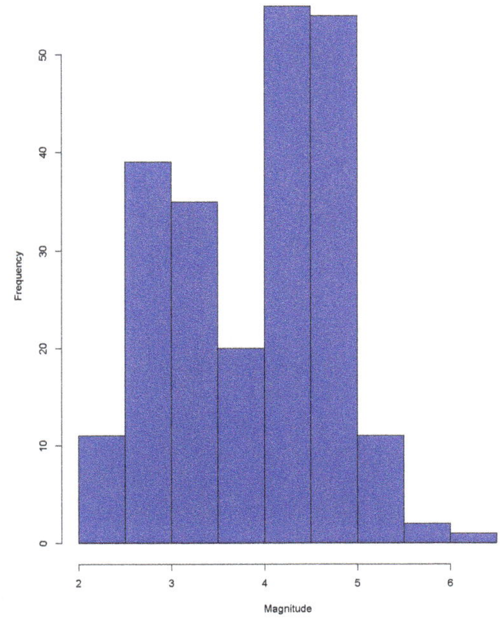

Figure 6. Histogram of US earthquake data collected in 2018.

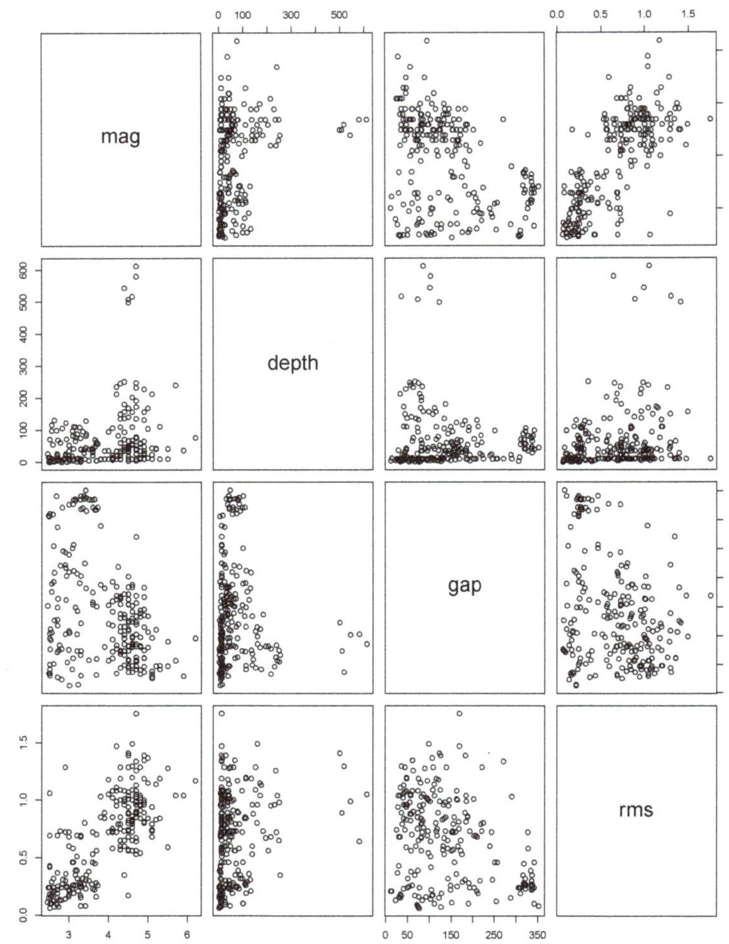

Figure 7. Scatter plot of earthquakes magnitudes, depth, gap and root-mean-square (RMS).

5.2. Analysis

We consider the model in Equation (7) and specify $\alpha_\beta = 10^5$ and $\kappa_\beta = 10^{-5}$. These choices lead to an MLG that approximates a multivariate normal distribution. This choice of hyper-parameters will give an approximately normal prior on $\boldsymbol{\beta}$. Inverse gamma priors are chosen for variance parameters σ_w^2 and σ^2, which is a usual choice of the variance parameters in Bayesian analysis. The full conditionals in the Appendix A are used to run a Gibbs sampler. We have seven candidate models in total, and $\boldsymbol{\beta} = (\beta_1, \beta_2, \beta_3)$ =(depth, gap, rms). The number of iterations of the Gibbs sampler is 15,000, and the number of burn-in iterations is 10,000. The trace plots of posterior samples are provided in the Appendix B to show the convergence of MCMC chain. We also compare to a model when $\boldsymbol{W}$ approximates to Normal. The "DIC_N" and "LPML_N" denote the DIC and LPML for a model when $\boldsymbol{W}$ approximates to normal respectively. Furthermore, we calculated the log probability density (LPD) for candidate models. Based on the results in Table 2, the three criteria selected the same model with β_1 and MLG spatial random effects. Our proposed criteria had consistent results with the LPD.

From Table 2, we know that the model with $\boldsymbol{\beta} = (\beta_1, 0, 0)$ has the smallest DIC and largest LPML. We also report the posterior estimates under the best model in Table 3 according to both DIC and LPML.

Table 2. Deviance information criterion (DIC) logarithm pseudo marginal likelihood (LPML), and log probability density (LPD) of candidate models.

Model	DIC	LPML	LPD	DIC_N	$LPML_N$	LPD_N
$\beta_1, \beta_2, \beta_3$	3058.71	−1535.68	−1528.58	3325.47	−1669.44	−1661.86
β_1, β_2	2936.72	−1472.54	−1469.38	3130.42	−1569.94	−1564.33
β_1, β_3	3037.96	−1522.69	−1516.97	3258.84	−1633.79	−1628.54
β_2, β_3	3056.02	−1533.71	−1526.38	3322.33	−1666.84	−1660.28
β_1	2890.80	−1446.60	−1445.789	2958.61	−1480.68	−1478.42
β_2	2908.10	−1457.28	−1452.35	3073.28	−1540.16	−1535.76
β_3	3034.67	−1519.84	−1518.62	3896.29	−1951.31	−1947.27

Table 3. Posterior estimation under the best model.

	Posterior Mean	Standard Error	95% Credible Interval
β_1	−0.00568	0.0009616	(−0.00763, −0.00389)
ϕ	24.8693	4.5693	(17.5827, 35.1427)
σ^2	2.1620	2.4563	(0.2642, 9.1086)
σ_w^2	4.9304	1.7632	(2.1670, 8.8958)

From these posterior estimates, the model we select just contains depth as the important covariates and 95% credible interval does not contain zero. We see that as the depth increases, the expected value of earthquakes magnitudes increases. The other two covariates, gap and RMS, have no significant effects on earthquake magnitudes. In other words, from these seven-day earthquake data, deep earthquakes will have bigger magnitudes than shallow earthquakes. From the posterior estimates of ϕ and σ_w^2, we can find that there exists spatial correlation of earthquake magnitudes between different locations. In addition, using MLG as spatial random effects increases the goodness of fit of regression model in this data. This result is consistent with the earthquake literature [2].

6. Discussion

In this paper, we propose a Bayesian variable selection criterion for a Bayesian spatial-temporal model for analyzing earthquake magnitudes. Our main methodological contributions are to use the multivariate log-gamma model for both the regression coefficients and spatial random effects and to do variable selection for regression covariates with spatial random effects. Both DIC and LPML have a good selection power to choose the true model. But Bayesian model assessment criteria such as DIC and LPML do not perform well in the high-dimensional case, because the number of candidate models is very large when the number of covariates increases a lot. Developing a high-dimensional variable selection procedure is one of the important future works. The other future work is to fit other earthquake magnitudes models such as the gamma model or the Weibull model. In addition, we need to propose some Bayesian model assessment criterion to select the true data model for earthquake magnitudes. For the nature hazards problem, we need to incorporate the temporal dependent structure of earthquakes. Recently, the ETAS model [24] (combining the Gutenberg–Richter law and the Omori law) has been widely studied. Modelling earthquake dynamics is an important approach for preventing economic loss caused by an earthquake. Incorporating self-exciting effects in our generalized linear model with spatial random effects is another important future work. Furthermore, we only consider earthquake information as the covariates in our model. It will increase the predictive accuracy for us to combine more geographical information such as fault line information or crustal movement in the future.

Author Contributions: Data curation, G.H.; Formal analysis, H.-C.Y.; Investigation, H.-C.Y.; Methodology, G.H.; Project administration, G.H.; Software, H.-C.Y.; Supervision, M.-H.C.; Visualization, H.-C.Y.; Writing—original draft, G.H.; Writing—review & editing, M.-H.C.

Funding: Chen's research was partially supported by NIH grants #GM70335 and #P01CA142538. Hu's research was supported by Dean's office of College of Liberal Arts and Sciences in University of Connecticut.

Conflicts of Interest: The authors declare no conflict of interest.

Appendix A. Full Conditionals Distributions for Pareto Data with Latent Multivariate Log-Gamma Process Models

From the hierarchical model in Equation (7), the full conditional distribution for $\boldsymbol{\beta}$ satisfies:

$$\begin{aligned} f(\boldsymbol{\beta}|\cdot) &\propto f(\boldsymbol{\beta}) \prod f(Z|\cdot) \\ &\propto \exp\left[\sum_i (\boldsymbol{X}(s_i)'\boldsymbol{\beta} + W(s_i)) - \sum_i (\log(Z(s_i) - \log(Z_m)) \exp(\boldsymbol{X}(s_i)'\boldsymbol{\beta} + W(s_i))\right] \\ &\times \exp\left\{\alpha_\beta \mathbf{1}_p' \Sigma_{\boldsymbol{\beta}}^{-1/2} \boldsymbol{\beta} - \kappa_\beta \mathbf{1}_p' \exp(\Sigma_{\boldsymbol{\beta}}^{-1/2} \boldsymbol{\beta})\right\}. \end{aligned} \tag{A1}$$

Rearranging terms we have

$$f(\boldsymbol{\beta}|\cdot) \propto \exp\left\{\boldsymbol{\alpha}_\beta' \boldsymbol{H}_\beta \boldsymbol{\beta} - \boldsymbol{\kappa}_\beta' \exp(\boldsymbol{H}_\beta \boldsymbol{\beta})\right\}, \tag{A2}$$

which implies that $f(\boldsymbol{\beta}|\cdot)$ is equal to $\text{cMLG}(\boldsymbol{H}_\beta, \boldsymbol{\alpha}_\beta, \boldsymbol{\kappa}_\beta)$, which is a shorthand for the conditional MLG distribution used in [20].

Similarly, the full conditional distribution for $\boldsymbol{W}$ satisfies:

$$\begin{aligned} f(\boldsymbol{W}|\cdot) &\propto f(\boldsymbol{W}) \prod f(Z|\cdot) \\ &\propto \exp\left[\sum_i (\boldsymbol{X}(s_i)'\boldsymbol{\beta} + W(s_i)) - \sum_i (\log(Z(s_i) - \log(Z_m)) \exp(\boldsymbol{X}(s_i)'\boldsymbol{\beta} + W(s_i))\right] \\ &\times \exp\left\{\alpha_W \mathbf{1}_n' \Sigma_W^{-1/2} \boldsymbol{W} - \kappa_W \mathbf{1}_n' \exp(\Sigma_W^{-1/2} \boldsymbol{W})\right\}. \end{aligned} \tag{A3}$$

Rearranging terms we have

$$f(\boldsymbol{W}|\cdot) \propto \exp\left\{\boldsymbol{\alpha}_W' \boldsymbol{H}_W \boldsymbol{W} - \boldsymbol{\kappa}_W' \exp(\boldsymbol{H}_W \boldsymbol{W})\right\}, \tag{A4}$$

which implies that $f(\boldsymbol{W}|\cdot)$ is equal to $\text{cMLG}(\boldsymbol{H}_W, \boldsymbol{\alpha}_W, \boldsymbol{\kappa}_W)$. Thus we obtain the following full-conditional distributions to be used within a Gibbs sampler:

$$\begin{aligned} \boldsymbol{\beta} &\sim \text{cMLG}(\boldsymbol{H}_\beta, \boldsymbol{\alpha}_\beta, \boldsymbol{\kappa}_\beta) \\ \boldsymbol{W} &\sim \text{cMLG}(\boldsymbol{H}_W, \boldsymbol{\alpha}_W, \boldsymbol{\kappa}_W) \\ \sigma^2 &\propto \text{MLG}(\mathbf{0}, \boldsymbol{\Sigma}_\beta^{1/2}, \alpha_\beta \mathbf{1}_p, \kappa_\beta \mathbf{1}_p) \times \text{IG}(a_1, b_1) \\ \sigma_w^2 &\propto \text{MLG}(\mathbf{0}, \boldsymbol{\Sigma}_W^{1/2}, \alpha_w \mathbf{1}_n, \kappa_w \mathbf{1}_n) \times \text{IG}(a_2, b_2) \\ \phi &\propto \text{MLG}(\mathbf{0}_n, \boldsymbol{\Sigma}_W^{1/2}, \alpha_w \mathbf{1}_n, \kappa_w \mathbf{1}_n) \times \text{IG}(a_3, b_3), \end{aligned} \tag{A5}$$

where "cMLG" is the conditional multivariate log gamma distribution from [20]. A motivating feature of this conjugate structure is that it is relatively straightforward to simulate from a cMLG. For σ^2, σ_w^2 and ϕ, we consider using a Metropolis–Hasting algorithm or slice sampling procedure [25].

The parameters of the conditional multivariate log gamma distribution are organized into in Table A1.

Table A1. Parameters of the full conditional distribution.

Parameter	Form
$\boldsymbol{H}_{\beta}$	$\begin{bmatrix} \boldsymbol{X} \\ \Sigma_{\beta}^{-1/2} \end{bmatrix}$
$\boldsymbol{\alpha}_{\beta}$	$\begin{bmatrix} \mathbf{1}_{n\times 1} \\ \alpha_{\beta}\mathbf{1}_{p\times 1} \end{bmatrix}$
$\boldsymbol{\kappa}_{\beta}^{(-1)}$	$\begin{bmatrix} (\log(Z(\mathbf{s})) - \log(Z_m))'\mathbf{1}_n e^{W'} \\ \frac{1}{\kappa_{\beta}}\mathbf{1}_p' \end{bmatrix}$
$\boldsymbol{H}_{W}$	$\begin{bmatrix} I_n \\ \boldsymbol{\Sigma}_{W}^{-1/2} \end{bmatrix}$
$\boldsymbol{\alpha}_{W}$	$\begin{bmatrix} \mathbf{1}_{n\times 1} \\ \alpha_{W}\mathbf{1}_{n\times 1} \end{bmatrix}$
$\boldsymbol{\kappa}_{W}^{(-1)}$	$\begin{bmatrix} (\log(Z(\mathbf{s})) - \log(Z_m))'\mathbf{1}_n (e^{X\beta})' \\ \frac{1}{\kappa_{W}}\mathbf{1}_n' \end{bmatrix}$

Appendix B. Trace Plot in Real Data Analysis

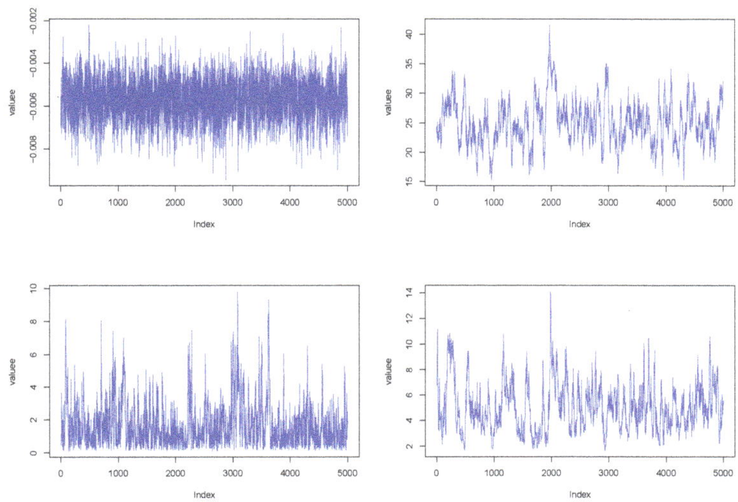

Figure A1. **(upper left)** Trace plot for β; **(upper right)** Trace plot for ϕ; **(lower left)** Trace plot for σ^2; **(lower right)** Trace plot for σ_w^2.

References

1. Mega, M.S.; Allegrini, P.; Grigolini, P.; Latora, V.; Palatella, L.; Rapisarda, A.; Vinciguerra, S. Power-law time distribution of large earthquakes. *Phys. Rev. Lett.* **2003**, *90*, 188501. [CrossRef] [PubMed]
2. Kijko, A. Estimation of the maximum earthquake magnitude, m max. *Pure Appl. Geophys.* **2004**, *161*, 1655–1681. [CrossRef]
3. Vere-Jones, D.; Robinson, R.; Yang, W. Remarks on the accelerated moment release model: Problems of model formulation, simulation and estimation. *Geophys. J. Int.* **2001**, *144*, 517–531. [CrossRef]
4. Charpentier, A.; Durand, M. Modeling earthquake dynamics. *J. Seismol.* **2015**, *19*, 721–739. [CrossRef]
5. Hu, G.; Bradley, J. A Bayesian spatial—Temporal model with latent multivariate log-gamma random effects with application to earthquake magnitudes. *Stat* **2018**, *7*, e179. [CrossRef]
6. Johnson, J.B.; Omland, K.S. Model selection in ecology and evolution. *Trends Ecol. Evol.* **2004**, *19*, 101–108. [CrossRef] [PubMed]

7. Cressie, N.; Calder, C.A.; Clark, J.S.; Hoef, J.M.V.; Wikle, C.K. Accounting for uncertainty in ecological analysis: The strengths and limitations of hierarchical statistical modeling. *Ecol. Appl.* **2009**, *19*, 553–570. [CrossRef] [PubMed]
8. Akaike, H. Information theory and an extension of the maximum likelihood principle. In *Selected Papers of Hirotugu Akaike*; Springer: Berlin, Germany, 1973; pp. 199–213.
9. Schwarz, G. Estimating the dimension of a model. *Ann. Stat.* **1978**, *6*, 461–464. [CrossRef]
10. Gelfand, A.E.; Dey, D.K. Bayesian model choice: Asymptotics and exact calculations. *J. R. Stat. Soc. Ser. B (Methodol.)* **1994**, *56*, 501–514. [CrossRef]
11. Geisser, S. *Predictive Inference*; Routledge: Abingdon, UK, 1993.
12. Ibrahim, J.G.; Laud, P.W. A predictive approach to the analysis of designed experiments. *J. Am. Stat. Assoc.* **1994**, *89*, 309–319. [CrossRef]
13. Spiegelhalter, D.J.; Best, N.G.; Carlin, B.P.; Van Der Linde, A. Bayesian measures of model complexity and fit. *J. R. Stat. Soc. Ser. B Stat. Methodol.* **2002**, *64*, 583–639. [CrossRef]
14. Chen, M.H.; Huang, L.; Ibrahim, J.G.; Kim, S. Bayesian variable selection and computation for generalized linear models with conjugate priors. *Bayesian Anal.* **2008**, *3*, 585. [CrossRef] [PubMed]
15. Bradley, J.R.; Holan, S.H.; Wikle, C.K. Bayesian Hierarchical Models with Conjugate Full-Conditional Distributions for Dependent Data from the Natural Exponential Family. *arXiv* **2017**, arXiv:1701.07506.
16. Chen, M.H.; Ibrahim, J.G. Conjugate priors for generalized linear models. *Stat. Sin.* **2003**, *13*, 461–476.
17. Geisser, S.; Eddy, W.F. A predictive approach to model selection. *J. Am. Stat. Assoc.* **1979**, *74*, 153–160. [CrossRef]
18. Tobler, W.R. A computer movie simulating urban growth in the Detroit region. *Econ. Geogr.* **1970**, *46*, 234–240. [CrossRef]
19. Gelfand, A.E.; Schliep, E.M. Spatial statistics and Gaussian processes: A beautiful marriage. *Spat. Stat.* **2016**, *18*, 86–104. [CrossRef]
20. Bradley, J.R.; Holan, S.H.; Wikle, C.K. Computationally Efficient Distribution Theory for Bayesian Inference of High-Dimensional Dependent Count-Valued Data. *arXiv* **2015**, arXiv:1512.07273.
21. Chen, M.H.; Shao, Q.M.; Ibrahim, J.G. *Monte Carlo Methods in Bayesian Computation*; Springer Science and Business Media: Berlin/Heidelberg, Germany, 2012.
22. Liang, H.; Wu, H.; Zou, G. A note on conditional AIC for linear mixed-effects models. *Biometrika* **2008**, *95*, 773–778. [CrossRef] [PubMed]
23. Baringhaus, L.; Franz, C. On a new multivariate two-sample test. *J. Multivar. Anal.* **2004**, *88*, 190–206. [CrossRef]
24. Ogata, Y. Statistical models for earthquake occurrences and residual analysis for point processes. *J. Am. Stat. Assoc.* **1988**, *83*, 9–27. [CrossRef]
25. Neal, R.M. Slice sampling. *Ann. Stat.* **2003**, *31*, 705–741. [CrossRef]

geosciences

MDPI

Article

Undrained Cyclic Laboratory Behavior of Sandy Soils

Francesco Castelli [1], Antonio Cavallaro [2], Salvatore Grasso [3,*] and Valentina Lentini [1]

[1] Faculty of Engineering and Architecture, University of Enna "Kore", 94100 Enna, Italy; francesco.castelli@unikore.it (F.C.); valentina.lentini@unikore.it (V.L.)

[2] National Research Council (CNR), Institute of Cultural Heritage Sciences (ISPC), 95124 Catania, Italy; a.cavallaro@ibam.cnr.it

[3] Department of Civil Engineering and Architecture, University of Catania, 95123 Catania, Italy

* Correspondence: sgrasso@dica.unict.it

Received: 1 August 2019; Accepted: 2 December 2019; Published: 11 December 2019

Abstract: The complex cyclic shear stress path experienced by the soil during an earthquake, which could also induce liquefaction phenomena, can be approximated in the laboratory only by using sophisticated testing apparatuses. Cyclic triaxial tests have been widely used, especially for coarse grained soils, as in this study. In the framework of the design for the seismic retrofitting of the "Ritiro viaduct" foundations along the A20 motorway connecting Messina with Palermo (Italy), a soil liquefaction study was also carried out. With this aim, a detailed geological and geotechnical characterization of the area was performed by in situ and laboratory tests, including seismic dilatometer Marchetti tests (SDMTs), the combined resonant column (RCT) and cyclic loading torsional shear tests (CLTSTs), and undrained cyclic loading triaxial tests (CLTxTs). In particular, the paper presents the results of cyclic triaxial tests carried out on isotropically consolidated specimens of a sandy soil. The seismic retrofitting works include the reinforcement of the foundation and replacement of the decks with newly designed type and structural schemes, mixed steel, and concrete with continuous girder. During the investigation, data were acquired for the characterization of materials, for the definition of degradation phenomena with the relative identification of possible causes, and for the estimation of the residual performance characteristics of the building. The structural campaign of investigations necessary to determine all of the key parameters useful for a correct definition of the residual performance capabilities of the work was divided into two phases: One in situ and one in the laboratory.

Keywords: in situ tests; laboratory tests; soil liquefaction; cyclic triaxial tests

1. Introduction

The present program of in situ investigations and laboratory tests originates from the static and seismic retrofitting works of the "Ritiro viaduct". The "Ritiro viaduct" (Figure 1), of the A20 Messina-Palermo (Italy) motorway, represents a vital node for the viability of Messina, the main connection to the motorway junctions. The works include the reinforcement of the foundations and replacement of the decks with newly designed type and structural schemes, mixed steel, and concrete with continuous girder.

During the investigation program, data were acquired for the characterization of materials, for the definition of degradation phenomena with the relative identification of possible causes, and for the estimation of the residual performance characteristics of the building. The structural campaign of investigations necessary to determine all of the parameters useful for a correct definition of the residual performance capabilities of the work has been useful to design the retrofitting works.

The campaign phase, in order to obtain as much data as possible without being extremely invasive towards the structural elements, was articulated through the execution of semi-destructive tests, such as the removal of concrete carrots, and non-destructive tests by means of execution of sclerometric tests, of pull-out with post-inserted grafts and in situ micro-seismic reliefs.

The area of the "Ritiro viaduct" was influenced by liquefaction phenomena during the 28 December 1908 Messina and Reggio Calabria historical earthquake. The evaluation of the liquefaction potential during earthquakes is an important subject in seismically active regions.

After the Niigata earthquake in 1964, which caused a lot of damage due to liquefaction, several studies [1–9] were performed in order to understand the cyclic behavior of sands.

During liquefaction, granular cohesionless saturated soil (gravel, sand, and low plasticity silt) loses its strength for a short interval of time, but long enough to cause significant failures. The liquefaction effects are usually evident on the ground surface (sand boils, large deformation, or fracture of the ground, etc.). The liquefaction susceptibility of soil deposits may be estimated by comparison between resistance profiles (i.e., Standard Penetration Test (SPT) blow count or Cone Penetration Test (CPT) tip resistance with depth) and critical values or estimation of a liquefaction safety factor (as a ratio of liquefaction resistance and action) function of depth.

The liquefaction potential can be obtained using either an estimation based on the maximum acceleration at the ground surface by a semi-empirical equation, or a dynamic calculation, including the reduction of soil stiffness due to built-up of pore pressure. Alternatively, the liquefaction resistance can be assessed by cyclic undrained laboratory tests on undisturbed or reconstituted specimens [10–12] or from correlations with the resistance measured by in situ tests (i.e., SPT or CPT).

In order to study the possible amplification phenomena of the "Ritiro viaduct" site, a comprehensive laboratory and in situ investigation was carried out to obtain a soil profile, with special attention being paid to the variation of the shear modulus (G) and damping ratio (D) with depth. This paper tries to summarize this information in a comprehensive way in order to provide a case record of site characterization for seismic response analysis.

Figure 1. The "Ritiro Viaduct" in Messina (Italy).

2. Geology and Seismicity of the Area

In Italy, recent strong earthquakes include the earthquake in central Italy that occurred on 24 August 2016 [13], the Emilia Romagna earthquake that occurred on 29 May 2012 [14–16], the L'Aquila earthquake that occurred on 6 April 2009 [17,18], the San Giuliano di Puglia earthquake that occurred on 31 October 2002 [19], the Umbria–Marche earthquake sequence of September–October 1997 [20], and the Catania earthquake that occurred on 13 December 1990 in southeastern Sicily.

The area under study is placed in the northeastern part of Sicily (Italy), at about 5 km from Messina old town. An example of its geological features is shown in Figure 2.

The area is covered by metamorphic complex in facies, with pan-African relicts in facies of granulitic deposits. The Messina Strait is located in the middle of the Calabrian Arc, one of the most seismically active areas of the Italian region and of the entire Mediterranean basin. The Strait's area itself has been struck several times in the past, though not with the same violence as in 1908.

Among Italian regions, Sicily is one of the most seismically active areas. In the past, strong earthquakes occurred in southeastern Sicily (1169, 1693) and in northeastern Sicily (1908). The M_W 7.1, 28 December 1908, Messina Straits earthquake (Figure 3) was the deadliest earthquake in recent European history, and also one of the first to be investigated with modern instrumental data [21]. The shaking was distinctly felt in Albania, Montenegro, and the Greek Ionian islands, about 400 km to the east and northeast of the Strait; and in Malta, about 250 km to the south. The earthquake was catastrophic in the epicentral area and was immediately followed by fires and by a large tsunami. Messina and Reggio Calabria were almost completely destroyed, buildings were severely damaged (Figure 3) over an area in excess of 6000 km^2 [21], and liquefaction phenomena occurred in the area.

According to [22] and [23] the main events reported in the historical catalogues and measured by Mercalli- Cancani- Sieberg scale (MCS) start around 91 BC (Reggio Calabria, IX–X MCS), 17 AD (Reggio Calabria, VIII–IX MCS), 361–363 (Messina Strait, X MCS), 853 (Messina, IX–X MCS), 1172 (Messina, VIII MCS), 1494 (Messina, VII–VIII MCS), 1509 sequence (area of Reggio Calabria, VIII MCS), 1659 (Southern Calabria, X MCS), 1783 sequence (Southwestern Calabria, X–XI MCS), 1894 (Southern Calabria, IX MCS), 1905 (Southwestern Calabria, XI MCS), and 1907 (Southern Calabria, VIII–IX MCS).

As well known, the characteristics of soil shaking are strongly influenced by local geological, geomorphological, and geotechnical conditions. These can modify significantly the amplitude, frequency, and duration of the seismic motion corresponding to the seismic hazard with reference to outcropping bedrock with horizontal topographical surface.

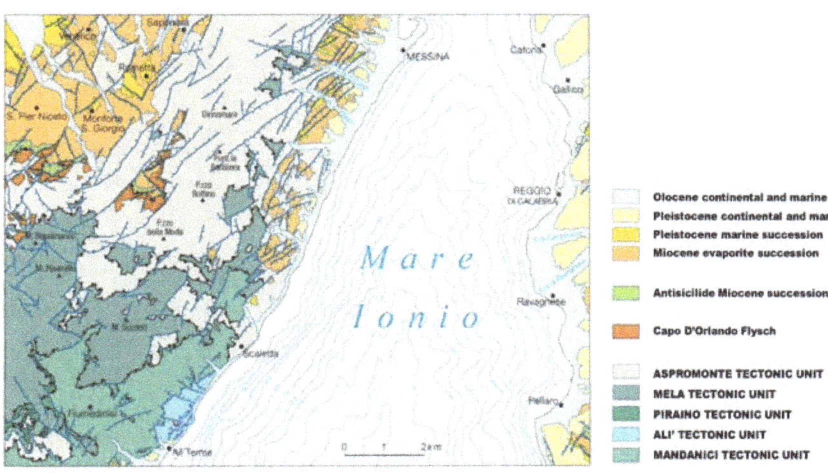

Figure 2. Geological synthesis of Messina area [24].

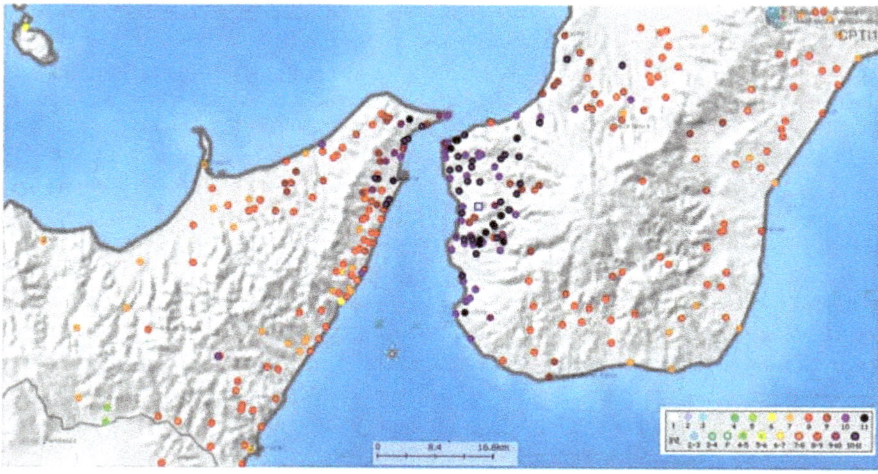

Figure 3. Messina Strait region with the 1908 earthquake epicenter and shocked localities with different colors according to Mercalli-Cancani-Sieberb scale [25].

To this aim, seismic microzonation defines the local seismic hazard through the identification of zones characterized by homogeneous seismic behavior that also includes the earthquake-induced effects, such as the slope instability, liquefaction in saturated granular soil, etc.

According to the Italian Guidelines for the Seismic Microzonation of the Department of Civil Defence, it is possible to distinguish three susceptibility classes: (1) Stable zones where no local effects occur (outcropping bedrock with a low steep morphological surface–slope angle\15°); (2) stable zones but susceptible of local seismic amplification, where seismic input motion amplifications are expected, as a result of the local litho-stratigraphic and morphological structure; (3) unstable zones, where the expected seismic effects can be attributed to irreversible soil deformations (slope instability, liquefaction, etc.).

A zoom of the Seismic Microzonation map of the area, and particularly of the "Ritiro Viaduct", is shown in Figure 4. In the map, the zones identified as homogeneous are characterized by similar parameters as lithological and litho-technical characteristics, depth of bedrock, geomorphological conditions, etc. The area of the "Ritiro Viaduct" is within the local amplification stable zones (Zone 0, Zone 1, and Zone 5) of the Messina microzonation map.

Therefore the seismicity that affected the area, but also the effects induced by earthquakes [26,27], constitute a hazard that should not be ignored in the design or in the seismic retrofitting works of infrastructure.

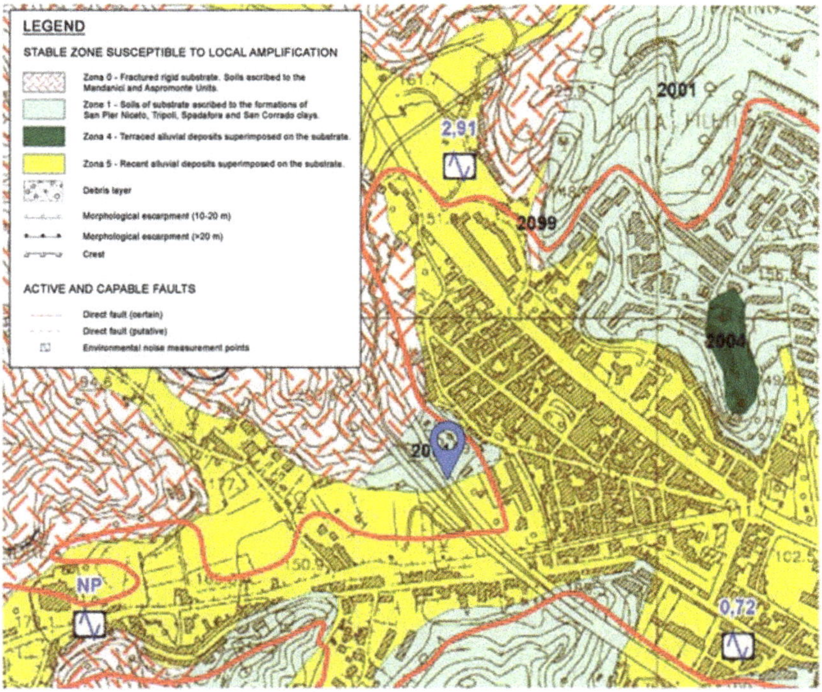

Figure 4. Microzonation map of Messina in the area of the "Ritiro viaduct".

3. Site Characterization Program and Basic Geotechnical Soil Properties

The site investigation was performed within the area of the "Ritiro viaduct" and reached a maximum depth of 35.00 m. Laboratory tests were performed on nine disturbed samples retrieved by means of a 101 mm tube sampler. To evaluate the geotechnical characteristics, the following in situ and laboratory tests were performed in the foundation soil located in the area of the "Ritiro viaduct": 28 Boreholes, 7 seismic dilatometer Marchetti tests (SDMTs) [28–31]; 9 particle size analysis, 25 direct shear tests (DST), 1 undrained triaxial test (UTxT), 8 cyclic loading torsional shear tests (CLTSTs) [32,33], 8 resonant column tests (RCTs) [34–36], and 3 cyclic loading triaxial tests (CLTxTs). The investigation program follows the approach and the methodology used in other test sites in the seismic areas of Catania and Messina [37–41]. On the basis of laboratory tests, the "Ritiro viaduct" deposits mainly consist of a normal consolidated grey/dark grey or dark brown sand with silty gravel or gravelly silt.

The results obtained by particle size analysis are shown in Figure 5 and Table 1. In Table 1, the soil samples were divided into Type A and Type B according to the depth and the presence of minor or major gravel. The uniformity coefficient (C_u), defined as the ratio of D_{60} to D_{10}, varies in the range of 5.48–52.24 and it points out the considerable non-homogeneity of the particle size. The maximum dry density was evaluated using a vibrating table, available at the geotechnical laboratory of the University Kore of Enna, according to ASTM (4253-83) standard.

Physical parameters were derived from standard classification tests performed on the samples retrieved by geotechnical survey. The values of soil unit weight and the minimum (e_{min}) and maximum (e_{max}) void ratios are summarized in Table 2. Most of the samples are coarse-grained soils, classifiable as silty sands to gravelly sands, showing a lower percentage of clayey material.

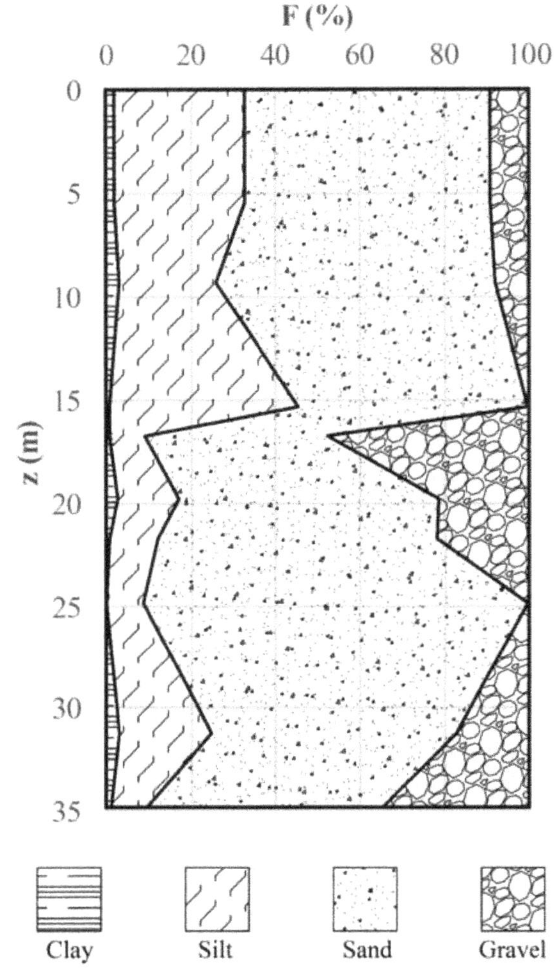

Figure 5. Profile of grain size distribution of the "Ritiro viaduct" area.

Table 1. Particle size analysis for the "Ritiro viaduct" area.

Boreholes	Type of Soil	H [m]	Gravel [%]	Sand [%]	Silt [%]	Clay [%]	d_{max} [mm]	D_{60} [mm]	D_{10} [mm]	C_u [-]	
S15C1	A	5.45	11.09	56.16	30.71	2.03	<9.5	0.1898	0.0124	15.31	Sand and gravelly silt of light gray color
S28C1	A	9.32	7.82	66.06	23.31	2.80	<19.0	0.1120	0.0180	6.22	Silty sand of light brown color
S28C2	A	15.30	0.35	54.35	44.67	0.63	<9.5	0.0699	0.0115	6.08	Sand and silt of dark gray color
S17C3	B	16.75	47.88	43.04	8.33	0.76	<9.5	3.5520	0.0680	52.24	Gravel and silty sand of dark brown color
S7C2	B	19.75	21.33	61.52	14.68	2.47	<9.5	0.5730	0.0171	33.51	Sand and silty gravel of dark gray color
S17C4	B	21.75	21.68	66.34	11.17	0.81	<19.0	0.7360	0.0354	20.79	Sand and silty gravel of dark brown color

Table 1. *Cont.*

Boreholes	Type of Soil	H [m]	Gravel [%]	Sand [%]	Silt [%]	Clay [%]	d_{max} [mm]	D_{60} [mm]	D_{10} [mm]	C_u [-]	
S28C4	B	24.80	0.24	91.00	8.76	0.00	<4.75	0.4957	0.0905	5.48	Sand lightly silty of light gray color
S7C4	B	31.25	17.18	58.22	21.64	2.96	<9.5	0.3532	0.0102	34.63	Sand and silty of light gray color
S17C7	B	34.75	34.41	55.56	8.97	1.06	<9.5	1.6010	0.0630	25.41	Sand and silty gravel of dark brown color

Type of soil A = soil with minor gravel; Type of soil B = soil with major gravel; H = depth; d_{max} = maximum diameter; D_{60} = the particle-size diameter for which 60% of the sample was finer; D_{10} = the particle-size diameter for which 10% of the sample was finer; C_u = uniformity coefficient.

Table 2. Mechanical characteristics for the "Ritiro viaduct" area.

Boreholes	Type of Soil	H [m]	γ [kN/m³]	D_{60} [mm]	C_u[-]	G_s[-]	e_o [-]	e_{min} [-]	e_{max} [-]	D_r [%]	CF [%]
S15C1	A	5.45	16.08	0.19	15.3	2.67	0.63	0.39	0.68	17.24	2.03
S28C1	A	9.32	18.24	0.11	6.2	2.80	0.57	0.44	0.69	48.00	2.80
S28C2	A	15.30	20.71	0.07	6.1	2.62	0.59	0.49	0.71	54.55	0.63
S17C3	B	16.75	15.18	3.55	52.2	2.49	0.64	0.52	0.72	40.00	0.76
S7C2	B	19.75	16.43	0.57	33.5	2.67	0.63	0.41	0.84	48.84	2.47
S17C4	B	21.75	16.80	0.74	20.8	2.75	0.64	0.51	0.70	31.58	0.81
S28C4	B	24.80	16.50	0.49	5.5	2.60	0.64	0.53	0.80	59.26	0.00
S7C4	B	31.25	16.97	0.35	34.6	2.69	0.67	0.37	0.79	28.57	2.96
S17C7	B	34.75	17.03	1.60	25.4	2.69	0.58	0.49	0.65	43.75	1.06

Type of soil A = soil with minor gravel; Type of soil B = soil with major gravel; H = depth; γ = total unit weight; D_{60} = the particle-size diameter for which 60% of the sample was finer; Cu = uniformity coefficient; G_s = specific gravity; e_o = initial void ratio; e_{min} = void ratio of soil in densest condition; e_{max} = void ratio of soil in loosest condition; D_r = relative density; CF = clay fraction.

The value of the natural moisture content w_n prevalently ranges from between 22 to 35% (Figure 6). Characteristic values of strength parameters are (cohesion) c′ = 5–24 kPa and (angle of shear resistance)′ = 23–40°, obtained by direct shear test, (undrained cohesion) cu = 228 kPa obtained by a undrained triaxial test performed on a cohesive sample, with a water content of 17%, retrieved at the depth of 14.65 m; G_s (specific gravity) ranged between 2.65 and 2.75, while e_o ranged between 0.29 and 0.67 (Table 2). Another stratum with a water content of 12% was found at the depth of about 19.25 m. Figure 6 shows index properties of the "Ritiro viaduct" area.

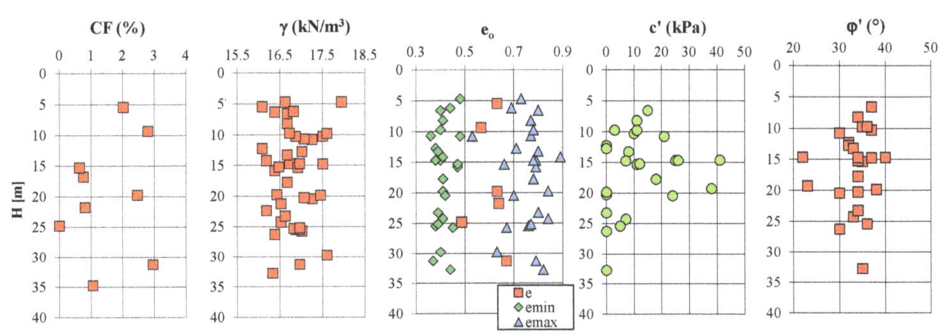

Figure 6. Particle size analysis of the "Ritiro viaduct" area.

4. Soil Properties by Laboratory Tests

The equivalent shear modulus G_{eq} and damping ratio D of "Ritiro viaduct" deposits were determined in the laboratory by means of a Resonant Column/Cyclic Loading Torsional Shear apparatus [42–44]. These dynamic parameters represent the basic data for the studies of local seismic response [45–48]. This apparatus was supplied at the Geotechnical Laboratory of the University Kore of Enna (Figure 7). A resonant column test consists of exciting one end of a confined solid or hollow cylindrical soil specimen. The specimen is fixed at the bottom (fixed-free test) and it is excited in torsion or flexure at the top by means of an electromagnetic drive system. Once the fundamental resonant frequency is established from measuring the motion of the free end, the velocity of the propagating wave and the degree of material damping are derived. The shear modulus is then obtained from the derived velocity V_s (in case of torsion) and the density of the sample.

Figure 7. Resonant column/cyclic torsional shear apparatus.

The equivalent shear modulus G_{eq} is the unload–reload shear modulus that is evaluated from RCT in function of velocity V_s and density ρ of the sample, while G_o is the maximum value or also "plateau" value as observed in the G-log(γ) plot. G is the secant modulus. Generally, G is constant until a certain strain limit is exceeded. This limit is called the elastic threshold shear strain (γ_t^e) and it is believed that soils behave elastically at strains smaller than (γ_t^e). The elastic stiffness at $\gamma < \gamma_t^e$ is thus already defined as G_o. Damping ratio D is defined by Equation (1):

$$D = \frac{\Delta W}{4\pi W}, \tag{1}$$

where ΔW is the area enclosed by the unloading–reloading loop and represents the total energy loss during the cycle, where W is the elastic stored energy. The RCT apparatus used is a fixed-free resonant column apparatus [49]. It enables the specimen consolidation under both isotropic and anisotropic stresses. It is composed of a drive system, a support system, and a base plate. The solid or hallow cylinder specimen is fixed at the bottom and its constraint at the base is due to the friction existing between the specimen and the porous synthesized bronze stone [50].

Torsional forces are applied at the top of the drive system, realized in aluminum. It is an electrical motor constituted of four magnets connected with the top of the sample and eight coils placed on the inox steel annular base, which is strictly linked to the support system. The weight of the motor is counterbalanced by a spring. A programmable function generator (PGF) excites the electrical motor of [51].

The support system, in addition to permitting the placement of the drive system, may possibly put the proximity transducers in and the filling in of water for saturated specimen tests. It is realized a Plexiglas cell pressure, to permit the isotropic consolidation using an air pressure source controlled with a manual pressure regulator. The base and the top plates are connected by three vertical rods inside the cell. In the resonant column test (RCT), the function generator produces a sinusoidal electric signal which is increased from an amplifier and transformed into torsional mechanical stress by an electromagnetic motor, consisting of eight coils and four magnets, connected to the head of the specimen. The magnetic field of the coils interacts with the magnets connected to the plate that transmits the torsional oscillation to the head of the specimen. Because the frequency of excitation is varied, the dynamic response of the sample varies in terms of amplitude. The latter is recorded by a accelerometer connected to the guide plate and proximity transducers measuring the relative movement between the plate and coils. Resonance frequency values were used for the calculation of the cutting module and the deformations. The decay curve, produced by interrupting the torsion excitation in resonance, allowed to evaluate the damping of the material following the amplitude decay method during the decrement of free vibration [42].

In the cyclic loading torsional shear test (CLTST), a cyclic torque is applied to the specimen by means of a torsional motor, to which a sinusoidal signal is sent at a fixed frequency of 0.5 Hz, continuously monitoring the torsion and angular deformation. The rotation of the specimen is measured thanks to the use of two proximity transducers. The data of the signal sent (proportional to the shear stress) and the corresponding torsion (proportional to the cutting deformation) are acquired simultaneously. The specimen shear module is then determined based on the average slope of the stress–strain hysteresis, while the damping is related to the cycle area of hysteresis as the ratio between the area enclosed by the unloading–reloading loop, and represents the total energy loss during the cycle and W is the elastic stored energy.

In the present work, solid cylindrical specimens were reconstituted by using tapping [50], in order to obtain the required relative and a good uniformity during the deposition.

The mold was assembled and a little depression was applied to let the membrane adhere to the inside surfaces. The material was placed in the mold using a funnel-pouring device. The soil was placed as loosely as possible in the mold by leaving the soil from the spout in a steady stream, holding the pouring device upright and vertical, and maintaining constant the fall height. It was possible to obtain different values of relative density changing the height of deposition. In order to realize high values of relative density, it could be necessary to beat delicately the mold surface during the deposition. Each sample was reconstituted with fresh sand. Each specimen was subjected to an isotropic load achieved in a Plexiglas pressure cell, using an air pressure source. The axial strain was measured by using a high-resolution proximity transducer, which monitors the aluminum top-cap displacement. Shear strain was measured by monitoring the top rotation with a couple of high-resolution proximity transducers. During a resonant column test, the proximity transducers were not able to appraise the value of the targets displacements, because of the high frequency of the oscillations. Then rotation on the top of the specimen was measured by means of an accelerometer.

The laboratory test conditions and the obtained small strain shear modulus G_0 are listed in Table 3. After the saturation phase, obtained by applying an appropriate back-pressure value, the undisturbed specimens were isotropically reconsolidated to the best estimate of the in situ mean effective stress. The same specimen was first subject to RCT, then to CLTST after a rest period of 24 h with opened drainage. CLTST was performed under stress control condition by applying a torque, with triangular time history, at a frequency of 0.1 Hz. The size of solid cylindrical specimens were radius = 25 mm and height = 100 mm.

Table 3. Test condition for the "Ritiro viaduct" area.

Boreholes	Type of Soil	H [m]	σ'_{vc} [kPa]	Test	G_o (1) [MPa]	G_o (2) [MPa]	G_o (3) [MPa]	Eo [MPa]	Δu_{max} [kPa]
S15C1	A	5.45	100	RCT–CLTST	51	54	21	-	-
S28C1	A	9.32	180	RCT–CLTST	105	83	324	-	-
S28C1	A	9.32	400	CLTxT	-	-	-	102	134
S28C2	B	15.30	180	RCT–CLTST	95	87	161	-	-
S17C3	B	16.75	300	RCT–CLTST	123	107	163	-	-
S7C2	B	19.75	300	RCT–CLTST	149	121	269	-	-
S17C4	B	21.75	350	CLTxT	-	-	-	209	344
S28C4	B	24.80	200	RCT–CLTST	109	99	378	-	-
S7C4	B	31.25	600	RCT–CLTST	160	135	-	-	-
S17C7	B	34.75	400	RCT–CLTST	158	142	-	-	-
S17C7	B	34.75	400	CLTxT	-	-	-	74	381

Type of soil A = soil with minor gravel; Type of soil B = soil with major gravel; H = depth; U = undrained; RCT = resonant column test; CLTST = cyclic loading torsional shear tests; CLTxT = cyclic loading triaxial test; G_o (1) from RCT, G_o (2) from CLTST after 24 h, G_o (3) from seismic dilatometer Marchetti test (SDMT).

The initial shear modulus G_o of soil is mainly influenced by the state of the soil, expressed by a combination of the void index e (or by the relative density D_r) and by the soil structure that reflects the deposition and the subsequent structural transformation processes such as aging, diagenesis, and cementation. For higher strain levels, the shear modulus G depends also, and especially on the strain level, on the stress–strain history and on the strain rate [52].

The G_o values (G_o (1) and G_o (2)), reported in Table 3, indicate moderate influence of strain rate, even at very small strain where the soil behavior is supposed to be elastic. Values of shear modulus G (MPa) and damping ratio D (%) versus from RCT and CLTST tests are reported in Figures 8–11, respectively. The initial shear modulus obtained during CLTST shows the effect of the soil degradation because of RCT; this effect tends to become negligible with the shear strain build up (Figure 12). Meanwhile, the D values obtained during RCT and CLTST follow the same trend and are thus comparable (Figure 13). The damping ratio values obtained from RCT by amplitude decay and CLTST method are quasi constant until a strain level of about 0.01%, higher values of D have been obtained from strain level higher than 0.01%. It is possible to see that the damping ratio from RCT and CLTST, at very small strains, is so equal to about 2%. Greater values of D are obtained from RCT for the strain level of about 0.1%.

It is supposed that RCT provides larger values of D than CLTST because of the rate (frequency) effect, in agreement with data shown by [53,54]. According to these researchers, the nature of soil damping in soils can be linked to the following phenomena:

— Non-linearity which governs the so-called hysteretic damping controlled by the current shear strain level. This kind of material damping is absent or negligible at very small strains.
— Viscosity of the soil skeleton (creep), which is relevant at very small strain rates.
— Viscosity of the pore fluid, which is relevant at very high frequencies.

Finally, higher values of the initial shear modulus (G_o (3)) were obtained from SDMTs. Generally, the small strain stiffness, determined in the laboratory on high quality reconstructed sample using appropriate apparatuses and procedures, is very close to that obtained in situ from seismic tests. Probably, in the case of the "Ritiro viaduct", disturbance phenomena occurred during reconstruction operations and differences in stress conditions determined lower values of the initial shear modulo in the laboratory.

The experimental results were used to determine the empirical parameters of the equation proposed by [55] to describe the shear modulus decay with shear strain level (Figure 14a,b and Figure 15a,b):

$$\frac{G(\gamma)}{G_o} = \frac{1}{1 + \alpha\gamma(\%)^{\beta}}, \tag{2}$$

where G(γ) = strain dependent shear modulus; γ = shear strain; α, β = soil constants.

Equation (2) allows the complete shear modulus degradation to be considered with strain level. The values of soil constants α and β obtained from RCTs and CLTSTs for soil type A and B are listed in Table 4.

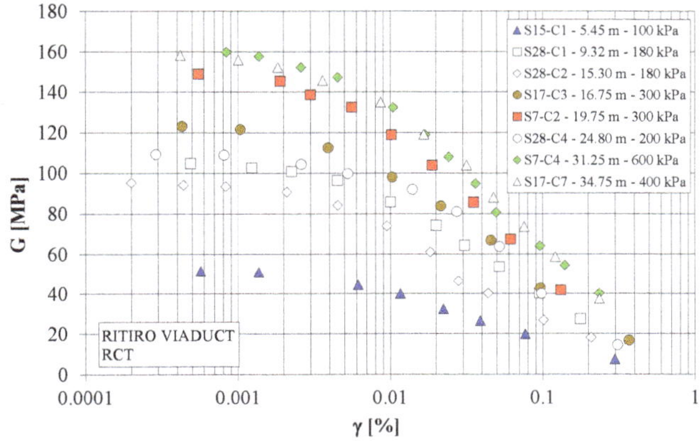

Figure 8. G–γ curves from RCTs.

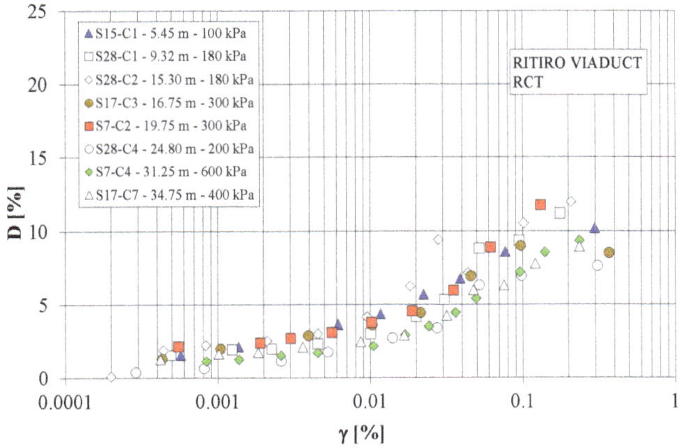

Figure 9. D–γ curves from RCTs.

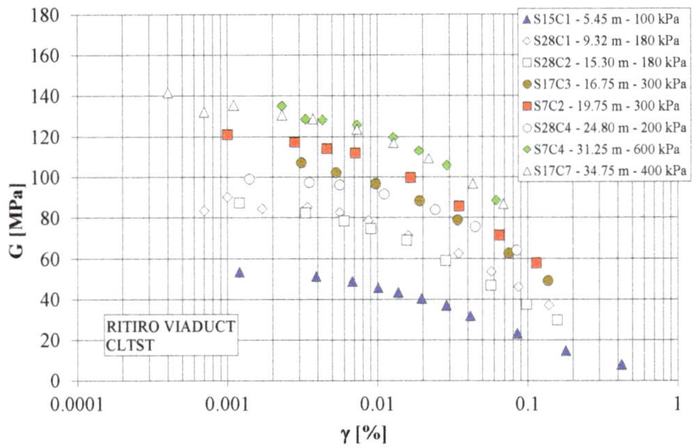

Figure 10. G–γ curves from CLTSTs.

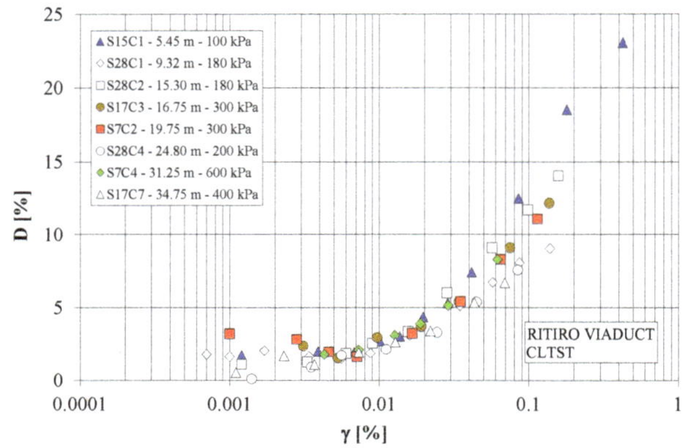

Figure 11. D–γ curves from CLTSTs.

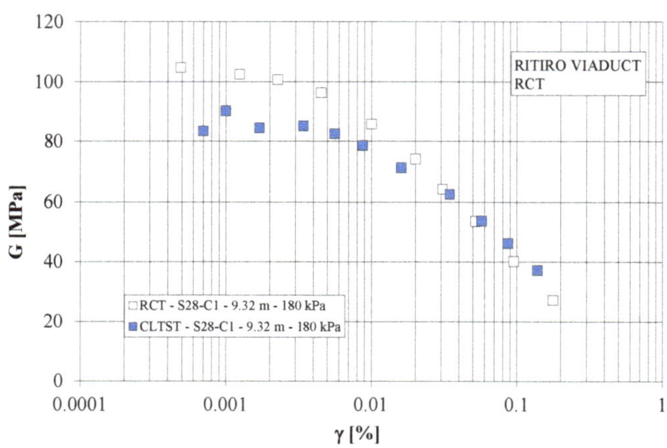

Figure 12. G–γ curves from RCTs and CLTSTs.

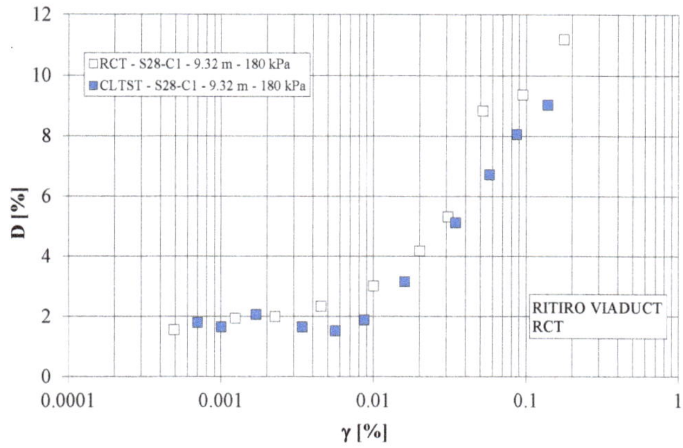

Figure 13. D–γ curves from RCTs and CLTSTs.

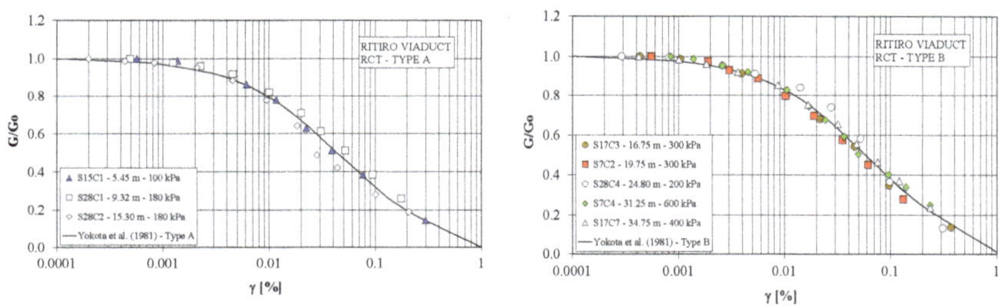

(**a**) (**b**)

Figure 14. G/G_o–γ curves from RCTs. (**a**) Type A; (**b**) Type B.

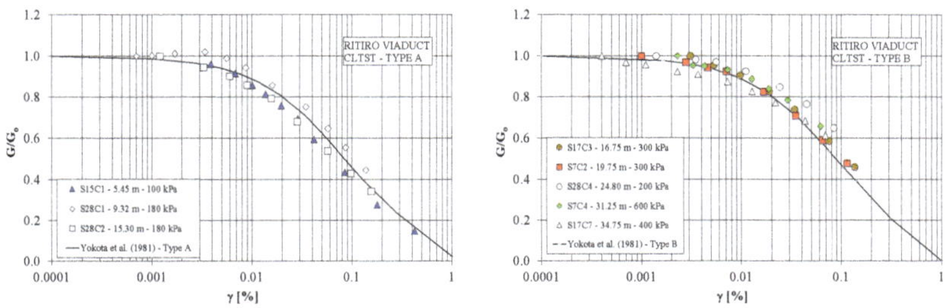

(**a**) (**b**)

Figure 15. G/G_o–γ curves from CLTSTs. (**a**) Type A; (**b**) Type B.

Table 4. Soil constants for the "Ritiro viaduct" area from RCTs and CLTSTs.

Test	Type of Soil	α [-]	β [-]	η [-]	λ [-]
RCT	A	18	0.92	14.5	1.7
RCT	B	13	0.90	18	2.2
CLTST	A	11	0.97	45	3.1
CLTST	B	10	0.95	65	3.5

As suggested by [55], the inverse variation of damping ratio with respect to the normalized shear modulus has an exponential form as that reported in Figure 16a,b and Figure 17a,b for the "Ritiro viaduct" area:

$$D(\gamma)(\%) = \eta \cdot \exp\left[-\lambda \cdot \frac{G(\gamma)}{G_o}\right], \tag{3}$$

where $D(\gamma)$ = strain dependent damping ratio; γ = shear strain; η, λ = soil constants.

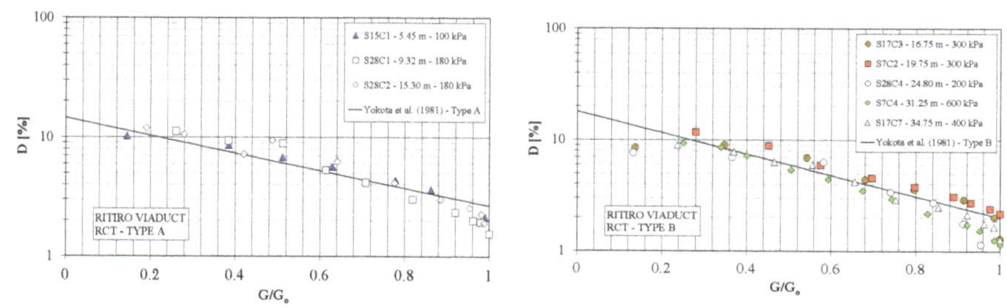

(a) (b)

Figure 16. D–G/G$_o$ curves from RCTs. (**a**) Type A; (**b**) Type B.

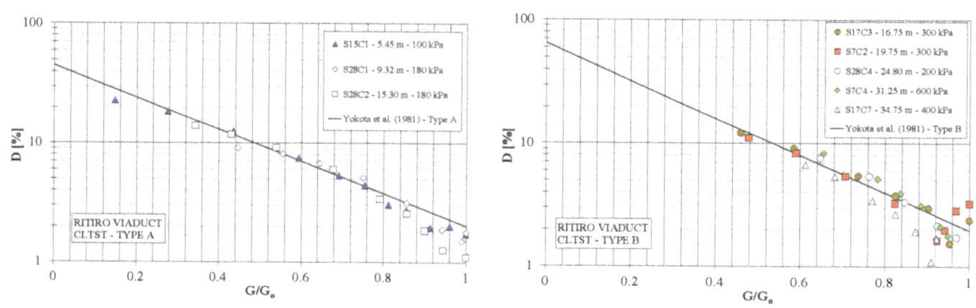

(a) (b)

Figure 17. D–G/G$_o$ curves from CLTSTs. (**a**) Type A; (**b**) Type B.

The values of soil constants η and λ obtained from RCTs and CLTSTs for soil type A and B are listed in Table 4.

Equation (3) assumes maximum value $D_{max} = 45\%$ for $G(\gamma)/G_o = 0$ and minimum value $D_{min} = 3.18\%$ for $G(\gamma)/G_o = 1$. Therefore, Equation (3) can be re-written in the following normalized form:

$$\frac{D(\gamma)}{D(\gamma)_{max}} = \exp\left[-\lambda \cdot \frac{G(\gamma)}{G_o}\right] \tag{4}$$

Figure 18a,b and Figure 19a,b show a comparison between the enveloping curves of the experimental data obtained during the RCTs and CLTSTs. It is possible to observe how, in general, the CLTSTs, with respect to the RCTs, determine a behavior of the soil characterized by a wider elastic field with the same level of deformation reached. This phenomenon may be due to the intergranular reassembly of the sandy soil due to the effect of the RCTS. Moreover, overall, in the case of CLTSTs higher values of D are observed.

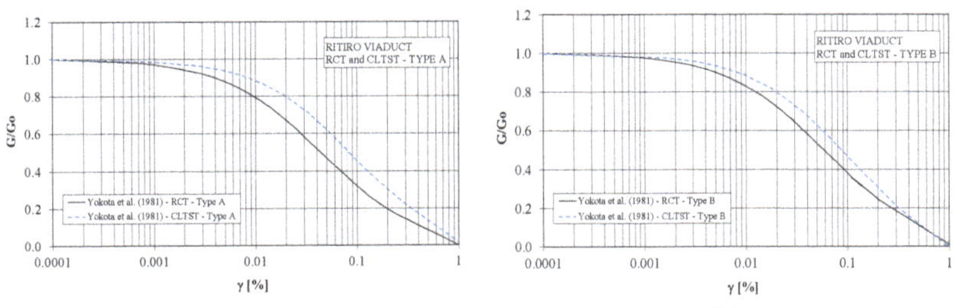

(a) (b)

Figure 18. G/G_o–γ curves from RCTS and CLTSTs by [55]. (**a**) Type A; (**b**) Type B.

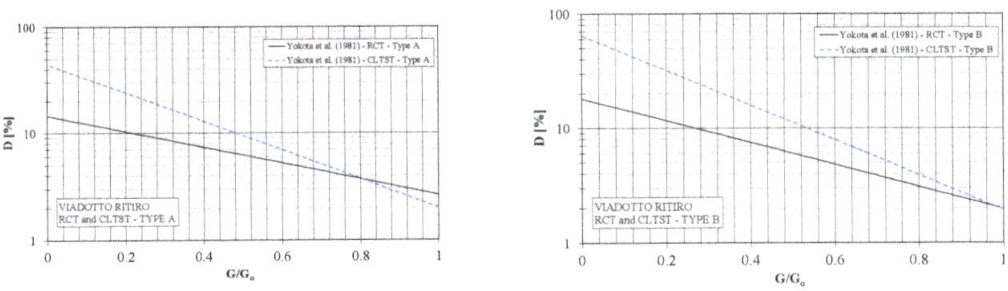

(a) (b)

Figure 19. D–G/G_o curves from RCTS and CLTSTs by [55]. (**a**) Type A; (**b**) Type B.

To perform triaxial tests, triaxial cells consisting of a structure were used for stainless steel and a Plexiglas cylinder (Figure 20). The maximum isotropic operating pressure was 1 MPa. For fluid of confinement, water was used. Sample drainage was allowed, when necessary, through porous stones placed on the two vertical load distribution bases. The specimen was subsequently placed in saturation by applying a back-pressure under an effective pressure isotropic, enough to prevent swelling. The test was then saturated by performing one measurement of parameter B. The value of approximately 0.95 was taken as an indirect measure of the complete saturation of the material. In the case of a value that is too low, saturation was prolonged for a further period of time, in some cases increasing counter-pressure again until a satisfactory value of B was reached and the specimen was brought, in several steps, to the final effective consolidation tension.

All test operations during saturation (until the Skempton B parameter reached at least 0.98) and isotropic consolidation were controlled by a panel that adjusts the confinement pressure and the counter-pressure and allows measurement of pressures and pressure volume variations of the specimen by means of a pressure transducer and a volumometer. The height variations of the specimen were detected by means of a displacement transducer. The application of cyclic loads took place by means of a contrast structure equipped with an electro-pneumatic system, which allowed to apply to the specimen a constant sinusoidal load of constant amplitude. The size of solid cylindrical specimens were radius = 35 mm and height = 140 mm. The laboratory test conditions and the obtained small strain shear modulus E_o are also listed in Table 3. During the cyclic triaxial test, the load sequence was characterized by steps of 40 strain controlled load cycles.

During the cyclic loading triaxial tests (CLTxTs) (Figure 21), the soil sample showed a rapid decrease of its mechanical characteristics at strain levels of about 10^{-2}%. It seems that it is not possible to investigate the values of the modulus of normal elasticity (Young's modulus) at very low strain levels (less than 10^{-3}%) due to undesired deformations caused by the deformability of the mechanical structure (system compliance) of the triaxial apparatus.

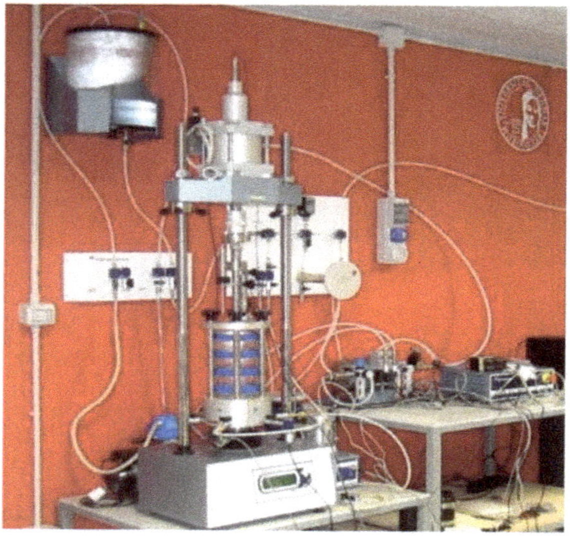

Figure 20. Cyclic triaxial apparatus.

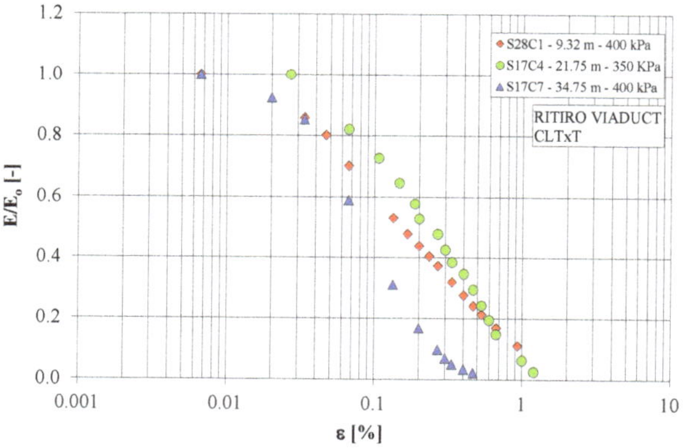

Figure 21. E/E_o–ε curves from CLTxTs.

Low values of the Young's modulus were obtained in correspondence with the test performed on the S17C7 sample, probably due to the low initial value of D_r.

The initial damping values are around 1%, while the maximum values are between 6 and 8% at a strain level of about 1% (Figure 22).

During the cyclic loading triaxial tests (CLTxT), unload–reload cycles became unstable and degradation phenomena of material occured when a certain limit strain was exceeded (Figures 23 and 24). This limit strain is defined as volumetric threshold shear strain and is rate-dependent. The degradation caused a decrease of stiffness, an increase of D and pore pressure build-up with the increase of N because of cyclic material degradation, as obtained from a CLTST and CLTxT on "Ritiro viaduct" soil [56].

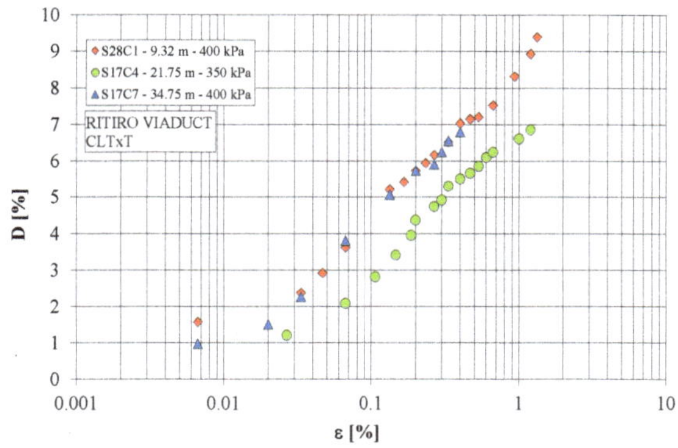

Figure 22. D–ε curves from CLTxTs.

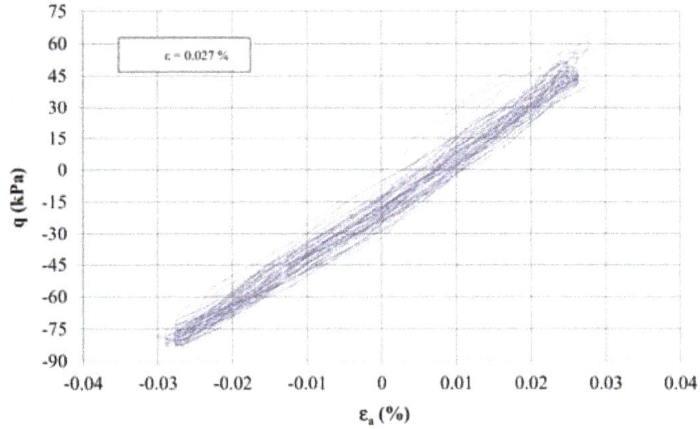

Figure 23. Stable cycles from CLTxT.

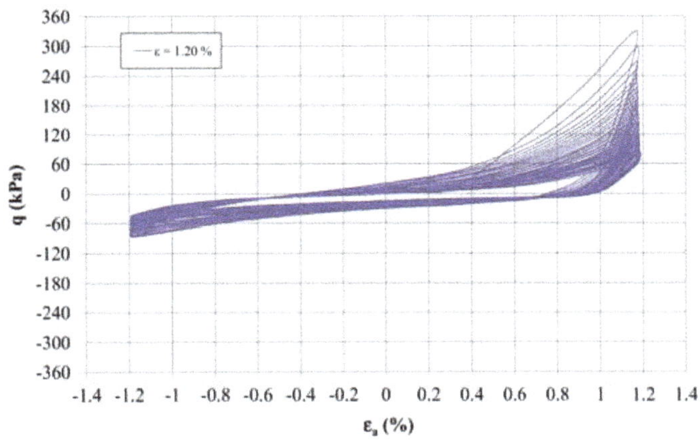

Figure 24. Unstable cycles from CLTxT.

Figure 25 shows the pore pressure build up during CLTxT. The pore pressure build up during CLTxT was so negligible at low strain. On the contrary, at strain level of about 0.15%, it is possible to observe an important increase of pore pressure due to the degradation phenomenon.

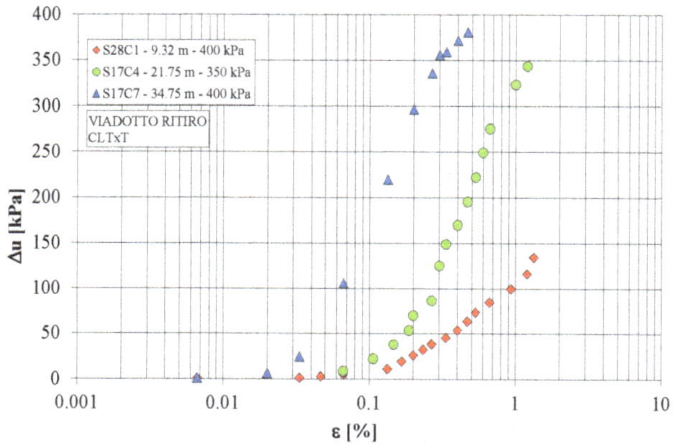

Figure 25. Δu–ε curves from CLTxTs.

Values of shear modulus G (MPa) and damping ratio D (%) versus γ (%) from CLTST and CLTxT are reported in Figures 26 and 27, respectively. In the case of CLTST, higher values of G were always obtained, compared to those of the CLTxT, also as a function of lower initial investigated strain levels. The trend of the G modulus seems to align only for strain levels higher than 0.1%, even if the results of the two types of tests are comparable only for a strain interval between 0.01 and 0.1%. This difference on G can probably be attributed to the high interstitial pressure values obtained during the CLTxT tests (Figure 25) due to the low initial value of D_r. Higher values of D were obtained from strain levels higher than 0.01%. Moreover, at the same strain level, in the case of CLTSTs, higher values of D were observed.

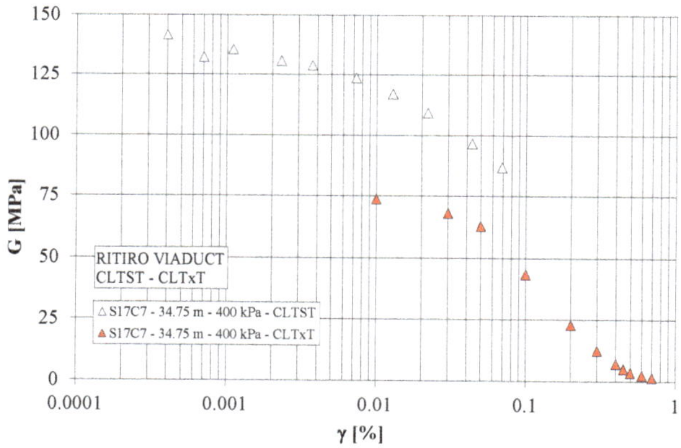

Figure 26. G–γ curves from CLTST and CLTxT.

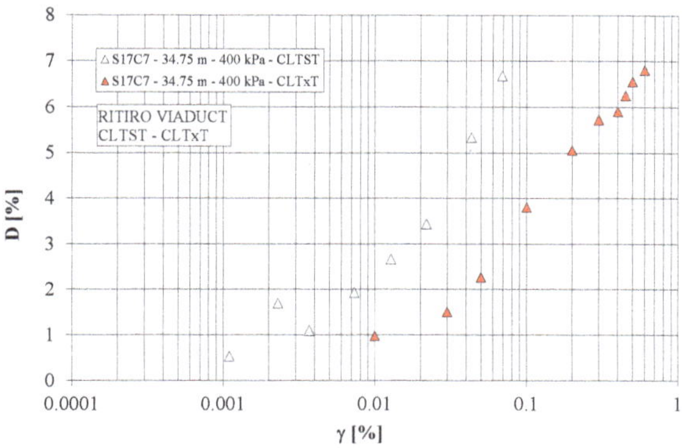

Figure 27. D–γ curves from CLTST and CLTxT.

5. Soil Properties by in Situ Tests

The use of in situ tests as complementary experimental techniques to laboratory experiments is now commonly recognized [57]. The small strain ($\gamma \leq 0.001\%$) shear modulus, G_o, was thus determined from SDMT. The SDMT provides a simple means for determining the initial elastic stiffness at very small strains and in situ shear strength parameters at high strains in natural soil deposits. Moreover, it was attempted to assess G_o by means of empirical correlations, based either on penetration test results or on laboratory test results [57]. The SDMT [17,58] provides a simple means for determining the initial elastic stiffness at very small strains and in situ shear strength parameters at high strains in natural soil deposits [32,59]. This apparatus was also used in offshore conditions by [60,61]. The test is conceptually similar to the seismic cone (SCPT). First introduced by [62], the SDMT was subsequently improved at Georgia Tech, Atlanta, USA [63–65]. A new SDMT system has recently been developed in Italy. The seismic modulus is a cylindrical instrumented tube, located above the DMT blade [66], housing two receivers at a distance of 0.50 m (see Figure 28). The test configuration "two receivers"/"true interval" avoids the problem connected with the possible inaccurate determination of the "first arrival" time sometimes met with the "pseudo interval" configuration (just one receiver). Moreover, the pair of seismograms recorded by the two receivers at a given test depth correspond to the same hammer blow and not to different blows in sequence, which are not necessarily identical. The adoption of the "true interval" configuration considerably enhances the repeatability in the V_s measurement (observed repeatability $V_s \approx 1$–2%). V_S is obtained as the ratio between the difference in distance between the source and the two receivers (S2–S1) and the delay of the arrival of the impulse from the first to the second receiver (Δt). V_s measurements are obtained every 0.5 m of depth. The shear wave source at the surface is a pendulum hammer ($\approx$10 kg), which hits horizontally a steel rectangular base pressed vertically against the soil (by the weight of the truck) and oriented with its long axis parallel to the axis of the receivers, so that they can offer the highest sensitivity to the generated shear wave.

Source waves are generated by striking a horizontal plank at the surface that is oriented parallel to the axis of a geophone connects by a co-axial cable with an oscilloscope [63,64]. The measured arrival times at successive depths provide pseudo interval V_s profiles for horizontally polarized vertically propagating shear waves. In Figure 28, the SDMT scheme for the measure of V_s is shown, while Figure 29 shows an example of seismograms obtained by SDMT at various test depths at the site of the "Ritiro viaduct" (it is a good practice to plot side-by-side the seismograms as recorded and re-phased according to the calculated delay). V_s may be converted into the initial shear modulus G_o by the theory of elasticity.

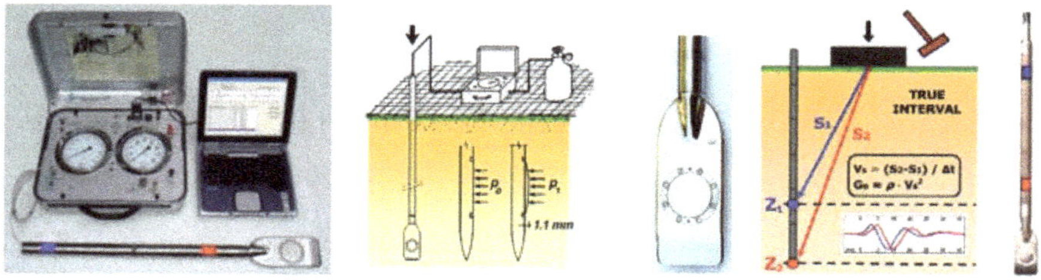

(a) (b) (c)

Figure 28. Seismic dilatometer equipment (**a**). Schematic layout of the flat dilatometer test (**b**) and of the seismic dilatometer test (**c**).

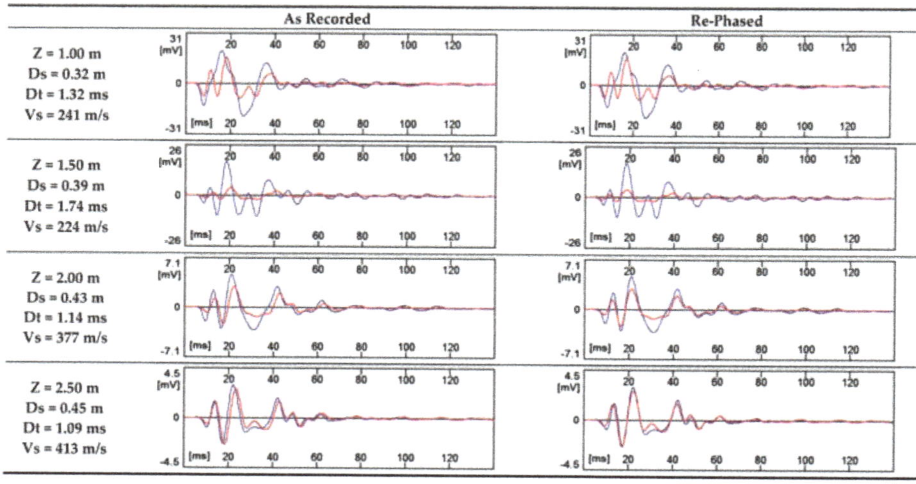

Figure 29. Example of seismograms obtained by SDMT at the site of the "Ritiro viaduct".

In three SDMT test verticals, only the seismic measurements were carried out, in pre-holes performed by means of a probe and filled with gravel (with grains of a diameter strictly between 5 and 15 mm), statically advancing with a penetrometer having a maximum thrust capacity equal to 20 tons. The noticeable difference between the density of the in situ material and the filling material of the pre-hole made the interpretation of the results particularly difficult. The combined knowledge of G_0 and of the one dimensional modulus M (from DMT) may be helpful in the construction of the G–γ modulus degradation curves [67–71].

A summary of SDMT parameters is shown in Figure 30, where:

- Id: Material index; gives information on soil type (sand, silt, clay).
- M: Vertical drained constrained modulus.
- Cu: Undrained shear strength.
- Phi: Angle of shear resistance.
- Kd: Horizontal stress index; the profile of Kd is similar in shape to the profile of the overconsolidation ratio OCR. Kd = 2 indicates in clays OCR = 1, Kd > 2 indicates overconsolidation. A first glance at the Kd profile is helpful to "understand" the deposit.
- V_S: Shear wave velocity.

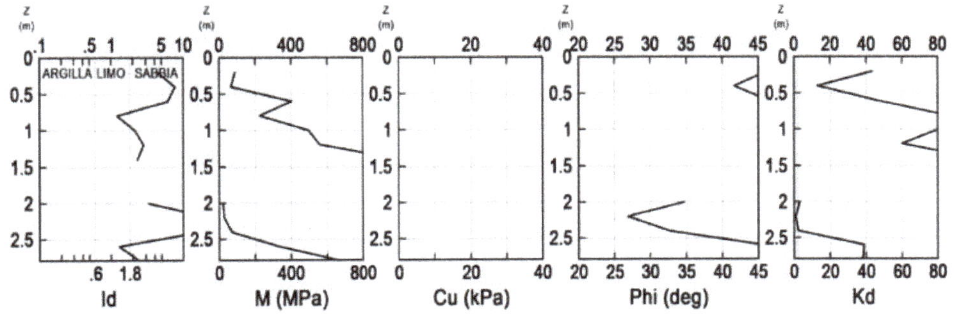

Figure 30. Results of the SDMTs (SDMT 4a) in terms of geotechnical parameters.

Figure 31 shows the values of G_o obtained in situ from SDMT and those measured in the laboratory from RCT performed on reconstructed solid cylindrical specimens, which were isotropically reconsolidated to the best estimate of the in situ mean effective stress. The G_o values are plotted in Figure 31 against depth. In the case of laboratory tests, the G_o values are determined at shear strain levels of less than 0.001%. A comparability exists between the laboratory and in situ test results. On average, the ratio of G_o (Lab) to G_o (Field) by RCT and SDMT was equal to about 1.80 at the depth of 19.75 m.

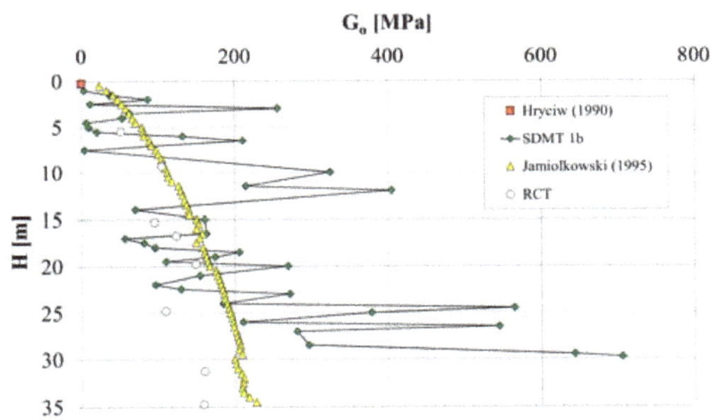

Figure 31. G_o obtained from SDMT, RCT, and empirical correlations.

It was also attempted to evaluate the small strain shear modulus by means of the following empirical correlations based on penetration tests results or laboratory results available in literature.

(a) [72]

$$G_o = \frac{530}{\left(\sigma'_v/p_a\right)^{0.25}} \frac{\gamma_D/\gamma_w - 1}{2.7 - \gamma_D/\gamma_w} K_o^{0.25} \cdot \left(\sigma'_v \cdot p_a\right)^{0.5}, \tag{5}$$

where G_o, σ'_v and p_a are expressed in the same unit; p_a = 1 bar is a reference pressure; γ_D and K_o are, respectively, the unit weight and the coefficient of earth pressure at rest, as inferred from SDMT results according to [66];

(b) [57]

$$G_o = \frac{600 \cdot \sigma_m'^{0.5} p_a^{0.5}}{e^{1.3}}, \tag{6}$$

where $\sigma'_m = (\sigma'_v + 2 \cdot \sigma'_h)/3$; $p_a = 1$ bar is a reference pressure; G_o, σ'_m, and p_a are expressed in the same unit. The values for parameters which appear in Equation (6) are equal to the average values that result from laboratory tests performed on quaternary Italian clays and reconstituted sands. A similar equation was proposed by [73] for Holocene clay deposits.

Equation (6) incorporates a term which expresses the void ratio; the coefficient of earth pressure at rest only appears in Equation (5). However only Equation (5) tries to obtain all the input data from the SDMT results. The G_o values obtained with the methods above are also plotted against depth in Figure 28. The method by [57] was applied considering a given profile of void ratio. The coefficient of earth pressure at rest was inferred from SDMT.

Since the purely dilatometric data only investigate the first meters of depth, the use of Equation (5) is limited. On the whole, Equation (6) seems to provide the most accurate trend of G_o with depth, but is not able to analyze stratigraphic variation along the depth, as can be seen in Figure 28. The results obtained by SDMT are comparable with the data of the RCT tests and they are able to identify the stratigraphic variations.

6. Analysis of the Effects on the Physical Environment

The results of cyclic triaxial tests carried out can be used for the seismic retrofitting works, which include also the reinforcement of the foundation of buildings. Data include characterization of materials, definition of degradation phenomena with the relative identification of possible causes, and estimation of the residual performance characteristics of the building. To take into account the analysis of the effects on the natural environment, according to the *"Manual for Zonation on Seismic Geotechnical Hazards"*, local seismic response, slope instability, and liquefaction of the area were analyzed using the results of cyclic triaxial tests and the results of other laboratory tests, including the combined resonant column tests (RCT) and cyclic loading torsional shear tests (CLTST). Local seismic response analysis using ONDA and DEEPSOIL computer codes was performed on the Ritiro Viaduct [74]. Results of the numerical analyses are presented as comparisons in terms of maximum acceleration profiles, maximum shear strain profiles, response spectra, surface response seismograms, Fourier spectra, and amplification ratios. The local seismic response (LSR) analysis was performed by using seismograms obtained for the 28 December 1908 Messina and Reggio Calabria earthquake scenario. The results of cyclic triaxial tests performed on samples were also used for the evaluation of the liquefaction resistance of soils of the Ritiro Viaduct [12]. The complex cyclic shear stress path experienced by the soil during an earthquake can be reproduced in the laboratory only by using sophisticated testing apparatuses. Cyclic triaxial tests have been widely used to assess soil liquefaction potential, especially for coarse-grained soils, as in this study.

7. Conclusions

In the framework of the design for the seismic retrofitting of the "Ritiro Viaduct" foundations along the A20 motorway, connecting the cities of Messina and Palermo, located in one of the most hazardous Italian seismic areas, a detailed geotechnical characterization was carried out. Indeed, the seismic effects induced by earthquakes play an important role in the planning and construction, as well as in the seismic retrofitting works of important infrastructures, such as the "Ritiro viaduct".

This paper focuses on a comprehensive laboratory and in situ investigations carried out to obtain a soil profile, with special attention to the variation of the shear modulus (G) and damping ratio (D) with depth. A detailed geological and geotechnical characterization of the area was performed by in situ and laboratory tests, including seismic dilatometer Marchetti tests (SDMTs), the combined resonant column (RCT) and cyclic loading torsional shear tests (CLTSTs), and undrained cyclic loading triaxial tests (CLTxTs).

The RCT and CLTST show a moderate influence of strain rate, even at very small strain, where the soil behavior is supposed to be elastic, while the D values obtained during RCT and CLTST follow the same trend and are thus comparable. It is possible to see that the damping ratio from RCT and

CLTST, at very small strains, is equal to about 2%. Moreover, higher values of the initial shear modulus G_0 were obtained from SDMTs. Probably, in the case of the "Ritiro viaduct", disturbance phenomena occurred during reconstruction operations and differences in stress conditions determined lower values of the initial shear modulus in the laboratory.

The experimental results were used to design the seismic retrofitting work and, in particular, to determine the empirical parameters of the proposed equation to describe the shear modulus decay and damping ratio build-up with shear strain level.

During the cyclic loading triaxial tests (CLTxTs), the soil sample showed a rapid decrease of its mechanical characteristics at strain levels of about 10^{-2}%. It seems that it is not possible to investigate the values of the modulus of normal elasticity (Young's modulus) at very low strain levels (less than 10^{-3}%) due to undesired deformations caused by the deformability of the mechanical structure (system compliance) of the triaxial apparatus. Low values of the Young's modulus were obtained in correspondence with the test performed on one sample, probably due to the low initial value of D_r. The initial damping values were around 1%. During CLTxT, unload–reload cycles became unstable and degradation phenomena of material occurred when a certain limit strain was exceeded. This limit strain is called volumetric threshold shear strain and it is rate-dependent. The degradation caused a decrease of stiffness, an increase of D, and pore pressure build-up with the increase of N because of cyclic material degradation. Finally, the in situ obtained results by SDMT, though higher, are comparable with the data of the RCT tests and they are able to identify the stratigraphic variations.

Author Contributions: F.C., A.C., S.G. and V.L. carried out the investigation and prepared the original manuscript according to the following percentages: 25% F.C., 25% A.C., 25% S.G. and 25% V.L.

Funding: This research received no external funding.

Conflicts of Interest: The authors declare no conflict of interest.

References

1. Seed, H.B.; Lee, K.L. Liquefaction of Saturated Sands During Cyclic Loading. *J. Soil Mech. Found.* **1966**, *92*, 105–134.
2. Yoshimi, Y.; Tokimatsu, K.; Kaneko, O.; Makihara, Y. Undrained Cyclic Shear Strength of Dense Nigata Sand. *Soils Found.* **1984**, *24*, 131–145. [CrossRef]
3. Sladen, J.A.; D'Hollander, R.D.; Krahn, J. The Liquefaction of Sands, a Collapse Surface Approach. *Can. Geotech. J.* **1985**, *22*, 564–578. [CrossRef]
4. Vaid, Y.P.; Chern, J.; Tumi, H. Confining Pressure, Grain Angularity and Liquefaction. *J. Geotech. Eng.* **1985**, *111*, 1229–1235. [CrossRef]
5. Michetti, A.M.; Esposito, E.; Guerrieri, L.; Porfido, S.; Serva, L.; Tatevossian, R.; Vittori, E.; Audemard, F.; Azuma, T.; Clague, J.; et al. Intensity Scale ESI 2007. *Mem. Descr. Della Carta Geol. D'italia* **2007**, *74*, 11–20.
6. Giona-Bucci, M.; Villamor, P.; Almond, P.; Tuttle, M.; Stringer, M.; Ries, W.; Smith, C.; Hodge, M.; Watson, M. Associations between Sediment Architecture and Liquefaction Susceptibility in Fluvial Settings: The 2010–2011 Canterbury Earthquake Sequence, New Zealand. *Eng. Geol.* **2018**, *237*, 181–197. [CrossRef]
7. Serva, L. History of the Environmental Seismic Intensity Scale ESI-07. *Geosciences* **2019**, *9*, 210. [CrossRef]
8. Tuttle, M.P.; Hartleb, R.; Wolf, L.; Mayne, P.W. Paleoliquefaction Studies and the Evaluation of Seismic Hazard. *Geosciences* **2019**, *9*, 311. [CrossRef]
9. Naik, S.P.; Kim, Y.; Kim, T.; Su-Ho, J. Geological and Structural Control on Localized Ground Effects within the Heunghae Basin during the Pohang Earthquake (MW 5.4, 15th November 2017), South Korea. *Geosciences* **2019**, *9*, 173. [CrossRef]
10. Toki, S.; Tatsukoa, F.; Miura, S.; Yoshimi, Y.; Yasuda, S.; Makihara, Y. Cyclic Undrained Triaxial Strength of Sand by a Cooperative Test Program. *Soils Found.* **1986**, *26*, 117–128. [CrossRef]
11. Flora, A.; Lirer, S.; Silvestri, F. Undrained Cyclic Resistance of Undisturbed Gravelly Soil. *Soil Dyn. Earthq. Eng.* **2012**, *43*, 366–379. [CrossRef]
12. Lentini, V.; Castelli, F. Liquefaction Resistance of Sandy Soils from Undrained Cyclic Triaxial Tests. *Geotech. Geol. Eng.* **2019**, *37*, 201. [CrossRef]

13. De Silva, F.; Sica, S.; Silvestri, F.; Aversa, S. Estimation of the ground shaking from the response of rigid bodies. *Ann. Geophy.* **2016**, *59*, 5. [CrossRef]
14. Facciorusso, J.; Madiai, C.; Vannucchi, G. The 2012 Emilia earthquake (Italy): Geotechnical characterization and ground response analyses of the paleo-Reno river levees. *Soil Dyn. Earthq. Eng.* **2016**, *86*, 71–88. [CrossRef]
15. Alessio, G.; Alfonsi, L.; Brunori, C.A.; Burrato, P.; Casula, G.; Cinti, F.R.; Civico, R.; Colini, L.; Cucci, L.; De Martini, P.M.; et al. Technologies and New Approaches Used by the INGV EMERGEO Working Group for Real-Time Data Sourcing and Processing During the Emilia Romagna (Northern Italy) 2012 Earthquake Sequence. *Ann. Geophys.* **2012**, *55*. [CrossRef]
16. Emergeo Working Group. Liquefaction Phenomena Associated with the Emilia Earthquake Sequence of May–June 2012 (Northern Italy). *Nat. Hazards Earth Syst. Sci.* **2012**, *13*, 935–947.
17. Monaco, P.; Marchetti, S.; Totani, G.; Marchetti, D. Interrelationship Beetween Small Strain Modulus G_0 and Operative Modulus. In Proceedings of the International Conference on Performance-Based Design in Earthquake Geotechnical Engineering (IS-Tokyo 2009), Tsukuba, Japan, 15–17 June 2009; Kokusho, T., Ed.; Taylor & Francis Group: London, UK, 2009; pp. 1315–1323.
18. Santucci de Magistris, F.; d'Onofrio, A.; Evangelista, L.; Foti, S.; Maraschini, M.; Monaco, P.; Amoroso, S.; Totani, G.; Lanzo, G.; Pagliaroli, A.; et al. Geotechnical characterization of the Aterno valley for site response analyses. *Rivista Italiana di Geotecnica* **2013**, *3*, 23–43.
19. Cavallaro, A.; Grasso, S.; Maugeri, M. Site Response Analysis for Tito Scalo Area (PZ) in the Basilicata Region, Italy. In Proceedings of the 4th Geotechnical Earthquake Engineering and Soil Dynamics Conference, Sacramento, CA, USA, 18–22 May 2008.
20. Cavallaro, A.; Maugeri, M. Dynamic Characterization of Soils at Sellano for Seismic Microzonation. *Riv. Ital. Geotec.* **2001**, *4*, 70–78.
21. Pino, N.A.; Piatanesi, A.; Valensise, G.; Boschi, E. The 28 December 1908, Messina Straits Earthquake (MW 7.1): A Great Earthquake through a Century of Seismology. *Seismol. Res. Lett.* **2008**, *80*, 243–259. [CrossRef]
22. Comerci, V.; Blumetti, A.M.; Brustia, E.; Di Manna, P.; Guerrieri, L.; Lucarini, M.; Vittori, E. Landslides Induced by the 1908 Southern Calabria: Messina Earthquake (Southern Italy). In *Landslide Science and Practice*; Margottini, C., Canuti, P., Sassa, K., Eds.; Springer: Berlin, Germany, 2013; Volume 5.
23. Comerci, V.; Vittori, E.; Blumetti, A.M.; Brustia, E.; Di Manna, P.; Guerrieri, L.; Lucarini, M.; Serva, L. Environmental Effects of the December 28, 1908, Southern Calabria–Messina (Southern Italy) Earthquake. *Nat. Hazards* **2015**, *76*, 1849–1891. [CrossRef]
24. ISPRA Istituto Superiore per la Protezione e la Ricerca Ambientale. Available online: http://www.isprambiente.gov.it/Media/carg/sicilia.html (accessed on 8 December 2019).
25. Istituto Nazionale di Geofisica e Vulcanologia INGV. Available online: https://emidius.mi.ingv.it/CPTI15-DBMI15/ (accessed on 8 December 2019).
26. Guerrieri, L.; Tatevossian, R.; Vittori, E.; Comerci, V.; Esposito, E.; Michetti, A.M.; Porfido, S.; Serva, L. Earthquake Environmental Effects (EEE) and Intensity Assessment: The Inqua Scale Project. *Boll. Soc. Geol. Ital.* **2007**, *126*, 375–386.
27. Meschis, M.; Roberts, G.P.; Mildon, Z.K.; Robertson JMichetti, A.M.; Faure Walker, J.P. Slip on a Mapped Normal Fault for the 28th December 1908 Messina Earthquake (Mw 7.1) in Italy. *Sci. Rep.* **2019**, *9*, 6481.
28. Castelli, F.; Cavallaro, A.; Grasso, S.; Ferraro, A. In Situ and Laboratory Tests for Site Response Analysis in the Ancient City of Noto (Italy). In Proceedings of the 1st IMEKO TC4 International Workshop on Metrology for Geotechnics, Benevento, Italy, 17–18 March 2016; pp. 85–90.
29. Castelli, F.; Cavallaro, A.; Grasso, S. SDMT Soil Testing for the Local Site Response Analysis. In Proceedings of the 1st IMEKO TC4 International Workshop on Metrology for Geotechnics, Benevento, Italy, 17–18 March 2016; pp. 143–148.
30. Cavallaro, A.; Maugeri, M.; Ragusa, A. In Situ Tests for the Geotechnical Characterization of Airship Hangar Soils in the City of Augusta. In Proceedings of the 2nd International Conference on Geotechnical Site Characterization, Porto, Portugal, 20–22 September 2004; pp. 1053–1059.
31. Cavallaro, A.; Capilleri, P.; Maugeri, M. Soil Characterisation of Catania Harbour by the Seismic Dilatometer Marchetti Test (SDMT). In Proceedings of the 3rd International Conference on the Flat Dilatometer, Roma, Italy, 15–17 June 2015.

32. Cavallaro, A.; Maugeri, M.; Lo Presti, D.C.F.; Pallara, O. Characterising Shear Modulus and Damping from in Situ and Laboratory Tests for the Seismic Area of Catania. In Proceedings of the 2nd International Symposium on Pre-Failure Deformation Characteristics of Geomaterials, Torino, Italy, 28–30 September 1999; pp. 51–58.
33. Lo Presti, D.C.F.; Jamiolkowski, M.; Cavallaro, A.; Pallara, O. Anisotropy of Small Strain Stiffness in Undisturbed and Reconstituted Clays. In Proceedings of the 2nd International Symposium on Pre-failure Deformation Characteristics of Geomaterials, Torino, Italy, 28–30 September 1999; pp. 3–10.
34. Cavallaro, A.; Maugeri, M.; Ragusa, A. Design Parameters of a Cohesive Soil from Laboratory Tests. In Proceedings of the A. W. Skempton Memorial Conference, London, UK, 29–31 March 2004; pp. 381–392.
35. Cavallaro, A.; Grasso, S.; Maugeri, M. Dynamic Clay Soils Behaviour by Different Laboratory and in Situ Tests. In Proceedings of the Geotechnical Symposium on Soil Stress-Strain Behavior: Measurement, Modelling and Analysis to Celebrate Prof. Tatsuoka's 60th Birthday, Rome, Italy, 16–17 March 2006; Volume 146, pp. 583–594.
36. Maugeri, M.; Cavallaro, A. Dynamic Characterisation of Soils from Laboratory Tests. In Proceedings of the 2nd International Conference on Earthquake Geotechnical Engineering, Lisbon, Portugal, 21–25 June 1999; pp. 863–868.
37. Castelli, F.; Cavallaro, A.; Ferraro, A.; Grasso, S.; Lentini, V.; Massimino, M.R. Static and Dynamic Properties of Soils in Catania City (Italy). *Ann. Geophys.* **2018**, *61*, 221.
38. Castelli, F.; Cavallaro, A.; Ferraro, A.; Grasso, S.; Lentini, V.; Massimino, M.R. Dynamic Characterisation of a Test Site in Messina (Italy). *Ann. Geophys.* **2018**, *61*, 222. [CrossRef]
39. Grasso, S.; Castelli, F.; Massimino, M.R.; Lentini, V. In Situ Measurements for Evaluating Liquefaction Potential under Cyclic Loading. In Proceedings of the 1st IMEKO TC-4 International Workshop on Metrology for Geotechnics, Benevento, Italy, 17–18 March 2016; pp. 79–84.
40. Castelli, F.; Lentini, V. Bearing capacity of shallow foundations on slopes: experimental analysis on reduced scale models. Poceedings of the 7th International Conference on Physical Modelling in Geotechnics, Zurich, Switzerland, 28 June–1 July 2010; Springman, L., Springman, S., Eds.; Taylor & Francis Group: London, UK; pp. 685–690, ISBN 978-0-415-59288-8.
41. Castelli, F.; Maugeri, M. Experimental Analysis of Waste Compressibility. In *GeoCongress 2008: Geotechnics of Waste Management and Remediation*; GeoCongress: New Orleans, LA, USA, GSP 177.
42. Capilleri, P.; Cavallaro, A.; Maugeri, M. Static and Dynamic Characterization of Soils at Roio Piano (AQ). *Ital. Geotech. J.* **2014**, *35*, 38–52.
43. Cavallaro, A.; Massimino, M.R.; Maugeri, M. Noto Cathedral: Soil and Foundation Investigation. *Constr. Build. Mater.* **2003**, *17*, 533–541. [CrossRef]
44. Cavallaro, A.; Cessari, L.; Gigliarelli, E. Site Characterization by in Situ and Laboratory Tests for the Structural & Architectural Restoration of Saint Nicholas Church, Nicosia, Cyprus. In Proceedings of the 2nd International Symposium on Geotechnical Engineering for the Preservation of Monuments and Historic Sites, Napoli, Italy, 30–31 May 2013; pp. 241–247.
45. Castelli, F.; Cavallaro, A.; Ferraro, A.; Grasso, S.; Lentini, V. A Seismic Geotechnical Hazard Study in the Ancient City of Noto (Italy). In Proceedings of the 6th Italian Conference of Researchers in Geotechnical Engineering (CNRIG), Bologna, Italy, 22–23 September 2016; Volume 158, pp. 535–540.
46. Cavallaro, A.; Grasso, S.; Maugeri, M. Volcanic Soil Characterisation and Site Response Analysis in the City of Catania. In Proceedings of the 8th National Conference on Earthquake Engineering, San Francisco, CA, USA, 18–22 April 2006; Volume 1290, pp. 835–844.
47. Cavallaro, A.; Castelli, F.; Ferraro, A.; Grasso, S.; Lentini, V. Site Response Analysis for the Seismic Improvement of a Historical and Monumental Building: The Case Study of Augusta Hangar. *Bull. Eng. Geol. Environ* **2018**, *77*, 1217–1248. [CrossRef]
48. Silvestri, F. Looking for objective Criteria in the Interpretation of Laboratory Stress-Strain Tests. In Proceedings of the 2nd International Symposium on Pre-failure Deformation Characteristics of Geomaterials, Torino, Italy, 28–30 September 1999; pp. 1305–1316.
49. Hall, J.R., Jr.; Richart, F.E., Jr. Dissipation of Elastic Wave Energy in Granular Soils. *J. Soil Mech. Found. Div.* **1936**, *89*, 27–56.
50. Drnevich, V.P.; Hardin, B.O.; Shippy, D.J. Modulus and Damping of Soils by Resonant Column Method. In *Dynamic Geotechnical Testing*; ASTM STP: West Conshohocken, PA, USA, 1978; Volume 654, pp. 91–125.

51. Stokoe, K.H., II; Hoar, R.J. Variable Affecting in Situ Seismic Measurement. Proceeding of the Conference on Earthquake Engineering and Soil Dynamics, Pasadena, CA, USA, 19–21 1978; ASCE: Pasadena, CA, USA, 1978; Volume 2, pp. 919–939.
52. Lo Presti, D.C.F. General Report: Measurement of Shear Deformation of Geomaterials in the Laboratory. In Proceedings of the Pre-failure Deformation of Geomaterials, Sapporo, Japan, 12–14 September 1994; pp. 1067–1088.
53. Shibuya, S.; Mitachi, T.; Fukuda, F.; Degoshi, T. Strain Rate Effect on Shear Modulus and Damping of Normally Consolidated Clay. *Geotech. Test. J.* **1995**, *18*, 365–375.
54. Tatsuoka, F.; Lo Presti, D.C.F.; Kohata, Y. Deformation Characteristics of Soils and Soft Rocks Under Monotonic and Cyclic Loads and Their Relations. In Proceedings of the 3rd International Conference on Recent Advances in Geotechnical Earthquake Engineering and Soil Dynamic, State of the Art 1, St. Louis, Missouri, 2–7 April 1995; Volume 2, pp. 851–879.
55. Yokota, K.; Imai, T.; Konno, M. Dynamic Deformation Characteristics of Soils Determined by Laboratory Tests. *Oyo Tec. Rep.* **1981**, *3*, 13–37.
56. Cavallaro, A.; Grasso, S.; Ferraro, A. A Geotechnical Engineering Study for the Safeguard, Restoration and Strengthening of Historical Heritage. In Proceedings of the 6th Italian Conference of Researchers in Geotechnical Engineering (CNRIG), Bologna, Italy, 22–23 September 2016; Volume 158, pp. 134–139.
57. Jamiolkowski, M.; Lo Presti, D.C.F.; Pallara, O. Role of In-Situ Testing in Geotechnical earthquake Engineering. In Proceedings of the 3rd International Conference on Recent Advances in Geotechnical Earthquake Engineering and Soil Dynamic, State of the Art 7, St. Louis, MO, USA, 2–7 April 1995; Volume II, pp. 1523–1546.
58. Marchetti, S.; Monaco, P.; Totani, G.; Marchetti, D. In Situ Tests by Seismic Dilatometer (SDMT). In *From Research to Practice in Geotechnical Engineering*; American Society of Civil Engineers: Reston, VA, USA, 2008; GSP 180. [CrossRef]
59. Cavallaro, A.; Ferraro, A.; Grasso, S.; Maugeri, M. Topographic Effects of the Monte Po Hill in Catania. *Soil Dyn. Earthq. Eng.* **2012**, *43*, 97–113. [CrossRef]
60. Cavallaro, A.; Grasso, S.; Maugeri, M.; Motta, E. An Innovative Low-Cost SDMT Marine Investigation for the Evaluation of the Liquefaction Potential in the Genova Harbour (Italy). In Proceedings of the 4th International Conference on Geotechnical and Geophysical Site Characterization (ISC'4), Porto de Galinhas, Brazil, 18–21 September 2012; Volume 1, pp. 415–422.
61. Cavallaro, A.; Grasso, S.; Maugeri, M.; Motta, E. Site Characterisation by in Situ and Laboratory Tests of the Sea Bed in the Genova Harbour (Italy). In Proceedings of the 4th International Conference on Geotechnical and Geophysical Site Characterization (ISC'4), Porto de Galinhas, Brazilian, 18–21 September 2012; Volume 1, pp. 637–644.
62. Hepton, P. Shear Wave Velocity Measurements during Penetration Testing. In Proceedings of the Geotechnology Conference Organized by the Institution of Civil Engineers, Birmingham, UK, 6–8 July 1988; pp. 275–278.
63. Martin, G.K.; Mayne, P.W. Seismic Flat Dilatometers Tests in Connecticut Valley Vaeved Clay. *Astm Geotech. Test. J.* **1997**, *20*, 357–361.
64. Martin, G.K.; Mayne, P.W. Seismic Flat Dilatometers Tests in Piedmont Residual Soils. *Geotech. Site Charact.* **1998**, *2*, 837–843.
65. Mayne, P.W.; Schneider, J.A.; Martin, G.K. Small and Large Strain Soil Properties from Seismic Flat Dilatometer Tests. In Proceedings of the 2nd International Symposium on Pre-Failure Deformation Characteristics of Geomaterials, Torino, Italy, 28–30 September 1999; Volume 1, pp. 419–427.
66. Marchetti, S. In Situ Tests by Flat Dilatometer. *J. Geotech. Eng. Div.* **1980**, *106*, 299–321.
67. Cavallaro, A.; Grasso, S.; Maugeri, M. Dynamic Geotechnical Characterization of Sangiuliano di Puglia Seismic Area. In Proceedings of the 6th International Conference on Case Histories in Geotechnical Engineering, Arlington, WV, USA, 11–16 August 2008.
68. Cavallaro, A.; Ferraro, A.; Grasso, S.; Maugeri, M. Site Response Analysis of the Monte Po Hill in the City of Catania. In Proceedings of the 2008 Seismic Engineering International Conference Commemorating the 1908 Messina and Reggio Calabria Earthquake MERCEA'08, Reggio Calabria, Italy, 8–11 July 2008; AIP Conference Proceedings. American Institute of Physics: College Park, MD, USA, 2008; Volume 1020, pp. 240–594.

69. Cavallaro, A.; Grasso, S.; Ferraro, A. Study on Seismic Response Analysis in "Vincenzo Bellini" Garden Area by Seismic Dilatometer Marchetti Tests. In Proceedings of the 5th International Conference on Geotechnical and Geophysical Site Characterization (ISC'5), Queensland, Australian, 5–9 September 2016; Volume 2, pp. 1309–1314.
70. Castelli, F.; Cavallaro, A.; Grasso, S.; Lentini, V. Seismic Microzoning from Synthetic Ground Motion Earthquake Scenarios Parameters: The Case Study of the City of Catania (Italy). *Soil Dyn. Earthq. Eng.* **2016**, *88*, 307–327. [CrossRef]
71. Cavallaro, A.; Capilleri, P.; Grasso, S. Site Characterization by in Situ and Laboratory Tests for Liquefaction Potential Evaluation during Emilia Romagna Earthquake. *Geosciences* **2018**, *8*, 242. [CrossRef]
72. Hryciw, R.D. Small Strain Shear Modulus of Soil by Dilatometer. *JGED* **1990**, *116*, 1700–1715.
73. Shibuya, S.; Tanaka, H. Estimate of Elastic Shear Modulus in Holocene Soil Deposits. *Soils Found.* **1996**, *36*, 45–55. [CrossRef]
74. Grasso, S.; Di Benedetto, C.; Ferraro, A.; Massimino, M.R.; Cavallaro, A. A comparative study on seismic response analysis of soils using different numerical codes. In Proceedings of the XVII ECSMGE-2019. Geotechnical Engineering Foundation of the Future, Reykjavík, Iceland, 1–6 September 2019; ISBN 978-9935-9436-1-3.

geosciences

Article

Machine Learning Methods for Seismic Hazards Forecast

Valeri G. Gitis and Alexander B. Derendyaev *

The Institute for Information Transmission Problems, Moscow 127051, Russia
* Correspondence: wintsa@gmail.com

Received: 25 April 2019; Accepted: 10 July 2019; Published: 12 July 2019

Abstract: In this paper, we suggest two machine learning methods for seismic hazard forecast. The first method is used for spatial forecasting of maximum possible earthquake magnitudes (M_{max}), whereas the second is used for spatio-temporal forecasting of strong earthquakes. The first method, the method of approximation of interval expert estimates, is based on a regression approach in which values of M_{max} at the points of the training sample are estimated by experts. The method allows one to formalize the knowledge of experts, to find the dependence of M_{max} on the properties of the geological environment, and to construct a map of the spatial forecast. The second method, the method of minimum area of alarm, uses retrospective data to identify the alarm area in which the epicenters of strong (target) earthquakes are expected at a certain time interval. This method is the basis of an automatic web-based platform that systematically forecasts target earthquakes. The results of testing the approach to earthquake prediction in the Mediterranean and Californian regions are presented. For the tests, well known parameters of earthquake catalogs were used. The method showed a satisfactory forecast quality.

Keywords: machine learning; expert estimate; maximum possible magnitudes of earthquakes; one class classification; seismic hazard; seismic zoning; earthquake forecasting

---

1. Introduction

Tectonic earthquakes are invariably preceded by a period when stresses increase in the Earth. This process forms anomalous changes in the geological environment near the source of the expected earthquake [1–3]. To describe the seismotectonic properties of the geological environment, various types of data are used: Earthquake catalogs, time series of geodetic [4], geophysical [5] and geochemical measurements [6], and aerospace observations [7]. The success of seismic hazard forecast is largely influenced by both completeness of the description of the spatial and spatio-temporal properties of the seismic process, and the possibility of their joint analysis. In our approach to joint analysis, all available data on the properties of the process are converted into grid fields [8].

Seismic zoning is prerequisite for seismic hazard assessment [9]. The most important and complex problem of seismic zoning is to map the maximum possible magnitudes of earthquakes (M_{max}). The values of M_{max} cannot be measured instrumentally. Two assumptions are used to construct a digital map of M_{max}: (1) The assumption of large earthquake repetition [10] and (2) the assumption that the values of M_{max} depend on the properties of the geological environment [11,12].

The statistical approach uses only the first assumption. This means that the M_{max} map is calculated using only those earthquakes whose epicenters fall into a sliding spatial window. The methods of extreme statistics are used for the estimation of M_{max} [13–16]. These methods require a sufficiently large number of observations, which may be unavailable for some zones in the region. To improve this method, [17] used the second assumption, and estimated M_{max} by earthquake epicenters within geologically homogeneous zones identified by an expert geologist. As in the previous approach, this method does not provide for extrapolating M_{max} values from zones in which there are many epicenters of strong earthquakes and therefore estimates of M_{max} are fairly accurate, to zones with similar seismotectonic properties, but with a small number of strong earthquakes.

The history of seismic observations is very short in relation to the speed of tectonic processes, and earthquakes with magnitudes close to maximum occur relatively rarely. To compensate for this effect, attempts are made to extrapolate reliable estimates to areas with similar seismotectonic properties of the geological environment. Reference [18] used the M_{max} mapping method based on a solution of a group of experts. One of the possible algorithmic approaches applies cluster analysis [19]. A cluster analysis program divides the region into zones with similar values of geological and geophysical characteristics. These zones may consist of several isolated areas. Next, the maximum magnitudes of earthquakes recorded in one area of the zone are extrapolated to all other areas. The disadvantages of this approach are related to the fact that the zoning of a region into quasi-homogeneous zones is largely determined a set of selected features, the method of measuring the similarity between clusters, the type of clustering algorithm and, finally, the criterion of stopping the clustering process.

We describe the method of approximation of interval expert estimates which is a regression approach to the construction of a forecast map of M_{max} [20–22]. To compile the map, dependence of M_{max} on properties of geological environments $\mathbf{x} = (x_1, \ldots, x_I)$ is used. The values at the training sample points are determined using expert knowledge. To this end, experts choose the most studied points of the region with different seismicity and geological conditions. The expert indicates the boundaries of the interval in which, in his opinion, lies the value of M_{max}, and evaluates the values of the confidence that M_{max} cannot exceed the lower and upper limits of the interval. In the assessment, the expert uses historical seismic data, instrumental data on the maximum magnitude of an earthquake in the vicinity of the point in question and data on the properties of the geological zone to which this point belongs. The algorithm generalizes the least squares approximation algorithm.

The task of predicting an earthquake is to determine the time, location, and magnitude of a future earthquake. Earthquake prediction studies are conducted in many directions. They include the study of the rock failure and earthquake precursor phenomena, the study of stochastic models for earthquake prediction, machine learning methods, and testing earthquake prediction algorithms [1–3,23–29]. At the same time, there are a number of works in which it is stated that earthquakes cannot be predicted [30].

Here we suggest a new method of machine learning, called the method of the minimum area of alarm, and describe a web-based platform that predicts earthquakes in automatic mode (http://distcomp.ru/geo/prognosis/). Our method solves the one-class classification problem (other methods can be found, for instance, in [31–33]). Our training sample set includes rare anomalous objects (the epicenters of target earthquakes) and grid fields of properties of the seismic process (field of features). The method allows one to detect the largest number of the target earthquakes for the training set, provided that the size of the spatio-temporal alarm area does not exceed a specified value. We present the results of testing the approach on the data of the Mediterranean and California regions.

2. The Method of Approximation of Interval Expert Estimates

Let the seismotectonic properties of the region under study be represented by a set of spatial grid fields of features $\mathbf{X}_1, \mathbf{X}_2, \ldots, \mathbf{X}_I$, and the values of the maximum possible magnitudes of earthquakes (M_{max}) be represented by a sample set of expert estimates. The task is to find from these data the function $F(\mathbf{x})$, which approximates the values of M_{max} at the sample set, where $\mathbf{x} = (x_1, \ldots, x_I)$ is the vector with the values of the fields of features. The M_{max} map is the $F(\mathbf{x})$ values calculated for all grid nodes of a region.

The type of expert evaluation should be convenient and straightforward for unambiguous understanding by all participants of the expert survey and should enable the expert to formalize his knowledge about the value of the forecast fully. These requirements correspond to interval expert estimates:

$$Q_{qn} = (m_{qn}^{(1)}, m_{qn}^{(2)}, w_{qn}^{(1)}, w_{qn}^{(2)}), \tag{1}$$

where $m_{qn}^{(1)}, m_{qn}^{(2)}$ are the interval boundaries within which all the values of M_{max} at the point n are the most probable and equally possible, in the opinion of the q-th expert, $m_{qn}^{(1)} \leq m_{qn}^{(2)}$; $w_{qn}^{(1)} > 0, w_{qn}^{(2)} > 0$ are the weighs on which the q-th expert indicates the degree of his confidence in the possibility that the value of M_{max} may be less or greater than the corresponding interval boundary $m_{qn}^{(1)}$ or $m_{qn}^{(2)}$.

We can assume that the expert estimate Q corresponds to some function of the subjective probability density $f(Y, Q)$, which reflects the expert's opinion about the value of Y at a given sample point. This function takes a constant value within the interval $[m^{(1)}, m^{(2)}]$ and decreases with the weights $w^{(1)}$ and $w^{(2)}$ respectively to the left and right of the interval boundaries:

$$f(Y, Q) = C \cdot \exp\{-(w^{(1)} \frac{|m^{(1)} - Y| + m^{(1)} - Y}{2} + w^{(2)} \frac{|m^{(2)} - Y| - m^{(2)} + Y}{2})^{p}\}, \tag{2}$$

where $p \geq 1$, and C is defined by the condition $\int_{-\infty}^{\infty} f(Y, Q)dy = 1$.

Suppose that there is a training sample $\{Q_{qn}, \mathbf{x}_n\}$, where q and n represent the expert and sample number. It is required to approximate the function $Y(x)$ in a certain class of functions $F(\mathbf{x}, \boldsymbol{\theta}) : \boldsymbol{\theta} \in \boldsymbol{\Theta}$, where $\boldsymbol{\Theta}$ is the domain of admissible values of the vector $\boldsymbol{\theta}$.

Let's replace Y in (2) with the value of the forecast function $F(\mathbf{x}, \boldsymbol{\theta})$ and consider the function

$$r(\mathbf{x}, \boldsymbol{\theta}) = -\ln f(Y, Q) + \ln C = (w^{(1)} \frac{|m^{(1)} - Y| + m^{(1)} - Y}{2} + w^{(2)} \frac{|m^{(2)} - Y| - m^{(2)} + Y}{2})^{p}. \tag{3}$$

The function $r(\mathbf{x}, \boldsymbol{\theta})$ determines the penalty for the inaccuracy of the approximation of the expert judgment Q by the value F of the forecast function. To estimate $\boldsymbol{\theta}$, the average penalty on the set is minimized. The estimation has the form

$$\hat{\boldsymbol{\theta}} = \arg\min_{\boldsymbol{\theta} \in \boldsymbol{\Theta}} \sum_{n} \sum_{q} r(F(\mathbf{x}_n, \boldsymbol{\theta}), Q_{qn}). \tag{4}$$

It is obvious that if the forecast function $F(\mathbf{x}, \boldsymbol{\theta})$ is linear in the parameters, then the functional (4) is convex. If the domain $\boldsymbol{\Theta}$ of admissible values of the vector is also convex, then it is possible to use iterative gradient algorithms for estimation.

It is easy to see from (3) and (4) that in case of $m^{(1)} = m^{(2)}$ and $w^{(1)} = w^{(2)}$ for all expert estimates the estimation algorithm (4) coincides with the method of the least absolute errors for $p = 1$, and with the method of least squares for $p = 2$. It was shown in [22] that under certain assumptions, the estimate (4) is an estimate of the maximum likelihood.

The method of approximation of interval expert estimates was repeatedly used to construct the maps of M_{max} in a number of regions, in particular, Bulgaria [34], Caribbean and Middle America Region [35], Central Europe [12,36], Costa Rica [37], the Caucasus [21], and North Caucasus [38]. In these papers, the dependences $M_{max}(\mathbf{x})$ were always estimated in a class of the sum of piecewise linear functions of geological and geophysical features. This estimation allows one to interpret the M_{max} map as the sum of non-linearly transformed fields of features.

For each of the above regions, from 10 to 200 geological and geophysical fields were analyzed. 3–4 of the most informative fields were selected from this set using the stepwise regression method. Prediction functions are the sum of piecewise linear dependencies on the values of these fields. The sum of nonlinearly transformed fields defines the M_{max} field. This is convenient for the seismotectonic interpretation of the M_{max} map.

Interpretation of the M_{max} map by a specialist allows a qualitative assessment of the accuracy of determining the maximum possible seismic hazard but is not an assessment of its accuracy. By definition, M_{max} values cannot be measured instrumentally. Statistical estimates of M_{max} estimates are possible only in areas with sufficiently high seismic activity. In the considered method, the values of M_{max} are replaced by interval expert estimates. These estimates are approximated by a nonlinear function of geological and geophysical fields. The accuracy of the M_{max} forecast is determined by the deviations of the M_{max} forecast map values from expert estimates. For the above regions, from 100 to 400 expert evaluations were used. The number of parameters estimated during training ranged from eight to 14 in each region. For those regions, who did not participate in the training, the average approximation errors of expert estimates are in the range from 0.2 to 0.34.

3. Method of the Minimum Area of Alarm

Let the properties of the seismic process are described by the spatial and spatio-temporal fields of features in a single coordinate grid with a step $\Delta x \times \Delta y \times \Delta t$. The values of these fields at the nodes of the grid $n = 1, \ldots, N$ correspond to the vectors of the I-dimensional feature space $\mathbf{f}^{(n)} = \{f_i^{(n)}\}$. A spatio-temporal forecast field $\mathbf{\Phi}$ is a function of the fields of features. It is trained using retrospective data: (1) A sample set of target earthquakes $q = 1, \ldots, Q$ with the magnitudes $M \geq M^*$ and (2) a set of grid fields of features $\boldsymbol{F}_i, i = 1, \ldots, I$, which describes spatial (quasi-stationary) and spatio-temporal (dynamic) properties of the seismotectonic process.

The method of the minimum area of alarm uses the following data model.

1. The epicenters of earthquakes with magnitudes $M \geq M^*$ (target events) are preceded by the anomalous (low-probability) values of the fields of features. Let's consider the fields of features to be designed in such a way that for each anomaly, the values of some of these fields are close to their maximum or minimum. To simplify the explanation, we assume that the anomalies refer only to the largest values of the fields of features.
2. If the $\boldsymbol{f}^{(q)}$ is an anomaly vector, preceding the target event q, then any vector $\boldsymbol{f}$ with the components $f_i \geq f_i^{(q)}$ for all $i = 1, \ldots, I$ can also precede a similar target event (monotonicity condition).

We will call the base vectors of the feature space the vectors for which $\boldsymbol{f} \geq \boldsymbol{f}^{(q)}$ componentwise. The nodes of the grid of the forecast field $\mathbf{\Phi}$ with the values $\phi \geq \phi^{(q)}$ we will call the base nodes of the forecast field.

From the assumption that anomalous refers only to the largest values of the fields of features and the monotonicity condition, it follows that the earthquake forecast can be carried out using the simplest threshold decision rule. If the value of the forecast field $\phi^{(n)\geq\theta}$, then spatio-temporal alarm cylinders are created at all base nodes of the forecast field with the values $\phi \geq \phi^{(q)}$. The alarm cylinder of the grid node n with the coordinates $(x^{(n)}, y^{(n)}, t^{(n)})$ has the center of the base in the node $(x^{(n)}, y^{(n)}, t^{(n)})$, the base radius R, and the element $[(x^{(n)}, y^{(n)}, t^{(n)}), (x^{(n)}, y^{(n)}, t^{(n)})]$. From this, it follows that for a given value of the threshold θ an earthquake with the epicenter coordinates and time (x^*, y^*, t^*) will be detected if and only if the cylinder with the center of the base (x^*, y^*, t^*), the radius R, and the element $[(x^*, y^*, t^* - T), (x^*, y^*, t^*)]$ contains at least one grid node with the value $\phi^{(n)} \geq \theta$. This cylinder will be called a precursor cylinder.

The alarm field detects an earthquake if its epicenter falls within an area consisting of a combination of alarm cylinders (alarm area). The quality of the forecast field at threshold θ is determined by two indicators: (1) The fraction of correctly detected events $Q^*(\theta)$ from all Q events $U(\theta) = Q^*(\theta)/Q$ (probability of detection) and (2) the fraction of number of grid nodes, falling in the alarm area $L^*(\theta)$, from the number of all grid nodes L of the analyzed area $V(\theta) = L^*(\theta)/L$ (alarm volume).

For training, we have a set of target events with magnitudes $M \geq M^*$ and a set of fields. At the first step, the algorithm should move from a set of target earthquakes to a set of target earthquake precursors. A precursor of the earthquake q is the vector $\boldsymbol{f}^{(q)}$ of a feature space which has a minimum volume of alarm $v^{(q)} = L^{(q)}/L$ among all vectors corresponding to the grid nodes of the precursor cylinder of the event q, where L is the number of all grid nodes of the analyzed area, $L^{(q)}$ is the number of nodes in the grid of the alarm area generated by the base points of the vector $\boldsymbol{f}^{(q)}$.

The algorithm for constructing the forecast field is nonparametric. There are the three most important versions of the algorithm. The first version of the algorithm is to construct the forecast field so that when the threshold θ decreases, the training earthquakes are detected in the sequence in which the corresponding alarm volumes increase $v^{(Q)} \leq v^{(Q-1)} \leq \ldots \leq v^{(2)} \leq v^{(1)}$ (this version is selected for testing). The version consists of the following steps.

1. To generate a training set $\{\boldsymbol{f}^{(q)}, v^{(q)}\}$, which consists of earthquake precursors $\boldsymbol{f}^{(q)}$ and corresponding alarm volumes $v^{(q)}$.
2. To sort the precursors $\boldsymbol{f}^{(q)}$, $q = 1, \ldots, Q$, by the alarm volume $v^{(Q)} \leq v^{(Q-1)} \leq \ldots \leq v^{(1)}$ in ascending order.
3. To assign to the nodes of the grid of the forecast field $\boldsymbol{\Phi}$ a value of 0.
4. To replace the value of 0 by Q at the nodes of the grid of the forecast field, for which the monotonicity condition, $f_i^{(n)} \geq f_i^{(Q)}$ for all $i = 1, \ldots, I$, is satisfied in the feature space; to replace the value of 0 by $Q - 1$ at the nodes of the grid of the forecast field, for which the monotonicity condition, $f_i^{(n)} \geq f_i^{(Q-1)}$ for all $i = 1, \ldots, I$ is satisfied in feature space, and then, successively, in the same way, to replace the values 0 by $Q + 1 - q$.

Obviously, the choice of the order of the earthquake precursors at the 2nd step of the algorithm determines the dependence $U(V)$ obtained from the forecast field. The 2nd version of the algorithm makes it possible to optimize the forecast field so that when the next target earthquake is detected, the alarm volume increases by a minimum value. To do this, one should arrange the precursors so that, when changing from event detection $q + 1$ to event q, the increase in alarm volume is minimal. Here, at each transition from the previously selected event $q + 1$ to q, a small search through the remaining q events is required. The 3rd version of the algorithm allows one to optimize the forecast field so that it detects the maximum number of target earthquakes with a total alarm volume of less than or equal to the predetermined value. In this case, you need to perform a full search on the selected number of events. The 3rd version of the algorithm allows optimizing the forecast field so that it detects the maximum number of target earthquakes, provided that the total alarm volume does not exceed the specified value. In this case, you need to perform a full search for the selected number of events.

4. Testing

The purpose of testing is to verify the proposed method of the forecast. Testing is carried out in accordance with the known characteristics of the catalog of earthquakes. Exploring the possibility to improve the quality of the forecast using a wider set of characteristics of earthquake catalogs or by adding other sources of input data is beyond the scope of this work.

The method of minimum area of alarm was tested on the platform of automatic earthquake forecast (http://distcomp.ru/geo/prognosis). The system tests the data with a constant step Δt. On each step (at time t) the raster fields of features are computed, the alarm area is trained based on data before the time t, and the system tests for time since t till $t + \Delta t$ if the alarm area covers an epicenter of the target earthquake. Then at time $t + \Delta t$, the training time is increased by Δt, the alarm zone is updated and the test is repeated.

Testing of the forecast method should provide an opportunity to compare different methods of solving the problem on the same indicators of the forecast quality. In this method, we use two quality indicators: The probability of detecting the target events from the test interval $U = Q^*/Q$ and the volume of alarm $V = L^*/L$. The number of target events Q is determined by a set of test samples, the number of target events detected Q^* is determined by the results of the forecast, the analysis area and its size L is selected at the beginning of the test, the size of the alarm zone L^* is determined by the training data.

In the following test experiments, the area of analysis was constructed in the following way: Any point is included in the area if in a circle around this point with radius $R = 100$ km for the period 1984–1993 there are more than 300 earthquake epicenters. This condition allows one to select a seismically active area for analysis but does not ensure its seismic homogeneity. Therefore, the indicator of the volume of alarm obtained during testing should be considered only in the context of the selected area of analysis. At the same time, the choice of the field of analysis according to a formal rule makes it possible to compare the results of the forecast obtained using various methods and according to different data.

One way to assess the quality of a forecast is to compare a regular forecast obtained by the algorithm under analysis (regular forecast) with a random one. We will assume that the forecast is random if the values of the forecast field are selected from a segment in accordance with a uniform distribution. Obviously, for this probabilistic model, the alarm volume V_r is equal to the probability of a random prediction U_r. It follows from this that comparing the probability of a regular forecast U with the probability of a random forecast U_r for the same alarm volumes $V = V_r$ is equivalent to comparing U with the corresponding alarm volume V. If, at the same time, a sample of target events were cleaned of aftershocks and foreshocks, then by proposing the independence of target events and using the binomial distribution model for them, we could build a confidence interval for estimating U.

In the number of articles, the results of a regular forecast are compared with the results of a forecast by a stationary field. In papers [39–41] the regular forecast is compared with the forecast by the 2D field of seismic activity (or earthquake epicenter density). The result of the comparison makes it possible to evaluate the efficiency of a regular forecast in relation to the forecast by the field $\boldsymbol{F}$, which is based only on the spatial heterogeneity of the seismic process. Comparison of results can be done in two ways. In one method the probabilities of regular prediction of Q target earthquakes are compared with the results of predicting the same earthquakes by a stationary field $\boldsymbol{F}$ (for example, $\boldsymbol{F}$ is a 2D field of the earthquake epicenter density). Another method uses the Gutenberg–Richter model [42]. In the beginning, the catalog of earthquakes with the following conditions is constructed: (1) The epicenters of earthquakes are in the area of analysis, (2) the magnitudes exceed the representative, and (3) the depth of the epicenters does not exceed the values specified for the target earthquake. It is assumed that b-value is the same for the entire area of analysis and the earthquake catalog agree well with the spatial distribution of seismicity. The alarm field $V(\theta)$ is calculated by the field $\boldsymbol{F}$. Then, in accordance with the alarm field, the dependence $N^*(V)$, is calculated, where N^* is the number of epicenters in the alarm zone, and V is the alarm volume. The dependence $N^*(V)$ is normalized to the number of all N epicenters in the analyzed area. According to the Gutenberg–Richter law and the assumption b =const, we have $N^*(V) = C\exp(d^* - bm)$, $N(V) = C\exp(d - bm)$, and $\mu(V) = N^*/N = \exp(d^* - d)$. Thus, the value $\mu(V)$ does not depend on the magnitude of earthquakes. It shows the proportion of earthquakes with a magnitude higher than a given, which fall into the alarm zone. Consequently, the value of $\mu(V)$ in the scope of our model is equal to the probability of forecasting the target earthquakes using the stationary field $\boldsymbol{F}$. If the field $\boldsymbol{F}$ is the density field of the earthquake epicenters, then the field obtained according to the Gutenberg–Richter law is denoted by the letter $\boldsymbol{\mu}$.

Testing was performed for two regions: The Mediterranean and California. The Mediterranean region: 10°–30° E, 34°–47° N. Input data: Earthquakes for the period from 27.05.1983 till 14.02.2018 with magnitudes $M \geq 2.7$ and depths of hypocenters $H \leq 160$ km from the International Seismological Centre catalog (see Materials and Methods). Target earthquakes: Magnitudes $M \geq 6.0$ and hypocenter depths $H \leq 60$ km. California region: 126°–114° W, 32°–43° N. Input data: Earthquakes for the period of 01.01.1983–15.02.2018 with magnitudes $M \geq 2.0$ and depths of hypocenters $H \leq 160$ km from the NEIC USGS catalog (see Materials and Methods). For the forecast, the target earthquakes with magnitudes $M \geq 5.7$ have been selected.

The following six fields of features were analyzed for forecasting:

- F_1 is the 3D field of the density of all considering earthquakes in the region.
- F_2 is the 3D field of mean magnitudes among all considering earthquakes in the region.

The estimation of 3D fields of F_1 and F_2 is performed with the method of local kernel regression. The kernel function for the n-th earthquake has the form $K_n = [\cosh^2(r_n/R)^2 \cosh^2(t_n/T)]^{-1}$, where $r_n < R\epsilon$, $t_n < T\epsilon$ are the distance and time interval between the n-th epicenter of the earthquake and the node of the 3D grid of the field, $\epsilon = 2$, $R = 50$ km, $T = 100$ days for F_1 and $R = 100$ km, $T = 730$ days for F_2.

- F_3 is the 3D field of negative temporal anomalies of the density of earthquakes.
- F_4 is the 3D field of positive temporal anomalies of the density of earthquakes.
- F_5 is the 3D field of positive temporal anomalies of mean earthquake magnitude.

  To estimate the field of F_3, F_4, F_5, the Student's t-statistic was used, which is defined for each grid node as the ratio of the difference of average values of the current (196 days) and background (3650 days) intervals to the standard deviation of this difference. Positive t-statistics values correspond to higher values on the test interval.
- F_6 is the 2D field of the density of earthquake epicenters: Kernel smoothing in the interval 1988–2008 the parameter $R = 50$ km.

The grid fields for the Mediterranean were calculated in a grid step $\Delta x \times \Delta y \times \Delta t = 0.2^\circ \times 0.13^\circ \times 49$ days. The forecast field was trained from 1998 until the next step of the forecast after 2008. The radius of the alarm cylinder is $R = 20$ km, and the element is $T = 50$ days. Testing is performed in 2008–2019. There are 11 target earthquakes in the analysis area. We used the method of stepwise selection to find the most informative fields of features. The algorithm selected the F_3 and F_6 fields to construct the alarm field.

We compare the earthquake prediction probabilities obtained using different fields of features in Table 1: U_1 is the forecast probability using the earthquake density field 2D (F_6), U_2 is the probability using the 3D field of negative earthquake density anomalies (F_3), $\mu(V)$ is the probability of forecast by 2D field of earthquake epicenters density, obtained using the Gutenberg–Richter model, and U_3 is the probability using F_3 and F_6 fields. We can see that the highest probability of a successful forecast occurs when the fields F_3 and F_6 are used together. When $V = 0.2$ ($U_r = 0.2$), the ratios for the prediction probability obtained with F_3 and F_6 fields to the prediction probabilities obtained with 2D earthquake density field (F_6) and for the field calculated using the Gutenberg–Richter ratio, are equal respectively $U_3(0.2)/U_1(0.2) = 0.91/0.64 = 1.49$ and $U_3(0.2)/\mu(0.2) = 0.91/0.41 = 2.2$. Table 1 shows the values of two types of alarm volumes: V_{learn} is the alarm volume received in accordance with the training data, and V_{test} is the alarm volume corresponding to the alarm volume V_{learn} but observed on the test data. You can see that when testing in almost all cases, except for testing the 2D-field F_6, the volumes of V_{test} are greater than V_{learn}. This is explained by the fact that the number of recorded earthquakes in a region changes over time (Figure 1). The number of earthquakes is influenced by the development of a seismic network and natural changes in the seismic process. The Figure 1 shows that the number of earthquakes increases significantly in the test interval. The same anomaly appears on the plot of the time series of average values of the density of earthquake epicenters throughout the analysis area (Figure 2). An increase in the density of earthquake epicenters leads to an increase in the field values of the F_3 function, which ultimately leads to an increase in the volume of anxiety during testing.

The grid fields for California were calculated in a grid step $\Delta x \times \Delta y \times \Delta t = 0.125^\circ \times 0.11^\circ \times 49$ days. The radius of the alarm cylinder is $R = 14$ km, and the element is $T = 100$ days. Testing of the earthquake forecast was performed for the interval 2009–2018. There were nine target earthquakes. The algorithm selected tree fields of features for the construction of the alarm field: F_4, F_5, and F_6.

Table 2 shows the probabilities of earthquake forecast for California.

Figures 3 and 4 show the test results for both regions. They depicted polygons selected as the area of analysis, and circles are the target epicenters of earthquakes in 2009–2018 with $M \geq 6.0$ for Mediterranean and $M \geq 5.7$ for California.

Table 1. Comparison of the probabilities of earthquake forecast for the Mediterranean.

V_{learn}: **Alarm Volumes for Learning Interval**	**0.01**		**0.05**		**0.1**		**0.15**		**0.2**	
Test indicators	V_{test}	U_{test}	V_{test}	U_{test}	V_{test}	U_{test}	V_{test}	U_{test}	V_{test}	U_{test}
Field F_6	0.00	0.00	0.03	0.09	0.10	0.09	0.15	0.45	0.20	0.64
Field F_3	0.00	0.00	0.02	0.09	0.20	0.36	0.29	0.45	0.35	0.64
$\mu(V)$: probability for the field F_6 obtained by the model	-	0.00	-	0.1	-	0.26	-	0.35	-	0.41
Fields F_3 and F_6	0.03	0.00	0.08	0.36	0.15	0.45	0.24	0.55	0.32	0.91

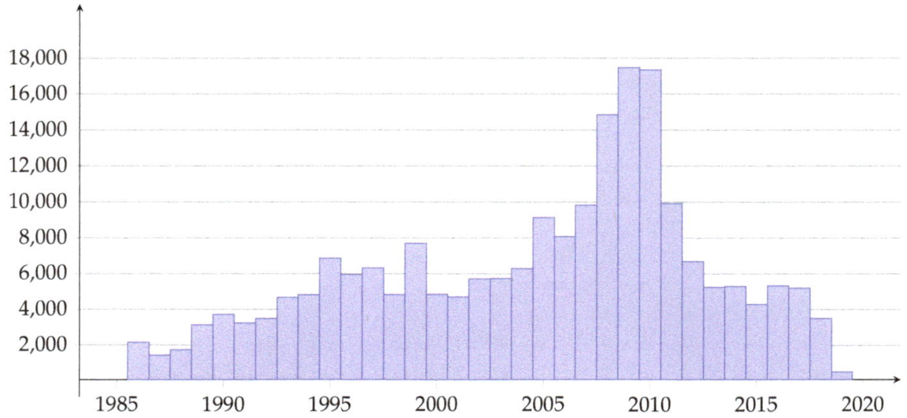

Figure 1. Histogram of the number of earthquakes with a magnitude greater than 2.7 and a depth of hypocenters less than 160 km.

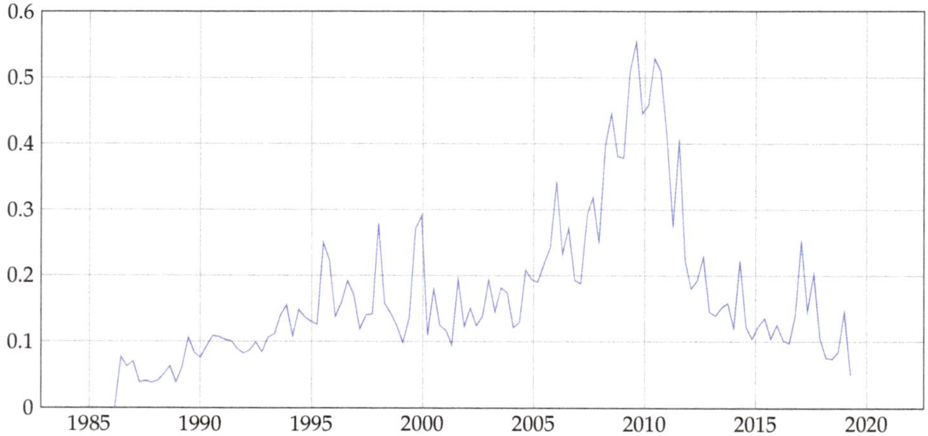

Figure 2. Time series of average values of the density of earthquake epicenters throughout the analysis area (trend line).

Table 2. Probabilities of earthquake forecast for California.

Volume of Alarm V_{learn}	Volume of Alarm V_{test}	Number of Correct Forecasts	Forecast Probability U
0.01	0.01	1	0.11
0.05	0.06	4	0.44
0.1	0.13	4	0.44
0.15	0.13	4	0.44
0.2	0.25	8	0.89

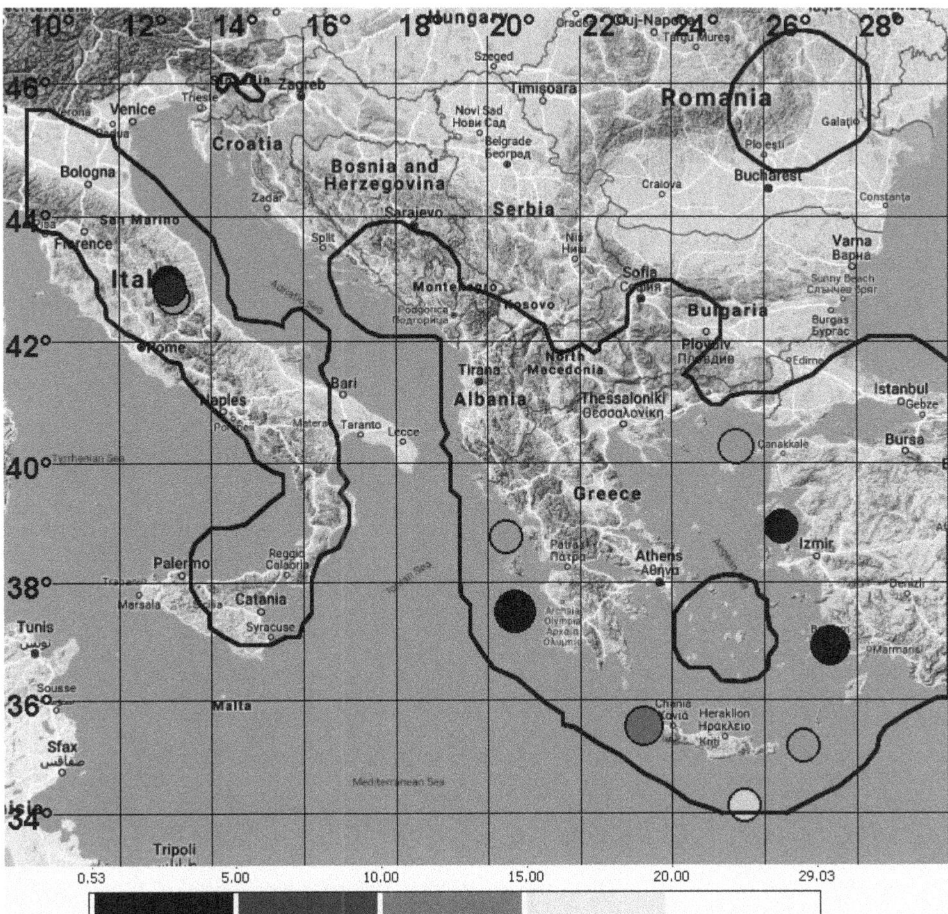

Figure 3. Area of analysis (marked with thick black line) and tested target epicenters of earthquakes in 2009–2018 in the Mediterranean region. Shades of grey indicates the minimum volume of alarm with which the epicenter was forecasted. Darkness of grey decreases in accordance with the volume of the alarm: 0.05, 0.1, 0.15, 0.2. A white color indicates that an earthquake is not forecasted with an alarm volume of less than 0.2.

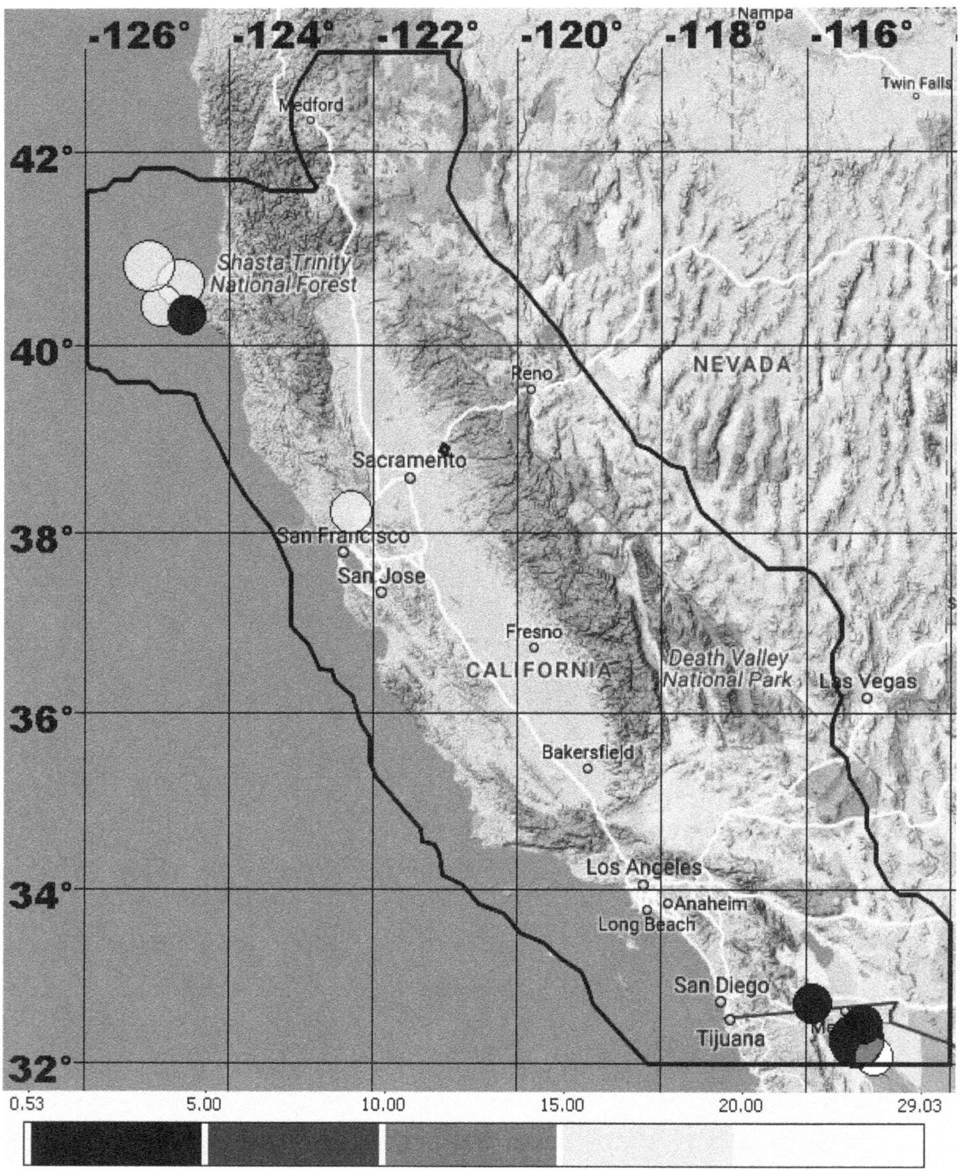

Figure 4. Area of analysis (marked with thick black line) and tested target epicenters of earthquakes in 2009–2018 in California. Shades of grey indicates the minimum volume of alarm with which the epicenter was forecasted. Darkness of grey decreases in accordance with the volume of the alarm: 0.05, 0.1, 0.15, 0.2. A white color indicates that an earthquake is not forecasted with an alarm volume of less than 0.2.

5. Discussion

The method of approximating the interval expert estimates compiles the M_{max} map, assuming a repetition of strong earthquakes and the existence of a relationship between M_{max} and the properties of the geological environment $x = (x_1, \ldots, x_I)$. At first, experts independently estimate M_{max} values in a set of the most studied points of a region. The algorithm approximates the dependence of M_{max} on geological and geophysical features by the function $F(x)$. The dependence $F(x)$ is defined as the sum of the non-linear functions of each of the feature. The M_{max} map is the $F(x)$ values calculated for the whole region. The presence of the formal forecast rule $F(x)$ allows the expert to study the contribution to the forecast of each feature and interpret the map as the sum of the nonlinearly transformed feature fields.

The method of the minimum area of alarm solves the problem of one-class classification. The method algorithm has two peculiar properties. The first relates to the data model. The model postulates two properties of anomalous objects: (1) Anomalous objects are unlikely, and some of their properties take values close to the maximum (or minimum) among the sample, and (2) the vectors of the space of features, which are componentwise larger (or smaller) of the vector corresponding to the anomalous object, can also be anomalous objects. Both these properties seem sufficiently natural. This model allows one to build a classification rule from a set of anomalous objects. In this case, normal objects are taken into account statistically through the probability of detecting anomalous objects by a random forecast. The second difference is that the algorithm allows constructing a forecast function that optimizes the probability of detecting anomalous objects in the training sample if the probability of a random forecast is not more than the predetermined value.

6. Conclusions

We considered two machine learning methods and their implementations to seismic hazard forecast. The method of approximation of interval expert estimates of M_{max} demonstrated good seismic zoning for many seismically active regions. The method of the minimum area of alarm is the basis of an automatic earthquake forecast system. The considered results of testing suggest that the method and the forecast system might contribute to advance in the problems of earthquake forecasting.

7. Supplement

The method of minimum area of alarm is the basis of an automated web-based platform that systematically forecasts target earthquakes. We presented the results of testing the approach to earthquake prediction in the Mediterranean and Californian regions. The goal of the test was to analyze the approach, the machine learning method, and the earthquake prediction platform. For the tests, ordinary parameters of earthquake catalogs were used. The testing was performed on data that did not participate in the training and showed a satisfactory forecast quality for both regions. The web-based platform has been launched and automatically calculates the seismic hazard fields from February 2018 [43]. During the time from 1 February 2018 to 8 July 2019, four target earthquakes occurred in these regions. In the Mediterranean region, two epicenters were predicted and fell into an area with an alarm volume of up to 15% (Figure 5), and in the California region, two epicenters fell into an area with an alarm volume greater than 20% and were not predicted (Figure 6).

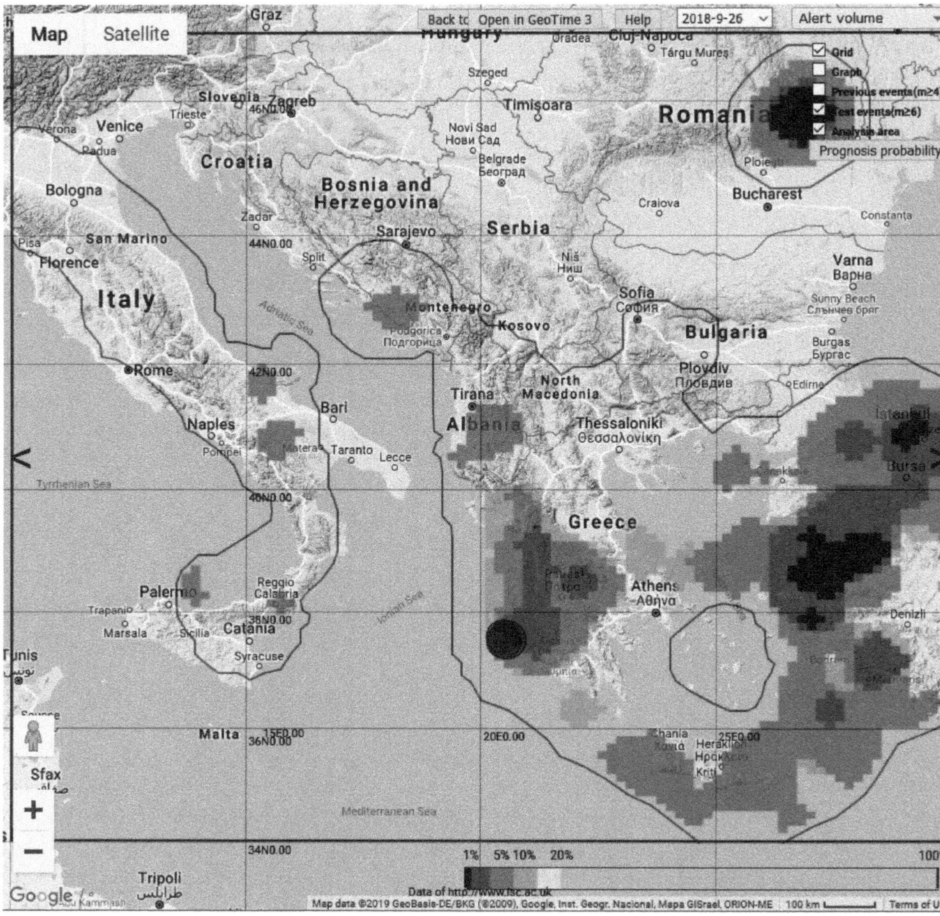

Figure 5. Screenshot of the web-based platform working window in the Mediterranean region: The map shows the alarm zone for earthquakes with a magnitude $M \geq 6.0$ and the predicted epicenters of earthquakes of 25 October 2018 with a magnitude of 6.6 and 30 October 2018 with a magnitude of 6.2 calculated for training according to data up to 26 September 2017. The palette shows areas with different alarm volumes in percent: 1%, 5%, 10%, 15%, 20%.

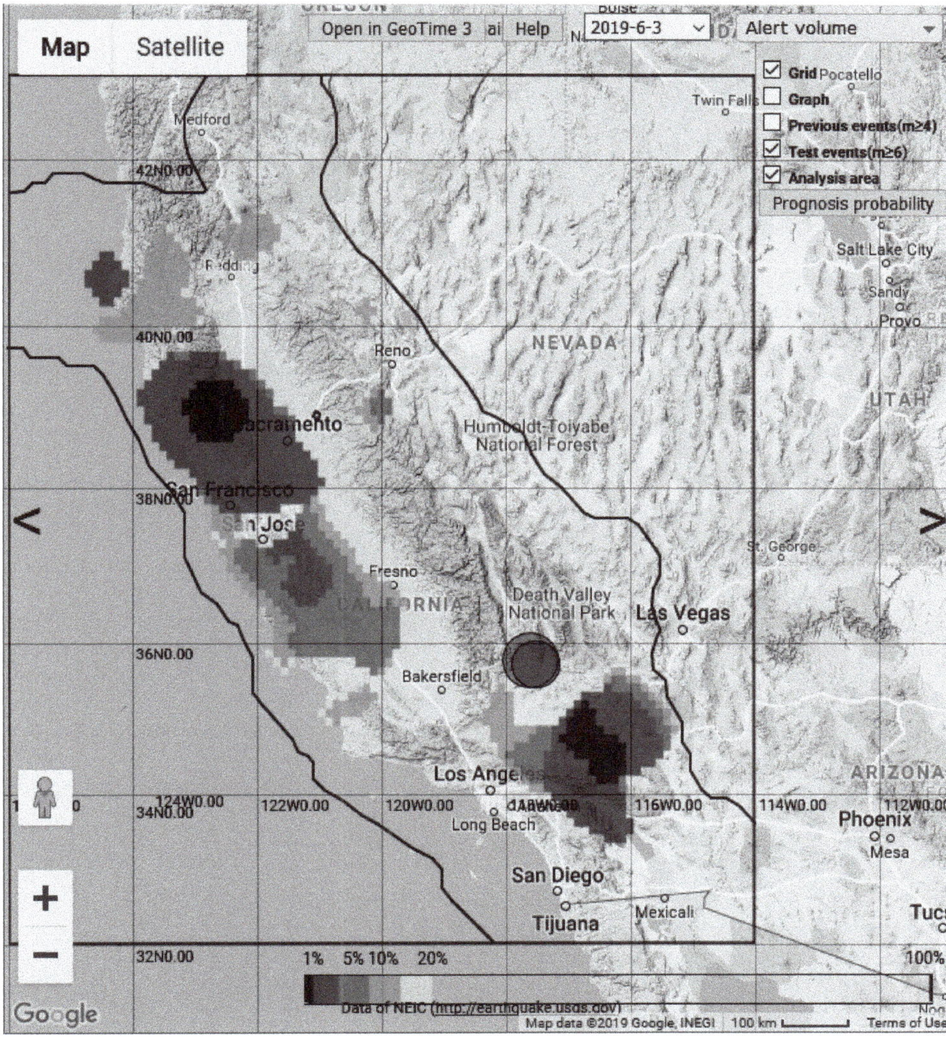

Figure 6. Screenshot of the web-based platform working window: The map shows the alarm zone for earthquakes with a magnitude $M \geq 5.7$ and the epicenters of earthquakes of 4 July 2019 with a magnitude of 6.4 and 6 July 2019 with a magnitude of 7.1 calculated during training according to data until 3 June 2019. The palette shows areas with different alarm volumes in percent: 1%, 5%, 10%, 15%, 20%.

8. Materials and Methods

International Seismological Centre [44] was searched using http://www.isc.ac.uk/iscbulletin/search/bulletin/ (last accessed on 14 February 2018). This online catalog was selected for its robustness and universality. It combines data from a lot of catalogs, and every earthquake with a magnitude more than three is manually checked. NEIC USGS catalog [45] was searched using https://earthquake.usgs.gov/earthquakes/feed/ (last accessed on 15 February 2018). This catalog was chosen because of numerous registered small earthquakes (magnitude of completeness less than 1.5) in California. Plots were made using the GeoTime 3 (www.geo.iitp.ru/GT3; [8]).

Author Contributions: Conceptualization, V.G.G.; Formal analysis, V.G.G. and A.B.D.; Investigation, V.G.G.; Methodology, A.B.D.; Project administration, V.G.G.; Visualization, A.B.D.; Software, A.B.D.; Writing—original draft, V.G.G.; Writing—review & editing, V.G.G. and A.B.D.

Funding: The paper is supported by the Russian Foundation for Basic Research, project 17-07-00494.

Conflicts of Interest: The authors declare no conflict of interest.

References

1. Sobolev, G. *Fundamentals of Earthquake Prediction*; Nauka: Moscow, Russia, 1993. (In Russian)
2. Sobolev, G.A.; Ponomarev, A.V. *Earthquake Physics and Precursors*; Nauka/Interperiodica: Moscow, Russia, 2003. (In Russian)
3. Keilis-Borok, V.I.; Soloviev, A.A. *Nonlinear Dynamics of the Lithosphere and Earthquake Prediction*; Springer Science & Business Media: Berlin/Heidelberg, Germany, 2011; p. 337.
4. Davis, J.L.; Wernicke, B.P.; Tamisiea, M.E. On seasonal signals in geodetic time series. *J. Geophys. Res. Solid Earth* **2012**, *117*. [CrossRef]
5. Silvia, M.T.; Robinson, E.A. *Deconvolution of Geophysical Time Series in the Exploration for Oil and Natural Gas*; Elsevier: Amsterdam, The Netherlands, 1979; Volume 10, .
6. Swarzenski, P.; Burnett, W.; Greenwood, W.; Herut, B.; Peterson, R.; Dimova, N.; Shalem, Y.; Yechieli, Y.; Weinstein, Y. Combined time-series resistivity and geochemical tracer techniques to examine submarine groundwater discharge at Dor Beach, Israel. *Geophys. Res. Lett.* **2006**, *33*. [CrossRef]
7. Collopy, P. Aerospace system value models: A survey and observations. In Proceedings of the AIAA Space 2009 Conference & Exposition, Pasadena, CA, USA, 14–17 September 2009; p. 6560.
8. Gitis, V.; Derendyaev, A.; Metrikov, P.; Shogin, A. Network geoinformation technology for seismic hazard research. *Nat. Hazards* **2012**, *62*, 1021–1036. [CrossRef]
9. Giardini, D.; Grünthal, G.; Shedlock, K.M.; Zhang, P. The GSHAP global seismic hazard map. *Ann. Geophys.* **1999**, *42*. [CrossRef]
10. Sieh, K. The repetition of large-earthquake ruptures. *Proc. Natl. Acad. Sci. USA* **1996**, *93*, 3764–3771. [CrossRef] [PubMed]
11. Anderson, J.G. Estimating the seismicity from geological structure for seismic-risk studies. *Bull. Seismol. Soc. Am.* **1979**, *69*, 135–158.
12. Schenk, V.; Schenkova, Z.; Gitis, V.G. Characteristic Features of Geonomic Forecasting Functions for the Maximum Possible Earthquake. *Nat. Hazards* **1994**, *10*, 97–115. [CrossRef]
13. Burton, P.W. Seismic risk in southern Europe through to India examined using Gumbel's third distribution of extreme values. *Geophys. J. Int.* **1979**, *59*, 249–280. [CrossRef]
14. Pisarenko, V. Statistical Evaluation of Maximum Possible Magnitude. *Earth Phys.* **1991**, *27*, 757–763.
15. Kijko, A. Estimation of the maximum earthquake magnitude, m max. *Pure Appl. Geophys.* **2004**, *161*, 1655–1681. [CrossRef]
16. Beirlant, J.; Kijko, A.; Reynkens, T.; Einmahl, J.H. Estimating the maximum possible earthquake magnitude using extreme value methodology: The Groningen case. *Nat. Hazards* **2018**, 1–23. . [CrossRef]
17. Molchan, G.; Kronrod, T. Calculation of seismic risk. In *Seismic Zoning of the Territory of the USSR*; Nauka: Moskow, Russia, 1980; pp. 69–82.
18. Bonchev, E.; Bune, V.; Christoskov, L.; Karagyuleva, J.; Kostadinov, V.; Reisner, G.; Rizikova, S.; Shebalin, N.; Sholpo, V.; Sokerova, D. A method for compilation of seismic zoning prognostic maps for the territory of Bulgaria. *Geol. Balc.* **1982**, *12*, 3–48.
19. Reisner, G.I.; Ioganson, L.I. Seismic potential of western Russia and other CIS and Baltic countries. In *Seismicity and Seismic Zonation of Northern Eurasia*; OIFZ RAN: Moscow, Russia, 1993; pp. 168–195.
20. Gitis, V.; Mironov, M.; Bune, V.; Vuchev, V. Mmax Earthquake Map Elaboration by the Method of Interval Expert Evaluations. *Phys. Earth* **1982**, *4*, 31–44.
21. Gitis, V.; Mironov, M.; Bune, V.; Kalenik, V.; Shchukin, Y.K. Mmax earthquake prediction on the basis of interval approximation of expert estimates. *Izv. AN SSSR Fizika Zemli* **1986**, *4*, 24–31.
22. Gitis, V.; Ermakov, B. *Fundamentals of Spatiotemporal Forecasting in Geoinformatics*; Fizmatgiz: Moscow, Russia, 2004; p. 256.

23. Kossobokov, V.G. User manual for M8. In *Algorithms for Earthquake Statistics and Prediction*; Seismological Society of America: Albany, CA, USA, 1997; pp. 167–222.
24. Zavyalov, A.D. *Intermediate Term Earthquake Prediction*; Nauka: Moscow, Russia, 2006. (In Russian)
25. Shebalin, P.N.; Narteau, C.; Zechar, J.D.; Holschneider, M. Combining earthquake forecasts using differential probability gains. *Earth Planets Space* **2014**, *66*. [CrossRef]
26. Amei, A.; Fu, W.; Ho, C.H. Time series analysis for predicting the occurrences of large scale earthquakes. *Int. J. Appl.* **2012**, *2*, 64–75.
27. Marzocchi, W.; Zechar, J.D. Earthquake Forecasting and Earthquake Prediction: Different Approaches for Obtaining the Best Model. *Seismol. Res. Lett.* **2011**, *82*, 442–448. [CrossRef]
28. Rhoades, D.A. Application of the EEPAS Model to Forecasting Earthquakes of Moderate Magnitude in Southern California. *Seismol. Res. Lett.* **2007**, *78*, 110–115. [CrossRef]
29. Rhoades, D.A. Mixture Models for Improved Earthquake Forecasting with Short-to-Medium Time Horizons. *Bull. Seismol. Soc. Am.* **2013**, *103*, 2203–2215. [CrossRef]
30. Geller, R.J.; Jackson, D.D.; Kagan, Y.Y.; Mulargia, F. Earthquakes cannot be predicted. *Science* **1997**, *275*, 1616. [CrossRef]
31. Bishop, C.M. *Pattern Recognition and Machine Learning*; Springer: Berlin, Germany, 2006.
32. Kotsiantis, S.B. Supervised Machine Learning: A Review of Classification Techniques. In *Emerging Artificial Intelligence Applications in Computer Engineering: Real Word AI Systems with Applications in eHealth, HCI, Information Retrieval and Pervasive Technologies*; IOS Press: Amsterdam, The Netherlands, 2007; pp. 3–24.
33. Khan, S.S.; Madden, M. A survey of recent trends in one class classification. In Proceedings of the Irish Conference on Artificial Intelligence and Cognitive Science, Dublin, Ireland, 19–21 August 2009; pp. 188–197.
34. Bune, B.; Vuchev, V.; Gitis, V.; Mironov, M. Application of the interval expert evaluation approximation method to compilation of M max maps of Bulgarian earthquakes. *Geol. Balc.* **1982**, *12*, 49–70.
35. Gitis, V.; Alvarez, L.; Chuy, T.; Cotilla, M.; Makarov, V.; Schukin Yu, K. Estimation of Mmax of the earthquakes of Caribbean and Middle America region with the help of GEO expert system. In Proceedings of the 4th International Symposium on the Analysis of Seismicity and on Seismic Risk, Bechyne, Czechoslovakia, 4–9 September 1989; pp. 128–137.
36. Schenk, V.; Gitis, V.G.; Schenkova, Z.; Mantlik, F.; Kotnauer, P.; Jurkov, E.F.; Shchukin, Y.K. Maximum Earthquake Prediction in Central Europe Given by the GEO 1.2 Expert System. In Proceedings of the 4th International Conference on Seismic Zonation, Stanford, CA, USA, 25–29 August 1991; Volume III, pp. 83–91.
37. Colombo, D.; Gitis, V.; De Franco, R. Application of pattern recognition techniques to long-term earthquake prediction in central Costa Rica. *Eng. Geol.* **1997**, *1*, 7–18. [CrossRef]
38. Gitis, V.; Tatevossian, R.; Vainshtock, A. Maximum expected magnitude assessment in a GEO computer environment: Case study. *Nat. Hazards* **1998**, *17*, 225–250. [CrossRef]
39. Molchan, G.M. Earthquake prediction as a decision-making problem. *Pure Appl. Geophys. PAGEOPH* **1997**, *149*, 233–247. [CrossRef]
40. Kossobokov, V.; Romashkova, L.; Keilis-Borok, V.; Healy, J. Testing earthquake prediction algorithms: Statistically significant advance prediction of the largest earthquakes in the Circum-Pacific, 1992–1997. *Phys. Earth Planet. Inter.* **1999**, *111*, 187–196. [CrossRef]
41. Kossobokov, V.; Shebalin, P. Earthquake prediction. In *Nonlinear Dynamics of the Lithosphere and Earthquake Prediction*; Springer: Berlin/Heidelberg, Germany, 2003; pp. 141–207.
42. Gutenberg, B.; Richter, C.F. Frequency of earthquakes in California. *Bull. Seismol. Soc. Am.* **1944**, *34*, 185–188.
43. Gitis, V.G.; Derendyaev, A.B. Web-Based GIS Platform for Automatic Prediction of Earthquakes. In Proceedings of the International Conference on Computational Science and Its Applications, Melbourne, Australia, 2–5 July 2018; pp. 268–283.
44. International Seismological Centre *On-Line Bulletin*; International Seismological Centre: Thatcham, UK, 2016.
45. Masse, R.; Needham, R. NEIC; the National Earthquake Information Center. *Earthq. Volcanoes USGS* **1989**, *21*, 4–45.

geosciences

Article

Suitability Analysis for the Emergency Shelters Allocation after an Earthquake in Japan

Tetsuya Akamatsu * and Kayoko Yamamoto

Graduate School of Informatics of Engineering, University of Electro-Communications, Chofu, Tokyo 182-8585, Japan

* Correspondence: a1930002@edu.cc.uec.ac.jp; Tel.: +81-42-443-5728

Received: 12 June 2019; Accepted: 24 July 2019; Published: 30 July 2019

Abstract: Japan is exposed to several natural phenomena; the damages caused by earthquakes were enormous in particular. An emergency shelter is a place for people to temporarily live when they cannot remain in their previous homes, and it is necessary for each community to, respectively, allocate such facilities in Japan. There are the districts that are short of such facilities, especially in rural and suburban areas, because emergency shelters mainly concentrate near large-scale stations and city centres in Japan. Against these backdrops, using geographic information systems (GIS), an applied statistical method and public open data related to population and emergency shelters, the present research aims to quantitatively conduct a suitability analysis for the emergency shelters allocation after an earthquake in Japan. Based on the results, the present research grasps the districts that are short of emergency shelters, and visually shows the places where such facilities should be newly established on the digital map of GIS. Additionally, the assessment method is reproducible in the spatial and temporal dimension. It is necessary to create an original data related to emergency shelters to raise the reliability of the results, as the present research has the limitation of data availability.

Keywords: emergency shelter; earthquake; disaster; weighted coefficient; suitability analysis; geographic information systems (GIS)

---

1. Introduction

In Japan, various disasters have been frequently occurred, and the damages especially caused by earthquakes were enormous. According to Yamaga (2019) [1], the earthquake catalog search conducted by the United States (U.S.) Geological Survey (USGS), 14.8% of the earthquakes larger than magnitude 5.0, which occurred worldwide from 2010 to 2015 concentrated in Japan. Therefore, it can be said that Japan is one of the few earthquake-ridden countries.

Meanwhile, due to the serious damages that are caused by disasters, victims who cannot remain in their homes stay in emergency shelters until they can settle in new places. The maximum distance elderly people and children can evacuate on foot is called as the maximum evacuation distance, and it is estimated to be around 1.5 to 2.0 km (Dictionary of Housing Terms, 2019) [2]. However, the distance that elderly people over 70 years old can walk is 879 m, which is significantly shorter when compared with those of the other age groups (Cabinet Office, 2019) [3]. The physical abilities of elderly people over 70 years old drastically decrease, which makes long-distance evacuations extremely difficult. Additionally, in Japan, as emergency shelters mainly concentrate near large-scale stations and city centres, there are districts that are short of such facilities, especially in rural and suburban areas. Additionally, focusing on the communities, Civiletti et al. (2016) [4] identified the institutional and social decisions that increase the resilience of the communities that are exposed at risk, by analyzing observations during seismic sequences that occurred in Italy in the last decades, and Cerchiello et al. (2018) [5] addressed the assessment of the social vulnerability and resilience level of the city of Nablus,

Palestine. Mojica et al. (2010) [6] and Awaotona (2012) [7] pointed out that it is necessary to take the countermeasures against disasters while considering vulnerable people (elderly and disabled people). However, in Japan, due to the Basic Act on Disaster Control Measures, it is possible to already take the measures for disabled people requiring assistance at the time of disasters in each municipality (Cabinet Office, 2013) [8]. Additionally, in Japan, the aging rate of elderly population who are over 65 years old is 26.7% in 2015, which is the highest in the world. For reference, the above rates are 22.4% in Italy, 21.2% in German, 19.9% in Sweden, and 19.1% in France (Ministry of Health, Labour and Welfare, 2016) [9]. Therefore, when considering the regional characteristics, such as the ratio of elderly people and population distribution, it is necessary to evaluate not only facility location, but also capacity, this is because elderly people have special needs. Additionally, it is also necessary to grasp the districts (made up of streets and towns, and the minimum level of Japanese communities), with a lack of emergency shelters.

Based on the social and academic background mentioned above, using Geographic Information Systems (GIS), which is an applied statistical method and public open data related to population and emergency shelters, and while considering regional characteristics, such as the ratio of elderly people and population distribution, the present research aims to conduct a suitability analysis for the emergency shelters allocation after an earthquake on a district scale in Japan. In the present research, the analysis will be conducted to examine whether emergency shelters are sufficient or not in the unit of districts, and where such facilities should be newly established. Additionally, referring the results of the preceding research studies both within and outside Japan introduced in the next section, the present research will develop a method for the suitability analysis for the emergency shelters allocation. Based on the results, the present research provides effective information that can be referred to determine the locations to establish new emergency shelters.

2. Literature Review

As mentioned in the previous section, Japan is exposed to several natural phenomena, and there is an accumulation of many research studies in the field related to emergency shelters. The present literature review is related to (1) research studies that are related to the suitability analysis of emergency shelters location, and (2) research studies related to the establishment of new emergency shelters. The following will introduce the major preceding research studies in the above two study areas, and discuss the originality of the present research in comparison with the others.

Regarding the representative research studies that are related to the suitability analysis of emergency shelters location, especially in Japan, Furihata et al. (1994) [10] evaluated the location of evacuation facilities taking the spatial patterns of distributed users into consideration and using GIS. Takeuchi et al. (2002) [11] proposed a shared area in the taking emergency shelter to have considered the approaching direction of the tsunami that is generated by a huge earthquake. Kongsomsaksakul et al. (2005) [12] proposed the optimal locations of emergency shelters for the flood evacuation planning, while using the combined distribution and assignment (CDA) model and genetic algorism (GA) in the U.S. Asou et al. (2007) [13] clarified evacuation behavior while using GA to propose the optimal arrangements of emergency shelters. Wei et al. (2008) [14] presented a diagnosis model for emergency shelter planning from the viewpoint of local people using GIS. Ng et al. (2010) [15] presented a hybrid bi-level model for the optimal shelter assignment in emergency evacuations in the U.S. Tai et al. (2010) [16] used six indicators to evaluate shelter and applied a spatial statistic model with local indicators of spatial association (LISA) to the evacuation choice of residents in the case of an earthquake in Taiwan. Kitajima (2013) [17] evaluated the location of emergency shelters using network analysis. Yu et al. (2016) [18] introduced a framework for the multi-criteria satisfaction assessment of the spatial distribution of urban emergency shelters while using a GIS-based analytic hierarchy process approach in Shanghai City, China. Vecere et al. (2016, 2017) [19,20] focused on a critical review of currently available methodologies and corresponding software packages that were specifically developed for estimating the number of displaced people and those who need public sheltering and temporary housing at the time of the 2011 Christchurch earthquake in New Zealand. Ashish et al. (2017) [21] presented a hybrid algorithm for efficiently managing location and relocation projects, by proposing a hybrid multi-objective decision model that is based on analytic hierarchy process (AHP), fuzzy set theory, and goal programming approach, referring two real case studies of Nepal earthquake. Xu et al. (2018) [22] developed a multi-objective mathematical model with four groups of the objectives, allied with a modified particle swarm optimization algorithm to solve the location-allocation problem for earthquake shelter in Beijing City, China. Nozaki et al. (2019) [23] conducted the location analysis of tsunami emergency shelter while considering inhabitants' preparedness for the coming Nankai Trough Earthquake.

Regarding the representative research studies that were related to the establishment of new emergency shelters, especially in Japan, Yamada et al. (2004) [24] proposed the planning support system for the locations of emergency shelters from the viewpoint of residents. Notsuda et al. (2005) [25] considered the optimum location of new emergency shelters and the placement of evacuees to each facility that is based on the location and capacity of existing emergency shelters, as well as the population distribution of evacuees. Nakai et al. (2012) [26] and Ikenaga et al. (2017) [27] considered the possibility of using vacant houses as emergency shelters and temporary evacuation shelters in the time of accidents. Asano et al. (2013) [28] and Miyoshi et al. (2017) [29] considered the possibility of using public and private lands as emergency shelters and safety evacuation areas. Araki et al. (2017) [30] examined setting up the patterns of non-designated emergency shelters at the time of the Great East Japan Earthquake (2011), which were based on GIS analysis and interview surveys. Sasaki et al. (2018) [31] examined the possibility of approximately 70,000 temples nationwide, complementing emergency shelters and safety evacuation areas that are expected to be in shortage when many disaster-affected residents appear during widespread disasters. Umeki et al. (2019) [32] proposed a method to determine the location of emergency shelters to aim at the reduction of the evacuation time of all victims.

In other countries, Kar et al. (2008) [33] developed a GIS-based model to determine the site suitability of emergency shelters for hurricane events, and proposed the candidate places to arrange new ones in the U.S. Alçada-Almeida et al. (2009) [34] incorporated multi-objective model into a GIS-based decision support system to locate emergency shelters during major fires in Portugal. Park et al. (2012) [35] developed a method that applied genetic optimization to determine optimal tsunami shelter locations with the goal of reducing evacuation time, thereby maximizing the probability of survival for the population in a coastal community in the U.S. Li et al. (2012) [36] developed a scenario-based bi-level programming model to optimize the selection of shelter locations, with explicit consideration of a range of possible hurricane events and the evacuation needs under each of those events in the U.S. Anhorn et al. (2015) [37] proposed a methodology to rank the suitability of open spaces for contingency planning and the placement of emergency shelter while using GIS in the immediate aftermath of a disaster in Turkey. Bayram et al. (2018) [38] proposed a scenario-based two-stage stochastic evacuation planning model that optimally locates shelter sites and assigns evacuees to nearest shelters and to shortest paths within a tolerance degree to minimize the expected total evacuation time in Turkey. Xu et al. (2016) [39] proposed a multi-criteria constraint location model to select and analyze the candidate for emergency shelters, and to determine the location of new ones while using GIS in China.

From the above, especially in Japan, research studies that are related to the suitability analysis of emergency shelters location increased after the Great Hanshin earthquake (1995), and research studies that are related to the establishment of new emergency shelters remarkably increased after the Great East Japan Earthquake. Because, at the time of the Great East Earthquake, large earthquake, tsunami, and an accident at the nuclear power station occurred at the almost same time, and it was essential to keep sufficient emergency shelters for many victims.

The present research will reveal its effectiveness by newly proposing the use of facilities, such as temporary evacuation shelters in the time of accidents, safety evacuation areas, and large-scale retail stores that are not set up to be emergency shelters as new ones, unlike the preceding research studies that were related to the establishment of new emergency shelters, as discussed in detail in Section 3. Additionally, though the preceding research studies that are related to the suitability analysis of emergency shelters location derived the optimum locations for emergency shelters, the present research will demonstrate the originality by conducting a quantitative assessment of current location of emergency shelters, and proposing an assessment method to consider regional characteristics, such as the ratio of elderly population and population distribution. Therefore, it is possible to conduct a suitability analysis for the emergency shelters allocation that appropriately reflect current conditions of each district, and decide on the location of new emergency shelters by suggesting specific sites on a small spatial scale as the unit of districts while using GIS.

3. Methods

From the results of the research studies that were related to the suitability analysis of emergency shelters location in the previous section, for the suitability analysis for the emergency shelters allocation, it is necessary to consider regional characteristics, such as location and capacity of existing emergency shelters, as well as the population distribution of evacuees. It is evident that GIS-based analysis, spatial data analysis, and an applied statistical method are effective for the above purpose. Additionally, from the results of the research studies related to the establishment of new emergency shelters in Japan, it is necessary to set new candidate facilities in both public and private lands, especially in the districts that are short of emergency shelters. On the other hand, from the results of the above research studies in other countries, it is evident that GIS-based methods are effective in conducting a suitability analysis for the emergency shelters allocation. Referring these results, in this section, the method of the present study will be proposed.

3.1. Framework and Process of Analysis

GIS will be used for the method of suitability analysis for the emergency shelter allocation after an earthquake. The framework and process of analysis in the present research are as shown below.

(1) Calculation of weighted coefficients and creation of the distribution map of emergency shelters in each district

The weighted coefficient for each district will be calculated. A distribution map for all the emergency shelters will be created in the form of the digital map of GIS. Additionally, the facility scale (area) will be added to the data of each emergency shelter.

(2) Calculation of the linear distance between each district and each emergency shelter

The linear distance between the center of each district and each emergency shelter will be calculated while using the distribution map of emergency shelters in (1).

(3) Suitability analysis for the emergency shelters allocation using weighted coefficients

Using the weighted coefficients and distribution maps of emergency shelters in (1) as well as the linear distance between each district and each emergency shelter in (2), the assessment value for each district will be calculated. In the present research, analysis will be separately conducted using the four types of weighted coefficients related to the regional characteristics that are described in detail in the next section. When determining the locations to establish new emergency shelters, the important regional characteristics might vary with areas. Based on the results, the present research provides the effective information to assist the policy and decision makers in planning new emergency shelters.

3.2. Assessment Method

3.2.1. Calculation of Weighted Coefficients and Creation of the Distribution Map of Emergency Shelters in Each District

(1) Distribution map of different types of weighted coefficients for each district

In the present research, the degree of importance is indicated by the weighted coefficient and an assessment method will be proposed. There are four types of weighted coefficients, including (a) the specialization coefficient of elderly population, (b) the ratio of permanent population, (c) the ratio of elderly population, and (d) the specialization coefficient of population density, indicating various regional characteristics. Regarding (a) (d), the specialization coefficients are the values that reveals which area has the highest against the average rates of elderly population and population density in the city.

Regarding (a), Saino (1992) [40] set a "specialization coefficient of elderly people" which divided the rate of people who are over 65 years old in each district by the same rate in the entire area. However, the health span in Japan for 2016 was 72.14 for men and 74.79 for woman and both greatly exceed the age of 70, and it is increasing every year (the 11th Japan 21 (secondary) Health Promotion Committee, 2018) [41]. Therefore, the present research sets the specialization coefficient of elderly population who are over 75 years old, instead of those who are 65 years old as the target. The specialization coefficient of elderly population is a value that reveals which area has the highest aging rate of elderly population in the city. The aging rate of elderly population can be calculated with Equation (1). Additionally, the specialization coefficient of elderly population can be calculated while adopting this in Equation (2).

$$A_i = \frac{p_{75i}}{p_i} \tag{1}$$

A_i: Aging rate of elderly population in district i
p_{75i}: Elderly population in district i (persons)
p_i: Population of district i (persons)

$$B_i = \frac{A_i}{A} \tag{2}$$

B_i: Specialization coefficient of elderly population in district i
A_i: Aging rate of elderly population in district i (%)
A: Aging rate of elderly population in the City (%)

Regarding (b), the ratio of permanent population for each district is important, as there is a greater need for emergency shelters in districts with a higher population. Equation (3) is adopted to calculate the ratio of permanent population.

$$C_i = \frac{p_i}{p} \tag{3}$$

C_i: Ratio of permanent population in district i
p_i: Population of district i (persons)
p: Population of the City (persons)

Regarding (c), the population distribution of elderly group is not reflected in this weighted coefficients, as (a) the specialization coefficient of elderly population only indicates which district has the highest aging rate of elderly population. Therefore, it is weighted with the ratio of elderly population in order for the weighting to reflect the population distribution of elderly group. The ratio of elderly population can be calculated with Equation (4).

$$D_i = \frac{p_{75i}}{p_{75}} \tag{4}$$

D_i: Ratio of elderly population in district i
p_{75i}: Elderly population in district i (persons)
p_{75}: Elderly population in the City (persons)

Regarding (d), districts that are larger than others and are around large-scale stations and city centres generally have larger populations. Therefore, it is important to take the population density into consideration, as it is necessary to prioritize the establishment of new emergency shelters in districts that are short of such facilities due to the high population density. Thus, new emergency shelters can be established and victims can stay in such facilities near their homes even in the districts with high population density. The specialization coefficient of population density is a value that reveals which area has the highest aging rate of population density in the city. Population density can be calculated with Equation (5), and adopting this in Equation (6), the specialization coefficient of population density can be calculated.

$$F_i = \frac{p_i}{a_i} \tag{5}$$

F_i: Population density in district i ($persons/km^2$)
p_i: Population of district i ($persons$)
a_i: Area of district i (km^2)

$$G_i = \frac{F_i}{F} \tag{6}$$

G_i: Specialization coefficient of population density in district i
F_i: Population density in district i ($persons/km^2$)
F: Population density in the City ($persons/km^2$)

(2) Creating distribution maps of emergency shelters

The distribution of emergency shelters will be displayed on the digital map of GIS. In the present research, the facilities, such as temporary evacuation shelters in the time of accidents and safety evacuation areas, are newly added to emergency shelters. These facilities were actually utilized as emergency shelters after the occurrence of disasters in the past. However, it is necessary to assume that an emergency shelter should be established as a temporary house in safety evacuation areas that are located outdoors. Based on Shigenobu et al. (2013) [42], the total size (area) of temporary houses can be calculated by multiplying the size of safety evacuation areas by the area ratio that is available for temporary houses. As there are multiple types of safety evacuation areas, the area ratio according to each type is set, as shown in Table 1. According to the Basic Act on Disaster Control Measures in Japan, it is possible to establish temporary houses in safety evacuation areas, such as open spaces and urban greens spaces that are shown in Table 1. Additionally, temporary houses were actually established in the above safety evacuation areas after the occurrence of disasters in the past. Additionally, after the Great East Japan Earthquake, over 2,000 people evacuated and stayed at large-scale retail stores in Ishinomaki City of Miyagi Prefecture. Therefore, since such large-scale retail stores are extremely wide and have a high capacity, they are also added as new emergency shelters.

Table 1. Area ratio available for temporary houses of safety evacuation areas.

Type of Safety Evacuation Area	Area Ratio Available for Temporary Houses (%)
Open Spaces	95.0
Urban green spaces	50.0

3.2.2. Calculation of the Linear Distance between Each District and Each Emergency Shelter

In the present research, the linear distance between each district and each emergency shelter, and the scale of each emergency shelter will be used when conducting analyses. Regarding the former, the evacuations can be quickly completed in areas that are close to emergency shelters. Additionally, regarding the latter, the larger an emergency shelter is, the more people it can contain. In order to calculate the former, one of the ArcGIS Pro analysis tools, called the "Generate Near Table", is used.

Using this tool, on the digital map of GIS, a proximity feature (emergency shelter) that is within 879 m from one or more features (center of each district) is selected, and n value is obtained. For each district, the n value will be the average of the number of emergency shelters within 879 m, which is the average walking distance for those in their 70's. Subsequently, using this tool again, the maximum number of features (n value) is set for each district, and the linear distance between each district and each emergency shelter is calculated.

3.2.3. Suitability Analysis for the Emergency Shelters Allocation

The present research develops an assessment method that is based on the p-median problem. The p-median problem, which is one of the facility location problems, places facilities by minimizing the total sum of the linear distance from users to their nearest facility, and it can be modeled, as shown in Equation (7). This model derives the optimum location that lessens the load for users in all districts as much as possible by changing X_{ij}. Equation (7) is changed to Equation (8) in order to respond the purpose of the present research.

$$\min_{X_{ij}} Z = \sum_i \sum_j w_i d_{ij} X_{ij} \tag{7}$$

$X_{ij} \in \{0, 1\}$: allocation to facility j in district i
w_i: Demand in district i
d_{ij}: Linear distance from district i to facility j

$$\text{minimize } H_i = w_i \times \frac{1}{n} \sum_{j=1}^{n} \frac{d_{ij}}{\sqrt{s_{ij}}} \tag{8}$$

H_i: Assessment value of district i
w_i: Weighted coefficient of district i
d_{ij}: Linear distance to the No. j closest emergency shelter from district i (m)
s_{ij}: Facility scale of the No. j closest emergency shelter from district i $\left(m^2\right)$

Specifically, X_{ij} of Equation (7) is removed in order to fix the locations of emergency shelters. Additionally, by calculating the assessment value for each district, it is possible to quantitatively grasp the emergency shelters allocation in each district. Additionally, as far as the originality of the present research, the average of the linear distance to the No. i closest emergency shelter in each district d_{ij}, divided by the square root of the facility scale of the emergency shelter s_{ij}, will be weighted with the coefficient w_i, and this will be the assessment value for that district.

As shown in the previous section, the four types of weighted coefficients, the linear distance between each district and each emergency shelter, and the facility scale of emergency shelters will be applied to Equation (8), the assessment value of each district will be calculated, and the results will be displayed on the digital maps of GIS. When not using a weighted coefficient, analyses are conducted based on the linear distance between each district and each emergency shelter, and the scale of such facilities.

3.2.4. Application of Assessment Method

The following two types of comparisons will be conducted in the discussion section in order to verify the validity of the assessment method proposed in the present research (Section 6).

(1) Comparison of the results between multiple areas with different regional characteristics

In the present research, multiple areas with different regional characteristics will be selected as target areas, and the suitability of emergency shelters allocation will be analyzed. Therefore, it is possible to verify the validity of the assessment method by comparing results of the selected areas with and without the four types of weighted coefficients that are related to the regional characteristics.

(2) Comparison of the results with and without weighted coefficients

The assessment method in the present research will adopt the four types of weighted coefficients as introduced in Section 3.2.1. Therefore, it is also possible to verify the validity of assessment method by comparing results with and without the four types of weighted coefficients that are related to the regional characteristics, focusing on the increase and decrease of the assessment values for both cases.

4. Selection of Target Areas and the Data Processing

4.1. Selection of Target Areas

For the analysis in the present research, it is essential to obtain the public open data that are related to population and emergency shelters, which will be introduced in the next section. However, in present Japan, very few cities make public the data related to emergency shelters. Among the cities that make public the above public open data, the four target areas selected for the present research are Chofu and Fuchu Cities in Tokyo metropolis, Toda City in Saitama Prefecture, and Shobara City in Hiroshima Prefecture. Table 2 describes the outlines of target areas. As described below, there are differences concerning the four types of weighted coefficients that are related to the regional characteristics introduced in Section 3.2.1 and the distribution of emergency shelters among these four cities.

Specifically, while the aging rate of elderly population (population aged over 75 years old) in Shobara City is significantly high, the rates in Toda City re fairly lower than the national average (11.2%). Additionally, the total populations of these two cities are 35,575 and 139,616, respectively, which makes them medium and small-scale cities. On the other hand, Chofu and Fuchu Cities, which are residential areas in Tokyo Metropolis, are large-scale cities with a population of 230,303 and 260,116, respectively. However, the distribution conditions of emergency shelters in these two cities are extremely different. Specifically, though emergency shelters are evenly distributed in the entire Chofu City, they are mostly located in the central part of Fuchu City.

According to the "Ordinance Covering Measures for Stranded Persons" in Tokyo Metropolis (2013), in the time of accidents, it is necessary for the companies in central Tokyo to keep employees at the workplaces. Therefore, central Tokyo, which has a concentration of companies as well as an extremely high daytime population, is not considered to be a target area in the present research. Additionally, according to the Basic Act on Disaster Control Measures in Japan, it is preferable for victims to stay in emergency shelters near their homes at the time of earthquakes. In the case of earthquakes, aftershocks follow for a while and additional damages might be generated, and it is dangerous for victims to still stay in their homes, even in good maintenance conditions that were constructed according to the Building Standards Act in Japan. Additionally, if victims stay together in emergency shelters near their homes, it is appropriate for rescue parties and government administrators to accurately confirm their safety and efficiently deliver relief goods to them.

Table 2. Outlines of target areas.

City	Population (Persons)	Area (km^2)	Aging Rate of Elderly Population (%)	Number of District	Number of Emergency Shelters
Chofu City	230,303	21.6	8.7	105	68
Fuchu City	260,116	29.4	8.3	148	73
Toda City	35,575	18.2	5.6	60	37
Shobara City	139,616	1247.0	23.7	120	75

4.2. Data Processing

Table 3 shows the utilized data and sources, and the utilization method of data in the present research.

Table 3. List of utilized data.

Utilized Data and Sources	Utilization Method of Data
Population by age (National senses 2010) (Source: Statistics Bureau, Ministry of Internal Affairs and Communications)	Creating distribution maps for the specialization coefficient of elderly population, the ratio of permanent population, the ratio of elderly population, and the specialization coefficient of population density
Data related to emergency shelters (Digital national land information) (Source: National Land Policy Bureau, Ministry of Land, Infrastructure, Transport and Tourism) Information related to emergency shelters (Source: Official websites of Chofu, Fuchu, Toda and Shobara Cities)	Creating a distribution map of emergency shelters

5. Results

The section will, respectively, describe the distinctive features of the result for each target area, referring Figures 1–20. These figures show the results with and without the four types of weighted coefficients that are related to the regional characteristics described in Section 3.2.1. These results are based on the suitability analysis for the emergency shelters allocation in the unit of districts of each target area mentioned in Section 3.2.3. Referring to these figures, the comparison of the results with and without the four types of weighted coefficients will be easily conducted. Furthermore, in these figures, all of the values have no units. As clearly shown in Table 2, the area of Shobara City is tremendously larger than those of other three cities. Therefore, the scale of results for Shobara City is different from those for other three cities on the digital maps of GIS.

5.1. Results for Chofu City

Figures 1–5 shows the results and n value is 6 for Chofu City. As mentioned in Section 3.2.2, for each district, the n value is the average of the number of emergency shelters within 879 m. From these figures, it is evident that there is a large difference between districts when not using a weighted coefficient (Figure 1), and when using the ratio of permanent population as a weighted coefficient (Figure 3). Without a weighted coefficient, there are districts with extremely high assessment values in the northern part, which are far from emergency shelters. With the ratio of permanent population as a weighted coefficient, there are districts with extremely high assessment values in the southeastern and southwestern parts, which have a large number of populations.

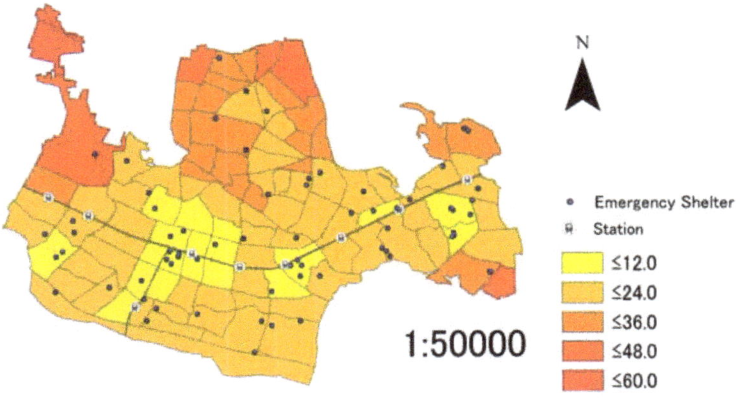

Figure 1. Result for Chofu City without a weighted coefficient.

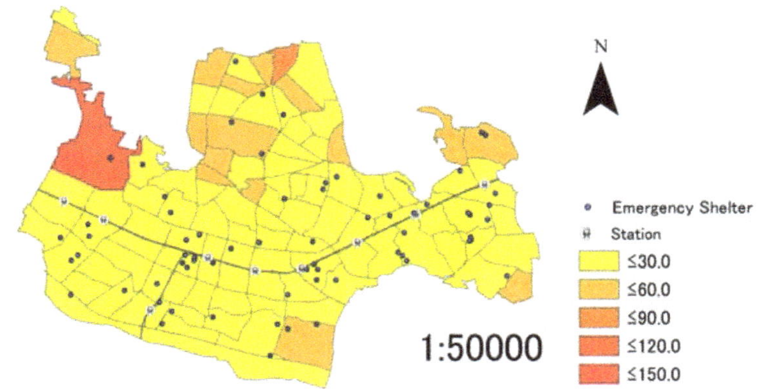

Figure 2. Result for Chofu City with the specialization coefficient of elderly population as a weighted coefficient.

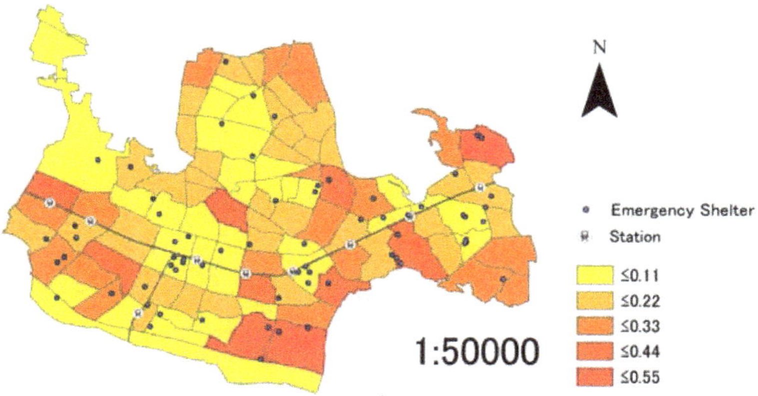

Figure 3. Result for Chofu City with the Ratio of permanent population as a weighted coefficient.

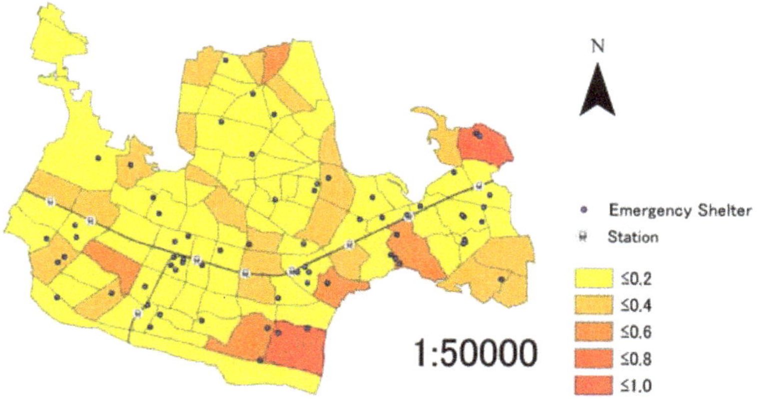

Figure 4. Result for Chofu City with the ratio of elderly population as a weighted coefficient.

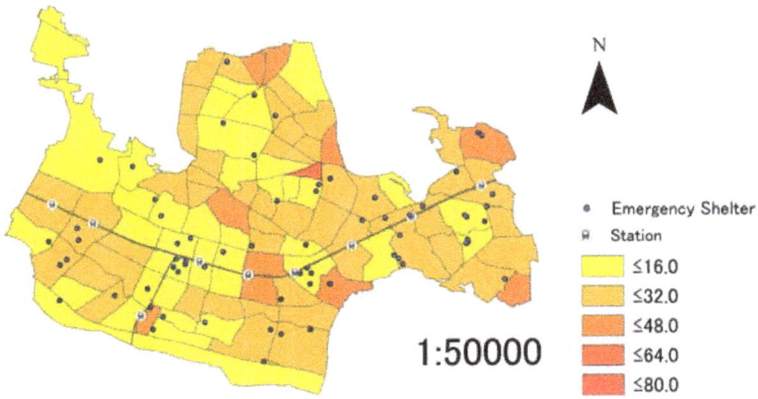

Figure 5. Result for Chofu City with the specialization coefficient of population density as a weighted coefficient.

5.2. Results for Fuchu City

Figures 6–10 shows the results and n value is 6 for Fuchu City. From these figures, it is evident that there is a large difference between districts when not using a weighted coefficient (Figure 6), and when using the ratio of permanent population (Figure 8). Without a weighted coefficient, the emergency shelters concentrated in the central part that has a low assessment value. With the ratio of permanent population as a weighted coefficient, there are districts with extremely high assessment values in the southeastern part, which are far from emergency shelters. Additionally, there are districts with low assessment values in central part, which have a large number of populations and emergency shelters.

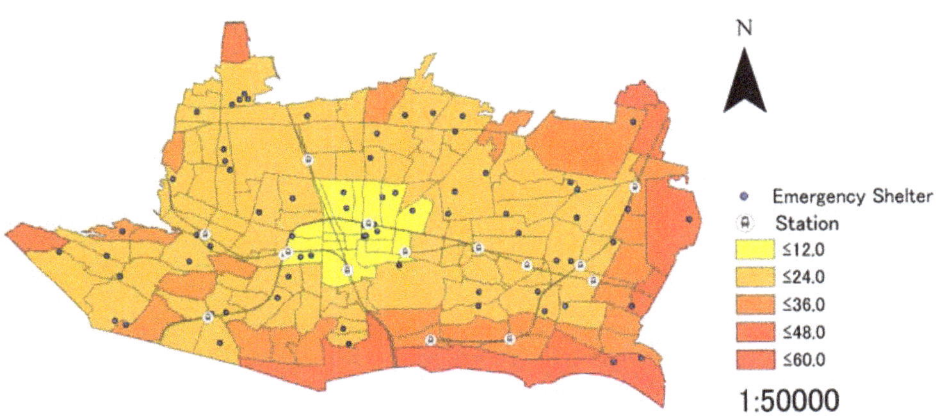

Figure 6. Result for Fuchu City without a weighted coefficient.

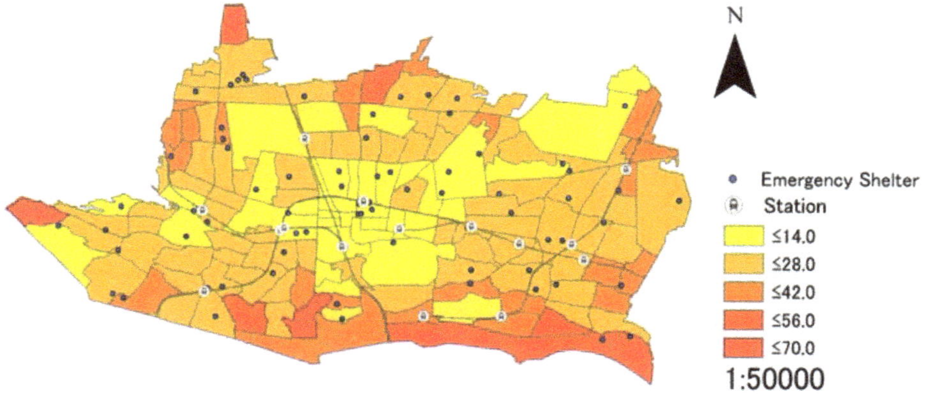

Figure 7. Result for Fuchu City with the specialization coefficient of elderly population as a weighted coefficient.

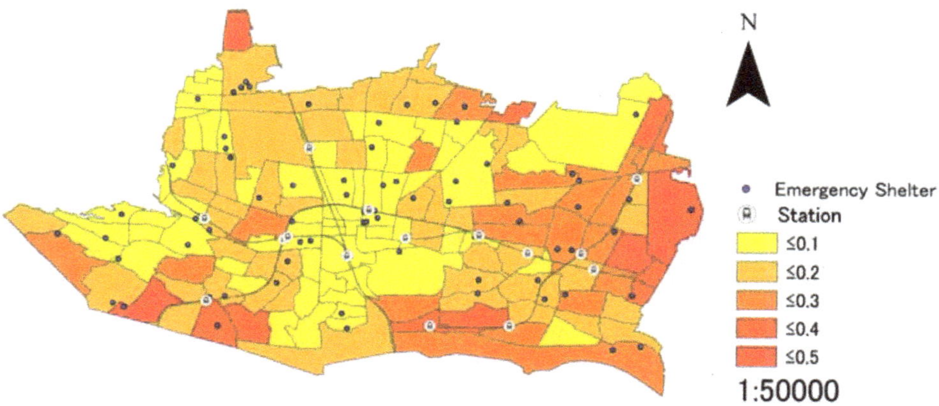

Figure 8. Result for Fuchu City with the ratio of permanent population as a weighted coefficient.

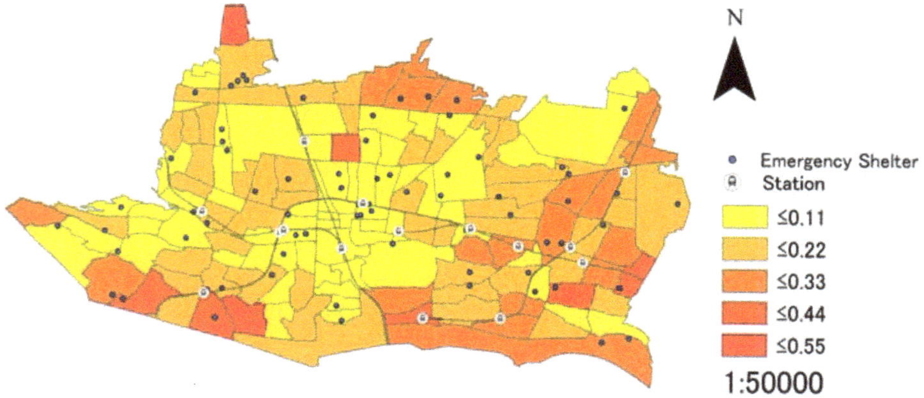

Figure 9. Result for Fuchu City with the ratio of elderly population as a weighted coefficient.

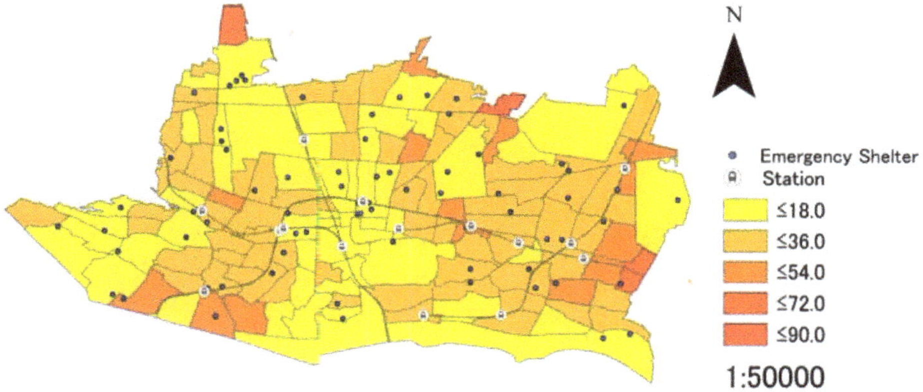

Figure 10. Result for Fuchu City with the specialization coefficient of population density as a weighted coefficient.

5.3. Results for Toda City

Figures 11–15 shows the results and n value is 5 for Toda City. From these figures, it is evident that there is a large difference between districts when not using a weighted coefficient (Figure 11), and when using the ratio of permanent population and the ratio of elderly population as the weighted coefficients (Figures 13 and 14). Without a weighted coefficient, emergency shelters concentrated in the districts that have low assessment values. With the ratio of permanent population and the ratio of elderly population as weighted coefficients, there are districts with extremely high assessment values in central part, which have a large number of populations and elderly populations.

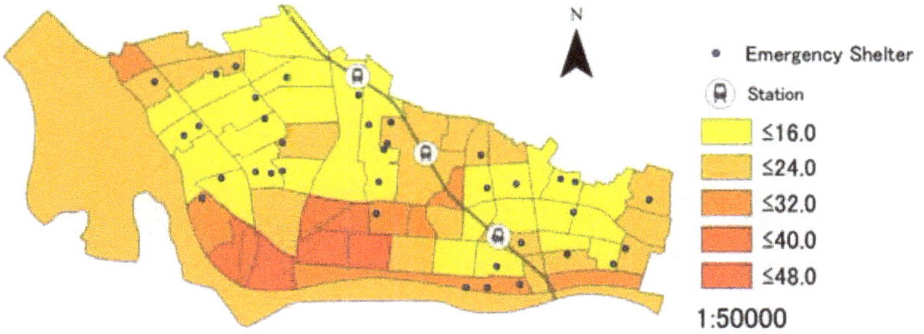

Figure 11. Result for Toda City without a weighted coefficient was not used.

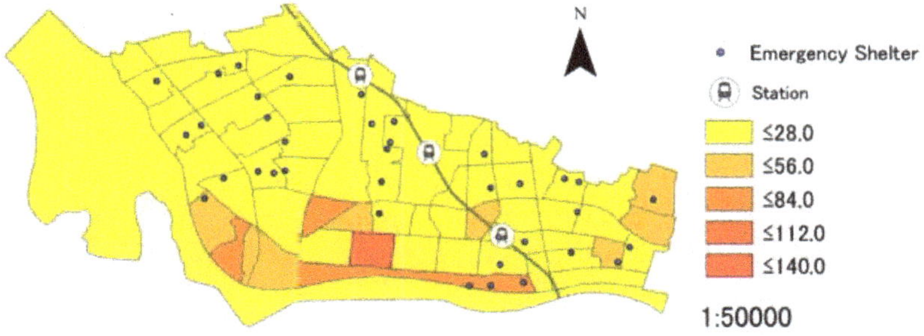

Figure 12. Result for Toda City with the specialization coefficient of elderly population as a weighted coefficient.

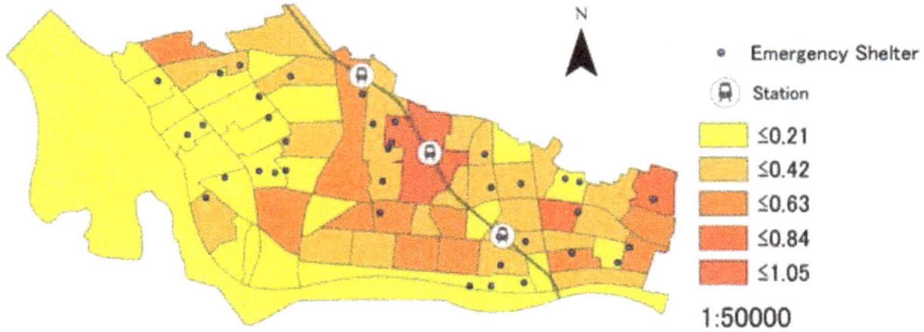

Figure 13. Result for Toda City with the ratio of permanent population as a weighted coefficient.

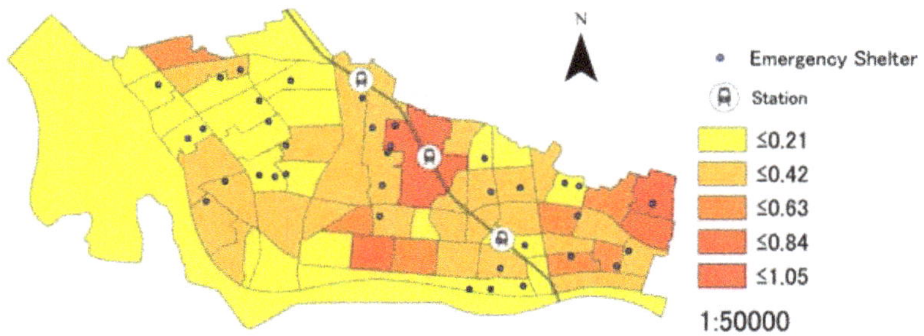

Figure 14. Result for Toda City with the ratio of elderly population as a weighted coefficient.

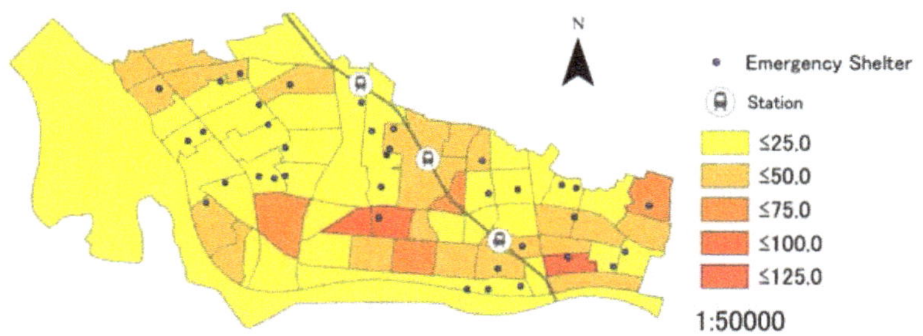

Figure 15. Result for Toda City with the specialization coefficient of population density as a weighted coefficient.

5.4. Results for Shobara City

Figures 16–20 shows the results and the n value is 1 for Shobara City. From these figures, it is evident that there is a large difference between districts when not using a weighted coefficient (Figure 16), and when using the specialization coefficient of elder population as the weighted coefficients (Figure 17). In both these two cases, districts with extremely high assessment values are distributed in the entire Shobara City, and they are far from emergency shelters. However, most of districts have low assessment values when using the specialization coefficient of population density as weighted coefficients (Figure 20).

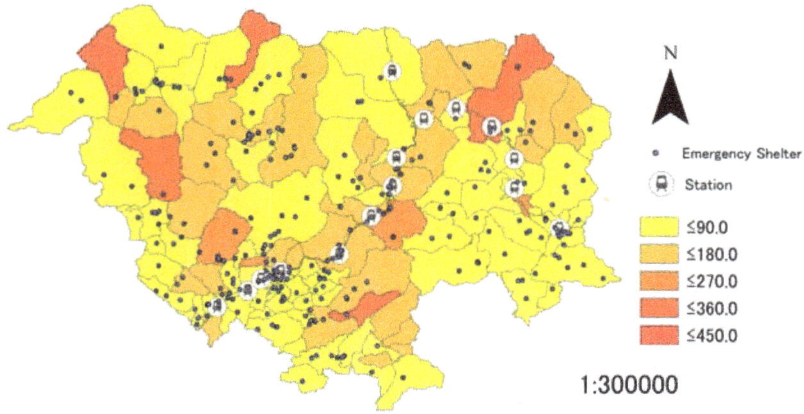

Figure 16. Result for Shobara City without a weighted coefficient.

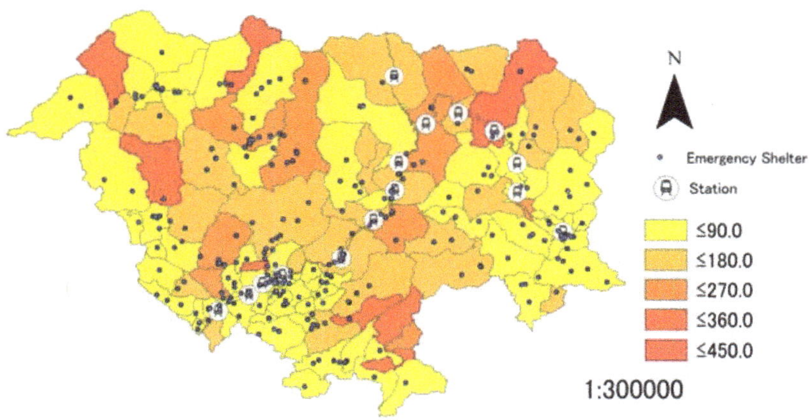

Figure 17. Result for Shobara City with the specialization coefficient of elderly population as a weighted coefficient.

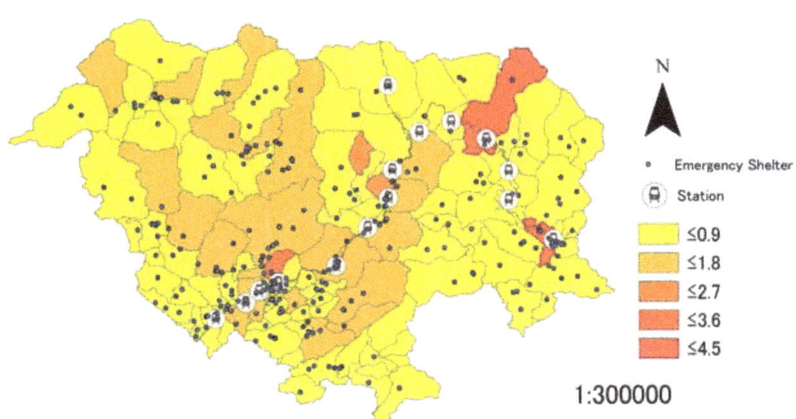

Figure 18. Result for Shobara City with the ratio of permanent population as a weighted coefficient.

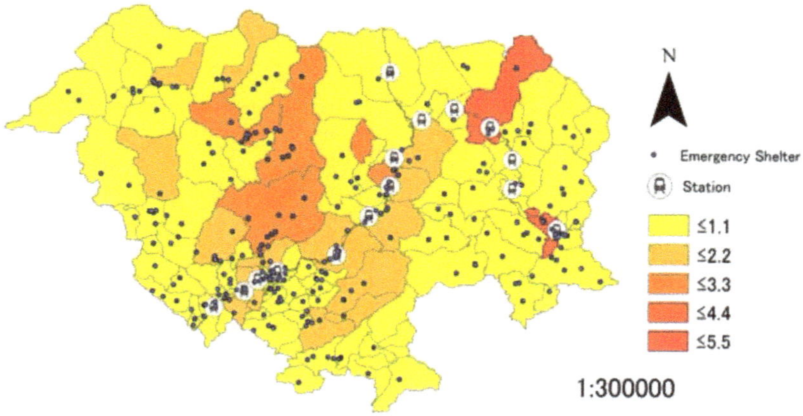

Figure 19. Result for Shobara City with the ratio of elderly population as a weighted coefficient.

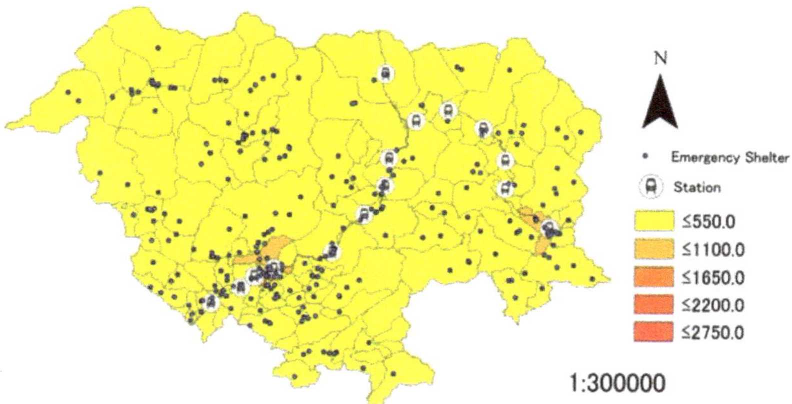

Figure 20. Result for Shobara City with the specialization coefficient of population density as a weighted coefficient.

6. Discussion

As mentioned in Section 3.2.4, this section will compare the results between multiple areas (Chofu, Fuchu, Toda and Shobara Cities) as well as the results with and without the four types of weighted coefficients that were related to the regional characteristics based on Figures 1–20 in order to verify the validity of the assessment method. Specifically, it is necessary to compare the results between multiple areas with different regional characteristics in order to verify the validity of the assessment method. Additionally, it is also necessary to compare the results with and without weighted coefficients, focusing on the increase and decrease of the assessment values for both cases. Furthermore, by verifying the assessment method, it is capable of proving the reliability of the results for the above four cities. Based on these two types of comparisons, focusing on regional characteristics, it is also possible to respectively grasp the suitability of emergency shelters allocation in the above four cities.

6.1. Comparison of the Results between Multiple Areas with Ddifferent Regional Characteristics

In this section, focusing on the maximum and minimum assessment values that are shown in Figures 1–20, the comparison of the results between target areas will be conducted. These assessment values are based on the suitability analysis for the emergency shelters allocation in the unit of districts of each target area mentioned in Section 3.2.3.

6.1.1. Results without a Weighted Coefficient

As mentioned in Section 3.2.3, when not using a weighted coefficient, analyses are conducted based on the linear distance between each district and each emergency shelter, and the scale of such facilities. Table 4 indicates the maximum (highest results) and minimum (lowest results) assessment values without a weighted coefficient. As shown in Table 2, though Shobara City has a large number of emergency shelters, because the area is extremely large, it is difficult to evenly place emergency shelters in the entire city. Therefore, there are some districts with extremely high assessment values in Shobara City. In contrast, because the areas of Chofu, Fuchu, and Toda Cities are relatively small, and the maximum assessment values in these three cities are not so high as that in Shobara City. Additionally, regarding the minimum assessment values, as all four cities have some districts with a concentration of emergency shelters, there is no noticeable difference between the four cities.

Table 4. Results without a weighted coefficient.

City	Chofu City	Fuchu City	Toda City	Shobara City
Maximum assessment value	59.81633	53.65750	47.21871	429.94276
Minimum assessment value	2.25878	1.872332	7.614647	2.11606

6.1.2. Results with the Specialization Coefficient of Elderly Population as a Weighted Coefficient

Table 5 indicates the results with the specialization coefficient of elderly population as a weighted coefficient. Districts where the specialization coefficient of elderly population is 0 are not considered. Regarding Shobara City, as the maximum assessment value without a weighted coefficient is significantly higher in comparison with the other three cities, the maximum assessment value remains high, regardless of the specialization coefficient of elderly population. Additionally, regarding Chofu, Fuchu, and Toda Cities, the maximum values of the specialization coefficient of elderly population are directly proportional to the maximum assessment values. However, regarding Fuchu City, as the minimum assessment value when not using a weighted coefficient is low, and the minimum value of the specialization coefficient of elderly population is also low, the minimum assessment value is lower in comparison with the other three cities. When comparing Chofu, Fuchu, and Toda Cities, which have almost the same areas as described in Table 2, when using the specialization coefficient of elderly population as a weighted coefficient, the values are directly proportional to the assessment values in these three cities.

Table 5. Results with the specialization coefficient of elderly population as a weighted coefficient.

City	Chofu City	Fuchu City	Toda City	Shobara City
Maximum value of specialization coefficient of elderly population	3.82633	2.02867	3.12336	2.35177
Minimum value of specialization coefficient of elderly population	0.26281	0.04191	0.14825	0.36911
Maximum assessment value	142.22402	67.88536	136.41473	448.82895
Minimum assessment value	2.24888	0.82389	2.23980	1.48857

6.1.3. Results with the Ratio of Permanent Population as a Weighted Coefficient

Table 6 indicates the results with the ratio of permanent population as a weighted coefficient. Districts where the ratio is 0 are not considered. The maximum and minimum values of the ratio of permanent population are directly proportional to the maximum and minimum assessment values when using the ratio of permanent population as a weighted coefficient, as it is made clear in Table 6.

Table 6. Results with the ratio of permanent population as a weighted coefficient.

City	Chofu City	Fuchu City	Toda City	Shobara City
Maximum value of ratio of permanent population	0.02835	0.01799	0.05574	0.06203
Minimum value of ratio of permanent population	0.00042	0.00009	0.00004	0.00116
Maximum assessment value	0.54371	0.47256	1.00212	5.68238
Minimum assessment value	0.01916	0.00166	0.00037	0.03124

6.1.4. Results with the Ratio of Elderly Population as a Weighted Coefficient

Table 7 indicates the results with the ratio of elderly population as a weighted coefficient. Districts where the ratio of elderly population is 0 are not considered. In a similar manner as the previous section, the maximum and minimum values of the ratio of elderly population are directly proportional to the maximum and minimum assessment values when using the ratio of elderly population as the weighted coefficient.

Table 7. Results with the ratio of elderly population as a weighted coefficient.

City	Chofu City	Fuchu City	Toda City	Shobara City
Maximum value of the ratio of elderly population	0.04965	0.02430	0.04953	0.06573
Minimum value of the ratio of elderly population	0.00030	0.00004	0.00011	0.00153
Maximum assessment value	0.93303	0.54209	1.04558	5.68238
Minimum assessment value	0.01359	0.00075	0.00202	0.03124

6.1.5. Results with the Specialization Coefficient of Population Density as a Weighted Coefficient

Table 8 indicates the results with the specialization coefficient of population density as a weighted coefficient. Districts where the specialization coefficient of population density is 0 are not considered. The maximum values of the specialization coefficient of population density are directly proportional to the maximum assessment values in all four cities. Additionally, regarding Toda City, as the minimum value of the specialization coefficient of population density is extremely low, the minimum assessment value is also low. On the other hand, regarding Shobara City, as the minimum value of the specialization coefficient of population density is extremely high, the minimum assessment value is also high. When comparing Chofu and Fuchu Cities, as there is a small difference between each minimum value of the specialization coefficient of population density, there is likewise a small difference between each minimum assessment value.

Table 8. Results with the specialization coefficient of population density as a weighted coefficient.

City	Chofu City	Fuchu City	Toda City	Shobara City
Maximum value of specialization coefficient of population density	2.88567	4.87911	5.85714	107.82974
Minimum value of t specialization coefficient of population density	0.02029	0.02364	0.00113	0.09622
Maximum assessment value	79.00819	86.45790	123.35271	2,691.62524
Minimum assessment value	0.75399	0.44378	0.00932	3.23330

6.1.6. Summary

This section will summarize the comparison results between multiple areas, while focusing on the maximum and minimum assessment values.

(1) Maximum assessment value

Regarding Shobara City, the maximum assessment value when not using a weighted coefficient is extremely high, and all of the values with different types of weighted coefficients are also significantly higher in comparison with the other three cities. Additionally, when comparing Chofu, Fuchu, and Toda Cities, as there is a small difference between the maximum assessment values when not using a weighted coefficient, the maximum values of weighted coefficients tend to be directly proportional to the maximum assessment values. Therefore, it is possible to verify the validity of the assessment method in the present research.

(2) Minimum assessment value

For all cities, as there is a small difference between the minimum assessment values when not using a weighted coefficient, the minimum values of weighted coefficients tend to be directly proportional to the minimum assessment values. Therefore, similar to when focusing on the maximum assessment value, it is possible to verify the validity of the assessment method in the present research.

6.2. *Comparison of the Results with and without Weighted Coefficients*

Based on the results of Section 5, this section will compare the results with and without the weighted coefficients, and verify the validity of the assessment method by focusing on the increase and decrease of assessment values in these two cases.

6.2.1. Comparison Concerning the Specialization Coefficient of Elderly Population

Districts with high assessment values also have high assessment values when not using a weighted coefficient, when using the specialization coefficient of the ratio of elderly population as a weighted coefficient. The same trend is seen between districts with low assessment values. Additionally, it is difficult to see any significant difference between the specialization coefficient of elderly population for each district, because the value is derived by dividing the rate of the population who are over 75 years old in each district by the same rate in the entire target area.

6.2.2. Comparison Concerning the Ratio of Permanent Population

When using the ratio of permanent population as a weighted coefficient, many districts with high assessment values also have high values, while many districts with low assessment values also have low values. When not using a weighted coefficient, many districts with high assessment values are far from any stations and city centres, while many districts with low assessment values are close to stations and the city centre. Therefore, when using the ratio of elderly population, which greatly differs between the districts as a weighted coefficient, the assessment values are reversed in the cases without a weighted coefficient and with the ratio of elderly population as the weighted coefficient.

6.2.3. Comparison Concerning the Ratio of Elderly Population

The trend is similar to that of the previous section when using the ratio of elderly population as a weighted coefficient. Specifically, many districts with high assessment values also have high values, while many districts with low assessment values also have low values. When not using a weighted coefficient, many districts with high values are far from any stations and city centres, while the districts with low assessment values are close to stations and city centres. Therefore, when using the ratio of elderly population that greatly differs between the districts as a weighted coefficient, the assessment values are reversed in the cases without a weighted coefficient and with the ratio of elderly population as a weighted coefficient.

6.2.4. Comparison Concerning the Specialization Coefficient of Population Density

When using the specialization coefficient of population density as the weighted coefficient, districts with high assessment values also have high values, while districts with low assessment values also have low values. The specialization coefficient of population density is high around stations and city centre, and it is lower as the distance from stations and city centers increases. When not using a weighted coefficient, the assessment values are high in districts that are far from any stations and city centre, while the assessment values are low in districts close to stations and city centre. Therefore, when using the specialization coefficient of population density, which greatly differs between the districts as a weighted coefficient, the assessment values are reversed in the cases without a weighted coefficient and with the ratio of elderly population as a weighted coefficient.

6.2.5. Summary

This section will summarize the comparison results with and without weighted coefficients. There is no significant difference between the assessment values with and without this weighted coefficient because there is a small difference between the specialization coefficient of elderly population in each district. However, as there is a significant difference between the other 3 weighted coefficients in each district, the weighed coefficients are directly proportional to the assessment values. In other words, when not using a weighted coefficient, districts with high assessment values have low weighted coefficient values, while districts with low assessment values have high weighted coefficient values. Therefore, the assessment values are reversed, depending on whether the other three weighted coefficients are used or not. Additionally, based on the above comparison results, it is possible to verify the validity of the assessment method in the present research.

7. Conclusions

The conclusion of the present research can be summarized in the following four points.

(1) The method in the present research modifies the p-median model that derives the best facility location and conducts a suitability analysis for the emergency shelters allocation in each district. The weighted coefficients such as the specialization coefficient of elderly population, the ratio of permanent population, the ratio of elderly population, and the specialization coefficient of population density that is related to the regional characteristics are integrated into the suitability analysis for the emergency shelters allocation in Japan, using the linear distance between each district and each emergency shelter as well as the coverage of such emergency shelters.

(2) As the quantitative data related to the above 4 types of weighted coefficients, the linear distance between each district and each emergency shelter, and the facility scale of emergency shelters are used to conduct a suitability analysis, the results are also quantitative, making it a useful indicator to analyze the suitability of emergency shelters allocation. Additionally, while there are only four types of weighted coefficients that are adopted in the present research, other regional characteristics can be adopted as weighed coefficients to expand the assessment method. Furthermore, it is possible to compare the sufficiency levels of emergency shelters between districts and point out the specific districts that are short of emergency shelters, as the suitability of emergency shelters allocation are analyzed by each district. Additionally, it is also possible to visually understand the suitability of emergency shelters allocation on a small spatial scale as the unit of districts, as the results are displayed on the digital maps of GIS.

(3) In the present research, the above four types of weighted coefficients, the linear distance between each district and each emergency shelter as well as the facility scale of emergency shelters are calculated while using open data, such as the National Census and the National Land Numerical Information. As analyses in the present study are conducted based on public information, by obtaining population data and geospatial data that are similar to the present research, analyses can be conducted while using data in other areas, as well as for the past and future. Therefore, the assessment method in

the present research has a high temporal reproducibility as well as spatial reproducibility. For example, by using the "future population estimate by region in Japan" of the National Institute of Population and Social Security Research [43] as future data, the shortage or overage of emergency shelters in the future can be evaluated.

(4) The present research has the limitation of data availability. Specifically, the data that are related to emergency shelters of contiguous cities could not be used, as the necessary data concerning the facility scale of emergency shelters were not available. Therefore, the assessment values of districts that were located near the administrative boundaries between the target areas and neighboring cities could not be considered to be accurate. For this reason, it is necessary to create original data related to emergency shelters.

The following are two issues that can be considered as future research topics.

(1) Calculation of the distance between each district and each emergency shelter

In the present research, as the linear distance between each district and each emergency shelter was adopted, such a distance differs from the actual road distance. Therefore, it is possible to increase the accuracy of the results by adopting the road distance. Manrique (2013) [44] demonstrated that the road distance of narrow roads that are only passable for those on foot is 1.271 times longer than the linear distance, and the road distance of wide roads that are drivable is 1.415 times longer than the linear distance. Referring to the results, it is possible to calculate road distance based on the linear distance.

(2) Application to other facilities in the time of accidents

While the present research focused on emergency shelters, the assessment method of the present research may also be applied to the suitability analysis of temporary evacuation shelters in the time of accidents.

Author Contributions: T.A. collected and processed the data used for variables, and conducted a suitability analysis for the emergency shelters allocation. K.Y. carried out background work, and developed the framework and process of analysis. She also initially drafted the paper. T.A. and K.Y. contributed to write up and review, and approved the paper manuscript.

Funding: This research received no external funding.

Conflicts of Interest: The authors declare no conflict of interest.

References

1. Susumu, Y. Magnitude and Earthquake. (In Japanese). Available online: https://www.s-yamaga.jp/nanimono/chikyu/magnitude-kazu.htm (accessed on 27 January 2019).
2. Recruit Sumai Company Ltd. Dictionary of Housing Terms. (In Japanese). Available online: https://suumo.jp/yougo/h/hinanngennkaikyori/ (accessed on 27 January 2019).
3. Cabinet Office. Awareness Related to the Development of Walkable Cities. (In Japanese). Available online: https://survey.gov-online.go.jp/h21/h21-aruite/2-2.html (accessed on 27 January 2019).
4. Civiletti, G.; Camassi, R.; Monteiro, R. Italian Seismic Sequences: Year 2000, the Emergency Phase in Romagn. *Procedia Eng.* **2016**, *161*, 2088–2092. [CrossRef]
5. Cerchiello, V.; Ceresa, P.; Monteiro, R.; Komendantova, N. Assessment of Social Vulnerability to Seismic Hazard in Nablus, Palestine. *Int. J. Disaster Risk Reduct.* **2018**, *28*, 491–506. [CrossRef]
6. Contreras, D.; Kienberger, S. GIS in the Vulnerability Assessment and Recovery Process in a Community with Elderly and Disabled People after a Disaster. In Proceedings of the International Conference on Rebuilding Sustainable Communities with the Elderly and Disabled People after Disasters, Boston, MA, USA, 12–15 July 2010; p. 20.
7. Awotona, A. *Rebuilding Sustainable Communities with Vulnerable Populations after the Cameras Have Gone: A Worldwide Study*; Cambridge Scholars Publishing: Cambridge, UK, 2012.
8. Cabinet Office. Measures for People Requiring Assistance During a Disaster. (In Japanese). Available online: http://www.bousai.go.jp/taisaku/hisaisyagyousei/youengosya/ (accessed on 11 July 2019).

9. Ministry of Health, Labour and Welfare. White Paper on Aging Society. (In Japanese). Available online: https://www8.cao.go.jp/kourei/whitepaper/w-2016/html/zenbun/index.html (accessed on 11 July 2019).
10. Furihata, T.; Matsuoka, K. A Facility Location Analysis Considering Spatial Patterns of Distributed Users. *Theory Appl. GIS* **1994**, *2*, 91–99. (In Japanese) [CrossRef]
11. Takeuchi, T.; Kondo, A. A Study about the Selection of the Place of Refuge at the Time of the Earthquake Tsunami: A Case Study in Susaki. *Infrastruct. Plan. Rev.* **2002**, *19*, 297–304. (In Japanese) [CrossRef]
12. Kongsomsaksakul, S.; Yang, C.; Chen, A. Shelter Location-Allocation Model for Flood Evacuation Planning. *J. East. Asia Soc. Transp. Stud.* **2005**, *6*, 4237–4252.
13. Aso, T.; Matsumoto, Y.; Morishita, K. A Study on the Optical Arrangement of Shelters Using Genetic Algorithm. *Dep. Bull. Pap. Yamaguchi Univ.* **2007**, *58*, 31–36. (In Japanese)
14. Wei, X.; Okada, N.; Takeuchi, Y.; Kajitani, Y. A Diagnosis Model for Disaster Shelter Planning from the Viewpoint of Local People-Case Study of Nagata Ward in Kobe City, Hyogo Prefecture, Japan. *Disaster Prev. Res. Inst. Annu.* **2008**, *50*, 233–239.
15. Ng, M.; Park, J. A Hybrid Bilevel Model for the Optimal Shelter Assignment in Emergency Evacuations. *Comput. -Aided Civ. Infrastruct. Eng.* **2010**, *25*, 547–556. [CrossRef]
16. Tai, C.; Lee, Y.; Yuan, C. Urban Disaster Prevention Shelter Location and Evacuation Behavior Analysis. *J. Archit. Build. Eng.* **2010**, *9*, 215–220. [CrossRef]
17. Kitajima, S. A study of Evacuation Shelter Location Using Network analysis. *Ann. Jpn. Assoc. Urban Sociol.* **2013**, *47*, 211–216. (In Japanese)
18. Yu, J.; Wen, J. Multi-criteria Satisfaction Assessment of the Spatial Distribution of Urban Emergency Shelters Based on High-Precision Population Estimation. *Int. J. Disaster Risk Sci.* **2016**, *7*, 413–429. [CrossRef]
19. Vecere, A.; Monteiro, R.; Ammann, W.J. Comparative Analysis of Existing Tools for Assessment of Post-Earthquake Short-Term Lodging Needs. *Procedia Eng.* **2016**, *161*, 2217–2221. [CrossRef]
20. Vecere, A.; Monteiro, R.; Ammann, W.J.; Giovinazzi, S.; Santos, R.H.M. Predictive Models for Post Disaster Shelter Needs Assessment. *Int. J. Disaster Risk Sci.* **2017**, *21*, 44–62. [CrossRef]
21. Trivedi, A.; Singh, A. A Hybrid Multi-Objective Decision Model for Emergency Shelter Location-Relocation Projects Using Fuzzy Analytic Hierarchy Process and Goal Programming Approach. *Int. J. Proj. Manag.* **2017**, *35*, 827–840. [CrossRef]
22. Xu, W.; Zhao, X.; Ma, Y.; Li, Y.; Qin, L.; Wang, Y.; Du, J. A Multi-Objective Optimization Based Method for Evaluating Earthquake Shelter Location–Allocation. *Geomat Nat. Hazards Risk* **2018**, *9*, 662–677. [CrossRef]
23. Nozaki, K.; Murakami, H. Study on Location Analysis of Tsunami Emergency Shelter and Inhabitants' Preparedness for the Coming Nankai Trough Earthquake: A case of Aoki Area, Miyazaki City. In Proceedings of the AIJ Chugoku Chapter Architectural Research Meeting, Yamaguchi, Japan, 2–3 March 2019; pp. 959–962. (In Japanese)
24. Yamada, T.; Takagi, A. A Planning Support System of Shelter Location from the Viewpoint of Resident. *Infrastruct. Plan. Rev.* **2004**, *21*, 325–334. (In Japanese) [CrossRef]
25. Notsuda, M.; Kishimoto, T. A study on Allocation and Location Model of Regional Evacuation Facilities: Studies on Location Planning of Evacuation Facilities No.1. *J. Archit. Plan.* **2005**, *70*, 115–122. (In Japanese) [CrossRef]
26. Nakai, S.; Kana, K.; Sakuma, Y. Research on Actual Condition of Existing Vacant Houses and Study on Possibility of Utilizing Them as "Spaces for Disaster Prevention" in the Densely Built-up Area: The Case of Tsuruhashi Area in Osaka City. *J. City Plan. Inst. Jpn.* **2012**, *47*, 1063–1068. (In Japanese)
27. Ikenaga, T.; Gokon, H.; Meguro, K. Feasibility Study on Temporary Housing Supply System Using Vacant Houses: A Case Study in Wakayama Prefecture. *J. Soc. Saf. Sci.* **2017**, *30/31*, 13–23. (In Japanese)
28. Asano, S.; Hirohata, D. Examination of the Guideline about the Choice of the Site Proposed for Construction on Emergency Temporary Housing for Public Lands: A Case Study in Shima-City. *J. City Plan. Inst. Jpn.* **2013**, *48*, 801–806. (In Japanese)
29. Miyoshi, S.; Kana, K.; Sakuma, Y. Evaluation Framework of Project for Making Open Space by Temporary Use of Private Land in Densely Built-Up Areas: A Case of Kobe Project. *J. City Plan. Inst. Jpn.* **2017**, *52*, 293–300. (In Japanese)
30. Araki, Y.; Tsuboi, S.; Hokojo, A. Setting up Patterns of Non-Designated Emergency Shelters after the Tsunami Evacuation: A Case Study in Kamaishi City after the Great East Japan Earthquake and Tsunami. *J. Archit. Plan.* **2017**, *741*, 2885–2895. (In Japanese) [CrossRef]

31. Sasaki, T.; Sasaki, N.; Katsumata, H. Study on the Status of Designation of Emergency Place and/or Evacuation Shelters, and the Possibility of Buddhist Temple Utilization When Widspreaded Disasters Occur: Based on a Questionnaire Survey of Local Governments Nationwide. *J. Archit. Plan.* **2018**, *747*, 877–884. (In Japanese) [CrossRef]
32. Umeki, H.; Nakamura, Y.; Fujimoto, M.; Mizumoto, T.; Suwa, H.; Arakawa, Y.; Yasumoto, K. Evacuation Center Determination Method Considering Bias of Congestion Degree on Areas. *J. Inf. Process.* **2019**, *60*, 608–616. (In Japanese)
33. Kar, B.; Hodgson, M.E. A GIS Based Model to Determine Site Suitability of Emergency Evacuation Shelters. *Trans. GIS* **2008**, *12*, 227–248. [CrossRef]
34. Alçada-Almeida, L.; Tralhão, L.; Santos, L.; Coutinho-Rodrigues, J. A Multiobjective Approach to Locate Emergency Shelters and Identify Evacuation Routes in Urban Areas. *Geogr. Anal.* **2009**, *41*, 9–29. [CrossRef]
35. Park, S.; van de Lindt, J.W.; Gupta, R.; Cox, D. Method to Determine the Locations of Tsunami Vertical Evacuation Shelters. *Nat. Hazards* **2012**, *63*, 891–908. [CrossRef]
36. Li, A.C.Y.; Nozick, L.; Xu, N.; Davidson, R. Shelter Location and Transportation Planning under Hurricane Conditions. *Transp. Res. Part E Logist. Transp. Rev.* **2012**, *48*, 715–729. [CrossRef]
37. Anhorn, J.; Khazai, B. Open Space Suitability Analysis for Emergency Shelter after an Earthquake. *Nat. Hazard Earth Sci. Syst.* **2015**, *15*, 789–803. [CrossRef]
38. Bayram, V.; Yaman, H. A Stochastic Programming Approach for Shelter Location and Evacuation Planning. *RAIRO Oper. Res.* **2018**, *52*, 779–805. [CrossRef]
39. Xu, J.; Yina, X.; Chen, D.; An, J.; Nie, G. Multi-Criteria Location Model of Earthquake Evacuation Shelters to Aid in Urban Planning. *Int. J. Disaster Risk Reduct.* **2016**, *20*, 51–62. [CrossRef]
40. Saino, T. Some Geographical Considerations on Aging of Population in Fukuoka City. *Kikan Chirigaku* **1992**, *44*, 29–36. (In Japanese) [CrossRef]
41. Special Committee for Promotion of National Health Promotion Exercise (Health Japan 21) in the 21st Century, Ministry of Health, Labour and Welfare, Evaluation Sheet. (In Japanese). Available online: https://www.mhlw.go.jp/file/05-Shingikai-10601000-Daijinkanboukouseikagakuka-Kouseikagakuka/0000166296_7.pdf (accessed on 27 January 2019).
42. Shigenobu, A.; Higaki, Y.; Ohtani, H. Calculation of Doorstop Make shift House Ground Product and the Rate of the Makeshift House Housing Complex Area Ratio Classified by Open Space: From the Example of the Urgent Makeshift House Housing Complex of the Great East Japan Earthquake (Miyagi Prefecture). In Proceedings of the AIJ Shikoku Chapter Architectural Research Meeting, Kami, Japan, 11 May 2013; pp. 115–116. (In Japanese)
43. National Institute of Population and Social Security Research, Population Projections of Each Region for Japan. 2018 March; (In Japanese). Available online: http://www.ipss.go.jp/pp-shicyoson/j/shicyoson18/2gaiyo_hyo/gaiyo.asp (accessed on 27 January 2019).
44. Manrique, R.; Luis, C. Proposal of an Evaluation Method of a Compact City Model. Ph.D. Thesis, University of Electro-Communications, Tokyo, Japan, 2014.

geosciences

MDPI

Article

Use of Macroseismic Intensity Data to Validate a Regionally Adjustable Ground Motion Prediction Model

Yuxiang Tang [1,*], Nelson Lam [1,2], Hing-Ho Tsang [2,3] and Elisa Lumantarna [1,2]

[1] Department of Infrastructure Engineering, The University of Melbourne, Parkville, VIC 3010, Australia
[2] Bushfire and Natural Hazards Cooperative Research Centre, Melbourne, VIC 3002, Australia
[3] Centre for Sustainable Infrastructure, Swinburne University of Technology, Melbourne, VIC 3122, Australia
* Correspondence: yuxiang.tang@unimelb.edu.au

Received: 2 September 2019; Accepted: 27 September 2019; Published: 30 September 2019

Abstract: In low-to-moderate seismicity (intraplate) regions where locally recorded strong motion data are too scare for conventional regression analysis, stochastic simulations based on seismological modelling have often been used to predict ground motions of future earthquakes. This modelling methodology has been practised in Central and Eastern North America (CENA) for decades. It is cautioned that ground motion prediction equations (GMPE) that have been developed for use in CENA might not always be suited for use in another intraplate region because of differences in the crustal structure. This paper introduces a regionally adjustable GMPE, known as the component attenuation model (CAM), by which a diversity of crustal conditions can be covered in one model. Input parameters into CAM have been configured in the same manner as a seismological model, as both types of models are based on decoupling the spectral properties of earthquake ground motions into a generic source factor and a regionally specific path factor (including anelastic and geometric attenuation factors) along with a crustal factor. Unlike seismological modelling, CAM is essentially a GMPE that can be adapted readily for use in different regions (or different areas within a region) without the need of undertaking any stochastic simulations, providing that parameters characterising the crustal structure have been identified. In addressing the challenge of validating a GMPE for use in an area where instrumental data are scarce, modified Mercalli intensity (MMI) data inferred from peak ground velocity values predicted by CAM are compared with records of MMI of past earthquake events, as reported in historical archives. South-Eastern Australia (SEA) and South-Eastern China (SEC) are the two study regions used in this article for demonstrating the viability of CAM as a ground motion prediction tool in an intraplate environment.

Keywords: seismic hazard; attenuation; GMPE; crustal model; MMI; intraplate region

1. Introduction

A ground motion prediction equation (GMPE) is a set of algebraic functions of earthquake magnitude, distance, and a site parameter, and may also include parameters to identify the style of faulting [1–5]. GMPEs are used to define the characteristics of ground motions for specific regions and earthquake scenarios (expressed in terms of magnitude-distance, or M-R combinations). In tectonically active regions (e.g., Western North America) where plenty of strong motion data can be captured by a network of densely distributed recording instruments, GMPEs are typically developed from regression analysis of recorded strong motion data. Empirical GMPEs, which were derived mainly from field recordings, should be capable of capturing regional specific earthquake ground motion characteristics [6,7]. However, empirical GMPEs may give results that are very sensitive to data from isolated records and the type of regression techniques adopted, and more so when data are scarce [8].

For tectonically stable regions of low-to-moderate seismicity, the conventional approach based on the regression of empirical data may not be feasible for developing GMPEs. To overcome the challenge of paucity in strong motion data, stochastic simulations of a seismological model expressed in the form of Fourier amplitude spectrum (FAS) can be undertaken to generate synthetic accelerograms [9,10]. This approach basically makes use of a seismological model to define the amplitude of the individual sinusoids constituting the acceleration time histories on the ground surface, whilst assigning random phase angles to the sinusoids. This approach of synthesizing ground motions is known as stochastic simulations. Standard calculation procedures may then be applied to derive the intensity of ground motion for any given earthquake scenario. A set of GMPE for the target regions can then be developed using the synthetic data obtained from simulations applied in a repetitive fashion to cover for different M-R combinations.

Complexity in the earthquake generation process and the associated uncertainties that are embodied in empirical GMPEs cannot possibly be captured completely by seismological modelling, nor by stochastic simulations. To address these intrinsic deficiencies of seismological modelling, a new class of (semiempirical) modelling approaches, namely the hybrid empirical method (HEM) [11,12] and the referenced empirical method (REM) [13,14], have been developed. However, the application of either HEM or REM requires a suitable "host region" containing abundant strong motion recordings and well-developed GMPEs.

In intraplate regions like South-Eastern Australia (SEA) and South-Eastern China (SEC), it is difficult to find a suitable "host region", or alternative well-established GMPEs, for applying HEM or REM. However, very useful macroseismic intensity information expressed in terms of the modified Mercalli intensity (MMI) has been recorded on isoseismal maps for earthquakes that occurred in this region for over hundreds of years. This type of data has been used in investigations for studying the attenuation behaviour of earthquake ground motions in Australia [15,16] and in many other regions of low-to-moderate seismicity [17]. The analysis of MMI data can result in the development of a GMPE that can be representative of the strong motion transmission properties of earthquake-affected areas based on observations over a long-time span (without requiring recording instruments to be placed close to the epicentre of any earthquake). For this reason, MMI data that have been recorded to date have much better coverage of strong motion conditions than instrumented data in regions of low-to-moderate seismicity, both in terms of time and space. Many current GMPEs developed in regions of low-to-moderate seismicity for the modelling of seismic hazards were derived from datasets of small magnitude earthquake events (for example, GMPEs developed in SEA were developed from datasets with a maximum moment magnitude M = 5.4 [18,19]). These GMPEs can be heavily biased to ground motion behaviour typifying small magnitude earthquakes. Their applicability in modelling large magnitude earthquake events occurring in the future is in doubt.

Another modelling challenge is reconciling modern GMPEs, which typically make use of peak ground acceleration (PGA), peak ground velocity (PGV), or response spectral accelerations (RSA) as intensity measures, with a GMPE derived from MMI data. Converting macroseismic intensity MMI data into ground motion parameters such as PGA or PGV has been studied by many scholars in the past few decades. A simple, and well-known, MMI-PGV conversion relationship was recommended by Newmark and Rosenblueth [20], Gaull et al. [21], and Lam et al. [15]. More rigorous conversion relationships have also been developed more recently (e.g., [22–24]). Regression analysis of MMI data, as obtained from historical archives, can therefore be transformed into attenuation relationships that are expressed in terms of PGA, PGV, or RSA.

In the absence of detailed spectral properties of a historical earthquake (which could only be derived from analyses of instrumented records), PGV is the preferred ground motion parameter, as opposed to PGA or RSA, for characterising the intensity of the earthquake for reasons outlined by Bommer and Alarcon [25]. First, PGV is a reliable, simple, and measurable intensity metric, which has a good correlation with damage distribution [26]. The accuracy of "shake maps" relies on good correlations between PGV and MMI values [22]. Second, PGV data can be employed for estimating the risk of damage to buried pipelines because of the good correlations between horizontal PGV and material strains [27]. Thus, fragility functions for buried pipelines expressed in terms of PGV can be found in open sources [28]. Third, PGV data can also be used for assessing the risks of liquefaction [29–31]. Fourth, PGV is one of the three parameters for scaling an elastic response spectrum model for engineering applications [32,33]. The seismic action model stipulated by the current Australian standard is based on scaling design PGV values that were derived from probabilistic seismic hazard analysis (PSHA), employing GMPEs found on MMI data [34]. This is evident of PGV being recognised as the key ground motion intensity measure.

This paper is aimed at introducing a generic, regionally adjustable GMPE (known as the component attenuation model with acronym: CAM) for the prediction of PGV for given earthquake scenarios in low-to-moderate seismicity regions. The use of macroseismic historical MMI data collected from SEA and SEC regions (the distribution of magnitude and distance for collected data can be found in Figure 1) to demonstrate the accuracy of predictions by CAM is also presented.

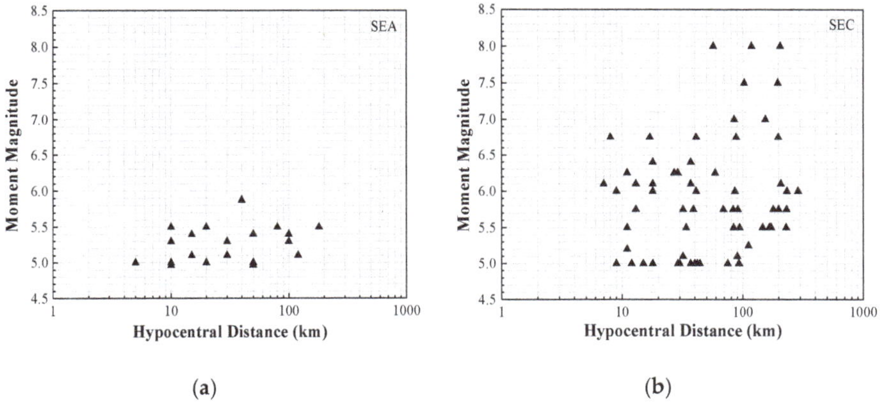

Figure 1. M-R combinations for selected regions. (**a**) South-Eastern Australia (SEA); (**b**) South-Eastern China (SEC).

2. The Component Attenuation Model (CAM) of PGV

The framework of CAM as a GMPE is defined in a generic functional form, which decouples the source and attenuation effects into different components, as shown by Equation (1).

$$\mathrm{PGV} = \Delta \times \alpha \times \beta \times G \times \gamma_c \times C \tag{1}$$

Equation (1) can also be presented in the logarithmic (base 10) format as shown by Equation (2):

$$\log\mathrm{PGV} = \log\Delta + \log\alpha + \log\beta + \log G + \log\gamma_c + \log C, \tag{2}$$

where Δ is the referenced PGV measured on reference hard rock sites (typifying Eastern North America) for the reference scenario of M = 6 and R = 30; α is the source factor, which is a function of the moment magnitude (M) and Brune stress drop ($\Delta\sigma$); β is the regional anelastic attenuation factor, which is responsible for the attenuation effects, excluding geometric attenuation (G); G is the geometric attenuation factor; γ_c is the crustal modification factor, which accounts for the amplification and attenuation phenomena within the upper 4 km of the rock crust, as well as amplification at the "mid-crust", which is in the surrounding of the location of the fault plane; and C is the calibration factor (which is used to minimise discrepancies between the model predictions and empirical recordings). The value of α is set as unity (and value of β and G close to unity) for a stress parameter value of 200 bars at M = 6 and R = 30 km. Seismological parameter values of stochastic simulations that CAM is based upon are summarised in Table 1. The listing of the values of model coefficients associated with each component factor in CAM are presented in the next section.

Table 1. Parameter values used in stochastic simulations for component attenuation model (CAM) modelling.

Parameter	Input Value
Ref. source shear wave velocity for hard rock (of Eastern North America)	β_{0sim} = 3.8 km/s [1]
Ref. source density for hard rock	ρ_{sim} = 2.8 g/cm^3 [1]
Source model	Generalised additive double-corner frequency model [35]
Spectral sag	$\varepsilon = 10^{0.605 - 0.255M}$
Distance	R = Hypocentral distance
Geometrical attenuation	Variable function (refer Table 2)
Stress drop, $\Delta\sigma$	$\Delta\sigma$ = 200 bars (default for intraplate regions)
Wave transmission quality factor	Q_0 = 120, 150, 200, 300, 400, 500, 600, 680, 800.
Exponential factor	$n = 0.0000008Q_0^2 - 0.0014Q_0 + 0.93$ [1] [36]
Time-averaged shear wave velocity for the top 30 m depth, V_{S30}	V_{S30} = 0.618, 0.76, 1.0, 1.2, 1.4, 1.6, 1.8, 2.0, 2.2, 2.4, 2.6, 2.78 km/s
Kappa factor	κ_0 = 0.001, 0.0025, 0.005, 0.0075, 0.01, 0.015, 0.02, 0.025, 0.03, 0.035, 0.04, 0.045, 0.05, 0.055, 0.06, 0.065, 0.07, 0.075, 0.08, 0.085, 0.09, 0.095, 0.1 s
Source duration	$0.5/f_a + 0.5/f_b$, where f_a and f_b are corner frequencies [35]
Path duration	0.05 × R, where R is the hypocentral distance [1]
Time step	dt = 0.002 s

[1] The equation is only used when the exponent value is unknown.

CAM, as introduced in the foregoing, is similar in methodology and philosophy to a generic ground motion prediction model (GMPE) introduced in reference [37]. A unique feature of CAM, which is not found in any GMPE, is the use of a geology-based crustal modelling approach, wherein crustal shear wave velocity profiles of bedrock to depths of tens of kilometres are used to derive modification factors of the upper crust (not to be confused with the site factor for modelling the effects of surficial sediments down to tens of metres only). Description of the crustal modelling methodology, complete with case studies, can be found throughout the rest of the article.

2.1. Generic Source Factor

The generic source factor is expressed in the form of Equation (3).

$$\log\alpha = a_1 \times M^{a_2} \times \Delta\sigma^{a_3} + a_4, \tag{3}$$

where a_1–a_4 are model coefficients. Equation (3), which was derived from the simulated data, has been normalised at M = 6. The model covers the range of M4–M8 and $\Delta\sigma$ = 30–300 bar.

2.2. Regional Whole Path Anelastic Attenuation Factor

The regional anelastic attenuation factor (β) is used to account for the attenuation of ground motions along the entire transmission path of the seismic wave from source to site. Results derived from stochastic simulations of the seismological model with increasing distance have been normalised at R = 30 km and the effects of geometric attenuation effect (G) have been removed (and accounted for separately). The anelastic attenuation factor so derived is expressed in the form of Equation (4):

$$\log\beta = (b_1 \times M + b_2) \times Q_0^{b_3} \times (\log R)^{b_4} + b_5, \tag{4}$$

where b_1–b_5 are model coefficients; M is moment magnitude, Q_0 is the regional dependent quality factor for wave transmission, and R is the hypocentral distance.

Higher frequency waves (with a larger number of wave cycles for a given distance) are more susceptible to anelastic attenuation along the wave travel path than lower frequency waves. Given that the frequency characteristics of an earthquake are magnitude-dependent, the extent of anelastic attenuation is accordingly magnitude-dependent. Hence, earthquake magnitude is a parameter in the predictive expression of Equation (4). The reliability (ability) of this expression to reproduce results of stochastic simulations of the upper crustal amplification phenomenon reasonably accurately is demonstrated in Section 3.

Predictions derived from Equations (3) and (4) have been subject to residual analysis, wherein the residual value is defined as $\delta = \log\left(Y_{sim}/Y_{pred}\right)$, where Y_{sim} and Y_{pred} are PGV measurements obtained from stochastic simulations and CAM predictions, respectively. $\delta > 0$ refers to underestimation of CAM and $\delta < 0$ refers to overestimation of CAM. A fourth order polynomial expression (Equation (5)) for defining the value of the adjustment factor $\beta_{adjustment\_factor}$ is used to adjust, through multiplication, predicted values of β from Equation (4) to minimise the values of the residuals, along with any other systematic modelling errors.

$$\beta_{adjustment\_factor} = b_6 R^4 + b_7 R^3 + b_8 R^2 + b_9 R + b_{10}, \quad (5)$$

where b_6–b_{10} are model coefficients.

2.3. Crustal Modification Factor

The crustal factor γ_c is to account for the combined effects of the amplification and attenuation of the upper earth crust, which is mainly dependent on the shear wave velocity profile in the upper 3–4 km of the earth crust. The literature refers to those phenomena as crustal modifications. The modification factor can be resolved into three components: (i) amplification of the upper crust; (ii) attenuation of the upper crust; and (iii) modification of the mid-crust. The respective factors representing each of these components are combined in a multiplicative manner, as represented by Equation (6).

$$\gamma_c = \gamma_{am} \times \gamma_{an} \times \gamma_{mc}, \quad (6)$$

where γ_{am}, γ_{an}, and γ_{mc} are factors representing amplifications of the upper crust, attenuation of the upper crust, and modification of the mid-crust, respectively.

Details of the derivation of each of these component factors contributing to crustal modifications are described in the rest of this section under separate subheadings.

2.3.1. Upper-Crustal Amplification

In modelling upper crustal amplification, information presented in reference [38], which is abbreviated herein as BJ97, and more recent updates of the model [39], have been incorporated for constructing the shear wave velocity profiles, which have also been used to infer the crustal density profiles [40]. The two profiles were then called up jointly to derive the upper crustal amplification factors by use of the square-root-impedance (SRI) method. The shear wave velocity profiles and the corresponding frequency-dependent amplification factors, so derived for the study regions, are presented in Figures 2 and 3.

The upper-crustal amplification factor γ_{am} was derived accordingly as a function of V_{S30} (time-averaged shear wave velocity in the upper 30 m of the earth crusts [41]) by curve-fitting simulated data. The general form of the function is shown by Equation (7):

$$\log \gamma_{am} = \gamma_1 \times M^{\gamma_2} \times V_{S30}^{\gamma_3} + \gamma_4 \times V_{S30}, \quad (7)$$

where γ_1–γ_4 are the regression coefficients.

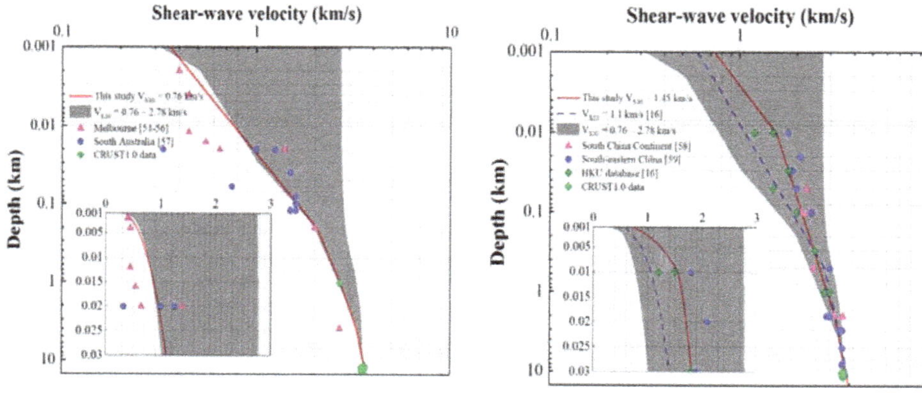

(**a**) (**b**)

Figure 2. Shear-wave velocity profile: (**a**) SEA; (**b**) SEC, in which CRUST1.0 is a global crustal database (https://igppweb.ucsd.edu/~gabi/crust1.html, last accessed in January 2019), the shear wave velocity data can be found in the "Supplementary Materials".

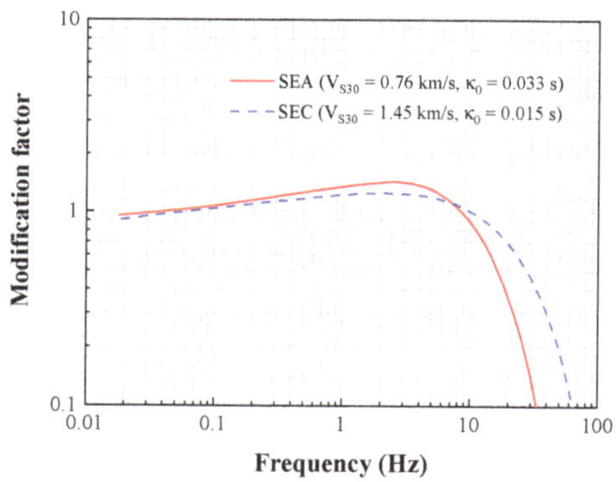

Figure 3. Frequency-dependent crustal modification factor. For both conditions, $\kappa_0 = 0.057/V_{s0.03}{}^{0.8} - 0.02$ [42,43].

Upper crustal amplification of seismic waves is also frequency-dependent, given that the amplification of higher frequency waves (of shorter wave lengths) are more sensitive to the shear wave velocity gradient of the earth crust than lower frequency waves (of longer wave lengths). This phenomenon can be explained by reference to the quarter wavelength principles in the analysis for crustal amplification. Further explanations can be found in reference [38]. The extent of upper crustal amplification is accordingly dependent on the frequency characteristics of the upward propagating seismic waves. Given that the frequency characteristics of an earthquake are magnitude dependent, the extent of upper crustal amplification is, therefore, also magnitude dependent. Hence, earthquake magnitude is a parameter in the predictive expression of Equation (7). The validity of using V_{S30} to characterise upper crustal conditions has been demonstrated in reference [42]. The reliability (ability) of Equation (7) to reproduce results of stochastic simulations of the upper crustal amplification phenomenon reasonably accurately is demonstrated in Section 3.

2.3.2. Upper Crustal Attenuation

The Kappa factor (κ_0) is another key parameter to be considered in upper crustal modelling. The attenuation behaviour of the ground motion in the upper crust at high frequencies (in addition to the whole path anelastic attenuation covered in Section 2.2) is controlled by the value of κ_0. Numerous studies that were targeted at modelling the value of κ_0 have been reported in the literature [41–46]. In simulations undertaken in this study, the value of κ_0 accordingly ranges from 0.001 s to 0.1 s. The functional form for the expression for determining the values of γ_{an} to account for the effects of upper crustal attenuation is represented by Equation (8).

$$\log\gamma_{an} = \gamma_5 \times \mathrm{M}^{\gamma_6}\kappa_0^{\gamma_7} + \gamma_8, \tag{8}$$

where γ_5–γ_8 are regression coefficients.

Table 2. Summary of the selected NGA-East seismological models.

Seismological Model	AB95 [1]	SGD02 [47]	A04 [48]	BS11 [49]
Source model [1]	$\frac{1-\varepsilon}{1+(f/f_a)^2} + \frac{\varepsilon}{1+(f/f_b)^2}\log\varepsilon = 2.52 - 0.637\mathrm{M}\log f_a = 2.41 - 0.533\mathrm{M}\ f_b = \sqrt{\left(f_c^2 - (1-\varepsilon)f_a^2\right)/\varepsilon}\, f_c = 4.9\times10^6\beta_0\left(\frac{\Delta\sigma}{M_0}\right)^{1/3}\Delta\sigma = 200\ \mathrm{bar}$			
Shear Wave Velocity at Source βos (km/s)	3.8	3.52	3.7	3.5
Geometrical Factor G (R in km) [2]	$R \le 70$: R^{-1} $70 < R \le 130$: R^0 $130 < R$: $R^{-0.5}$	$R \le 80$: $R^{-(1.0296-0.0422(M-6.5))}$ $80 < R$: $R^{-0.5(1.0296+0.0422(M-6.5))}$	$R \le 70$: $R^{-1.3}$ $70 < R \le 140$: $R^{0.2}$ $140 < R$: $R^{-0.5}$	$R \le 50$: R^{-1} $50 < R$: $R^{-0.5}$
Quality Factor Q	$680 f^{0.36}$	$351 f^{0.84}$	$\max(1000, 893 f^{0.32})$	$410 f^{0.5}$
Upper Crustal Amplification Parameter	$V_{S30} = 0.76$ km/s	$V_{S30} = 0.76$ km/s	$V_{S30} = 0.76$ km/s	$V_{S30} = 0.76$ km/s
Upper Crustal Attenuation Parameter	$\kappa_0 = 0.025$ s	$\kappa_0 = 0.025$ s	$\kappa_0 = 0.025$ s	$\kappa_0 = 0.025$ s

[1] source models presented in the original references have been replaced by the more updated source model presented in the table based on recommendations in reference [35]. [2] R is hypocentral distance in km.

As for whole path anelastic attenuation, upper crustal attenuation is also magnitude-dependent. Thus, earthquake magnitude is also a parameter in Equation (8) for predicting the extent of upper crustal attenuation. The significance of the magnitude term is well demonstrated in Section 3.

Residual analysis has been conducted for CAM in totality, incorporating the generic source, regional path (including anelastic and geometric attenuation factors), and crustal modification factor. An apparent trend with increasing distance R has been identified. A fourth order polynomial expression (Equation (9)) for defining the value of the adjustment factor $\gamma_{\mathrm{adjustment\_factor}}$ is used to adjust, through multiplication, predicted values of PGV from Equation (3), (4), (5), (7), and (8) to minimise values of the residuals, along with other systematic modelling errors.

$$\gamma_{\mathrm{adjustment\_factor}} = \gamma_9\mathrm{R}^4 + \gamma_{10}\mathrm{R}^3 + \gamma_{11}\mathrm{R}^2 + \gamma_{12}\mathrm{R} + \gamma_{13}, \tag{9}$$

where γ_9–γ_{13} are regression coefficients.

2.3.3. Mid-Crustal Modification

Another crustal factor to consider is the mid-crustal modification factor γ_{mc}, which is used to account for the effects of the density and shear wave velocity of the earth crust at the depth of the source of the earthquake. Mid-crustal amplification is purely a source phenomenon, as it represents the increase in the amplitude of shear waves (generated at the source of the earthquake) with decreasing shear wave velocity and density of the earth crust at the depth of the source from where seismic waves are emitted. In calculating the amplification factor a high shear wave velocity value of 3.8 km/s and a crustal density value of 2.8 g/cm^3, both of which are characteristics of hard rock conditions in the shield regions of Eastern North America, are used as the "benchmark" conditions, for which the amplification factor is set as unity. These benchmark parameters for shear wave velocity and crustal density are denoted as β_{0sim} and $\rho_{sim,}$ respectively.

The value of γ_{mc} can be found using Equation (10), which has been derived by the authors based on the source model defined in reference [35].

$$\gamma_{mc} = \frac{\rho_{sim}}{\rho_S} \times \left(\frac{\beta_{0sim}}{\beta_{0S}}\right)^k \tag{10}$$

where $k = -0.273 \times M + 3.278$, $\rho_{sim} = 2.8$ g/cm^3, $\beta_{0sim} = 3.8$ km/s, and ρ_S and β_{0S} are the density and shear wave velocity of the earth crust at the depth of the source (i.e., the mid-crust).

2.4. Verification of CAM Using Various Seismological Models

To verify CAM as a tool for translating seismological models into GMPEs for predictions of PGV, four NGA(Next Generation Attenuation)-East seismological models (as listed in Table 2), namely Atkinson and Boore (1995) [1], Silva, Gregor, and Darragh (2002) [47], Atkinson (2004) [48], and Boatwright and Seekins (2011) [49], have been used to verify CAM. The acronyms of these four models are AB95, SGD02, A04, and BS11. The PGV values, so derived from stochastic simulation of the (selected) seismological models, were compared with PGV values predicted by use of CAM for the respective model. Results of the verification analyses are presented in Section 3.2.

2.5. Validation of CAM Using MMI Data

Historical macroseismic MMI data collected from South-Eastern Australia (SEA) and South-Eastern China (SEC) has also been adopted to validate CAM as a regionally adjustable generic GMPE. The magnitude and distance distributions are shown in Figure 1 for the two selected study regions. The well-known MMI-PGV correlation proposed by Atkinson and Kaka [23] was adopted in this study (with residual corrections) to transfer the predicted PGV values (as derived from CAM) into MMI values for comparison purposes. The correlation relationship without residual corrections is expressed by Equation (11), and that with residual corrections is expressed by Equation (12). The historical MMI data can be found in this paper in the section presenting "Supplementary Materials".

$$\text{MMI} = \begin{cases} 4.37 + 1.32 \times \log\text{PGV} & \log\text{PGV} \le 0.48 \\ 3.54 + 3.03 \times \log\text{PGV} & \log\text{PGV} > 0.48 \end{cases}, \tag{11}$$

$$\text{MMI} = \begin{cases} 4.37 + 1.32 \times \log\text{PGV} + 0.47 - 0.19 \times \text{M} + 0.26 \times \log\text{R} & \log\text{PGV} \le 0.48 \\ 3.54 + 3.03 \times \log\text{PGV} + 0.47 - 0.19 \times \text{M} + 0.26 \times \log\text{R} & \log\text{PGV} > 0.48 \end{cases}, \tag{12}$$

where M and R are the moment magnitude and hypocentral distance, respectively, and the unit of PGV is cm/s.

For recordings collected from SEA, some of the events were recorded in terms of local magnitude (M_L). The conversion between M (moment magnitude, same as M_W) and M_L would need to be undertaken in the first place to obtain correct estimates of the event magnitude. In this study, the bilinear conversion relationship developed for Australian conditions [50] as defined by Equation (13) was adopted.

$$\text{M} = \begin{cases} 2/3\text{M}_L + 1.2, & \text{M}_L \le 4.5 \\ \text{M}_L - 0.3, & \text{M}_L > 4.5 \end{cases} \tag{13}$$

As the earthquake recordings for SEC were derived originally from the ancient yearbook, some records dated back to as early as the 15th century [51,52]. Thus, magnitude conversion is filled with uncertainties. The magnitude recordings compiled in this study (which can be found in the section titled "Supplementary Materials") are expressed in terms of moment magnitude.

Another important component in CAM is the upper crustal modification factor. Regional shear wave velocity (SWV) profiles can be used for deriving the upper crustal amplification factor. The shear wave velocity profiles in this study were constructed based on the use of a geology-based modelling approach, as recommended in reference [53]. With this modelling approach, a compressional (P) wave velocity profile is converted into a shear (S) wave velocity profile using relationships that have been developed by regression analysis of recorded data collated from multiple sources. More details in relation to the process of modelling the shear wave velocity profile for the two study regions can be found in the "Supplementary Materials" section. The parameter values used for constructing SWV profiles are listed in Table 3. In Table 3, the values of Z_S (depth of the upper sedimentary crustal layer) and Z_C (combined thickness of the soft and hard sedimentary crustal rock layers) for each region were obtained from the average estimates of the thickness of soft sediment layer and total sediment layers, respectively, in CRUST1.0 database (https://igppweb.ucsd.edu/~gabi/crust1.html, last accessed in January 2019). $V_{S0.03}$ (shear-wave velocity values at the depth of 0.03 km) values were identified by curve-fitting to minimise discrepancies (defined as sum of squares errors) between the modelled and recorded SWV values. The SWV value at 8 km depth (V_{S8}), representing conditions at the source of the earthquake, has also been determined for the two study regions based on information of the regional crustal structure. The detailed information about the recorded SWV value can be found the references [54–59]. The velocity profile showing the upper bound of 2.78 km/s in Figure 2 is the shear wave velocity profile for generic hard rock conditions, as recommended in reference [38]. The velocity profile showing the lower bound of 0.76 km/s in Figure 2 was derived from interpolation between the shear wave velocity profiles for generic rock and generic hard rock conditions, as recommended in reference [38]. The modelling approach introduced in reference [39] has been adopted. The frequency-dependent modification factors for the study regions are shown in Figure 3.

Table 3. Parameter values for modelling shear wave velocity (SWV) profiles for South-Eastern Australia (SEA) and South-Eastern China (SEC).

Parameter	SEA	SEC
Z_S (km)	1.0	0.01
Z_C (km)	4.0	2.0
$V_{S0.03}$ (km/s) [1]	1.1	1.81
V_{S8} (km/s) [2]	3.5	3.6 [60]
n	0.141	0.136
function form	$Z \le 0.2$, $V_{SZ} = V_{S0.03}(Z/0.03)^{0.3297}$; $0.2 < Z \le Z_S$ [4], $V_{SZ} = V_{S0.2}(Z/0.2)^{0.1732}$ [3]; $Z_S < Z \le Z_C$ [5], $V_{SZ} = V_{SZC}(Z/Z_C)^n$; $Z_C < Z$, $V_{SZ} = V_{S8}(Z/8)^{0.0833}$.	$0 < Z \le Z_S$, $V_{SZ} = V_{SZI}(Z/Z_I)^{0.3297}$ ($Z_I = \min(Z_S, 0.03)$); $Z_S < Z \le Z_C$, $V_{SZ} = V_{SZC}(Z/Z_C)^n$; $Z_C < Z$, $V_{SZ} = V_{S8}(Z/8)^{0.0833}$.

1. $V_{S0.03}$ is the shear wave velocity value at the depth of 0.03 km; 2. V_{S8} is the shear wave velocity value at the depth of 8 km; 3. $V_{S0.2}$ is the shear wave velocity value at the depth of 200 m; 4. Z_S is the depth of the upper sedimentary crustal layer; 5. Z_C is the combined thickness of the soft and hard sedimentary crustal rock layers.

The parameter values used in CAM and the selected GMPEs for comparison purposes for SEA and SEC regions are summarised in Table 4.

Table 4. Parameter values used in CAM for SEA and SEC alongside the selected GMPEs.

Parameter	SEA	SEC
Source Shear Wave Velocity (km/s)	3.5	3.6 [60]
Source Density (g/cm^3)	2.8 [18]	2.9 [61,62]
Stress Drop (bar)	200	200
Δ (cm/s)	3.9 [1]	3.9 [1]
Geometric Attenuation Factor (G) [1]	$0 \le R \le 70$, $R^{-1}70 < R \le 130$, $R^0 130 < R$, $R^{-0.5}$	$0 \le R \le 70$, $R^{-1}70 < R \le 130$, R^0 $130 < R$, $R^{-0.5}$
Quality Factor (Q_0)	200 (New South Wales) [15] 100 (Victoria) [15] 300 (South Australia) [15]	320 [16]
V_{S30} (km/s)	0.76 [2]	1.45 [2]
κ_0 (s)	0.03 [42]	0.02 [42]
Conversion Factor (PGV_S/PGV_R)	1.5 [63,64][3]	1.5 [16]
Source Factor (α) at M6R30	1.00	1.00
Anelastic Attenuation Factor (β) at M6R30	0.95	0.95
Path Adjustment Factor ($\beta_{adjustment\_factor}$) at M6R30	0.98	0.98
Upper Crustal Amplification Factor (γ_{am}) at M6R30	2.22	1.52
Upper Crustal Attenuation Factor (γ_{an}) at M6R30	0.48	0.6
Crustal Adjustment Factor ($\gamma_{adjustment\_factor}$) at M6R30	1.16	1.16
Mid-crustal Modification Factor (γ_{mc}) at M6R30	1.1	1.06
Selected GMPEs	SGC09 [65] [4] A12 [18] [5]	CB08 [66] [6] CY08 [2] [7]

[1] Both Δ values for SEA and SEC were determined from the calculated PGV at M6R30 from program GENQKE for hard rock conditions [9,15]; [2] V_{S30} values are based on Figure 2. [3] PGV_S and PGV_R refer to PGV value on an average soil site and rock site, respectively, the value of 1.5 for the rock to the soil site conversion is based on stipulations by the earthquake loading standard for sites on shallow soil sediments [63] as explained in reference [64]; [4] SGC refers to Somerville et al. (2009) [65]; [5] A12 refers to Allen (2012) [18]; [6] CB08 refers to Campbell and Bozorgnia (2008) [66]; [7] CY08 refers to Chiou and Youngs (2008) [2].

3. Results of Verification Analyses

3.1. PGV Modelling

Values for each of the component factors in CAM are presented in, Figures 4–8, alongside results generated from stochastic simulations of the respective seismological model. Figure 4 shows the source factor (α) as a function of M and Δσ.

Figure 5 shows the path factor (β) as a function of moment magnitude M, R, and wave transmission quality factor (Q_0). Figure 6 shows the upper crustal amplification factor (γ_{am}) as a function of M and V_{S30}. Figure 7 shows the upper crustal attenuation factor (γ_{an}) as a function of M and κ_0. Figure 8 shows the overall PGV obtained both from stochastic simulations and CAM predictions. The regression coefficients together with the regression goodness (R^2) values for different component factors demonstrating excellent agreement between the two sets of results are summarised in Table 5.

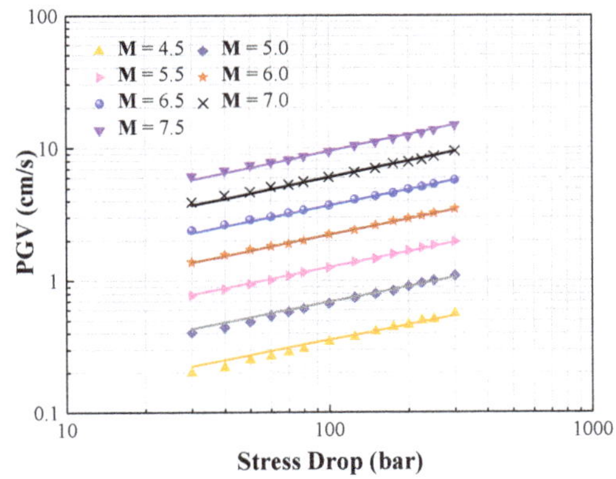

Figure 4. Simulated values of peak ground velocity (PGV) (symbols) alongside predictions by CAM (lines) for varying moment magnitudes (M4.5–M7.5) and stress drops (30–300 bar), at R = 30 km.

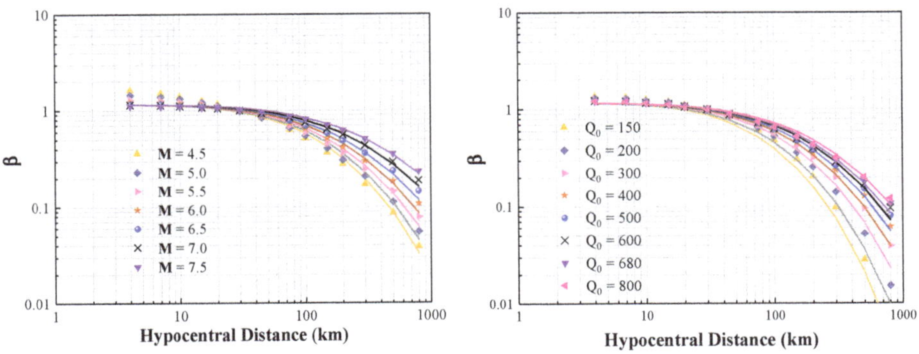

(a) (b)

Figure 5. Anelastic attenuation factor (β). (**a**) Simulated PGV values normalised at Q_0 = 680 for varying M values (M4.5–M7.5) and distances (4–800 km) shown alongside predictions by CAM (lines); (**b**) simulated PGV values normalised at M = 6 for varying Q_0 values (150–800) and distances (4–800 km) shown alongside predictions by CAM (lines).

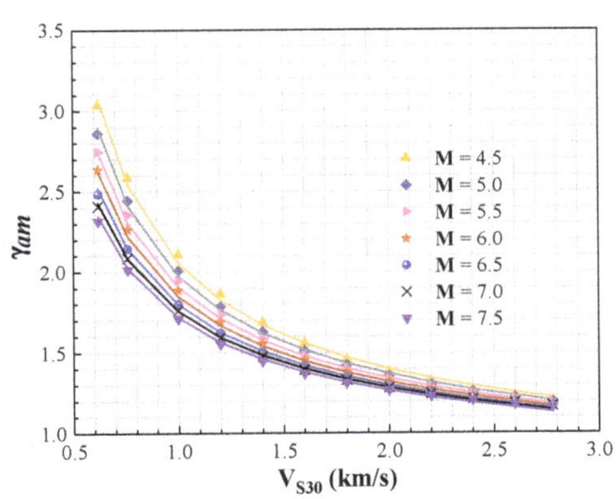

Figure 6. Regression analysis results of upper crustal amplification factor (γ_{am}).

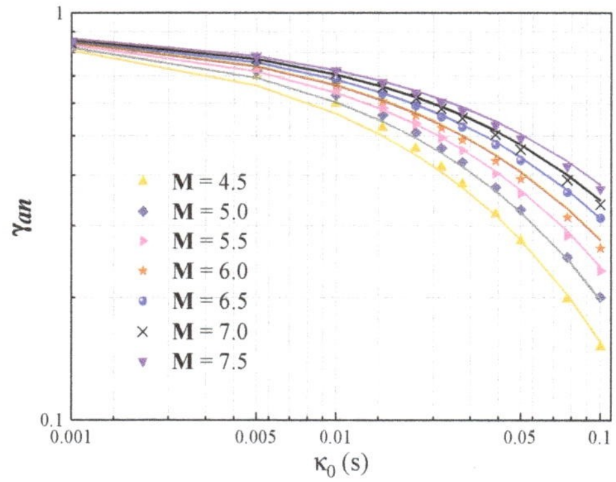

Figure 7. Regression analysis results of upper crustal attenuation factor (γ_{an}).

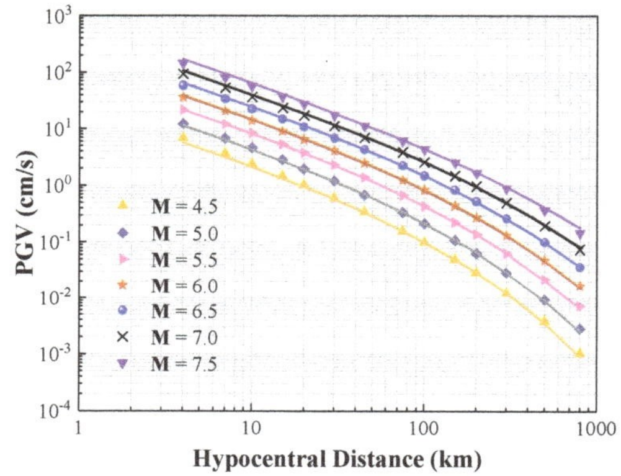

Figure 8. Simulated PGV values (symbols) for varying M values (M4.5–M7.5) and distances (4–800 km) shown alongside predictions by CAM (lines) for $\Delta\sigma$ = 200 bar, V_{S30} = 0.76 km/s, and κ_0 = 0.025 s.

Table 5. Coefficients of CAM for modelling PGV.

$\log\alpha$	Δ	a_1	a_2	a_3	a_4	R^2
	2.952	27.797	0.0841	0.0059	−33.35	0.9994
$\log\beta$	b_1	b_2	b_3	b_4	b_5	R^2
	0.06287	−0.6326	−0.4963	4.431	0.06135	0.9807
$\log\beta_{adjustment}$	b_6	b_7	b_8	b_9	b_{10}	R^2
	0.01714	−0.06931	0.08404	−0.09224	0.1389	0.9975
$\log\gamma_{am}$	γ_1	γ_2	γ_3	γ_4		R^2
	0.7334	−0.5251	−0.8479	−0.019		0.9953
$\log\gamma_{an}$	γ_5	γ_6	γ_7	γ_8		R^2
	−21.35	−1.351	0.5584	−0.03336		0.9978
$\log\gamma_{adjustment}$	γ_9	γ_{10}	γ_{11}	γ_{12}	γ_{13}	R^2
	−0.01333	0.07378	−0.1294	0.1046	0.01838	0.9948

This paper focuses on the use of CAM for predictions of PGV. CAM also provides predictions for response spectral accelerations. Refer to "Supplementary Materials", which provides the link to access CAM for response spectrum modelling.

3.2. Translating Seismological Models into GMPEs in Terms of PGV

Four well-known NGA-East seismological models (as listed in Table 2) have been used as examples to verify the accuracy of CAM. The selected seismological models are namely Atkinson and Boore (1995) [1], Silva, Gregor, and Darragh (2002) [47], Atkinson (2004) [48], and Boatwright and Seekins (2011) [49], with acronyms: AB95, SGD02, A04, and BS11. Each of these models has its own attenuation properties (encompassing geometric attenuation and whole path anelastic attenuation). In this study, the same generic source factor (of the generalised additive double-corner frequency form with stress drop of 200 bar) and site conditions (V_{S30} = 0.76 km/s and κ_0 = 0.025 s) have been input into the seismological models for defining the Fourier amplitude spectrum (FAS) and for making predictions of PGV through stochastic simulations. All the selected seismological models have been translated into PGV predictive models with a reasonable level of accuracy (as shown in Figure 9).

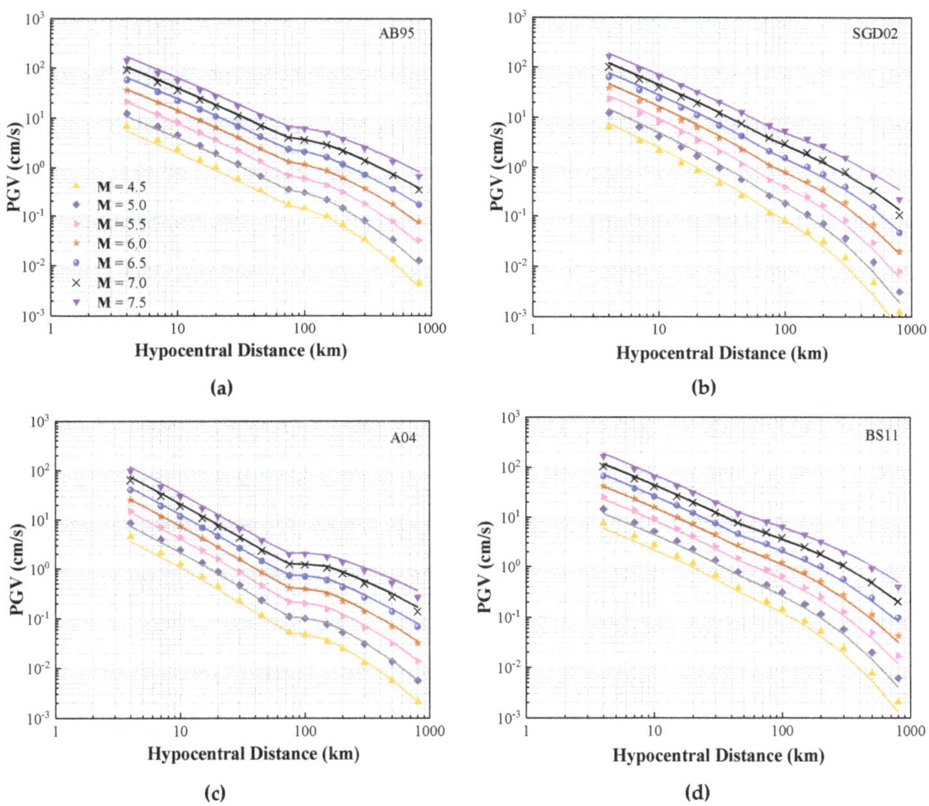

Figure 9. Comparison between predictions by CAM (lines) and simulations of seismological models (symbols). (**a**) AB95 model, with trilinear geometric spreading and Q_0 = 680; (**b**) SGD02 model, with magnitude-dependent geometric spreading and Q_0 = 351; (**c**) A04 model, with trilinear geometric spreading and Q_0 = 893; (**d**) BS11 model, with bilinear geometric spreading and Q_0 = 410.

The process of transforming a seismological model into a ground motion prediction model is summarised as follows: (a) stochastic simulations of a seismological model based on a given set of seismological parameters for generating an ensemble of accelerograms, each of which has its own array of random phase angles; (b) calculation of the response spectrum for every accelerogram that has been generated; (c) statistical analysis of the calculated response spectral ordinates for determining their mean values; and (d) developing a GMPE for providing median ground motion predictions by collation of the mean response spectral values (across the natural period range of engineering interests) for different combinations of seismological parameters including M and R.

3.3. Comparing with Historical MMI Data and Existing GMPEs

In regions of low-to-moderate seismicity where instrumented strong motion data are lacking, ground motion models can be verified by use of macroseismic data. The most commonly used metrics of this type are modified Mercalli intensity (MMI) data. A more recently developed alternative to the MMI scale is the environmental seismic intensity (ESI) scale. Further descriptions of the ESI scale can be found in Section 4. At present, only data presented in the MMI scale are currently available in the two study regions for verifying CAM.

Four candidate GMPEs have been selected for use in the comparative study for evaluating the accuracies of GMPEs in terms of their level of agreement with MMI data. For modelling PGVs in SEA, GMPEs that have been compared are namely: (i) SGC09 (non-cratonic condition) [65], (ii) A12 (shallow earthquake) [18], and (iii) CAM-SEA (this study). For modelling PGVs in SEC, GMPEs that have been compared are namely: (i) CB08 [66], (ii) CY08 [2], and (iii) CAM-SEC (this study). CB08 model and CY08 model are selected because that PGV predictions by these two models have been suggested in the literature to be appropriate for use in South China [61,62]. Details of seismological parameters that have been identified for the two regions for input to CAM can be found in Table 4. Results of evaluations for the two regions are presented in Figures 10 and 11, respectively. In both figures the x-axis is the historical recorded MMI values based on observations on average soil sites (as presented in "Supplementary Materials"). The y-axis shows the MMI values that are predicted from the four candidate GMPEs and have incorporated an average site factor of 1.5 for transforming predictions from rock sites to average soil sites. The key reference for sourcing intensity information from isoseismal maps was AGSO (1995) [67]. Information presented in website: https://earthquakes.ga.gov.au/ has also been used to identify locations of epicentres of the historical earthquake events.

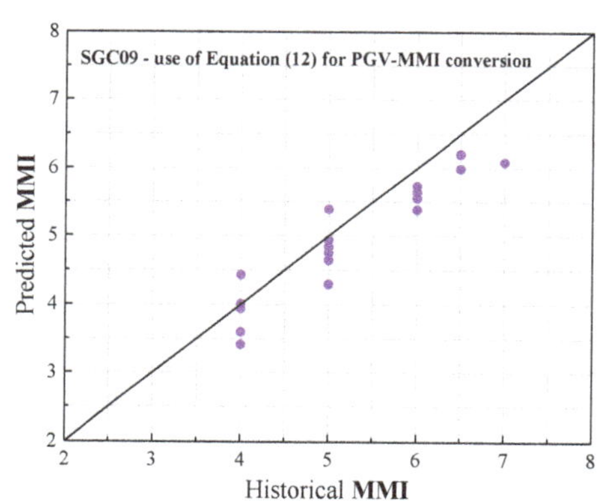

(a) SGC09

Figure 10. *Cont.*

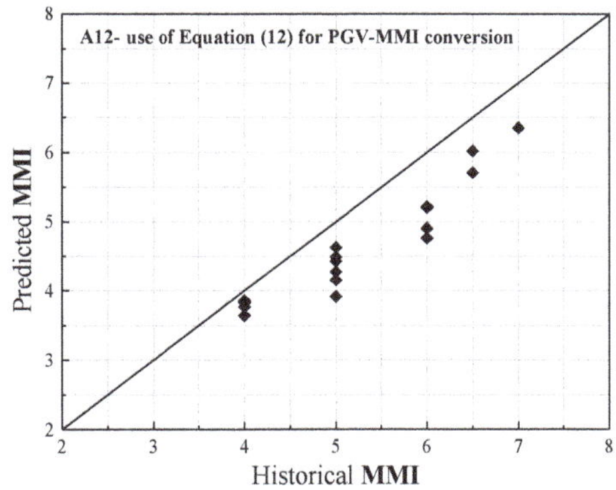

(b) A12

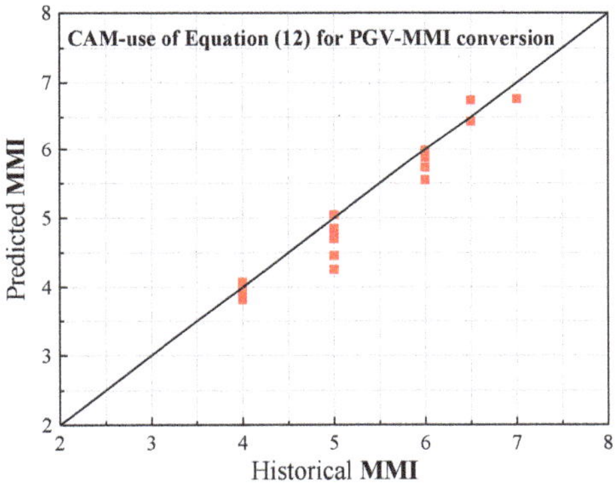

(c) CAM (this study)

Figure 10. Comparison between historical recorded and model-predicted modified Mercalli intensity (MMI) for SEA.

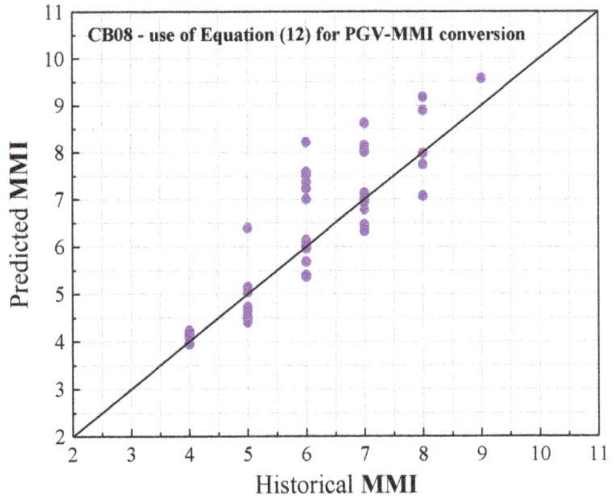

(a) CB08

Figure 11. *Cont.*

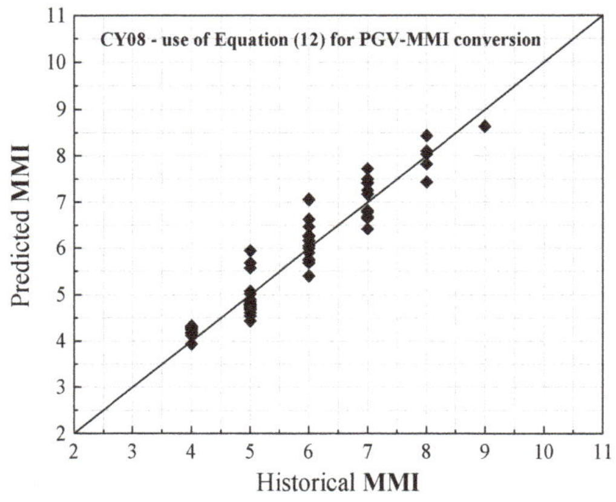

(b) CY08

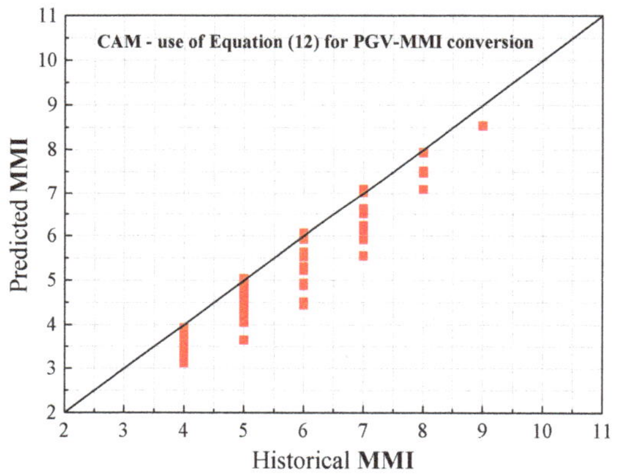

(c) CAM (this study)

Figure 11. Comparison between historical recorded and model-predicted MMI for SEC.

4. Discussion

CAM essentially decouples the effect of earthquake source, path, and crust into separate components, thereby enabling it to be regionally adjustable in order that predictions for SEA and SEC can be covered in one model.

In Figure 9, all the simulated results (symbols) and predictions by CAM (lines) are shown to match well for the selected magnitude ranges, indicating that CAM can accurately represent the four selected seismological models: AB95, SGD02, A04, and BS11. The minor mismatch, as displayed in the figures, can be explained by the different relationships that have been used to determine the value of the exponent factor n in the Q_0 function ($Q = Q_0 f^n$), as recommended in the literature; refer to reference [36]. A prominent feature of the A04 model is that the function for determining the Q factor is not of the typical form involving parameters Q_0 and n, but is instead defined as: $Q(f) = \max(1000,\ 893f^{0.32})$. The strategy of setting $Q_0 = 893$ for all $T_n < 1.0$ s, and $Q_0 = 1000$ for all $T_n \geq 1.0$ s has been adopted in predicting response spectral values. According to Bommer and Alarcon [25], the ratio between response spectral acceleration (RSA) and PGV (RSA/PGV) is nearly constant at $T_n = 0.5$ s, thus, $Q_0 =$ 893 was adopted in CAM for predicting the value of PGV in this study.

By a rough glance at Figure 10, predictions by CAM for SEA (Figure 10c) are shown to be in better agreement with the recorded values for MMI than the other two candidate GMPEs (Figure 10a,b). Adopting the geometric attenuation factor of $R^{-1.33}$, as per recommendations by A12 [18], at short distance as opposed to the conventional factor of R^{-1} is controversial, whilst good match between the model predictions and field recorded data has been demonstrated in references [18,19]. It is noted that when calibrating seismological parameters (to achieve agreement between predictions from a seismological model and field recorded data) there are trade-offs of the assumed stress drop values with the assumed rate of geometrical attenuation. Stress drop behaviour of earthquakes, as assumed by different groups of investigators (based on calibration), can be very inconsistent. The geometrical factors adopted by the two study groups can accordingly be very inconsistent too ($R^{-1.33}$ versus R^{-1}), whilst achieving good agreement between predictions and recorded data in their respective studies.

Specific studies on ground motion characteristics of the SEC region have not been reported in the literature. In this study, the geometric attenuation form was adopted from AB95 model to represent the attenuation characteristics in the SEC region. CB08 [66] and CY08 [2] models were selected for comparison purpose, given that PGV predictions by these two models have been suggested in the literature to be appropriate for use in South China [61,62]. There appears to be under-predictions by CAM for SEC (Figure 11c), warranting further investigations whilst overpredictions by the other two models are also shown (Figure 11a,b).

Discrepancies of results presented in this study are considered to have been resulted from the following causes:

1. Uncertainties with the relationship for conversion from MMI to PGV: Although the adopted MMI–PGV conversion function is recommended by many studies, there are still significant variances when applying the function to a diversity of regions, which is demonstrated by the discrepancies between not using residual corrections (Equation (11)) and using residual corrections (Equation (12)) in the relationship functions shown in Figures 10 and 11.
2. Uncertainties with the modification factor for conversion from MMI on a soil site to MMI on a rock site: a factor of 1.5 was adopted for both SEA (recommended by AS1170.4-2007 [64]) and SEC region (recommended by Tsang et al. [16]). CAM can only give predictions on rock sites and thus the conversion between soil sites and rock sites is essential. Uncertainties with magnitude conversion: in SEA, local magnitude (M_L) has been converted into moment magnitude (M) based on studies conducted by Geoscience Australia [50]. However, the magnitude of ancient recordings in SEC has not been assured (the magnitude identified with individual recordings is assumed to be in moment magnitude).
3. Uncertainties with shear wave velocity profiling: a geology-based approach for constructing SWV profile was adopted. This approach can make the best use of local recording data, thereby minimising inter-regional variability when calculating the upper crustal modification factor. For SEA region, the proposed SWV profile resulted in a V_{S30} value that is the same as previous study (0.76 km/s) [68]. However, for SEC region, the SWV profile obtained from this study (V_{S30} = 1.45 km/s) is different from that presented by Tsang et al. [16] (V_{S30} = 1.1 km/s, which is different from $V_{S0.03}$). More local data for accurate SWV profile modelling is required in future studies.
4. Uncertainties with the seismological parameters: no complete seismological model has been developed specifically for the SEC region. The parameter values (including stress drop value and geometric attenuation factor) used in CAM are mainly default values that are expected for a typical intraplate region.
5. Another intrinsic limitation of CAM is that it has not taken earthquake duration effects into account in a comprehensive manner. Incorporating an adjustment factor for earthquake duration effects into CAM is recommended for its future development.

Another matter of consideration is the selection of the best metrics for quantifying the intensity of earthquake ground shaking (and ground deformation) in historical, and prehistorical, earthquake events. The most commonly used macroseismic intensity metrics is the MMI scale that has been used for verifying the accuracy of CAM, as described in the previous section of the article. A review of metrics presented in the MMI scale for quantifying historical earthquake hazards can be found in Ref. [24]. The alternative environmental seismic intensity scale (ESI) [69] is a new macroseismic metric that is based on traces left on the landscape that were caused by earthquake activities in the natural environment. ESI allows the intensity of ground shaking in an earthquake event to be post-dicted in situations where no information on damage to buildings is available or when diagnostic damage-based elements have saturated [70]. Thus, prehistorical events that are not within the scope of any historical archives can be covered. The basic idea of the ESI is to make use of traces of geological and/or geomorphological nature that have been left behind by primary and secondary surface ruptures and mass movements, so generated by large magnitude earthquakes to post-estimate the intensity of the hazard and the magnitude of the event [71]. Example applications of the ESI scale can be found in references [71–75]. Results reported in reference [73] indicate that incorporating ESI data into probabilistic/deterministic seismic hazard analysis can result in significant changes to the modelled PGA values. Another type of macroseismic information is paleo liquefaction data, which can also be employed to provide very valuable information in support of seismic hazard analyses [76].

5. Conclusions

CAM was introduced as an engineering tool for providing realistic ground motion predictions for intraplate regions. The modelling principle is based on utilising up-to-date (and generic) knowledge on the frequency behaviour of seismic waves that radiated from the source of a local (small and medium magnitude) earthquake in combination with knowledge on local geophysical and crustal conditions, which control frequency modifications along the wave travel path. The seismological model provides the framework for integrating both knowledge bases into the predictions. This decoupling approach to modelling is more scientific, and rational, than simply applying the logic tree procedure to assign weighing factors on existing ground motion models (that are typically biased to conditions and areas where instrumented data are abundant).

This article focused on PGV as a ground motion intensity metric that can be translated into MMI data (which can be compared with data extracted from isoseismal maps of historical earthquake events). The use of algebraic expressions to make predictions of the PGV is an important feature in CAM. Essentially, an adaptable seismological model is presented as a GMPE, thereby waiving away the need for undertaking stochastic simulations along with response spectral computations. An important, and significant, achievement of this article was to have the algebraic expressions in CAM verified. The outcome from the MMI comparative study undertaken for SEA also shows a great deal of promise for CAM, but there are also scopes of improving the match with SEC data in a future study.

The user-friendly setting of CAM (featuring the use of algebraic expressions) serves to facilitate engineering professionals to become more involved with ground motion modelling, thereby gaining a good perspective of the modelling rationale and the underlying assumptions. In summary, this article represents a contribution towards improving transparencies in seismic hazard modelling and in the selection, and scaling of, accelerograms for engineering applications.

The success of CAM in the future relies on the investment of resources into studying crustal and geophysical conditions in intraplate regions for the users of CAM become better informed. This is important given that the quality of the output from any predictive model can only be as good as the quality of input into the model. Further investigations on the MMI-PGV conversion relationship are also warranted for comparison across GMPE models involving the use of MMI data to become more robust.

Supplementary Materials: Supplementary materials can be found at http://www.mdpi.com/2076-3263/9/10/422/s1. The detailed information about the proposed geology-based shear wave modelling approach is available online at: https://www.dropbox.com/s/jp4o7j08fe3gqxn/Geology-based-SWV-modelling.pdf?dl=0. The **MMI** recordings for the SEA region can be found online at: https://www.dropbox.com/s/v3eqh3w686ho97j/MMI%

20events%20SEA.xlsx?dl=0. The **MMI** recordings for the SEC region can be found online at: https://www.dropbox.com/s/8vr4wq9mbsee5c0/MMI%20Events%20SEC.xlsx?dl=0. The vs. data can be found online at: https://www.dropbox.com/s/u3v127wjt10cwwk/Shear%20wave%20velocity%20data.xlsx?dl=0. The link to access **CAM** for response spectrum modelling: https://www.dropbox.com/s/jjfbfc8cm2srub3/CAM-Response-spectral-acceleration.pdf?dl=0.

Author Contributions: Conceptualization, N.L., H.-H.T., and Y.T.; methodology, Y.T., H.-H.T.; software, Y.T.; validation, Y.T. and N.L.; formal analysis, N.L., Y.T.; investigation, Y.T., N.L.; resources, Y.T.; data curation, N.L., H.-H.T., and Y.T.; writing—original draft preparation, Y.T.; writing—review and editing, N.L., H.-H.T., and E.L.; visualization, Y.T.; supervision, N.L., H.-H.T., and E.L.; project administration, N.L.; funding acquisition, N.L.

Funding: The first author was funded by China Scholarship Council, CSC grant number 201506030102. The project was funded by Commonwealth of Australia through the Cooperative Research Centre program.

Acknowledgments: The extensive technical support given by Yiwei Hu at the University of Melbourne is gratefully acknowledged.

Conflicts of Interest: The authors declare no conflict of interest.

References

1. Atkinson, G.M.; Boore, D.M. Ground-motion relations for Eastern North America. *Bull. Seism. Soc. Am.* **1995**, *85*, 17–30.
2. Chiou, B.J.; Youngs, R.R. An NGA model for the average horizontal component of peak ground motion and response spectra. *Earthq. Spectra* **2008**, *24*, 173–215. [CrossRef]
3. Brune, J.N. Tectonic stress and the spectra of seismic shear waves from earthquakes. *J. Geophys. Res.* **1970**, *75*, 4997–5009. [CrossRef]
4. Atkinson, G.M.; Boore, D.M. Earthquake ground-motion prediction equations for Eastern North America. *Bull. Seism. Soc. Am.* **2006**, *96*, 2181–2205. [CrossRef]
5. Hassani, B.; Atkinson, G.M. Adjustable generic ground-motion prediction equation based on equivalent point-source simulations: Accounting for kappa effects. *Bull. Seism. Soc. Am.* **2018**, *108*, 913–928. [CrossRef]
6. Boore, D.M.; Stewart, J.P.; Seyhan, E.; Atkinson, G.M. NGA-West2 equations for predicting PGA, PGV, and 5% damped PSA for shallow crustal earthquakes. *Earthq. Spectra* **2014**, *30*, 1057–1085. [CrossRef]
7. Campbell, K.W.; Bozorgnia, Y. NGA-West2 ground motion model for the average horizontal components of PGA, PGV, and 5% damped linear acceleration response spectra. *Earthq. Spectra* **2014**, *30*, 1087–1115. [CrossRef]
8. Zafarani, H.; Mousavi, M.; Noozad, A.; Ansari, A. Calibration of the specific barrier model to Iranian plateau earthquakes and development of physically based attenuation relationships for Iran. *Soil Dyn. Earthq. Eng.* **2008**, *28*, 550–576. [CrossRef]
9. Lam, N.T.K.; Wilson, J.L.; Hutchinson, G.L. Generation of synthetic earthquake accelerograms using seismological modelling: A review. *J. Earthq. Eng.* **2000**, *4*, 321–354. [CrossRef]
10. Boore, D.M. Simulation of ground motion using the stochastic method. *Pure Appl. Geophys.* **2003**, *160*, 635–676. [CrossRef]
11. Campbell, K.W. Prediction of strong ground motion using the hybrid empirical method and its use in the development of ground-motion (attenuation) relations in Eastern North America. *Bull. Seism. Soc. Am.* **2003**, *93*, 1012–1033. [CrossRef]
12. Pezeshk, S.; Zandieh, A.; Campbell, K.W.; Tavakoli, B. *Ground Motion Prediction Equations for CENA Using the Hybrid Empirical Method in Conjunction with NGA-West2 Empirical Ground Motion Models, in NGA-East: Median Ground Motion Models for the Central and Eastern North America Region*; PEER Rept. No. 2015/04, Chapter 5; Pacific Earthquake Engineering Research Center: Berkeley, CA, USA, 2015; pp. 119–147.
13. Atkinson, G.M. Ground-motion prediction equations for Eastern North America from a referenced empirical approach: Implications for epistemic uncertainty. *Bull. Seismol. Soc. Am.* **2008**, *98*, 1304–1318. [CrossRef]
14. Hassani, B.; Atkinson, G.M. Referenced empirical ground-motion model for Eastern North America. *Seism. Res. Lett.* **2015**, *86*, 477–491. [CrossRef]
15. Lam, N.T.K.; Sinadinovski, C.; Koo, R.; Wilson, J.L. Peak ground velocity modelling for Australian intraplate Earthquakes. *J. Seismol. Earthq. Eng.* **2003**, *5*, 11–21.
16. Tsang, H.H.; Sheikh, M.N.; Lam, N.T.K.; Chandler, A.M.; Lo, S.H. Regional differences in attenuation modelling for Eastern China. *J. Asian Earth Sci.* **2010**, *39*, 441–459. [CrossRef]

17. Noman, M.N.A.; Cramer, C.H. *Empirical Ground-Motion Prediction Equations for Eastern North America*; PEER Rept. No. 2015/04; Chapter 8; Pacific Earthquake Engineering Research Center, University of California: Berkeley, CA, USA, 2015; pp. 193–212.
18. Allen, T.I. *Stochastic Ground-Motion Prediction Equations for Southeastern Australian Earthquakes Using Updated Source and Attenuation Parameters*; Geoscience Australia: Canberra, ACT, Australia, 2012.
19. Allen, T.I.; Dhu, T.I.; Cummins, P.R.; Schneider, J.F. Empirical attenuation of ground-motion spectral amplitudes in Southwestern Western Australia. *Bull. Seismol. Soc. Am.* **2006**, *96*, 572–585. [CrossRef]
20. Newmark, N.M.; Rosenblueth, E. *Fundamentals of Earthquake Engineering*; Prentice Hall Inc.: New Jersey, NJ, USA, 1971.
21. Gaull, B.A.; Michael-Leiba, M.O.; Rynn, J.M.W. Probabilistic earthquake risk maps of Australia. *Aust. J. Earth Sci.* **1990**, *37*, 169–187. [CrossRef]
22. Kaka, S.I.; Atkinson, G.M. Relationships between instrumental ground-motion parameters and Modified Mercalli Intensity in eastern North America. *Bull. Seism. Soc. Am.* **2004**, *94*, 1728–1736. [CrossRef]
23. Atkinson, G.M.; Kaka, S.I. Relationship between felt intensity and instrumental ground motion in the Central United States and California. *Bull. Seism. Soc. Am.* **2007**, *97*, 497–510. [CrossRef]
24. Yaghmaei-Sabegh, S.; Tsang, H.H.; Lam, N.T.K. Conversion between Peak Ground Motion Parameters and Modified Mercalli Intensity Values. *J. Earthq. Eng.* **2011**, *15*, 1138–1155. [CrossRef]
25. Bommer, J.J.; Alarcon, J.E. The prediction and use of peak ground velocity. *J. Earthq. Eng.* **2006**, *10*, 1–31. [CrossRef]
26. Akkar, S.; Ozen, O. Effect of peak ground velocity on deformation demands for SDOF systems. *Earthq. Eng. Struct. Dyn.* **2005**, *34*, 1551–1571. [CrossRef]
27. Todorovska, M.I.; Trifunac, M.D. Hazard mapping of normalized peak strain in soils during earthquakes: Mircozonation of a metropolitan area. *Soil Dyn. Earthq. Eng.* **1996**, *15*, 321–329. [CrossRef]
28. FEMA. *FEMA's Software Program for Estimating Potential Losses from Disasters*; FEMA: Washington, DC, USA, 2003.
29. Trifunac, M.D. Empirical criteria for liquefaction in sands via standard penetration tests and sesimic wave energy. *Soil Dyn. Earthq. Eng.* **1995**, *14*, 419–426. [CrossRef]
30. Kostadinov, M.V.; Towhata, I. Assessment of liquefaction-inducing peak ground velocity and frequency of horizontal ground shaking at onset of phenomenon. *Soil Dyn. Earthq. Eng.* **2002**, *22*, 309–322. [CrossRef]
31. Orense, R.P. Assessment of liquefaction potential based on peak ground motion parameters. *Soil Dyn. Earthq. Eng.* **2005**, *25*, 225–240. [CrossRef]
32. Newmark, N.M.; Blume, J.A.; Kapur, K.K. Seismic design spectra for nuclear power plants. *J. Power Div. ASCE* **1973**, *99*, 287–303.
33. Newmark, N.M.; Hall, W.J. *Earthquake Spectra and Design*; Earthquake Engineering Research Institute: El Cerrito, CA, USA, 1982.
34. Wilson, J.L.; Lam, N.T.K.; Australian Earthquake Engineering Society. *AS 1170.4-2007 Commentary: Structural Design Actions*; Part 4, Earthquake Actions in Australia; Australian Earthquake Engineering Society: McKinnon, VIC, Australia, 2007.
35. Boore, D.M.; Alessandro, C.D.; Abrahamson, N.A. A generalization of the double-corner-frequency source spectral model and its use in the SCEC BBP validation exercise. *Bull. Seismol. Soc. Am.* **2014**, *104*, 2387–2398. [CrossRef]
36. Mak, S.; Chan, L.S.; Chandler, A.M.; Koo, R. Coda Q estimates in the Hong Kong Region. *J. Asian Earth Sci.* **2004**, *24*, 127–136. [CrossRef]
37. Yenier, E.; Atkinson, G.M. Regional adjustable generic ground-motion prediction equation based on equivalent point-source simulations: Application to Central and Eastern North America. *Bull. Seismol. Soc. Am.* **2015**, *105*, 1989–2009. [CrossRef]
38. Boore, D.M.; Joyner, W.B. Site amplification for generic rock sites. *Bull. Seismol. Soc. Am.* **1997**, *87*, 327–341.
39. Boore, D.M. Short Note: Determining generic velocity and density models for crustal amplification calculations, with an update of the Boore and Joyner (1997) generic amplification for $V_{S(Z)}$ = 760 m/s. *Bull. Seism. Soc. Am.* **2016**, *106*, 316–320. [CrossRef]
40. Brocher, T.M. Empirical relations between elastic wavespeeds and density in the earth's crust. *Bull. Seismol. Soc. Am.* **2005**, *95*, 2081–2092. [CrossRef]

41. Silva, W.J.; Darragh, R.B.; Gregor, N.N.; Martin, G.; Abragamson, N.A.; Kircher, C. *Reassessment of Site Coefficients and Near-Fault Factors for Building Code Provisions*; Final Technical Report; N.E.R. Program: El Cerrito, CA, USA, 1999.
42. Chandler, A.M.; Lam, N.T.K.; Tsang, H.H. Near-surface attenuation modelling based on rock shear-wave velocity profile. *Soil Dyn. Earthq. Eng.* **2006**, *26*, 1004–1014. [CrossRef]
43. Chandler, A.M.; Lam, N.T.K.; Tsang, H.H. Regional and Local Factors in Attenuation Modelling: Hong Kong Case Study. *J. Asian Earth Sci.* **2006**, *27*, 892–906. [CrossRef]
44. Drouet, S.; Cotton, F.; Guéguen, P. V_{S30}, κ, regional attenuation and M_W from accelerograms: Application to magnitude 3–5 French earthquakes. *Geophys. J. Inter.* **2010**, *182*, 880–898. [CrossRef]
45. Edwards, B.; Fäh, D.; Giardini, D. Attenuation of seismic shear wave energy in Switzerland. *Geophys. J.Inter* **2011**, *185*, 967–984. [CrossRef]
46. Van Houtte, C.; Drouet, S.; Cotton, F. Analysis of the origins of κ (kappa) to compute hard rock to rock adjustment factors for GMPEs. *Bull. Seismol. Soc. Am.* **2011**, *101*, 2926–2941. [CrossRef]
47. Silva, W.J.; Gregor, N.N.; Darragh, R.B. *Development of Regional Hard Rock Attenuation Relations for Central and Eastern North America*; Pacific Engineering and Analysis: El Cerrito, CA, USA, 2002.
48. Atkinson, G.M. Empirical attenuation of ground motion spectral amplitudes in Southeastern Canada and the Northeastern United States. *Bull. Seismol. Soc. Am.* **2004**, *94*, 1079–1095. [CrossRef]
49. Boatwright, J.; Seekins, L. Regional spectral analysis of three moderate earthquakes in Northeastern North America. *Bull. Seismol. Soc. Am.* **2011**, *101*, 1769–1782. [CrossRef]
50. Ghasemi, H.; Allen, T.I. Testing the sensitivity of seismic hazard in Australia to new empirical magnitude conversion equations. In Proceedings of the AEES, Canberra, ACT, Australia, 24–26 November 2017.
51. Gao, W.X. *China Earthquake Yearbook*; Seismological Press: Beijing, China, 1990.
52. Gu, G.X. *Catalogue of Chinese Earthquake (1831 B.C.–1969 A.D.)*; Science Press: Beijing, China, 1989.
53. Chandler, A.M.; Lam, N.T.K.; Tsang, H.H. Shear wave velocity modelling in crustal rock for seismic hazard analysis. *Soil Dyn. Earthq. Eng.* **2006**, *25*, 167–185. [CrossRef]
54. Lam, N.T.K.; Asten, M.W.; Chandler, A.M.; Tsang, H.H.; Venkatesan, S.; Wilson, J.L. Seismic attenuation modelling for Melbourne based on the SPAC-CAM Procedure. In Proceedings of the AEES, Mount Gambier, SA, Australia, 5–7 November 2004.
55. Roberts, J.; Asten, M.; Tsang, H.H.; Venkatesan, S.; Lam, N.T.K. Shear Wave Velocity Profiling in Melbourne Silurian Mudestone Using the SPAC Method. In Proceedings of the AEES, Mount Grambier, SA, Australia, 5–7 November 2004.
56. Lam, N.T.K.; Venkatesan, S.; Wilson, J.; Asten, M.; Roberts, J.; Chandler, A.; Tsang, H.H. Generic Approach for Modelling Earthquake Hazard. *Adv. Struct. Eng.* **2006**, *9*, 67–82. [CrossRef]
57. Collins, C.; Kayen, R.; Carkin, B.; Allen, T.I.; Cummins, P.R.; McPherson, A. *Shear Wave Velocity Measurement at Australian Ground Motion Seismometer Sites by the Spectral Analysis of Surface Waves (SASW) Method*; Earthquake Engineering in Australia: Canberra, ACT, Australia, 2006.
58. Zhao, B.; Zhang, Z.; Bai, Z.; Badal, J.; Zhang, Z. Shear Velocity and Vp/Vs Ratio Structure of the Crust beneath the Southern Margin of South China Continent. *J. Asian Earth Sci.* **2013**, *62*, 167–179. [CrossRef]
59. Zhao, M.; Qiu, X.; Xia, S.; Xu, H.; Wang, P.; Tan, K.; Lee, C.; Xia, K. Seismic Structure in the Northeastern South China Sea: S-wave Velocity and Vp/Vs Ratios Derived from Three-component OBS Data. *Tectonophysics* **2010**, *480*, 183–197. [CrossRef]
60. Zhang, Z.; Yang, L.; Teng, J.; Badal, J. An Overview of the Earth Crust Under China. *Earth Sci. Rev.* **2011**, *104*, 143–166. [CrossRef]
61. Xie, J.; Li, X.; Wen, Z.; Wu, C. Near-Source Vertical and Horizontal Strong Ground Motion from the 20 April 2013 Mw 6.8 Lushan Earthquake in China. *Seism. Res. Lett.* **2014**, *85*, 23–33. [CrossRef]
62. Megawati, K. Hybrid Simulations of Ground Motions from Local Earthquakes Affecting Hong Kong. *Bull. Seismol. Soc. Am.* **2007**, *97*, 1293–1307. [CrossRef]
63. Standards Australia. Earthquake actions in Australia. In *AS 1170.4-2007: Structural Design Actions*; Standards Australia: Sidney, NSW, Australia, 2007.
64. Lam, N.T.K.; Wilson, J.L. The new response spectrum model for Australia. *Electron. J. Struct. Eng.* **2008**, *8*, 6–24.
65. Somerville, P.; Graves, R.; Collins, N.; Song, S.G.; Ni, S.; Cummins, P. Source and Ground Motion Models for Australian Earthquakes. In Proceedings of the AEES, Newcastle, NSW, Australia, 11–13 December 2009.

66. Campbell, K.W.; Bozorgnia, Y. NGA ground motion model for the geometric mean horizontal component of PGA, PGV, PGD and 5% damped linear elastic response spectra for periods ranging from 0.01 s to 10 s. *Earthq. Spectra* **2008**, *24*, 139–171. [CrossRef]
67. AGSO (Australian Geological Survey Organisation). *Atlas of Iso-Seismal Maps of Australian Earthquakes*; McCue, K., Ed.; Australian Geological Survey Organisation: Canberra, ACT, Australia, 1995.
68. Allen, T.I.; Leonard, M.; Clark, D.; Ghasemi, H. *The 2018 National Seismic Hazard Assessment for Australia-Model Overview*; Clark, D., Ed.; Geoscience Australia: Canberra, ACT, Australia, 2018.
69. Michetti, A.M.; Esposito, E.; Guerrieri, L.; Porfido, S.; Serva, L.; Tatevossian, R.; Vittori, E.; Audemard, F.; Azuma, T.; Clague, J.; et al. Intensity Scale ESI 2007. *Mem. Descr. Carta Geol.* **2007**, *74*, 11–20.
70. Serva, L. History of the Environmental Seismic Intensity Scale ESI-07. *Geoscience* **2019**, *9*, 210. [CrossRef]
71. Grützner, C.; Walker, R.; Ainscoe, E.; Elliott, A.; Abdrakhmatov, K. Earthquake Environmental Effects of the 1992 Ms7.3 Suusamyr Earthquake, Kyrgyzstan, and Their Implications for Paleo-Earthquake Studies. *Geoscience* **2019**, *9*, 271. [CrossRef]
72. Chunga, K.; Livio, F.A.; Martillo, C.; Lara-Saavedra, H.; Ferrario, M.F.; Zevallos, I.; Michetti, A.M. Landslides Triggered by the 2016 Mw 7.8 Pedernales, Ecuador Earthquake: Correlations with ESI-07 Intensity, Lithology, Slope and PGA-h. *Geoscience* **2019**, *9*, 371. [CrossRef]
73. Caccavale, M.; Sacchi, M.; Spiga, E.; Porfido, S. The 1976 Guatemala Earthquake: ESI Scale and Probabilistic/Deterministic Seismic Hazard Analysis Approaches. *Geoscience* **2019**, *9*, 403. [CrossRef]
74. King, T.R.; Quigley, M.; Clark, D. Surface-Rupturing Historical Earthquakes in Australia and Their Environmental Effects: New Insights from Re-Analyses of Observational Data. *Geoscience* **2019**, *9*, 408. [CrossRef]
75. Silva, P.G.; Rodríguez-Pascua, M.A.; Robles, J.L.G.; Élez, J.; Pérez-López, R.; Davila, M.B.B. Catalogue of the Geological Effects of Earthquakes in Spain Based on the ESI-07 Macroseismic Scale: A New Database for Seismic Hazard Analysis. *Geoscience* **2019**, *9*, 334. [CrossRef]
76. Tuttle, M.P.; Hartleb, R.; Wolf, L.; Mayne, P.W. Paleo liquefaction Studies and the Evaluation of Seismic Hazard. *Geoscience* **2019**, *311*, 1–61.

geosciences

MDPI

Article

Comparison of Earthquake-Triggered Landslide Inventories: A Case Study of the 2015 Gorkha Earthquake, Nepal

Sansar Raj Meena * and Sepideh Tavakkoli Piralilou

Department of Geoinformatics—Z_GIS, University of Salzburg, 5020 Salzburg, Austria; sepideh.tavakkoli-piralilou@stud.sbg.ac.at

* Correspondence: sansarraj.meena@sbg.ac.at

Received: 13 September 2019; Accepted: 9 October 2019; Published: 10 October 2019

Abstract: Despite landslide inventories being compiled throughout the world every year at different scales, limited efforts have been made to critically compare them using various techniques or by different investigators. Event-based landslide inventories indicate the location, distribution, and detected boundaries of landslides caused by a single event, such as an earthquake or a rainstorm. Event-based landslide inventories are essential for landslide susceptibility mapping, hazard modeling, and further management of risk mitigation. In Nepal, there were several attempts to map landslides in detail after the Gorkha earthquake. Particularly after the main event on 25 April 2015, researchers around the world mapped the landslides induced by this earthquake. In this research, we compared four of these published inventories qualitatively and quantitatively using different techniques. Two principal methodologies, namely the cartographical degree of matching and frequency area distribution (FAD), were optimized and applied to evaluate inventory maps. We also showed the impact of using satellite imagery with different spatial resolutions on the landslide inventory generation by analyzing matches and mismatches between the inventories. The results of our work give an overview of the impact of methodology selection and outline the limitations and advantages of different remote sensing and mapping techniques for landslide inventorying.

Keywords: mass movements; inventory map; amalgamation; earth observation (EO); spatial resolution

1. Introduction

Landslides are the most frequent hazards of mountain regions throughout the world [1]. Given landslides' variable characteristics, they cause enormous damage to human life and infrastructure [2]. Landslides are usually caused by a trigger, like an earthquake or rainfall, and these two phenomena are considered to be the common physical triggers for event-based landslides [3]. In the Himalayan region, rainfall in the monsoon period triggers several massive and small landslides every year [4]. However, landslides triggered by earthquakes are severely destructive. For instance, the recent global landslide triggering events of the Wenchuan earthquake (2008), China [5], the Gorkha earthquake (2015), Nepal [6], and the Bihar earthquake (2002), India [7], resulted in a large number of casualties and severe damage to private and public infrastructures. However, there are several reasons that make it difficult to extract information about the exact location of landslides in an area, such as difficulty in accessing the hazard-affected remote areas [1]. A landslide inventory map, including the exact location and the exact boundaries along with the distribution, is the prerequisite for landslide analysis, susceptibility assessment, and mapping [8,9]. Furthermore, for the case of event-based landslides, detailed and state-of-the-art information about the landslides is critical. To better understand the triggering factors in an event-based landslide, different aspects of tracking, recording, and analysing data must be considered [10]. Several definitions are available regarding landslide inventories in the literature. According to [11], landslide inventory maps are the basis for obtaining records about the date of occurrence, location, and type of slope movements. A dataset of a landslide inventory can provide information on the time of the event, the location of occurrence, the type of landslide, and the extent of the landslide [12].

Landslide inventory maps are the basis of determining hazard and risk assessment [13,14]. Golovko et al. [15] analyzed multiple sources of slope failures for the establishment of a comprehensive multitemporal landslide inventory. They also described landslide inventories as the prerequisite to enable landslide hazard assessment. Landslide inventories are also crucial in carrying out a risk analysis by analyzing the impacts of past landslide events and relating them to the present criteria to predict future landslide-prone areas [16]. Thus, it is crucial to record required information on landslide occurrence to link this with triggering factors.

Landslide inventory maps can give information about probable threatened areas to disaster management authorities whcih can be used for reconstruction planning after an earthquake event [6]. Inventories are also a basis for training and validating various knowledge-based, machine learning, and deep learning methodologies related to automatic landslide detection [1,14].

Event-based landslide inventory maps can be prepared from various sources. Recently, the availability and use of high-resolution remote sensing optical images has been very useful in the identification of landslides [14]. For instance, triggering events like earthquakes trigger thousands of landslides in remote areas. Therefore, remote sensing and earth observation (EO) data play a significant role in mapping and analyzing inventories. The EO data of pre- and post-landslide events are required for conducting classification and interpretation of the hazard-affected area. There have been several attempts to map landslides using expert-based approaches, such as manual rule-based, automatic, and semi-automatic classification techniques. Two main approaches for the classification and extraction of landslides from the EO data, namely object-based and pixel-based, were distinguished [1]. However, the spatial resolution of the available EO data plays a critical role in the quality of the resulting landslide inventory maps [17,18]. Although most studies so far relied on EO data with a single scale for landslide extraction, some works considered multi-resolution satellite imageries and EO data [18]. Even using a single satellite image frame, some studies performed multi-scale methodologies and observed higher landslide extraction accuracies compared to single-scale performances [17]. Scientific progress toward pixel-based automatic identification of landslides was made using remote sensing imageries with different resolutions [19]. Considerable progress toward pixel-based automatic identification of landslides was achieved using deep-learning convolution neural networks (CNN) [1,14]. Ghorbanzadeh et al. [1] applied CNN and other machine learning models to identify landslides in RapidEye data from the Rasuwa district in Nepal. The extracted landslides from different models were then tested using Global Position System (GPS) data along with a manually detected landslide inventory of image spectral features from RapidEye data; topographic input was also used, including digital elevation models (DEM) from Advanced Land Observing Satellite (ALOS) data. In another pixel-based study [20], unsupervised classification resulted in the detection of about 60% of manually extracted and mapped landslides. There were also several object-based studies for landslide extraction, such as [21] and [22], which compared their results with manually extracted results. Therefore, manual landslide extraction and mapping is considered a standard technique for receiving the most detailed and accurate inventory. As manual detection and extraction is the most reliable technique for inventory generation, it is preferable for testing state-of-the-art models, such as deep-learning CNN models. Therefore, the quality of the manually detected landslide inventory is critical, as it usually considered to be the ground truth [14]. There are a limited number of landslide and mass movement detection studies evaluating the quality of their applied inventory dataset. In a specific study [14], the frequency area distribution (FAD) method was applied to evaluate the three available landslide inventory maps, with one of them being selected for validation of the results.

On one hand, the quality of a landslide inventory map easily affects the overall accuracy of any landslide detection, susceptibility assessment, or hazard- and risk- mapping study. On the other hand, there are a limited number of landslide studies that tackled the problem of comparison of two or more inventories [3,23]. Pellicani and Spilotro [23] compared archives and surveyed inventories for the Daunia region in Italy. They compared two landslide inventory maps to determine the corresponding quality through direct comparison. Another work by [3] compared photo-interpreted and semi-automatic landslide inventory maps in the Pogliaschina catchment, Italy. They compared the quality of rainfall event-based inventories cartographically and statistically.

Quality and completeness levels of a landslide inventory depend upon the accuracy, certainty, and type of information included in the map [24]. Criteria for the assessment of inventories are lacking from previous landslide research [3,25,26]. For the comparison of inventories, two or more inventories are needed to compare the quality of landslide maps. Despite the great importance of establishing the quality of a landslide inventory for scientific investigations, the number of such studies is limited. Comparing two or more inventories does not occurs often due to the limited availability of two or more inventories. However, there are some studies that were carried out regarding landslide mapping, and also several databases of landslides were compiled earlier by [27,28]. After the 1989 Loma Prieta, California, earthquake (Mw 6.9) a total of 1046 landslides were mapped using field investigations and aerial photographs in an area of about 15,000 km^2 in central California. The spatial distribution of the landslides was investigated statistically using one-way analysis of variance (ANOVA) and regression techniques. Correlations of landslide occurrence with distance from the earthquake source, slope steepness, and rock type were determined [29]. A comprehensive database of devastating landslides caused by catastrophic earthquakes that took place all over the world was compiled by Rodríguez et al. [28], covering the period 1980–1997. Another work by Esposito et al. [30] discussed and described the ground effects and landslides triggered by the 1997 Umbria–Marche seismic sequence in an area of 700 km^2. The environmental, seismic intensity (ESI) scale, and earthquake hazard were studied by [31,32]. The ESI scale is a measurement which defines the earthquake intensity by considering the size and spatial distribution of earthquake environmental effects. Lekkas et al. [33] used the ESI scale and its correlation with geological structures for seismic hazard estimation of the 2008 Mw 7.9 Wenchuan, China, earthquake. In another study by [34], correlations between ESI-07 intensity, slope, and lithology were discussed regarding landslides triggered by the 2016 Mw 7.8 Pedernales, Ecuador, earthquake.

Furthermore, Ferrario et al. [35] investigated the role of earthquake environmental effects within seismic sequences from the 2018 Lombok (Indonesia). Statistical analysis was carried out for three nearly complete landslide inventories triggered by the 12 May 2008, Wenchuan Mw 7.9 earthquake of China [36]. Correlations of landslide occurrence with topographical factors and seismic parameters were studied for three inventories. This literature review shows that, over the past decades, there was an improving trend regarding both the documentation and statistical evaluation of the earthquake-induced landslides in some important works.

Several of the nowadays commonly applied methodologies were mostly developed by the Italian groups for the census of the effects induced by earthquakes of moderate magnitude. During the Emilia Romagna (northern Italy) 2012 earthquake sequence, for processing and real-time data sourcing, new approaches and technologies were developed by the INGV EMERGEO working group [37]. Just after the earthquake event, the EMERGEO working group surveyed the epicentral area for co-seismic geological hazards. Later, they organized the records and processed them with the EMERGEO Information System (siE). The EMERGEO working team of Civico et al. [38] presented a 1:25,000 scale map of the surface ruptures after the 30 October 2016 Mw 6.5 Norcia earthquake, central Italy. They used 11,000 oblique photographs taken from helicopter flights, which were verified with field datasets. They also provided the datasets through a database of the co-seismic effects following the Norcia earthquake [39].

In this study, we evaluated four manually extracted landslide inventory maps of the Gorkha earthquake 2015, Nepal. Four different research teams from different parts of the world carried out landslide inventories related to this earthquake and published their resulting maps. These studies used various EO data sources, such as very high-resolution WorldView imagery and coarser-resolution imageries. We compared these inventory maps quantitatively and qualitatively using the cartographical degree of matching and frequency area distribution (FAD) methods [40]. The reasons for the differences are discussed here, outlining the limitations and advantages of different mapping techniques.

2. Study Area

The epicentre of the earthquake was in the Gorkha district and its aftershock was about 140 km in the Dholaka district [41]. However, landslides triggered by the earthquake and several aftershocks were scattered around large areas. Seven districts that were severely affected by the earthquake were selected for inventory mapping by different researchers. The study area covered a 14,502 km^2 region, which spread over most of central Nepal. The area had two primary drainage systems, namely Narayani and Saptakoshi [42]. Our case study area lay in central Nepal country within the fold and thrust zone of Himalaya. This zone was caused by the collision of the Eurasian plate with the Indian Plate [43] (see Figure 1). The collision resulted in extensive crustal shortening and upheaval, leading to the formation of the quintessential collided orogen, Himalaya. All evaluations of the present study were done based on an overlapping region of four considered inventory maps.

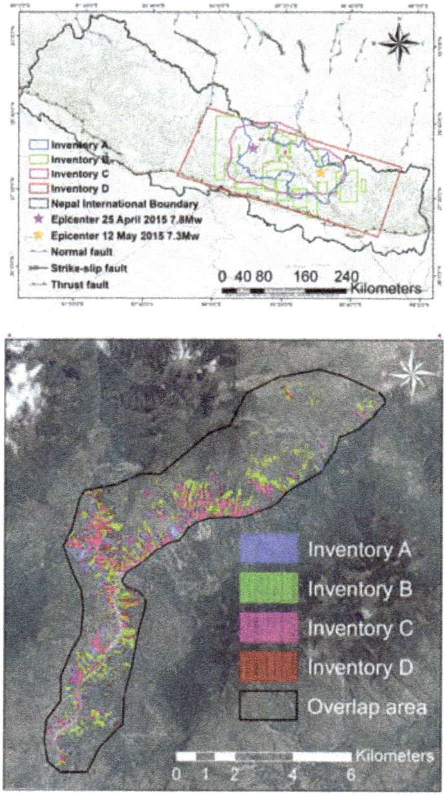

Figure 1. Area covered by the different investigators during mapping landslide inventories of the 2015 Gorkha earthquake.

3. Landslide Inventory Maps of Gorkha Earthquake (National Scale)

3.1. Inventory A

Authors carried out a field survey and manual interpretation of Landsat-8 Enhanced Thematic Mapper (ETM) images with a resolution of 15 m, Gaofen-1 (GF-1) images with resolution of 2 m, GF-2 images with resolution of 0.8 m, and Google Earth imagery to prepare the polygon-based inventory [42]. A total of 15,456 km^2 area was selected for investigation based on previous studies and updates during the digitation process. A total of 3716 co-seismic landslides were mapped. The largest landslide mapped was 9983 m^2 and the smallest landslide was less than 50 m^2. The co-seismic landslides covered a total area of 15.93 km^2.

3.2. Inventory B

Earthquake-induced landslides were mapped by comparing pre- and post-event very high-resolution satellite imageries. DigitalGlobe WorldView-2 and -3 imagery, along with Pleiades satellite data, were used [44]. The spatial resolution of imagery used in mapping landslides varied from 30–50 cm in most of the areas. Images were acquired from 26 April to 15 June 2015, with most images collected between 2 May and 8 May 2015. This team was the only group working on Gorkha earthquake who differentiated between landslide source and deposits areas. They were able to map more than 25,000 landslides in the area affected by the earthquake.

3.3. Inventory C

Authors used Google Earth imagery, which was updated after the Gorkha earthquake, to prepare a polygon-based inventory. A total of 4000 km^2 area was selected for investigation based on previous studies and updates during the digitizing process [45]. In most of the region, the satellite imagery of pre-earthquake was from December 2014 and post-earthquake imagery from 2–4 May, which was around one week after the main earthquake. A total of 17,000 co-seismic landslides were mapped by [45]. Results also showed the spatial correlation of topographical parameters with landslide occurrence.

3.4. Inventory D

The landslide inventory map was manually mapped by the authors in the immediate aftermath of the earthquake using a range of EO data sources, including web-hosted high-resolution optical data in Google™ Crisis Response (e.g., United Kingdom - Disaster Monitoring Constellation-2 (UK-DMC2), Disaster Monitoring Constellation for the International Charter (DMCii), Worldview, Digital Globe Inc., SPOT National Centre for Space Studies (CNES), imagery accessed via the Disaster Charter, imagery available from United States Geological Survey (USGS) Hazards Data Distribution System (HDDS Explorer), and imagery specifically tasked over regions of interest (e.g., Pleiades CNES). A total of 2117 co-seismic landslides were mapped by [46] (see Figure 2).

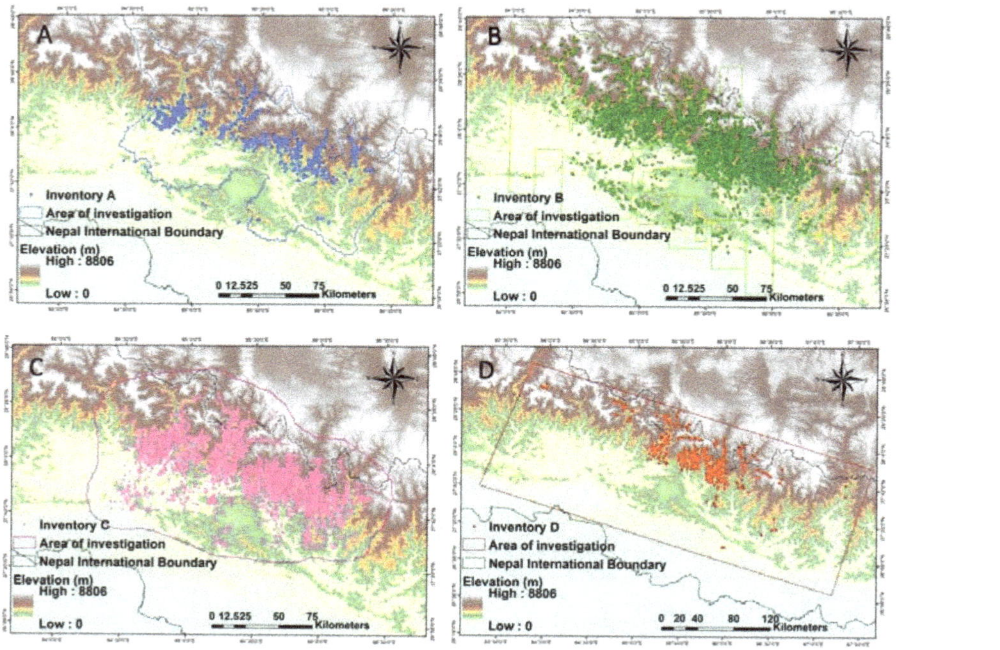

Figure 2. Landslide mapped after the 2015 Gorkha earthquake for the affected region. (**A**) Inventory A, (**B**) inventory B, (**C**) inventory C, and (**D**) inventory D.

4. Landslide Inventory Comparison Methodologies

4.1. Spatial Distribution

Landslide inventories for the Gorkha event were collected from various sources. Detailed information on inventories is given in Table 1. Out of the inventories, we used four that were available and prepared by research teams from various parts of the world. To analyze the landslide density per square kilometre, point-based inventory data were used. Figure 3 shows the landslide density distribution of selected inventories for the Gorkha event; all of the inventories were co-seismic and prepared using post-earthquake satellite imageries.

Information regarding the data used to prepare the inventories is given in Table 1. There was a significant variation in the number of landslides mapped for the Gorkha event, ranging from $N_{LT} = 2117$ to $N_{LT} = 24,915$. None of these inventories classified the types of landslides as proposed by [47]. The data sources used for mapping varied from high-resolution imagery of about 30 cm resolution to Google Earth images. The landslide density analysis results, illustrated in Figure 3, showed that most of the landslides were concentrated in the Gorkha and Sindhupalchock districts for most of the inventories; the southern part of the Rasuwa district was also severely hit, as can be interpreted from the Figure 3. Differences in the density of landslides in an area depended on factors such as the effect of amalgamation while mapping, the purpose of the mapping, and the data sources used for mapping. Areas near the Dholaka and Sindhupalchok districts showed high landslide density ias a result of the aftershock on 12 May 2015 in the Dholaka district. The higher density of landslides showed that inventories were prepared after the aftershock of 12 May 2015 and hence had a larger number of landslides compared to inventories prepared for the first earthquake event. Another factor that affected the density of landslides was the different coverage of the mapping area. Comparing the different inventories at larger scale is not an easy task as their coverage areas are different. To compare landslide inventories prepared by different interpreter, we chose a commonly mapped area that was mapped by most of the inventories. For the analysis, we chose four inventories from various sources which were polygon-based, and compared them statistically for smaller regions.

Table 1. Landslide inventories prepared after the 2015 Gorkha earthquake, Nepal.

#	Landslide Inventory	No. of Landslides (N_{LT})	Geometry Type	Area Coverage	Produced by
1.	Valagussa et al. (2016)	4300	Polygon	Central Nepal	[48]
2.	Roback et al. (2018)	24,915	Polygon	Central Nepal	[44]
3.	Martha et al. (2017)	15,551	Polygon	Central Nepal	[49]
4.	Regmi et al. (2016)	2645	Polygon	Central Nepal	[42]
5.	Meena, Mavrouli, and Westen (2018)	2513	Polygon	Central Nepal	[50]
6.	Kargel et al. (2016)	4312	Polygon	Central Nepal	[51]
7.	Gnyawali et al. (2016)	19,332	Point	Central Nepal	[45]
8.	Robinson et al. (2017)	2117	Polygon	Central Nepal	[46]

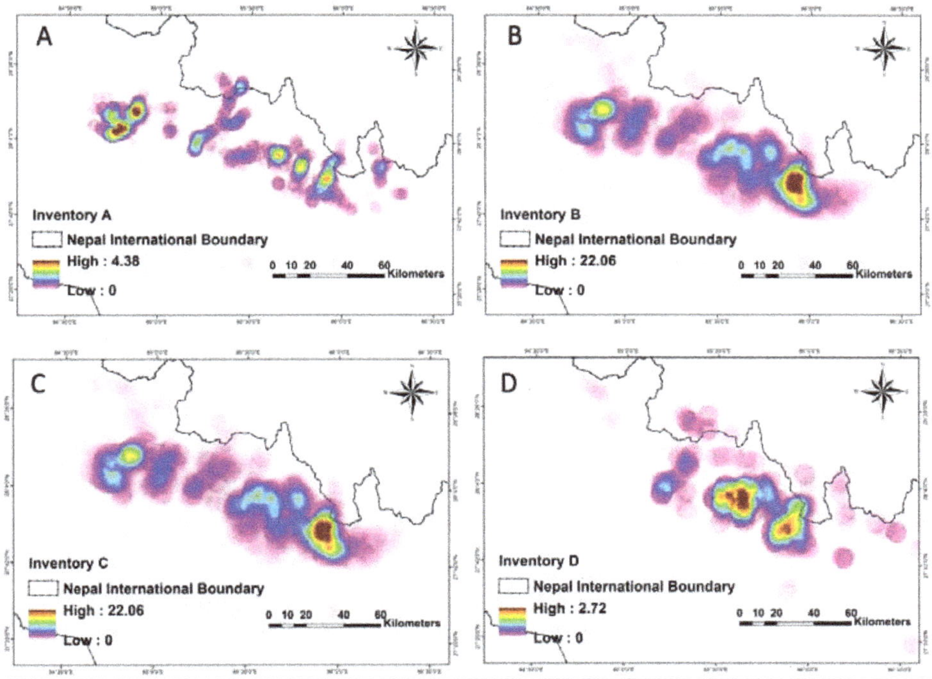

Figure 3. Landslide inventory densities per km^2 for the Gorkha event. (**A**) Inventory A, (**B**) inventory B, (**C**) inventory C, and (**D**) inventory D.

To compare the inventories for the Gorkha earthquake, we chose four polygon-based inventories that were available. The commonly mapped area was covered in all four inventories, which allowed for the statistical comparison of the inventories. We compared the four inventories and examined the differences and similarities of the total area covered, the number of landslides mapped, and the size of the landslides.

4.2. Cartographical Degree of Matching

For the comparison of four inventories available for the common mapped area in the Rasuwa district, an attempt was made to determine the cartographic matching and mismatch (see Figure 4).

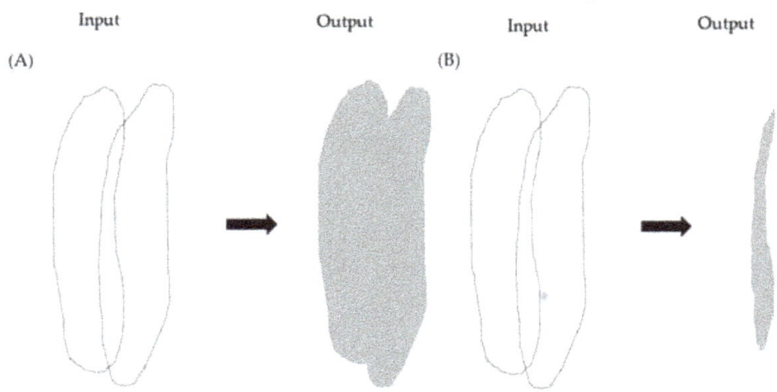

Figure 4. (**A**) Example of a union with features within a feature class that overlaps. (**B**) Example of an intersection with a feature within a feature class that overlaps.

Carrara et al. (1993) [52] provided a method to evaluate the degree of matching and mismatch between two inventory maps. The mismatch index, E, was given by

$$E = \frac{(A_1 \cup A_2) - (A_1 \cap A_2)}{(A_1 \cup A_2)}, \ 0 \le E \le 1 \tag{1}$$

$$M = 1 - E, 0 \le M \le 1 \tag{2}$$

where E is the mismatch and M is the matching between the inventories. Union and intersection are used in Equation (1) to observe the matching and mismatch between the inventories.

This equation was valid for up to two inventories. However, in our case, we wanted to compare four inventories. Thus, we formulated an equation for the comparison of four inventories (see Equation (3)).

$$E = 1 - \frac{(A \cap B) \cup (A \cap C) \cup (A \cap D) \cup (B \cap C) \cup (B \cap D) \cup (C \cap D)}{(A \cup B \cup C \cup D)} \tag{3}$$

The statistics of the four landslide inventories can be seen in Table 2. The total number of landslides varied from 33 to 49 for the same area. Moreover, there was a significant difference in the total area of landslides. Cartographical mapping differences in a single landslide for the same event can be seen in Figure 5. Interpreters mapped the landslide boundary differently, and the causes of such differences are discussed in Section 5. Based on Equation (3), we observed cartographical match and mapping errors, as presented in Table 3. The pairwise comparison of inventories and comparison of inventories mapped by different interpreters are represented in Figures 6 and 7, respectively.

Table 2. Statistics for the four event-based landslide inventories.

	A	B	C	D
Number of mapped landslides #	144	498	197	336
Minimum landslide area (m^2)	500.53	35.50	95.49	112.22
Maximum landslide area (m^2)	157,265.25	118,805.11	764,038	151,708.90
Mean landslide area (m^2)	9990.18	6520.23	25,599.34	8816.13
Standard deviation of landslide area (m^2)	19,203.38	11,595.57	74,255.39	17,832.65
Total landslide area (m^2)	1,438,587.08	3,247,077.22	5,043,071.10	2,962,222.12

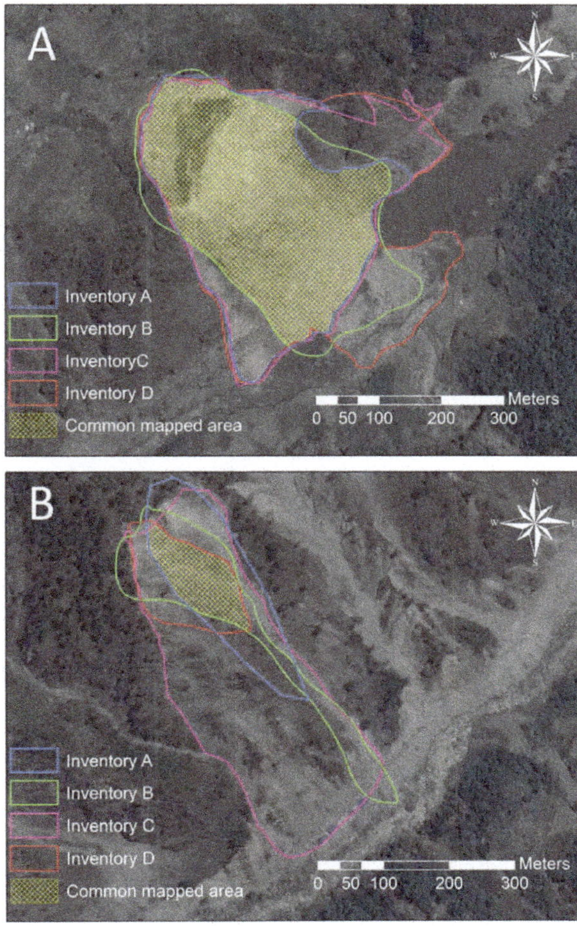

Figure 5. Mapping by different interpreters for the Gorkha earthquake of the same landslides. (**A**,**B**) represent examples of the mapped landslides.

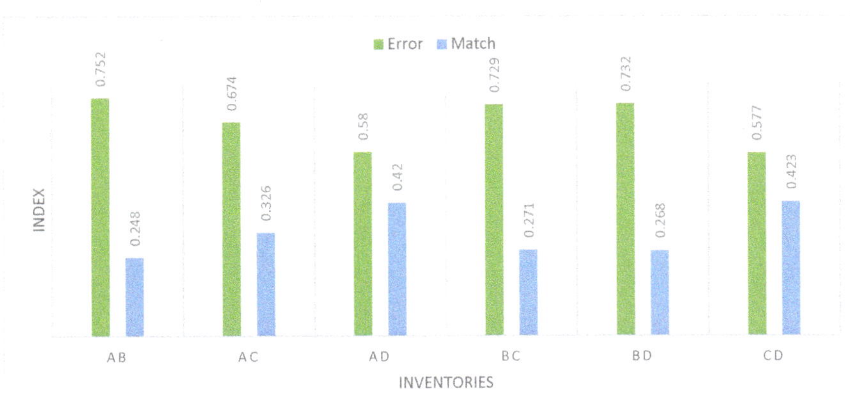

Figure 6. Pairwise comparison of inventories mapped by different interpreters.

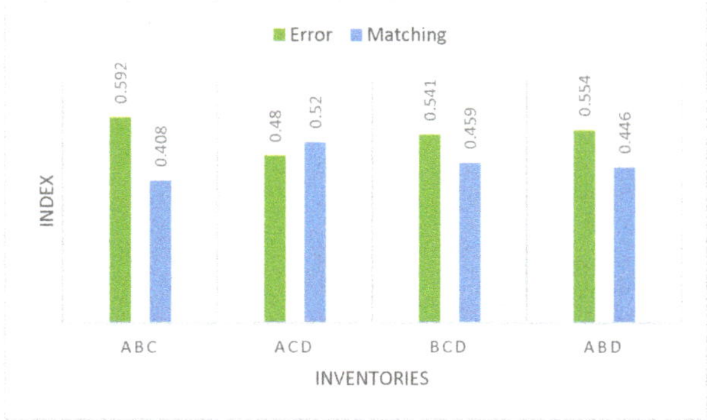

Figure 7. Comparison of inventories mapped by different interpreters.

Table 3. Cartographical matching and mismatch between inventories.

Inventories	Area m^2	Percentage of the Area Covered Relative to the Total Study Area	Mapping Error, E	Mapping Match, M
Landslide area Inventory A	294,0746.32	6.49		
Landslide area Inventory B	3,256,245.33	7.18		
Inventory A ∪ Inventory B	4,966,842.26	10.96		
Inventory A ∩ Inventory B	1,231,685.97	2.72	0.752	0.248
Landslide area Inventory A	2,940,746.32	6.49		
Landslide area Inventory C	5,043,072.67	11.12		
Inventory A ∪ Inventory C	6,005,657.34	13.25		
Inventory A ∩ Inventory C	1,960,725.08	4.32	0.674	0.326
Landslide area Inventory A	2,940,746.32	6.49		
Landslide area Inventory D	4,230,476.82	9.33		
Inventory A ∪ Inventory D	5,051,960.15	11.14		
Inventory A ∩ Inventory D	2,122,932.49	4.68	0.58	0.42
Landslide area Inventory B	3,256,245.33	7.18		
Landslide area Inventory C	5,043,072.67	11.12		
Inventory B ∪ Inventory C	6,518,452.96	14.38		
Inventory B ∩ Inventory C	1,765,857.66	3.90	0.729	0.271
Landslide area Inventory B	3,256,245.33	7.18		
Landslide area Inventory D	4,230,476.82	9.33		
Inventory B ∪ Inventory D	5,909,070.8	13.03		
Inventory B ∩ Inventory D	1,582,813.94	3.49	0.732	0.268
Landslide area Inventory C	5,043,072.67	11.12		
Landslide area Inventory D	4,230,476.82	9.33		
Inventory C ∪ Inventory D	6,530,387	14.40		
Inventory C ∩ Inventory D	2,764,641.79	6.10	0.577	0.423

4.3. *Frequency Area Distribution (FAD)*

Landslide inventories were statistically analyzed using frequency area distribution (FAD) curves, in which the landslide areas were plotted versus the cumulative and non-cumulative landslide frequencies. In the study by [53], observations showed that the power law was valid for medium and large landslides. The probability of occurrence of landslide size can be given by the power-law equation.

$$p(x) = cX^{-\beta}, \tag{4}$$

where X is the observed values, c is a normalization constant, and β is the power-law exponent.

Figure 8 shows the power-law distribution for medium to massive landslides and divergence from the power-law toward lower frequencies with a rollover point, where frequency decreased for smaller landslides. The trend of the FAD of most landslide inventories diverged from the power-law for small landslides [53–56]. The point where this divergence began was defined as the cut-off point [56,57]. For non-cumulative probability density distributions of landslide areas, the peak point of the probability distribution curve, after which the probability value began to decrease for smaller landslides following a positive power-law decay, was referred to as the rollover point [58]. According to [58], in a power-law distribution, the slope of the distribution was defined with a power-law exponent. The part that was represented by large events was referred to as the power-law tail, as shown in Figure 9 (with a scaling parameter, β). Malamud et al. [53] investigated four well-documented landslide events and concluded that rollover was a real phenomenon for landslide-event inventories, depending upon the bias and under-sampling of the smaller landslides. They modeled the FAD for these four inventories and established theoretical curves to estimate the total landslide area triggered by an earthquake or rainfall event.

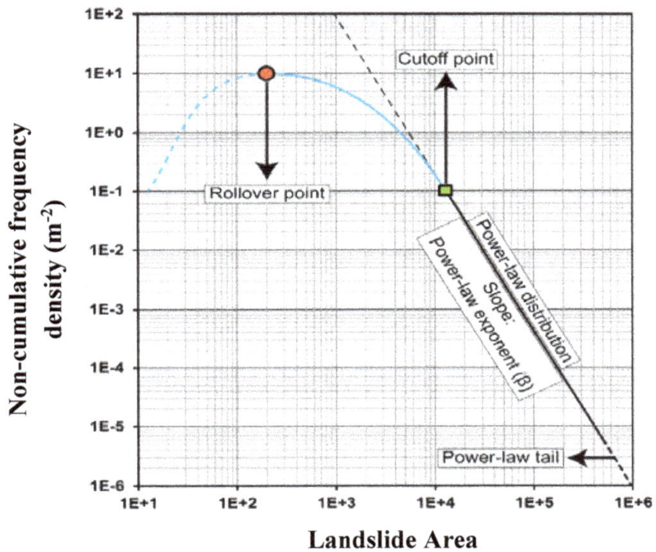

Figure 8. Schematic representation of the main components of a non-cumulative frequency area distribution (FAD) for a landslide inventory.

Further distributions were used to fit the frequency area distribution of landslides. The double Pareto model described the majority of the data well, but [59] indicated that the same model was less good at the tails of the distribution. Another method was proposed by [60], who showed that the entire FAD of landslides could be explained by a three-parameter inverse-gamma distribution (equation). This approach also described a way to estimate the landslide event magnitude (*mLs*). The *mLS* is the indication of the size of the landslide triggering event and gives an indication of the severity of the event in terms of landslide occurrence in a particular area for an event.

$$p(A_L; \rho, a, s) = \frac{1}{a\Gamma(\rho)}\left[\frac{a}{A_L - s}\right]^{\rho+1} exp\left[-\frac{a}{A_L - s}\right] \quad (5)$$

where ρ is the parameter primarily controlling power-law decay for medium and large values, $\Gamma(\rho)$ is the gamma function of ρ, A_L is landslide area, a is the location of rollover point, s is the exponential decay for small landslide areas, and $-(\rho+1)$ is the power-law exponent. Malamud et al. [60] provided a best fit for the power-law exponent and showed that $-(\rho+1) = 2.4$.

Table 4 shows that the power-law exponent of four analysed inventories ranged from 2.27 to 2.48, which was consistent with the literature describing an interval having a central tendency of around 2.3–2.5 [55,58]. The minimum landslide area mapped ranges from 35.50 m^2 to 500.53 m^2 for the inventories. Also, the largest landslide mapped ranges from 764,038 m^2 to 157,265.25 m^2. The rollover points ranged from 256.74 m^2 to 1258 m^2 for the inventories.

Table 4. Comparison of the frequency area statistics of the landslide area.

Inventories	Total number of Landslides N_{LT}	Total Area of Landslides A_L m^2	Minimum Area of Landslides $minA_L$ m^2	Maximum Area of Landslides $maxA_L$ m^2	Power Law Exponent (β)	Rollover Point (m^2)
Inventory 1	144	1,438,587.08	500.53	157,265.25	2.48	1411.43
Inventory 2	498	3,247,077.22	35.50	118,805.11	2.30	85.13
Inventory 3	197	5,043,071.10	95.49	764,038	2.27	223.22
Inventory 4	336	2,962,222.12	112.22	151,708.90	2.37	289.99

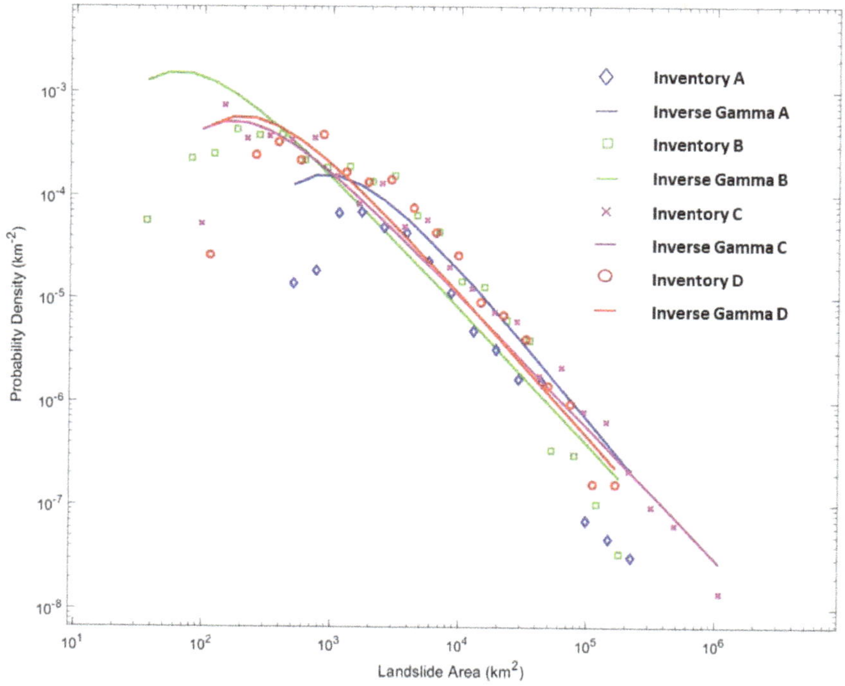

Figure 9. Landslide frequency size distribution, representing the dependence of landslide probability density p on the landslide area.

There was a zigzag pattern in the plotted figure of landslide probability density against the inverse gamma fit. The differences in the probability distribution and inverse gamma fit may have been the result of gaps regarding mapped landslides for given inventories, indicating that some landslides were missing or not mapped by interpreters due to various reasons, such as rapid mapping or the amalgamation of smaller landslides into single landslide features. The rollover points for all inventories differed to each other. For the inventory by interpreter B, the rollover point toward the smaller landslides was 85.13, which was smaller in comparison to other inventories, as there was a wider distribution of small landslides in this inventory.

5. Discussion and Conclusions

The availability of four independent landslide inventories for the same triggering event in the same geographical area allowed us to quantitatively compare and outline strengths and weaknesses of the methods used to prepare the inventories. The inventory by interpreter B (Figure 10b) portrayed more landslides (498) than the other inventories. This difference was significant, as the other interpreters could not detect most of the smaller landslides visually using Google Earth imagery, which was used to prepare the three inventories by interpreters A, C, and D. The more significant number of landslides in the inventory by interpreter B with the very high-resolution Worldview imagery used for the visual identification of the landslides, compared to the spatial resolution of the satellite imagery used for the other inventories, is explained here. The amalgamation of the landslide information merged multiple very small landslides into a single larger landslide, reducing the total number of the mapped event landslides. Considering only the number of landslides, quantitative identification could not be performed because of both the amalgamation of adjacent landslides and the subjectivity of the landslide extraction process. Many factors caused the amalgamation of landslides in a landslide inventory map. The manual extraction of landslide borders and representation with polygons is a subjective process that is affected by the applied method, the preferences of the experts and interpreters, and how much time and effort are invested into the inventory generating process [61]. Amalgamation is the mapping of nearby smaller landslides as a single polygon, which may lead to possibly severe distortion of the statistical analysis of these inventories. An adjacent landslide polygon is usually described as a single polygon if the runouts or scars overlap in areas. Thus, differentiating between them is difficult. Low image resolution, the working scale, and the contrast between affected and unaffected areas are all other reasons for amalgamation [62]. In some regions, landslides can be very condensed, and several contiguous landslides may join runout areas. Amalgamation is often due to errors resulting from a lack of expertise of the interpreter. It may also happen when landslide inventory mapping is carried out using (semi)-automated classification and change detection based on optical satellite images, e.g., [63].

There are different numbers of mapped landslides for the same area interpreted by different interpreters, which related to the scale of working, applied EO spatial resolution, personal perspective, and the method of mapping that was adopted. Inventory C (Figure 10C) and inventory D (Figure 10D) landslides were mapped using Google Earth imagery; 197 and 336 landslides were mapped, respectively. In Figure 10A, inventory A had 144 landslides for the same common area using Google Earth imagery. Figure 10B shows 498 landslides in inventory B, which was the result of using high-resolution Digital Globe WorldView 2–3 satellite imagery by the author. However, it should be mentioned that very high resolution (VHR) optical satellite imagery can have significant distortions and georeferencing errors in high-relief areas if they are orthorectified using only provided rational polynomial coefficient (RPC) models and not by using manual ground control points (GCPs). Also, this can be more severe, as many of the VHR rapid emergency acquisitions over a disaster area-of-interest can have small incidence angles, thus increasing automatic orthorectification errors. Differences in the mapped outlines of landslides can be due to this. The offset from automatic RPC processing is no less than 5 m for the newer sensors (WorldView 2- 3, Pleiades) and even less for older satellites (GeoEye-1, etc). High mountain relief and low-resolution digital elvetaion model (DEM) used for orthorectification (e.g., Shuttle Radar Topography Mission (SRTM), Advanced Spaceborne Thermal Emission and Reflection Radiometer (ASTER) or Advanced Land Observing Satellite (ALOS) may add more possible errors to these offsets [64].

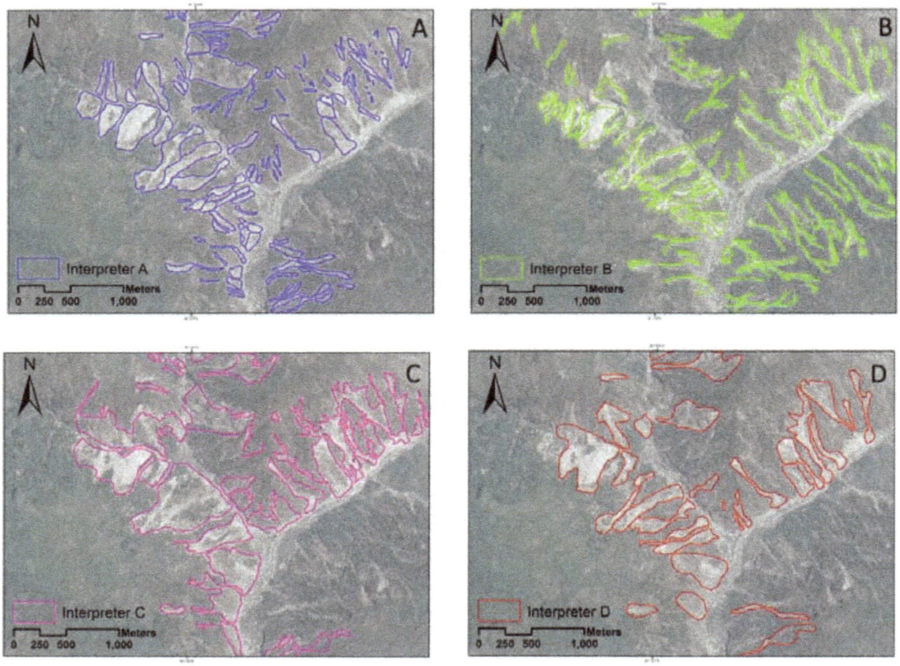

Figure 10. Effect of amalgamation for the same area mapped by four different people in the area near the Trishuli river, (**A**) Inventory A, (**B**) inventory B, (**C**) inventory C, and (**D**) inventory D.

The results of our mismatch index in this work confirm that the difference between the four-event inventories was significant. However, the mismatch index was in the range of the differences measured by other investigators in previous studies that compared landslide inventories in similar physiographical settings (e.g., [25,52]). In addition to the causes for the mismatch discussed by these investigators, in our test case, the difference was also the result of the period of landslides mapped and of the different spatial resolutions of the satellite imagery. Despite the mismatch index, E, visual inspection of the four inventories revealed a similar spatial distribution of the event landslides in the four landslide maps. This matter was confirmed (i) by the spatial correlation between the four inventories and (ii) by the similarities of the double Pareto density functions for the four inventories (see Figure 10). The results of the power-law exponent of the four analysed inventory ranged from 2.27 to 2.48, which was consistent with the literature that showed that the power-law exponent interval had a central tendency around 2.3–2.5 [55,58].

According to the results of the present study, we conclude that both comparison methods, i.e., the cartographical degree of matching and FAD, are essential for evaluating the quality of landslide inventory productions in terms of their similarities in the total number, area size, and spatial density of landslides. However, the cartographical degree of matching method is suitable more for the evaluation and validation of the location and boundaries of the landslide affected areas. Therefore, the cartographical degree of matching method is essential for landslide extraction/annotation studies, especially those using machine learning models for this aim. The accuracy of the resulting landslide extraction from remote sensing data by using the machine-learning models is directly related to the quality of the training data of landslide inventory datasets [1]. Thus, the cartographical degree of the cartographical matching method plays a vital role in enhancing the accuracy of any landslide extracting studies using machine learning models.

On the other hand, although the FAD is also advantageous for landslide detection studies, it is a crucial method for studies which analyse the landslide areas and their distribution in a case study area. Landslide susceptibility mapping, landslide-prone region analysis, and landslide modeling and risk assessment are considered to be studies, where the FAD method should be considered to evaluate the applied inventory datasets.

For the Gorkha earthquake-affected region in central Nepal, we compared four independent event landslide inventories showing landslides triggered by a high-intensity earthquake that hit the area on 25 April 2015. The first inventory was obtained through the visual interpretation of Google Earth imagery. The second inventory was obtained by exploiting a semi-automatic procedure applied to a VHR resolution worldview satellite imagery. We compared the four inventories by exploiting methods already present in the literature and by proposing new qualitative and quantitative criteria. Comparison of the four independent event inventory maps led to the conclusion that the mismatch between the four inventories was significant but consistent with differences measured by other investigators in similar physiographical areas. The mismatch was attributed to (i) different spatial resolutions of the satellite images, (ii) the amalgamation of smaller landslides into larger landslides while delineating the boundaries of landslides; the minimum landslide area was 35.50 m^2, which was mapped using VHR resolution worldview satellite imagery, whereas, the landslide mapped using Landsat 8 ETM imagery was 500.53 m^2, and (iii) the inability of the operator to recognize landslides in shadowed areas. For our future work, we will apply these inventories along with some data-driven models to generate landslide susceptibility maps and compare the resulting susceptibility accuracies.

Author Contributions: Conceptualization, S.R.M.; methodology, S.R.M., and S.T.P.; validation, S.R.M.; data curation, S.R.M.; writing—original draft preparation, S.R.M., and S.T.P.; writing—review and editing, S.R.M.; visualization, S.R.M. All authors read and approved the final manuscript.

Funding: This research is partly funded by the Open access fund by University of Salzburg

Acknowledgments: We would like to thank three anonymous reviewers for their valuable comments.

Conflicts of Interest: The authors declare no conflict of interest.

References

1. Ghorbanzadeh, O.; Blaschke, T.; Gholamnia, K.; Meena, S.R.; Tiede, D.; Aryal, J. Evaluation of Different Machine Learning Methods and Deep-Learning Convolutional Neural Networks for Landslide Detection. *Remote Sens.* **2019**, *11*, 196. [CrossRef]
2. Meena, S.R.; Mishra, B.K.; Tavakkoli Piralilou, S. A Hybrid Spatial Multi-Criteria Evaluation Method for Mapping Landslide Susceptible Areas in Kullu Valley, Himalayas. *Geosciences* **2019**, *9*, 156. [CrossRef]
3. Mondini, A.; Viero, A.; Cavalli, M.; Marchi, L.; Herrera, G.; Guzzetti, F. Comparison of event landslide inventories: the Pogliaschina catchment test case, Italy. *Nat. Hazards Earth Syst. Sci.* **2014**, *14*, 1749–1759. [CrossRef]
4. Dhital, M. Causes and consequences of the 1993 debris flows and landslides in the Kulekhani watershed, central Nepal. In Proceedings of the 3rd International Conference on Debris-Flow Hazards Mitigation: Mechanics, Prediction and Assessment, Davos, Switzerland, 10–12 September 2003; Rickenmann, D., Chen, C.-L., Eds.; Millpress: Rotterdam, The Netherlands, 2003.
5. Xu, Q.; Fan, X.M.; Huang, R.Q.; Westen, C.V. Landslide dams triggered by the Wenchuan Earthquake, Sichuan Province, south west China. *Bull. Eng. Geol. Environ.* **2009**, *68*, 373–386. [CrossRef]
6. Meena, S.R.; Ghorbanzadeh, O.; Blaschke, T. A Comparative Study of Statistics-Based Landslide Susceptibility Models: A Case Study of the Region Affected by the Gorkha Earthquake in Nepal. *ISPRS Int. J. Geo-Inf.* **2019**, *8*, 94. [CrossRef]
7. Keefer, D.K. Investigating landslides caused by earthquakes—A historical review. *Surv. Geophys.* **2002**, *23*, 473–510. [CrossRef]

8. Ghorbanzadeh, O.; Blaschke, T. Optimizing Sample Patches Selection of CNN to Improve the mIOU on Landslide Detection. In Proceedings of the 5th International Conference on Geographical Information Systems Theory, Applications and Management: GISTAM, Heraklion, Greece, 3–5 May 2019; Volume 1, p. 8. [CrossRef]
9. Harp, E.L.; Keefer, D.K.; Sato, H.P.; Yagi, H. Landslide inventories: The essential part of seismic landslide hazard analyses. *Eng. Geol.* **2011**, *122*, 9–21. [CrossRef]
10. Damm, B.; Klose, M. The landslide database for Germany: Closing the gap at national level. *Geomorphology* **2015**, *249*, 82–93. [CrossRef]
11. Guzzetti, F.; Mondini, A.C.; Cardinali, M.; Fiorucci, F.; Santangelo, M.; Chang, K.-T. Landslide inventory maps: New tools for an old problem. *Earth-Sci. Rev.* **2012**, *112*, 42–66. [CrossRef]
12. Ciampalini, A.; Raspini, F.; Bianchini, S.; Frodella, W.; Bardi, F.; Lagomarsino, D.; Di Traglia, F.; Moretti, S.; Proietti, C.; Pagliara, P. Remote sensing as tool for development of landslide databases: the case of the Messina Province (Italy) geodatabase. *Geomorphology* **2015**, *249*, 103–118. [CrossRef]
13. Van Westen, C.J.; Ghosh, S.; Jaiswal, P.; Martha, T.R.; Kuriakose, S.L. From landslide inventories to landslide risk assessment; an attempt to support methodological development in India. In *Landslide Science and Practice: Landslide Inventory and Susceptibility and Hazard Zoning*; Springer: Berlin/Heidelberg, Germany, 2013; Volume 1, pp. 3–20.
14. Ghorbanzadeh, O.; Meena, S.R.; Blaschke, T.; Aryal, J. UAV-Based Slope Failure Detection Using Deep-Learning Convolutional Neural Networks. *Remote Sens.* **2019**, *11*, 2046. [CrossRef]
15. Golovko, D.; Roessner, S.; Behling, R.; Wetzel, H.-U.; Kaufmann, H. GIS-based integration of heterogeneous data for a multi-temporal landslide inventory. In *Landslide Science for a Safer Geoenvironment*; Springer: Berlin/Heidelberg, Germany, 2014; pp. 799–804.
16. Ghorbanzadeh, O.; Feizizadeh, B.; Blaschke, T.; Khosravi, R. Spatially Explicit Sensitivity and Uncertainty Analysis for the landslide risk assessment of the Gas Pipeline Networks. In Proceedings of the 21st AGILE Conference on Geo-Information Science, Lund, Sweden, 12–15 June 2018; pp. 1–7.
17. Lahousse, T.; Chang, K.; Lin, Y. Landslide mapping with multi-scale object-based image analysis—A case study in the Baichi watershed, Taiwan. *Nat. Hazards Earth Syst. Sci.* **2011**, *11*, 2715–2726. [CrossRef]
18. Sansar Raj, M.; Thimmaiah, G.N. Impact of Spatial Resolution of Digital Elevation Model on Landslide Susceptibility Mapping: A case Study in Kullu Valley, Himalayas. *Geosciences* **2019**, *9*, 360. [CrossRef]
19. Schweigl, J.; Straka, W. Working with Landslide Inventories and Susceptibility Maps in Lower Austria. In *Landslide Science and Practice: Landslide Inventory and Susceptibility and Hazard Zoning*; Margottini, C., Canuti, P., Sassa, K., Eds.; Springer: Berlin/Heidelberg, Germany, 2013; Volume 1, pp. 43–50.
20. Borghuis, A.M.; Chang, K.; Lee, H.Y. Comparison Between Automated and Manual Mapping of Typhoon-triggered Landslides from SPOT-5 Imagery. *Int. J. Remote Sens.* **2007**, *28*, 1843–1856. [CrossRef]
21. Danneels, G.; Pirard, E.; Havenith, H.-B. Automatic landslide detection from remote sensing images using supervised classification methods. In Proceedings of the IEEE International Geoscience and Remote Sensing Symposium (IGARSS 2007), Barcelona, Spain, 23–28 July 2007; pp. 3014–3017.
22. Keyport, R.N.; Oommen, T.; Martha, T.R.; Sajinkumar, K.; Gierke, J.S. A comparative analysis of pixel-and object-based detection of landslides from very high-resolution images. *Int. J. Appl. Earth Obs. Geoinf.* **2018**, *64*, 1–11. [CrossRef]
23. Pellicani, R.; Spilotro, G. Evaluating the quality of landslide inventory maps: Comparison between archive and surveyed inventories for the Daunia region (Apulia, Southern Italy). *Bull. Eng. Geol. Environ.* **2015**, *74*, 357–367. [CrossRef]
24. Fan, X.; Scaringi, G.; Korup, O.; West, A.J.; van Westen, C.J.; Tanyas, H.; Hovius, N.; Hales, T.C.; Jibson, R.W.; Allstadt, K.E.; et al. Earthquake-induced chains of geologic hazards: Patterns, mechanisms, and impacts. *Rev. Geophys.* **2019**. [CrossRef]
25. Galli, M.; Ardizzone, F.; Cardinali, M.; Guzzetti, F.; Reichenbach, P. Comparing landslide inventory maps. *Geomorphology* **2008**, *94*, 268–289. [CrossRef]
26. Guzzetti, F.; Cardinali, M.; Reichenbach, P.; Carrara, A. Comparing landslide maps: A case study in the upper Tiber River basin, central Italy. *Environ. Manag.* **2000**, *25*, 247–263. [CrossRef]
27. Keefer, D.K. Landslides caused by earthquakes. *Geol. Soc. Am. Bull.* **1984**, *95*, 406–421. [CrossRef]
28. Rodríguez, C.E.; Bommer, J.J.; Chandler, R.J. Earthquake-induced landslides: 1980–1997. *Soil Dyn. Earthq. Eng.* **1999**, *18*, 325–346. [CrossRef]

29. Keefer, D.K. Statistical analysis of an earthquake-induced landslide distribution — the 1989 Loma Prieta, California event. *Eng. Geol.* **2000**, *58*, 231–249. [CrossRef]
30. Esposito, E.; Porfido, S.; Simonelli, A.L.; Mastrolorenzo, G.; Iaccarino, G. Landslides and other surface effects induced by the 1997 Umbria–Marche seismic sequence. *Eng. Geol.* **2000**, *58*, 353–376. [CrossRef]
31. Serva, L.; Vittori, E.; Comerci, V.; Esposito, E.; Guerrieri, L.; Michetti, A.M.; Mohammadioun, B.; Mohammadioun, G.C.; Porfido, S.; Tatevossian, R.E. Earthquake Hazard and the Environmental Seismic Intensity (ESI) Scale. *Pure Appl. Geophys.* **2016**, *173*, 1479–1515. [CrossRef]
32. Guerrieri, L.; Michetti, A.; Reicherter, K.; Serva, L.; Silva, P.; Audemard, F.; Azuma, T.; Baiocco, F.; Baize, S.; Blumetti, A. Earthquake environmental effect for seismic hazard assessment: The ESI intensity scale and the EEE catalogue. Mem. descr. *Carta Geol. D'italia* **2015**, *97*, 11–20.
33. Lekkas, E.L. The 12 May 2008 Mw 7.9 Wenchuan, China, Earthquake: Macroseismic Intensity Assessment Using the EMS-98 and ESI 2007 Scales and Their Correlation with the Geological Structure. *Bull. Seismol. Soc. Am.* **2010**, *100*, 2791–2804. [CrossRef]
34. Chunga, K.; Livio, F.A.; Martillo, C.; Lara-Saavedra, H.; Ferrario, M.F.; Zevallos, I.; Michetti, A.M. Landslides Triggered by the 2016 Mw 7.8 Pedernales, Ecuador Earthquake: Correlations with ESI-07 Intensity, Lithology, Slope and PGA-h. *Geosciences* **2019**, *9*, 371. [CrossRef]
35. Ferrario, M.F. Landslides triggered by multiple earthquakes: insights from the 2018 Lombok (Indonesia) events. *Nat. Hazards* **2019**, *98*, 575–592. [CrossRef]
36. Xu, C.; Xu, X.; Yao, X.; Dai, F. Three (nearly) complete inventories of landslides triggered by the May 12, 2008 Wenchuan Mw 7.9 earthquake of China and their spatial distribution statistical analysis. *Landslides* **2014**, *11*, 441–461. [CrossRef]
37. Alessio, G.; Alfonsi, L.; Brunori, C.A.; Burrato, P.; Casula, G.; Cinti, F.R.; Civico, R.; Colini, L.; Cucci, L.; De Martini, P.M.; et al. Technologies and new approaches used by the INGV EMERGEO Working Group for real-time data sourcing and processing during the Emilia Romagna (northern Italy) 2012 earthquake sequence. *Ann. Geophys.* **2012**, 55. [CrossRef]
38. Civico, R.; Pucci, S.; Villani, F.; Pizzimenti, L.; De Martini, P.M.; Nappi, R. Surface ruptures following the 30 October 2016 Mw 6.5 Norcia earthquake, central Italy. *J. Maps* **2018**, *14*, 151–160. [CrossRef]
39. Villani, F.; Civico, R.; Pucci, S.; Pizzimenti, L.; Nappi, R.; De Martini, P.M.; The Open EMERGEO Working Group. A database of the coseismic effects following the 30 October 2016 Norcia earthquake in Central Italy. *Sci. Data* **2018**, *5*, 180049. [CrossRef] [PubMed]
40. Tanyaş, H.; van Westen, C.J.; Allstadt, K.E.; Anna Nowicki Jessee, M.; Görüm, T.; Jibson, R.W.; Godt, J.W.; Sato, H.P.; Schmitt, R.G.; Marc, O.; et al. Presentation and Analysis of a Worldwide Database of Earthquake-Induced Landslide Inventories. *J. Geophys. Res. Earth Surf.* **2017**, *122*, 1991–2015. [CrossRef]
41. Tsou, C.-Y.; Chigira, M.; Higaki, D.; Sato, G.; Yagi, H.; Sato, H.P.; Wakai, A.; Dangol, V.; Amatya, S.C.; Yatagai, A. Topographic and geologic controls on landslides induced by the 2015 Gorkha earthquake and its aftershocks: an example from the Trishuli Valley, central Nepal. *Landslides* **2018**, *15*, 953–965. [CrossRef]
42. Regmi, A.D.; Dhital, M.R.; Zhang, J.-Q.; Su, L.-J.; Chen, X.-Q. Landslide susceptibility assessment of the region affected by the 25 April 2015 Gorkha earthquake of Nepal. *J. Mt. Sci.* **2016**, *13*, 1941–1957. [CrossRef]
43. Ni, J.; Barazangi, M. Seismotectonics of the Himalayan collision zone: Geometry of the underthrusting Indian plate beneath the Himalaya. *J. Geophys. Res. Solid Earth* **1984**, *89*, 1147–1163. [CrossRef]
44. Roback, K.; Clark, M.K.; West, A.J.; Zekkos, D.; Li, G.; Gallen, S.F.; Chamlagain, D.; Godt, J.W. The size, distribution, and mobility of landslides caused by the 2015 Mw7. 8 Gorkha earthquake, Nepal. *Geomorphology* **2018**, *301*, 121–138. [CrossRef]
45. Gnyawali, K.R.; Maka, S.; Adhikari, B.R.; Chamlagain, D.; Duwal, S.; Dhungana, A.R. Spatial implications of earthquake induced landslides triggered by the April 25 Gorkha earthquake Mw 7.8: preliminary analysis and findings. In Proceedings of the International Conference on Earthquake Engineering and Post Disastor Reconstruction Planning, Bhaktapur, Nepal, 24–26 April 2016.
46. Robinson, T.R.; Rosser, N.J.; Densmore, A.L.; Williams, J.G.; Kincey, M.E.; Benjamin, J.; Bell, H.J. Rapid post-earthquake modelling of coseismic landsliding intensity and distribution for emergency response decision support. *Nat. Hazards Earth Syst. Sci.* **2017**, *17*, 1521–1540. [CrossRef]
47. Cruden, D.M.; Varnes, D.J. Landslide Types and Processes. *Spec. Rep. Natl. Res. Counc. Transp. Res. Board* **1996**, *247*, 76.

48. Valagussa, A.; Frattini, P.; Crosta, G.; Valbuzzi, E. Pre and post 2015 Nepal earthquake landslide inventories. In *Landslides and Engineered Slopes. Experience, Theory and Practice;* CRC Press: Boca Raton, FL, USA, 2016; pp. 1957–1964.
49. Martha, T.R.; Roy, P.; Mazumdar, R.; Govindharaj, K.B.; Kumar, K.V. Spatial characteristics of landslides triggered by the 2015 M w 7.8 (Gorkha) and M w 7.3 (Dolakha) earthquakes in Nepal. *Landslides* **2017**, *14*, 697–704. [CrossRef]
50. Meena, S.R.; Mavrouli, O.; Westen, C.J. Web based landslide management system for Nepal. In Proceedings of the 33rd Himalaya-Karakorum-Tibet Workshop (HKT), Lausanne, Switzerland, 10–12 September 2018; pp. 109–110.
51. Kargel, J.; Leonard, G.; Shugar, D.H.; Haritashya, U.; Bevington, A.; Fielding, E.; Fujita, K.; Geertsema, M.; Miles, E.; Steiner, J. Geomorphic and geologic controls of geohazards induced by Nepal's 2015 Gorkha earthquake. *Science* **2016**, *351*, aac8353. [CrossRef] [PubMed]
52. Carrara, A. Uncertainty in evaluating landslide hazard and risk. In *Prediction and Perception of Natural Hazards;* Springer: Berlin/Heidelberg, Germany, 1993; pp. 101–109.
53. Malamud, B.D.; Turcotte, D.L.; Guzzetti, F.; Reichenbach, P. Landslide inventories and their statistical properties. *Earth Surf. Process. Landf.* **2004**, *29*, 687–711. [CrossRef]
54. Guzzetti, F.; Malamud, B.D.; Turcotte, D.L.; Reichenbach, P. Power-law correlations of landslide areas in central Italy. *Earth Planet. Sci. Lett.* **2002**, *195*, 169–183. [CrossRef]
55. Stark, C.P.; Guzzetti, F. Landslide rupture and the probability distribution of mobilized debris volumes. *J. Geophys. Res. Earth Surf.* **2009**, *114*, 1–16. [CrossRef]
56. Tanyaş, H.; van Westen, C.J.; Allstadt, K.E.; Jibson, R.W. Factors controlling landslide frequency—Area distributions. *Earth Surf. Process. Landf.* **2019**, *44*, 900–917. [CrossRef]
57. Stark, C.P.; Hovius, N. The characterization of landslide size distributions. *Geophys. Res. Lett.* **2001**, *28*, 1091–1094. [CrossRef]
58. Van Den Eeckhaut, M.; Poesen, J.; Govers, G.; Verstraeten, G.; Demoulin, A. Characteristics of the size distribution of recent and historical landslides in a populated hilly region. *Earth Planet. Sci. Lett.* **2007**, *256*, 588–603. [CrossRef]
59. Guthrie, R.H.; Evans, S.G. Analysis of landslide frequencies and characteristics in a natural system, coastal British Columbia. *Earth Surf. Process. Landf.* **2004**, *29*, 1321–1339. [CrossRef]
60. Malamud, B.D.; Turcotte, D.L.; Guzzetti, F.; Reichenbach, P. Landslides, earthquakes, and erosion. *Earth Planet. Sci. Lett.* **2004**, *229*, 45–59. [CrossRef]
61. Soeters, R.; van Westen, C.J. Slope instability recognition, analysis, and zonation. *Landslides Investig. Mitig.* **1996**, *247*, 129–177.
62. Marc, O.; Hovius, N. Amalgamation in landslide maps: effects and automatic detection. *Nat. Hazards Earth Syst. Sci.* **2015**, *15*, 723–733. [CrossRef]
63. Martha, T.R.; Kerle, N.; Jetten, V.; van Westen, C.J.; Kumar, K.V. Characterising spectral, spatial and morphometric properties of landslides for semi-automatic detection using object-oriented methods. *Geomorphology* **2010**, *116*, 24–36. [CrossRef]
64. Tang, C.; Van Westen, C.J.; Tanyas, H.; Jetten, V.G. Analysing post-earthquake landslide activity using multi-temporal landslide inventories near the epicentral area of the 2008 Wenchuan earthquake. *Nat. Hazards Earth Syst. Sci.* **2016**, *16*, 2641–2655. [CrossRef]

geosciences

MDPI

Article

Dynamics of the Zones of Strong Earthquake Epicenters in the Arctic–Asian Seismic Belt

Lyudmila P. Imaeva [1,2,*], Valery S. Imaev [1,3] and Boris M. Koz'min [3]

[1] Institute of the Earth's Crust SB RAS, 128 Lermontov Street, Irkutsk 664033, Russia; imaev@crust.irk.ru
[2] Academy of Sciences of the Sakha Republic (Yakutia), pr. Lenina, 33, Yakutsk 677007, Russia
[3] Institute of Geology of Diamonds and Precious Metals SB RAS, 39 Lenin Prospect, Yakutsk 677077, Russia; b.m.kozmin@diamond.ysn.ru
* Correspondence: imaeva@crust.irk.ru; Tel.: +79149441598

Received: 28 February 2019; Accepted: 9 April 2019; Published: 12 April 2019

Abstract: Our comprehensive study of the Russian Arctic region aims to clarify the features and types of seismotectonic deformation of the crust in the Arctic–Asian Seismic Belt, specifically in the zones of strong earthquakes in the Laptev Sea Segment, the Kharaulakh Segment, and the Chersky Seismotectonic Zone. We have analyzed modern tectonic structures and active fault systems, as well as tectonic stress fields reconstructed by tectonophysical analysis of the Late Cenozoic faults and folds. The investigated neotectonic structures are ranked with respect to the regional classification principles. Changes in the crustal stress–strain state in the lithospheric plate boundaries between the Eurasian, North American, and Okhotsk Sea Plates are analyzed, and regularities of such changes are discovered. A set of models has been constructed for the studied segments of plate boundaries with account of the dynamics of the regional geological structures. The models can give a framework for the assessment of potential seismic risks of seismogenerating structures in the Russian Arctic region.

Keywords: Arctic–Asian seismic belt; regional segment; active fault; paleoseismogenic structure; Late Cenozoic deformation; earthquake mechanism; seismotectonic deformation; potential seismicity

1. Introduction

The Russian Arctic region is covered by comprehensive studies combining geological, geophysical, and seismological methods that have identified the Arctic–Asian seismic belt, which includes the spreading Gakkel Ridge, the system of rift basins in the Laptev Sea shelf, and the seismogenerating structures of the continental crust in the Chersky Seismotectonic Zone [1]. The geodynamic processes in this belt are studied as indicators of types of seismotectonic deformations of the crust in the contact zone where the Eurasian, North American, and Okhotsk Sea lithospheric plates interact and move relative to each other (Figure 1).

This paper reviews published results regarding the Laptev Sea Segment, Kharaulakh Segment, and the Chersky Seismotectonic Zone [2,3], and presents new research results. It discusses the dynamics of the formation of neotectonic structures and the types of the crustal stress–strain state in the zones of strong earthquake epicenters in the Arctic–Asian seismic belt. In each of the above-mentioned segments, the structural–tectonic positions of the modern structures and the systems of active faults are identified and analyzed on the basis of the field geological and geostructural data collected by the authors and the literature data.

Our research provides data for discovering the tectonic positions and the structural dynamic pattern of the main fields of earthquake epicenters in the study area and allows identifying the blocks that acts as tectonic stress concentrators. Changes in the crustal stress–strain state in different tectonic settings at the margins of the Eurasian, North American, and Okhotsk Sea lithospheric plates are analyzed, and the regularities of such changes are discovered.

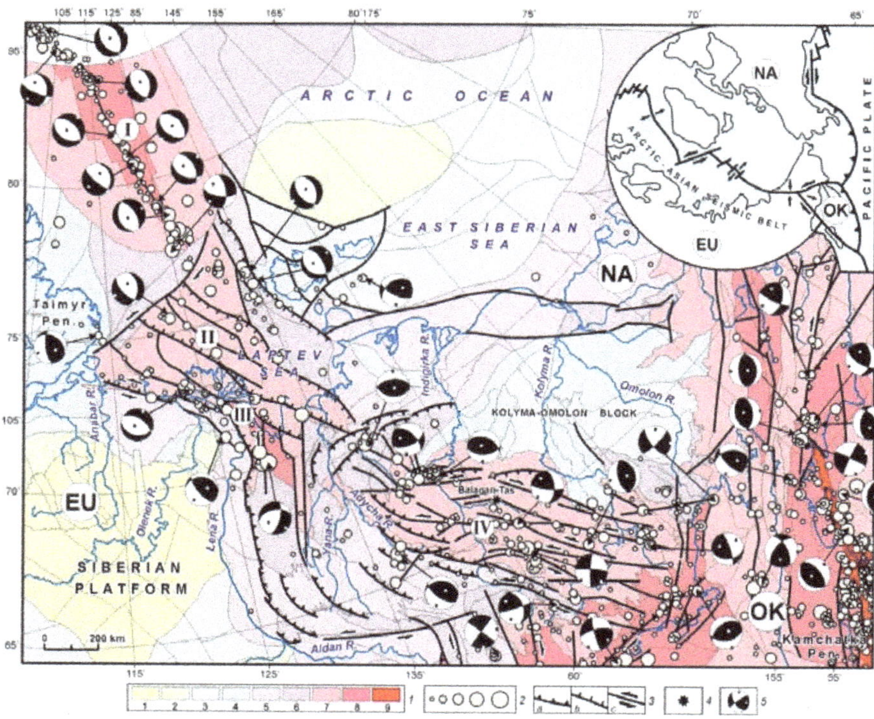

Figure 1. Geodynamic activity of neotectonic structures of the Arctic–Asian seismic belt (modified after [1]). 1—classes of geodynamic activity of the domains: 1–2—low activity, 3–5—medium activity, 6–9—high activity; 2—earthquake epicenters sized by magnitude: ≤ 4.0, 4.1–5.0, 5.1–6.0, 6.1–7.0, ≥ 7.0; 3—kinematics of active faults: (a) thrusts, (b) normal faults, (c) strike-slip faults; 4—Balagan–Tas volcano; 5—focal mechanisms of earthquakes (lower hemisphere; the principal stress axes of compression and extension are marked by black and white dots, respectively). Lithospheric plates: EU—Eurasian, NA—North American, OK—Okhotsk. Inset—location of the Arctic–Asian seismic belt. Segments: I—spreading Gakkel Ridge, II—Laptev Sea rift, III—Kharaulakh, IV—Chersky Seismotectonic Zone.

2. Materials and Methods

Seismotectonic studies are based on the concept of the structural and dynamic uniformity of the geophysical medium and regularities in the development of seismogeodynamic processes, which is elaborated by the Institute of Physics of the Earth of the Russian Academy of Sciences (IPE RAS) [4,5] and developed in [1–3,6,7]. The research method used in our studies of the Russian Arctic includes several stages. The first stage aims at establishing the general trends in the neotectonic development of the study area. Modern (Late Cenozoic) structures in the study area are analyzed considering the recent tectonics as a structural framework comprising active faults and other features of modern tectonic activity related to regional seismicity. Neotectonic structures of the Russian Arctic are ranked by their degree of activity, according to the regional principles of the classification described in [6,7].

In our study, a domain is a neotectonic geodynamic taxon of the territorial rank, which is considered to be a spatially localized integral object with a multifactorial interaction of its main components in the profile of the earth's crust. The classification of domains is a multi-level system, including nine levels (i.e., classes) of activity of modern geodynamic processes that lead to the formation of neotectonic structures. Each activity class is characterized by its specific set of primary and additional features pertaining to tectonics (geodynamic settings), geophysics (seismicity, heat flow, field gravity anomalies, and crustal thickness), morphostructure (terrain elevation, difference between the highest and lowest elevations, and rates of vertical and horizontal movement of the ground surface), material composition, deformation indicators, and GPS measurements. Additionally, the inherited dynamics of neotectonic structures are considered with respect to the conditions in the previous stages of the domain development. To assess the modern geodynamic activity and specify the class of each domain, we interpreted both its primary and additional features. The nine classes of domains are grouped by the degree of the modern geodynamic activity: low (classes 1–2), moderate (classes 4–6), and high (classes 7–9) activity. The characteristics of the classes and the data on active faults are shown in Figure 1.

The next stage of the study aims at identifying the most probable areas of the recent activity for a more detailed investigation and search for reference objects. Large-scale morphostructural and structural–dynamic mapping is carried out to provide the two components of morphotectonic analysis. The main concept of this analysis is consistency between terrain features and corresponding rates and types of endogenous processes. The relative movements of crustal blocks during neotectonic activation cycles create the main features of the terrain, specifically the morphostructures bordered by active faults. The types of endogenous geodynamics are reflected in the features of modern geodynamic activity in the blocks of different ranks and linear fault zones between the blocks.

The database, including geological, geophysical, and seismological data on the study area, provides the basis for investigating the structural–dynamic features of the main fields of earthquake epicenters characterized by the maximum seismic potential. To this end, this study stage includes the collection and interpretation of large-scale remote data and laser-scanning images of the isoseismal areas of strong earthquakes, mapping active faults, fault kinematic analysis, and selection of areas to be covered by detailed field surveys. Attention is given to additional indicators of recent fault activity, such as displacements identified by repeated geodetic surveys, earthquake epicenters confined to the zones of dynamic influence of faults, focal mechanisms of earthquakes as indicators of the dynamics and directions of crustal movements, geothermal and gas-hydrochemical anomalies that give evidence of an increased permeability of the crust, seismic profiling, seismic survey data, gravimetric data, and electrical survey data. Field studies collect and clarify information on the deformation and displacement of young relief elements and sediments and discover evidence of strong paleo- and modern earthquakes. Trenching is performed on seismogenic deformation sites. The field database helps to clarify the kinematic types of Late Cenozoic folds and faults and the structural parageneses of active faults.

Reference objects (i.e., zones of earthquake epicenters) are selected in the seismogeodynamic zones and studied in detail. In such zones, we identify linear fault zones and blocks that act as tectonic stress accumulators, which may have high seismic potential; determine the kinematic types of contact zones of the main seismogenerating structures; and create models showing regional structures and dynamics. The experience of some Russian and international joint research projects shows that the above-described sequence of data collection and processing ensures that the resultant datasets provide a complete and reliable investigation of modern seismogeodynamic processes.

The research results reported in this paper are based on an updated and more comprehensive regional database consolidated by the authors, as well as the data on geology, tectonics, geophysics, seismogeology, and hydrogeology of the study area from publications and sources provided by industrial companies and research organizations.

3. Results

3.1. Laptev Sea Segment

3.1.1. Structure and Tectonics of Laptev Sea Segment

The Laptev Sea shelf is located at the northern margin of the Eurasian Plate (Figure 1). In this area, the SE flank of the mid-oceanic Gakkel Ridge is traced orthogonally to the continental slope (Figure 2). The oceanic basin is in contact with the shelf areas of the East Arctic seas, which represent a system of structures between the continent and the ocean. This area experienced several phases of collisional deformation in the Late Paleozoic and Mesozoic [8,9].

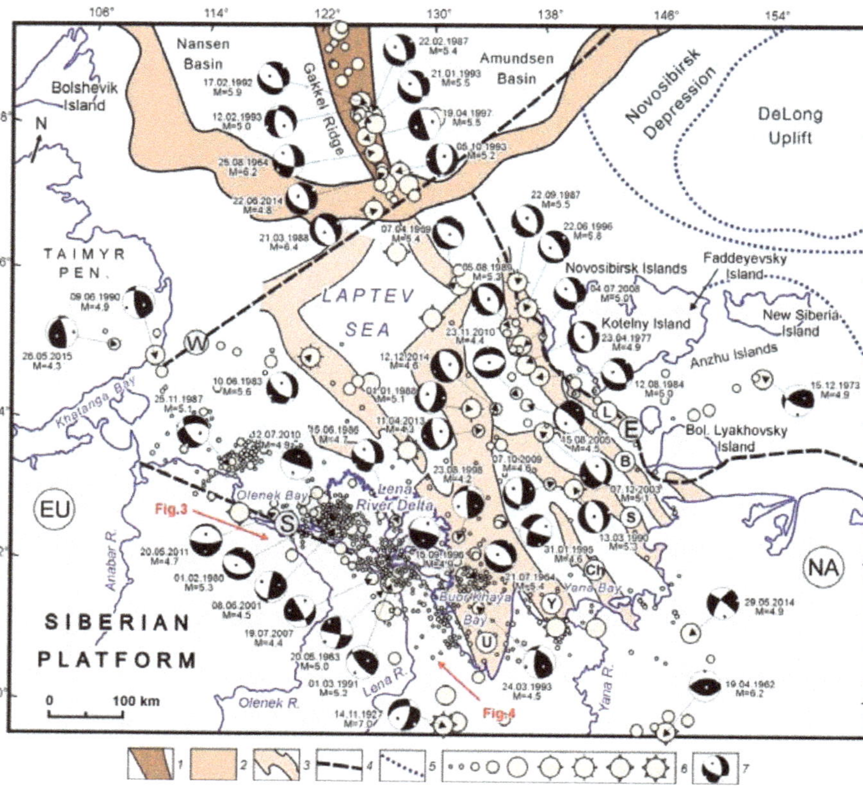

Figure 2. Seismotectonics of the Laptev Sea segment (modified after [7]). 1—Gakkel spreading ridge; 2—continental slope; 3—grabens at the bottom of the Laptev Sea: L—Lyakhov, B—Belkovsky–SvyatoyNos, S—Shirostonsky, Ch—Chondonsky, Y—Ust–Yana, U—Ust–Lena; 4—conventional boundaries of the Laptev Sea plate: S—southern, W—western, E—eastern; 5—boundaries of large troughs and uplifts; 6—earthquake epicenters sized by magnitude: ≤3.0, 3.5, 4.0, 4.5, 5.0, 5.5, 6.0, 6.5, 7.0; 7—focal mechanisms, dates and magnitudes of earthquakes (lower hemisphere; the principal stress axes of compression and extension are marked by black and white dots, respectively). Lithospheric plates: EU—Eurasian, NA—North American.

At the end of the Cretaceous and Cenozoic, intensive rifting took place due to opening of the Eurasian spreading basin [8,9]. The Laptev Sea Rift has been described in detail using the data from marine multichannel seismic profiling by reflected waves (CDP–SRM) [10–12]. According to these surveys, the entire Laptev Sea area is framed by the Mesozoic fold systems that are traced for considerable distances into the sea. In seismic profiles, the fold systems are recorded as an acoustic basement and can be viewed as a tectonic base of the Laptev Sea Rift. The basement top is disturbed only by normal faulting, which gives evidence of the post-folding age of this surface and suggests that the sedimentary cover may be dated to the Cenozoic or Late Cretaceous.

The large thickness of the sedimentary cover (to 10 km) is indicative of the presence of both Cenozoic and Cretaceous deposits in the Laptev Sea Rift. The basement includes the Paleozoic and Early Mesozoic formations of the Verkhoyansk–Kolyma fold system, wherein folding was completed by the Middle Cretaceous. During the Cenozoic, subsidence occurred on both the continental shelf and coastal lowlands. The profiles of the exhumed Cenozoic sediments on several coastal lowlands and islands have been studied in detail. On the Anzhu Islands, the Upper Cretaceous and Cenozoic deposits, and even the Miocene sediments, are disturbed by linear folds in thrusts that are non-conformably overlain with horizontally deposited Upper Pliocene sediments [13]. It can thus be suggested that Cenozoic extension was interrupted by a compression episode at the end of the Miocene. According to the seismic profiling, a system of narrow NW-striking grabens, troughs (Ust–Lena, Ust–Yana, Chondon, Bel'kovsky–Svyatoy Nos, etc.), and associated submarine horsts have been identified in this area (Figure 2) [11,14–16]. The troughs are up to 200–250 km long and 40–60 km wide. The most remarkable element of this rift system is the Ust–Lena graben, traced for a distance of 400–420 km north of the southern termination of Buor–Khaya Bay; its northern part is 150–170 km wide, and gradually narrows to 30–40 km towards the south. Faults bordering the grabens control the details of their interior structure. Two types of faults are typical of this area: sublatitudinal and NW-striking normal faults. Strike-slips faults oriented sub-orthogonally to the normal faults are also observed. Such faults form the E–NE, NE, and NW-trending fault systems that have horizontally displaced the deposits in the depressions by almost two kilometers.

3.1.2. Seismotectonics of Laptev Sea Segment

The Laptev Sea Segment is a zone of diffuse seismicity, which spatially corresponds to the system of rift basins in the Laptev Sea shelf and occupies the area between the Tajmyr Peninsula, Lena River delta, and Novosibirsk Islands (Figure 2). This zone includes several NW-striking subzones of more densely spaced earthquake hypocenters. Strong earthquakes mainly tend to occur in the zone that extends from the Gakkel Ridge to the Yana Bay of the Laptev Sea and follows the boundary between the Eurasian and North American lithospheric plates. In this zone, the earthquake sources are either concentrated in the basins of the Laptev Sea rift system or occur at the sides of the depressions. Furthermore, two zones without much seismic activity are identified along the boundaries of the Laptev Sea shelf. The Lena–Tajmyr zone stretches across the Lena River delta, along the coast of Olenek Bay of the Laptev Sea, and towards the Tajmyr Peninsula; and another zone extends sublongitudinally from the East Siberian Sea between the Faddeev and New Siberian islands and can be traced further northwards (Figure 2).

In our previous studies [1,7], seismotectonic deformation parameters were calculated from the focal mechanisms of local earthquakes recorded at the southeastern termination of the Gakkel Ridge, the continental slope and the rift depressions of the Laptev Sea shelf (Figure 2). Seismic moment tensors were taken from the Global Centroid Moment Tensor (CMT) Catalog and the International Seismological Centre (ISC) Bulletin [17]. According to our calculations, the seismotectonic setting of extension is evidently dominant in the above-mentioned areas of the Russian Arctic. The principal stress axis of extension is subhorizontal and oriented in a NE–SW direction across the strike of the main structures [7].

Considering the regular coincidence of earthquake epicenters with the tectonic structures of the Laptev Sea shelf, we distinguish the three most seismically active portions of the Laptev Sea Segment, which are termed 'eastern', 'southern', and 'western' (Figure 2). The eastern portion is traced along the Belkovsky–Svyatoy Nos and Lyakhovsky grabens and coincident with the northern flank of the Mesozoic fold zone belonging to the Verkhoyansk–Kolyma fold system. The southern portion is traced along the latitudinal Mesozoic fold branch of the Olenek sector in the portion of the Lena–Anabar regional suture (Figure 3). The western portion is traced from the Tajmyr fold system along and towards the abyssal basin. In plan, these seismogenerating structures contour the margins of the Laptev Sea Microplate [1,3,7]. According to the seismological data, the western and eastern boundaries of the Laptev Sea Microplate are subjected to compression (Figures 1 and 2). In this area, thrusting caused local earthquakes in the contact zones between the continental and rift structures.

From the south, the Laptev Sea Rift is separated from the Siberian platform by the Mesozoic faults of the Olenek section of the Lena–Anabar regional suture (Figure 3). In this area, a series of

W–NW-trending folds formed in the Mesozoic due to sublatitudinal left-lateral shearing along the northern margin of the Siberian platform. Such folds may continue on the Laptev Sea shelf, wherein their possible extension is limited by an extended zone of high-gradient gravity anomalies. The latter is detected as a sublatitudinal system of alternating small-size linear positive and negative anomalies of varying intensity [18]. In this zone, the seismic records show several areas of maximum seismic activity values, which correlate with the above-mentioned system of gravity anomalies (Figure 3).

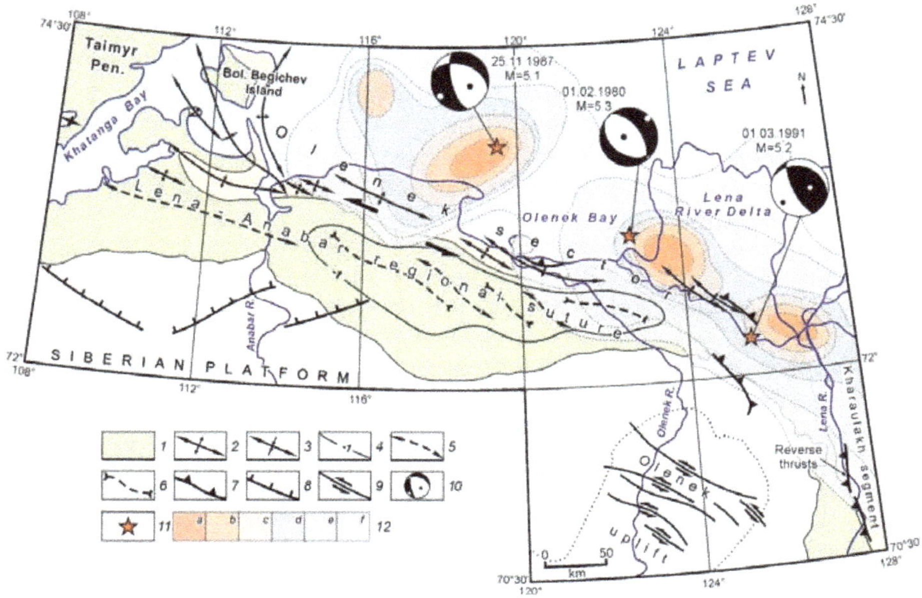

Figure 3. Seismotectonics of the Olenek sector of the Lena–Anabar regional suture (modified after [7]). 1—Lower Cretaceous continental deposits; 2—anticline axis; 3—syncline axis; 4—isoheights of the Lower Cretaceous bottom; 5—axis of consedimental bar; 6—axis of consedimental basin; 7–9—kinematics of active faults: 7—thrust, 8—normal fault, 9—strike-slip fault; 10—focal mechanisms, dates and magnitudes of earthquakes (lower hemisphere; the principal stress axes of compression and extension are marked by black and white dots, respectively); 11—earthquake epicenter; 12—density of earthquake epicenters (number of seismic events per 1° × 1°): (a) 26–30, (b) 21–25, (c) 16–20, (d) 11–15, (e) 6–10, (f) 1–5.

In the zone of the Lena–Anabar regional suture, the seismic process develops in conditions of both extension (Lena River delta, shores of Olenek, and Anabars bays) and compression (Taimyr Peninsula). Focal mechanism solutions for this zone are varying, as well as the fault kinematics, including normal, reverse, and strike-slip faults, and their modifications. The analysis of seismotectonic deformation structures shows the dominance of crustal stretching with a small shear component [6,7]. The extension stress azimuth in this zone differs in orientation from that of the Laptev Sea Rift. The principal stress axes are directed across the strike of the main tectonic elements; those with low dip angles are NW–SW-trending (Figures 1 and 3). Most focal mechanisms of the local earthquakes show normal faulting, with the exception of the 1990 M_S 4.9 Taimyr earthquake in the western margin of the Laptev Sea shelf and the 2015 M_S 4.3 earthquake in the NW part of the Taimyr Peninsula (Figure 2) [3,7]. Their focal mechanism solutions significantly differ from those of all other seismic events recorded in the area along the Laptev Sea coast (Figures 1 and 3).

The 1990 Taimyr earthquake, which occurred at depth of 15 km, caused thrust-and reverse-type displacements, along the gently sloping and subvertical planes of the NW-striking and submeridional faults, respectively. It should be noted that the conclusion on compression in the Taimyr Peninsula, which is based on the focal mechanisms, is consistent with the results of the geostructural studies based on the geological survey data [19]. It is also supported by the rates of visible uplifting of the Laptev Sea coast, which were estimated from long-term measurements of the global ocean level. Over a 10-year measurement period, the rates of modern tectonic uplifting in the Taimyr Peninsula were +(1–2) mm/yr on average [20].

Thus, the modern geodynamic setting of the Laptev Sea Segment is determined by the seismogenerating structures located in the area where the riftogenic zone of the Eurasian basin of the Arctic Ocean extends up to the shelf area. This predetermines the style of seismotectonic deformation, the locations of zones with specific tectonic settings, and the crustal stress-strain state, as well as the structural and dynamic features of the main fields of earthquake epicenters in the study area. In general, the seismotectonic deformation parameters reflect the trends in the distribution of crustal stress fields reconstructed from the structural data.

3.2. Kharaulakh Segment

3.2.1. Structure and Tectonics of Kharaulakh Segment

The Kharaulakh Segment is the northern flank of the Verkhoyansk fold–thrust belt (Figure 4). In the Neo-Proterozoic, this segment originated at the reworked margin of the Siberian platform and developed as a passive continental margin. Its evolution is reflected in the structures and kinematic types of the observed dislocations. During the Cenozoic, the rotation pole of the North American and Eurasian plates changed its position several times and thus caused alternating extension and compression periods, as confirmed by the structural tectonics of this area [7,9,21]. In the Kharaulakh Segment, the Cenozoic megacomplex is represented mainly by Paleocene-Eocene continental deposits that occur with a sharp angular unconformity on different horizons of the deformed Precambrian–Mesozoic megacomplex [6,7]. These deposits fill a series of sublongitudinally oriented depressions, e.g., Kengdei, Kunga, Sogin, Bykov, etc. (Figure 4), which formed in the Paleogene during the earliest rifting stage along the continental continuation of the spreading Gakkel Ridge. At some locations, the Paleogene deposits are folded and cut by thrusts and reverse faults.

All the above-mentioned observations provide evidence of compression in the Cenozoic. The structural studies of the Kharaulakh Segment (Figure 4) suggest sublatitudinal compression in the Middle Miocene, according to the Cenozoic profiles of this segment and adjacent areas [21].

The next episode in the Cenozoic evolution of the Kharaulakh Segment was extension in the Pliocene–Quaternary. Young normal faults that displaced the Neogene weathering crust are observed along the coast of Buor–Khaya Bay (Figure 4). The extension axis was either sublatitudinal or NE-trending. The Quaternary evolution of the near-coastal zone of the Kharaulakh Segment was sharply different from the development of its continental part. Normal faulting dominated in this area and defined the block structure of this territory, as shown by differences in the hypsometric positions of the Late Quaternary and Holocene deposits. The data on normal faults give evidence of the extension phase and suggest that the extension axes were sublatitudinal and NE-oriented [7]. Thus, the Kharaulakh Segment is a transition zone where the mid-oceanic and continental crustal structures are conjugated [9,16,21].

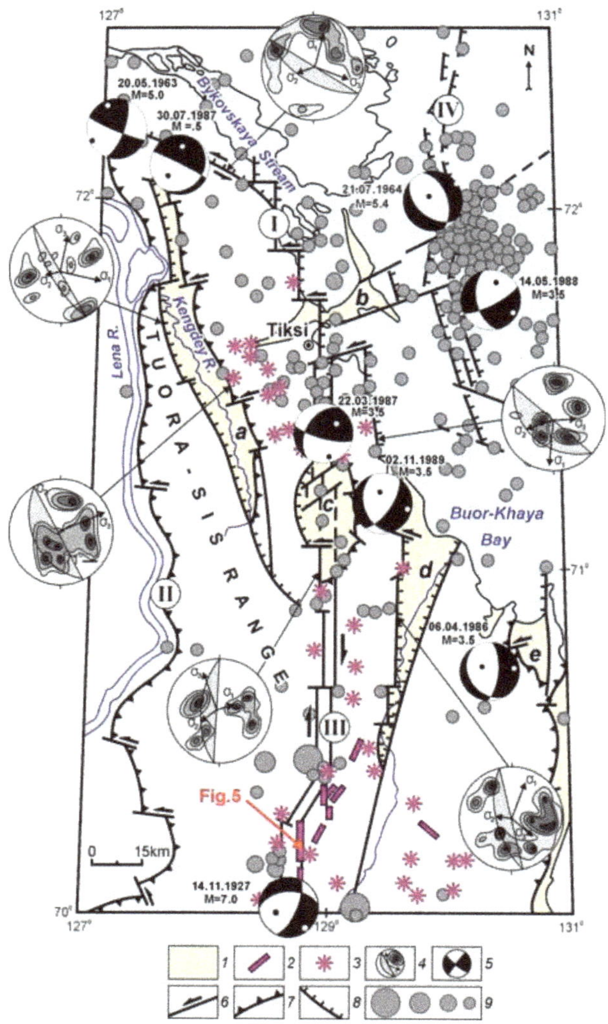

Figure 4. Seismotectonic map of the Kharaulakh segment (modified after [21]). 1—Cenozoic depression: (a) Kengdey, (b) Khorogor, (c) Kunga, (d) Kharaulakh, (e) Nyaybinskaya; 2—seismodislocations; 3—locations of seismogravity effects; 4—rock fracturing diagrams, positions of vectors of the principal stress axes and fault planes; 5—focal mechanisms, dates and magnitudes of earthquakes (lower hemisphere; the principal stress axes of compression and extension are marked by black and white dots, respectively); 6–8—kinematics of active faults: 6—strike-slip fault, 7—thrust, 8—normal fault; 9—earthquake epicenters sized by magnitude: ≥6.8, 6.7–5.0, 4.9–4.0, ≤3.9. Systems of active faults: I—Primorskaya, II—West Verkhoyansk, III—Kharaulakh, IV—Buor-Khaya.

3.2.2. Seismotectonics of Kharaulakh Segment

In our study, we use the data collected from seismic and geostructural surveys of the Kharaulakh Segment in combination with the information on tectonic stress fields reconstructed by tectonophysical analysis of the Late Cenozoic faults and folds in the study area. The available geological and geophysical datasets are also used to identify the systems of regional and local faults that were active in the Cenozoic. Their kinematics are confirmed by corresponding fracturing diagrams and earthquake focal mechanisms (Figure 4). In the zones of their dynamic influence, we identify seismogenerating structures of various sizes, which correlate with seismic events of M ≥ 7.5. Four main groups of fault systems are distinguished with respect to spatial locations, lengths, and kinematics: (I) Primorskaya (normal and strike-slip faults), (II) West Verkhoyansk (thrusts), (III) Kharaulakh (normal and strike-slip faults), and (IV) Buor–Khaya (normal faults).

The most active strike-slip faults are located in the central zone of the Kharaulakh Segment, which is traced as a system of closely spaced subparallel longitudinal faults arranged 'en echelon' to each other (Figure 4). In the map, this zone coincides with a field of minimum gravity values, which extends far to the south beyond the study area [7,21]. In the aerial photographs, morphological features show that the Kharaulakh faults are active in the Quaternary; numerous troughs, grabens, landslides, and avalanches are related to these faults. Furthermore, the available seismic records give evidence of active faulting in the Kharaulakh Segment. In 1927–1928, five strong earthquakes (M_s 5.8–7.0) took place near the village of Bulun (Yakutia) at the southern termination of the fault zone in the same crustal block.

In the southern (most active) flank of this area, the sublongitudinal faults cross the western slopes of the Kharaulakh Ridge and run parallel to its axial line for a distance of 15 km. In the aerial photograph (Figure 5), a straight fault line displaces numerous river channels and cuts the watershed ridges.

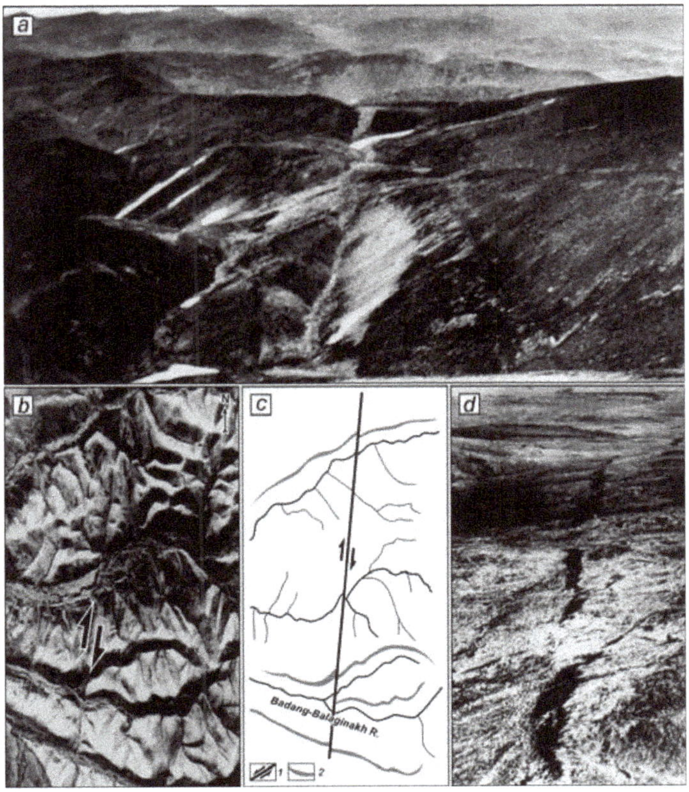

Figure 5. Seismodislocation Beris (after [21]). (**a**) Photograph of surface ruptures resulting from the Bulun earthquakes of 1927–1928 (M ≥7.0); (**b**) aerial photograph of the Beris fault scarp, arrows point to the fault; (**c**) fragment of the deciphered image (1—strike-slip fault; 2—watershed axis); (**d**) photograph of seismogenic extension ruptures in the Kharaulakh fault zone (original location of the Beris fault).

According to the field structural and geological observations in the zone of dynamic influence and the rock fracturing analysis, this is a right-lateral strike-slip fault with a normal component. These fault kinematics are confirmed by the focal mechanism solution of the Bulun earthquake (1927; M_s 7.0). The dip and strike of the fault plane are consistent in the reconstructions based on both the structural and seismic data. Recent activation features are clearly observed in the 'diagonal link' zone, which includes more than 20 paleo and modern seismodislocations caused by gravity and tectonic events. Some of the seismodislocations are marked by outcropped fault segments with horizontal displacement of 5–7 m (Figure 5).

Another seismically active area is the Buor–Khaya normal fault zone along the western coast of the Buor–Khaya Bay (Figure 4), which includes normal faults observed in the coastal outcrops from the Bykov Channel to the Kharaulakh depression for a distance of more than 160 km. The faults cut the basement of the rift structure, and many of them penetrate into the upper horizons of the sedimentary cover, which reveals their young age (Pliocene-Quaternary) [7,21]. The same age is determined from the field geological and geomorphological observation data [22] and multichannel seismic profiling [23]. Some of the faults are traced on land and to the Buor–Khaya Bay bottom and are clearly reflected in the sea bottom relief [11,12]. Their activity is evidenced by morphological features detected in the satellite images, as well as by their relationship to earthquakes (M_s 3.5–5.4) and seismodislocations.

Considering the structural dynamics of the Kharaulakh Segment, we distinguish several areas that differ in types of the crustal stress state (Figure 4). The zones of strong earthquake epicenters differ in seismic parameters from one area to another, as shown by the data from the field geological and structural observations and the seismotectonic crustal deformation regimes reconstructed from the seismic data. According to our analysis of the state of crustal stresses, the Kharaulakh Segment is a unique transition region where extension is replaced with compression [7,21]. Similar regions on the globe are the Afar Rift in East Africa and the Northern California region. In the Kharaulakh Segment, stresses are mainly concentrated in the systems of sublongitudinal strike-slip faults with a normal component, which compose a block-concentrator that has a maximum seismic potential. Faulting in this area was influenced by the zone of the left-lateral strike-slip displacements with a normal component in the Olenek sector of the Lena–Anabar regional suture, as well as by the dynamically conjugated northwestern system of faults traced from the Chersky mountains (Figures 1 and 3). This conclusion is confirmed by the general sublongitudinal strike of the tectonic structures in the Kharaulakh Segment. The flanks of the dome structures show the northwestward deviation. In this area, the Tuora–Sis Range (Figure 4) extends to the Lena River left bank (Chekanov Range). Other facts in support of the above findings are the Cenozoic basins bordered by the strike-slip faults with a normal component from the west and the thrust faults from the east. The structural and seismological data show that the Kharaulakh duplex, represented in the west by ramp anticlines of the Tuora–Sis Range, is the main seismogenerating structure in the Kharaulakh Segment [21].

3.3. *Segment of the Chersky Seismotectonic Zone*

3.3.1. Structure and Tectonics of the Segment

The Chersky Seismotectonic Zone belongs to a collage of terranes, which differ in structural and geological evolution features, comprising the Verkhoyansk fold–thrust belt [24]. In the Middle Jurassic, most of these terranes amalgamated through various episodes of subduction, collision and extension into a single larger tectonic block termed the Kolyma–Omolon Superterrane. The southeastern segment of the Chersky Seismotectonic Zone constitutes a continental fragment of the Okhotsk crustal plate. Its basement is composed of Archaean and Early Proterozoic crystalline rocks overlain with continental volcanic rocks of the Okhotsk–Chukotka belt [24,25]. The structure of the Chersky Seismotectonic Zone is complex and includes numerous linear fold zones and crustal blocks characterized by a high degree of seismic activity [26]. In this zone, the Yana–Indigirka and Indigirka–Kolyma regional sectors are recognized by various features, including specific parageneses of active structures, anomalies of potential fields, and seismicity patterns (Figure 6).

3.3.2. Seismotectonics of the Segment

Structural, morphotectonic, and seismic data were analyzed to investigate the kinematic features of active faults in the Chersky Zone (Figure 6). The detailed analysis shows that most of the active faults were formed due to horizontal compression and crustal shortening. This conclusion is supported by the structural observations of movements in the fault zones, field studies of tectonic fracturing, intense dislocations of Cenozoic sediments, as well as geological maps [26]. Compression is also confirmed by the focal mechanism solutions of strong earthquakes recorded in the zones of influence of active faults.

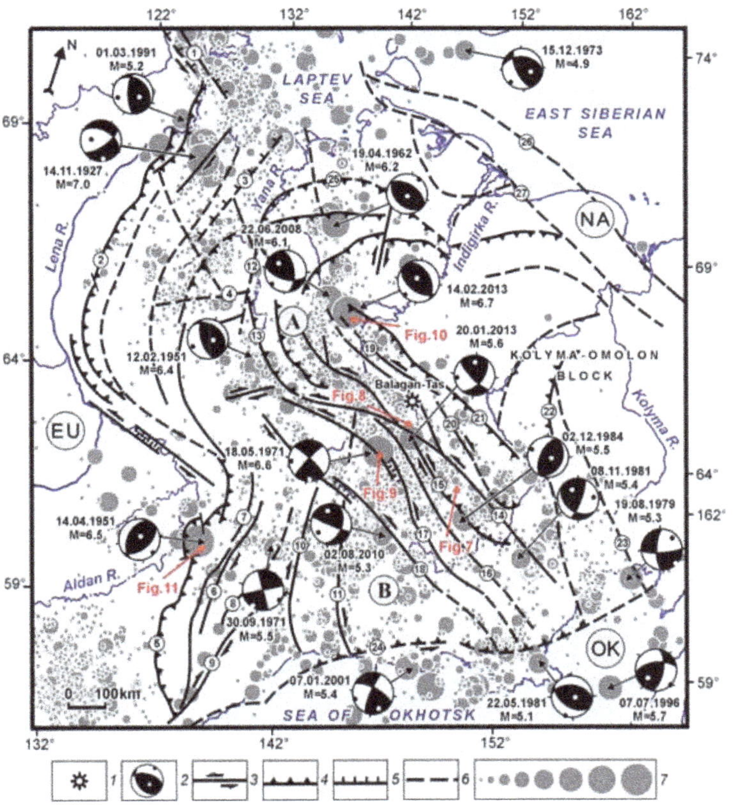

Figure 6. Seismicity and active faults of the Chersky Seismotectonic Zone (modified after [26]). 1—Balagan–Tas volcano; 2—focal mechanisms, dates and magnitudes of earthquakes (lower hemisphere; the principal stress axes of compression and extension are marked by black and white dots, respectively); 3–5—kinematics of active faults: 3—strike-slip fault, 4—thrust and reverse faults, 5—normal fault; 6—inferred fault; 7—earthquake epicenters sized by magnitude: ≤3.0, 3.5, 4.0, 4.5, 5.0, 5.5, 6.0, ≥6.5. Active faults: 1—Lena–Anabar regional suture, 2—Verkhoyansk marginal suture, 3—Omoloi, 4—Ege-Khaya, 5—Nelkan–Kyllakh regional suture, 6—Burkhala, 7—East Sette-Daban, 8—Yudoma, 9—Bilyakchan, 10—Ketanda, 11—Nyut-Ulbei, 12—Yana, 13—Adycha-Taryn, 14—Ulakhan, 15—Darpir, 16—Inyali-Debin, 17—Chai-Yureya, 18—Bryungade, 19—Ilin-Tas, 20—Arga-Tas, 21—Myatis, 22—Yarhodon, 23—Korkodon, 24—Chelomdzha-Yamsk, 25—Polousnensky, 26—Yarkan (South Anyui), 27—Khetachan. Lithospheric plates: EU—Eurasian, NA—North American, OK—Okhotsk. Sectors: A—Yana-Indigirka, B—Indigirka-Kolyma.

The Ulakhan strike-slip fault system is the largest in the NE regions of Russia (Figures 6 and 7). It has a complete set of characteristic of similar fault systems located in other regions of the world: a major shear suture represented by a straight-line fault, a system of echelon ruptures, an asymmetrical geometric pattern of folds and fractures feathering the major suture, an intrusion belt at a distance from the major suture, and a wide stratigraphic range of rocks [24]. Specific structural features of this fault system are an echelon series of left-lateral strike-slip faults and a chain of young pull-apart mini-troughs that formed in the crustal stretching segments of this fault. The Ulakhan fault crosses the Rassokha and Omulevka rivers. During the Middle Pleistocene–Holocene, the Ulakhan fault shifted the river channels to the left for about 24 km [26,27], which shows that an average rate of horizontal tectonic movements along the fault amounted to 5–7 mm/yr.

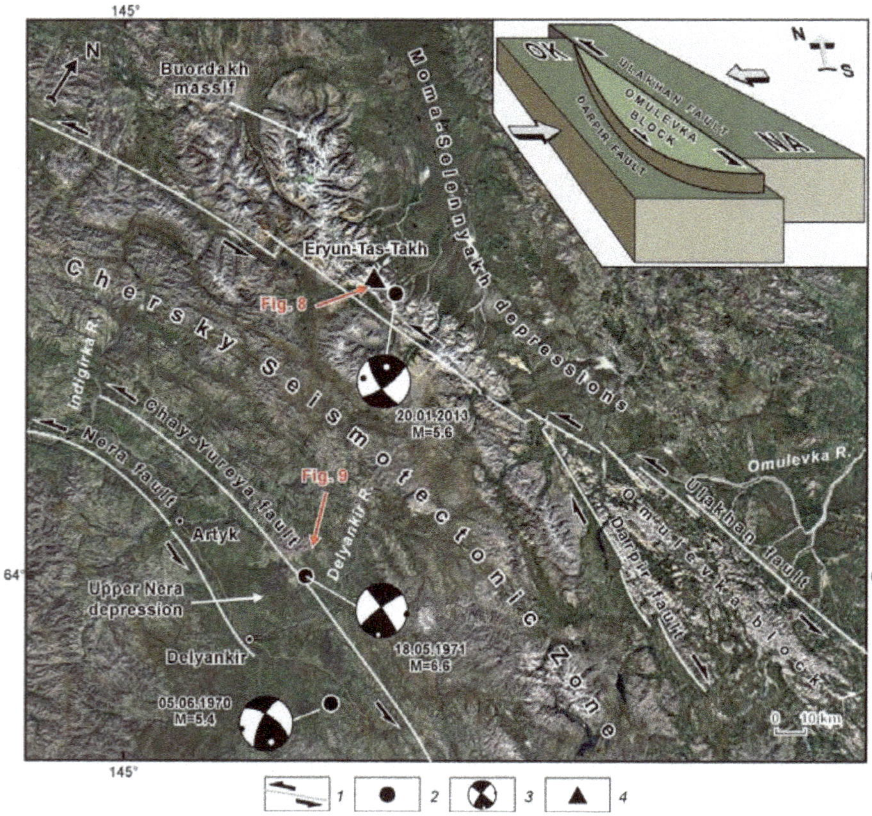

Figure 7. Structural dynamics of the Omulevka block. 1—active strike-slip fault; 2—earthquake epicenter; 3—focal mechanisms, dates and magnitudes of earthquakes (lower hemisphere; the principal stress axes of compression and extension are marked by black and white dots, respectively); 4—seismodislocation Eryun–Tas–Takh. Arrows show directions of plate movements. Lithospheric plates: NA—North American, OK—Okhotsk.

In the Ulakhan fault zone, many traces of ancient and modern seismodislocations are observed, including the most spectacular one—an almost 50 m high landslide dam in the upper reaches of the Tirekhtyakh River. This natural dam occurred due to a catastrophic earthquake hundreds or thousand years ago [28] (Figure 8).

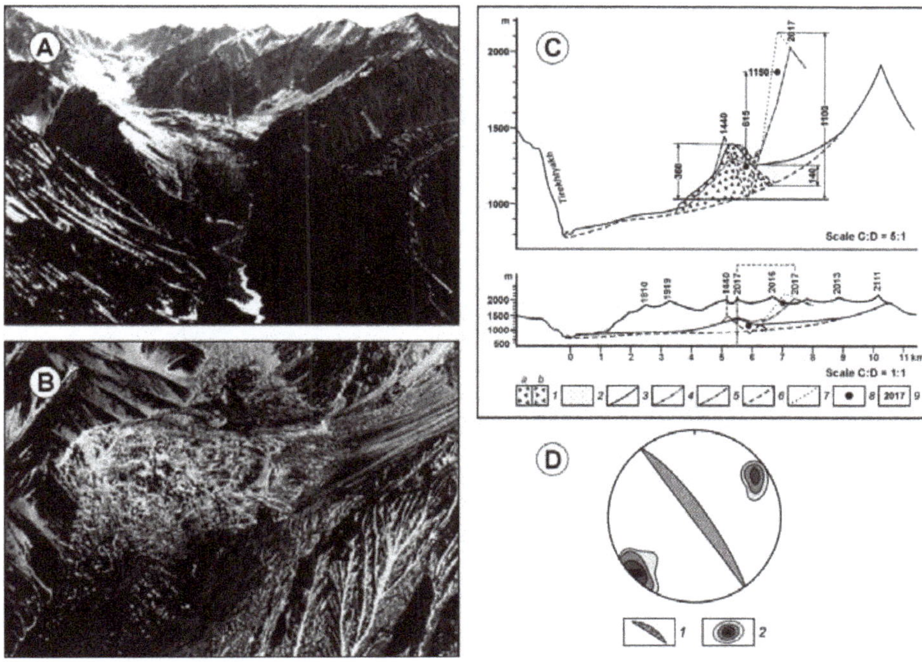

Figure 8. Dam on the Eryun–Tas–Takh River (after [25,26]). (**A**) front view (the headwater segment the Eryun–Tas–Takh River is blocked by the dam); (**B**) top view (the river valley and the detachment wall are blocked by collapsed rocks); (**C**) transverse profile across the seismogenic dam (after [28]): 1—coarse-grained material: (a) granite gneiss, (b) carbonates, 2—fine-grained sandstone material, 3—erosion line of the rock collapse, 4—line before the rock collapse, 5—reconstructed line of the rock collapse wall, 6—lower boundary of the dam lake, 7—hypsometric level before the rock collapse, 8—observation points, 9—absolute elevation; (**D**) rock fracturing diagram for the active fault zone: 1—active fault plane, 2—fracture density isolines.

The Ulakhan fault conjugates with the Darpir fault (Figure 7). This pair of active conjugated faults reflects the style of tectonic deformation at the SE boundary between the North American and Okhotsk plates. Both faults are clearly detectable on satellite images. In topographic maps, these faults are shown as lengthy lineaments of the NW strike, which converge at an angle of 20–25°. The faults border the uplifted Omulevsky block of Paleozoic rocks, which is elevated relative to the Mesozoic structures by 450–550 m. The Omulevsky block is viewed as a terrane that formed in the Mesozoic structural frame during collisional and post-collisional transformations of the Verkhoyansk–Kolyma belt. The background seismicity in the Omulevsky block suggests that this rootless terrane [24] was detached from the neighboring rocks as a result of horizontal displacements.

The Chai–Yureya fault is a significant structure in the Chersky system of active left-lateral strike-slip faults (hereafter the Chersky fault system), which is traced from the Okhotsk Sea coast to the Indigirka River as a series of separate fault segments typical of left-lateral strike-slip faults (Figures 6 and 9).

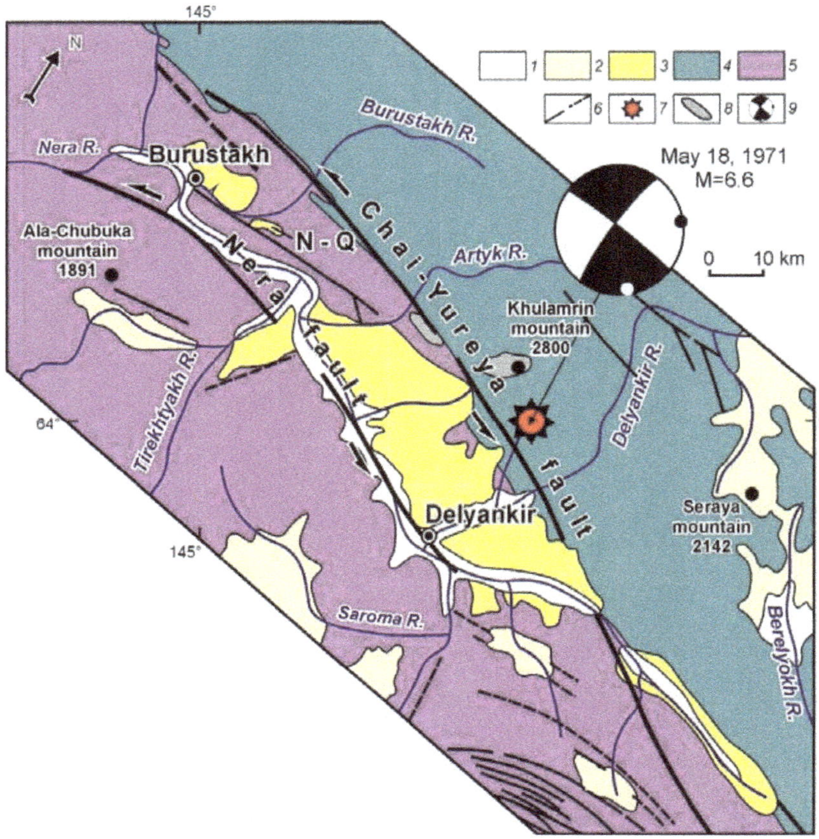

Figure 9. Geological model of the 1971 Artyk earthquake epicenter zone [29]. 1—modern alluvial deposits; 2—Middle Upper Quaternary water–glacial deposits; 3—Neogene–Quaternary deposits of the Verkhnener basins; 4—Jurassic sediments; 5—Triassic sediments; 6—lines of confirmed and inferred tectonic faults; 7—Artyk earthquake epicenter; 8—area of maximum values of ground surface deformation; 9—focal mechanism, dates and magnitudes of earthquakes (lower hemisphere; the principal stress axes of compression and extension are marked by black and white dots, respectively).

The entire length of this fault is detected as a zone of intense rock dislocations, which is marked by a linear magnetic anomaly and a sharp gravity gradient jump [29]. The Artyk earthquake (1971, M_S 6.6), which took place in the zone of the Chai–Yureya fault, was one of the strongest earthquakes in NE Asia. During 12 months after the main shock, more than 1500 aftershocks were recorded in a 60 km long area extending in the NW direction from the epicenter of the main seismic event along the Chai–Yureya fault. The distribution of the aftershocks with depth and focal solution of the main shock show strike-slip movements on the nearly vertical fault plane (Figure 9) [2].

In the north–west, the Chersky fault system is conjugated with the Polousnensky thrust zone of sublatitudinal strike. The latter is represented by a series of subparallel thrust and reverse faults with planes dipping to the south and south–west (Figure 6). The focal mechanism of the Irgichan earthquake (1962, M_S 6.2) also shows thrusting in the zone of the Polousnensky faults.

The Adycha–Taryn zone of active thrust faults borders the entire Chersky fault system from the south–west. This zone includes a series of 'en echelon' faults shifted by straight transverse strike-slip faults. Thrusting is confirmed by the focal mechanism solutions of the earthquakes (1951, M_S 5.2–6.4) that occurred near the Upper Adycha depression [29]. Several seismodislocations are located in the area where the Triassic sandstones are thrust onto the Holocene terrace. In the middle course of the Adycha River, the Triassic rocks are thrust onto the Middle Pleistocene and Middle Quaternary rocks. The planes of all the above-mentioned thrust faults dip to the northeast underneath the Chersky Ridge. The northeastern boundary of the Chersky Seismotectonic Zone is the Myatiss thrust fault that is traced over 700 km along the NE foot of the Moma Ridge in the zone of its junction with the Indigirka–Zyryansky trough (Figure 6). At the NW flank of this fault, there is the Andrey–Tas area of maximum seismicity, as evidenced by the Uyanda (1984, M_S 5.6), Andrey–Tas (2008, M_S 6.1), and Ilin–Tas (Abyi) (2013, M_S 6.9) earthquakes, as well as numerous minor seismic events (Figure 10). The main shock of the Ilin–Tas earthquake and its aftershocks occurred in the duplex compressional structure defined by the tectonic deformation data. Compression is confirmed by the focal mechanism solutions [30,31].

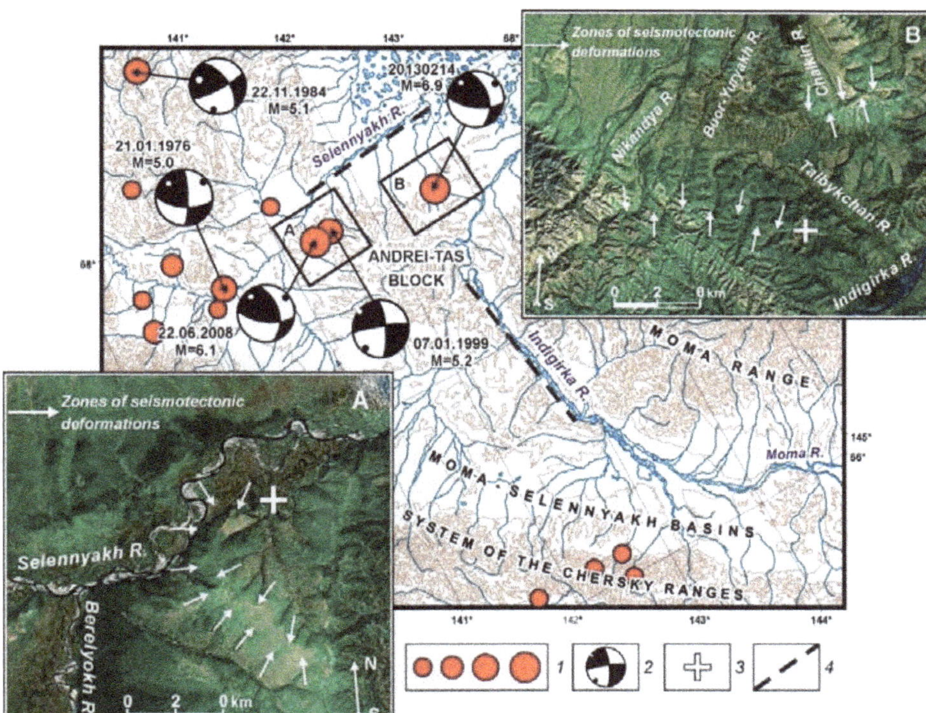

Figure 10. Strong earthquakes in the Andrei–Tas area of maximum seismic activity values (after [30,31]). 1—earthquake epicenters sized by magnitude: 3.0–3.9, 4.0–4.9, 5.0–5.9, 6.0–6.9; 2—focal mechanisms, dates and magnitudes of earthquakes (lower hemisphere; the principal stress axes of compression and extension are marked by black and white dots, respectively); 3—Andrei-Tas earthquake epicenter of 22 June 2008 (inset **A**) and Ilin–Tas (Abyi) earthquake epicenter of 14 February 2013 (inset **B**); 4—zones of seismogravity effects The insets show seismotectonic deformation structures resulting from the earthquakes.

The Andrey–Tas block bordered with thrust faults is a compact rectangular structure experiencing intense uplift (up to 2500 m high) and subjected to faulting of various ranks and kinematic types. Shearing is observed mainly in its central part, and a fan-shaped series of thrust faults develops in the NW flank. In the images from Roskosmos Geoportal, seismotectonic dislocations are detectable in the epicenter areas of the Andrey–Tas and Ylin–Tas earthquakes (Figure 10). The strike and morphological features suggest reverse faulting and shearing along this fault. The features detected from the seismotectonic and seismogravitational effects are possibly related to recent seismic events recorded in the Andrei–Tas block and the paleoearthquakes. It is noted that the areas with seismotectonic deformations and the epicenter zones of the seismic events are spatially coincident, and the shapes of the ground surface dislocations are consistent with the movements reconstructed from the corresponding focal mechanism solutions.

The Chersky fault system is sublongitudinally adjacent to the Ketandino–Ulbey system of right-lateral strike-slip faults (Figure 6). In the space images and aerial photographs, the latter are detected from the segments of straight fault lines that disturb the watershed parts of the ridges. Displaced granite intrusions and geological boundaries along these faults suggest that the right-lateral shear displacements amounted to 20 km [26,29]. This type of shearing is also confirmed by the focal mechanism solutions of strong earthquakes that occurred in the zones of dynamic influence of the faults. The Chelomdzha–Yamsky fault traced near the Okhotsk Sea coast is a boundary between the outer and inner zones of the Okhotsk–Chukotka volcanogenic belt (Figure 6). The magnetic and gravity fields are detected from magnetic anomalies and a sharp gravity gradient jump, respectively [29]. The fault is accompanied by a chain of intermountain depressions filled with Neogene and Quaternary sediments. The fault kinematics are defined by straightness and the presence of folded and thrust dislocations in the Cenozoic sediments of the depressions influenced by the fault, which is a thrust with a left-lateral strike-slip component.

Seismic activity increases at the sutures bordering the Chersky Seismotectonic Zone. The Sette–Daban earthquake (1951, M_S 6.5) occurred in the Kyllakh block, which is the area of dynamic influence of the Nelkano–Kyllakh marginal suture (Figures 6 and 11). In this area, the deformation field is complex due to overlapping of the seismogenerating structures that belong to the Arctic–Asian, Okhotsk–Chukotka, and Baikal–Stanovoy seismic belts. The structure is characterized by lystric thrust faults that steeply dip near the surface and continue eastwards as low-angle detachments at depth. In the western part of the Nelkano–Kyllakh area, the sediments are thrust onto the subhorizontal beds of the Jurassic and Cretaceous deposits of the Siberian platform [24,25]. Due to thrusting, the eastern wing of the anticline is overlain with the Vend–Cambrian deformed beds of the East Sette–Daban tectonic zone. The E–NE orientated compression is evidenced by two fault planes reconstructed from the seismological data for the Sette–Daban earthquake, the NE plane (right-lateral strike-slip), and the NW plane (left-lateral strike-slip) [27,32]. In general, the identified seismotectonic deformation parameters reflect the trends in the distribution of crustal stress fields reconstructed from the structural data.

In order to investigate the crustal stress state of the Chersky Seismotectonic Zone, we analyzed the focal mechanism solutions of earthquakes recorded in the frontal zone, where the Kolyma–Omolon superterrane interacts with the Eurasian plate (Figure 12). These seismic events took place in conditions of steady NE-oriented compression. The principal compressional axis is subhorizontal (dip from 3° to 44°) and acts across the strike of the structural elements in this zone (Figure 6). The extension stress axis is often coincident with the fault strike and oriented either horizontally or subvertically (dip from 2° to 85°). The spatial orientations of the intermediate stress axis are chaotic, and the dip angles vary in a wide range (from 0° to 82°). The above-described stress orientations reconstructed from the focal solutions are dominant across the entire Chersky zone.

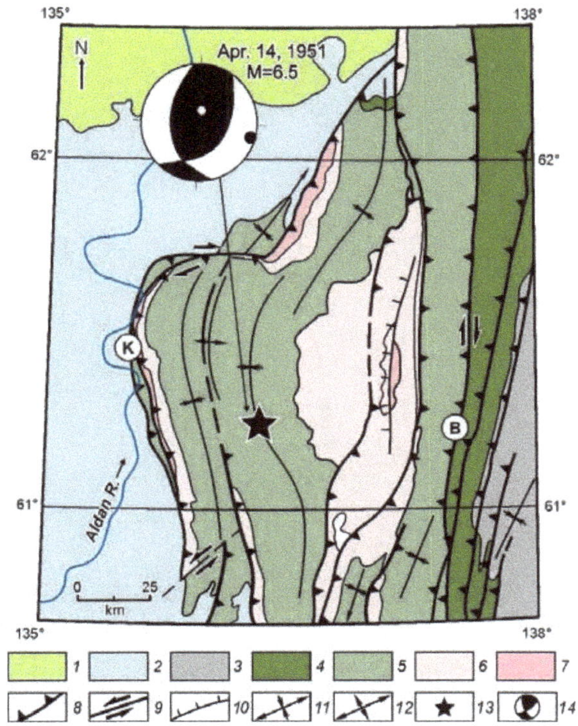

Figure 11. Structural dynamics of the Kyllakh block, South Verkhoyansk sector (modified after [27,32]). 1–7—ages of deposits: 1—Cretaceous, 2—Jurassic, 3—Carboniferous–Permian, 4—Ordovician–Silurian–Devonian, 5—Vendian–Cambrian, 6—Middle–Upper Riphean, 7—Lower Riphean; 8–10—kinematics of active faults: 8—thrust; 9—strike-slip; 10 – normal fault; 11—syncline axis; 12—anticline axis; 13—Kyllakh earthquake epicenter; 14—focal mechanism, dates and magnitudes of earthquakes (lower hemisphere; the principal stress axes of compression and extension are marked by black and white dots, respectively). Faults: B—Burkhala, K—Kyllakh.

According to the kinematics of the principal tectonic stresses inferred from the seismic data, most of the earthquake focal mechanisms in the Chersky Seismotectonic Zone yield reverse (40%), strike-slip (30%), and thrust (20%) solutions, and about 10% show a combination of strike-slip and normal faulting (the percentage is calculated for 24 solutions; see Figure 6). These data clearly demonstrate that in the zone of interaction between the Kolyma–Omolon superterrane and the Eurasian plate, the seismic process develops under compression along the system of major faults that are conjugated with the marginal thrust and reverse faults.

Based on the structural pattern of the major seismogenerating zones and their dynamics, we proposed a model for the entire Chersky Seismotectonic Zone (the inset in Figure 12). This model shows transpression settings initiated by the interacting frontal structures in the contact zone between the Eurasian and North American lithospheric plates moving relative to each other at different rates [7,29]. Such a setting is possible if the Kolyma–Omolon block (in the frontal part of the North American plate) operated as an active indenter during convergence of the lithospheric plates in the NE direction. Due to its impact, a fan-shaped set of NW-trending left-lateral and SE-trending right-lateral strike-slip faults was formed. Seismogenerating zones of reverse faulting and thrusting, which developed at the terminations of the above-mentioned faults, have high seismic potential. Specific features of the modern geodynamics of the Chersky Seismotectonic Zone are reflected in the regular pattern of the fields of local earthquake epicenters.

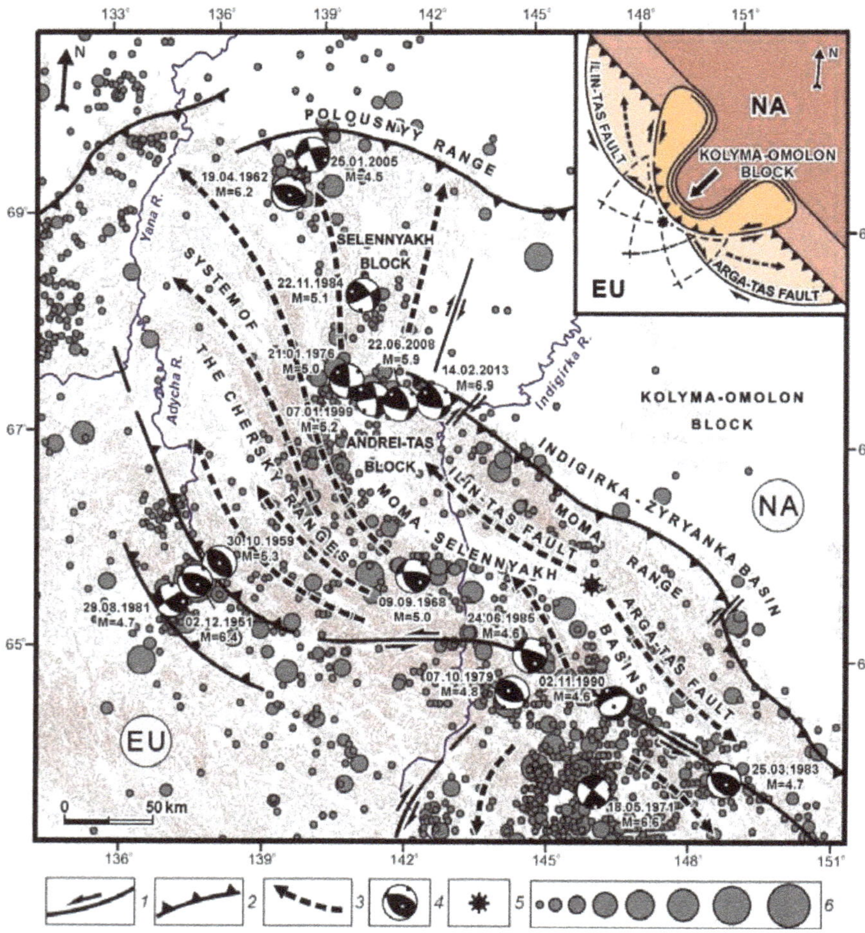

Figure 12. Scheme showing contemporary dynamics and earthquake epicenters in Yana–Indigirka segment of Chersky seismotectonic zone (after [26]). 1–2—kinematics of active faults: 1—strike-slip, 2—thrusts and reverse; 3—directions of movements of blocks; 4—focal mechanisms of earthquakes (lower hemisphere; the principal stress axes of compression and extension are marked by black and white dots, respectively); 5—Balagan–Tas volcano; 6—earthquake epicenters sized by magnitude: ≤3.0, 3.5, 4.0, 4.5, 5.0, 5.5, 6.0, ≥6.5.

Due to compression, the continental part of the Okhotsk plate is shifted in the E–SE direction, which contributes to the left-lateral shearing along the Ulakhan faults and right-lateral displacements in the Ketandino–Ulbey zone (Figure 6). This conclusion is supported by the development of the Pliocene–Quaternary extension zones along the northern coast of the Okhotsk Sea, which resulted from shifting of the Okhotsk Sea plate to the southeast. The E–SE thrusting of the continental part of the Okhotsk Seaplate is confirmed by the topographic and geomorphological observations that show uplifting in the mountainous area along the Okhotsk Sea coast. The watershed line between the Okhotsk Sea and the Arctic Ocean is disproportionately close to the Okhotsk Sea. In this region, shifting to the southeast relative to the Eurasian plate at a rate of 2–4 cm/yr is detected by satellite geodetic monitoring of the reference points installed by the US Geological Survey in Magadan and Petropavlovsk–Kamchatsky [26,29]. The model showing the structure and geodynamics of the Chersky Seismotectonic Zone allows us to discover the regularities in the occurrence of active structures of various types and identify the blocks that act as tectonic stress concentrators that have high seismic potential ($M_s \geq 6.5$).

4. Discussion and Conclusions

The territory of the Arctic–Asian seismic belt is influenced by a variety of geodynamic processes: spreading (Gakkel Ridge), rifting (Laptev Sea shelf), transtension (Kharaulakh Zone), and transpression (Chersky Seismotectonic Zone). The fault–block structures differ by kinematic types in the belt segments but reflect the regularities in the development of the major seismogenerating structures in the entire Arctic–Asian seismic belt.

Spreading takes place in the Gakkel Ridge that represents the Arctic segment of the boundary between the Eurasian and North American lithospheric plates. In this area, the dominating seismotectonic setting is crustal stretching, and the principal stress axes are sublatitudinal across the strike of the main structural elements. South of the continental slope is the part on the Laptev sea microplate (Figure 2). The seismotectonic deformation parameters in this area show transverse subhorizontal extension. According to the seismological data, compression takes place at the western and eastern margins of the Laptev Sea microplate, as well as in the Lena River delta, in response to rifting in the Laptev Sea shelf. The Olenek sector (a structural border between the Laptev Sea rift system and the Siberian platform) is also subjected to compression, with a small left-lateral shear component. Zones differing in the crustal stress state (stretching, compression, and a variety of stretching–compression combinations) are observed in the Kharaulakh Segment, where the mid-oceanic and continental structures are conjugated. Southeast of the transtension area, in the Chersky Seismotectonic Zone, the North American and Eurasian plates approach each other at an oblique angle. In the Chersky zone, the steady NE–trending compression / transpression is confirmed by the structural–tectonic and seismological data and evidenced by the seismogenerating structures of thrust, reverse and shear types.

A change in the geodynamic setting from tension to compression can be satisfactorily explained in the global plate tectonics theory, considering that an assumed pole of rotation of the North American and Eurasian plates is located near the Buor–Khaya Bay [21,29]. It can, therefore, be concluded that stretching (extension) is currently typical of the neotectonic structures located to the north of the rotation pole, while compression takes place in the structures located to the south and southeast of the rotation pole.

The geodynamic processes dominating in the regional segments of the Arctic–Asian seismic belt predetermine the types of focal zones of strong earthquakes. The transtension zones, where the mid-oceanic and continental crustal structures are conjugated, demonstrate the highest seismic potential. In the transpression zones, increased seismic potential is typical for the blocks that acts as tectonic stress concentrators. Such blocks are formed at the terminations of the wings of active shear systems and associated with 'pull-apart' and/or duplex compression structures. In remote images, these blocks are clearly detectable from the geomorphological and morphodynamic features of the modern terrain. The flanks of the Chersky zone, which are confined to the reactivated structures of the marginal sutures, are also characterized by an increased level of seismic activity. Strong earthquake focal zones are formed in accordance with the dynamics of the major seismogenerating structures located in the central zone of the Arctic–Asian seismic belt.

Our results of the seismogeodynamic analysis maybe used for assessments of potential seismic risks of seismogenerating, and potentially active, structures in the Russian Arctic region, which are located at the lithospheric plate boundaries between the Eurasian, North American and Okhotsk Sea Plates.

Author Contributions: L.P.I. carried out the preparation and analysis of the structural-geological and morphodynamic situation in the seismic zone of Chersky, and she also had the idea of determining the geodynamic activity of neotectonic structures. V.S.I. conducted field work on the study of the pleistoseist regions of strong earthquakes of the Chersky seismic belt. He was also responsible for determining the kinematics of active faults and the paleoseismological situation in the seismic zone. B.M.K. was engaged in catalogues of earthquake epicenters and focal mechanisms of earthquakes in the Chersky belt, as well as materials of seismological data.

Funding: This study was supported by the grant of the Russian Foundation for Basic Research (RFBR) No. 19-05-00062. The study was carried out at the financial support of RFBR in the framework of the scientific project No. 16-05-000224, a project of the Institute of the earth's crust Siberian branch of RAS (No. 346-2018-0001) and the Institute of diamond Geology and noble me-for metal Siberian branch of RAS (No. 0381-2616-0001) and programs of the Government of republic Yakutia for study of region in 2016-2020-glare of Sakha (Yakutia) on complex study territory of the Republic for 2016-2020.

Conflicts of Interest: The authors declare no conflict of interest.

References

1. Imaeva, L.P.; Imaev, V.S.; Mel'nikova, V.I. The stress–strain state of recent structures in the northeastern sector of the Russian Arctic Region. *Dokl. Earth Sci.* **2018**, *479*, 371–374. [CrossRef]
2. Imaeva, L.P.; Imaev, V.S.; Koz'min, B.M. Structural-dynamic model of the Cherskyseismotectonic zone (continental part of the Arctic-Asian seismic belt). *J. Asian Earth Sci.* **2016**, *116*, 59–68. [CrossRef]
3. Imaeva, L.P.; Gusev, G.S.; Imaev, V.S.; Mel'nikova, V.I. Neotectonic activity and parameters of seismotectonic deformations of seismic belts in the Northeast Asia. *J. Asian Earth Sci.* **2017**, *148*, 254–264. [CrossRef]
4. Shebalin, N.V.; Ulomov, V.I.; Tatevossian, R.E.; Trifonov, V.G.; Yoffe, A.I.; Kozhurin, A.I. *Unified Seismogeological Taxonomy of the Northern Eurasia*; IUGG-Abstracts. SB21C-14; IUGG: Boulder, CO, USA, 1995.
5. Ulomov, V.I.; Shumilina, L.S. A set of new general seismic zonation maps for the Russian Federation territory. *Seism. Resist. Constr.* **1998**, *4*, 30–34.
6. Imaeva, L.P.; Gusev, G.S.; Imaev, V.S.; Ashurkov, S.V.; Mel'nikova, V.I.; Seredkina, A.I. Geodynamic activity of modern structures and tectonic stress fields in Northeast Asia. *Geodyn. Tectonophys.* **2017**, *8*. [CrossRef]
7. Imaeva, L.P.; Imaev, V.S.; Mel'nikova, V.I.; Koz'min, B.M. Recent structures and tectonic regimes of the stress–strain state of the Earth's crust in the Northeastern Sector of the Russian Arctic Region. *Geotectonics* **2016**, *50*, 535–552. [CrossRef]
8. Grachev, A.F.; Demenitskaya, R.M.; Karasik, A.M. Problems of relationship between the Moma continental rift and the structure of the Gakkel mid-ocean ridge. *Geophys. Methods Surv. Arct. Reg.* **1973**, *8*, 56–75.
9. Zonenshain, L.P.; Kuz'min, M.I.; Natapov, L.M. *Plate Tectonics of the USSR Territory*; Nedra: Moscow, Russia, 1990; Volume 2, p. 334.
10. Gramberg, I.S.; Demenitskaya, R.M.; Sekretov, S.B. The system of rift grabens of the Laptev Sea shelf as the missing link in the rift belt of the Gakkel–Moma ridge. *Dokl. Akad. Nauk SSSR* **1990**, *311*, 689–694.
11. Drachev, S.S. Tectonics of the Laptev Sea rift system. *Geotectonics* **2000**, *34*, 467–481.
12. Drachev, S.S.; Mazur, S.; Campbell, S.; Green, C.; Tishchenko, A. Crustal architecture of the East Siberian Arctic Shelf and adjacent Arctic Ocean constrained by seismic data and gravity modeling results. *J. Geodyn.* **2018**, 1–78. [CrossRef]
13. Dorofeev, V.K.; Blagoveshchenskii, M.G.; Smirnov, A.N.; Ushakov, V.I. *New Siberian Islands: Geological Structure and Minerageny*; Institute of Oceanology: St. Petersburg, Russia, 1999; 130p.
14. Avetisov, G.P. *Seismoactive Zones of the Arctic Region*; Institute of Oceanology: St. Petersburg, Russia, 1996; 185p.
15. Anokhin, V.M.; Gusev, E.A.; Rekant, P.V. Character of synoceanic tectonics of the Laptev Sea oceanic margin. In *Tectonics and Geodynamics of the Continental Lithosphere*; GEOS: Moscow, Russia, 2003; Volume 1, pp. 10–12.
16. Bogdanov, N.A. Continental margins: General problems of the structure and tectonic evolution. In *Fundamental Problems of General Tectonic*; Nauchnyimir: Moscow, Russia, 2001; pp. 231–249.
17. Global Centroid Moment Tensor Project: On-Line Bull. Available online: http://www.globalcmt.org (accessed on 25 September 2015).
18. Grigor'ev, M.N.; Imaev, V.S.; Imaeva, L.P.; Koz'min, B.M.; Kunitskii, V.V.; Larionov, A.G.; Mikulenko, K.I.; Skryabin, R.M.; Timirshin, K.V. *Geology, Seismicity, and Permafrost Processes in the Arctic Areas of West Yakutia*; Yakutsk Scientific Center: Yakutsk, Russia, 1996; 84p.

19. Fedorov, G.B.; Antonov, O.M.; Bolshiyanov, D.Y. Features of the regime of modern tectonic movements of Central Taimyr. *News Russ. Geogr. Soc.* **2001**, *133*, 76–81.
20. Stovas, M.V. Young tectonic uplift in the coasts of the Kara, Laptev, Eastern Siberian, and Chukchi seas. *Dokl. Akad. Nauk SSSR* **1965**, *161*, 193–194.
21. Imaev, V.S.; Imaeva, L.P.; Smekalin, O.P.; Chipizubov, A.V.; Ovsyuchenko, A.N.; Kolodeznikov, I.I. Neotectonics of the Kharaulakh sector of the Laptev shelf. *Russ. Geol. Geophys.* **2018**, *59*, 831–844. [CrossRef]
22. Galabala, R.O. New data on the structure of the Lena River delta. In *The Quaternary of Northeast Asia*; North Eastern Complex Scientific Research Institute, Far East Branch of the USSR Academy of Sciences: Magadan, Russia, 1987; pp. 152–172.
23. Avetisov, G.P.; Guseva, Y.V. Deep structure of the Lena River delta area from seismological data. *Sov. Geol.* **1991**, *4*, 73–81.
24. Parfenov, L.M.; Oksman, V.S.; Prokop'ev, A.V.; Timofeev, V.F.; Tret'yakov, F.F.; Trunilina, V.A.; Deikunenko, A.V. Assemblage of terranes in the Verkhoyansk–Kolyma orogenic zone. In *Tectonics, Geodynamics, and Metallogeny of the Territory of the Sakha (Yakutia) Republic*; Nauka: Moscow, Russia, 2001; pp. 199–254.
25. Gusev, G.S. *Folded Structures and Faults of the Verkhoyansk–Kolyma System of Mesozoides*; Nauka: Moscow, Russia, 1979; 207p.
26. Imaeva, L.P.; Imaev, V.S.; Koz'min, B.M. Dynamics of seismogenerating structures in the frontal zone of the Kolyma–Omolon Superterrane. *Geotectonics* **2016**, *50*, 349–365. [CrossRef]
27. Fujita, K.; Kozmin, B.M.; Mackey, K.G.; Riegel, S.A.; Imaev, V.S.; McLean, M.S. Seismotectonics of the Chersky seismic belt, eastern Russia (Yakutia) and Magadandistrict, Russia. In *Geology, Geophysics and Tectonics of Northeastern Russia: A Tribute to Leonid Parfenov*; Stone, D.B., Fujita, K., Layer, P.W., Miller, E.L., Prokopiev, A.V., Toro, J., Eds.; Stephan Mueller Special Publication: 2009; EGU: Vienna, Austria, 19–24 April 2009; Series 4; pp. 117–145. [CrossRef]
28. Vazhenin, B.P. Paleoseismic dislocations in the Chersky seismic belt. In *Seismological and Petrophysical Studies in Northeastern Russia*; North Eastern Complex Scientific Research Institute, Far East Branch of the USSR Academy of Sciences: Magadan, Russia, 1992; pp. 79–102.
29. Imaeva, L.P.; Kolodeznikov, I.I. (Eds.) *Seismotectonics of the Northeastern Sector of Russian Arctic*; Institute of the Earth's Crust: Irkutsk, Russia; Institute of Diamond and Precious Metals Geology: Novosibirsk, Russia; Siberian Branch of RAS: Siberian, Russia, 2017; 136p.
30. Imaeva, L.P.; Koz'min, B.M.; Imaev, V.S.; Mackey, K.G. Structural-dynamic analysis of the epicentral zone of the Ilin-Tas earthquake (Feb 14, 2013, M_s 6.9). *J. Seismol.* **2015**, *19*, 341–353. [CrossRef]
31. Imaeva, L.P.; Koz'min, B.M.; Imaev, V.S.; Mackey, K.G. Seismotectonic Studies of the pleistoseist Area of the M_s 6.9 Ilin-Tass earthquake in northeast Yakutia. *Izv. Phys. Solid Earth* **2016**, *52*, 821–835. [CrossRef]
32. Imaeva, L.P.; Imaev, V.S.; Koz'min, B.M. Seismotectonic activation of modern structures of the Siberian Craton. *Geotectonics* **2018**, *52*, 618–633. [CrossRef]

MDPI
St. Alban-Anlage 66
4052 Basel
Switzerland
Tel. +41 61 683 77 34
Fax +41 61 302 89 18
www.mdpi.com

Geosciences Editorial Office
E-mail: geosciences@mdpi.com
www.mdpi.com/journal/geosciences

www.ingramcontent.com/pod-product-compliance
Lightning Source LLC
Chambersburg PA
CBHW041938260926
51850CB00018B/242